FreeCAD 0.18 Black Book (Colored)

By
Gaurav Verma
Matt Weber
(CADCAMCAE Works)

Edited by
Kristen

ISBN # 978-1-77459-026-3

DEDICATION

To teachers, who make it possible to disseminate knowledge
to enlighten the young and curious minds
of our future generations

To students, who are the future of the world

THANKS

To my friends and colleagues

To my family for their love and support

Training and Consultant Services

At CADCAMCAE WORKS, we provide effective and affordable one to one online training on various software packages in Computer Aided Design(CAD), Computer Aided Manufacturing(CAM), Computer Aided Engineering (CAE), Computer programming languages(C/C++, Java, .NET, Android, Javascript, HTML and so on). The training is delivered through remote access to your system and voice chat via Internet at any time, any place, and at any pace to individuals, groups, students of colleges/universities, and CAD/CAM/CAE training centers. The main features of this program are:

Training as per your need

Highly experienced Engineers and Technician conduct the classes on the software applications used in the industries. The methodology adopted to teach the software is totally practical based, so that the learner can adapt to the design and development industries in almost no time. The efforts are to make the training process cost effective and time saving while you have the comfort of your time and place, thereby relieving you from the hassles of traveling to training centers or rearranging your time table.

Software Packages on which we provide basic and advanced training are:

CAD/CAM/CAE: CATIA, Creo Parametric, Creo Direct, SolidWorks, Autodesk Inventor, Solid Edge, UG NX, AutoCAD, AutoCAD LT, EdgeCAM, MasterCAM, SolidCAM, DelCAM, BOBCAM, UG NX Manufacturing, UG Mold Wizard, UG Progressive Die, UG Die Design, SolidWorks Mold, Creo Manufacturing, Creo Expert Machinist, NX Nastran, Hypermesh, SolidWorks Simulation, Autodesk Simulation Mechanical, Creo Simulate, Gambit, ANSYS and many others.

Computer Programming Languages: C++, VB.NET, HTML, Android, Javascript and so on.

Game Designing: Unity.

Civil Engineering: AutoCAD MEP, Revit Structure, Revit Architecture, AutoCAD Map 3D and so on.

We also provide consultant services for Design and development on the above mentioned software packages

For more information you can mail us at:
cadcamcaeworks@gmail.com

Table of Contents

Chapter 2 : Sketching

Chapter 3 : Solid Modeling

Chapter 4 : Solid Modeling

Chapter 5 : Solid Modeling Practice

Chapter 6 : Drafting

Chapter 7 : Arch Modeling

Chapter 8 : Path Workbench

Chapter 9 : FEM Workbench

Chapter 10 : Tech Drawing and Image Workbench

Chapter 11 : Miscellaneous Workbench

Preface

FreeCAD is a parametric (history based) open source CAD/CAM/CAE software which finds its uses in various fields of designing. The software has a modular structure which allows to plugin extensions (modules) to add functionality to the core application. An extension can be as complex as a whole new application programmed in C++ or as simple as a Python script or self-recorded macro. You have complete access to almost any part of FreeCAD from the built-in Python interpreter, macros or external scripts, be it geometry creation and transformation, the 2D or 3D representation of that geometry (scenegraph) or even the FreeCAD interface. The software allows import/export to standard formats such as STEP, IGES, OBJ, STL, DXF, SVG, STL, DAE, IFC or OFF, NASTRAN, VRML in addition to FreeCAD's native FCStd file format. FreeCAD's installer allows flexible installations on Windows systems. Packages for Ubuntu systems are also available and updated regularly. There is a huge library of external workbenches available for FreeCAD on web. You can download free copy of software from https://www.freecadweb.org/ web link.

The **FreeCAD 0.18 Black Book** is the first edition of our series on FreeCAD. This book is written to help beginners in creating some of the most complex solid models. The book follows a step by step methodology. In this book, we have tried to give real-world examples with real challenges in designing. We have tried to cover most of the topics utilized in industries for designing. The book covers almost all the information required by a learner to master the FreeCAD. The book starts with sketching and ends at advanced topics like Path (CAM), and FEM (Simulation). Some of the salient features of this book are :

In-Depth explanation of concepts

Every new topic of this book starts with the explanation of the basic concepts. In this way, the user becomes capable of relating the things with real world.

Topics Covered

Every chapter starts with a list of topics being covered in that chapter. In this way, the user can easily find the topics of his/her interest easily.

Instruction through illustration

The instructions to perform any action are provided by maximum number of illustrations so that the user can perform the actions discussed in the book easily and effectively. There are about 1350 illustrations that make the learning process effective.

Tutorial point of view

At the end of concept's explanation, the tutorial make the understanding of users firm and long lasting. Almost each chapter of the book has tutorials that are real world projects. Moreover most of the tools in this book are discussed in the form of tutorials.

Project

Projects and exercises are provided to students for practicing.

For Faculty

If you are a faculty member, then you can ask for video tutorials on any of the topic, exercise, tutorial, or concept.

Formatting Conventions Used in the Text

All the key terms like name of button, tool, drop-down etc. are kept bold.

Free Resources

Link to the resources used in this book are provided to the users via email. To get the resources, mail us at ***cadcamcaeworks@gmail.com*** with your contact information. With your contact record with us, you will be provided latest updates and informations regarding various technologies. The format to write us mail for resources is as follows:

Subject of E-mail as ***Application for resources of......................... book.***
Also, given your information like
Name:
Course pursuing/Profession:
Contact Address:
E-mail ID:

Note: We respect your privacy and value it. If you do not want to give your personal informations then you can ask for resources without giving your information.

About Authors

The author of this book, Matt Weber, has written more than 16 books on CAD/CAM/CAE available in market. He has coauthored SolidWorks Simulation, SolidWorks Electrical, SolidWorks Flow Simulation, and SolidWorks CAM Black Books. The author has hands on experience on almost all the CAD/CAM/CAE packages. If you have any query/doubt in any CAD/CAM/CAE package, then you can contact the author by writing at cadcamcaeworks@gmail.com

The author of this book, Gaurav Verma, has written and assisted in more than 16 titles in CAD/CAM/CAE which are already available in market. He has authored Autodesk Fusion 360 Black Book, AutoCAD Electrical Black Book, Autodesk Revit Black Books, and so on. He has provided consultant services to many industries in US, Greece, Canada, and UK. He has assisted in preparing many Government aided skill development programs. He has been speaker for Autodesk University, Russia 2014. He has assisted in preparing AutoCAD Electrical course for Autodesk Design Academy. He has worked on Sheetmetal, Forging, Machining, and Casting designs in Design and Development departments of various manufacturing firms.

For Any query or suggestion

If you have any query or suggestion, please let us know by mailing us on ***cadcamcaeworks@gmail.com***. Your valuable constructive suggestions will be incorporated in our books and your name will be addressed in special thanks area of our books on your confirmation.

Page left blank intentionally

Chapter 1

Starting with FreeCAD

Topics Covered

The major topics covered in this chapter are:

- ***Overview of FreeCAD***
- ***Installing FreeCAD***
- ***Starting FreeCAD***
- ***File Menu***
- ***View Menu***
- ***Standard Views***
- ***Freeze Display***
- ***Draw Style***
- ***Tools Menu***
- ***Navigating in the 3D view***
- ***FreeCAD Interface***

OVERVIEW OF FREECAD

FreeCAD is an open-source parametric 3D modeling application, made primarily to design real-life objects. Parametric modeling describes a certain type of modeling, where the shape of the 3D objects you design are controlled by parameters. For example, the shape of a brick might be controlled by three parameters: height, width and length. In FreeCAD, as in other parametric modelers, these parameters are part of the object, and stay modifiable at any time, after the object has been created. It connects your entire product development process in a single cloud-based program that works on all the three major platforms viz. Windows, Mac OS, and Linux; refer to Figure-1.

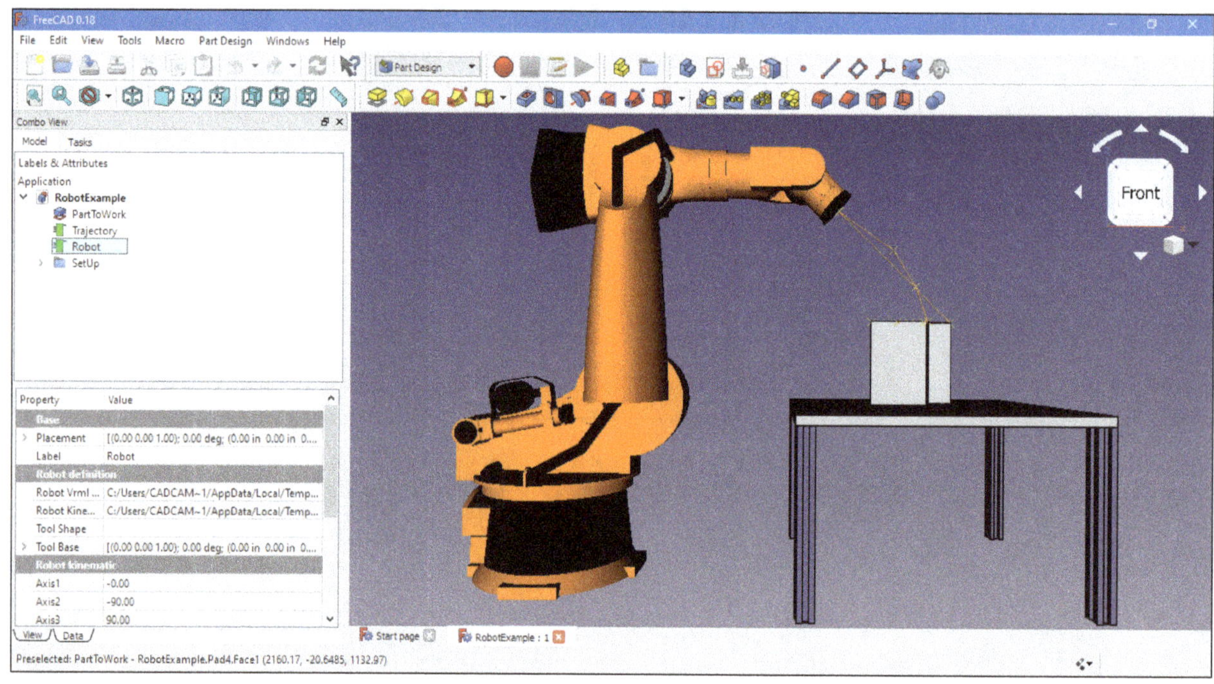

Figure-1. Application interface

FreeCAD is not designed for a particular kind of work or to make a certain kind of objects. Instead, it allows a wide range of uses and permits users to produce models of all sizes and purposes from small electronic components to 3D-printable pieces and all the way up to buildings. Each of these tasks have different dedicated sets of tools and workflows available. Being open-source, FreeCAD benefits from the contributions and efforts of a large community of programmers, enthusiasts, and users worldwide. FreeCAD is essentially an application built by the people who use it, instead of being made by a company trying to sell you a product. And of course, it also means that FreeCAD is free, not only to use but also to distribute, copy, modify, or even sell.

FreeCAD also benefits from the huge, accumulated experience of the open-source world. In its bowels, it includes several other open source components, as FreeCAD itself can be used as a component in other applications. It also possesses all kinds of features that have become a standard in the open-source world, such as supporting a wide range of file formats being hugely scriptable, customizable, and modifiable. All this is made possible through a dynamic and enthusiast community of users.

INSTALLING FREECAD

- Connect your pc with the internet connection and then go to **https://www.freecadweb.org/** as shown in Figure-2.

Figure-2. FreeCAD website

- Click on the **Download now** button, the **Downloads** tab of this website will be displayed; refer to Figure-3.

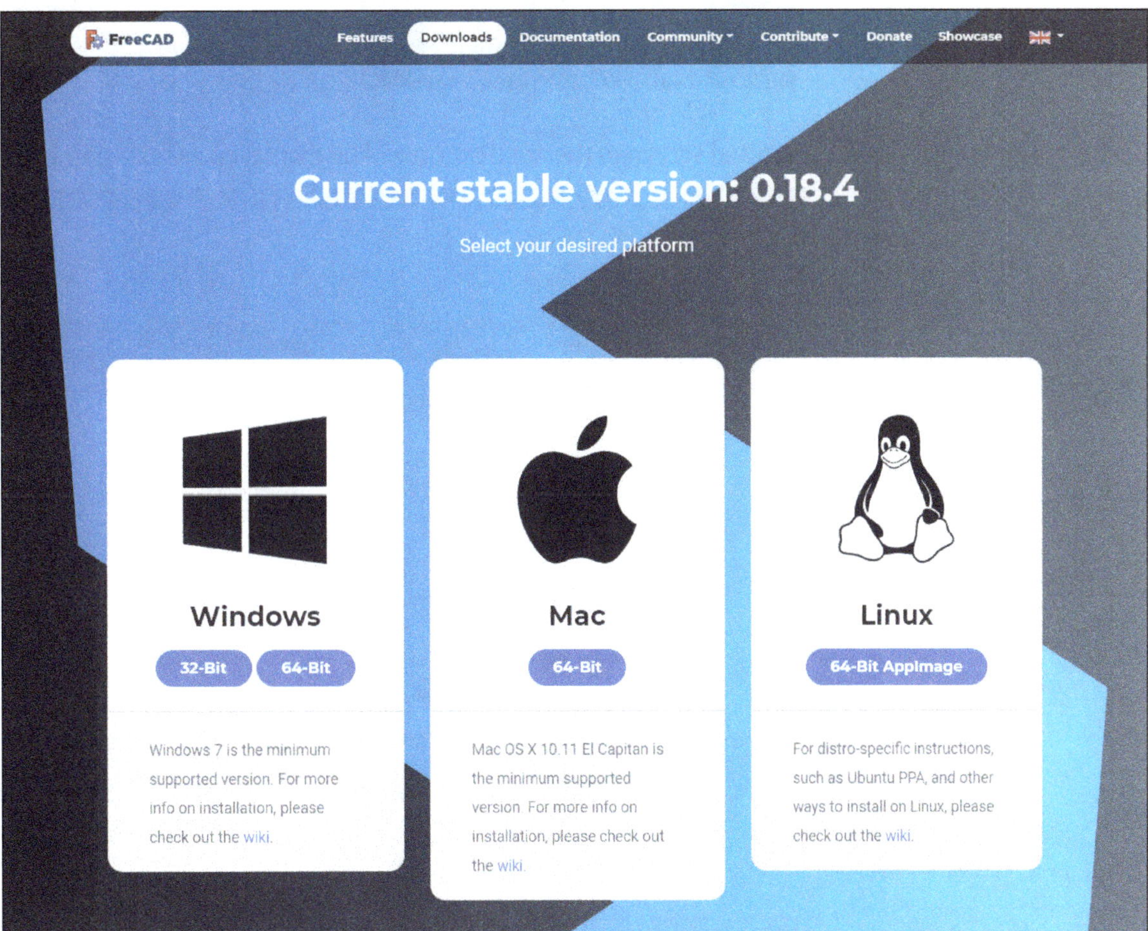

Figure-3. Downloads tab of FreeCAD website

- Select desired Operating System and the type of processor installed on your PC. The software will begin to download.
- Open the download setup file and follow the instructions as per the setup window. The software will be installed in a couple of minutes.

STARTING FREECAD

- To start **FreeCAD** from **Start** menu, in Windows 7, click on the **Start** button in the **Taskbar** at the bottom left corner, click on **All Programs** folder and then on **FreeCAD** folder. If you are using Windows 8 or later then direct click on **FreeCAD** folder in the **Start** menu and select the **FreeCAD** icon; refer to Figure-4. Click on the program name to start the application.

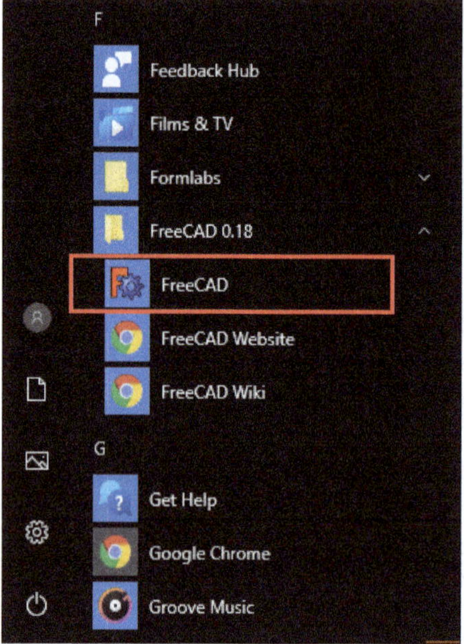

Figure-4. Start menu

- On starting the application, the initial screen of the application will be displayed as shown in Figure-5.

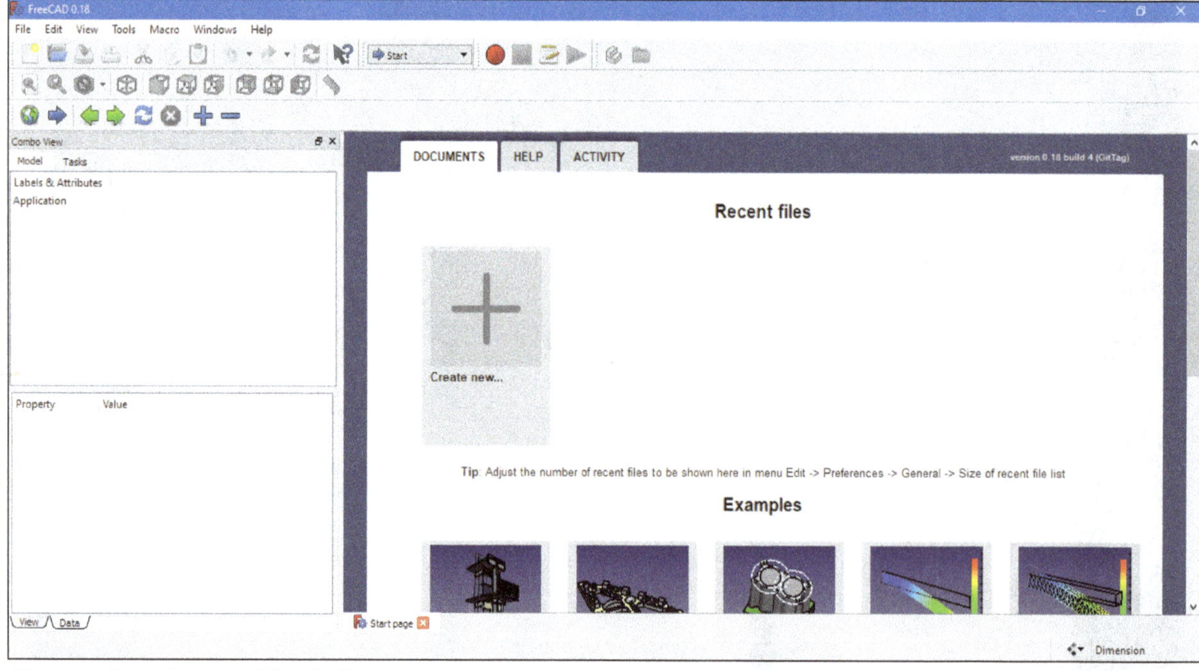

Figure-5. Initial screen of FreeCAD

FILE MENU

The options in the **File** menu are used to manage files and related parameters. Various tools of **File** menu are discussed next.

New

The **New** tool is used for initiating a new document file. The procedure to use this tool is discussed next.

- Click on the **New** tool from the **File** menu or **Ribbon** or press **Ctrl+N** key from keyboard; refer to Figure-6. The new document file will open; refer to Figure-7.

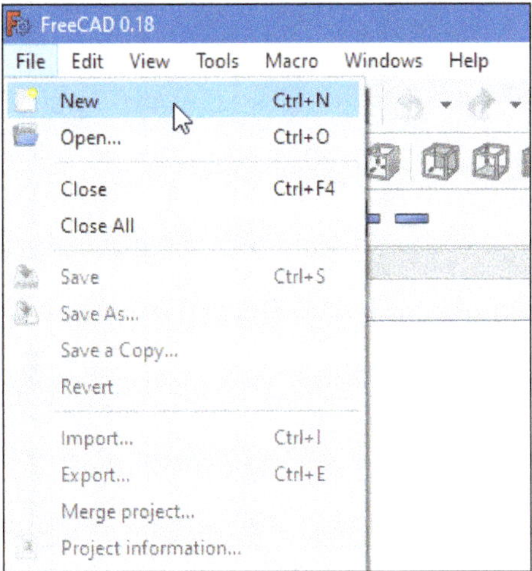

Figure-6. New tool

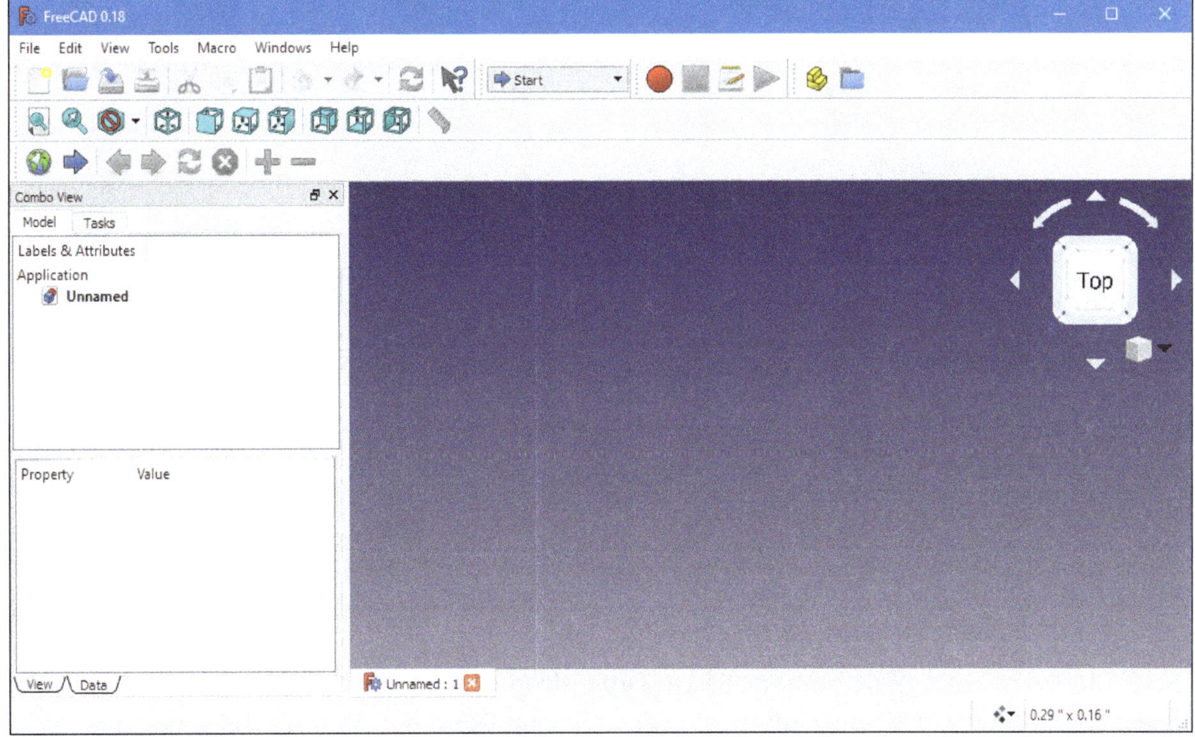

Figure-7. New document file

Open

The **Open** tool is used to open FreeCAD files and import files of other CAD applications. The procedure to use this tool is discussed next.

- Click on the **Open** tool from the **File** menu or **Ribbon** or press **CTRL+O** key from keyboard; refer to Figure-8. The **Open document** dialog box will be displayed; refer to Figure-9.

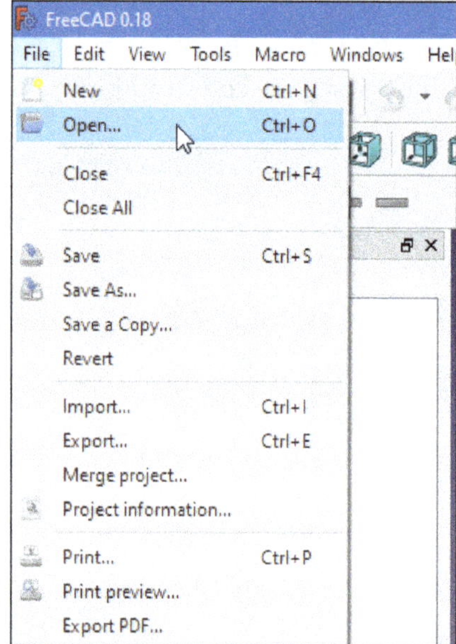

Figure-8. Open tool

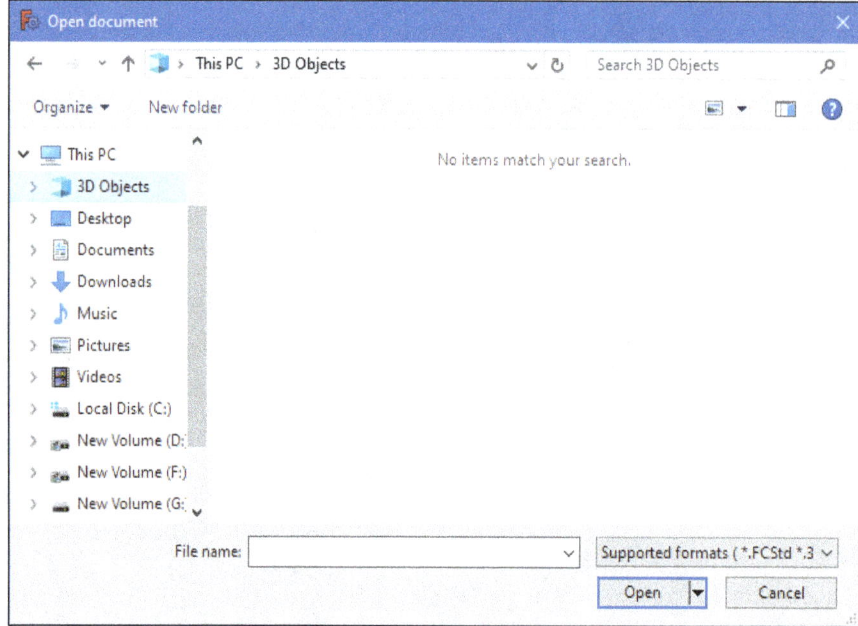

Figure-9. Open document dialog box

- Select desired format of file from the **File Type** drop-down to define format of file to be opened.
- Select desired file which you want to open and click on **Open** button from the dialog box. The file will open.

Close

The **Close** tool is used to close the current file without closing the application. The procedure to use this tool is discussed next.

- Click on the **Close** tool from **File** menu or press **Ctrl+F4** key from keyboard; refer to Figure-10. The current file will be closed.

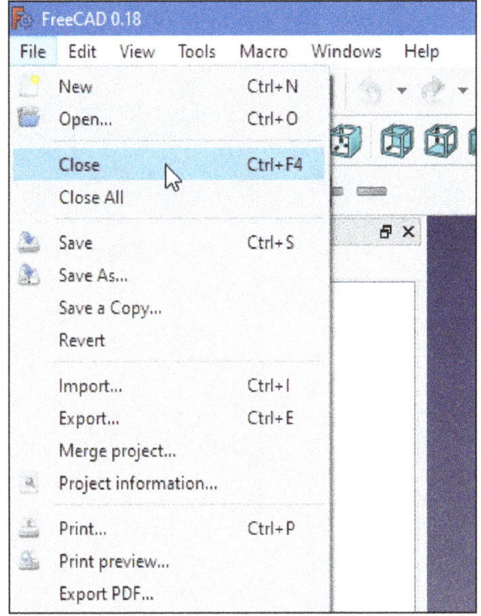

Figure-10. Close tool

Close All

The **Close All** tool is used to close all the files opened without closing the application. The procedure to use this tool is discussed next.

- Click on the **Close All** tool from the **File** menu; refer to Figure-11. All the files which are currently opened will be closed.

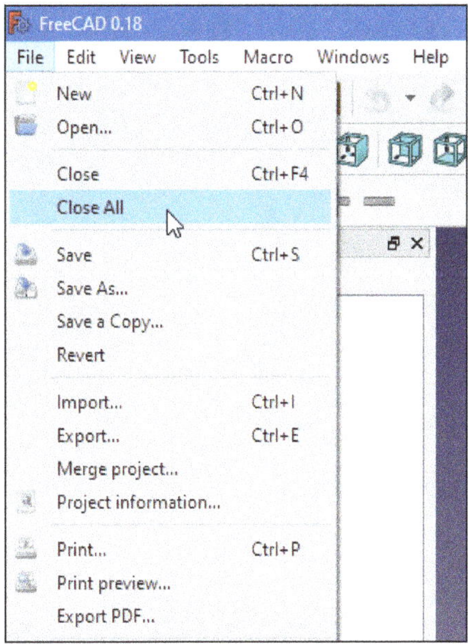

Figure-11. Close All tool

Save

The **Save** tool is used to save the current file opened in the application. The procedure to use this tool is discussed next.

- Click on the **Save** tool from **File** menu or press **Ctrl+S** key from keyboard or select the **Save** button from **Toolbar** as shown in Figure-12. The **Save FreeCAD Document** dialog box will be displayed; refer to Figure-13.

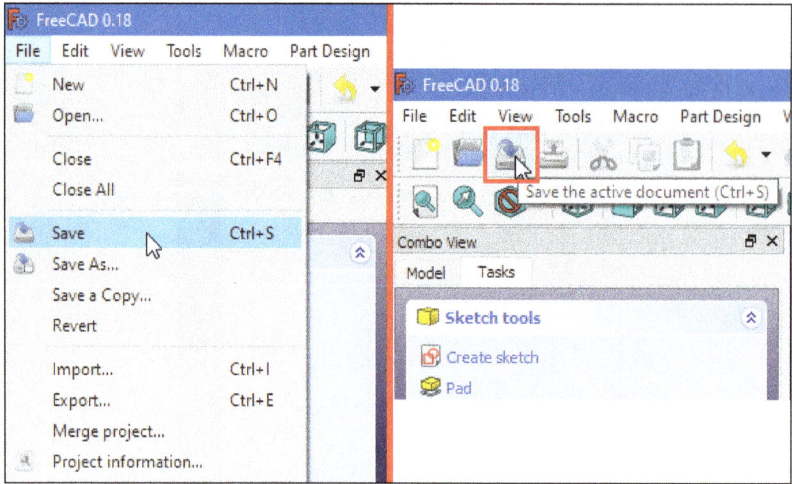

Figure-12. Save tool

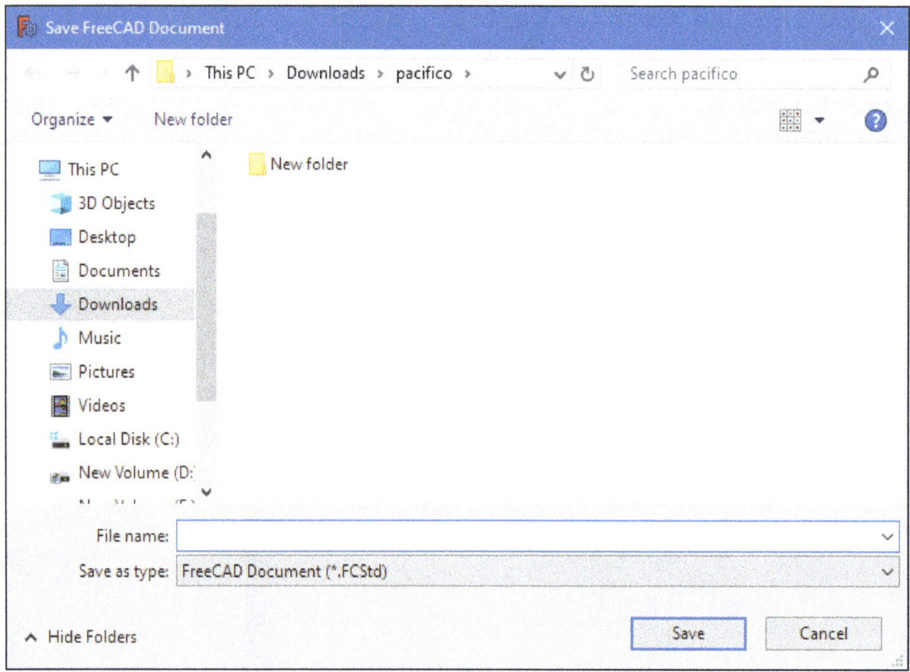

Figure-13. Save FreeCAD Document dialog box

- Specify desired location and name for the file, and click on the **Save** button from the dialog box to save the file.

Save As

Using the **Save As** tool, you can save the file with different name. The procedure to use this tool is discussed next.

- Click on **Save As** tool from **File** menu; refer to Figure-14. The **Save FreeCAD Document** dialog box will be displayed as shown in Figure-13.
- Rest of the procedure is same as discussed in the **Save** tool.

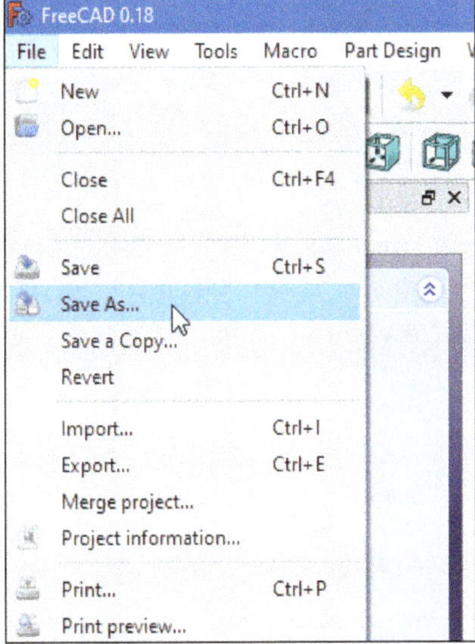

Figure-14. Save As tool

Save a Copy

The **Save a Copy** tool is used to create a copy of the file with new name. The procedure to use this tool is discussed next.

- Click on the **Save a Copy** tool from the **File** menu; refer to Figure-15. The **Save FreeCAD Document** dialog box will be displayed as shown in Figure-13.
- Rest of the procedure is same as we have discussed earlier.

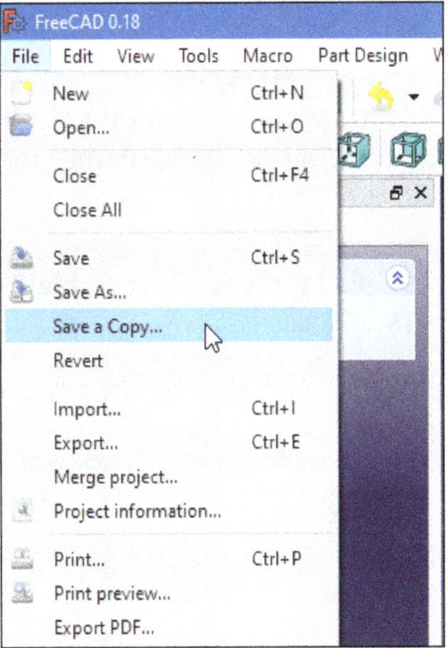

Figure-15. Save a Copy tool

Revert

The **Revert** tool is used to close the active document and reopens the last saved version of this document. Using this tool, you will lose all the changes you have made since the last file save. The procedure to use this tool is discussed next.

- Click on the **Revert** tool from the **File** menu; refer to Figure-16. The **Revert document** dialog box will be displayed asking you for the confirmation to discard all the changes; refer to Figure-17.

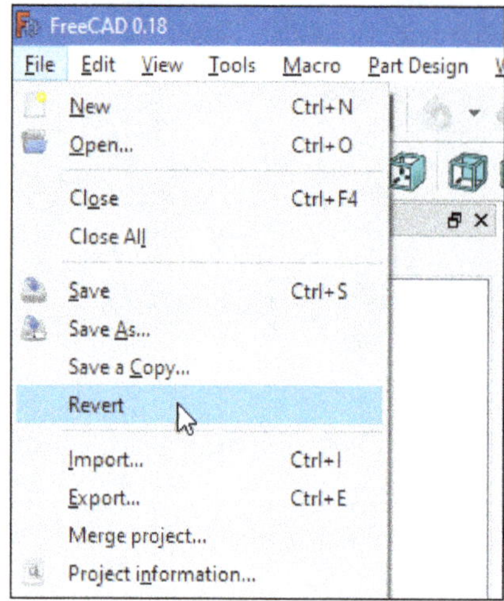

Figure-16. Revert tool

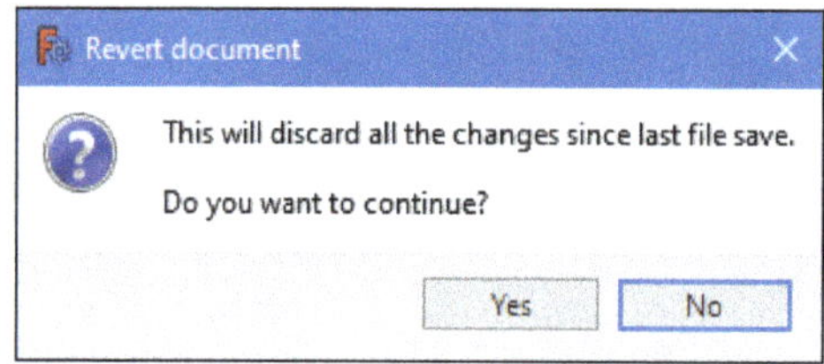

Figure-17. Revert document dialog box

- Click on the **Yes** button from the dialog box to discard all the changes.

Import

Many times, importing reduces lots of extra work of rebuilding the base sketch for models. In FreeCAD, we can directly use the CAD files for creating or manipulating the model. The procedure to use this tool is discussed next.

- Click on the **Import** tool from the **File** menu; refer to Figure-18. The **Import file** dialog box will be displayed; refer to Figure-19.

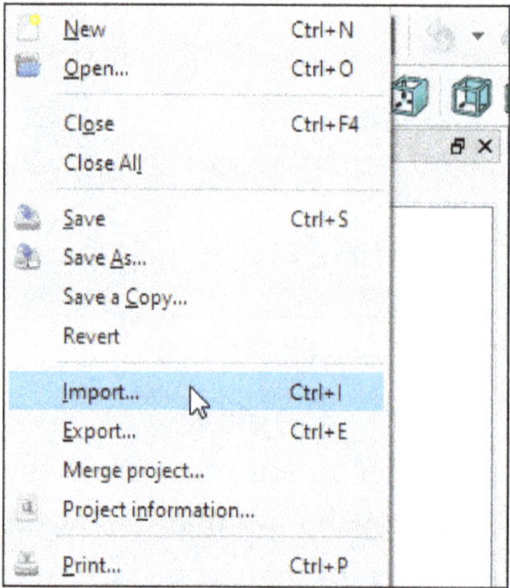

Figure-18. Import tool

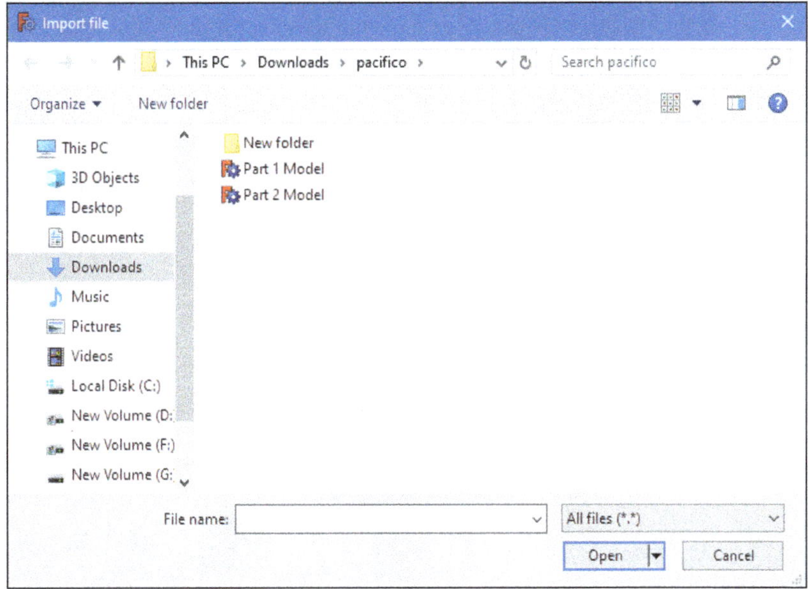

Figure-19. Import file dialog box

- Select desired format of non-native file you want to open from the **File Type** drop-down; refer to Figure-20.

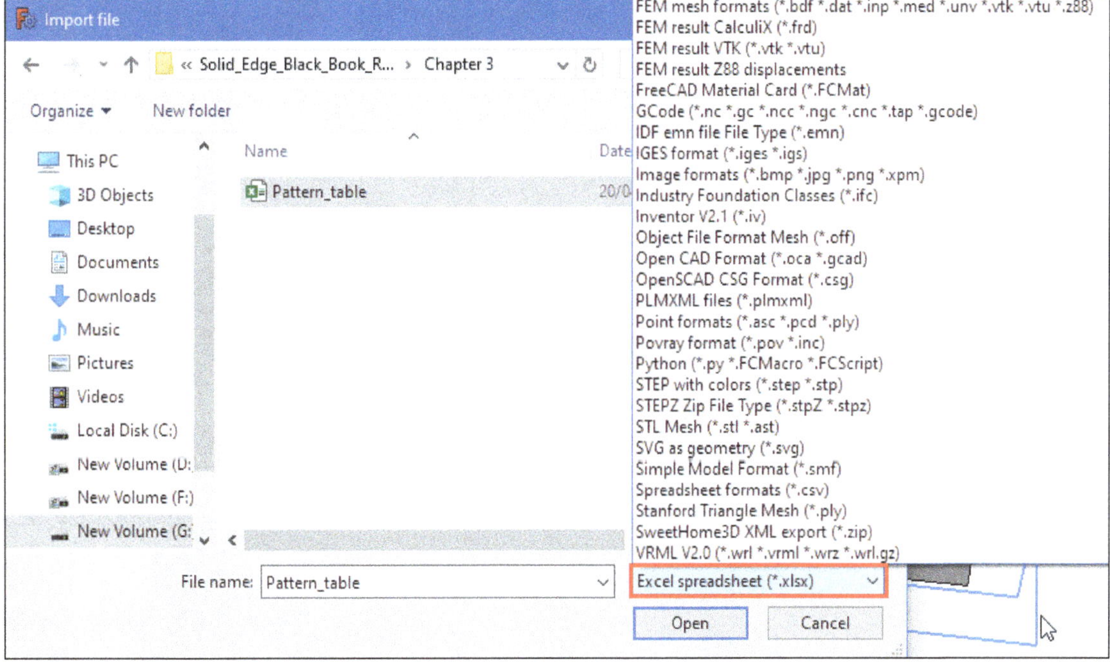

Figure-20. File type drop down

- Select desired file which you want to import and click on the **Open** button from the dialog box.

Export

The **Export** tool is used to export the existing file in various different formats. The procedure to use this tool is discussed next.

- Select the solid part or sketch which you want to export from the **Model** tab of the **Combo View** as shown in Figure-21.

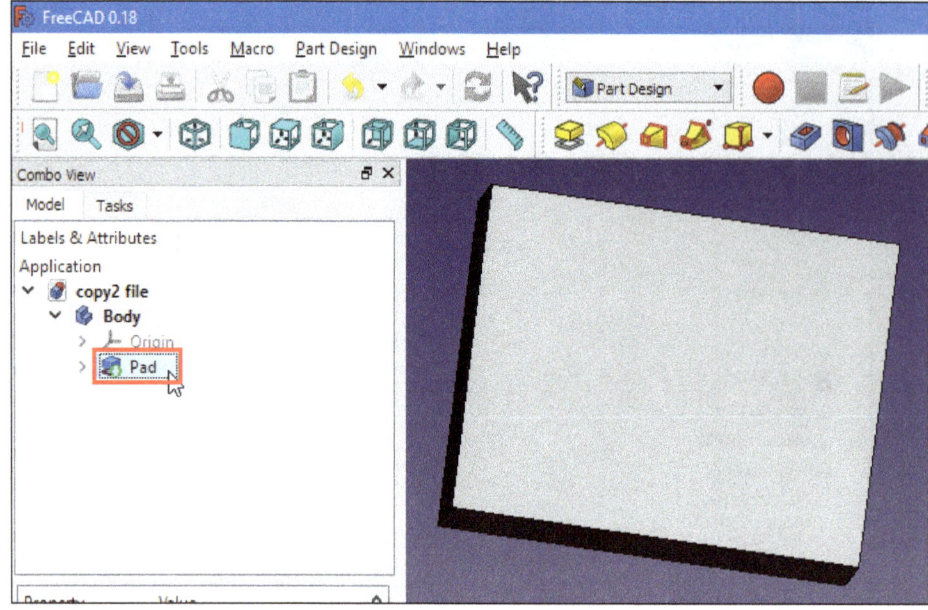

Figure-21. Solid model selected to export

- Click on the **Export** tool from **File** menu; refer to Figure-22. The **Export file** dialog box will be displayed; refer to Figure-23.

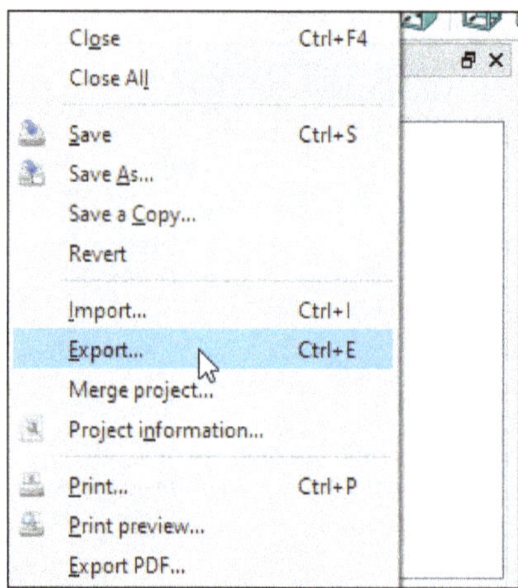

Figure-22. Export tool

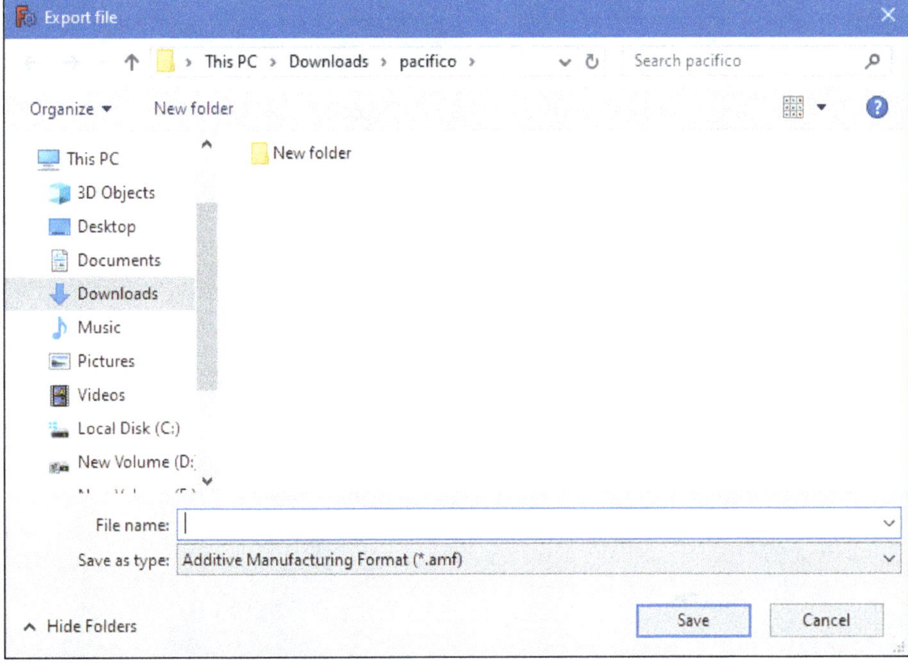

Figure-23. Export file dialog box

- Select desired format of file from the **Save as type** drop-down.
- Specify desired name for the file in the **File name** edit box and click on **Save** button from the dialog box to export the file.

Merge project

The **Merge project** tool is used to add the contents of a FreeCAD file into the active document. The procedure to use this tool is discussed next.

- Click on the **Merge project** tool from the **File** menu; refer to Figure-24. The **Merge project** dialog box will be displayed; refer to Figure-25.

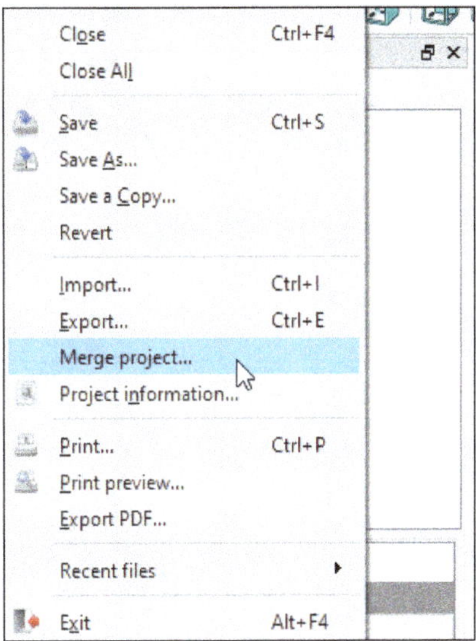

Figure-24. Merge project tool

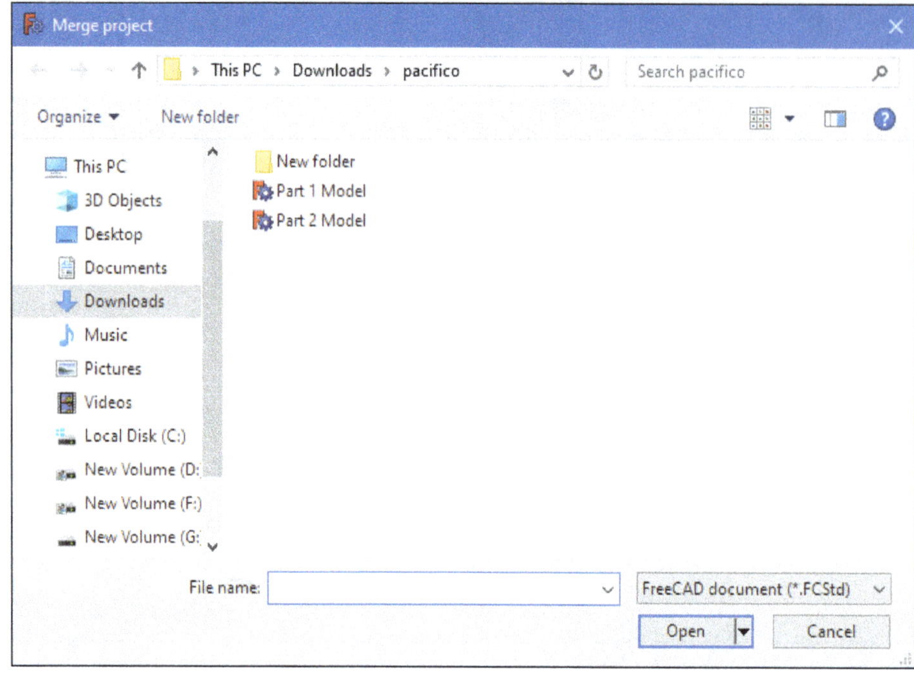

Figure-25. Merge project dialog box

- Select desired file which you want to add in the active document and click on **Open** button from the dialog box. The model of selected file will be added in the active document.

Project information

The **Project information** tool is used to show the project information in the dialog box belonging to the active document. Some of this information can be edited. The procedure to use this tool is discussed next.

- Click on the **Project information** tool from the **File** menu; refer to Figure-26. The **Project information** dialog box will be displayed; refer to Figure-27.

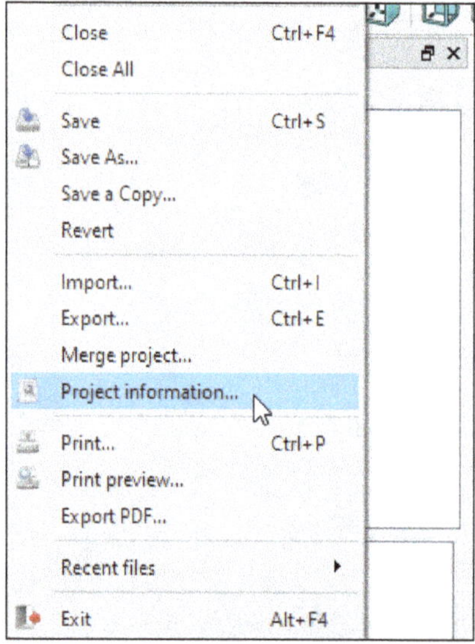

Figure-26. Project information tool

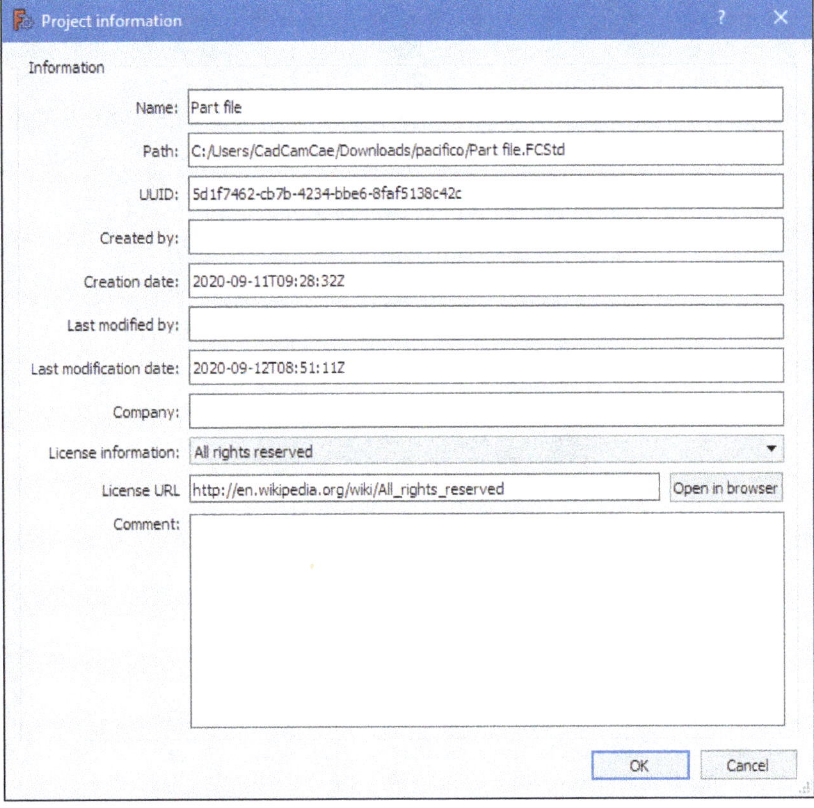

Figure-27. Project information dialog box

- Specify and edit the information about the active document as desired and click on **OK** button from the dialog box.

Print

As the name suggests, the **Print** tool is used to get print out of the model/drawing on paper using printer connected to the system.

Print Preview

The **Print Preview** tool is used to check the print before sending command to printer. The procedure to use this tool is discussed next.

- Click on the **Print preview** tool from the **File** menu; refer to Figure-28. The **Print Preview** dialog box will be displayed along with the preview of the model; refer to Figure-29.

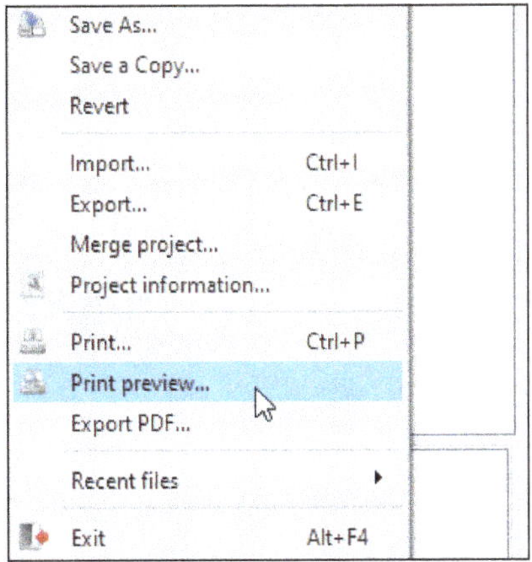

Figure-28. Print preview tool

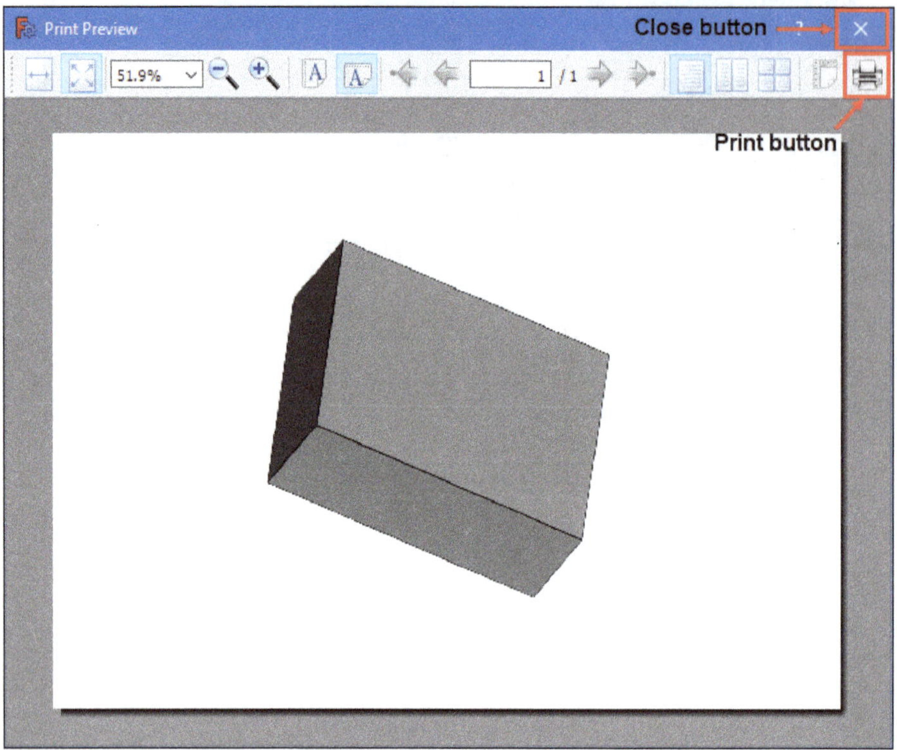

Figure-29. Print Preview dialog box

- If the preview is satisfactory then click on the **Print** button otherwise, click on the **Close** button from the toolbar displayed at the top of the modeling area and modify the model; refer to Figure-29.

Export PDF

The **Export PDF** tool is used to export or save the active document in a pdf file format. The procedure to use this tool is discussed next.

- Click on the **Export PDF** tool from **File** menu; refer to Figure-30. The **Export PDF** dialog box will be displayed; refer to Figure-31.

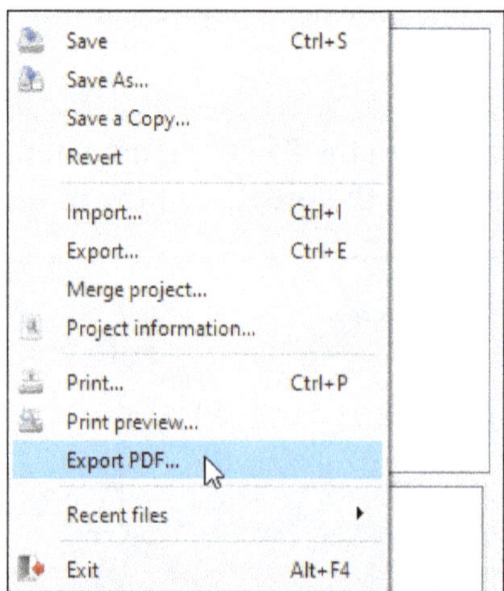

Figure-30. Export PDF tool

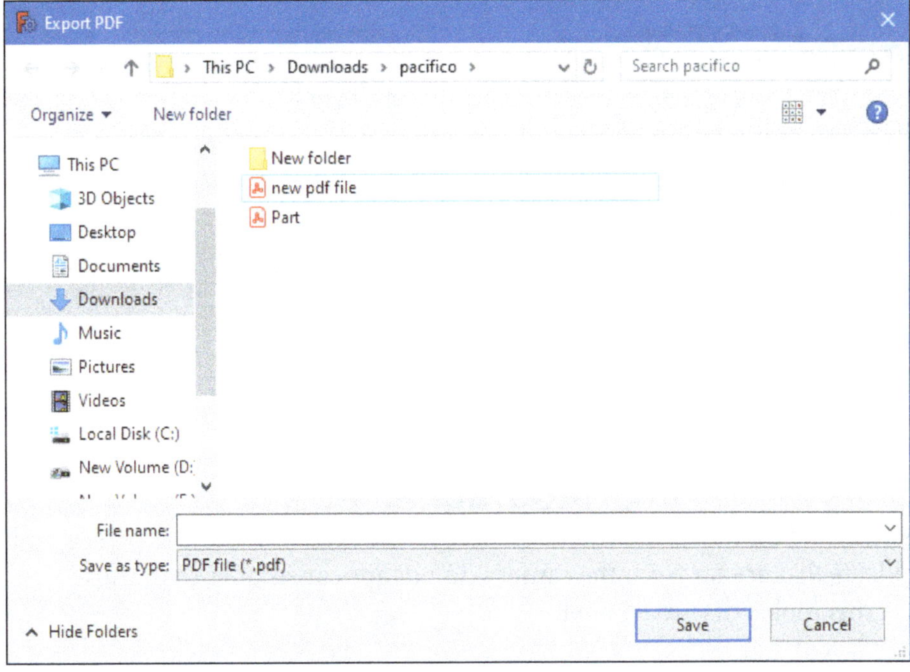

Figure-31. Export PDF dialog box

- Specify desired file name for saving the file as pdf and click on **Save** button from the dialog box.

VIEW MENU

The **View** menu provides tools to change the 3D view, the view properties of objects in the model, and tools related to the display of interface components. Various tools of **View** menu are discussed next.

Create new view

The **Create new view** tool is used to create a new 3D view for the active document. It can be useful if you want to inspect the model from multiple directions or at different zoom levels. The procedure to use this tool is discussed next.

- Click on the **Create new view** tool from the **View** menu; refer to Figure-32. A new window will be opened with the 3D view of an active model; refer to Figure-33.

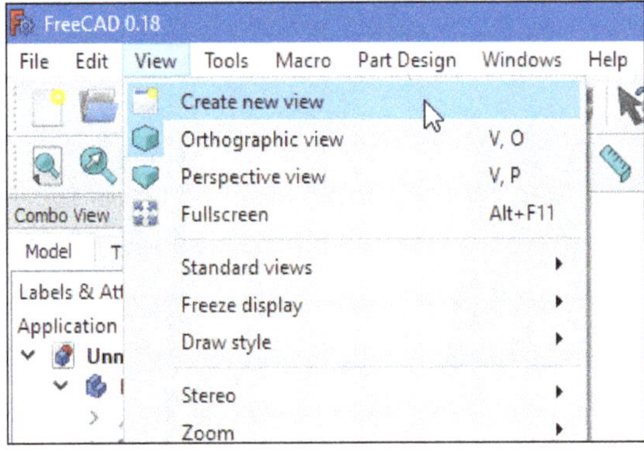

Figure-32. Create new view tool

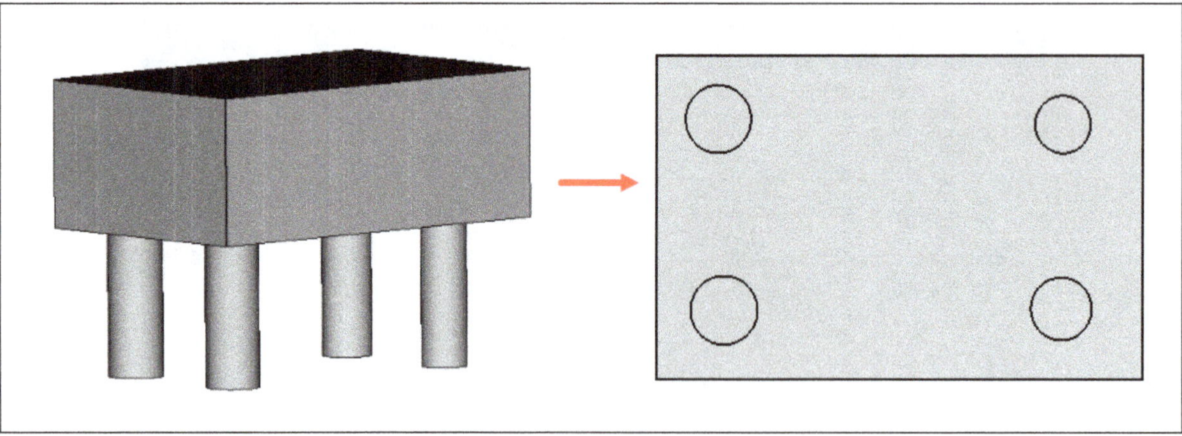

Figure-33. New 3D view of a model

Orthographic view

The **Orthographic view** tool switches the camera in the active 3D view to orthographic view mode. In this mode, objects that are far from the camera do not appear smaller than those that are closer. The procedure to use this tool is discussed next.

- Click on the **Orthographic view** tool from the **View** menu; refer to Figure-34. An orthographic view of the model will be displayed; refer to Figure-35.

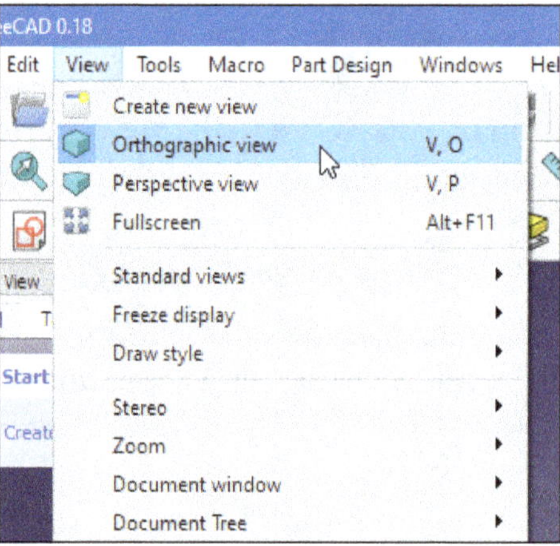

Figure-34. Orthographic view tool

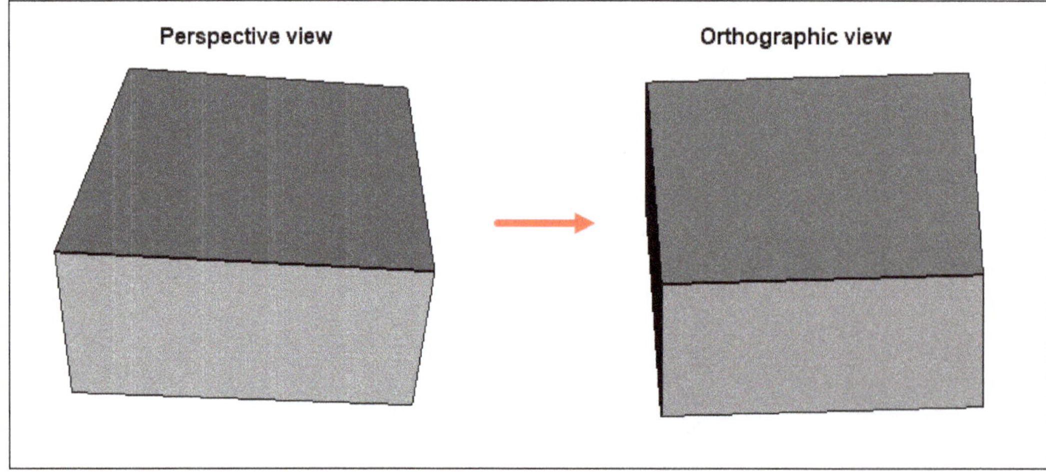

Figure-35. Orthographic view of the model

Perspective view

The **Perspective view** tool switches the camera in the active 3D view to perspective view mode. In this mode, objects that are far from the camera appear smaller than those that are closer. The procedure to use this tool is discussed next.

- Click on the **Perspective view** tool from the **View** menu; refer to Figure-36. A perspective view of the model will be displayed; refer to Figure-37.

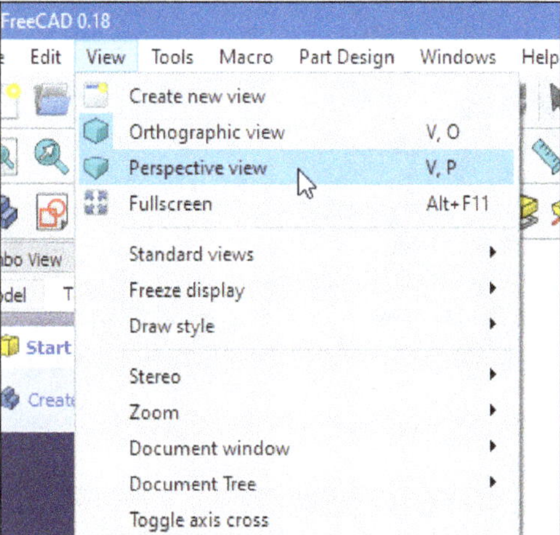

Figure-36. Perspective view tool

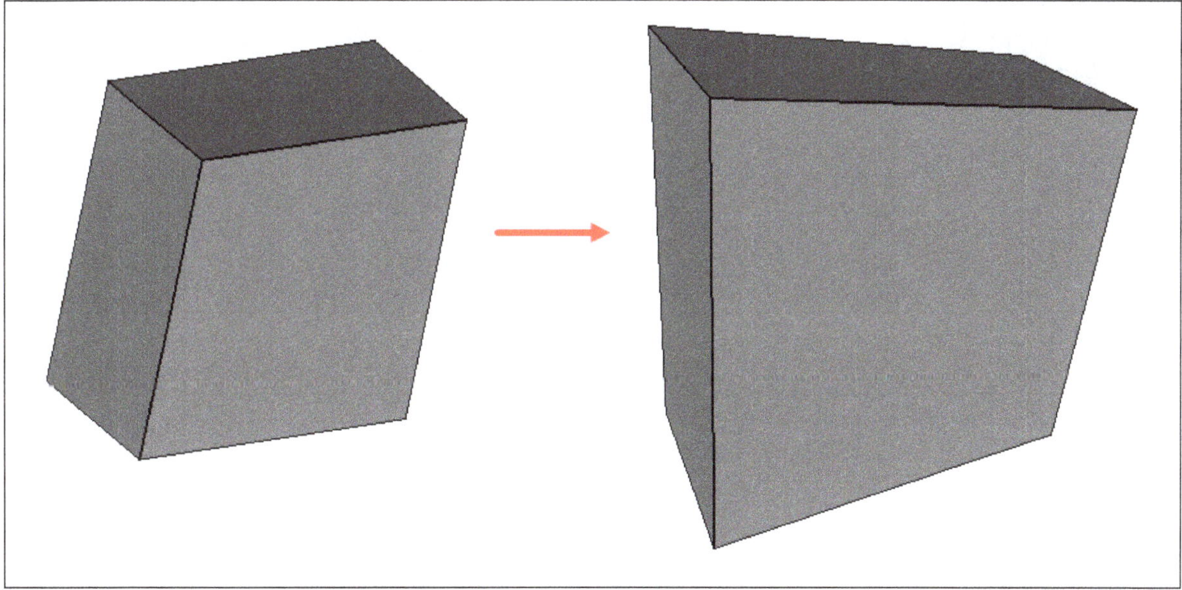

Figure-37. Perspective view of the model

Fullscreen

The **Fullscreen** tool switches the FreeCAD's main window in fullscreen mode. The procedure to use this tool is discussed next.

- Click on the **Fullscreen** tool from the **View** menu; refer to Figure-38. The view of the model will be displayed in Fullscreen window; refer to Figure-39.

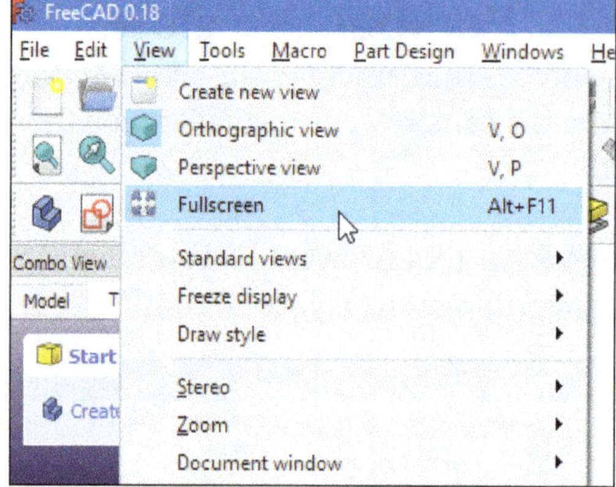

Figure-38. Fullscreen tool

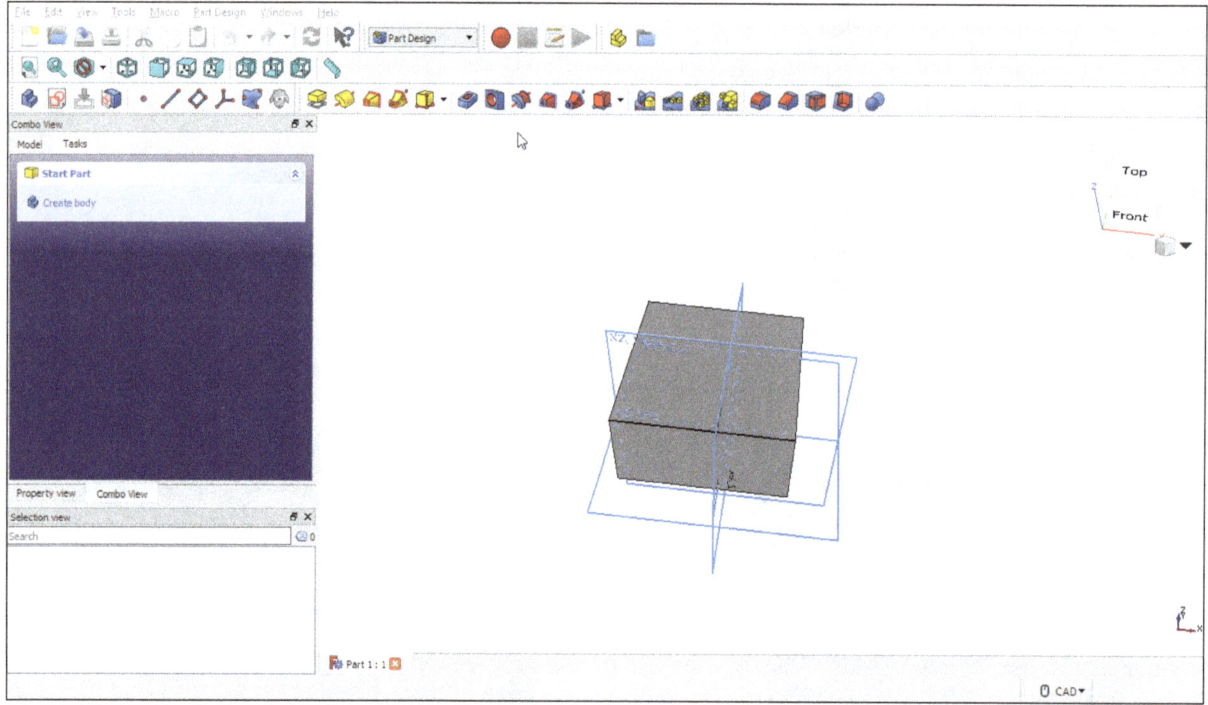

Figure-39. Fullscreen mode

STANDARD VIEWS

Standard views provide tools to orient the part, assembly, or sketch in one of the preset standard views. The standard views display the model or drawing through one, two, or four viewports.

Various tools available in the **Standard views** cascading menu from the **View** menu; refer to Figure-40, are discussed next.

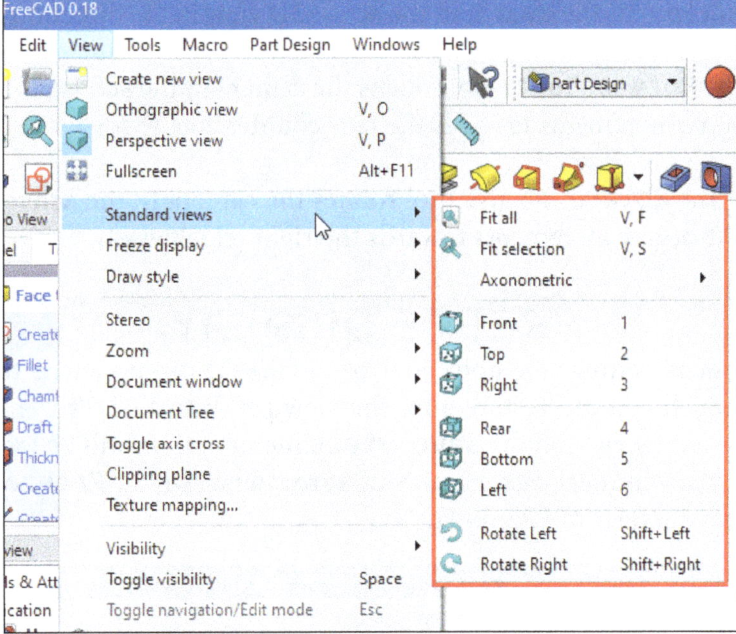

Figure-40. Standard views cascading menu

- **`Fit all`** : The **`Fit all`** tool zooms and pans the camera so that all visible objects fit inside the active 3D view.

- **`Fit selection`** : The **`Fit selection`** tool zooms and pans the camera so that all selected objects fit inside the active 3D view.

- **`Axonometric`** : The **`Axonometric`** cascading menu has three view tools as;

`Isometric View` : The **`Isometric`** tool realigns the camera in the active 3D view to obtain an isometric view. For a truly isometric view, the 3D view must be in orthographic mode but the tool also works, if the view is in perspective mode.

`Dimetric View` : The **`Dimetric`** tool realigns the camera in the active 3D view to obtain a dimetric view.

`Trimetric View` : The **`Trimetric`** tool realigns the camera in the active 3D view to obtain a trimetric view.

- **`Front`** : The **`Front`** tool points the camera in the active 3D view in the direction of the positive Y axis.

- **`Top`** : The **`Top`** tool points the camera in the active 3D view in the direction of the negative Z axis.

- **`Right`** : The **`Right`** tool points the camera in the active 3D view in the direction of the negative X axis.

- **`Rear`** : The **`Rear`** tool points the camera in the active 3D view in the direction of the negative Y axis.

- **`Bottom`** : The **`Bottom`** tool points the camera in the active 3D view in the direction of the positive Z axis.

- **Left** : The **Left** tool points the camera in the active 3D view in the direction of the positive X axis.

- **Rotate Left** : The **Rotate Left** tool rotates the camera in the active 3D view around the view direction in 90-degree increments towards the left (counterclockwise).

- **Rotate Right** : The **Rotate Right** tool rotates the camera in the active 3D view around the view direction in 90-degree increments towards the right (clockwise).

FREEZE DISPLAY

FreeCAD can store camera settings for upto 50 'frozen views'. Frozen view is the view with current specified orientation which you can restore after the view get disturbed. Frozen views are not stored in the document and if not saved with the **Save views** menu option, will be lost when the FreeCAD application closes. The tools that deal with frozen views are available in the **Freeze display** cascading menu from the **View** menu; refer to Figure-41.

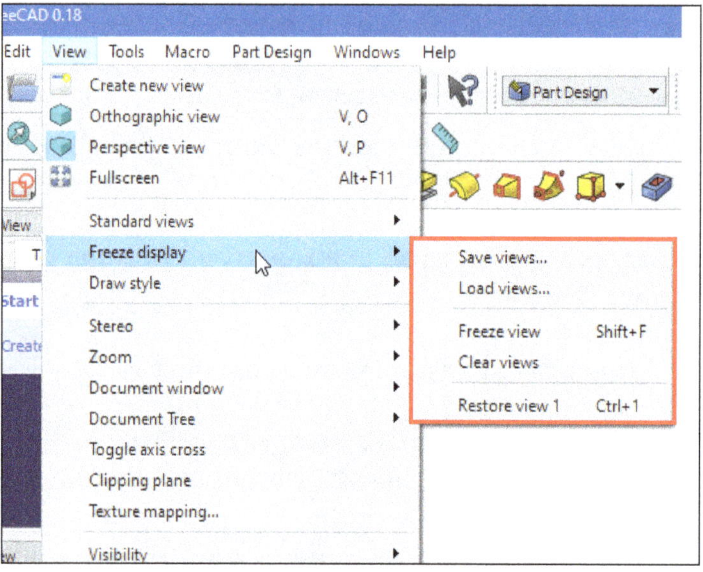

Figure-41. Freeze display cascading menu

Save views

The **Save views** tool saves all existing frozen views in a file with the ***.cam** extension. To use this tool, one or more frozen views must exist. A frozen view can be created with the **Freeze view** menu option which will be discussed later. The procedure to use this tool is discussed next.

- Click on the **Save views** tool from the **Freeze display** cascading menu. The **Save frozen views** dialog box will be displayed; refer to Figure-42.

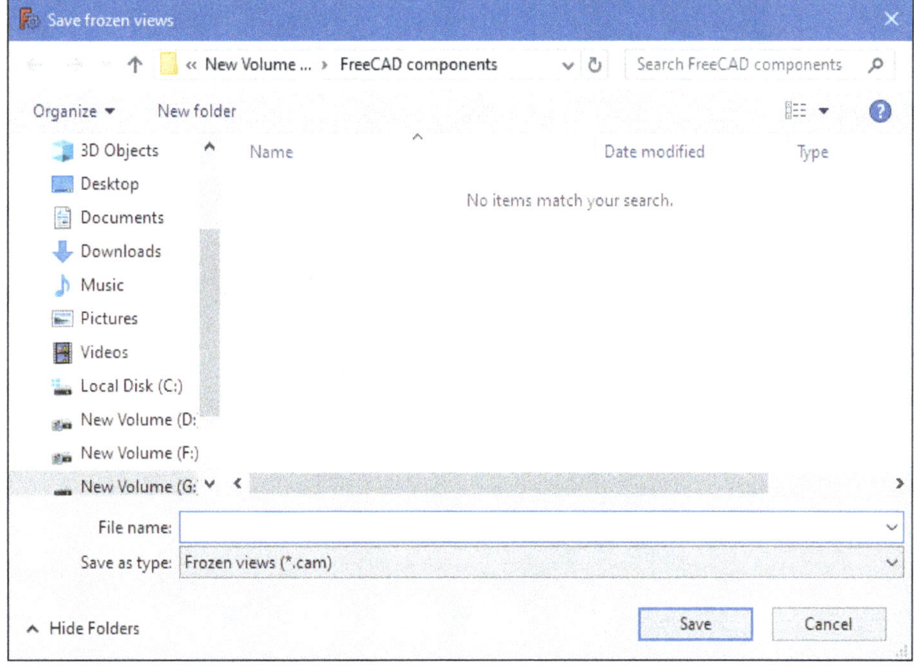

Figure-42. Save frozen views dialog box

- Specify desired name for the file in the **File name** edit box and click on **Save** button from the dialog box to save the frozen view.

Load views

The **Load views** tool loads the frozen views from a file with the ***.cam** extension. Note that when you load a frozen view then all existing frozen views will be deleted. The procedure to use this tool is discussed next.

- Click on the **Load views** tool from the **Freeze display** cascading menu.
- If the frozen views are saved earlier then the **Restore views** dialog box will be displayed; refer to Figure-43, asking you to confirm you want to lose all existing frozen views.
- Click on the **Yes** button from the dialog box, the **Restore frozen views** dialog box will be displayed; refer to Figure-44.
- If the frozen views are not saved earlier then also the **Restore frozen views** dialog box will be displayed.

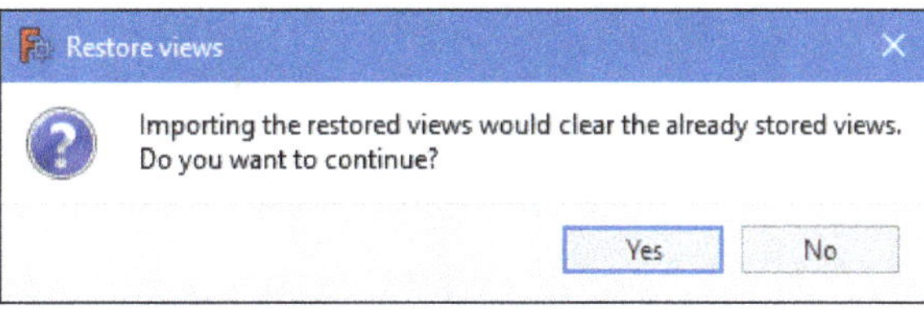

Figure-43. Restore views dialog box

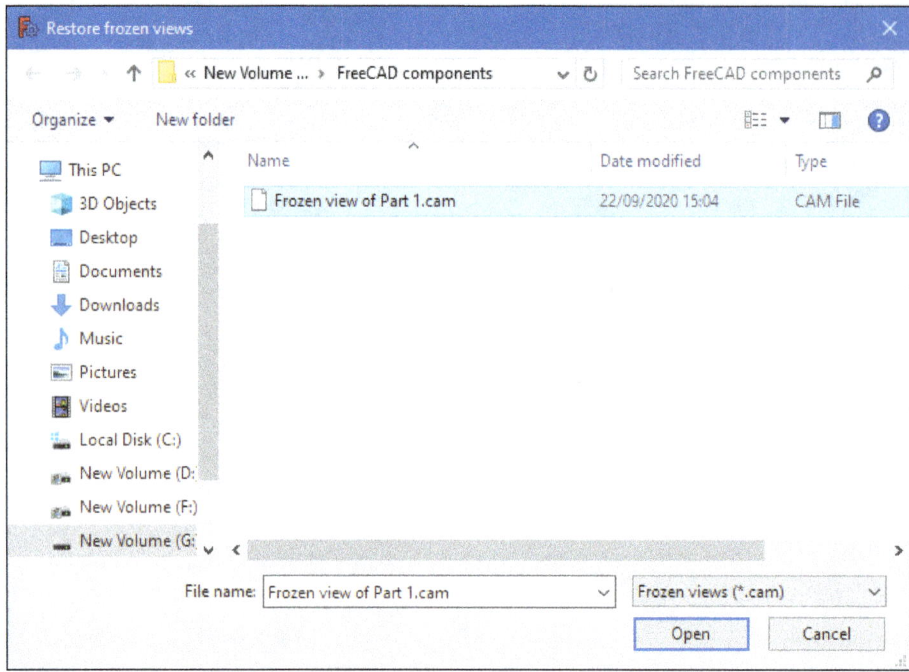

Figure-44. Restore frozen views dialog box

- Select the file which you want to restore and click on **Open** button from the dialog box.

Freeze view

The **Freeze view** tool stores the current camera settings in a new frozen view. You can create maximum 50 frozen views. The procedure to use this tool is discussed next.

- Click on the **Freeze view** tool from **Freeze display** cascading menu. The frozen view of the model will be created; refer to Figure-45.

Figure-45. Frozen view of the model

Clear views

The **Clear views** tool deletes all existing frozen views. Click on the **Clear views** tool from **Freeze display** cascading menu, all the existing frozen views will be deleted.

Restore view

For each frozen view, a **Restore view** option is added with which the frozen view can be restored. The procedure to use this tool is discussed next.

- Click on the **Restore view** tool from **Freeze display** cascading menu, the frozen view of the model will be restored; refer to Figure-46.

Figure-46. Restored frozen view of the model

DRAW STYLE

The **Draw style** tool is used to override the effect of the view display mode property of objects in a 3D view.

The tools available in the **Draw style** cascading menu of the **View** menu; refer to Figure-47, are used to display the model in different draw styles as shown in Figure-48.

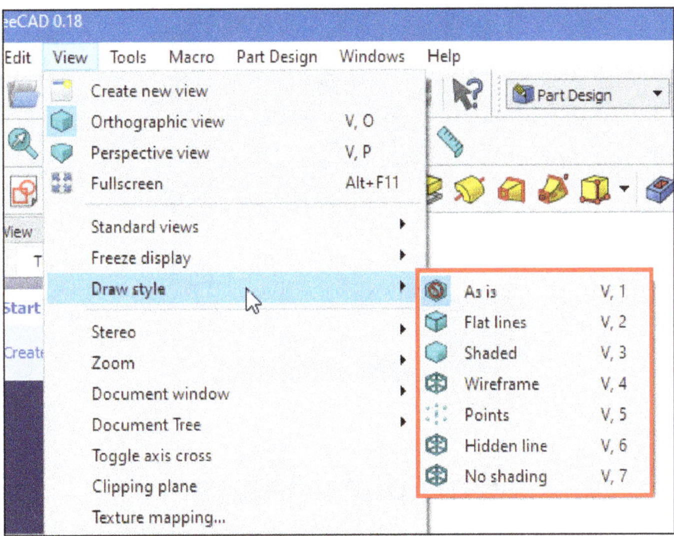

Figure-47. Draw style cascading menu

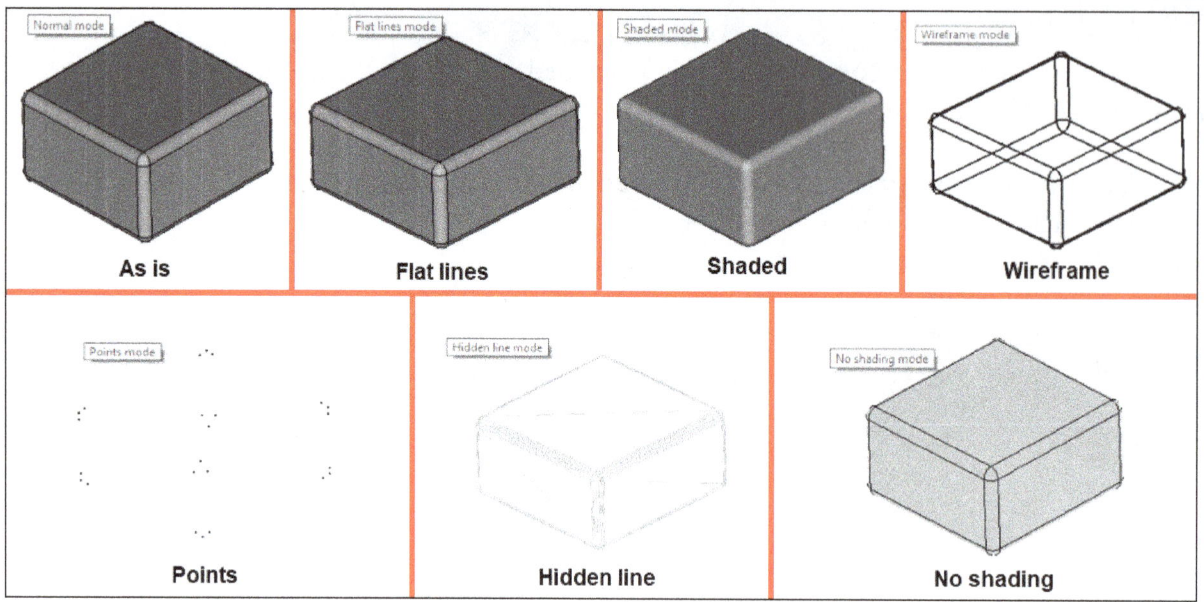

Figure-48. Various type of draw styles

You will learn about other tools of **View** menu later in this book.

TOOLS MENU

The **Tools** menu provides tools for debugging models, customizing FreeCAD behaviour as well as auxiliary tools; refer to Figure-49. Various tools available in the **Tools** menu are discussed next.

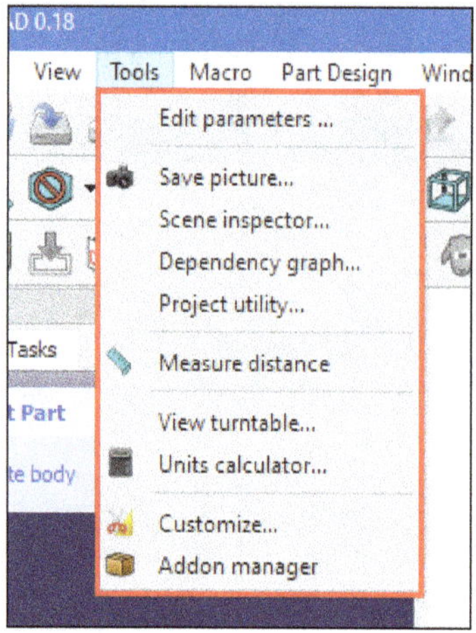

Figure-49. Tools menu

Edit parameters

The **Edit parameters** tool gives access to the parameters that control the program. By clicking on this tool, the **Parameter Editor** dialog box will be displayed as shown in Figure-50. The **Parameter Editor** controls the behaviour of FreeCAD and its workbenches. The parameters are stored in a file called **user.cfg**.

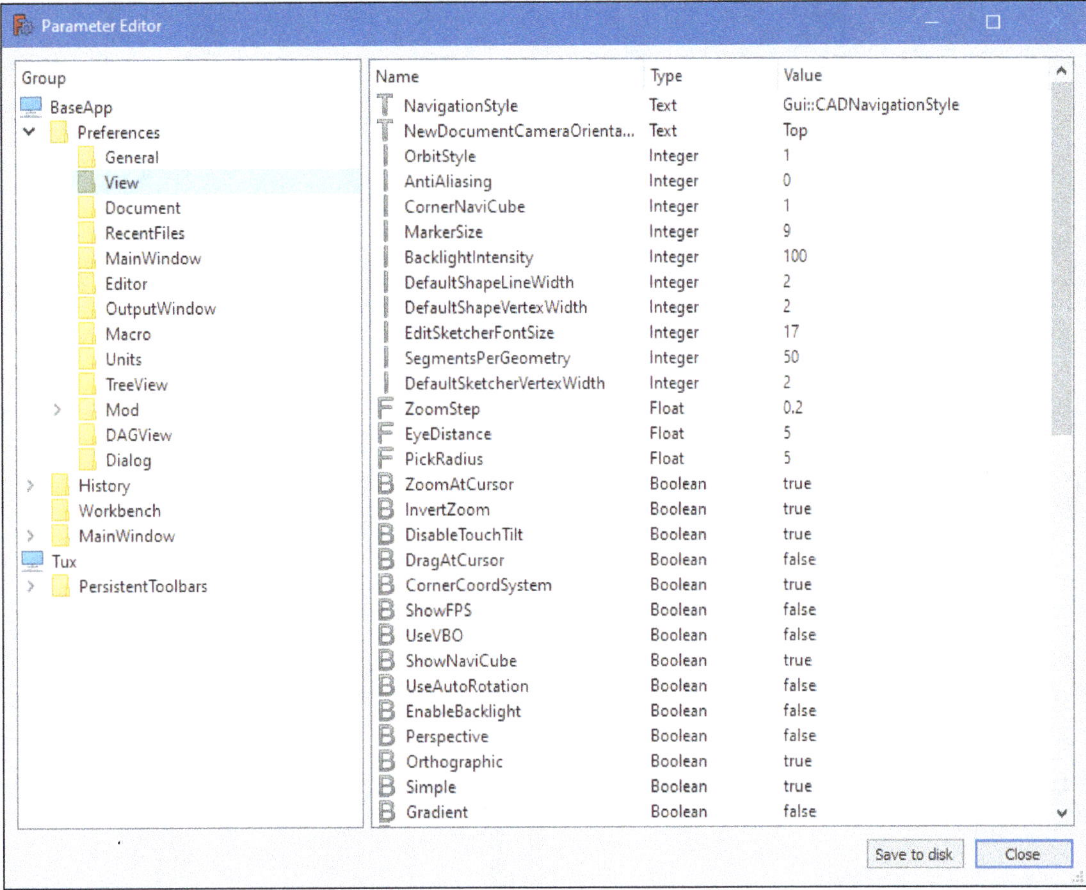

Figure-50. Parameter Editor dialog box

- Specify desired parameters and click on **Save to disk** button to update the **user.cfg** file.
- Click on the **Close** button to close the dialog box. You will learn more about the parameters in this dialog box later whenever needed.

Save picture

The **Save picture** tool is used to create an image file or a screenshot from the active 3D view. The procedure to use this tool is discussed next.

- Click on the **Save picture** tool from **Tools** menu. The **Save picture** dialog box will be displayed; refer to Figure-51.

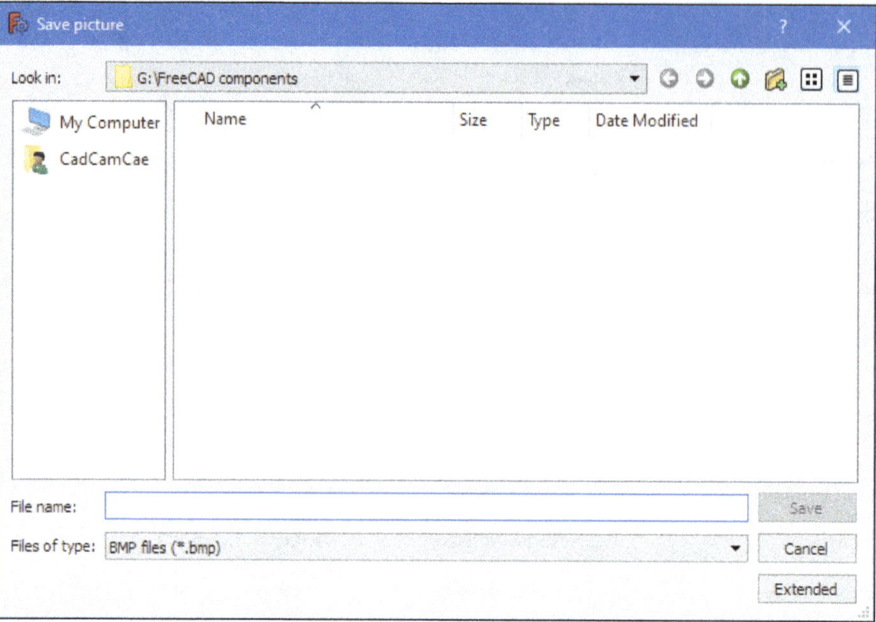

Figure-51. Save picture dialog box

- Click on the **Extended** button from the dialog box. An additional panel will be added as shown in Figure-52.

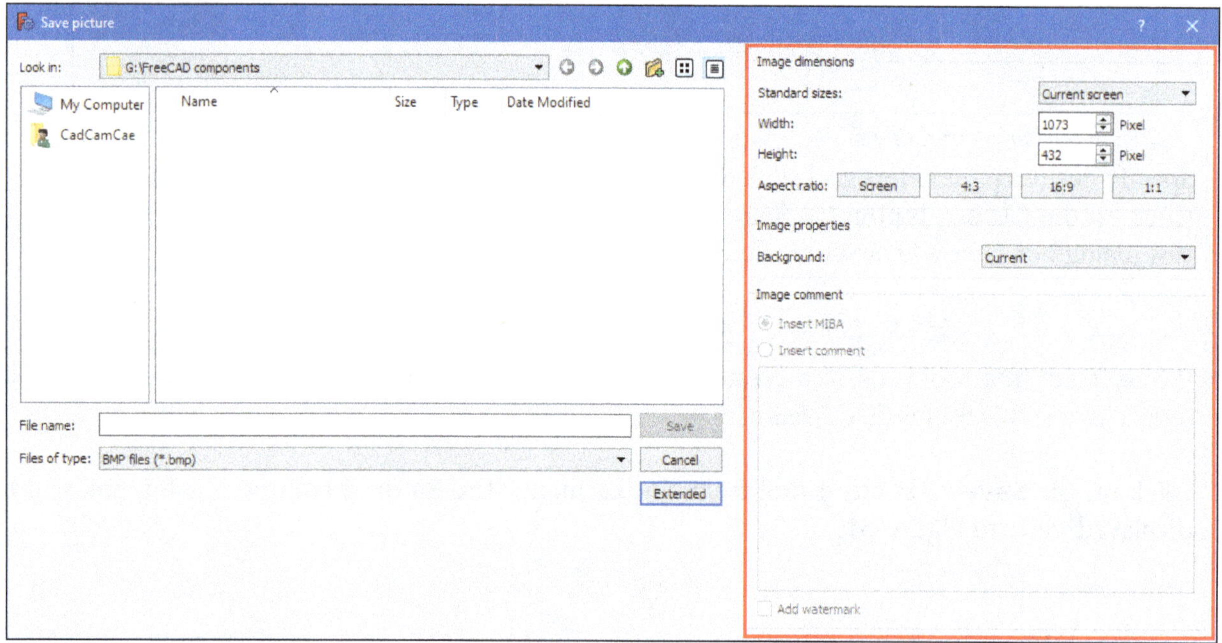

Figure-52. Expanded view of Save picture dialog box

- Select standard size for the image from **Standard sizes** drop-down list or specify width and height of the image in the **Width** and **Height** edit boxes respectively in the **Image dimensions** area of the dialog box.
- Select desired **Aspect ratio** button to set the width to height ratio of the image.
- Select desired option from **Background** drop down list in the **Image properties** area of the dialog box to set background for the image.
- Select **Insert MIBA** radio button from **Image comment** area to add MIBA information to the file. The options in the **Image comment** area of the dialog box will be activate only on selecting **JPG**, **JPEG**, and **PNG** image file types.
- Select **Insert comment** radio button to type a comment in the text field to embed a comment in the file.

- Select **Add watermark** check box to add a watermark of the FreeCAD logo in the image file.
- Specify desired file name and click on **Save** button from the dialog box.

Scene inspector

The **Scene inspector** tool display an overview of all nodes in the scenegraph of the active 3D view. It is more a utility for programmers than for average users. It can be used to find out why the rendering is slow or why something is not rendered properly.
The procedure to use this tool is discussed next.

- Click on the **Scene inspector** tool from **Tools** menu. The **Scene Inspector** dialog box will be displayed as shown in Figure-53.

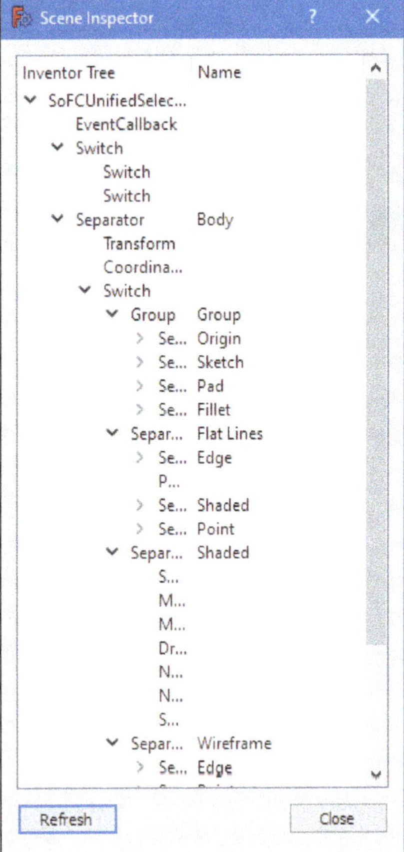

Figure-53. Scene Inspector dialog box

- This dialog box is modeless, it means it can stay open while you continue working in FreeCAD.
- Click on the **Refresh** button to update the overview or click on **Close** button to close the dialog box.

Dependency Graph

The **Dependency Graph** tool displays dependencies between objects of the active document in a dependency graph. In this graph, objects are listed in reverse chronological order, with the first created object at the bottom.
The dependency graph is purely a visualization tool therefore it cannot be edited. It automatically updates, if changes are made to the model.

Note that to use this tool, a third party software named **Graphviz** needs to be installed.

If the software is not pre-installed, install it after downloading from the **Graphviz** website. If the software is installed in an unconventional location, FreeCAD will ask you for software installation path.

The procedure to use this tool is discussed next.

- Click on the **Dependency graph** tool from **Tools** menu. The **Graphviz not found** dialog box will be displayed, if **Graphviz** is not installed in standard location; refer to Figure-54.

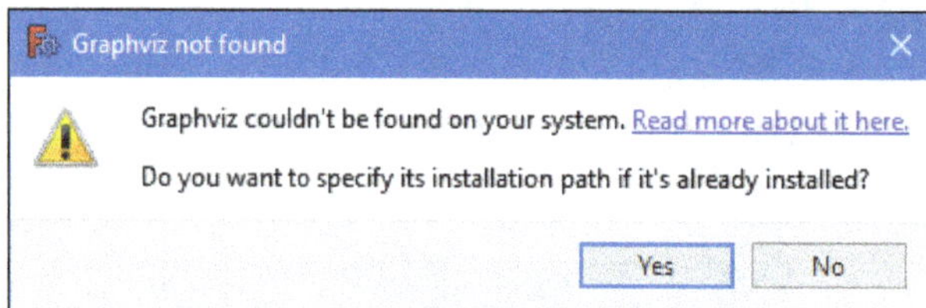

Figure-54. Graphviz not found dialog box

- Click on **Yes** button from the dialog box. The **Graphviz installation path** dialog box will be displayed asking for software installation folder; refer to Figure-55.

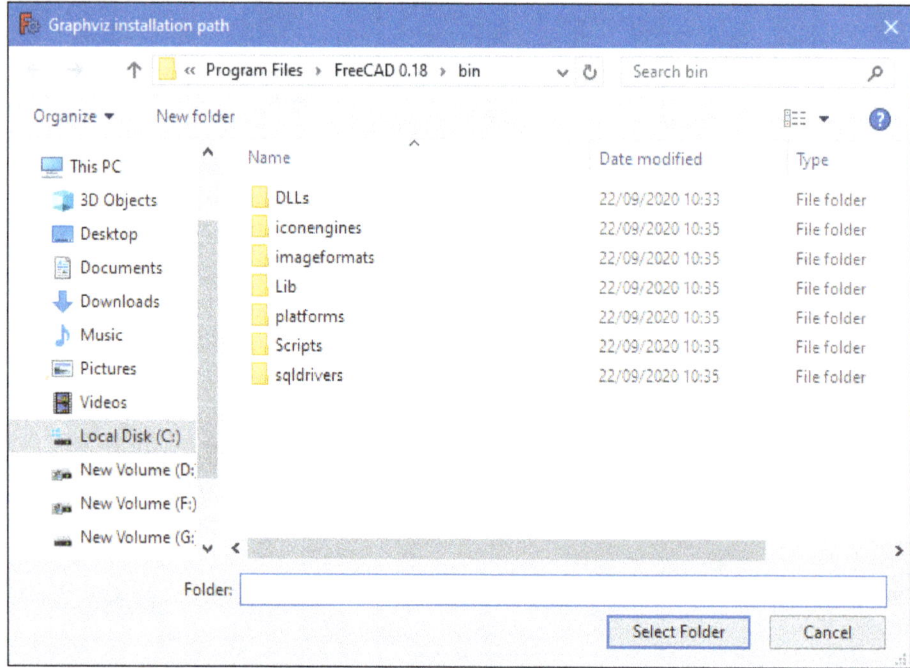

Figure-55. Graphviz installation path dialog box

- Browse to desired folder in your system and click on **Select Folder** button from the dialog box. A new window named **Dependency graph** will be opened displaying dependencies between objects; refer to Figure-56.

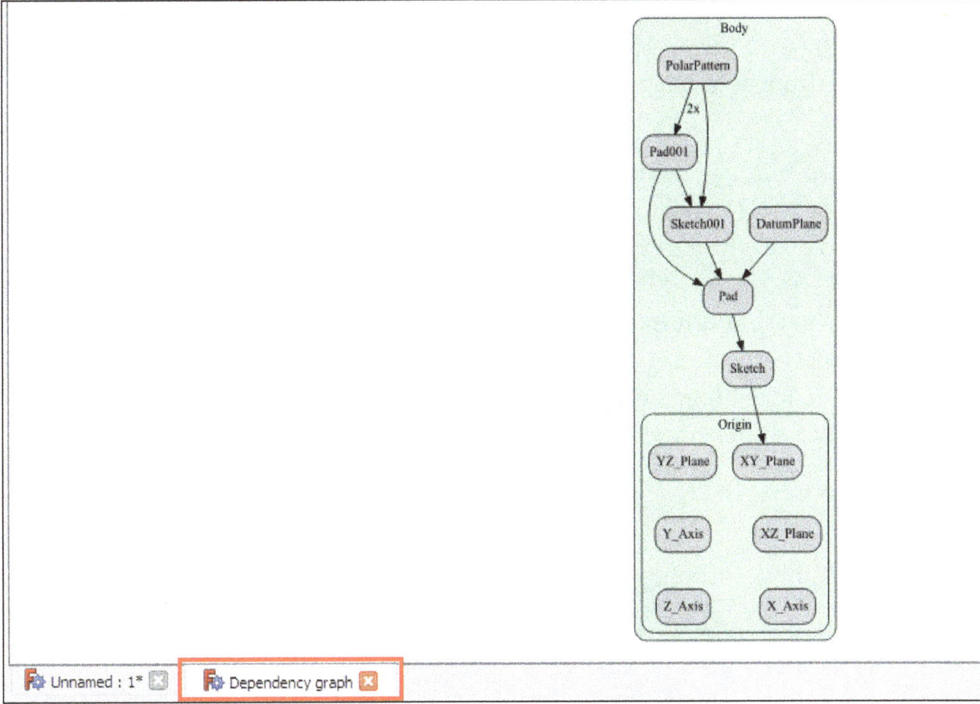

Figure-56. Dependency graph window

Project utility

The **Project utility** tool assists in repairing damaged project files.

By using this tool, you can extract files from a FreeCAD project file **(*.FCStd)**, which is in fact a **ZIP** file and after manual edits, create a new project file from them.

The procedure to use this tool is discussed next.

- Click on **Project utility** tool from **Tools** menu. The **Project utility** dialog box will be displayed; refer to Figure-57.

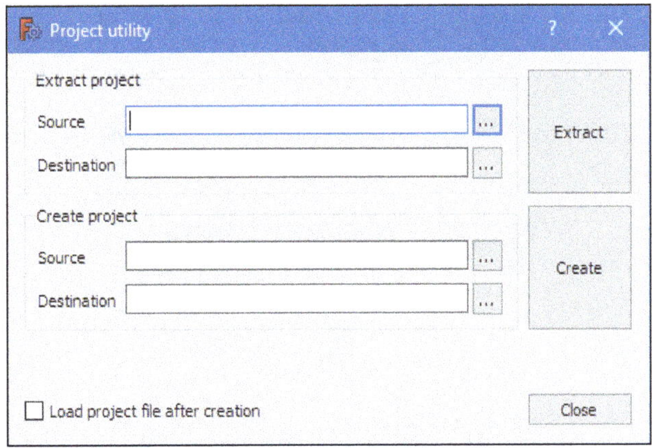

Figure-57. Project utility dialog box

- Click on the **Browse** button for **Source** edit box to define location of source file for extracting/creating the project in the **Extract project/Create project** area, respectively.
- Similarly, click on the **Browse** button for **Destination** edit box to specify the destination location for extracting/creating the project in the **Extract project/Create project** area, respectively.

- Click on the **Extract** button to create zip file and click on the **Create** button to save project file.
- Select the **Load project file after creation** check box to open the created file.
- Click on the **Close** button to close the dialog box.

Measure distance

The **Measure distance** tool is used to create a distance object that measures and displays the distance between two points. The procedure to use this tool is discussed next.

- Click on the **Measure distance** tool from **Tools** menu. The measurement scale symbol will get attached to the cursor.
- Select the first and second dimension points of object to measure the distance between the points. The measured distance dimension will be displayed; refer to Figure-58.

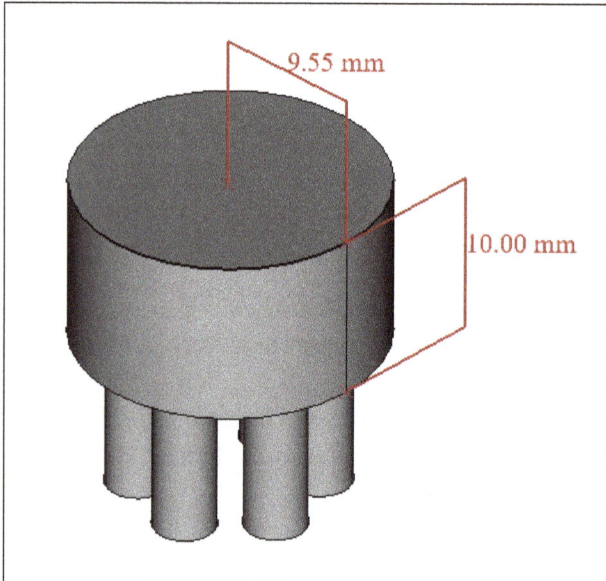

Figure-58. Dimension created on the object

View turntable

The **View turntable** tool continuously rotates the camera in a desired angle and speed for desired time. The procedure to use this tool is discussed next.

- Click on **View turntable** tool from **Tools** menu. The **View Turntable** dialog box will be displayed; refer to Figure-59.

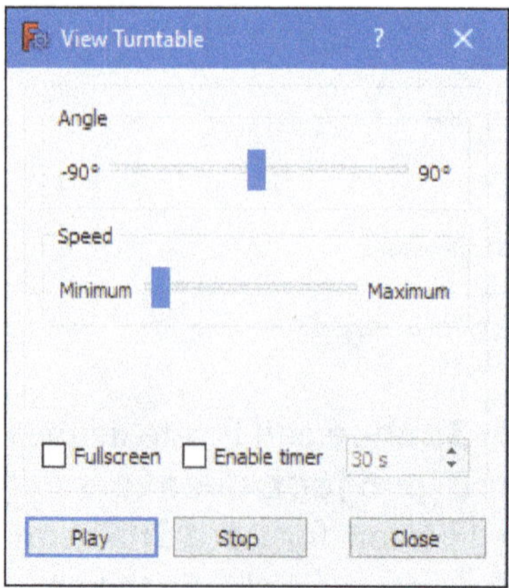

Figure-59. View Turntable dialog box

- Move the **Angle** slider to specify the angle on which you want to rotate the model.
- Move the **Speed** slider to adjust rotation speed of the model.
- Select **Fullscreen** check box to switch 3D view to fullscreen mode.
- Select the **Enable timer** check box to specify the duration of rotation.
- Click on **Play** button to start rotating the model and click on **Stop** button to stop the rotation.
- Click on **Close** button to close the dialog box.

Units calculator

The **Units calculator** tool can be used to convert values from one unit system to another. The procedure to use this tool is discussed next.

- Click on **Units calculator** tool from **Tools** menu. The **Units calculator** dialog box will be displayed; refer to Figure-60. This dialog box is modeless, it means it can stay open while you continue working in FreeCAD.

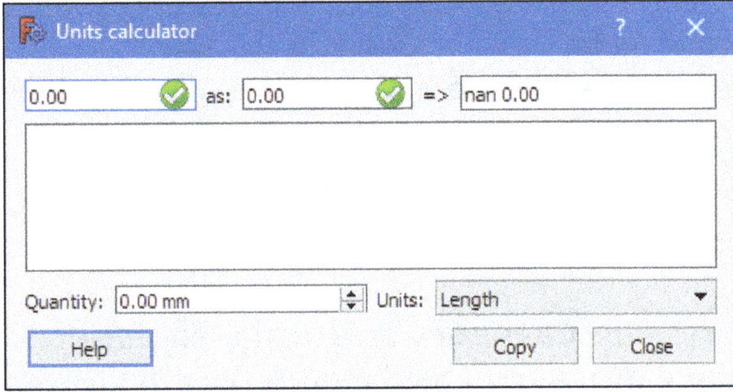

Figure-60. Units calculator dialog box

- Specify desired values in the first two edit boxes to perform conversion.
- In first edit box, specify the value to be converted and in the **as:** edit box, specify the unit value to be converted.
- Result will be displayed in **=>** edit box. For example, if you want to convert **55** mm to inch then specify **55** in first edit box and **25.4** in the **as:** edit box; refer to Figure-61.

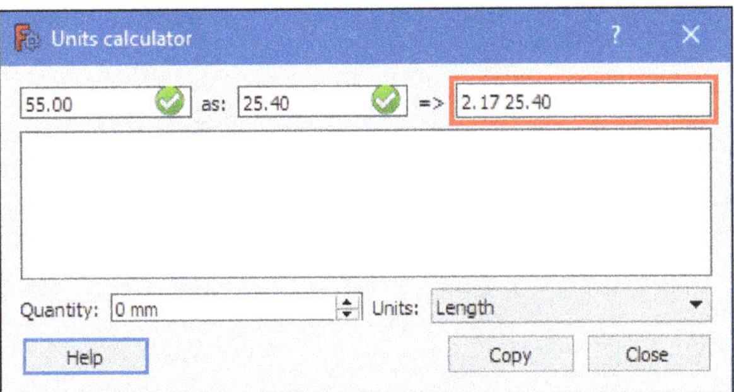

Figure-61. Result of conversion

Customize

Customize tool gives access to several customization options. The customization depends on the workbenches that have been loaded in the current FreeCAD session. The procedure to use this tool is discussed next.

- Click on the **Customize** tool from **Tools** menu. The **Customize** dialog box will be displayed; refer to Figure-62.

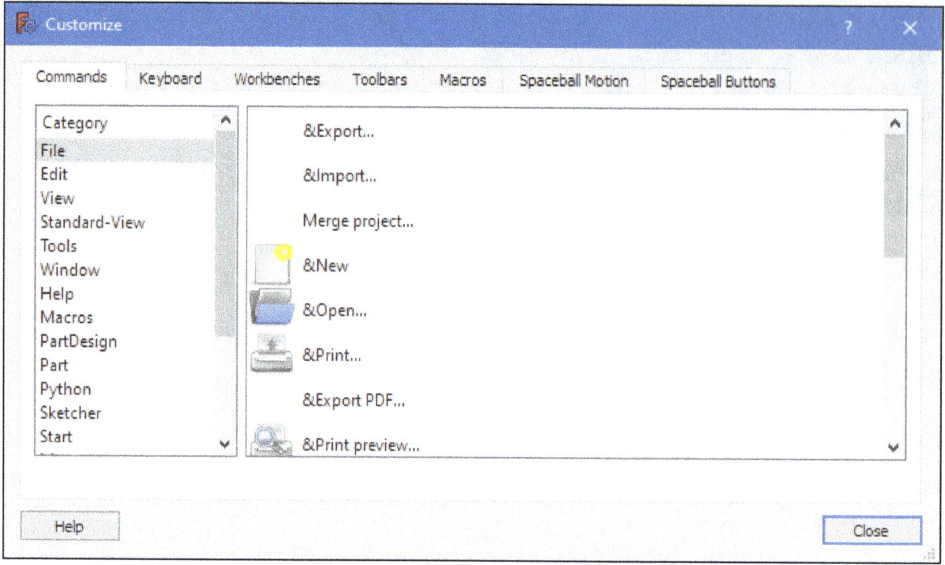

Figure-62. Customize dialog box

The options in various tabs of the **Customize** dialog box are discussed next.

Commands tab

In the **Commands** tab, you can browse all the available commands or tools in FreeCAD.

- Select a command category in the **Category** area on the left side of dialog box. The tools available in the selected category are shown in area on the right; refer to Figure-63.
- Hovering cursor on a tool will display its tooltip.
- Select desired tool to check information about it in the status bar of dialog box.

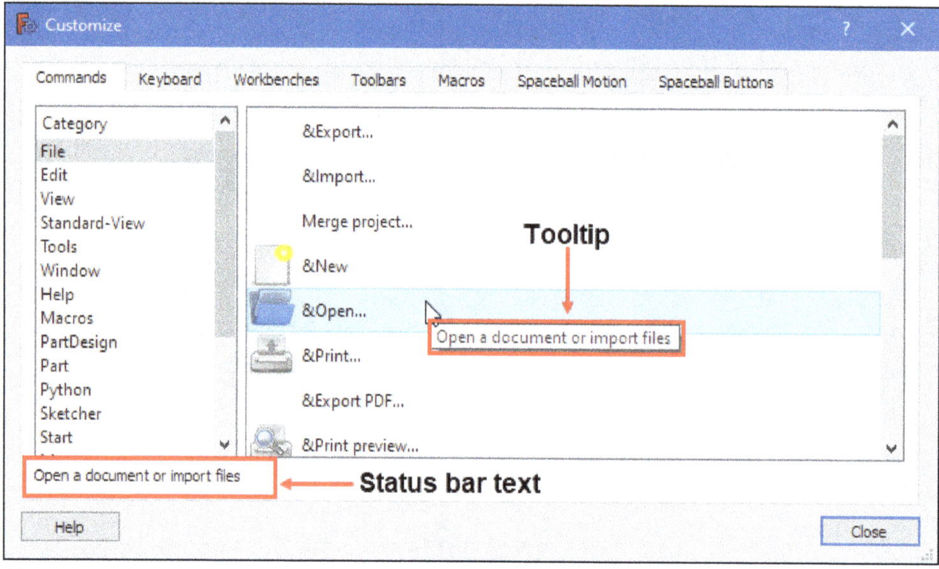

Figure-63. Commands tab

- Click on the **Close** button to exit the dialog box.

Keyboard tab

Using the options in **Keyboard** tab, you can define custom keyboard shortcuts for using the tool. The **Keyboard** tab in **Customize** dialog box is shown in Figure-64.

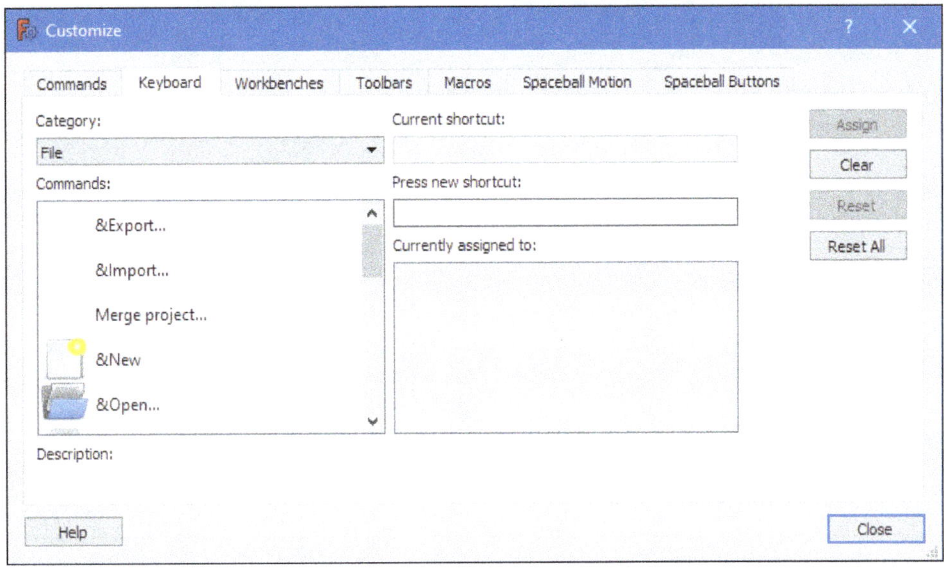

Figure-64. Keyboard tab

- Select desired command category from **Category** drop-down list. Commands of selected category will be displayed in the **Commands** area.
- Select desired command from the **Commands** area of the dialog box.
- The **Current shortcut** input box displays the pre-assigned shortcut for the selected tool.
- To specify a new shortcut for the tool, click in the **Press new shortcut** input box and press new shortcut keys from the keyboard. If the specified shortcut is already in use for other command then **Already defined shortcut** dialog box will be displayed as shown in Figure-65, asking you to override the shortcut.

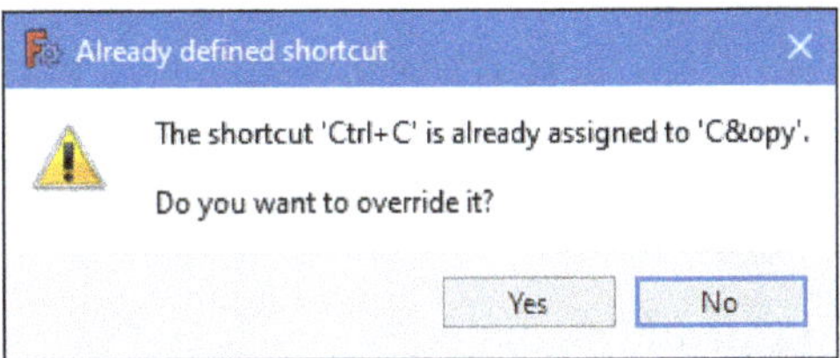

Figure-65. Already defined shortcut dialog box

- Click on **Yes** button from the **Already defined shortcut** dialog box and click on **Assign** button from **Customize** dialog box to assign the new shortcut.
- Click on **Clear** button to remove the specified shortcut for the selected command.
- Click on **Reset** button to remove a custom shortcut for the selected command.
- Click on **Reset All** button to remove all custom shortcuts specified for the commands.
- Click on the **Close** button to exit the dialog box.

Workbenches tab

In **Workbenches** tab, you can modify the workbench selector list. The **Enabled workbenches** list shows the workbenches as they will appear in the Workbench selector.

- To disable a workbench, select a workbench in the **Enabled workbenches** list and click on button. The workbench will be moved to the **Disabled workbenches** list; refer to Figure-66.

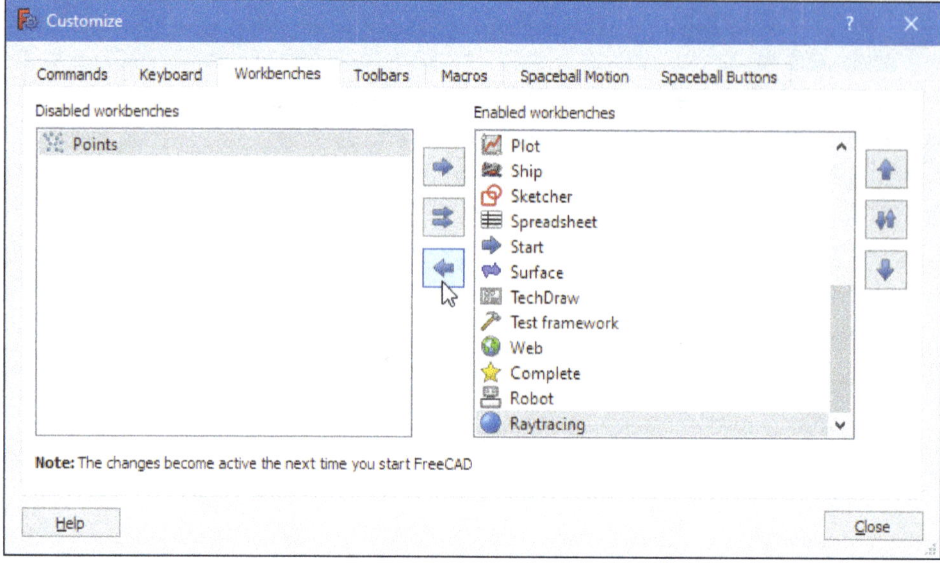

Figure-66. Workbenches tab

- To re-enable a workbench, select a workbench in the **Disabled workbenches** list and click on button, the workbench will be moved to the **Enabled workbenches** list.
- To re-enable all workbenches, click on button.
- Select a workbench in the **Enabled workbenches** list and click on button or button to change the selected workbench position.
- Click on button to sort all the workbenches, alphabetically.
- Click on the **Close** button to exit the dialog box.

Note that, the changes you made will become active after you restart the FreeCAD.

Toolbars tab

In the **Toolbars tab**, you can create custom toolbar and perform toolbar modification; refer to Figure-67.

- In the drop-down list in right side of the dialog box, select a workbench whose custom toolbar is to be modified.

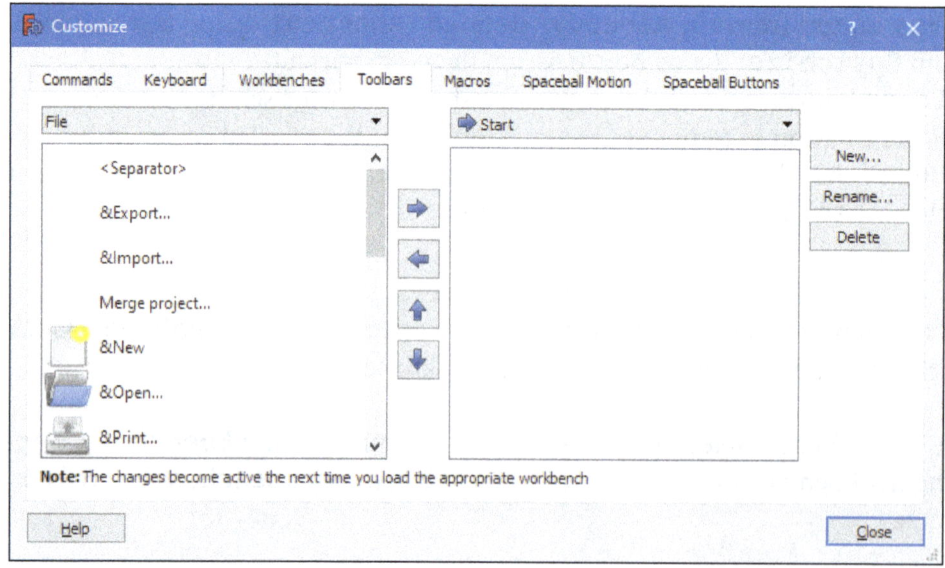

Figure-67. Toolbars tab

- Click on the **New** button from right side of the dialog box. The **New toolbar** dialog box will be displayed asking you to specify the toolbar name; refer to Figure-68.

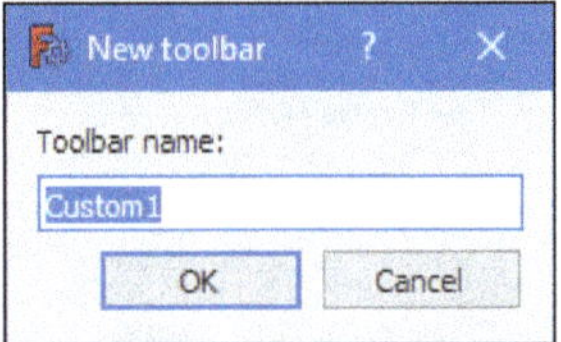

Figure-68. New toolbar dialog box

- Specify desired name in the **Toolbar name** edit box and click on **OK** button. The new toolbar will be created in the right area of the dialog box.
- Select the newly created toolbar from the right area to change its name and click on **Rename** button, the **Rename toolbar** dialog box will be displayed; refer to Figure-69.

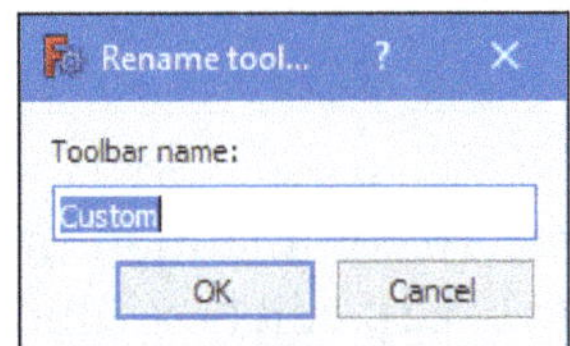

Figure-69. Rename toolbar dialog box

- Specify desired new name of the toolbar and click on **OK** button.
- If you want to delete the toolbar then select the toolbar from right area and click on **Delete** button.
- To disable a toolbar, deselect the checkbox in front of toolbar name in the right panel. The toolbar will be invisible in the FreeCAD interface.
- To add a command in the toolbar, select the created toolbar in the right area. If no toolbar is selected, the command will be added to the first toolbar available in the drop-down list in right side of the dialog box.
- Select desired command category from the drop-down list available in the left side of the dialog box. The list of related commands will be displayed.
- Select desired command from the left panel which you want to add to the toolbar and click on button.
- To remove a command, select it from the right panel and click on button.
- To change a command position, select it and click on button or button.
- Click on the **Close** button to exit the dialog box.

Macros tab

In the **Macros** tab, you can manage macros created earlier. FreeCAD uses a dedicated folder for user macros and only macros in that folder can be set up.

- Select desired macro from the **Macro** drop-down list. The parameters related to macro will be displayed; refer to Figure-70.

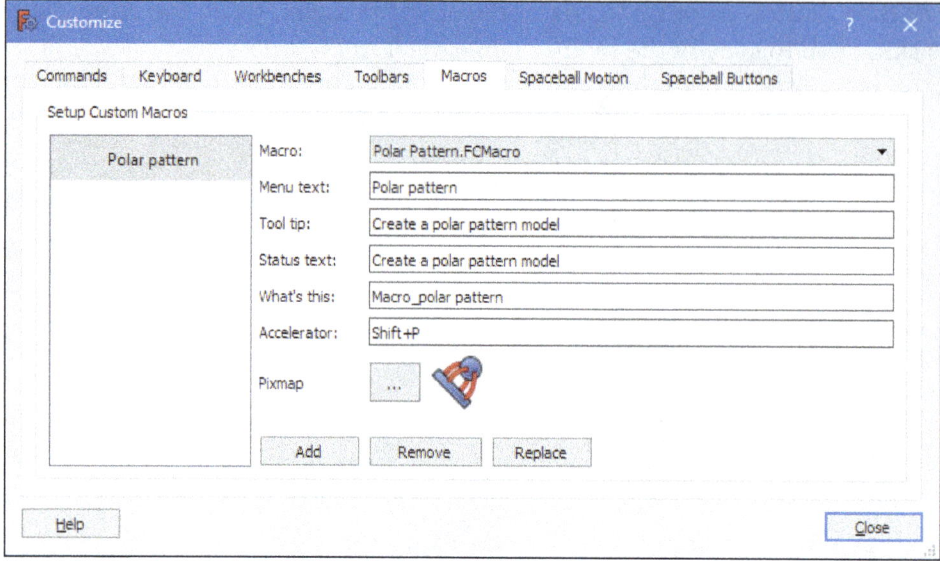

Figure-70. Macros tab

- Specify desired name to identify the macro command in the **Menu text** edit box.
- Specify a desired tooltip text in the **Tool tip** edit box which will display near the cursor when you hover the cursor on the macro icon.
- Specify desired text in the **Status text** edit box which will display in the status bar when you hover the cursor on the macro icon.
- Specify the wiki page name for the macro if available on the internet in the **What's this** edit box.
- Specify the keyboard shortcut for a macro in the **Accelerator** edit box.
- To add an icon for the macro command, click on **Pixmap** button. The **Choose Icon** dialog box will be displayed; refer to Figure-71.

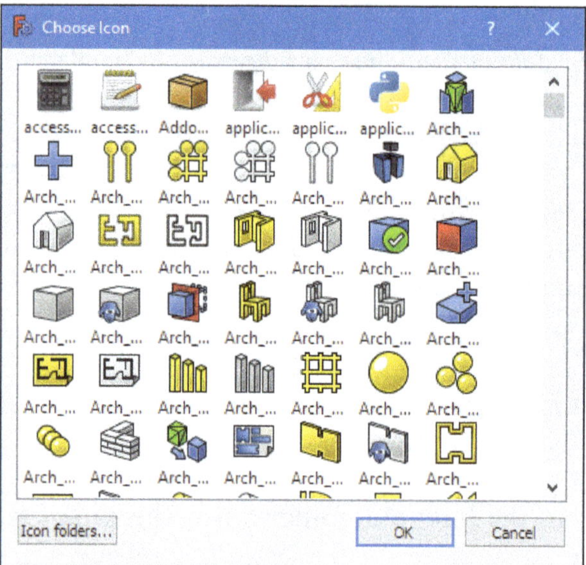

Figure-71. Choose Icon dialog box

- Select desired icon from the panel, the **Choose Icon** dialog box will be closed automatically. Click on **Icon folders** button from the **Choose Icon** dialog box if you want to choose an icon from the local drive.
- Click on the **Add** button from the **Customize** dialog box to add modified macro in the macro list available for document.
- To remove a macro, select it from the left area of the dialog box and click on **Remove** button.
- To change the macro command, double-click on the command from the left panel. Make desired changes and click on **Replace** button.

- Click on the **Close** button to exit the dialog box.

Similarly, you can use the **Spaceball Motion** and **Spaceball buttons** tabs to customize functions of mouse with spaceball feature.

Addon Manager

With the **Addon Manager** tool, you can install and manage external workbenches and macros provided by the FreeCAD community.

- Click on **Addon Manager** tool from **Tools** menu. The **Addon manager** dialog box will be displayed; refer to Figure-72.

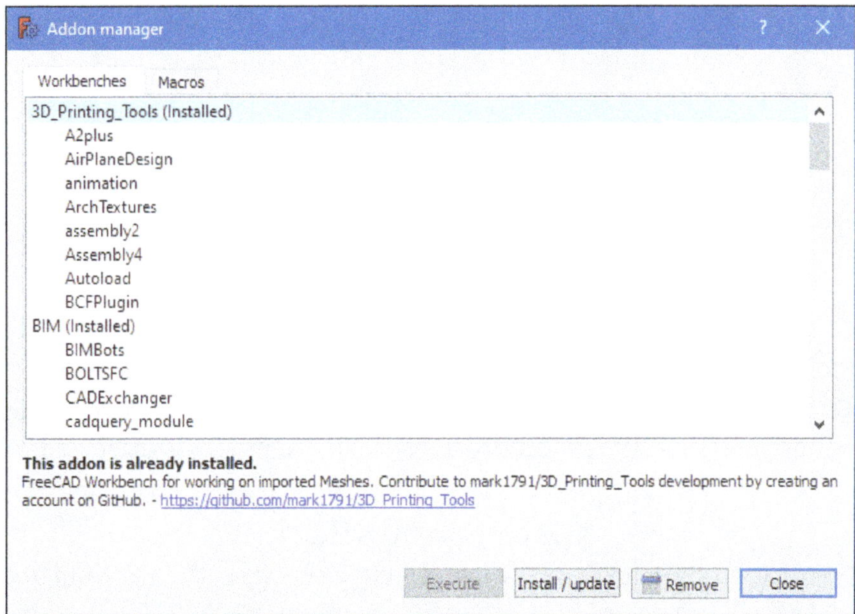

Figure-72. Addon manager dialog box

Note that you need an internet connection to download the addon resources.

- Select desired addon which you want to install from the list available in **Workbenches** tab.
- Click on **Install/update** button to install the addon.
- To uninstall the addon, select the addon from the list which you want to uninstall and click on **Remove** button from the dialog box.
- If you have installed or updated a workbench then by clicking on **Close** button, a new **Addon manager** dialog box will open informing you to restart FreeCAD for the changes to take effect.
- Click on **OK** button from the dialog box and restart FreeCAD.

NAVIGATING IN THE 3D VIEW

Navigating in the FreeCAD 3D view can be done with a mouse, a Space Navigator device, the keyboard, a touchpad, or a combination of these. FreeCAD can use several navigation modes which determine how the three basic view manipulation operations (pan, rotate, and zoom) are done as well as how to select objects on the screen. Navigation modes are accessed from the **Preferences screen** or directly by right-clicking anywhere on the 3D view; refer to Figure-73.

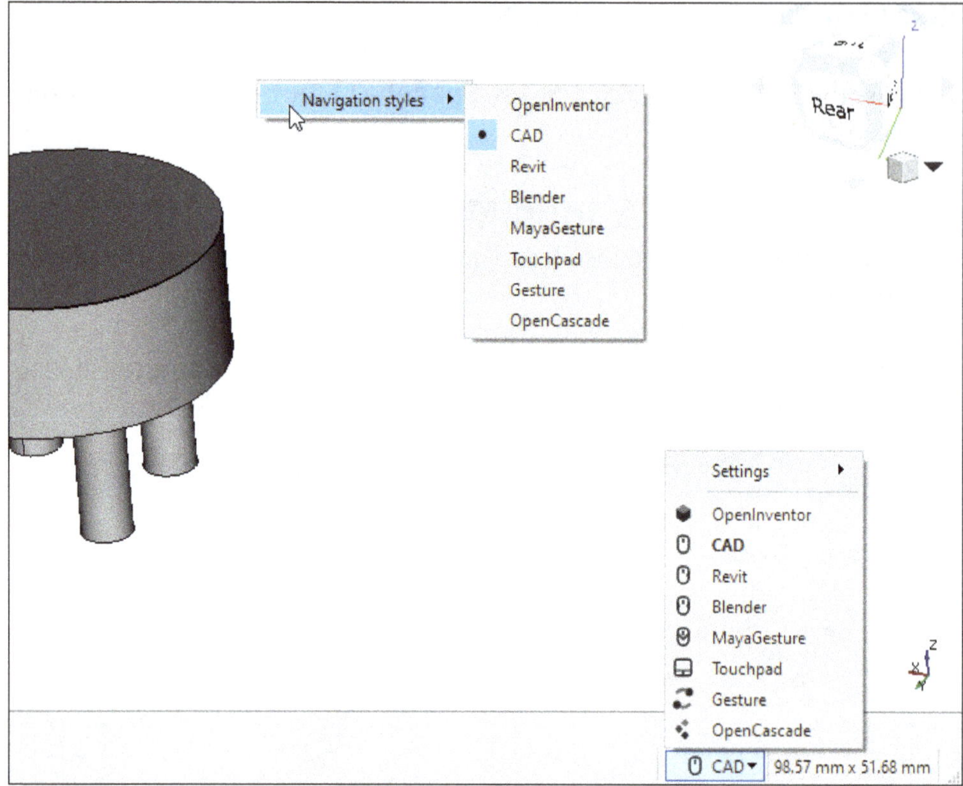

Figure-73. Navigation modes in 3D view

Each of these modes attributes different mouse buttons, mouse+keyboard combinations, or mouse gestures to these four operations. The procedure to use all these modes is discussed next.

CAD Navigation

This is the default navigation style. It allows the user a simple control of the view and does not require the use of keyboard keys except to make multi-selections. The mouse gestures used for object manipulation in **CAD** navigation style are shown in Figure-74.

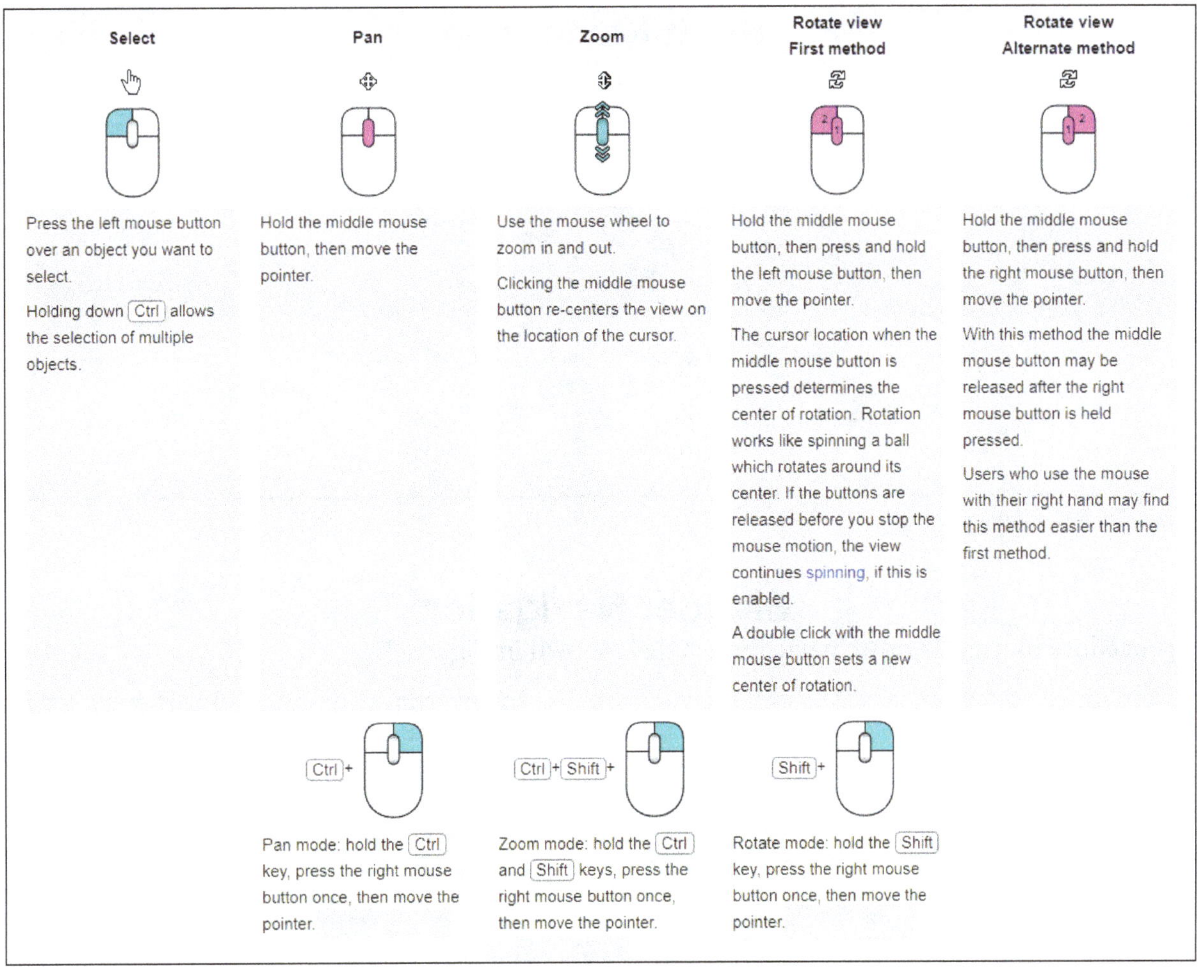

Select
Pan
Zoom
Rotate view
First method
Rotate view
Alternate method
Press the left mouse button over an object you want to select.
Holding down Ctrl allows the selection of multiple objects.
Hold the middle mouse button, then move the pointer.
Use the mouse wheel to zoom in and out.
Clicking the middle mouse button re-centers the view on the location of the cursor.
Hold the middle mouse button, then press and hold the left mouse button, then move the pointer.
The cursor location when the middle mouse button is pressed determines the center of rotation. Rotation works like spinning a ball which rotates around its center. If the buttons are released before you stop the mouse motion, the view continues spinning, if this is enabled.
A double click with the middle mouse button sets a new center of rotation.
Hold the middle mouse button, then press and hold the right mouse button, then move the pointer.
With this method the middle mouse button may be released after the right mouse button is held pressed.
Users who use the mouse with their right hand may find this method easier than the first method.
Ctrl +
Ctrl + Shift +
Shift +
Pan mode: hold the Ctrl key, press the right mouse button once, then move the pointer.
Zoom mode: hold the Ctrl and Shift keys, press the right mouse button once, then move the pointer.
Rotate mode: hold the Shift key, press the right mouse button once, then move the pointer.

Figure-74. CAD navigation style

OpenInventor Navigation

In **OpenInventor** navigation, you must hold down the **Ctrl** key in order to select objects. The procedure to use this navigation style is shown in Figure-75.

Select
Pan
Zoom
Rotate view
Ctrl +
Hold Ctrl, then press the left mouse button over an object you want to select.
Hold the middle mouse button, then move the pointer.
Use the mouse wheel to zoom in and out.
Alternatively, hold the middle mouse button, then press and hold the left mouse button, then move the pointer.
Hold the left mouse button, then move the pointer.

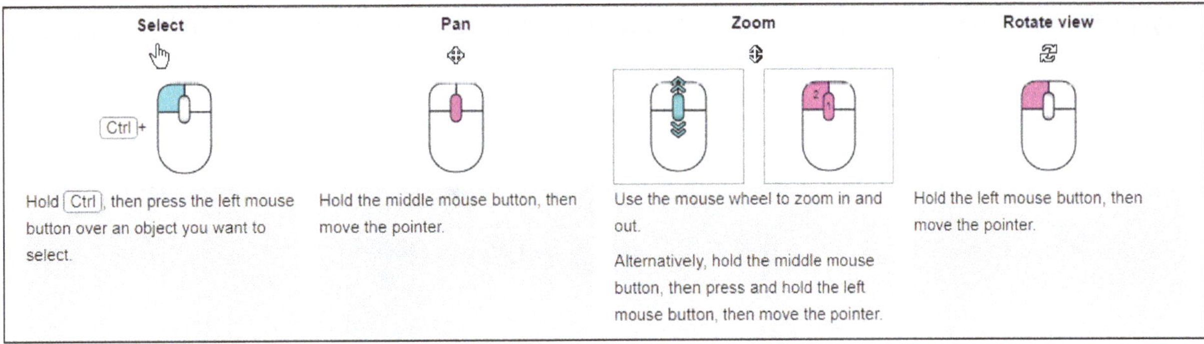

Figure-75. OpenInventor navigation style

Revit Navigation

The procedure to use **Revit** navigation style is shown in Figure-76.

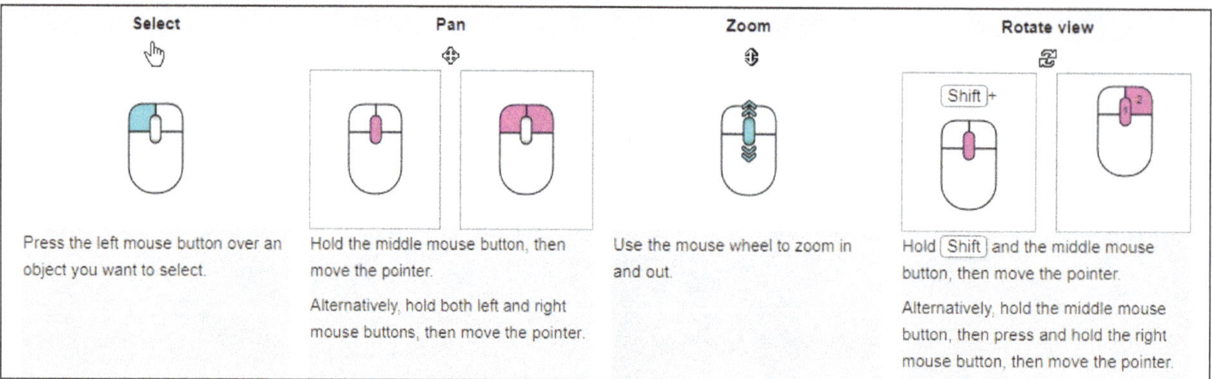

Figure-76. Revit navigation style

Blender Navigation

The procedure to use **Blender** navigation style is shown in Figure-77.

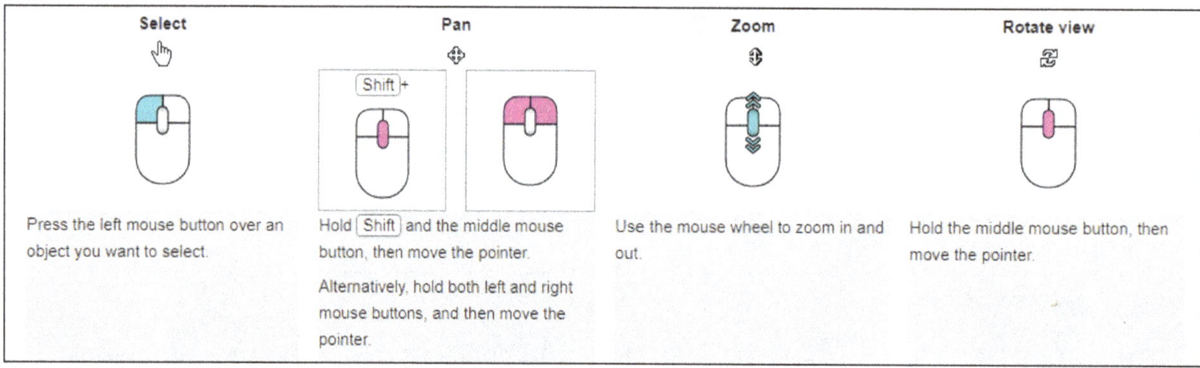

Figure-77. Blender navigation style

MayaGesture Navigation

In **Maya-Gesture** navigation, panning, zooming, and rotating the view require the **Alt** key together with a mouse button; therefore, a three-button mouse is required. The procedure to use this navigation is shown in Figure-78.

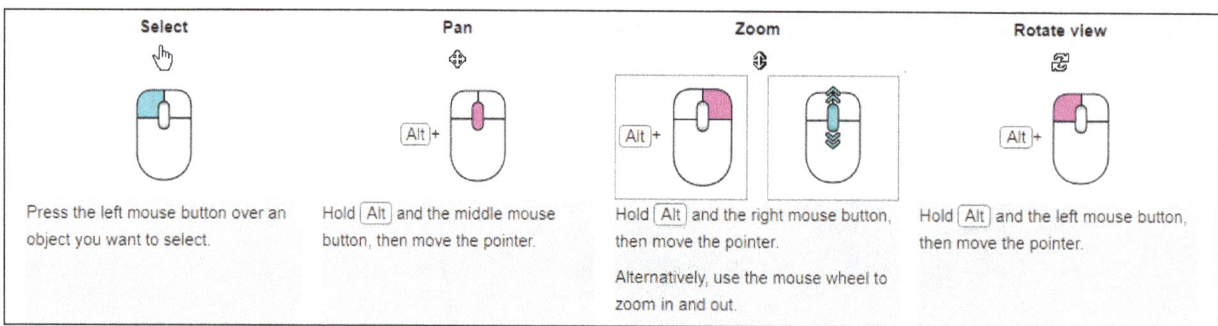

Figure-78. MayaGesture navigation style

Touchpad Navigation

In **Touchpad** navigation, panning, zooming, and rotating the view require a modifier key together with the touchpad. The procedure to use this navigation is shown in Figure-79.

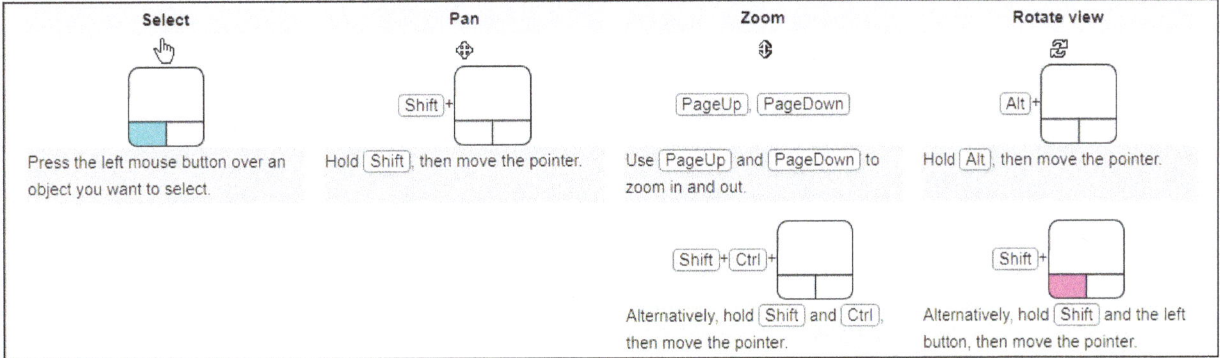

Figure-79. Touchpad navigation style

Gesture Navigation

The procedure to use **Gesture** navigation style is shown in Figure-80.

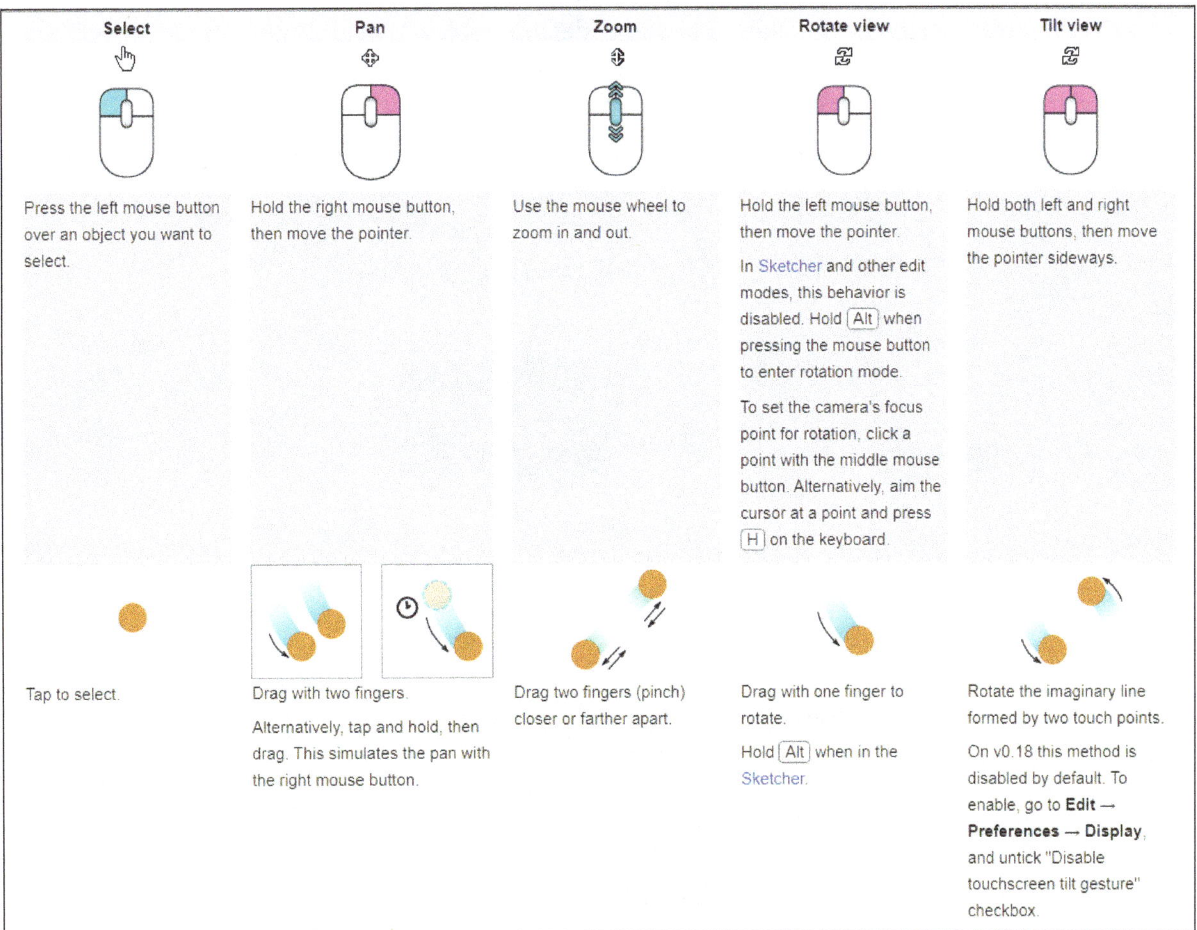

Figure-80. Gesture navigation style

OpenCascade Navigation

The procedure to use **OpenCascade** navigation is shown in Figure-81.

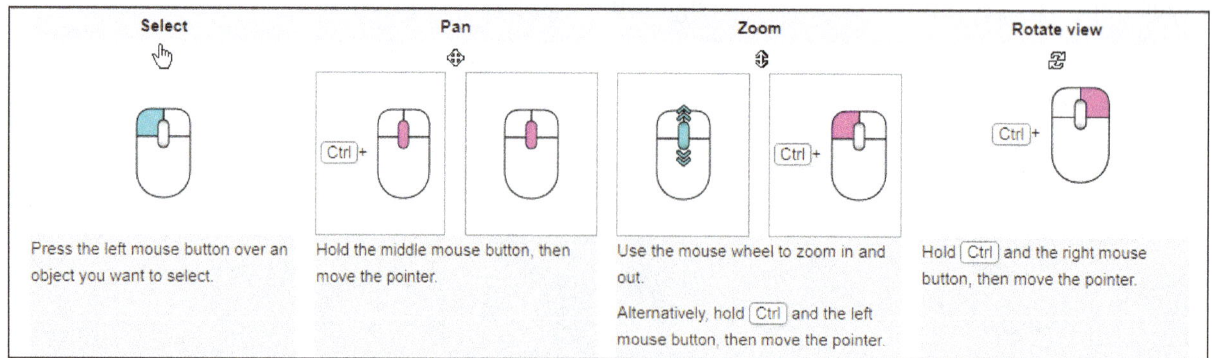

Figure-81. OpenCascade navigation style

FREECAD INTERFACE

The FreeCAD interface is based on the modern Qt, a well known graphical user interface **(GUI)** toolkit and has a state of the art organization. Some aspects of the interface can be customized. You can add custom toolbars with tools from several workbenches or tools defined in macros and you can create your own keyboard shortcuts. But the menus and default toolbars that come with FreeCAD and its workbenches cannot be changed. Different parts of the FreeCAD interface are discussed next; refer to Figure-82.

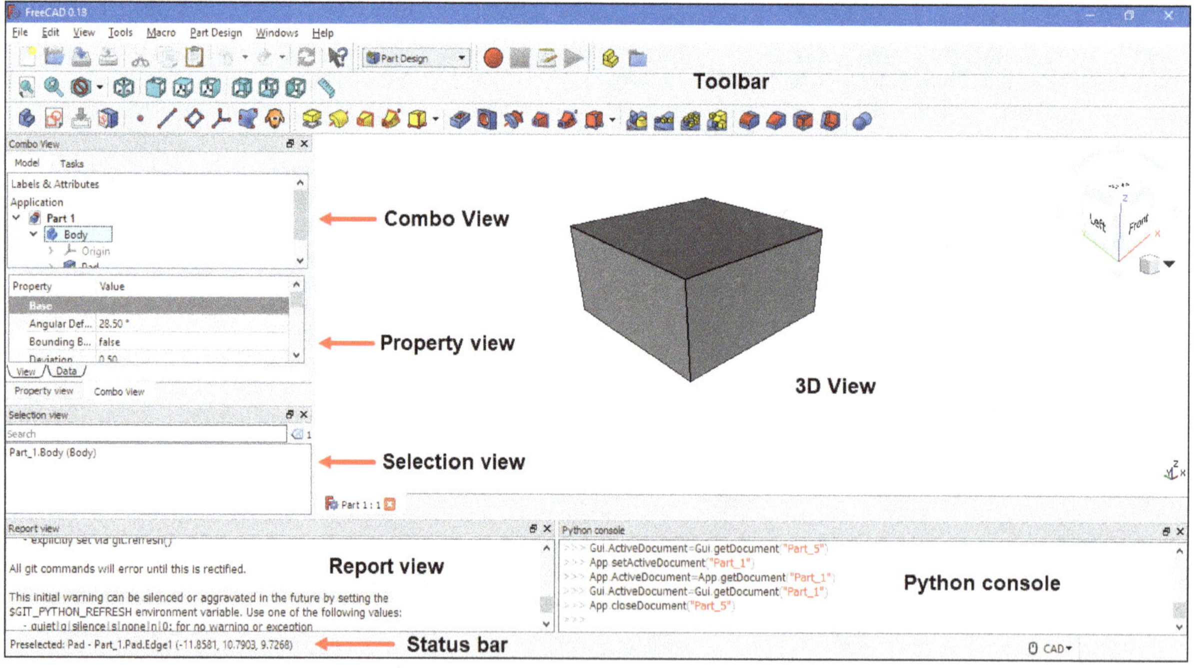

Figure-82. FreeCAD interface

- **Toolbars** : The standard toolbars that appear in the interface are:

- The tools in the **File** toolbar are used to work with files, open documents, copy, paste, undo, and redo actions.
- The **Workbench** toolbar contains a single widget to select the active workbench.
- The tools in **Macro** toolbar are used to record, edit, and execute macros.
- In the **View** toolbar, the tools are used to control how objects appear in the 3D view.

- The tools in the **Structure** toolbar are used to organize objects in the document and create links to additional documents.

- **3D View**: The **3D view** is the main component of the interface. It can be undocked out of the main window if you can have several views of the same document (or same objects) or several documents opened at the same time. You can select objects or parts of objects by clicking them and you can pan, zoom, and rotate the view with the mouse buttons.

- **Combo View** : The **Combo view** is one of the main panels in the FreeCAD interface. It is located on the left side of screen by default. It is composed of two tabs. The **Model** tab shows you the tree view which is a representation of the document's content including 2D and 3D geometry with their parametric history but also supporting objects that contain data saved in the document.

- The **Tasks** tab is where FreeCAD will prompt you for values specific to the tool you're currently using at the time—for example, entering a 'length' value when the **Line** tool is being used. It will close automatically after pressing the **OK** (or **Cancel**) button. Also, by double-clicking the related object in the combo view, most tools will allow you to reopen the task panel in order to modify the settings.

- **Property editor** : The **Property editor** appears when the **Model** tab of the **Combo View** is active in the interface. It allows managing the publicly exposed properties of the objects in the document. Generally, the property editor is intended to deal with just one object at one time. The values shown in the property editor belong to the selected object of the active document. Despite this, some properties like colors can be set for multiple selected objects. If there are no elements selected, the property editor will be empty.

- **Selection view** : The **Selection view** is a panel in the interface by default located below the combo view. Just like the property editor, it shows more information about the currently selected objects. A selection can be made by picking an object in the 3D view or in the tree view. Multiple bodies can be selected by holding **Ctrl** key.

- **Report view** : The **Report view** is a panel that shows text messages from FreeCAD processes and tools. The report view is normally hidden but it is a good idea to leave open as it will list any information, warnings, or errors to help you decipher (or debug) what you may have done wrong.

- **Python console** : The **Python console** is a panel that runs an instance of the Python interpreter. It can be used to control FreeCAD processes. It creates and modifies objects and their properties. The Python console in FreeCAD has basic syntax highlighting able to differentiate with various styles and colors, comments, strings, numeric values, built in functions, printed text output, and delimiters like parentheses and commas.

- **Status bar** : The **Status bar** is a simple information bar that appears at the bottom of the FreeCAD interface. When the mouse pointer is over a button or menu, the usage information of that command is displayed both in a textual pop-up and in the status bar. It shows the zoom level in the right corner next to the mouse navigation style. The status bar also shows the last pre-selected object or element of an object and the coordinates of the mouse pointer when the last pre-selection occurred; refer to Figure-83.

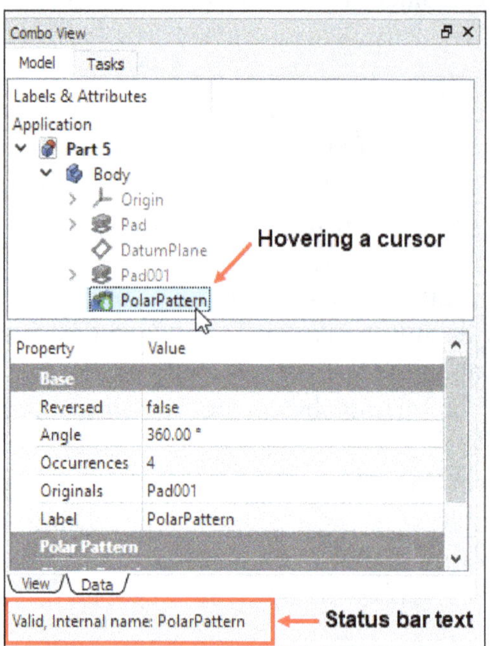

Figure-83. Valid Internal name Polar Pattern status bar text

Chapter 2

Sketching

Topics Covered

The major topics covered in this chapter are:

- ***Introduction***
- ***Starting Sketch***
- ***Sketch Creation Tools***
- ***Sketch Editing Tools***
- ***Sketcher Constraints***
- ***Sketcher Tools***
- ***Sketcher B-Spline Tools***
- ***Sketching Practical and Practice***

INTRODUCTION

In Engineering, sketches are based on real dimensions of real world objects. These sketches work as building blocks for various 3D operations. In this chapter, we will be working with sketch entities like; Line, Circle, Arc, Polygon, Ellipse, and so on to form base feature for various 3D operations. Note that the sketching environment is the base of 3D models so you should be proficient in sketching.

In this chapter, we will be working in **Sketcher Workbench**. We will learn about various tools of toolbar used in 2D sketching.

STARTING SKETCH

- To start a new sketch file, click on the **New** button from **File menu** drop-down and select **Sketcher** workbench from **Switch between workbenches** drop-down in the **Toolbar**; refer to Figure-1. The tools to create 2D sketching will be displayed in toolbar in inactive mode; refer to Figure-2.

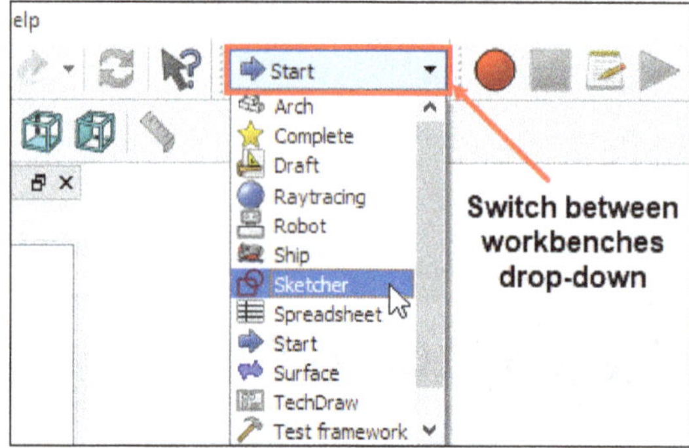

Figure-1. Sketcher workbench

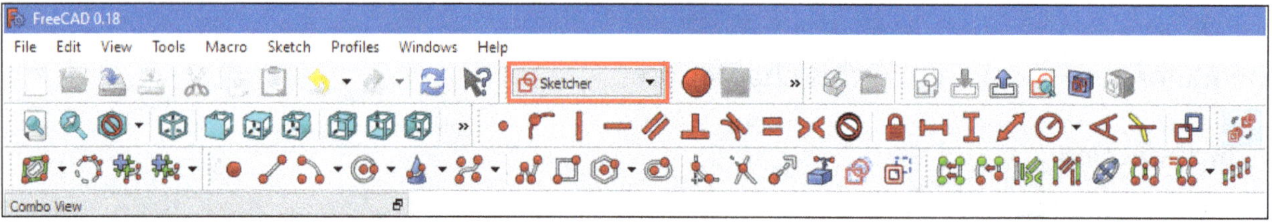

Figure-2. Sketching tools

- Click on **Create a new sketch** button from the toolbar. The **Choose orientation** dialog box will be displayed asking you to select the plane; refer to Figure-3.

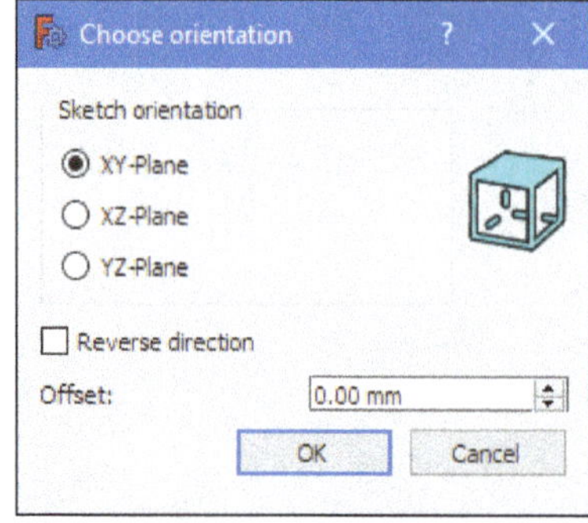

Figure-3. Choose orientation dialog box

- Select desired radio button from **Sketch orientation** area to define which plane will be used as sketching plane.
- Select the **Reverse direction** check box to reverse the plane direction.
- Specify the offset distance for plane in **Offset** edit box. The plane for sketching will be at specified

offset distance perpendicular to selected base plane.

- Click on the **OK** button to close the dialog box. The selected plane will become parallel to the screen and act as current sketching plane. Now, we are ready to draw sketch on the selected plane.

First, we will start with the sketch creation tools and later we will discuss the other tools.

SKETCH CREATION TOOLS

After selecting the plane, all the sketch creation tools will become active. These tools are discussed next.

Point

The **Point** tool creates a point in the current sketch which can be used for constructing geometry elements. The procedure to use this tool is discussed next.

- Click on **Create a Point** tool from **Toolbar** in the **Sketcher** workbench; refer to Figure-4.

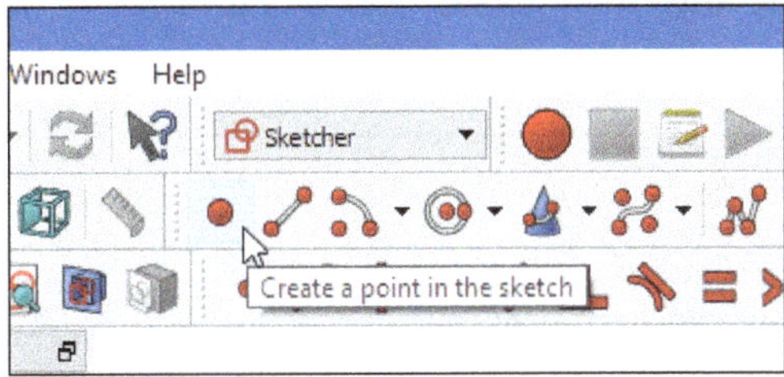

Figure-4. Point tool

- Click at desired locations in the 3D view area. The points will be created; refer to Figure-5.

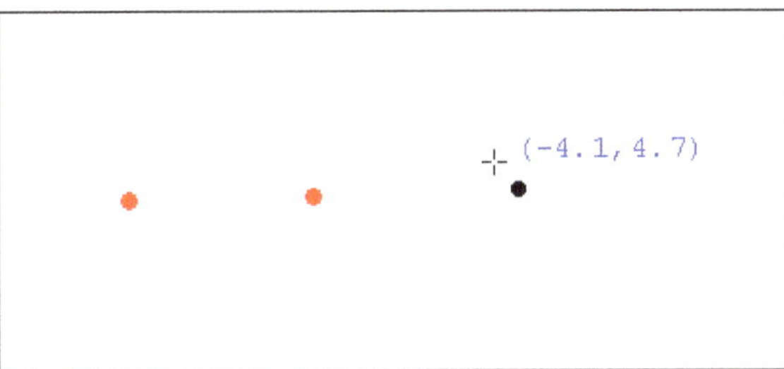

Figure-5. Creation of point

- Press **ESC** key from keyboard or press **RMB** to exit the tool.

Line

The **Line** tool is used to create a line by picking two points in the 3D view. The procedure to use this tool is discussed next.

- Click on **Create a Line** tool from **Toolbar** in **Sketcher** workbench; refer to Figure-6. You can also press **L** key from keyboard to select the **Line** tool. You will be asked to specify starting point of the line.

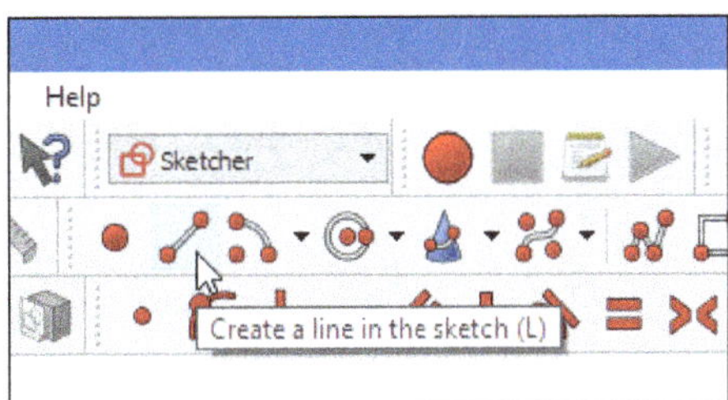

Figure-6. Line tool

- Click at desired location to specify start point and move the cursor away in desired direction in which you want to create the line; refer to Figure-7. You will be asked to specify location of end point of the line.

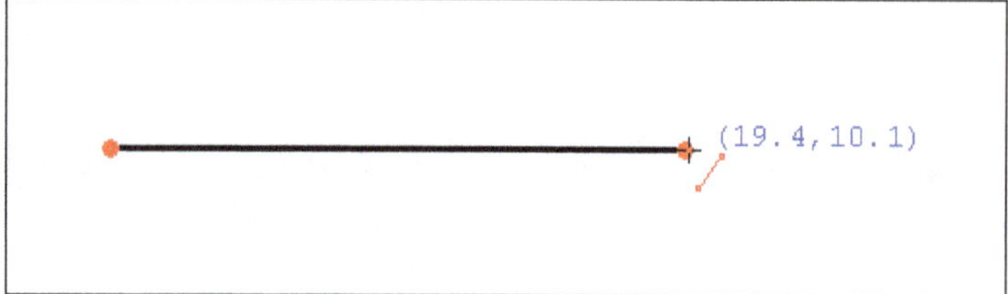

Figure-7. Creation of line

- Click at desired location to specify the end point and right-click to exit the tool. If you want to create a line for construction purpose then select the line you have created and click **RMB** on an empty area of the 3D view, a shortcut menu will be displayed.
- Click on the **Toggle construction geometry** option from the shortcut menu, the normal line will become construction line and vice-versa; refer to Figure-8. Note that construction geometry is displayed in blue color.

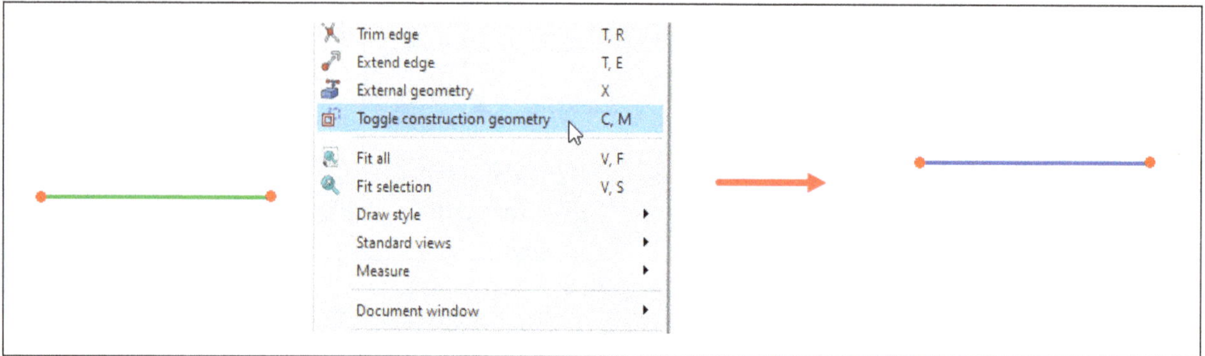

Figure-8. Creation of construction line

- Press **ESC** key or press **RMB** to exit the tool.

Arc

The tools available in **Create an arc** drop-down from **Toolbar** in the **Sketcher** workbench are used to create an arc using three points. There are two tools in **Create an arc** drop-down; **Center and end points** and **End points and rim point**; refer to Figure-9. The procedures to use these tools are discussed next.

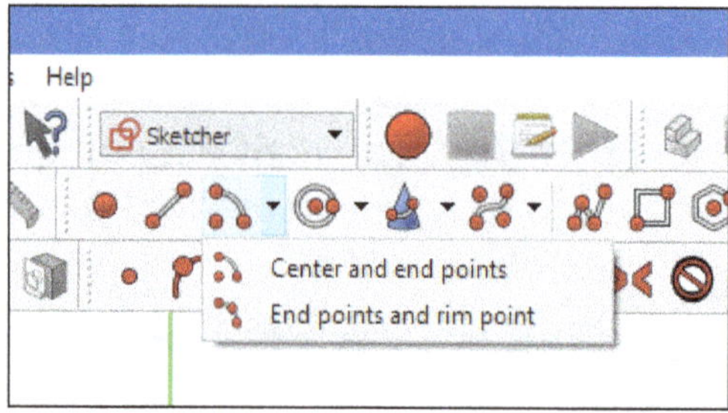

Figure-9. Create an arc drop-down

Creating center and end points arc

The **Center and end points arc** tool is used to create an arc using three points (the center point, the start angle along the radius, and the end angle). The procedure to use this tool is discussed next.

- Click on **Center and end points arc** tool from **Create an arc** drop-down. You will be asked to specify center point of the arc.
- Click at desired location in the 3D view area to specify center point of the arc; refer to Figure-10.

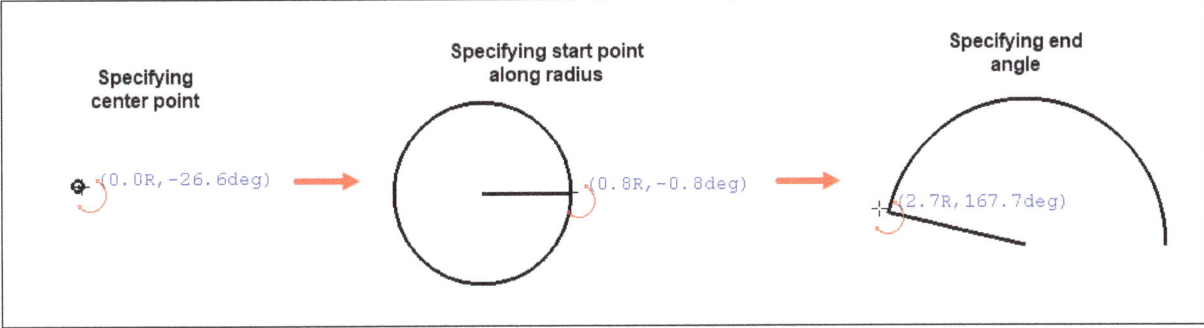

Figure-10. Creation of center and endpoints arc

- Move the cursor to desired location and click to specify starting point of the arc along the radius.
- Click on desired location to specify end point of the arc at a desired angle. The arc will be created.
- Press **ESC** key or press **RMB** to exit the tool.

Creating end points and rim point arc

The **End points and rim point arc** tool is used to create arc using three points (the start point, the end point, and a point on the arc). The procedure to use this tool is discussed next.

- Click in the **End points and rim point arc** tool from **Create an arc** drop-down. You will be asked to specify starting point of the arc.
- Click on the 3D view area to specify starting point of the arc; refer to Figure-11. You will be asked to specify end point of the arc.

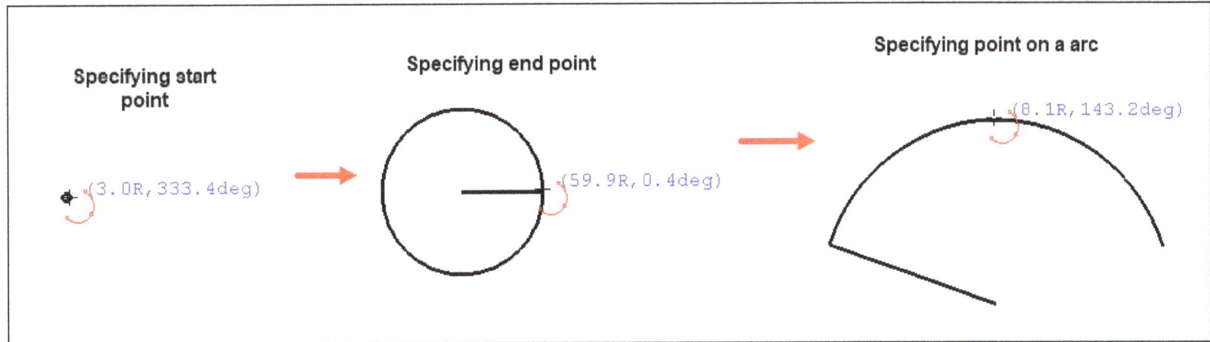

Figure-11. Creation of End points and rim point arc

- Move the cursor away from the point created and click at desired location to specify end point of the arc. You will be asked to specify rim point of the arc.
- Click at desired location to specify a point on the arc at a desired angle. The arc will be created.
- Press **ESC** key or press **RMB** to exit the tool.

Circle

The tools available in **Create a circle** drop-down from **Toolbar** in the **Sketcher** workbench are used to create circle using two and three points. There are two tools in **Create a circle** drop-down; **Center and rim point** and **3 rim points**; refer to Figure-12. The procedures to use these tools are discussed next.

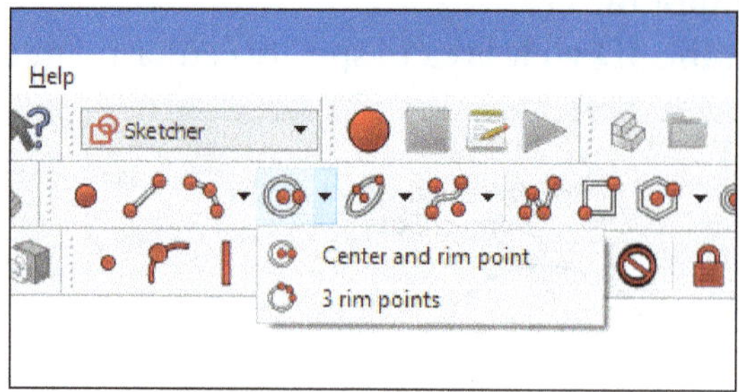

Figure-12. Create a circle drop-down

Center and rim point circle

The **Center and rim point** tool is used to create a circle using two points (the center and a point over the circumference of circle). The procedure to use this tool is discussed next.

- Click on **Center and rim point** tool from **Create a circle** drop-down. You will be asked to specify center point of the circle.
- Click in the 3D view area to specify center point of the circle; refer to Figure-13. You will be asked to specify rim point of the circle.
- Click on desired location to specify a point of circumference of the circle. The circle will be created.
- Press **ESC** key or press **RMB** to exit the tool.

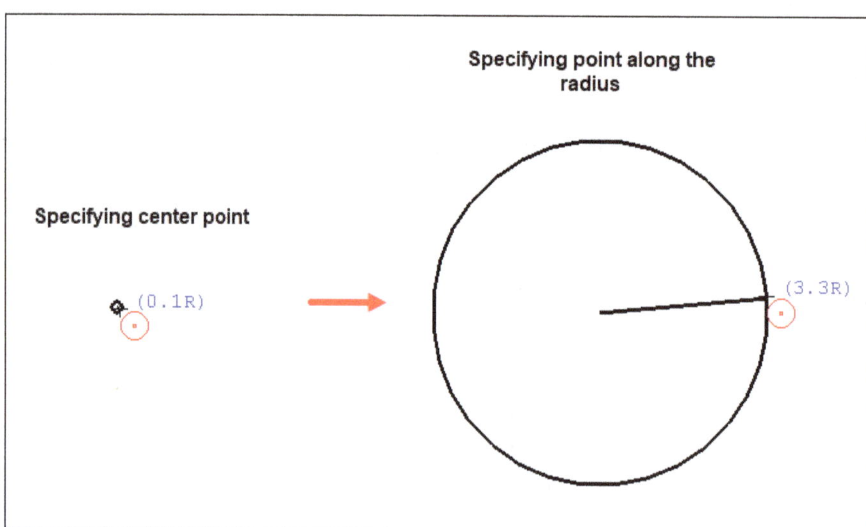

Figure-13. Creation of center and rim point circle

3 rim points circle

The **3 rim points** tool is used to create a circle using 3 rim points or circumferential points as reference. The procedure to use this tool is discussed next.

- Click on the **3 rim points** tool from **Create a circle** drop-down. You will be asked to specify first rim or circumferential point of a circle.
- Click in the 3D view area to specify first rim or circumferential point of a circle; refer to Figure-14. You will be asked to specify second rim point for circle.

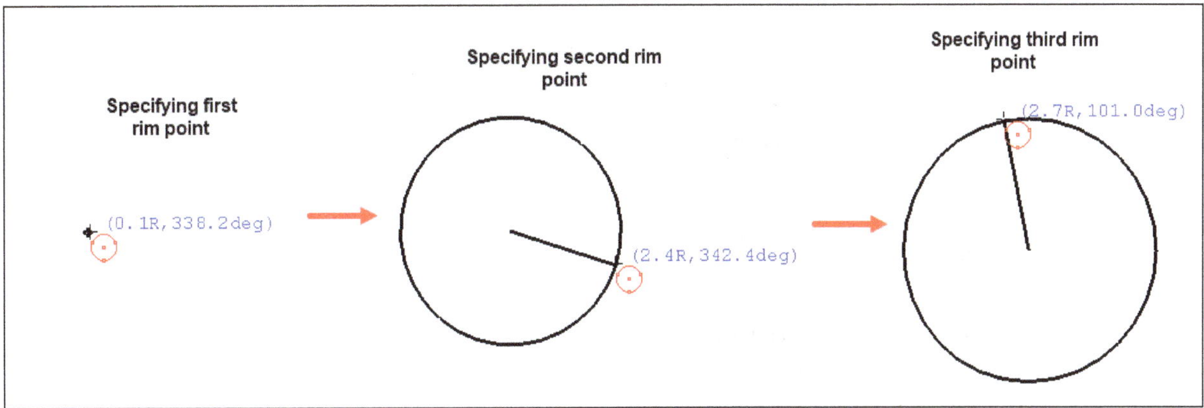

Figure-14. Creation of 3 rim points circle

- Click on desired locations to specify second and then third rim or circumferential point of a circle. The circle will be created.
- Press **ESC** key or press **RMB** to exit the tool.

Conic

There are five tools available in **Create a conic** drop-down from **Toolbar** in the **Sketcher** workbench to create ellipses and arcs; **Ellipse by center**, **Ellipse by 3 points**, **Arc of ellipse**, **Arc of hyperbola**, and **Arc of parabola**; refer to Figure-15. The procedures to use these tools are discussed next.

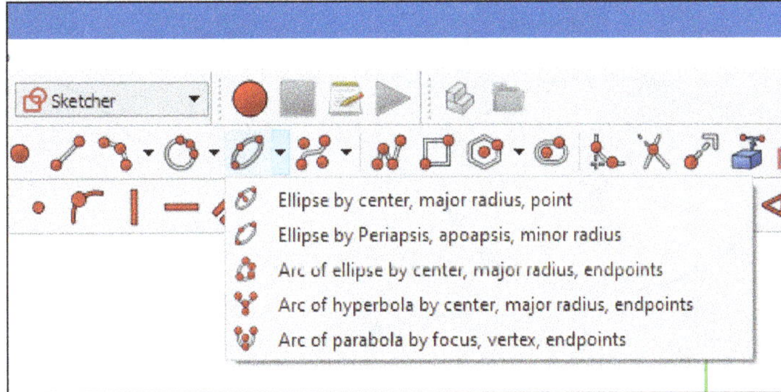

Figure-15. Create a conic drop-down

Ellipse by center

The **Ellipse by center** tool is used to create an ellipse using three points (the center, the end of major radius, and the minor radius). The procedure to use this tool is discussed next.

- Click on **Ellipse by center** tool from **Create a conic** drop-down. You will be asked to specify center point of the ellipse.
- Click in the 3D view area to specify center point of the ellipse; refer to Figure-16.

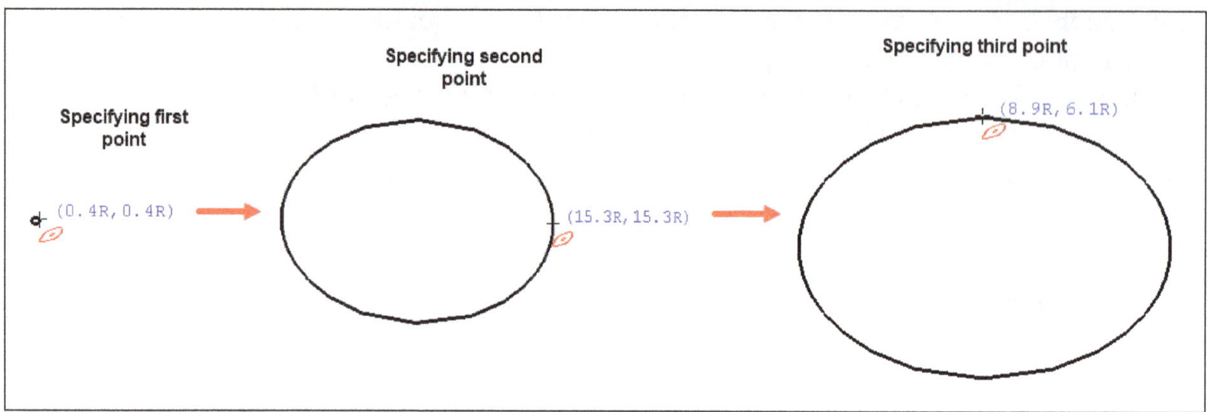

Figure-16. Creation of Ellipse by center

- Move the cursor away from specified point and click to specify second point of the ellipse.
- Click at desired location to specify third point of the ellipse. The ellipse will be created.
- Press **ESC** key or press **RMB** to exit the tool.

Ellipse by 3 points

The **Ellipse by 3 points** tool is used to create an ellipse using three points (the **Periapsis** - first crossing of major diameter with ellipse, the **Apoapsis** - second crossing of major diameter with ellipse, and third point defines the minor radius of ellipse). The procedure to use this tool is discussed next.

- Click on **Ellipse by 3 points** tool from **Create a conic** drop-down. You will be asked to specify first major diameter of ellipse.
- Click in the 3D view area to specify first point of major diameter of ellipse as shown in Figure-17. You will be asked to specify opposite point for major diameter.

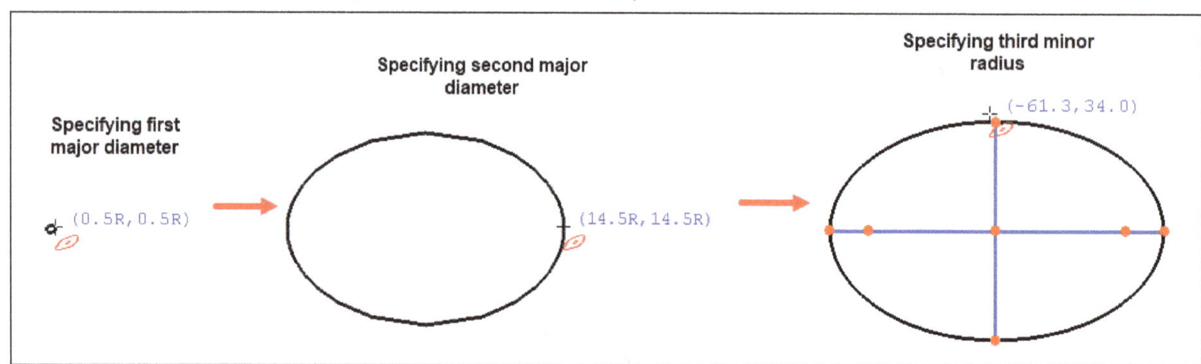

Figure-17. Creation of Ellipse by three points

- Click on the opposite side of the center point to specify second point of major diameter of ellipse. You will be asked to specify a circumferential point for minor diameter of ellipse.
- Click on desired location to specify third minor radius of the ellipse. An ellipse will be created.
- Press **ESC** key or press **RMB** to exit the tool.

Arc of ellipse

The **Arc of ellipse** tool is used to create an arc of ellipse using four points (the center, the end of major radius, the start point, and the end point). The procedure to use this tool is discussed next.

- Click on the **Arc of ellipse** tool from **Create a conic** drop-down. You will be asked to specify center point of the arc.
- Click in the 3D view area to specify center point of the arc; refer to Figure-18. You will be asked to specify major radius of ellipse.

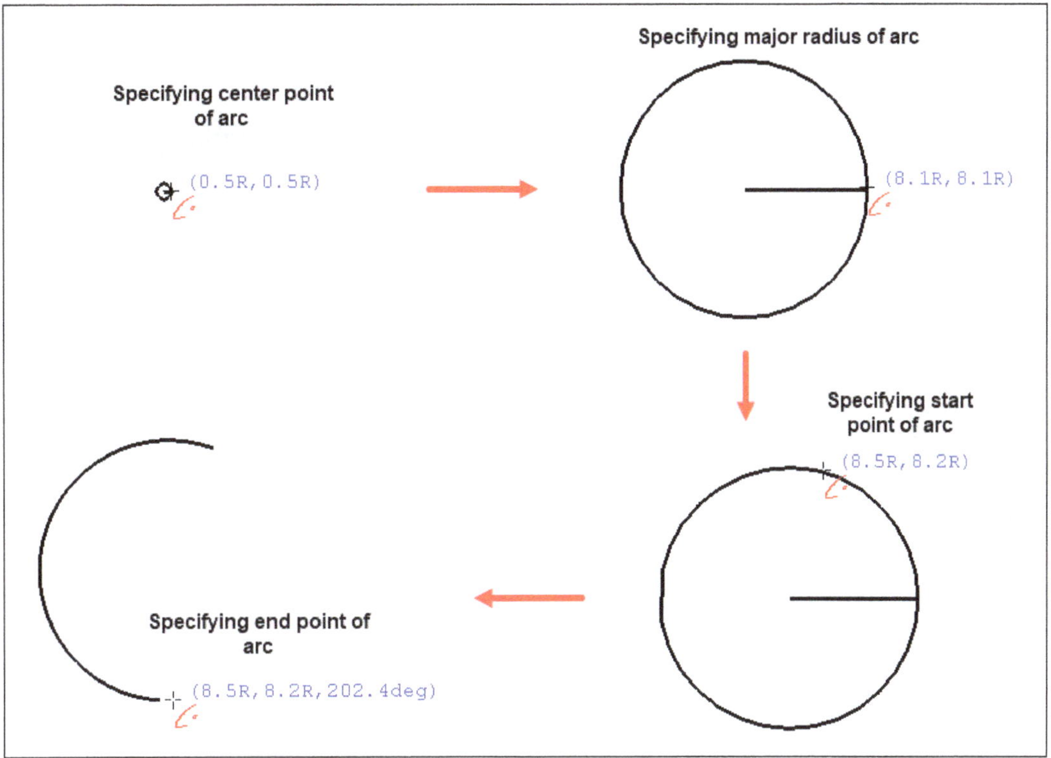

Figure-18. Creation of arc of ellipse

- Move the cursor away from the center point and click to specify major radius of the arc. You will be asked to specify start point of arc.
- Click on desired location to specify starting point and then end point of the arc. The arc of ellipse will be created.
- Press **ESC** key or press **RMB** to exit the tool.

Arc of hyperbola

The **Arc of hyperbola** tool is used to create an arc of hyperbola using four points (the center point, the major radius of arc, the start point, and the end point of arc). The procedure to use this tool is discussed next.

- Click on the **Arc of hyperbola** tool from **Create a conic** drop-down. You will be asked to specify center point of arc.
- Click in the 3D view area to specify center point of arc; refer to Figure-19. You will be asked to specify major radius of arc.

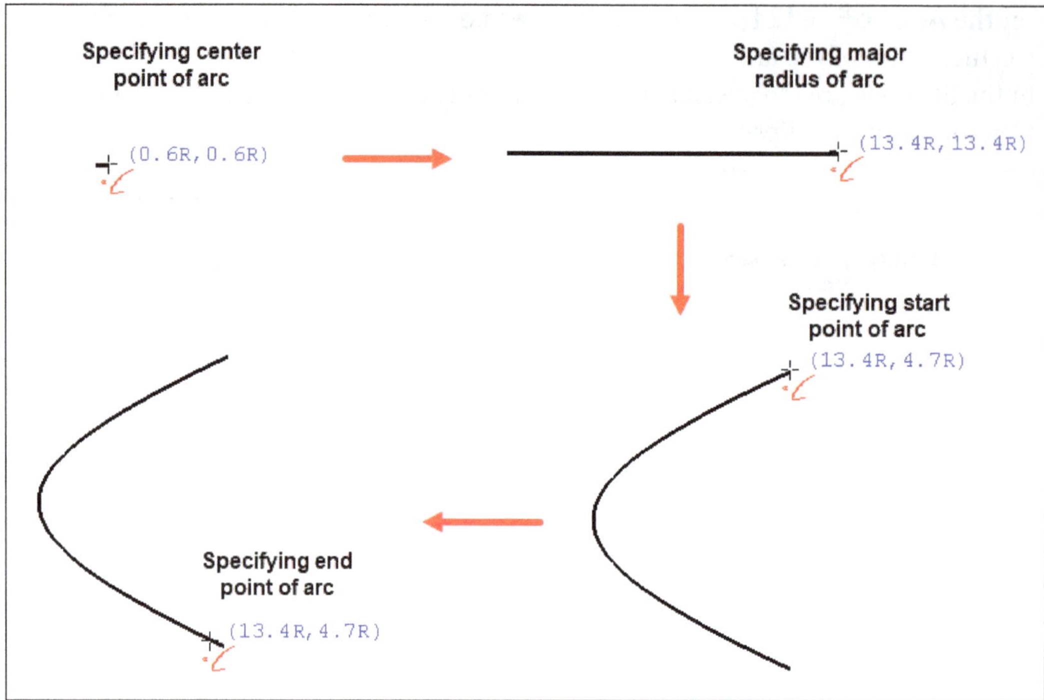

Figure-19. Creation of arc of hyperbola

- Move the cursor away from the center point and click to specify major radius of the arc. You will be asked to specify starting point of arc.
- Click on desired location to specify starting point of arc and then the end point of arc. The arc of hyperbola will be created.
- Press **ESC** key or press **RMB** to exit the tool.

Arc of parabola

The **Arc of parabola** tool is used to create an arc of parabola using four points (the focus of arc, the vertex of arc, the start point, and the end point of arc). The procedure to use this tool is discussed next.

- Click on the **Arc of parabola** tool from **Create a conic** drop-down. You will be asked to specify focus of the arc.
- Click in the 3D view area to specify focus point of arc; refer to Figure-20.

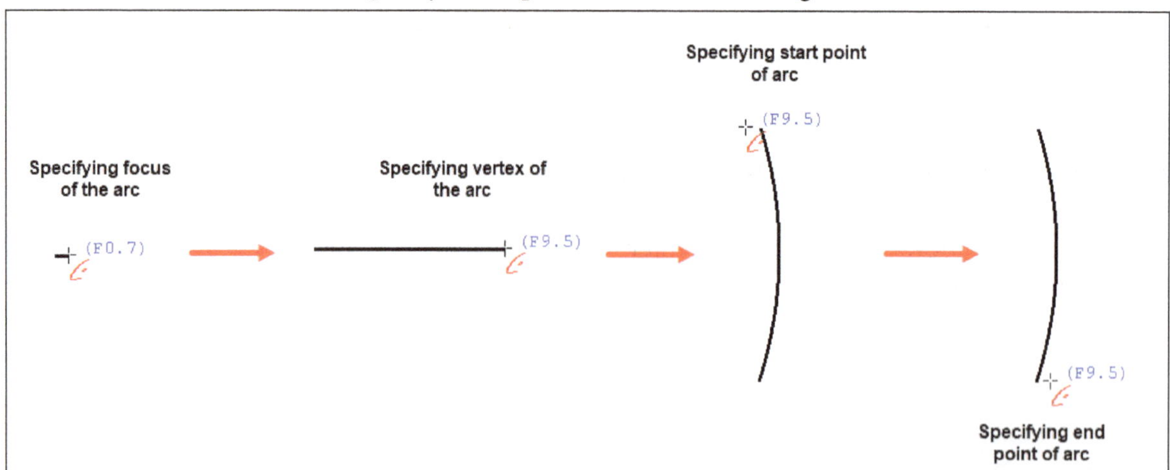

Figure-20. Creation of arc of parabola

- Move the cursor away and click to specify vertex of the arc.
- Click on desired location to specify starting point of arc and then the end point of arc. The arc of parabola will be created.

- Press **ESC** key or press **RMB** to exit the tool.

B-spline

There are two tools available in **Create a B-spline** drop-down from **Toolbar** in the **Sketcher** workbench to create B-spline curves; **B-spline by control points** and **Periodic B-spline by control points**; refer to Figure-21. The procedures to use these tools are discussed next.

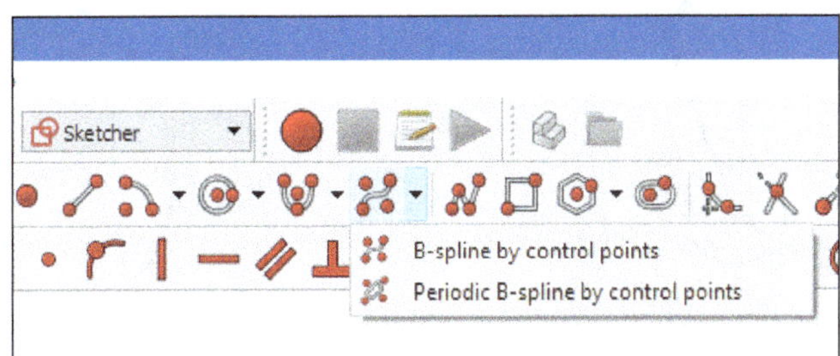

Figure-21. Create a B spline drop-down

B-spline by control points

The **B-spline by control points** tool is used to create open B-spline curves by tracing its control points. The procedure to use this tool is discussed next.

- Click on the **B-spline by control points** tool from **Create a B-spline** drop-down. You will be asked to create the series of points.
- Click in the 3D view area to specify a series of points to create spline; refer to Figure-22.

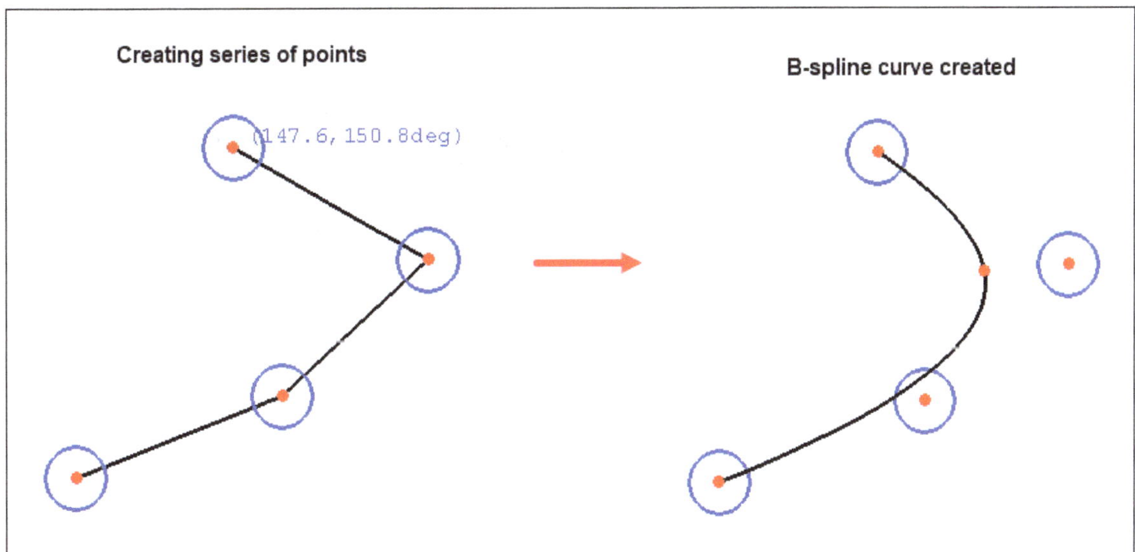

Figure-22. Creation of B spline

- Press **ESC** key or press **RMB** to exit the creation, the B-spline curve will be created.
- Press **ESC** key or press **RMB** again to exit the tool.

Periodic B-spline by control points

The **Periodic B-spline by control points** tool is used to create closed B-spline curves by tracing its control points. The procedure to use this tool is discussed next.

- Click on the **Periodic B-spline by control points** tool from **Create a B-spline** drop-down. You will be asked to create series of points.

- Click in the 3D view area to specify a series of points for creating spline; refer to Figure-23.

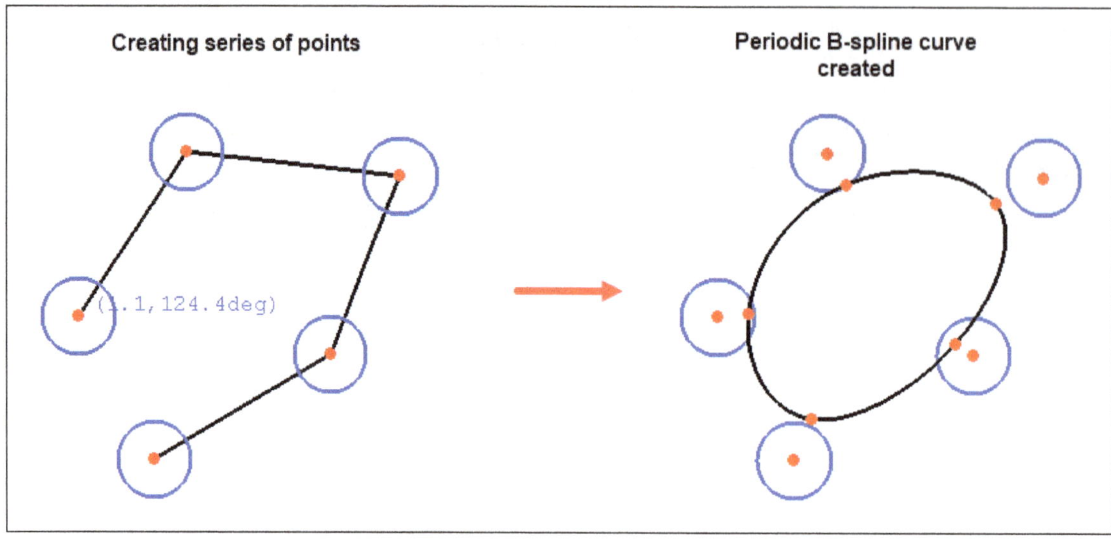

Figure-23. Creation of Periodic B spline

- Press **RMB** to exit the creation, the periodic B-spline curve will be created.
- Press **ESC** key or press **RMB** again to exit the tool.

Polyline

The **Create a polyline** tool is used to create continuous line and arc segments connected by their vertices. The polyline tool has multiple modes that can be toggled with the **M** key.

For example, you can draw tangent or perpendicular arcs following a line or arc segment. The procedure to use this tool is discussed next.

- Click on **Create a polyline** tool from **Toolbar** in the **Sketcher** workbench; refer to Figure-24. You will be asked to create a line segment.

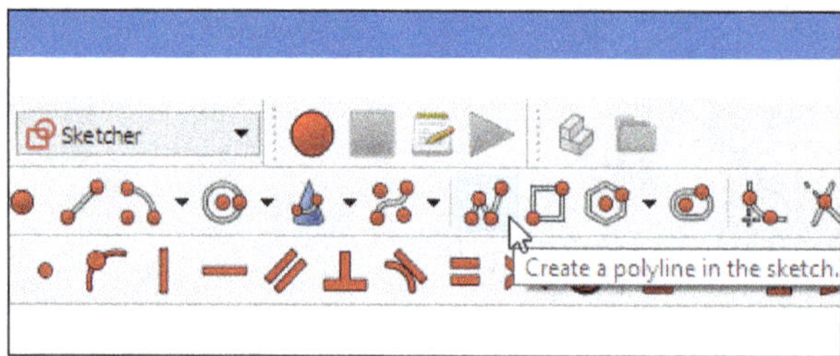

Figure-24. Polyline tool

- Click in the 3D view area and create a line segment. Press **M** key from the keyboard to toggle different modes of creating polyline; refer to Figure-25.

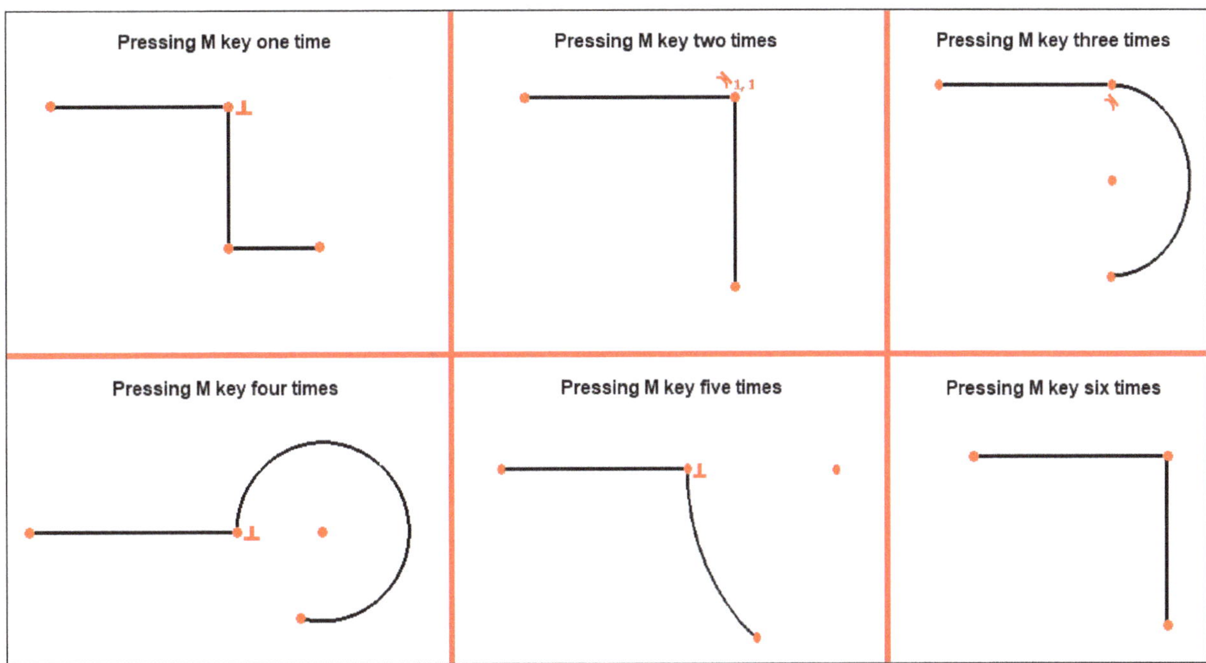

Figure-25. Creation of Polyline multiple modes with M key

- After creating a line segment, press **M** key for one time to create a line perpendicular to the previous entity.
- Press **M** key for two times to create a line tangential to the previous entity.
- Press **M** key for three times to create an arc tangential to the previous entity.
- Press **M** key for four times to create an arc perpendicular to the left of the previous entity.
- Press **M** key for five times to create an arc perpendicular to the right of the previous entity.
- Press **M** key for six times to create a line coincident to the previous entity as in the normal state.
- Press **ESC** key or press **RMB** to exit the creation and continue to create a new line.
- Press **ESC** key or press **RMB** again to exit the tool.

Rectangle

The **Create a rectangle** tool is used to create rectangle by using two opposite points. The procedure to use this tool is given next.

- Click on **Create a rectangle** tool from **Toolbar** in the **Sketcher** workbench; refer to Figure-26. You will be asked to specify first corner point of the rectangle.

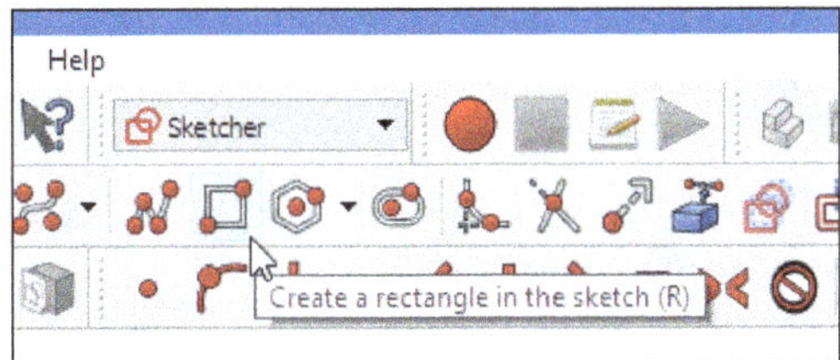

Figure-26. Rectangle tool

- Click in the 3D view area to specify first corner point of the rectangle; refer to Figure-27.

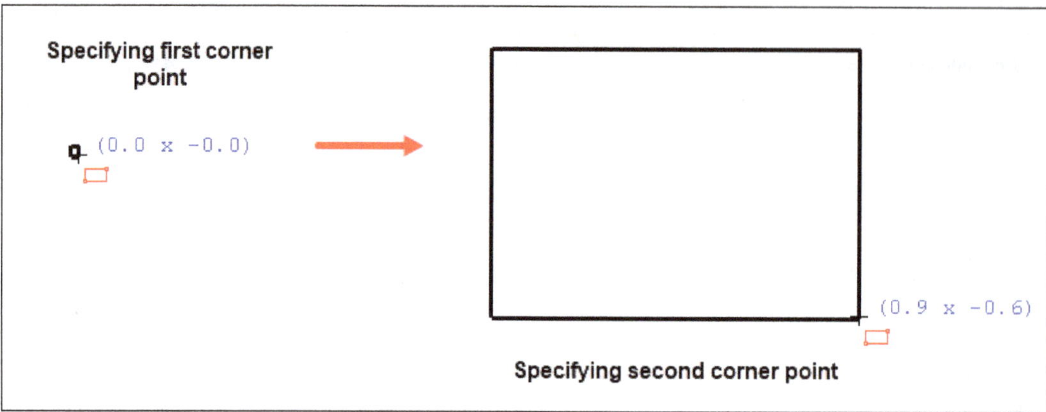

Figure-27. Creation of rectangle

- Move the cursor to desired location and click to specify second corner point of the rectangle. The rectangle will be created.
- Press **ESC** key or press **RMB** to exit the tool.

Polygon

There are seven tools available in **Create a regular polygon** drop-down from **Toolbar** in the **Sketcher** workbench to create a polygon; **Triangle**, **Square**, **Pentagon**, **Hexagon**, **Heptagon**, **Octagon**, and **Regular Polygon**; refer to Figure-28. The procedures to use these tools are discussed next.

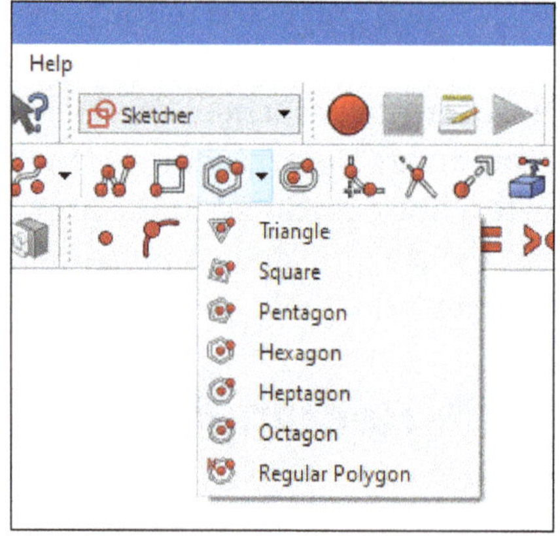

Figure-28. Create a regular polygon drop-down

Triangle

The **Triangle** tool is used to create an equilateral triangle inscribed in a construction geometry circle. The procedure to use this tool is discussed next.

- Click on the **Triangle** tool from the **Create a regular polygon** drop-down. You will be asked to specify center of the triangle.
- Click in the 3D view area to specify center of the triangle as shown in Figure-29.

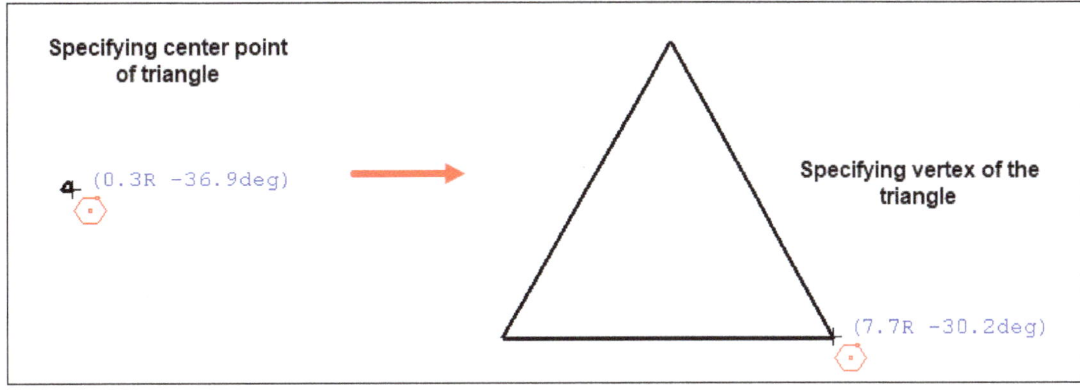

Figure-29. Creation of triangle

- Move the cursor away from the center point and click to specify a vertex of the triangle. The triangle will be created.
- Press **ESC** key or press **RMB** to exit the tool.

Square

The **Square** tool is used to create a square inscribed in a construction geometry circle. The procedure to use this tool is discussed next.

- Click on the **Square** tool from **Create a regular polygon** drop-down. You will be asked to specify center point of the square.
- Click in the 3D view area to specify center of the square; refer to Figure-30.

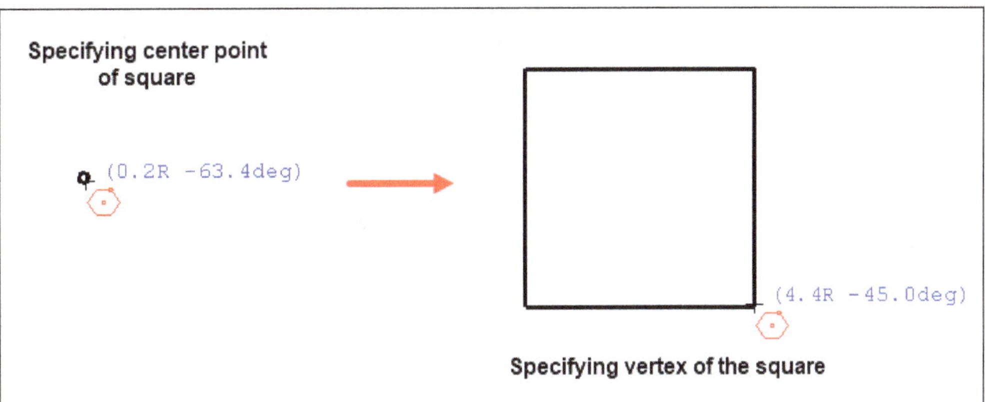

Figure-30. Creation of square

- Move the cursor away from center point and click to specify vertex of the square. The square will be created.
- Press **ESC** key or press **RMB** to exit the tool.

Pentagon

The **Pentagon** tool is used to create a pentagon inscribed in a construction geometry circle. The procedure to use this tool is discussed next.

- Click on the **Pentagon** tool from **Create a regular polygon** drop-down. You will be asked to specify center of the pentagon.
- Click in the 3D view area to specify center of the pentagon; refer to Figure-31.

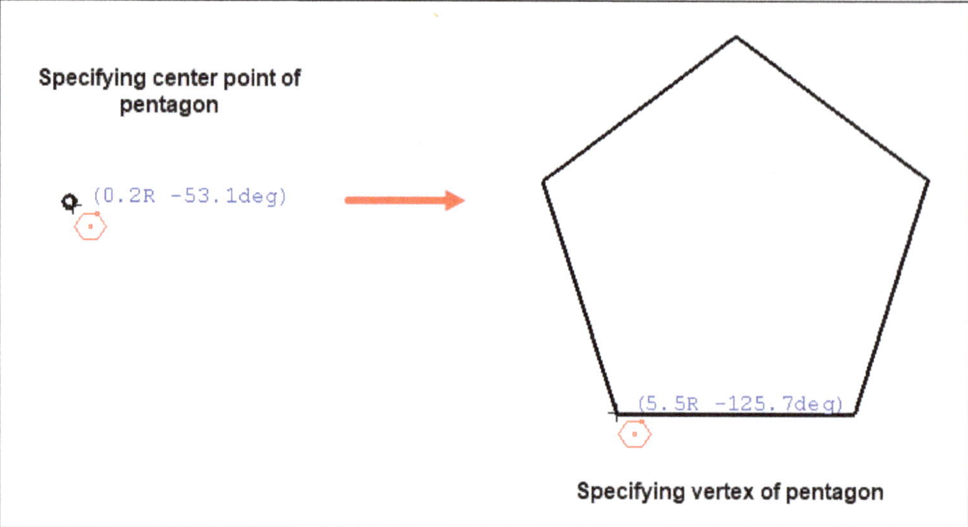

Figure-31. Creation of pentagon

- Move the cursor away from center point and click to specify vertex of pentagon. The Pentagon will be created.
- Press **ESC** key or press **RMB** to exit the tool.

Hexagon

The **Hexagon** tool is used to create a hexagon inscribed in a construction geometry circle. The procedure to use this tool is discussed next.

- Click on the **Hexagon** tool from **Create a regular polygon** drop-down. You will be asked to specify center of the hexagon.
- Click in the 3D view area to specify center of the hexagon; refer to Figure-32.

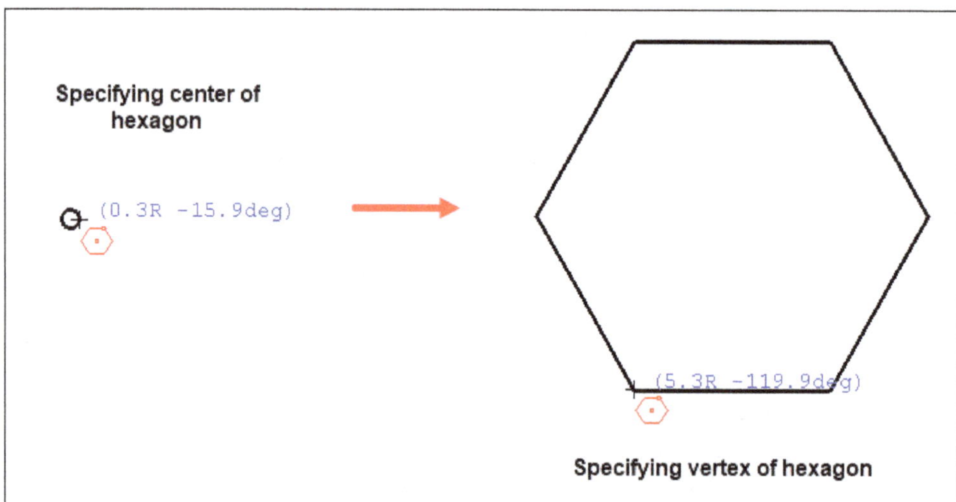

Figure-32. Creation of hexagon

- Move the cursor away from center point and click to specify vertex of the hexagon. The hexagon will be created.
- Press **ESC** key or press **RMB** to exit the tool.

Heptagon

The **Heptagon** tool is used to create a heptagon inscribed in a construction geometry circle. The procedure to use this tool is discussed next.

- Click on the **Heptagon** tool from **Create a regular polygon** drop-down. You will be asked to specify center of the heptagon.
- Click in the 3D view area to specify center point for heptagon; refer to Figure-33.

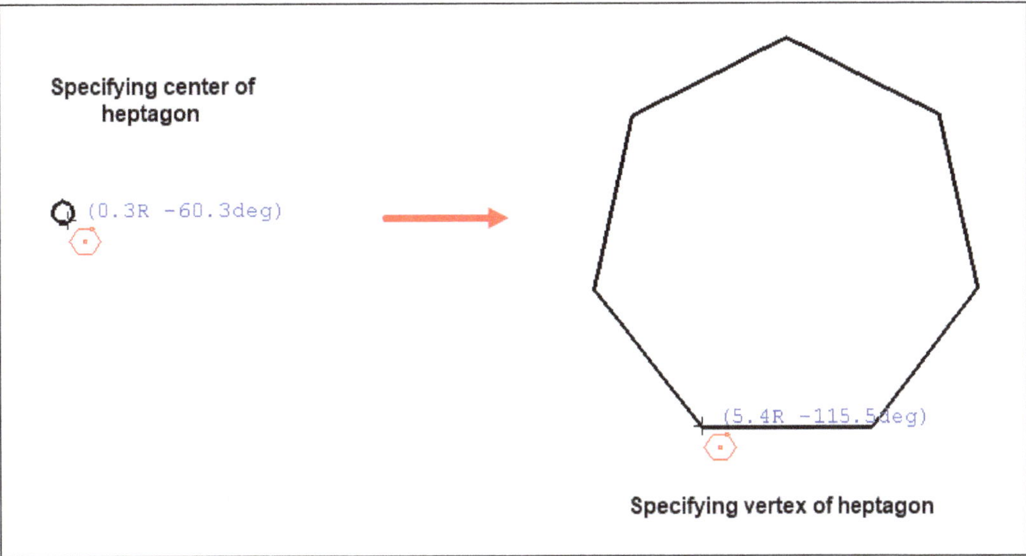

Figure-33. Creation of heptagon

- Move the cursor away from center point and click to specify vertex of heptagon. The heptagon will be created.
- Press **ESC** key or press **RMB** to exit the tool.

Octagon

The **Octagon** tool is used to create an octagon inscribed in a construction geometry circle. The procedure to use this tool is discussed next.

- Click on the **Octagon** tool from **Create a regular polygon** drop-down. You will be asked to specify center point of octagon.
- Click in the 3D view area to specify the center of octagon; refer to Figure-34.

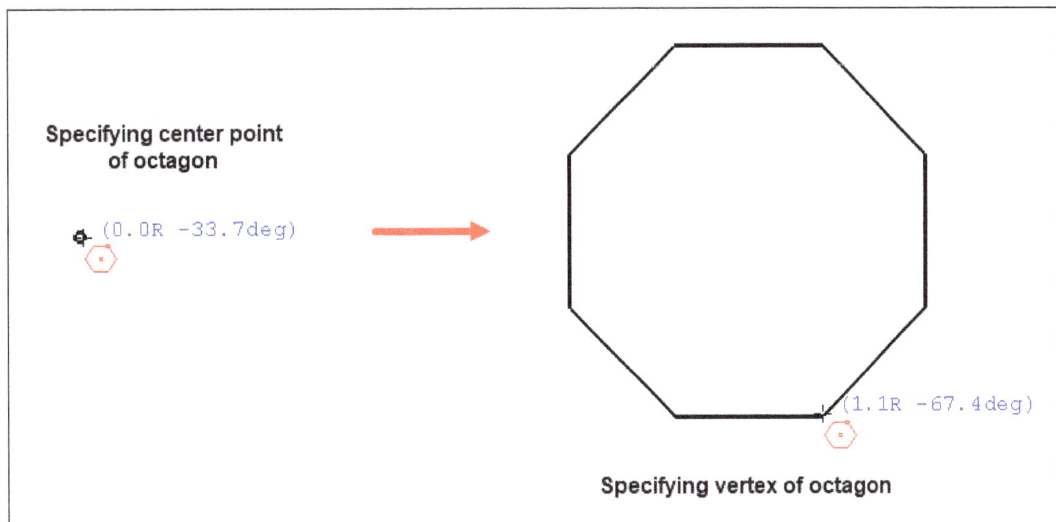

Figure-34. Creation of octagon

- Move the cursor away from center point and click to specify vertex of octagon. The octagon will be created.
- Press **ESC** key or press **RMB** to exit the tool.

Regular Polygon

A **Regular Polygon** tool is used to create a polygon inscribed in a construction geometry circle. The procedure to use this tool is discussed next.

- Click on the **Regular Polygon** tool from **Create a regular polygon** drop-down. The **Create array** dialog box will be displayed asking you to specify the number of sides; refer to Figure-35.

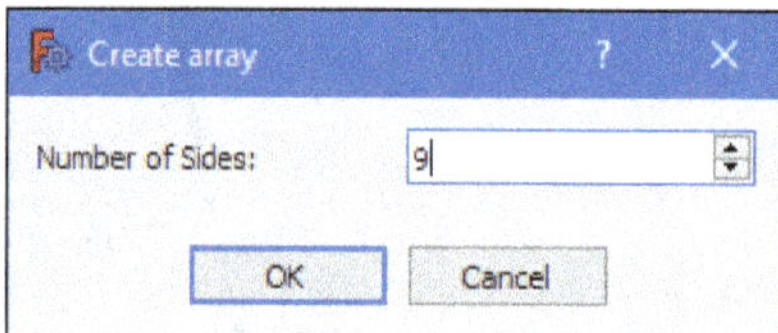

Figure-35. Create array dialog box

- Specify desired number of sides you want to have in a polygon in the **Number of Sides** edit box. Click on the **OK** button to close the dialog box. You will be asked to specify center point of polygon.
- Click in the 3D view area to specify the center point for polygon; refer to Figure-36.
- Move the cursor away from center point and click to specify the vertex of polygon. The regular polygon will be created.
- Press **ESC** key or press **RMB** to exit the tool.

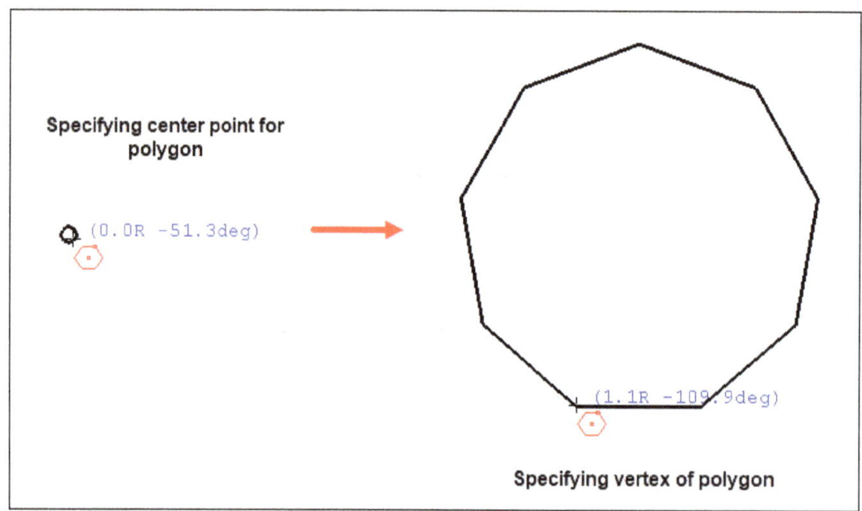

Figure-36. Creation of regular polygon

Slot

The **Slot** tool is used to create a slot by selecting the center of one semicircle and circumferential point of the other semicircle. The procedure to use this tool is discussed next.

- Click on the **Create a slot** tool from **Toolbar** in the **Sketcher** workbench; refer to Figure-37. You will be asked to specify center of one semicircle.

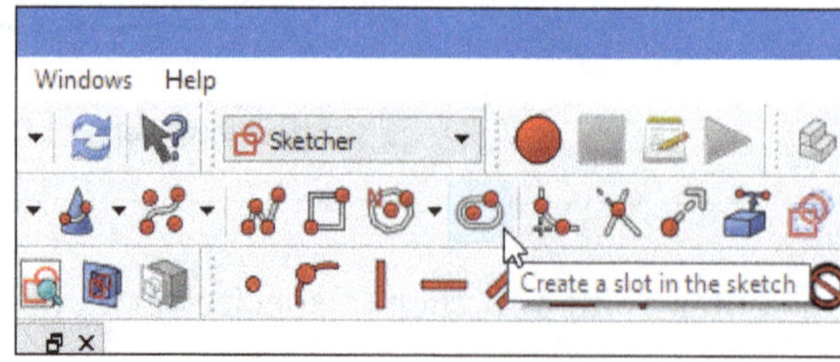

Figure-37. Slot tool

- Click in the 3D view area to specify center point of one semicircle; refer to Figure-38.

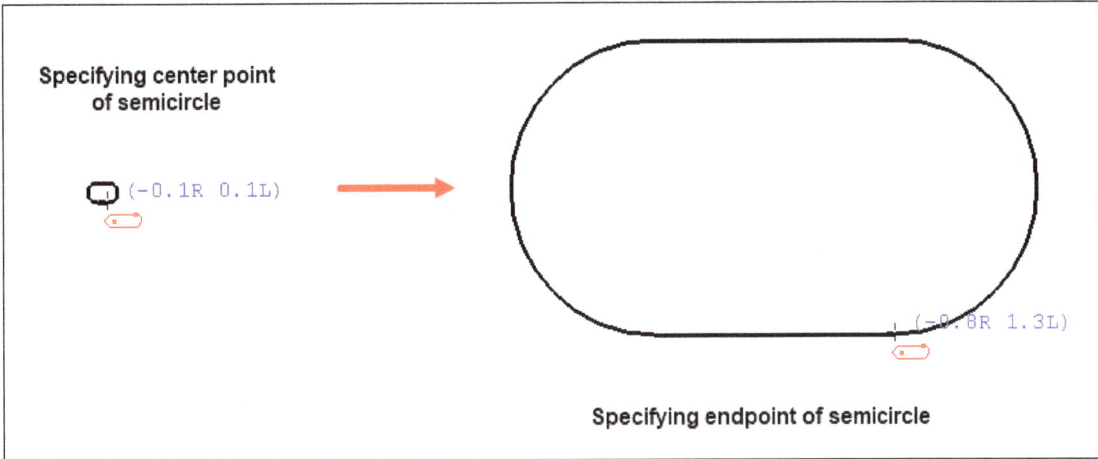

Figure-38. Creation of slot

- Move the cursor away and click at desired location to specify circumferential point of other side semicircle. The slot will be created.
- Press **ESC** key or press **RMB** to exit the tool.

SKETCH EDITING TOOLS

In **FreeCAD** software, the standard tools to edit sketch entities are available in the **Toolbar** of **Sketcher** workbench; refer to Figure-39.

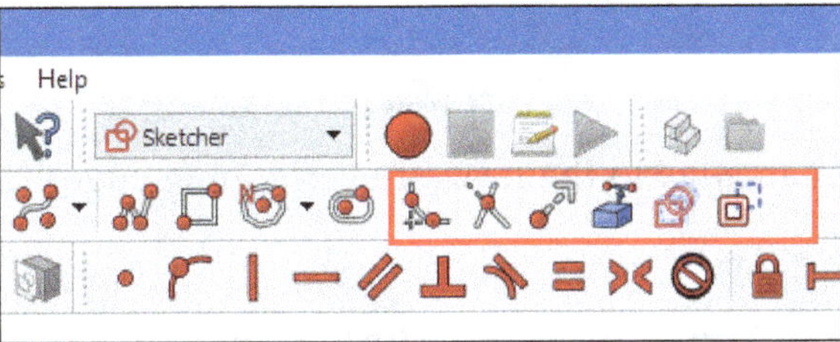

Figure-39. Sketch editing tools

Fillet

The **Fillet** tool is used to create fillet between two lines joined at one point. The procedure to use this tool is discussed next.

- Click on the **Fillet** tool from **Toolbar** in the **Sketcher** workbench; refer to Figure-40. You will be asked to select the vertex or lines of an entity.

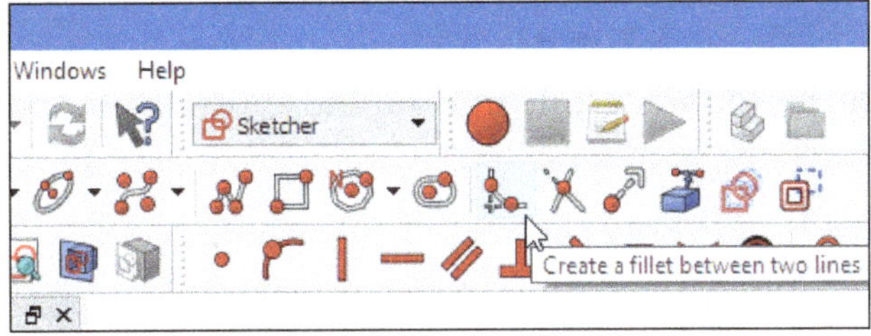

Figure-40. Fillet tool

- Select the vertex connecting two lines or select the two connected lines of an entity. The fillet will be created; refer to Figure-41.

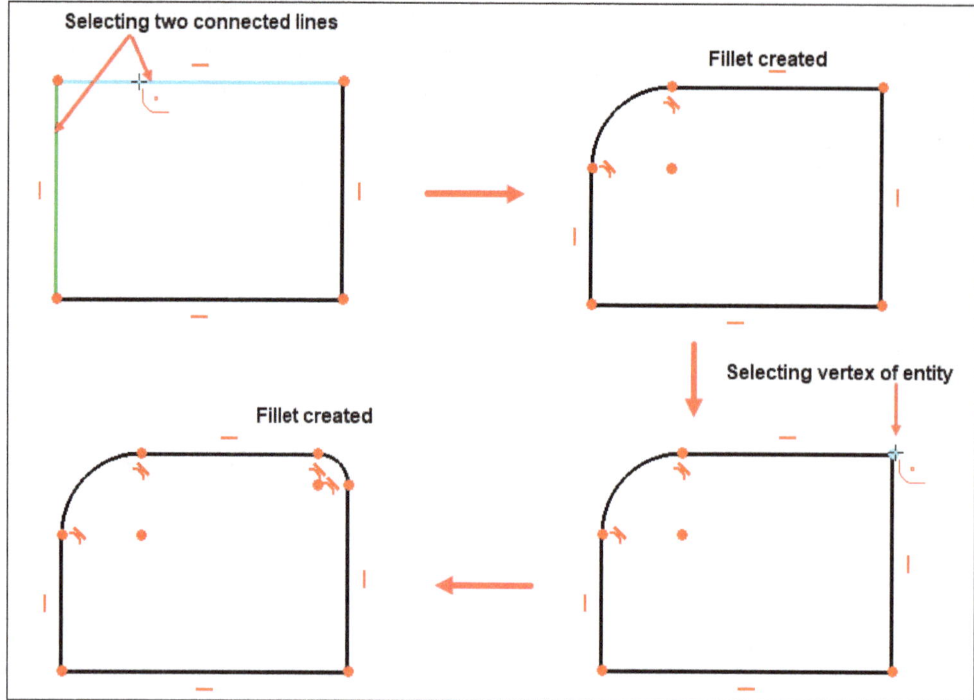

Figure-41. Fillet creation

Note that, the distance you click from the vertex will specify the fillet radius.

- Press **ESC** key or press **RMB** to exit the tool.

Trim

The **Trim** tool is used to trim an edge to the nearest overlapping edge. The procedure to use this tool is discussed next.

- Click on the **Trim** tool from **Toolbar** in the **Sketcher** workbench; refer to Figure-42. You will be asked to select the curve to be trimmed.

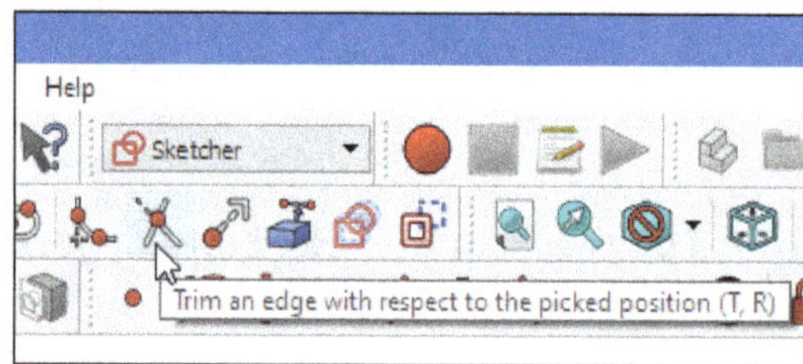

Figure-42. Trim tool

- Select an overlapping curve you want to trim from the entity. The curve will be trimmed; refer to Figure-43.

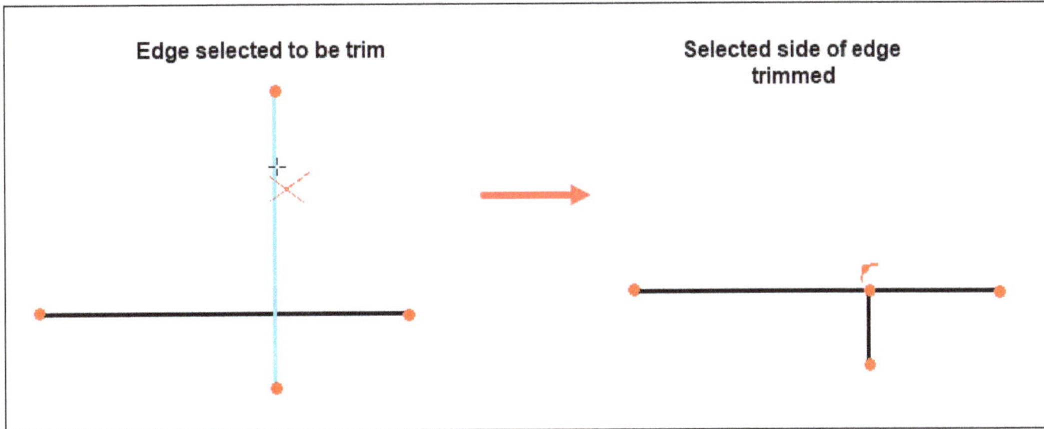

Figure-43. Edge trimmed

- If there are sketch on both sides of clicked position then the selected side of edge will be trimmed.
- Press **ESC** key or press **RMB** to exit the tool.

Extend

The **Extend** tool is used to extend a curve to a specifed location in the sketch or to another curve. The procedure to use this tool is discussed next.

- Click on the **Extend** tool from **Toolbar** in the **Sketcher** workbench; refer to Figure-44. You will be asked to select a line or an arc.

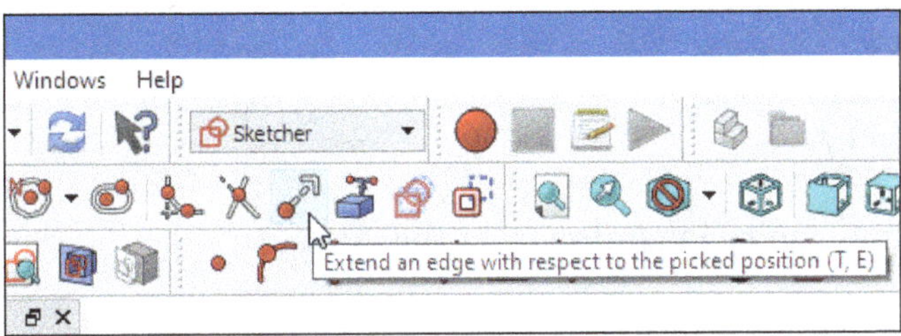

Figure-44. Extend tool

- Select a line or an arc you want to extend in a specified location or upto selected entity; refer to Figure-45.

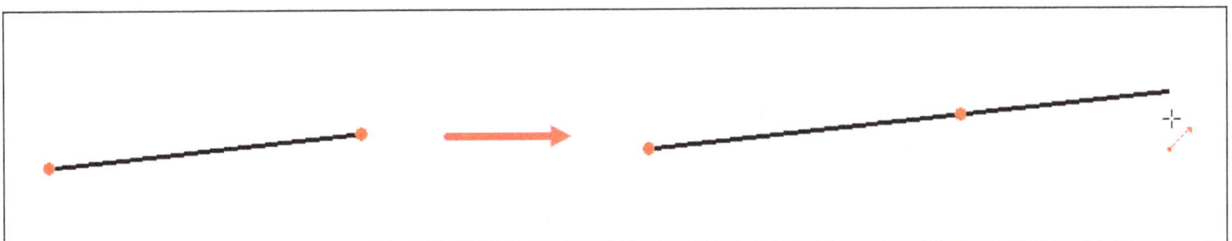

Figure-45. Line extended to an arbitrary location

- To extend a line/arc to an another curve, hover the cursor on the curve. When the curve is highlighted and the **Point on object** constraint icon appears then press **LMB** to extend the line. The line will be extended; refer to Figure-46.

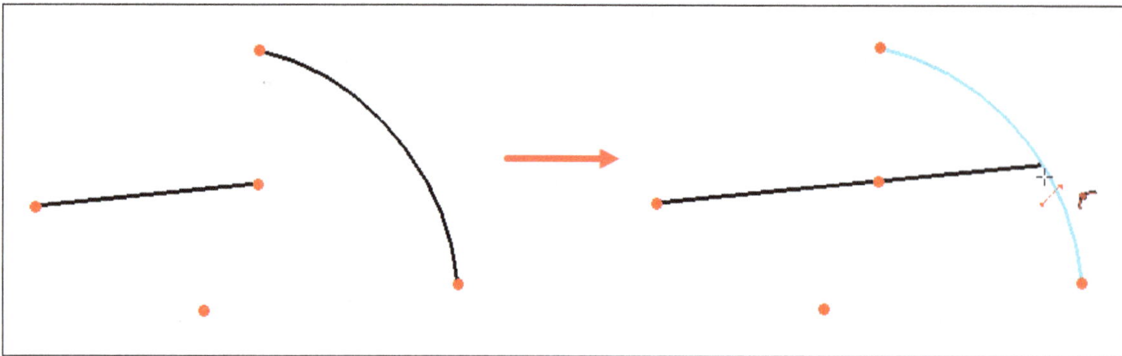

Figure-46. Line extended to an edge

- To extend an edge to a point in the sketch, hover the cursor over the point. When the point is highlighted and the **Coincident** constraint icon appears then press **LMB** to extend the line. The line will be extended; refer to Figure-47.

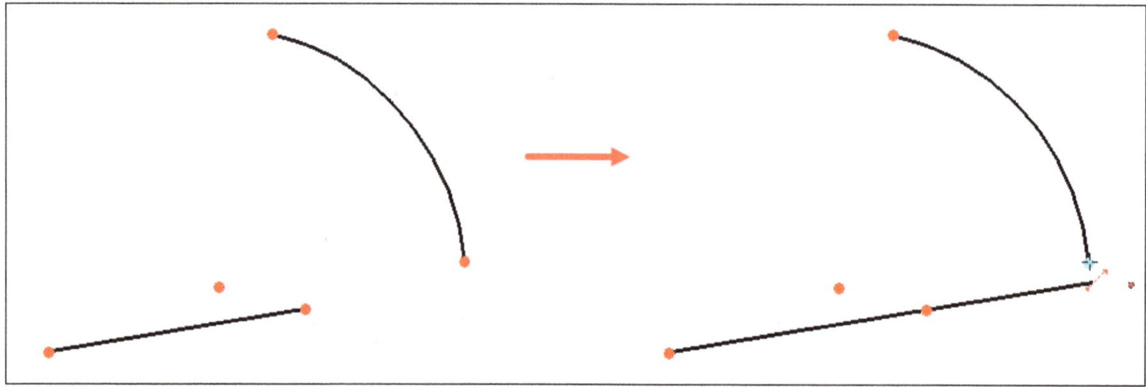

Figure-47. Line extended to a point

Note that only arcs and lines can be extended from this tool.

- Press **ESC** key or press **RMB** to exit the tool.

External Geometry

The **External Geometry** tool is used to apply a constraint between sketch geometry and geometry external to current sketch. It works by inserting a linked construction geometry in the sketch. The procedure to use this tool is discussed next.

- Click on the **External Geometry** tool from **Toolbar** in the **Sketcher** workbench; refer to Figure-48. You will be asked to select a line, arc, or curve.

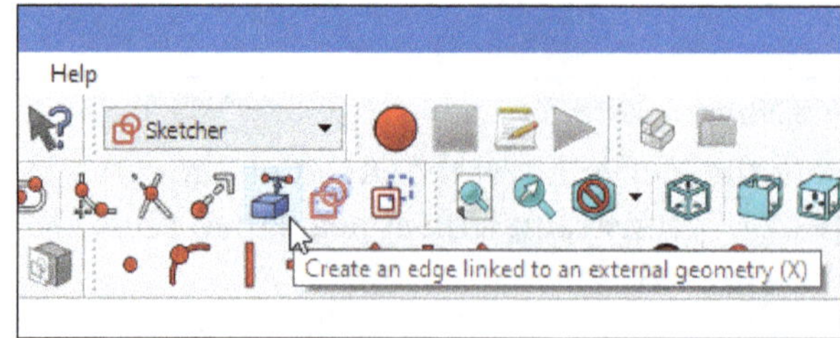

Figure-48. External Geometry tool

- Select a curve that you want to link to the sketch; refer to Figure-49. The selected curve will become part of current sketch. Now, you can apply constraints between new sketch curves and previous selected sketch curves.

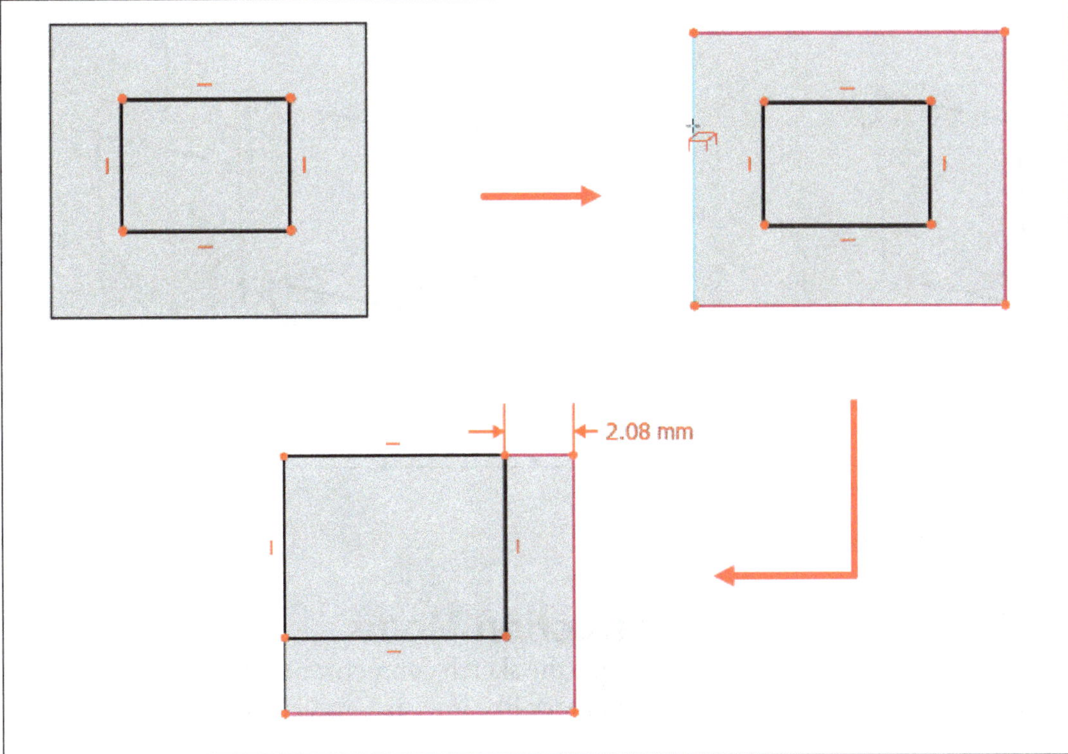

Figure-49. External geometry creation

- Press **ESC** key or press **RMB** to exit the tool.

Carbon Copy

The **Carbon Copy** tool copies all the geometry and constraints from another sketch into the active sketch. The procedure to use this tool is discussed next.

- Click on the **Carbon Copy** tool from **Toolbar** in the **Sketcher** workbench; refer to Figure-50. You will be asked to select curves from another sketch.

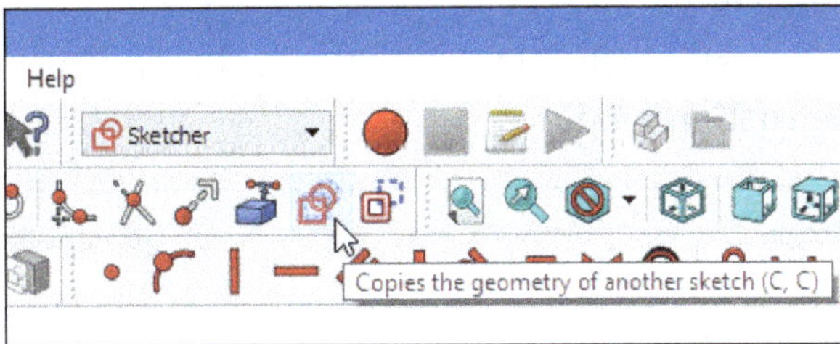

Figure-50. Carbon Copy tool

- Click on curves from the existing sketch, the geometry elements as well as constraints are copied into the active sketch; refer to Figure-51.

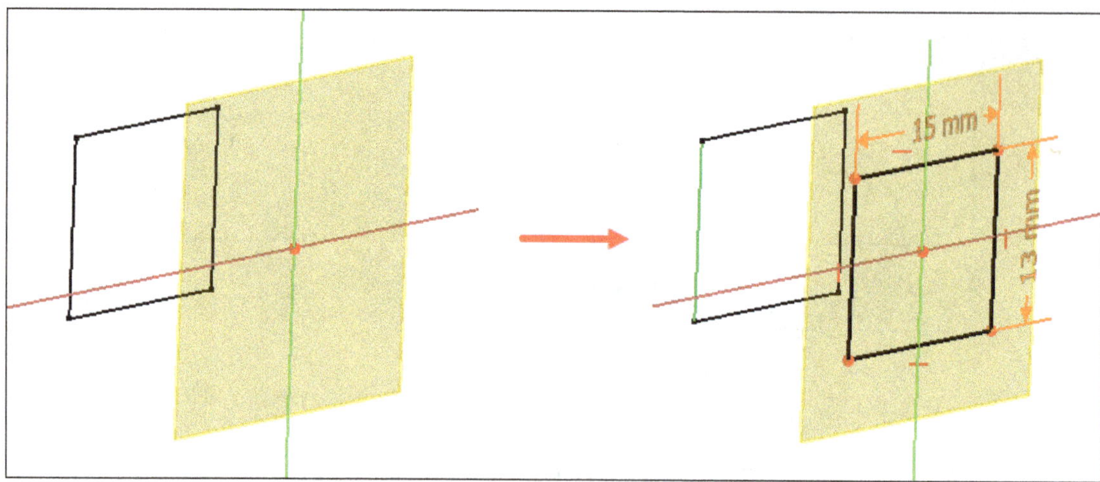

Figure-51. Carbon copy of sketch geometry

- Press **ESC** key or press **RMB** to exit the tool.

Construction Mode

The **Construction Mode** tool is used to toggle the sketch geometry between construction mode and sketch mode. It can be used on any type of geometry like line, arc, and circle. The procedure to use this tool is discussed next.

- Click on the **Construction Mode** tool from **Toolbar** in the **Sketcher** workbench; refer to Figure-52. You will be asked to select the sketch.

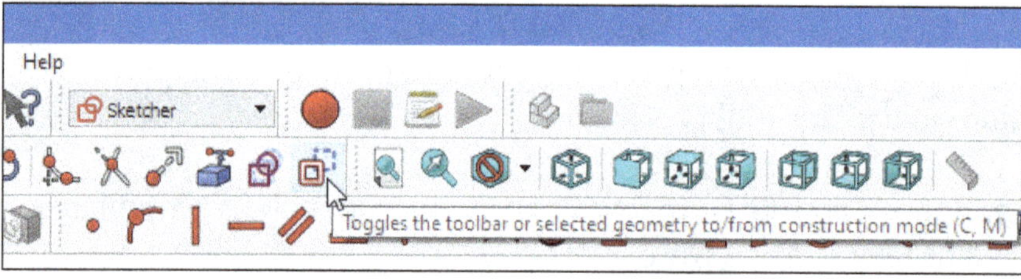

Figure-52. Construction mode tool

- Select one or more sketch geometry in the 3D view to toggle the construction mode; refer to Figure-53. The sketch will be in construction mode.

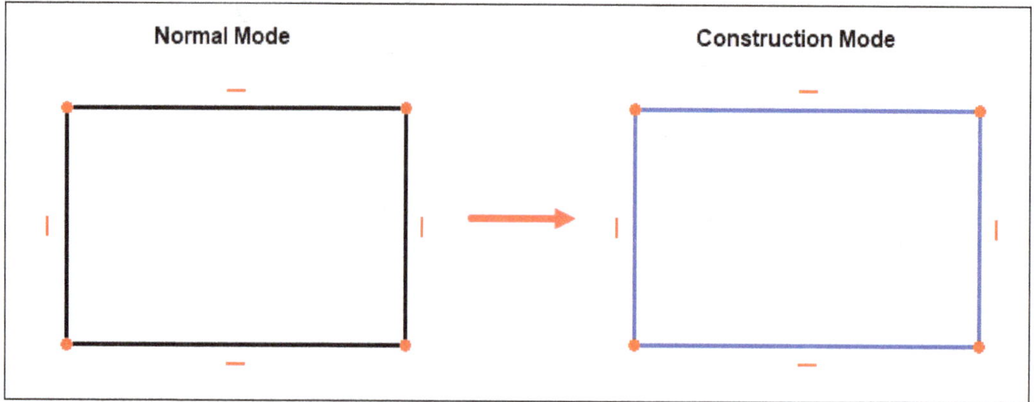

Figure-53. Construction mode sketch

Note that if there is not any sketch selected earlier and if you select **Construction Mode** tool then all the sketches will be created in construction mode until you click the button again.

- Press **ESC** key or press **RMB** to exit the tool.

SKETCHER CONSTRAINTS

Constraints are used to define lengths, set rules between sketch elements, and to lock the sketch along the vertical and horizontal axes.

Geometric constraints

Geometric constraints are not associated with numeric data. The tools to apply geometric constraints are available in the **Toolbar** of **Sketcher** workbench; refer to Figure-54. The procedures to use these tools are discussed next.

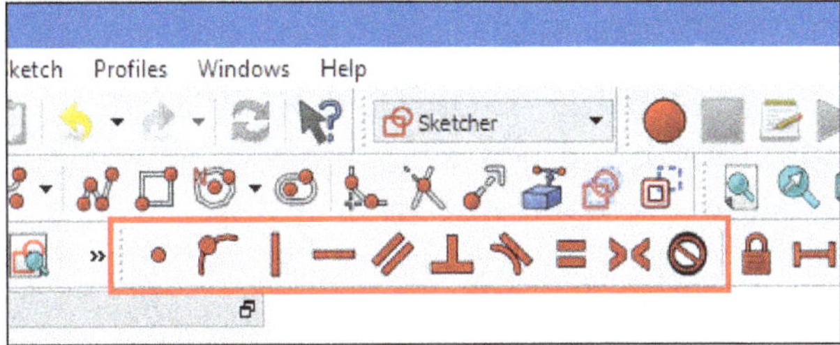

Figure-54. Geometric constraints tools

Coincident constraint

The **Coincident constraint** tool is useful when there is a break in a profile, for example: where two lines end near each other and need to be joined. A coincident constraint on their endpoints will close the gap and make two selected points coincident. The procedure to use this tool is discussed next.

- Click on the **Coincident constraint** tool from **Toolbar** in the **Sketcher** workbench; refer to Figure-55. You will be asked to select the endpoints.
- Select the endpoints of the sketch which you want to join or make coincident of each other. The endpoints of the sketch will become coincident to each other; refer to Figure-56.
- Press **ESC** key or press **RMB** to exit the tool.

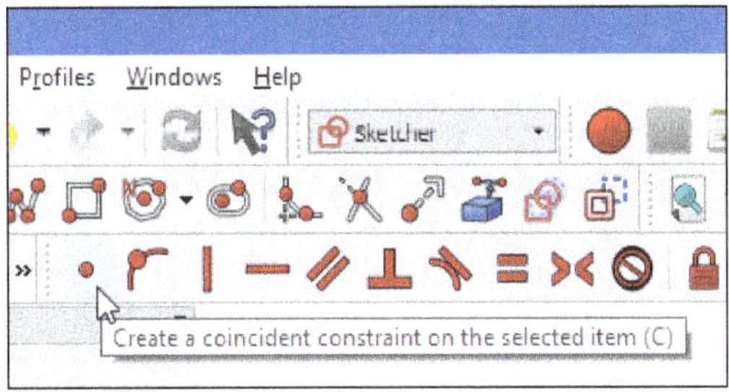

Figure-55. Coincident constraint tool

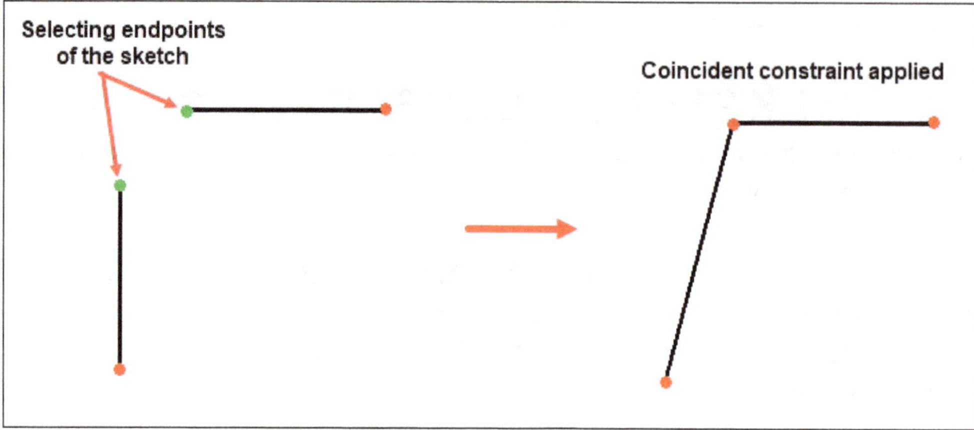

Figure-56. Applying coincident constraint

Point On Object constraint

The **Point On Object constraint** tool affixes the selected point onto another object such as a line, arc, or curve. The procedure to use this tool is discussed next.

- Click on the **Point On Object constraint** tool from **Toolbar** in the **Sketcher** workbench; refer to Figure-57. You will be asked to select the point.

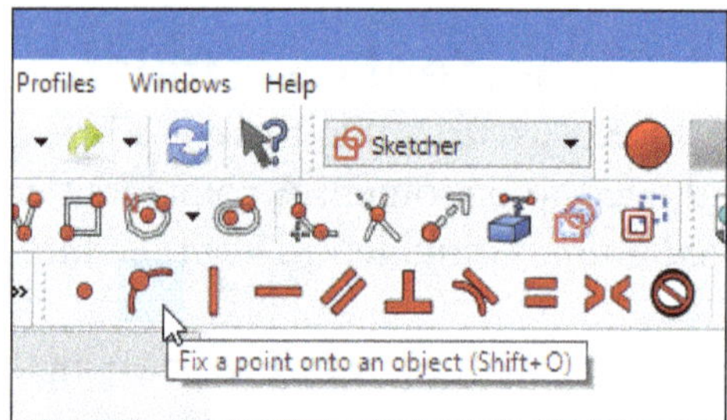

Figure-57. Point On Object constraint tool

- Select the point you want to affix on the line, arc, or sketch curve.
- Select the curve on which you want to affix the point. The point will be affixed on the line and the constraint will be applied; refer to Figure-58.
- Press **ESC** key or press **RMB** to exit the tool.

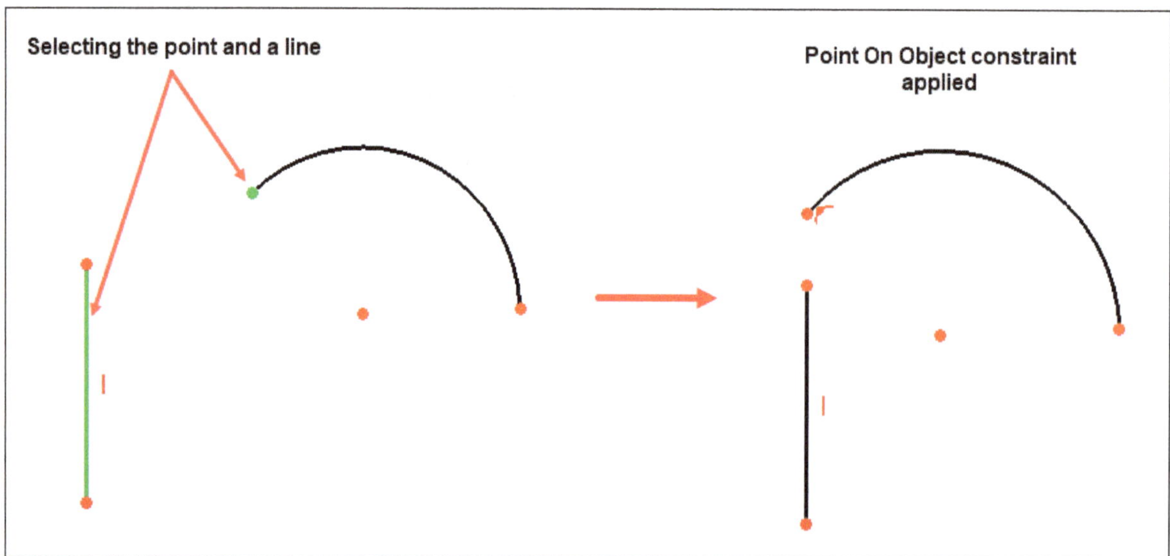

Figure-58. Applying point on object constraint

Vertical constraint

The **Vertical constraint** tool is used to constrain the selected lines or polyline elements to a true vertical orientation. The procedure to use this tool is discussed next. Note that only lines will be selectable from this tool.

- Click on the **Vertical constraint** tool from **Toolbar** in the **Sketcher** workbench; refer to Figure-59. You will be asked to select the lines or vertices.

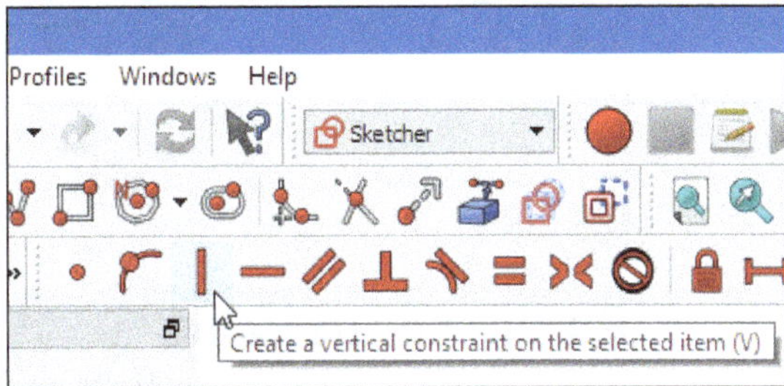

Figure-59. Vertical constraint tool

- Select the lines to be constrained vertically. The selected lines will become vertically constrained; refer to Figure-60.
- Press **ESC** key or press **RMB** to exit the tool.

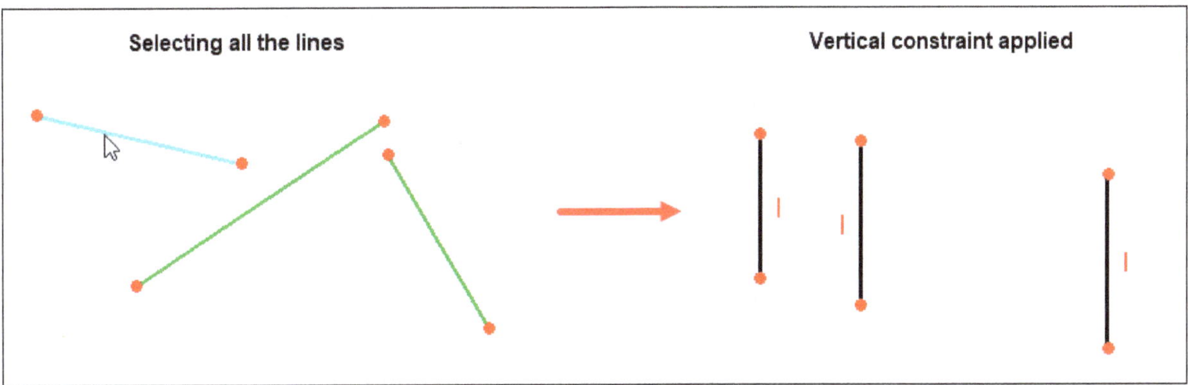

Figure-60. Applying vertical constraint

Horizontal constraint

The **Horizontal constraint** tool forces a selected line or lines in the sketch to be parallel to the horizontal axis of the sketch. The procedure to use this tool is discussed next.

- Click on the **Horizontal constraint** tool from **Toolbar** in the **Sketcher** workbench; refer to Figure-61. You will be asked to select the lines.

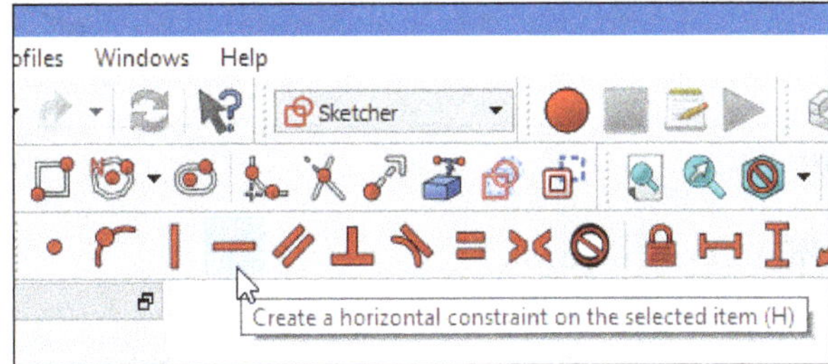

Figure-61. Horizontal constraint tool

- Select the lines which you want to constrain horizontally. The selected lines will become horizontally constrained; refer to Figure-62.

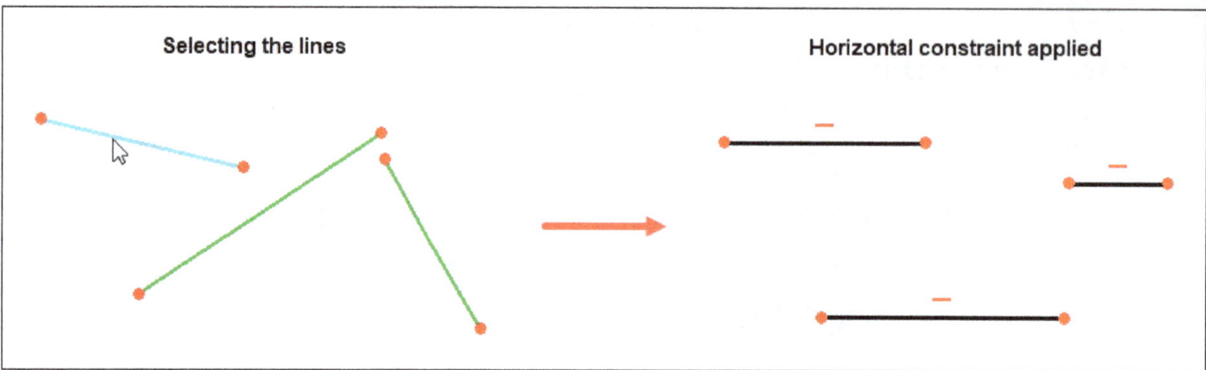

Figure-62. Applying horizontal constraint

- Press **ESC** key or press **RMB** to exit the tool.

Parallel constraint

The **Parallel constraint** tool forces two selected straight lines to be parallel to each other. The procedure to use this tool is discussed next.

- Click on the **Parallel constraint** tool from **Toolbar** in the **Sketcher** workbench; refer to Figure-63. You will be asked to select the lines.

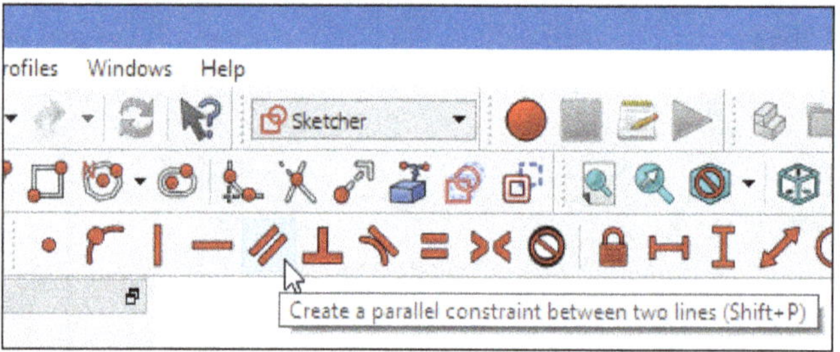

Figure-63. Parallel constraint tool

- Select the lines or line and axis which you want to make parallel.
- The selected lines will become parallel to each other; refer to Figure-64.

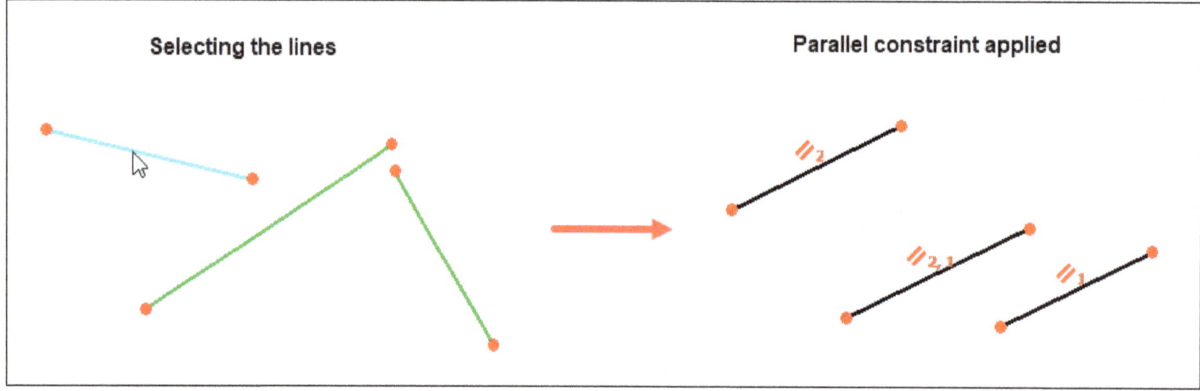

Figure-64. Applying parallel constraint

- Press **ESC** key or press **RMB** to exit the tool.

Perpendicular constraint

The **Perpendicular constraint** tool makes two lines perpendicular to each other or two curves to be perpendicular at their intersection point. The procedure to use this tool is discussed next.

- Click on the **Perpendicular constraint** tool from **Toolbar** in the **Sketcher** workbench; refer to Figure-65. You will be asked to select desired entities.

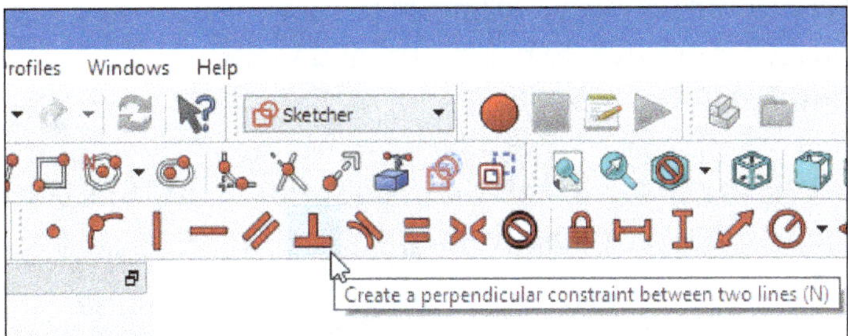

Figure-65. Perpendicular constraint tool

- Select the lines, curves, arcs, circles, ellipses, etc. which you want to make perpendicular. The entities selected will become perpendicular to each other; refer to Figure-66.
- Press **ESC** key or press **RMB** to exit the tool.

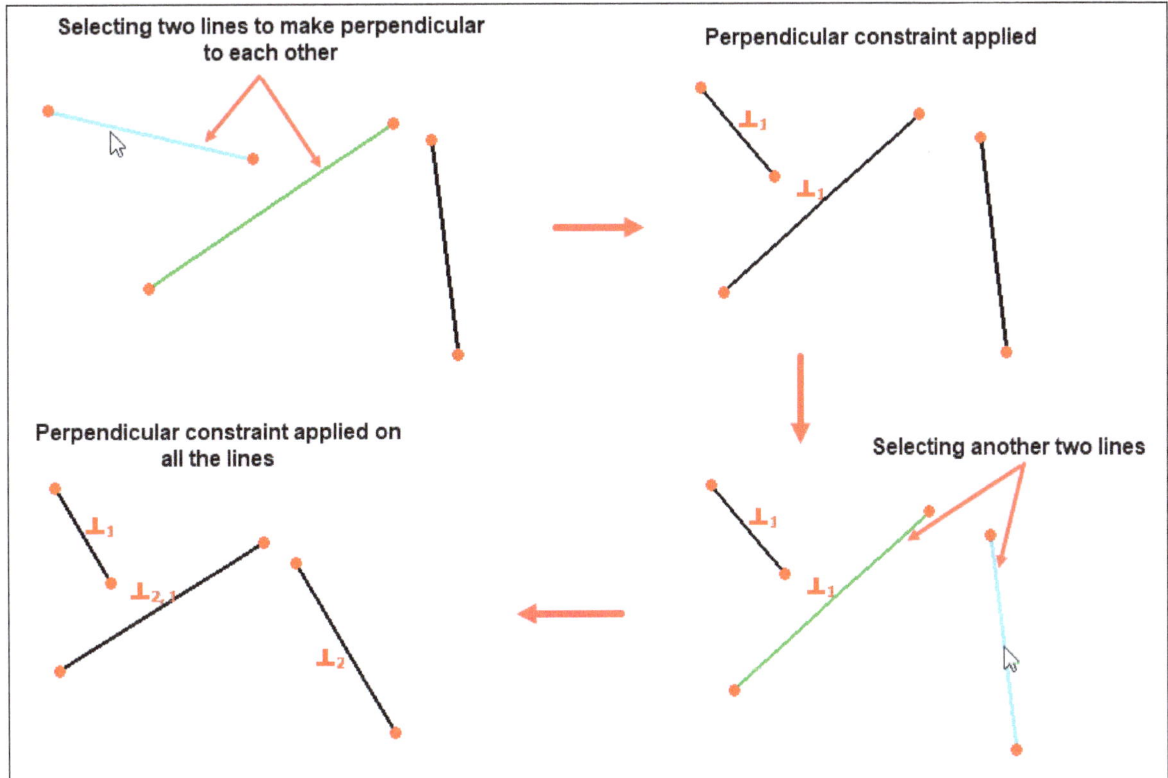

Figure-66. Applying perpendicular constraint

Tangent constraint

The **Tangent constraint** tool applies a tangent constraint between arcs, circles, curves, and lines. If you are applying tangent constraint between two lines then they will become collinear. A line segment does not have to lie directly on an arc or circle to be constrained tangent to that arc or circle. The procedure to use this tool is discussed next.

- Click on the **Tangent constraint** tool from **Toolbar** in the **Sketcher** workbench; refer to Figure-67. You will be asked to select the entities.

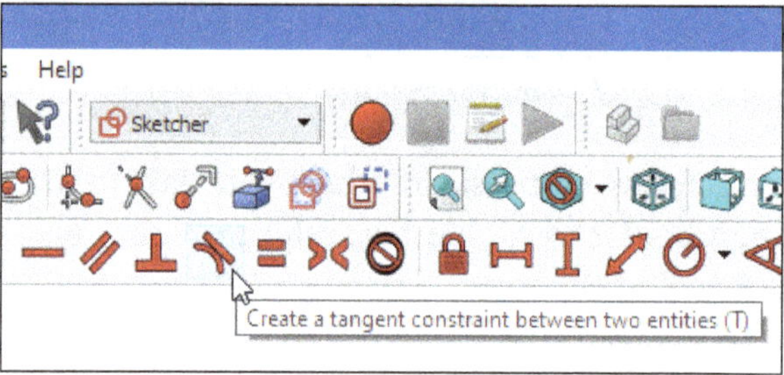

Figure-67. Tangent constraint tool

- Select the entities on which you want to apply tangent constraint. The entities will become tangentially constraint to each other; refer to Figure-68.

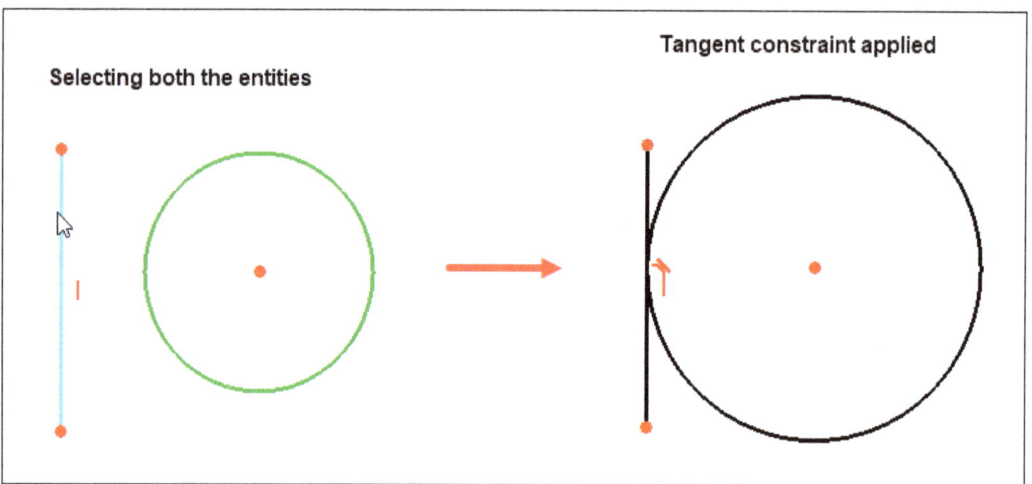

Figure-68. Applying tangent constraint

- Press **ESC** key or press **RMB** to exit the tool.

Equal constraint

The **Equal constraint** tool forces two or more sketch segments to have equal length. If applied to arcs or circles, the radii are constrained to be equal. The procedure to use this tool is discussed next.

Note that **Equal constraint** cannot be applied to geometric entities which are not of the same type.

- Click on the **Equal constraint** tool from **Toolbar** in the **Sketcher** workbench; refer to Figure-69. You will be asked to select the entities of same geometry type.

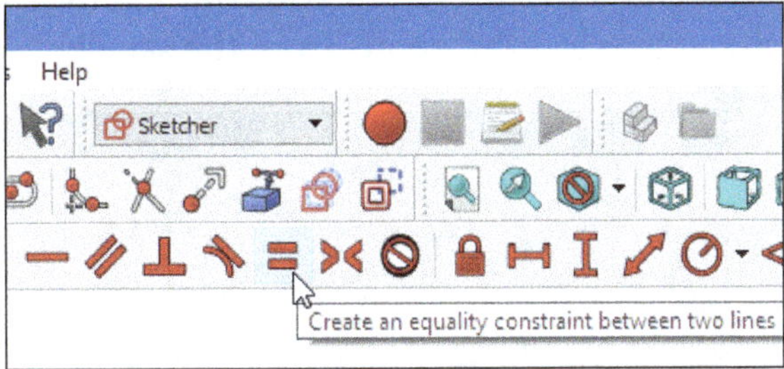

Figure-69. Equal constraint tool

- Select the entities of same geometry on which you want to apply equal constraint. The selected entities will become equally constrained; refer to Figure-70.
- Press **ESC** key or press **RMB** to exit the tool.

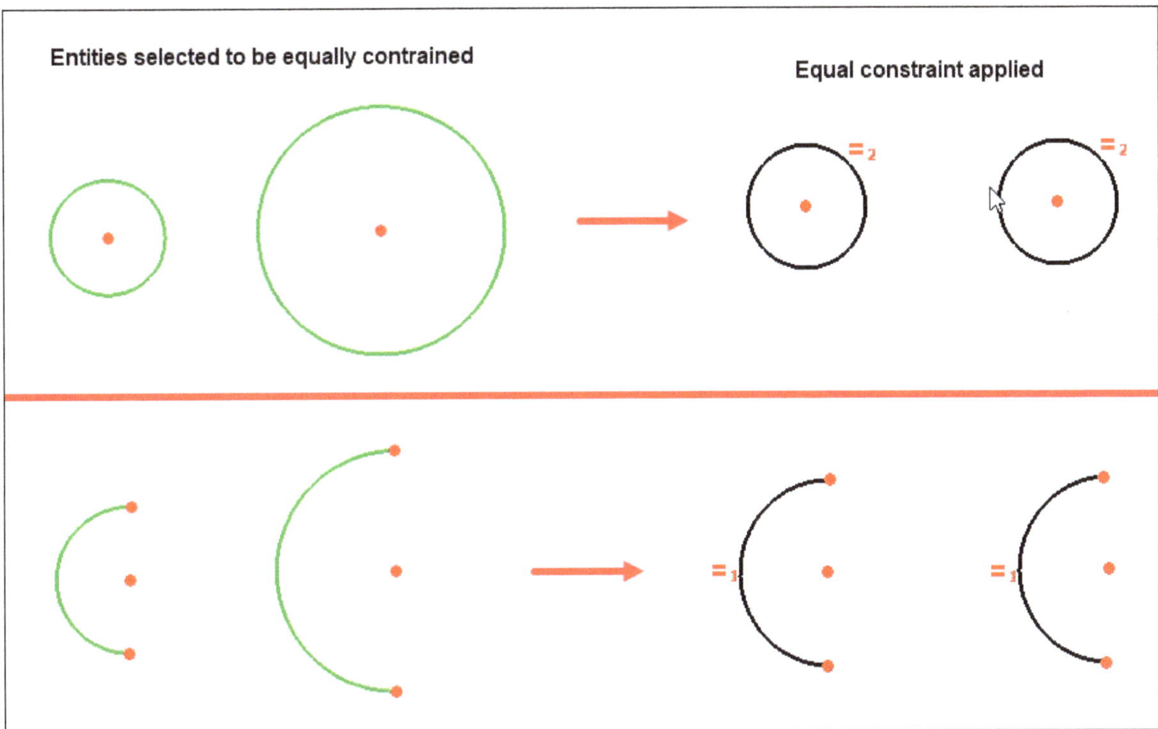

Figure-70. Applying equal constraint

Symmetric constraint

The **Symmetric constraint** tool is used to constrain the two selected points to be symmetrical about a given line, i.e., both selected points are constrained to lie on a normal to the line through both points and are constrained to be equidistant from the line. The procedure to use this tool is discussed next.

- Click on the **Symmetric constraint** tool from **Toolbar** in the **Sketcher** workbench; refer to Figure-71. You will be asked to select the points from the sketch.

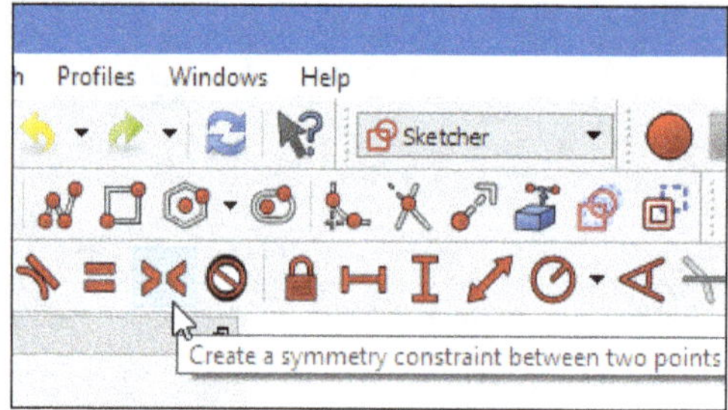

Figure-71. Symmetric constraint tool

- Select points and a line from the sketch. The selected entities will be constrained symmetrically; refer to Figure-72.

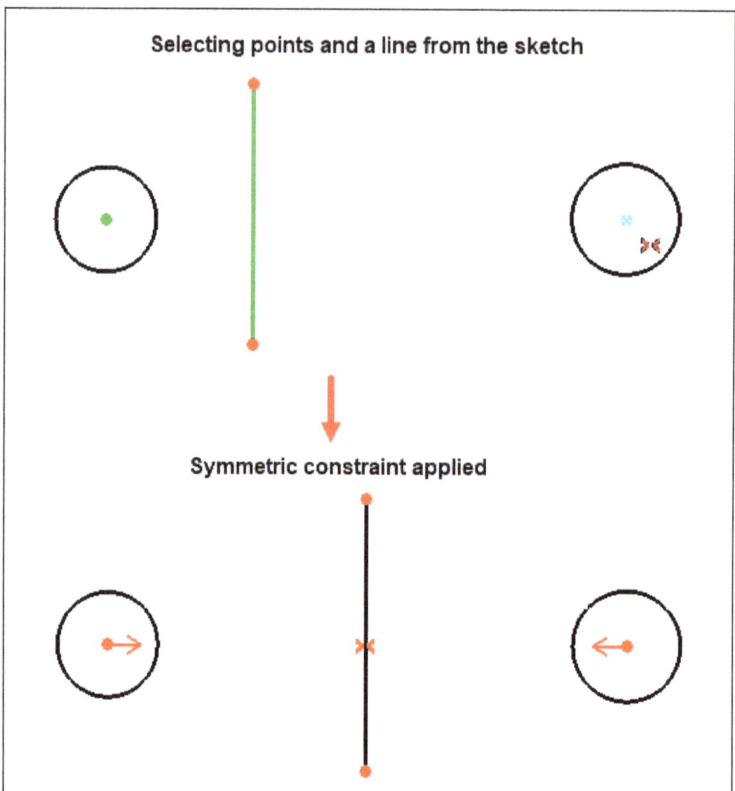

Figure-72. Applying symmetric constraint

- Press **ESC** key or press **RMB** to exit the tool.

Block constraint

The **Block constraint** tool is used to block a sketch entity from moving, that is, it prevents its vertices from changing their current positions. It is mainly useful to fix the position of B-splines, otherwise it is difficult to fully constrain it. The procedure to use this tool is discussed next.

- Click on the **Block constraint** tool from **Toolbar** in the **Sketcher** workbench; refer to Figure-73. You will be asked to select the entity.

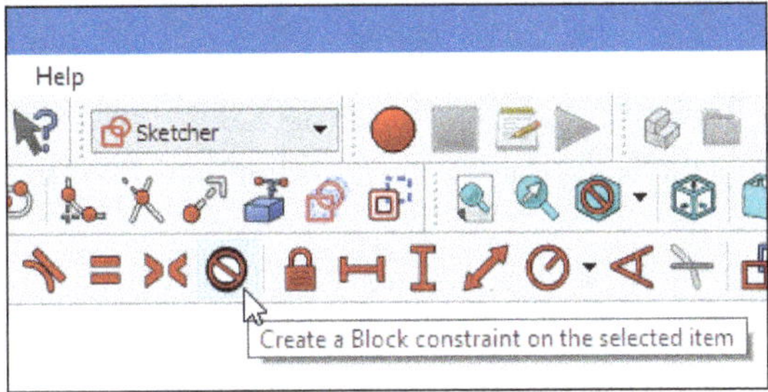

Figure-73. Block constraint tool

- Select the entity or entities whose position you want to block. Now, the position of entity will be blocked; refer to Figure-74.

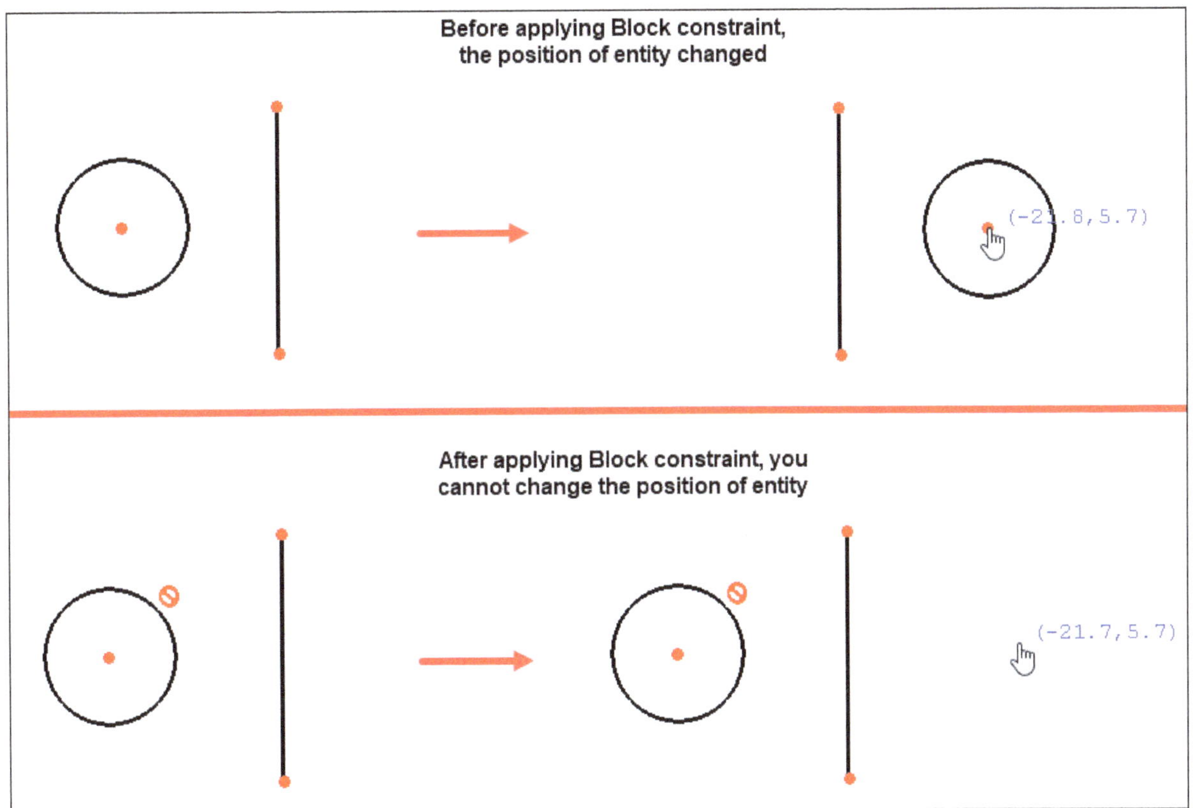

Figure-74. Applying block constraint

- Press **ESC** key or press **RMB** to exit the tool.

Dimensional constraints

Dimensional constraints are associated with numeric data for which you can use the expressions. The tools to apply dimensional constraints are available in the **Toolbar** of **Sketcher** workbench; refer to Figure-75. The procedures to use these tools are discussed next.

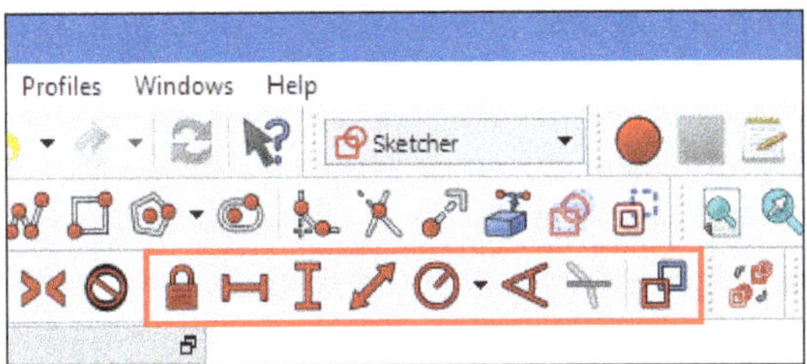

Figure-75. Dimensional constraint tools

Lock constraint

Lock constraint tool applies **Horizontal distance** and **Vertical distance** dimensions to selected vertices (points) in the sketch.

If a single vertex is selected, the horizontal and vertical distance constraints will refer to the sketch origin point. If two or more points are selected, horizontal and vertical distance constraints will be added for each pair of points.

The procedure to use this tool is discussed next.

- Click on the **Lock constraint** tool from **Toolbar** in the **Sketcher** workbench; refer to Figure-76. You will be asked to select the vertices.

Figure-76. Lock constraint tool

- Select one or more vertices from the sketch on which you want to apply lock constraint. The lock constraint will be applied to the sketch entity; refer to Figure-77.

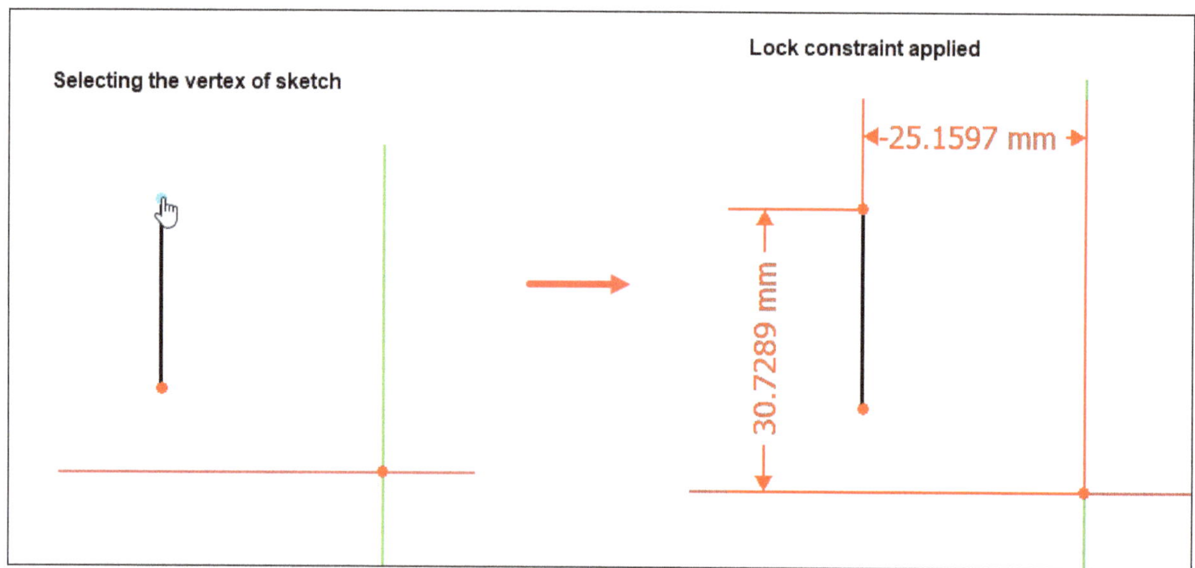

Figure-77. Applying lock constraint

- Double click on the dimension value in the 3D view and modify the value as desired in the edit box displayed.
- Press **ESC** key or press **RMB** to exit the tool.

Horizontal distance constraint

The **Horizontal distance constraint** fixes the horizontal distance between two points or line endpoints. If only one point is selected, the distance is set to the sketch origin. The procedure to use this tool is discussed next.

- Click on the **Horizontal distance constraint** tool from **Toolbar** in the **Sketcher** workbench; refer to Figure-78. You will be asked to select endpoints or a horizontal line.

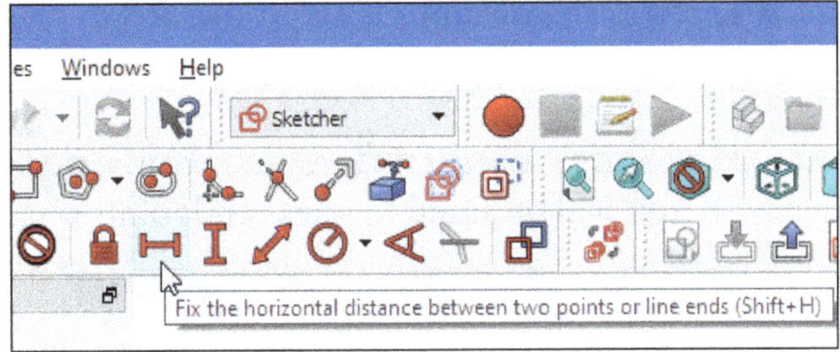

Figure-78. Horizontal distance constraint tool

- Select endpoints of a sketch or select a line to fix the horizontal distance or length. The **Insert length** dialog box will be displayed; refer to Figure-79, asking you to specify the horizontal length of entity.

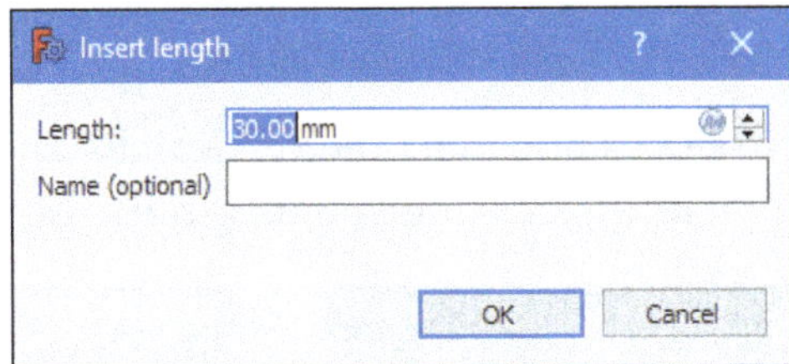

Figure-79. Insert length dialog box

- Specify desired value in the **Length** edit box of the dialog box.
- You can also specify name of the dimension parameter in the **Name** edit box of the **Insert Length** dialog box.
- Click on **OK** button from the dialog box. The horizontal length or distance of the line/point will be modified; refer to Figure-80.

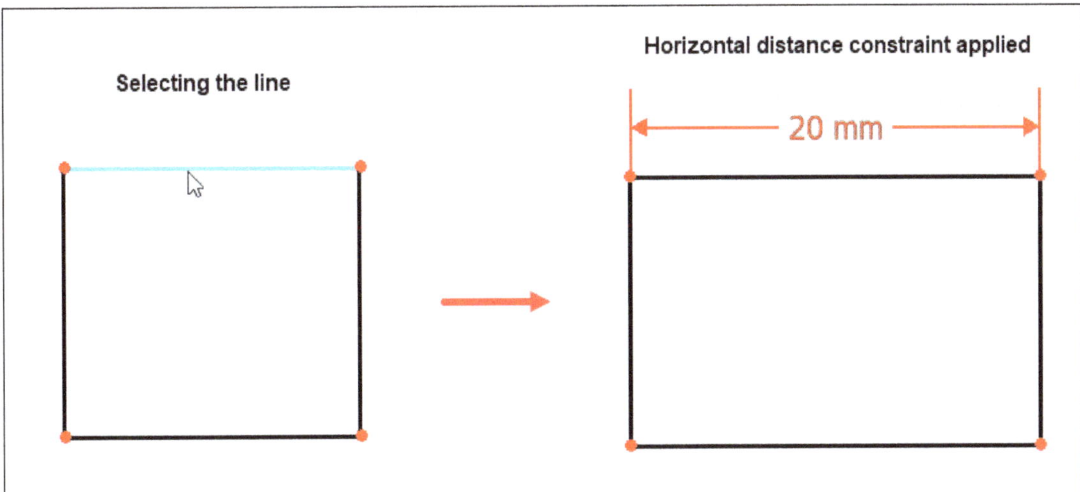

Figure-80. Applying horizontal distance constraint

- Press **ESC** key or press **RMB** to exit the tool.

Vertical distance constraint

The **Vertical distance constraint** tool fixes the vertical distance between two points or line endpoints. If only one point is selected, the distance is set to the sketch origin. The procedure to use this tool is discussed next.

- Click on the **Vertical distance constraint** tool from **Toolbar** in the **Sketcher** workbench; refer to Figure-81. You will be asked to select endpoints or a line.

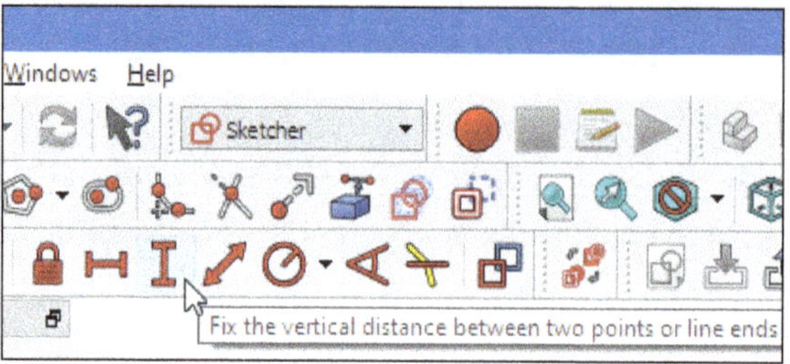

Figure-81. Vertical distance constraint tool

- Select endpoints of a sketch or select a line to fix the vertical distance. The **Insert length** dialog box will be displayed; refer to Figure-82, asking you to specify the vertical length of entity.

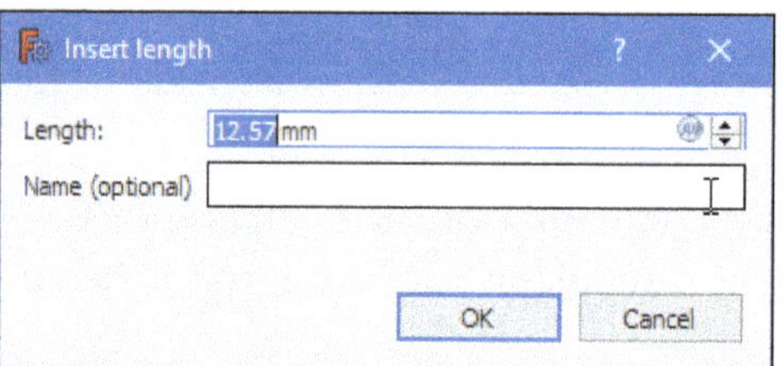

Figure-82. Insert length dialog box

- Specify desired value in the **Length** edit box and click on **OK** button from the dialog box.
- The vertical length or distance of the sketch will be modified; refer to Figure-83.

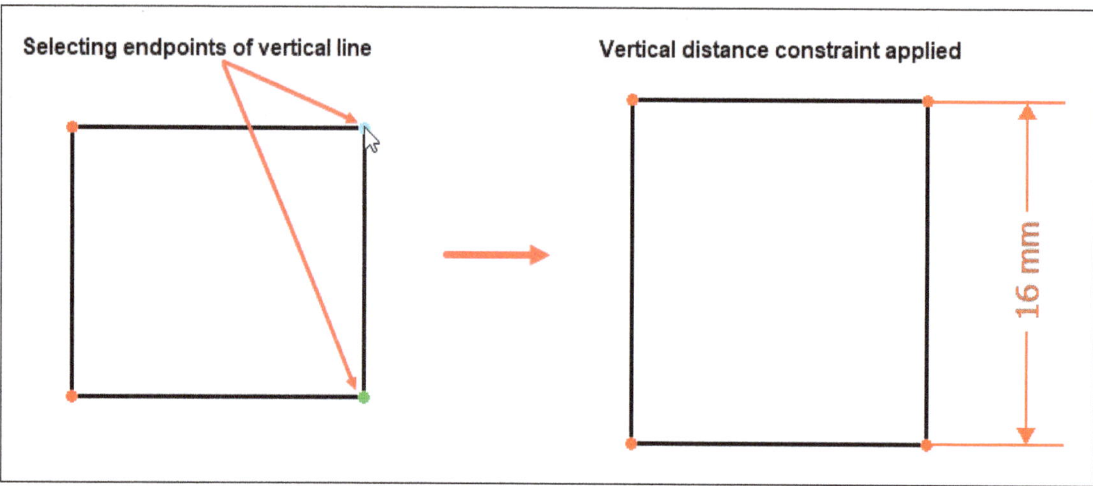

Figure-83. Applying vertical distance constraint

- Press **ESC** key or press **RMB** to exit the tool.

Distance constraint

The **Distance constraint** tool defines the distance of a selected line by constraining its length or defines the distance between two points by constraining the distance between them. Using this tool, you can specify aligned distance of selected line. The procedure to use this tool is discussed next.

- Click on the **Distance constraint** tool from **Toolbar** in the **Sketcher** workbench; refer to Figure-84. You will be asked to select the entities.

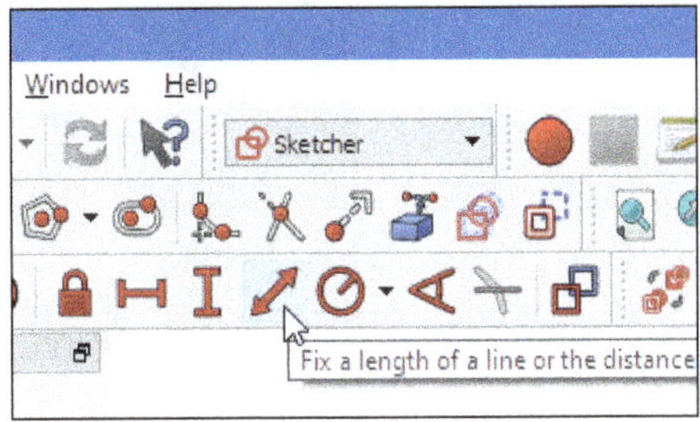

Figure-84. Distance constraint tool

- Select two points/a line/a point and a line to apply the distance constraint in the sketch. The **Insert length** dialog box will be displayed as discussed earlier asking you to specify the distance of selected entity.
- Specify desired distance value in the **Length** edit box and click on **OK** button from the dialog box. The length or distance of the selected entity will be modified; refer to Figure-85.

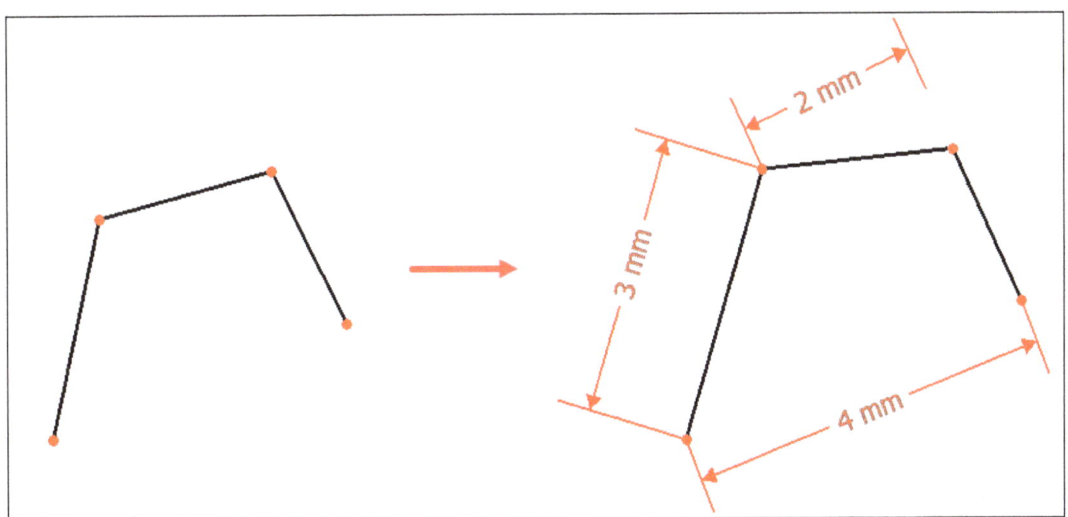

Figure-85. Applying distance constraint

- Press **ESC** key or press **RMB** to exit the tool.

Constrain radius

The **Constrain radius** tool defines the radius of a selected arc or circle by constraining the radius. The procedure to use this tool is discussed next.

- Click on the **Constrain radius** tool from **Constrain an arc or a circle** drop-down in the **Toolbar** of **Sketcher** workbench; refer to Figure-86. You will be asked to select a circle or arc.

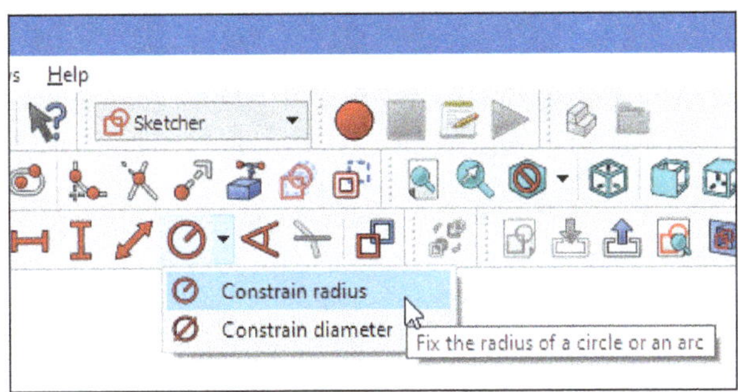

Figure-86. Constrain radius tool

- Select one or more circles and arcs to define the radius. The **Change radius** dialog box will be displayed; refer to Figure-87.

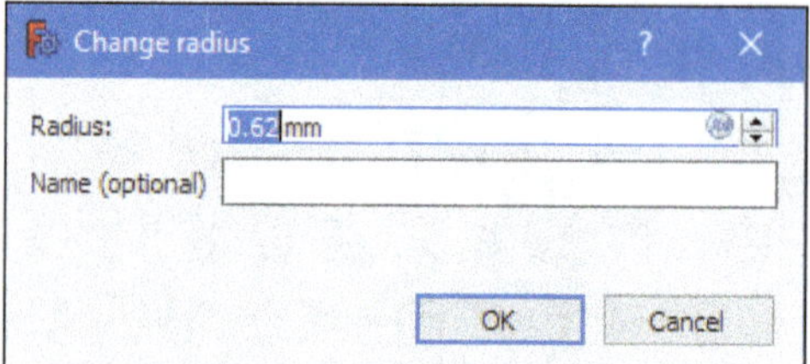

Figure-87. Change radius dialog box

- Specify desired value in **Radius** edit box and click on **OK** button from the dialog box. The radius of selected entity will be modified; refer to Figure-88.

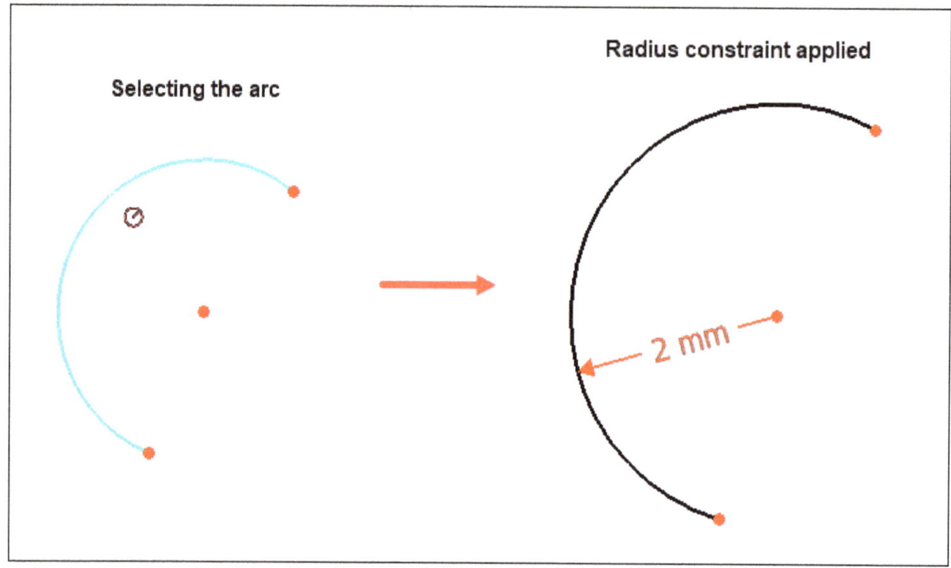

Figure-88. Applying radius constraint

- Press **ESC** key or press **RMB** to exit the tool.

Constrain diameter

The **Constrain diameter** tool defines the diameter of selected arc or circle. The procedure to use this tool is discussed next.

- Click on the **Constrain diameter** tool from **Constrain an arc or circle** drop-down in the **Toolbar** of **Sketcher** workbench; refer to Figure-89. You will be asked to select an arc or circle.

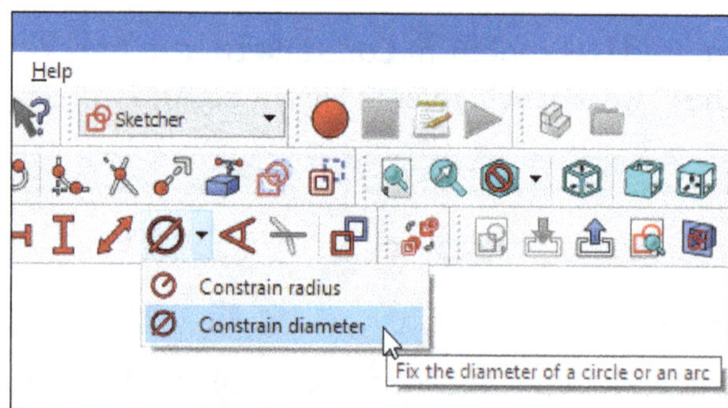

Figure-89. Constrain diameter tool

- Select one or more circles or arcs to define the diameter. The **Change diameter** dialog box will be displayed; refer to Figure-90.

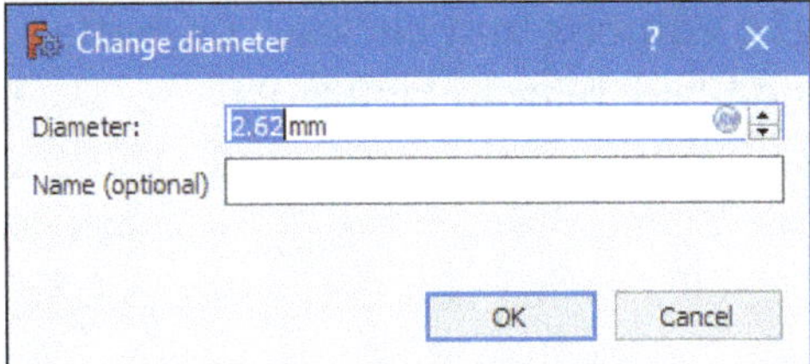

Figure-90. Change diameter dialog box

- Specify desired value in **Diameter** edit box and click on **OK** button from the dialog box. The diameter of selected entity will be modified; refer to Figure-91.

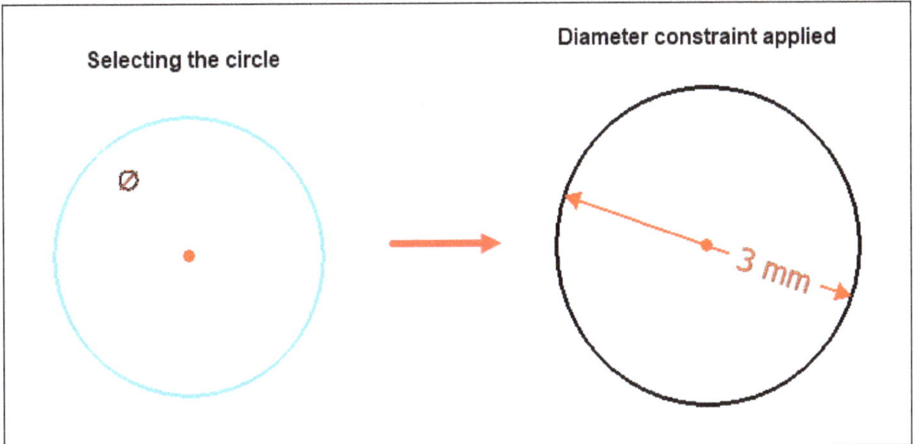

Figure-91. Applying diameter constraint

- Press **ESC** key or press **RMB** to exit the tool.

Angle constraint

The **Angle constraint** tool defines the internal angle between two selected lines. It is capable of setting slopes of individual lines, angles between lines, angles of intersections of curves, and angle spans of circular arcs. The procedure to use this tool is discussed next.

- Click on the **Angle constraint** tool from **Toolbar** in the **Sketcher** workbench; refer to Figure-92. You will be asked to select one or multiple entities.

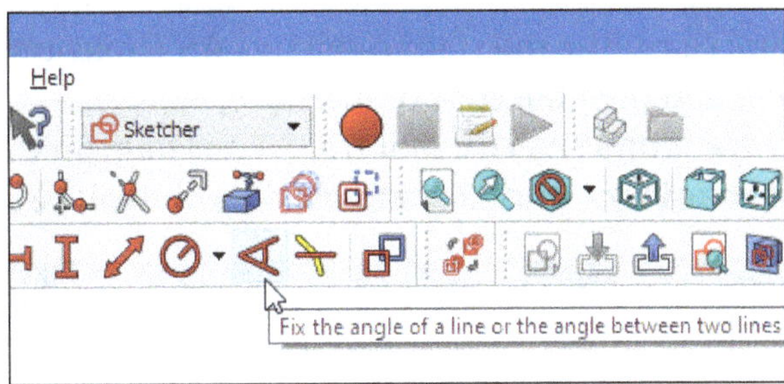

Figure-92. Angle constraint tool

- Select the entities from the sketch, the **Insert angle** dialog box will be displayed; refer to Figure-93.

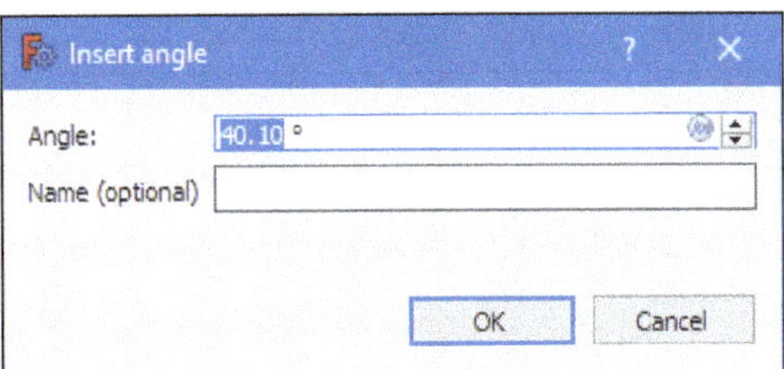

Figure-93. Insert angle dialog box

- Specify desired value in **Angle** edit box and click on **OK** button from the dialog box. The angle between the entities will be modified; refer to Figure-94.

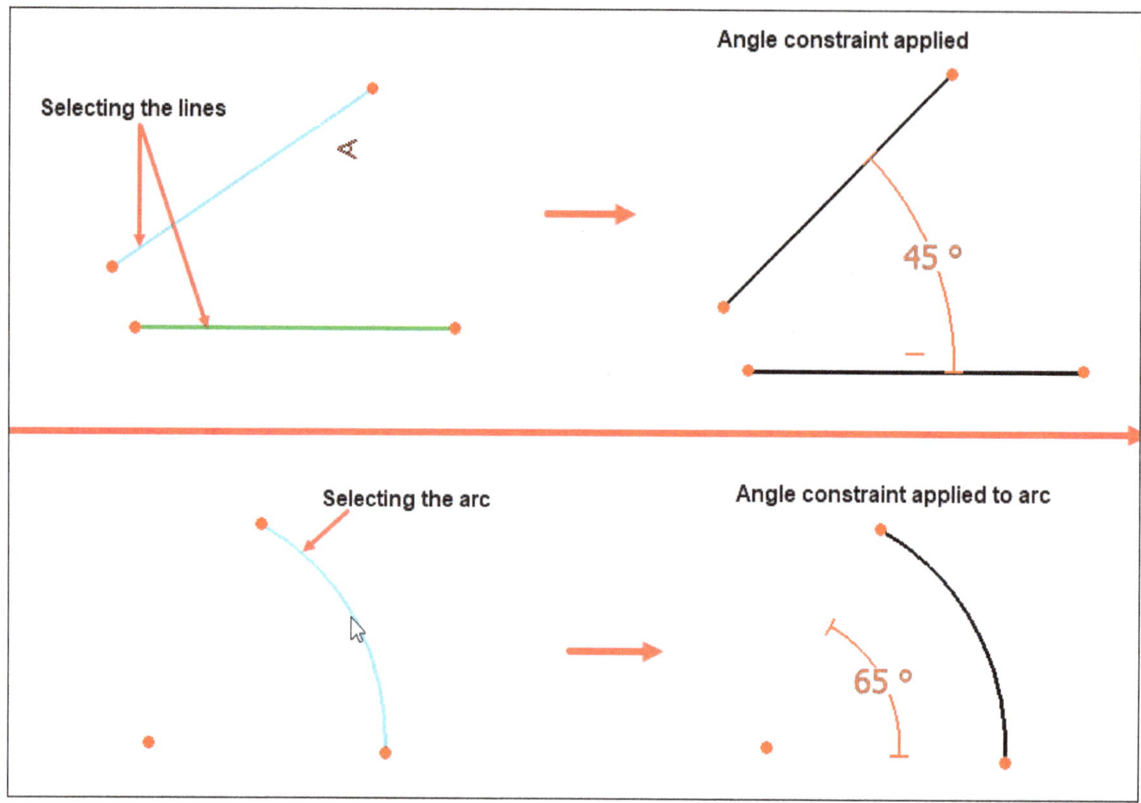

Figure-94. Applying angle constraint

- Press **ESC** key or press **RMB** to exit the tool.

Snell's Law constraint

The **Snell's Law constraint** tool constrains two lines to follow the law of refraction of light as it penetrates through an interface, where two materials of different refraction indices meet. The procedure to use this tool is discussed next.

- You need two lines to follow a beam of light and a curve to act as an interface. The lines should be on different sides of the interface. The interface can be a line, circle/arc, or ellipse/arc of ellipse.
- Select the endpoint of lines and then the edge of interface.
- Click on the **Snell's Law constraint** tool from **Toolbar** in the **Sketcher** workbench; refer to Figure-95. The **Refractive index ratio** dialog box will be displayed; refer to Figure-96.

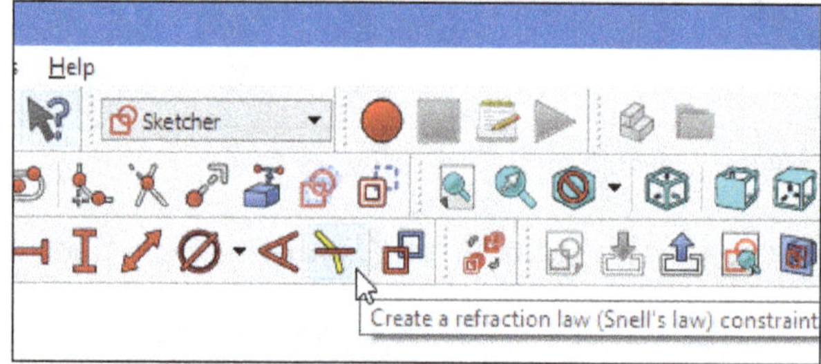

Figure-95. Snell's Law constraint tool

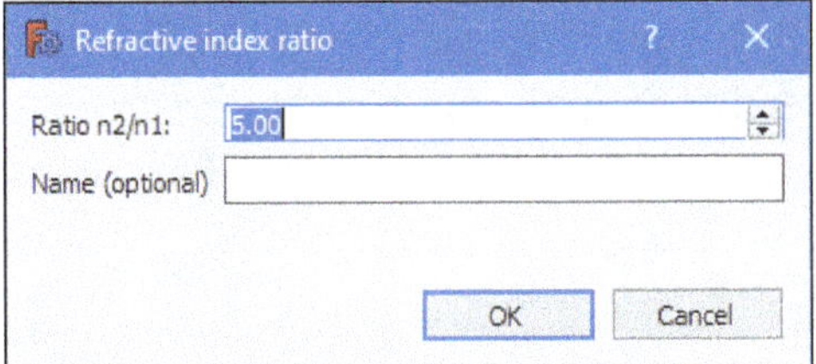

Figure-96. Refractive index ratio dialog box

- Specify desired value in **Ratio n2/n1** edit box and click on **OK** button from the dialog box. The endpoints will be constrained on the interface and the Snell's law will become constrained; refer to Figure-97.

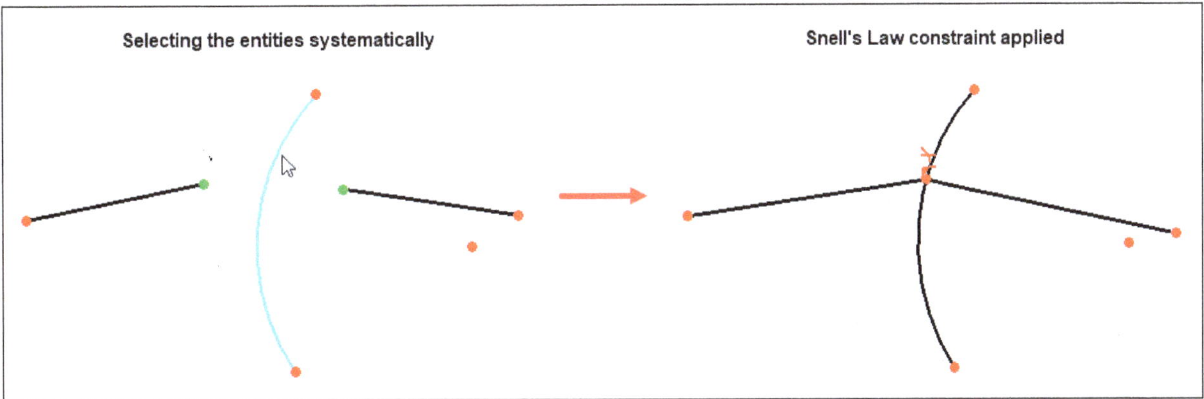

Figure-97. Applying snell's law constraint

- Press **ESC** key or press **RMB** to exit the tool.

Toggle Driving constraint

The **Toggle Driving constraint** tool is used to switch between driving and driven dimension constraints (except **Snell's Law constraint**). The icons in the toolbar turns blue and rather than dimensional constraints, reference dimensions are created; refer to Figure-98.

Figure-98. Dimensional constraints icons turns blue

Reference dimensions do not constrain the sketch. The procedure to use this tool is discussed next.

- Click on the **Toggle driving constraint** tool from **Toolbar** in the **Sketcher** workbench; refer to Figure-99. The dimensional constraint icons turn from red to blue.

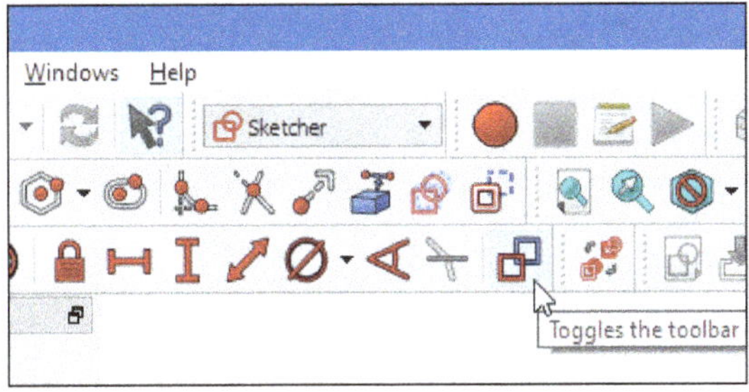

Figure-99. Toggles driving constraint tool

- Click on the **Vertical constraint** tool from **Toolbar** in the **Sketcher** workbench. You will be asked to select the entity from sketch to create reference dimension.
- Select the entity from the sketch of which you want to create reference dimension. The reference dimension will be applied to the selected entity; refer to Figure-100.

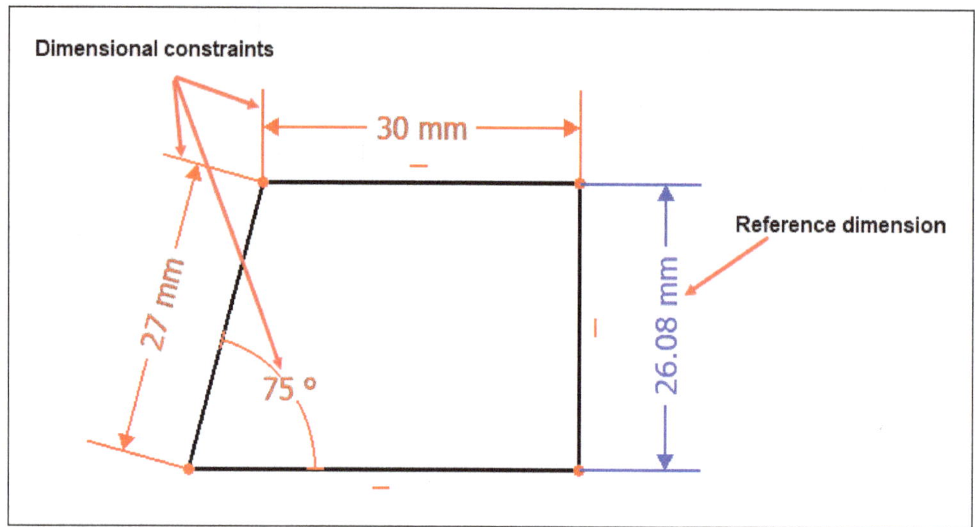

Figure-100. Applying reference dimension

- Press **ESC** key or press **RMB** to exit the tool.

SKETCHER TOOLS

The **Sketcher** tools in **FreeCAD** are available in the **Sketcher** workbench of **toolbar**; refer to Figure-101. You can also access the sketcher tools from the **Sketcher tools** cascading menu in the **Sketch** menu; refer to Figure-102. The procedure to use these tools are discussed next.

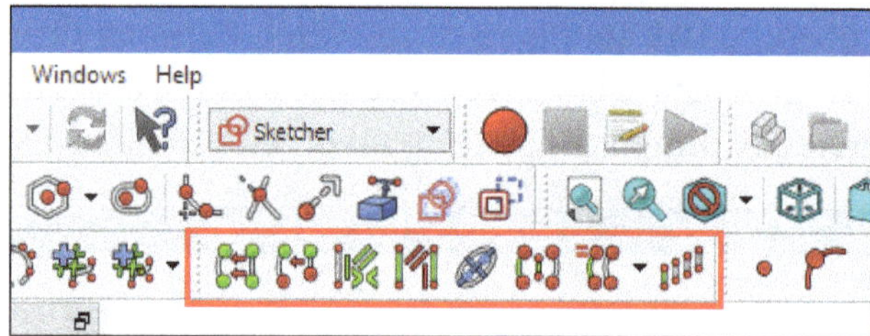

Figure-101. Sketcher tools

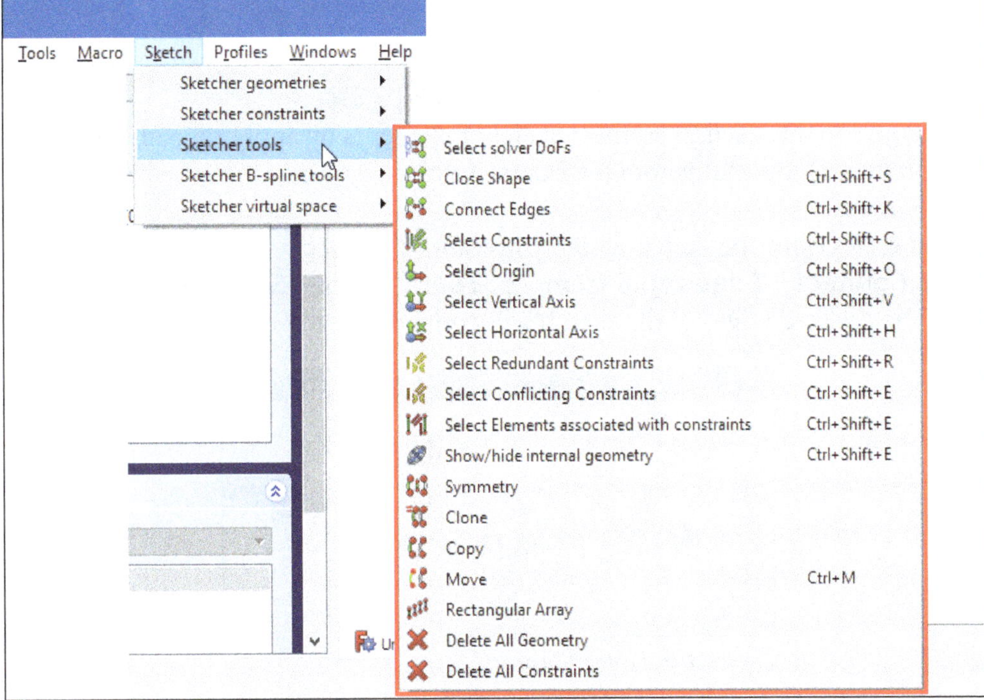

Figure-102. Sketcher tools in sketch menu

Close Shape

The **Close Shape** tool is used to create a closed shape by applying coincident constraints to endpoints. The procedure to use this tool is discussed next.

Note that the tool will connect the elements in the order of their selection. While editing a sketch, select two or more sketcher elements from the **Model** panel of **Combo View** or in the 3D view area.

- Select desired entities from the sketch which you want to connect.
- Click on the **Close Shape** tool from **Toolbar** in the **Sketcher** workbench; refer to Figure-103. The selected entities will connect to each other and become a closed shape entity; refer to Figure-104.

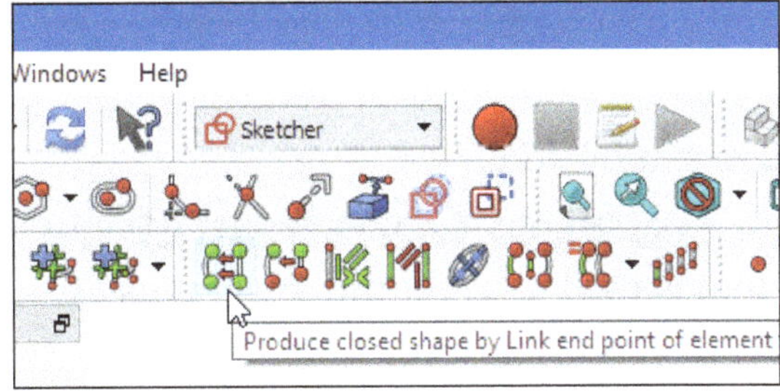

Figure-103. Close Shape tool

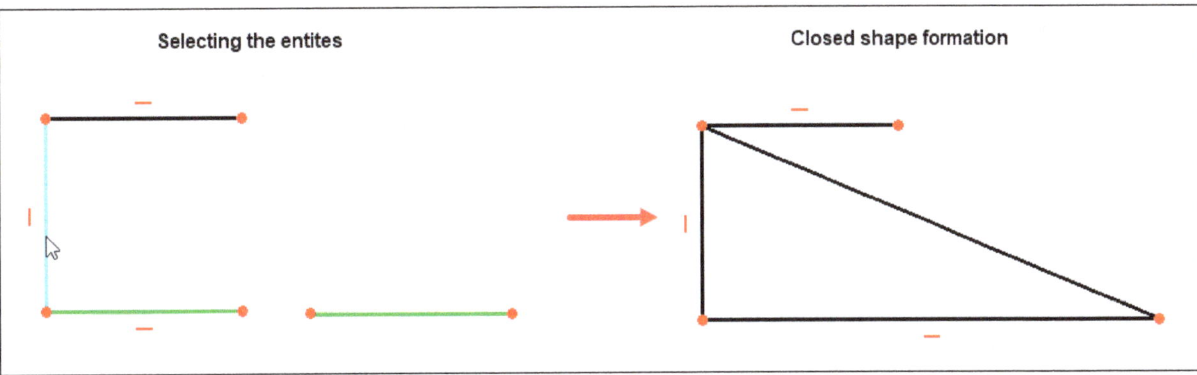

Figure-104. Creation of closed shape

- Press **ESC** key or press **RMB** to exit the tool.

Connect Edges

The **Connect Edges** tool is used to connect sketcher elements by applying coincidents constraints to endpoints. The procedure to use this tool is discussed next.

- Select desired edges from the sketch which you want to connect.
- Click on the **Connect Edges** tool from **Toolbar** in the **Sketcher** workbench; refer to Figure-105. The selected edges of the sketch will connect to each other; refer to Figure-106.

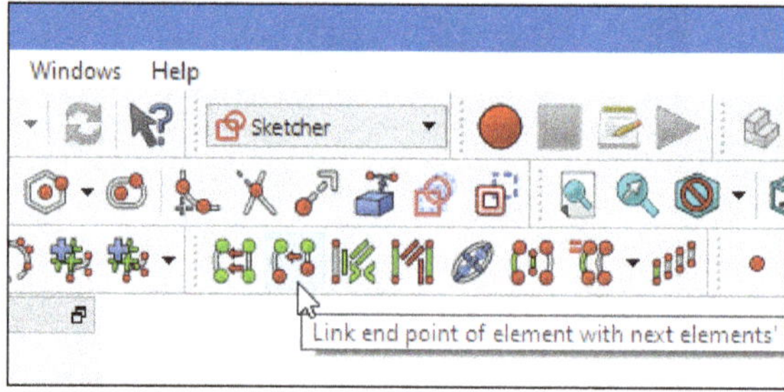

Figure-105. Connect edges tool

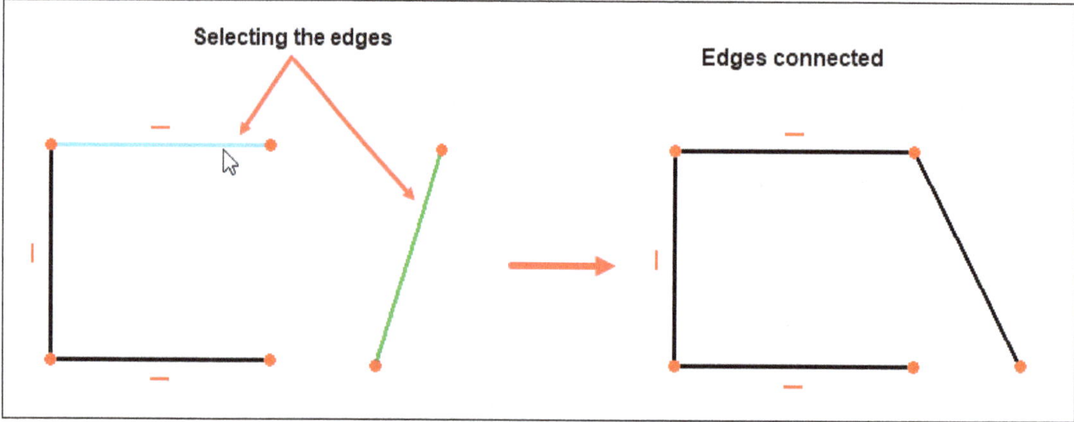

Figure-106. Connecting the edges

- Press **ESC** key or press **RMB** to exit the tool.

Select Constraints

The **Select Constraints** tool is used to select the constraints associated to the selected sketcher elements. The procedure to use this tool is discussed next.

- Select the entities of the sketch whose constraints are to be selected.
- Click on the **Select Constraints** tool from **Toolbar** in the **Sketcher** workbench; refer to Figure-107. The constraints associated with selected entities will be selected; refer to Figure-108.

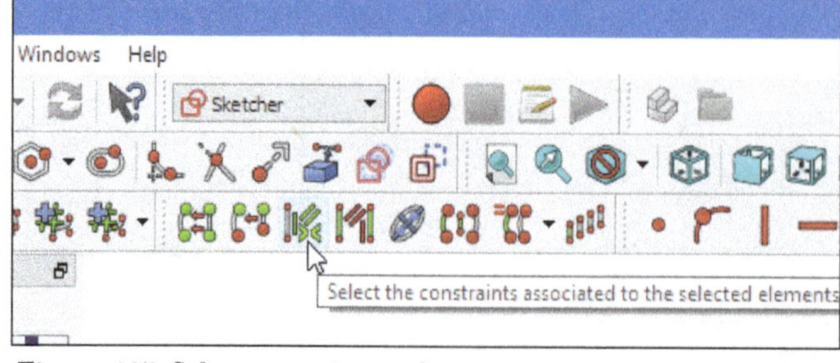

Figure-107. Select constraints tool

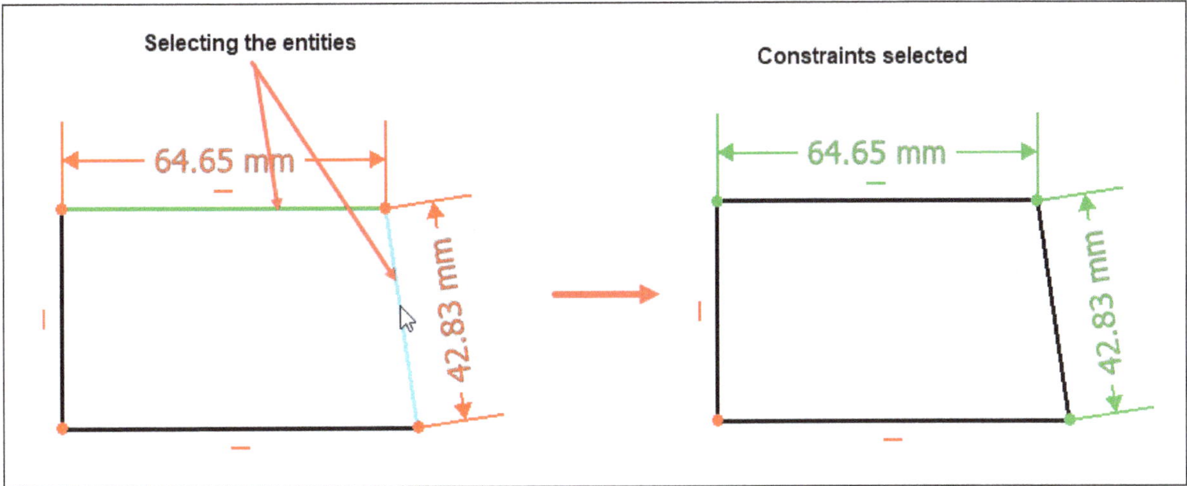

Figure-108. Selection of constraints

- Press **ESC** key or press **RMB** to exit the tool.

Select Elements

The **Select Elements** tool is used to select sketcher elements associated with constraints. The procedure to use this tool is discussed next.

- Select the constraints of those entities which you want to select.
- Click on the **Select Elements** tool from **Toolbar** in the **Sketcher** workbench; refer to Figure-109. The entities associated with selected constraints will be selected; refer to Figure-110.

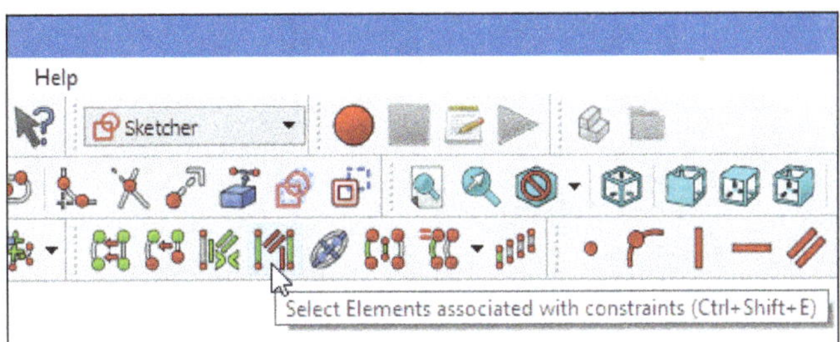

Figure-109. Select elements tool

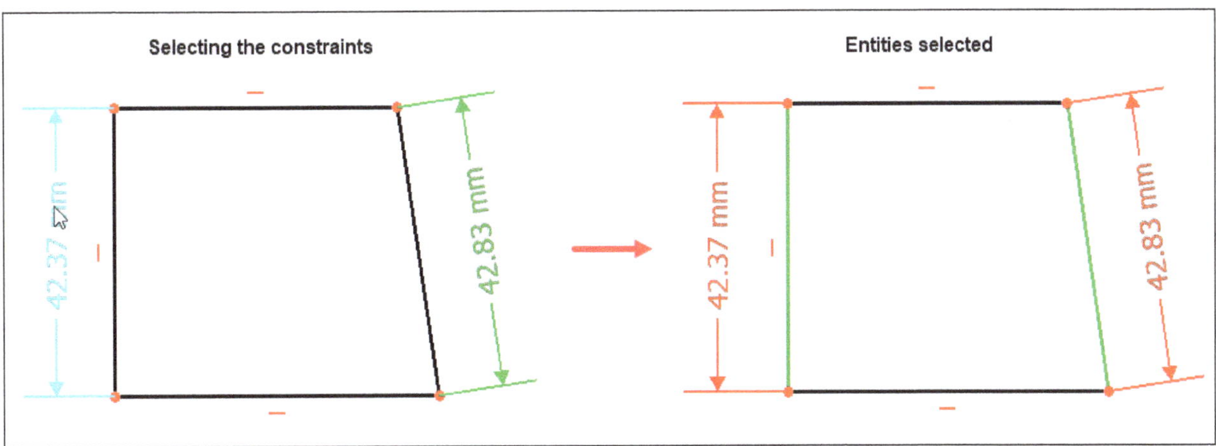

Figure-110. Selection of elements

- Press **ESC** key or press **RMB** to exit the tool.

Show/Hide internal geometry

The **Show/Hide internal geometry** tool is used to show or hide unnecessary internal geometry or recreates missing geometry of a selected ellipse, arc of ellipse/hyperbola/parabola, or B-spline. The procedure to use this tool is discussed next.

- Select desired entity of a sketch of which you want to show or hide the unnecessary geometry.
- Click on the **Show/Hide internal geometry** tool from **Toolbar** in the **Sketcher** workbench; refer to Figure-111. The geometry of selected entity will be hidden from the sketch; refer to Figure-112.

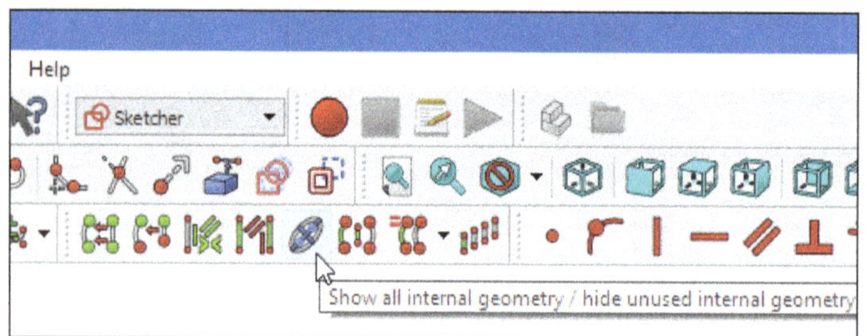

Figure-111. Show hide internal geometry tool

- Press **ESC** key or press **RMB** to exit the tool.

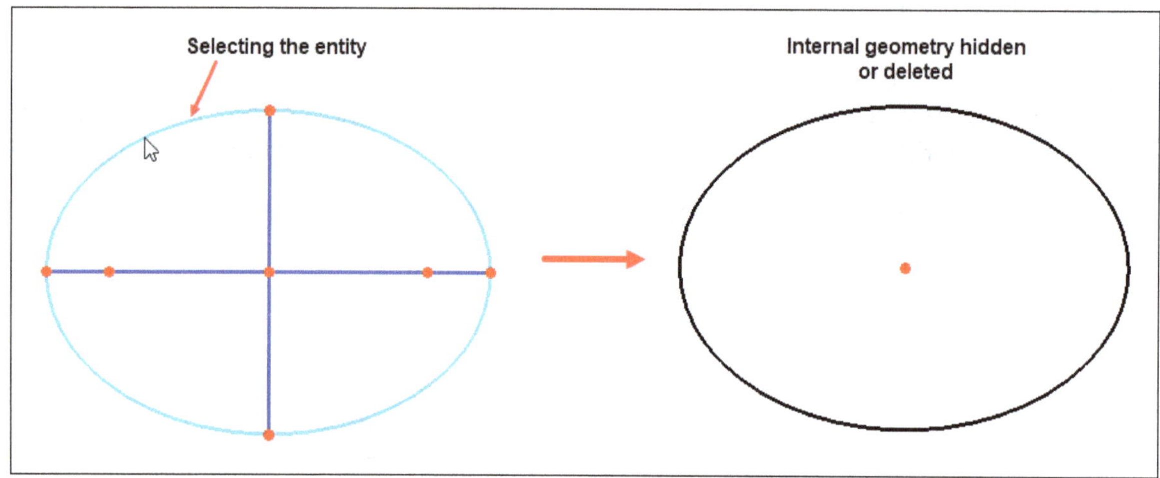

Figure-112. Internal geometry deleted

Symmetry

The **Symmetry** tool is used to mirror sketcher geometry in reference to a chosen line or sketch axis. Note that last selected entity will be used as symmetry axis. The procedure to use this tool is discussed next.

- Select the geometry you want to copy and then a line or sketch axis to be used as symmetry axis.
- Click on the **Symmetry** tool from **Toolbar** in the **Sketcher** workbench; refer to Figure-113. The selected geometry will be copied symmetrically to the selected line or sketch axis; refer to Figure-114.

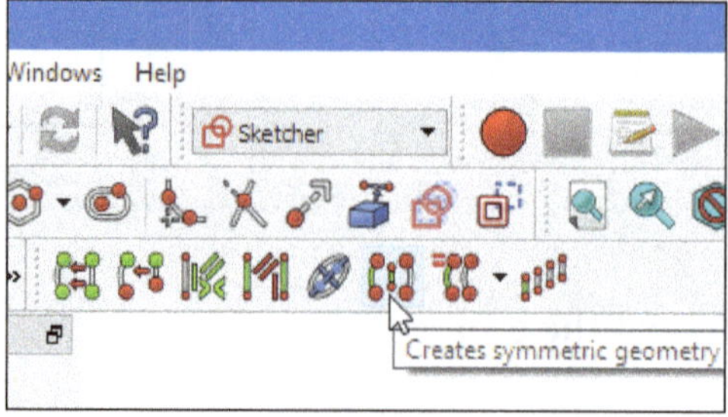

Figure-113. Symmetry tool

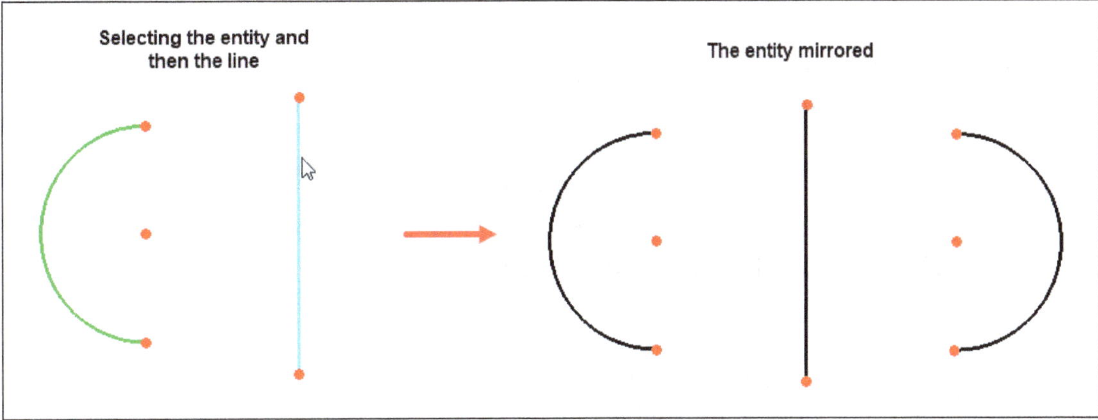

Figure-114. Symmetry entity creation

- Press **ESC** key or press **RMB** to exit the tool.

Clone

The **Clone** tool is used to clone the selected sketch elements from one point to another using the last selected point as reference. The procedure to use this tool is discussed next.

- Select the sketch entity which you want to clone.
- Click on the **Clone** tool from a drop-down available in the **Toolbar** of **Sketcher** workbench; refer to Figure-115. A line with clone icon will be attached to the cursor asking you to place the clone entity.

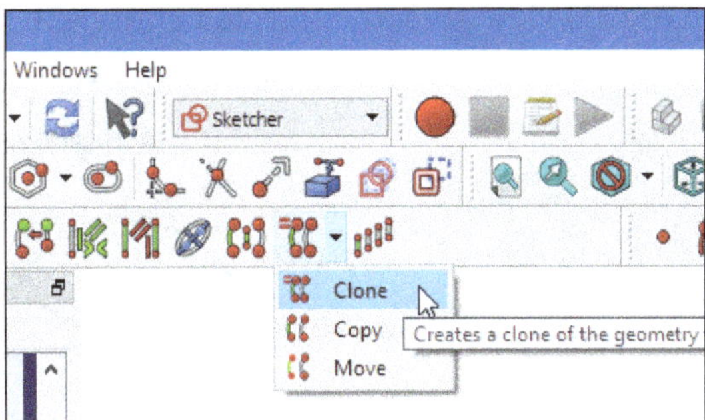

Figure-115. Clone tool

- Click on desired location in the 3D view area, the clone of selected entity will be placed; refer to Figure-116.

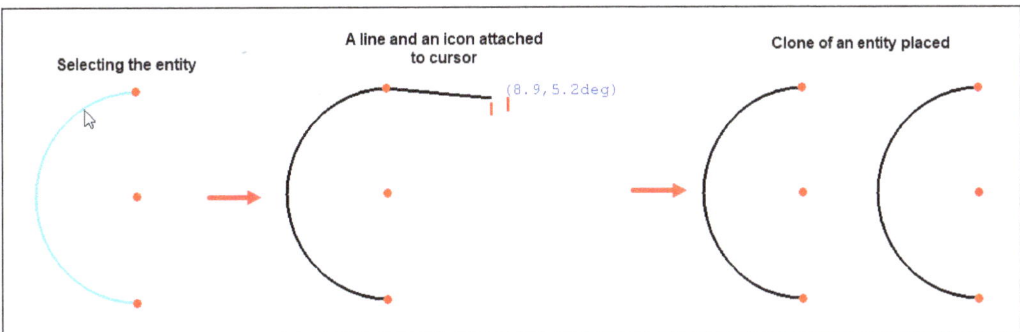

Figure-116. Creation of clone

- Press **ESC** key or press **RMB** to exit the tool.

Copy

The **Copy** tool is used to copy selected sketch elements from one point to another using the last selected point as reference. The procedure to use this tool is same as discussed for the **Clone** tool.

Move

The **Move** tool moves the selected sketch elements from one point to another using the last selected point as reference. The procedure to use this tool is discussed next.

- Select the entity which you want to move from one place to another.
- Click on the **Move** tool from the drop-down in the **Toolbar** of **Sketcher** workbench; refer to Figure-117. A line and a move icon will be attached to the cursor asking you to move the entity.

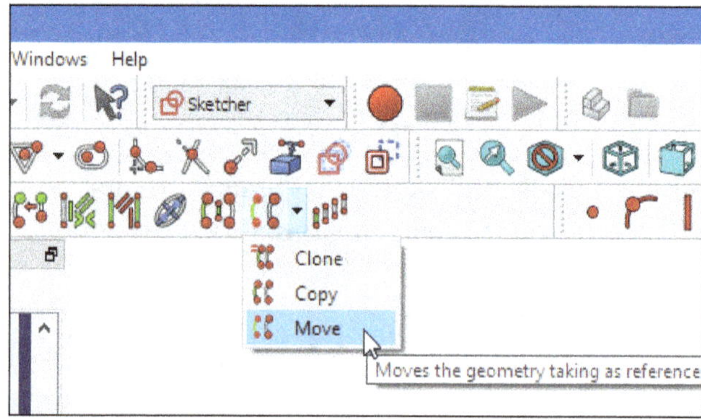

Figure-117. Move tool

- Click on desired location in the 3D view area, the selected entity will be moved; refer to Figure-118.

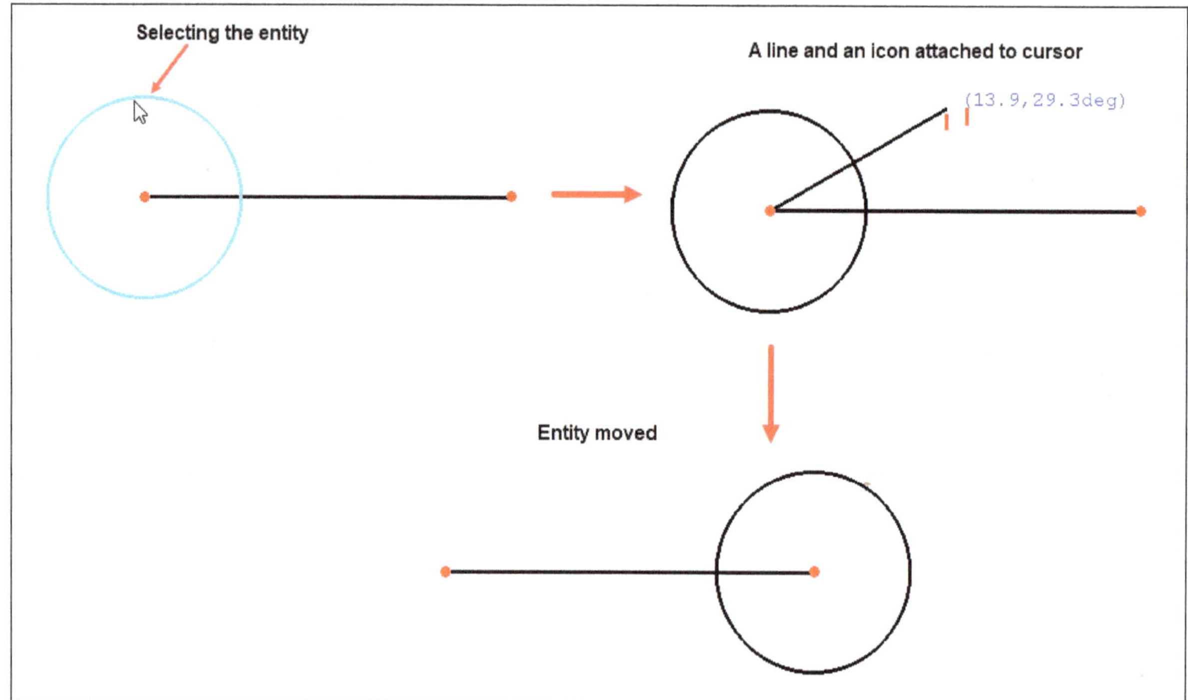

Figure-118. Entity moved

- Press **ESC** key or press **RMB** to exit the tool.

Rectangular Array

The **Rectangular Array** tool creates a rectangular array pattern of the geometry taking last selected point as reference. The procedure to use this tool is discussed next.

- Select the entity of which you want to create a rectangular array pattern.
- Click on the **Rectangular Array** tool from **Toolbar** in the **Sketcher** workbench; refert to Figure-119. The **Create array** dialog box will be displayed; refer to Figure-120, asking you to specify the parameters.

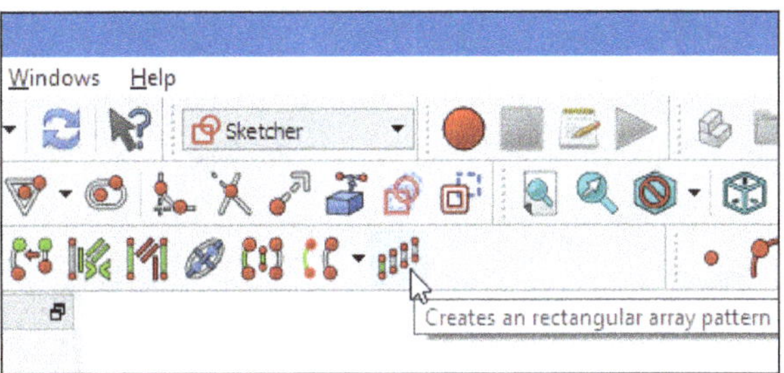

Figure-119. Rectangular array tool

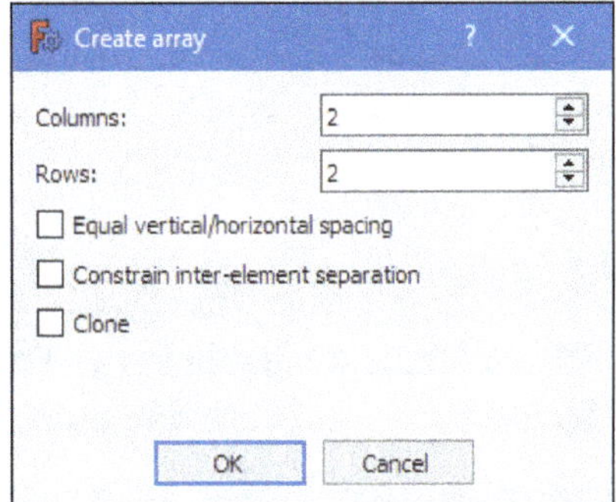

Figure-120. Create array dialog box

- Specify the number of columns and rows in the **Columns** and **Rows** edit boxes, respectively.
- Specify the other parameters as desired and click on **OK** button from the dialog box. A line and an icon will be attached to the cursor.
- Click on desired location in the 3D view area to specify the pattern. The rectangular array pattern will be created; refer to Figure-121.
- Press **ESC** key or press **RMB** to exit the tool.

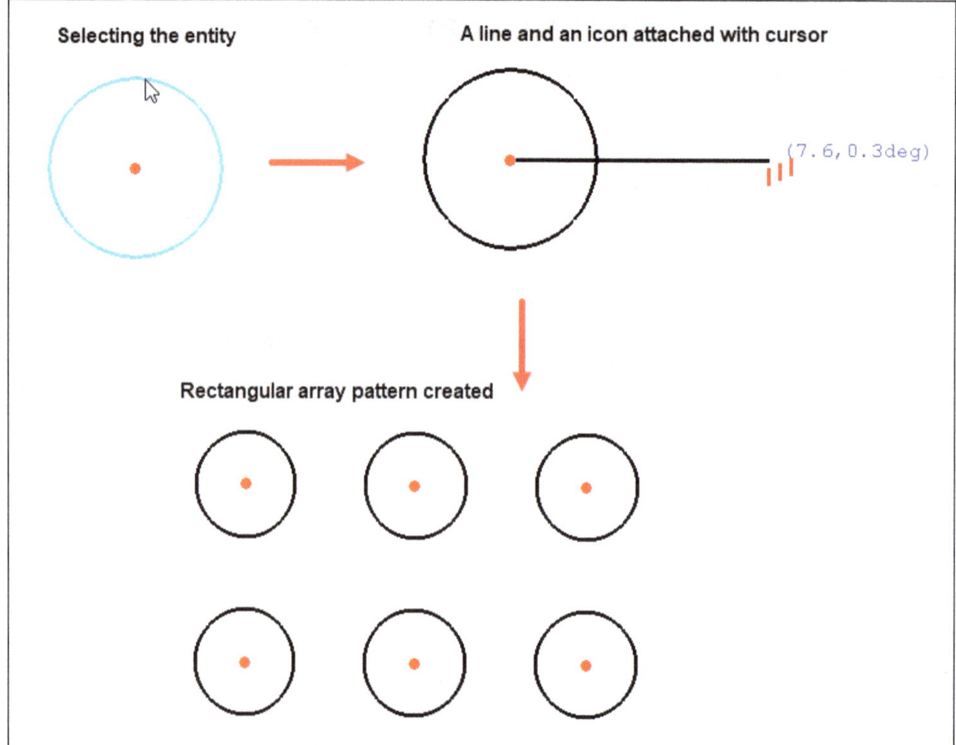

Figure-121. Creation of rectangular array pattern

SKETCHER B-SPLINE TOOLS

The **Sketcher B-spline** tools in **FreeCAD** are available in the **Toolbar** of **Sketcher** workbench; refer to Figure-122. The procedures to use these tools are discussed next.

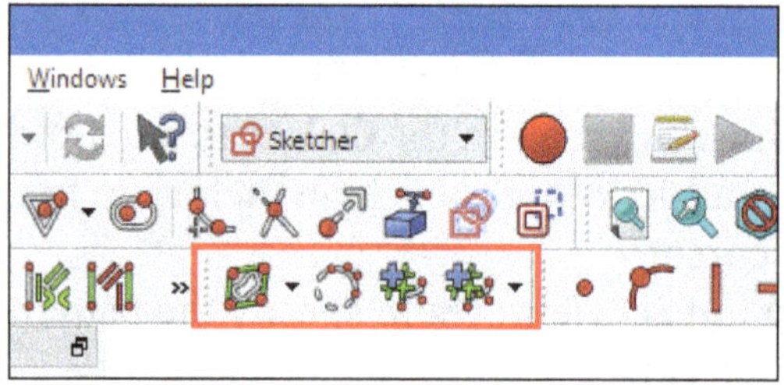

Figure-122. Sketcher B spline tools

Show/Hide B-spline degree

The **Show/Hide B-spline degree** tool shows or hides the display of the degree of a B-spline curve. The procedure to use this tool is discussed next.

- Select the B-spline entity whose degree you want to show or hide.
- Click on the **Show/Hide B-spline degree** tool from **Show/hide B-spline information layer** drop-down in the **Toolbar** of **Sketcher** workbench; refer to Figure-123. The degree of B-spline will be displayed as shown in Figure-124.

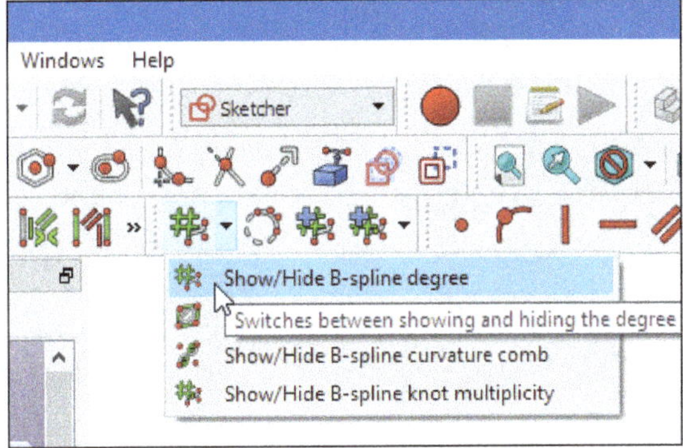

Figure-123. Show hide B spline degree tool

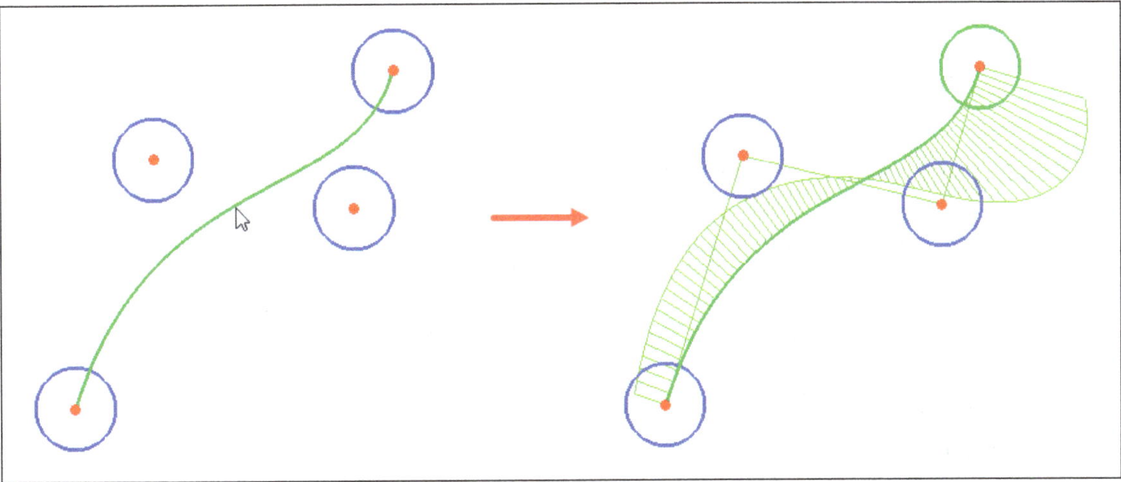

Figure-124. Degree of B spline displayed

- Press **ESC** key or press **RMB** to exit the tool.

Show/Hide B-spline control polygon

The **Show/Hide B-spline control polygon** tool shows or hides the display of the defining polygon of a B-spline curve. The procedure to use this tool is discussed next.

- Select a B-spline entity whose polygon you want to show or hide.
- Click on the **Show/Hide B-spline control polygon** tool from **Show/hide B-spline information layer** drop-down in the **Toolbar** of **Sketcher** workbench; refer to Figure-125. The polygon of selected B-spline entity will be shown or hidden; refer to Figure-126.

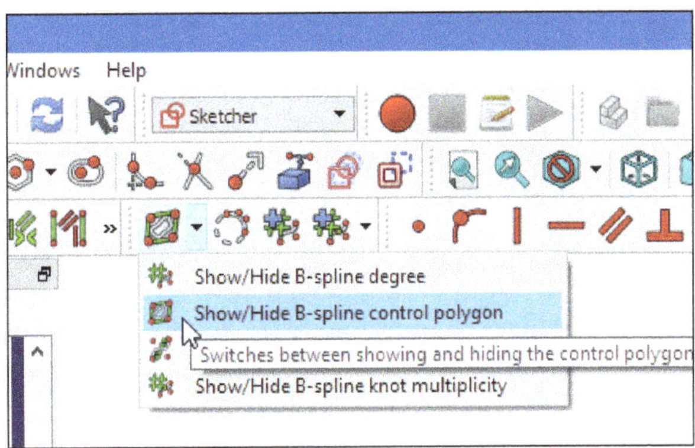

Figure-125. Show Hide B spline control polygon tool

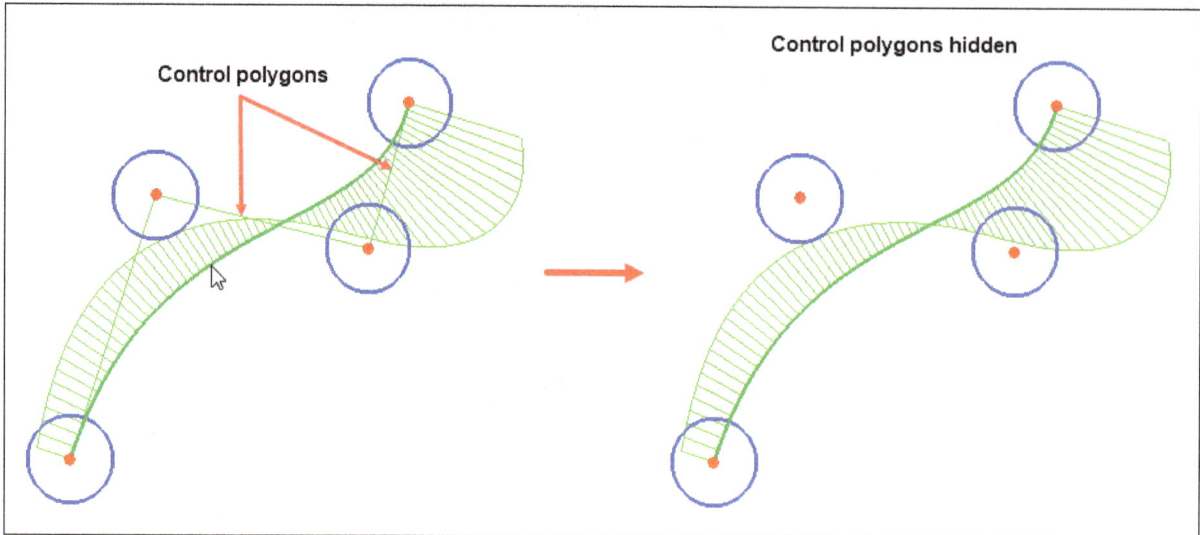

Figure-126. B spline control polygons hidden

- Press **ESC** key or press **RMB** to exit the tool.

Show/Hide B-spline curvature comb

The **Show/Hide B-spline curvature comb** tool shows or hides the display of the curvature comb of a B-spline curve. The procedure to use this tool is discussed next.

- Select the B-spline entity whose curvature comb you want to show or hide.
- Click on the **Show/Hide B-spline curvature comb** tool from **Show/hide B-spline information layer** drop-down in the **Toolbar** of **Sketcher** workbench; refer to Figure-127. The curvature comb of selected B-spline entity will be shown or hidden as shown in Figure-128.

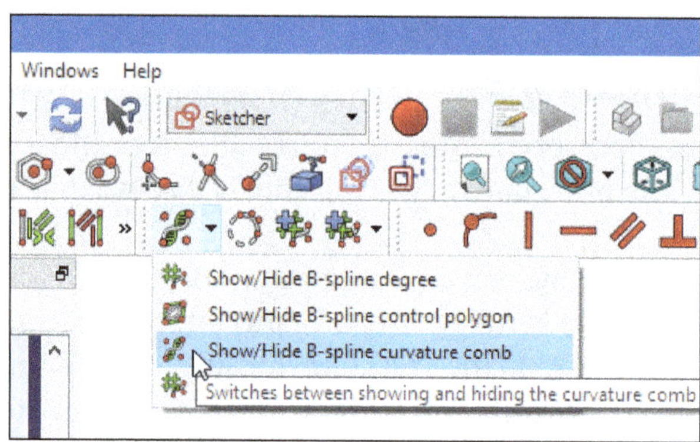

Figure-127. Show hide B spline curvature comb tool

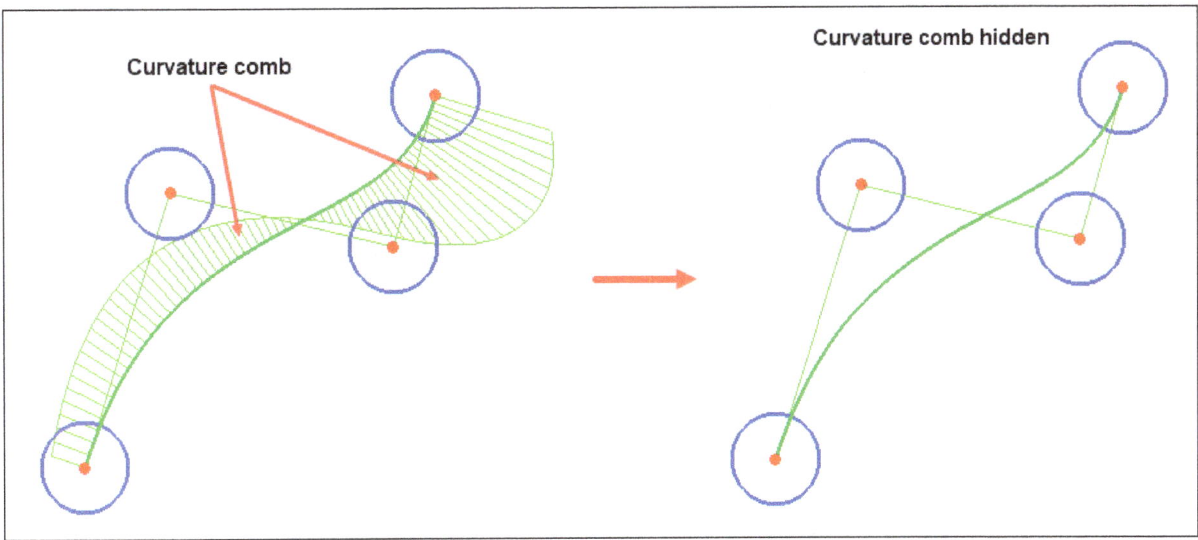

Figure-128. B spline curvature comb hidden

- Press **ESC** key or press **RMB** to exit the tool.

Show/Hide B-spline knot multiplicity

The **Show/Hide B-spline knot multiplicity** tool shows or hides the display of the knot multiplicity of a B-spline curve. The procedure to use this tool is discussed next.

- Select the B-spline entity whose knot multiplicity you want to show or hide.
- Click on the **Show/Hide B-spline knot multiplicity** tool from **Show/hide B-spline information layer** drop-down in the **Toolbar** of **Sketcher** workbench; refer to Figure-129. The knot numbers of spline vertices will be displayed; refer to Figure-130.
- Press **ESC** key or press **RMB** to exit the tool.

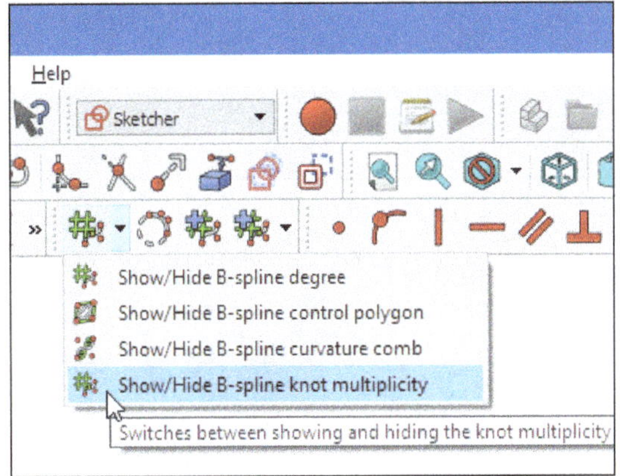

Figure-129. Show hide B spline knot multiplicity tool

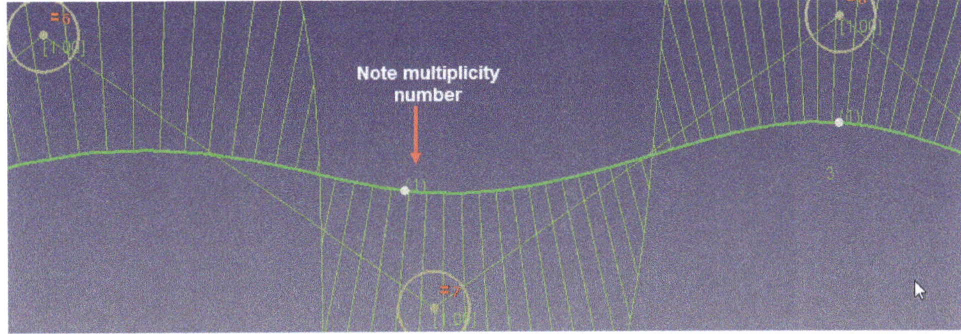

Figure-130. Showing_knot_multiplicity_number.png

Convert Geometry to B-spline

The **Convert Geometry to B-spline** tool converts compatible geometry, edges, and curves into a B-spline. The procedure to use this tool is discussed next.

- Select the geometry you want to convert into a B-spline.
- Click on the **Convert Geometry to B-spline** tool from **Toolbar** in the **Sketcher** workbench; refer to Figure-131. The selected geometry will be converted into a B-spline as shown in Figure-132.

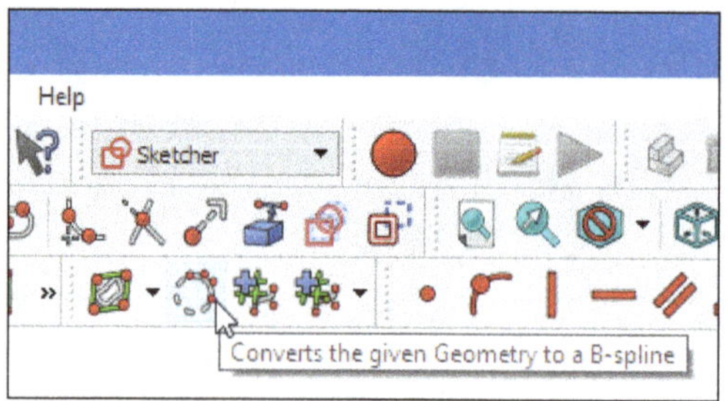

Figure-131. Convert geometry to B spline tool

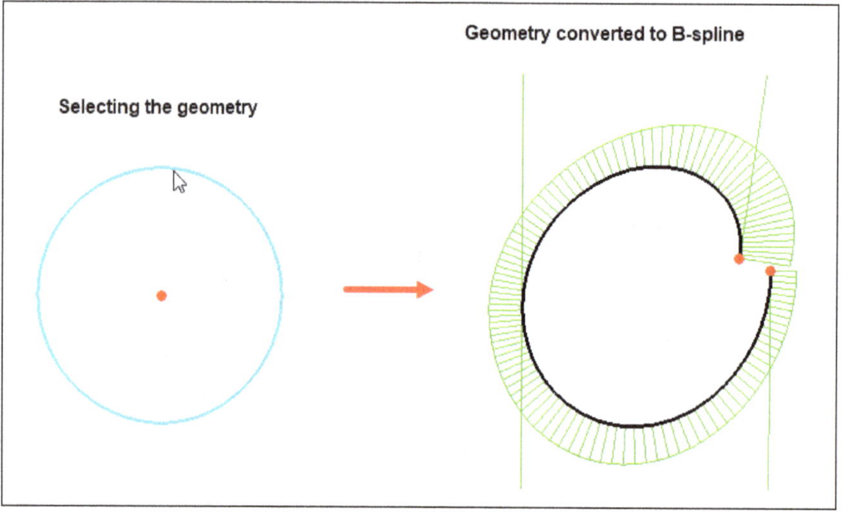

Figure-132. Geometry converted to B spline

- Press **ESC** key or press **RMB** to exit the tool.

Increase degree

The **Increase degree** tool increases the degree of the B-spline degrees define the number of vertices of spline that can be modified. The procedure to use this tool is discussed next.

- Select the B-spline entity whose degree you want to increase.
- Click on the **Increase degree** tool from **Toolbar** in the **Sketcher** workbench; refer to Figure-133. The degree of B-spline entity will be increased; refer to Figure-134.

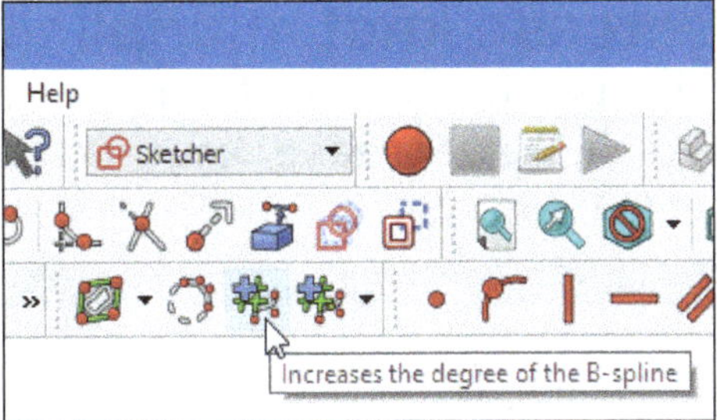

Figure-133. Increase degree tool

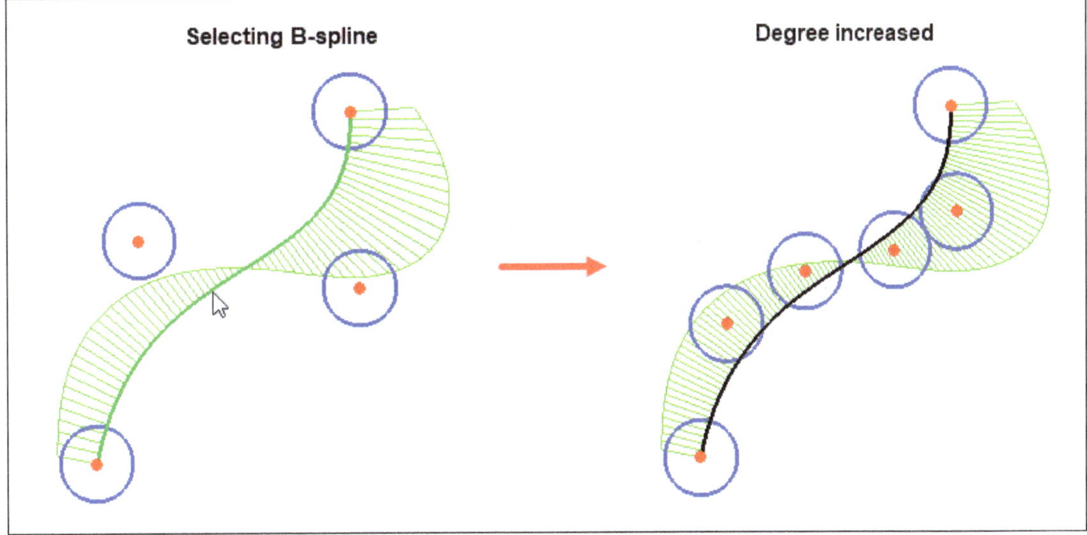

Figure-134. Degree increased of a B spline

- Press **ESC** key or press **RMB** to exit the tool.

Increase knot multiplicity

The **Increase knot multiplicity** tool increases the knot multiplicity of a B-spline curve knot. The number of times a knot value is duplicated is called the knot multiplicity. A knot value is said to be a full-multiplicity knot if it is duplicated degree many times. Select the knot from spline and click on the **Increase knot multiplicity** tool from **Toolbar** in the **Sketcher** workbench; refer to Figure-135.

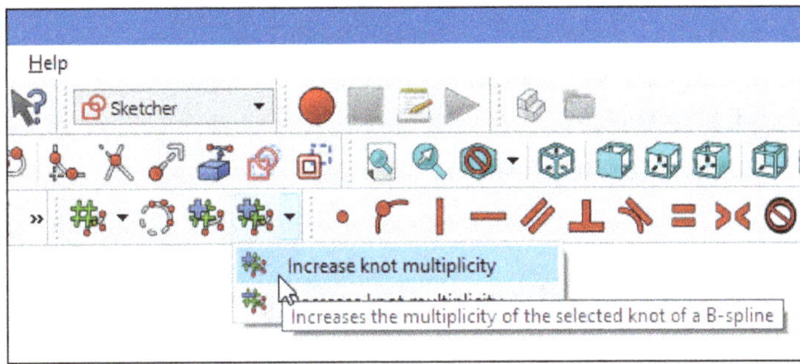

Figure-135. Increase knot multiplicity tool

Decrease knot multiplicity

The **Decrease knot multiplicity** tool decreases the knot multiplicity of a B-spline curve knot. Select the knot from spline and click on the **Decrease knot multiplicity** tool from **Toolbar** in the **Sketcher** workbench; refer to Figure-136.

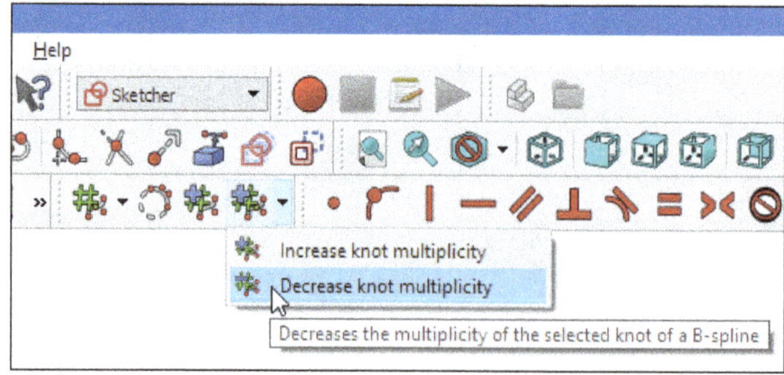

Figure-136. Decrease knot multiplicity tool

PRACTICAL 1

Create the sketch as shown in Figure-137. Also, dimension the sketch as per the figure.

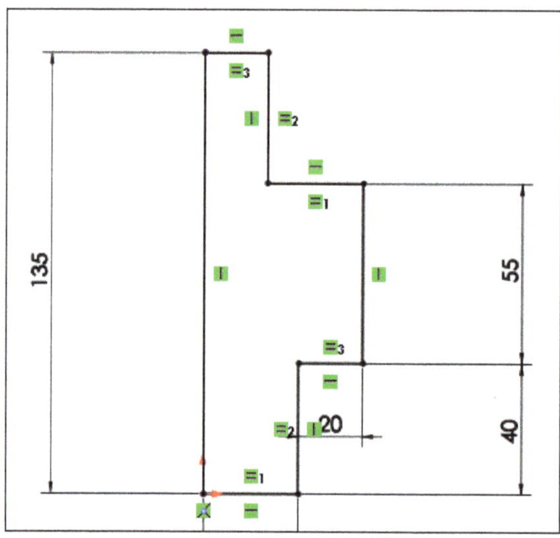

Figure-137. Practical 1

Steps to be performed:

- Start a new part file.
- Select a sketching plane and activate sketching mode.
- Create the sketch using **Line** tool.
- Apply the dimensions using dimension tools.
- Save the file.

Starting a New Sketch

- Start FreeCAD if not started yet.
- Select the **Sketcher** option from the **Switch between workbenches** drop-down in the **Toolbar**. The sketcher environment will be displayed.
- Click on the **Create a new sketch** tool from the **Toolbar** or **Sketch** menu. The **Choose orientation** dialog box will be displayed; refer to Figure-138.

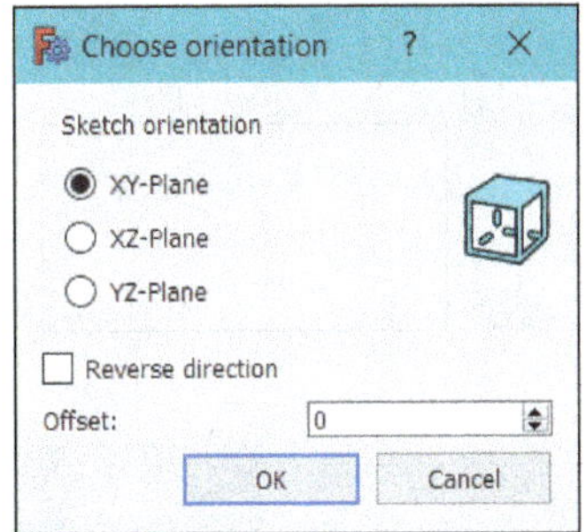

Figure-138. Choose orientation dialog box

- Select the **XY-Plane** tool from the **Sketch orientation** area of dialog box to create sketch on XY plane.
- Click on the **OK** button from the dialog box. The tools to create sketch will become active.

Creating Line Sketch

- Click on the **Create Line** tool from the **Toolbar** or **Sketcher geometries** cascading menu of **Sketch** menu to activate line tool. You will be asked to specify start point of line.
- Click at the origin to specify start point of line and then create a straight horizontal line of approximate length **100** mm; refer to Figure-139. (Make sure to select the **Grid snap** check box from **Edit controls** dialog in **Combo View** to snap cursor on origin). The process to set unit to mm has been discussed earlier.

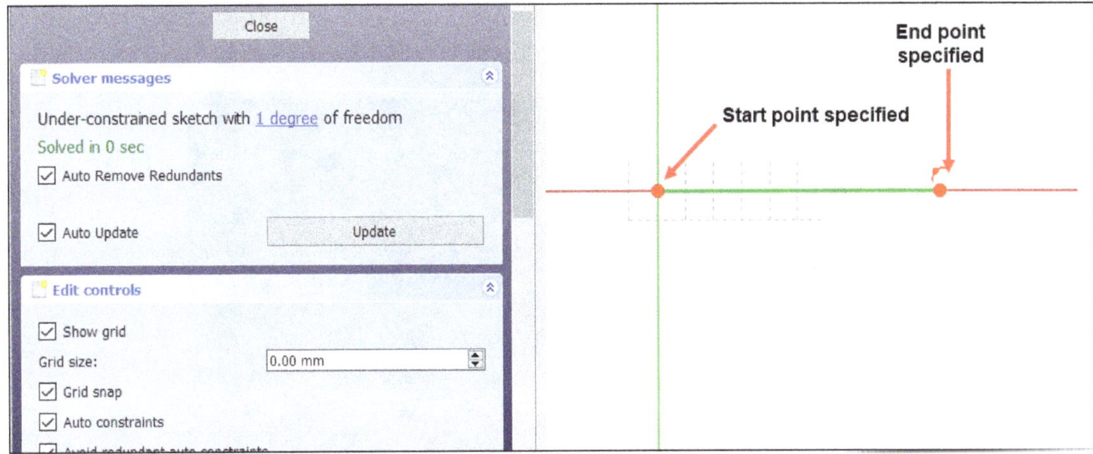

Figure-139. Horizontal line created

- Click on the end point of newly created line and create a line of length approximately **20** upward; refer to Figure-140.
- Create other lines of the sketch as discussed earlier; refer to Figure-141.

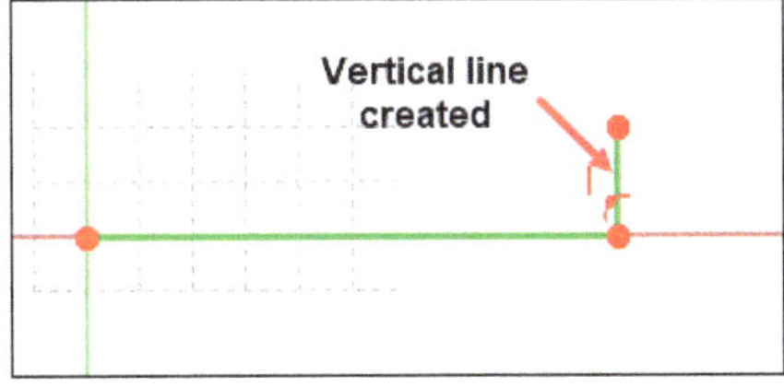

Figure-140. Vertical line created

Applying Dimensions

- Click on the **Constrain distance** tool from the **Toolbar** or **Sketcher constraints** cascading menu of **Sketch** menu. You can also use **SHIFT+D** shortcut key to activate this tool. On activating this tool, you will be asked to select a line.
- Select the base line of sketch. The **Insert Length** dialog box will be displayed with length of line; refer to Figure-142.

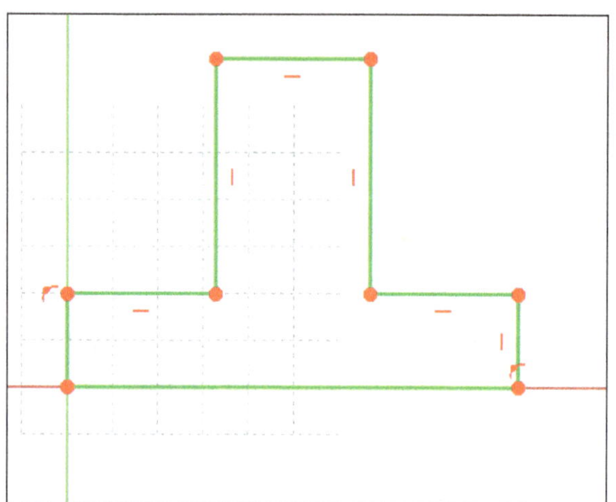

Figure-141. Sketch created

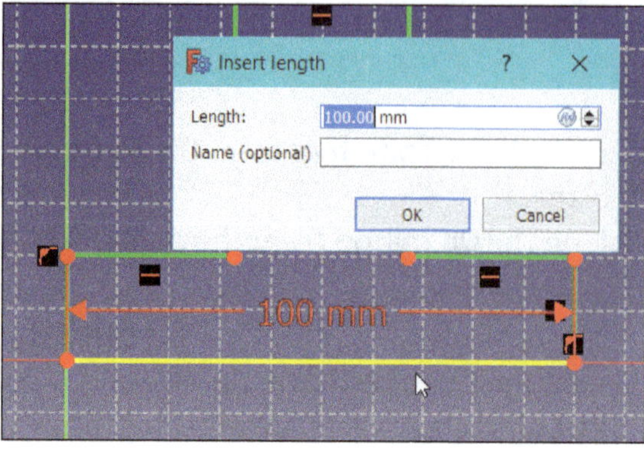

Figure-142. Insert length dialog box

- Specify desired value in the **Length** edit box and click on the **OK** button to create the dimension.
- Repeat the steps to create all the dimensions of sketch; refer to Figure-143.

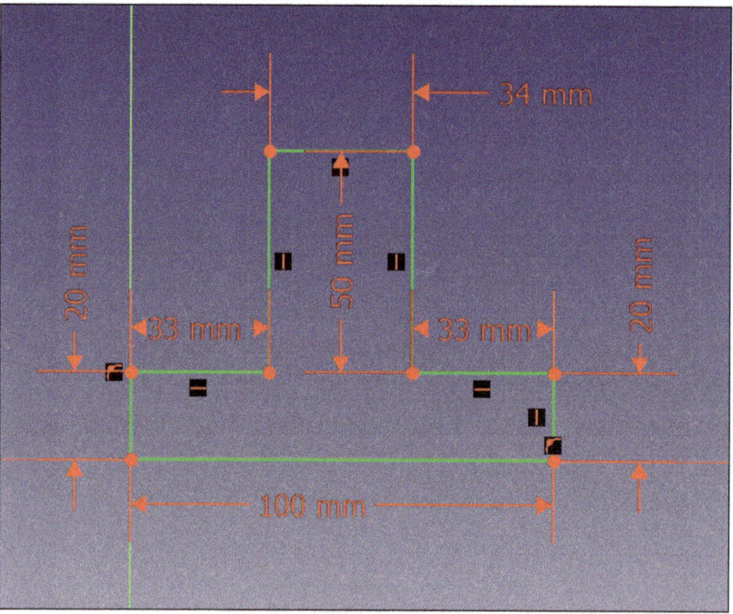

Figure-143. Dimensions created for sketch

- After specifying all the constraints and dimensions, the sketch should be displayed as fully constrained; refer to Figure-144.

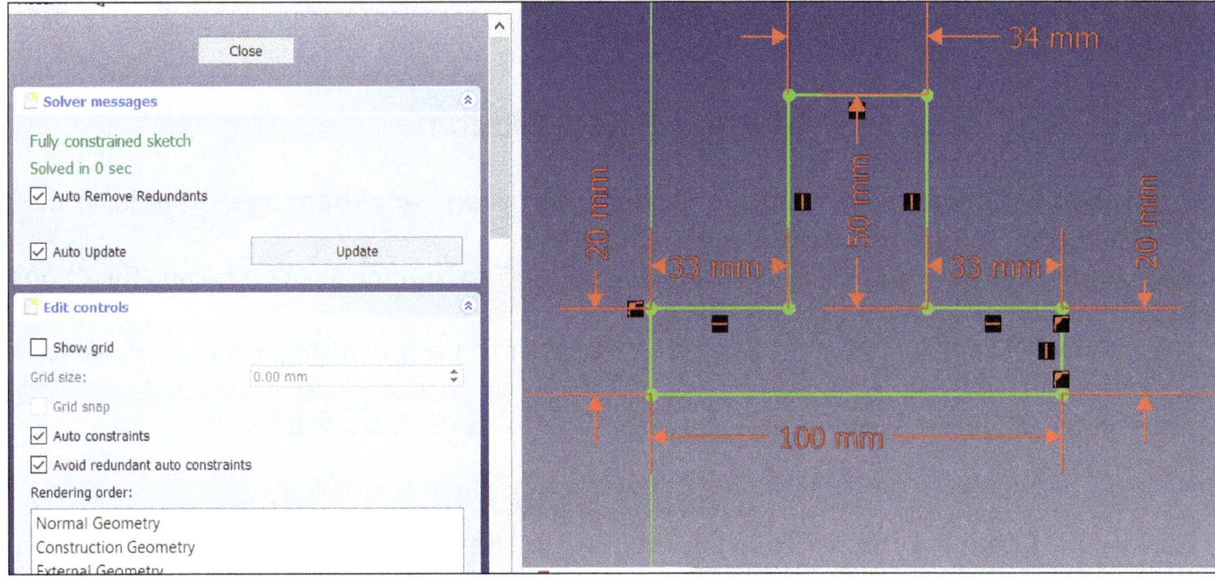

Figure-144. Fully constrained sketch

- Click on the **Close** button from **Tasks** panel of **Combo View** to exit sketch mode.
- Click on the **Save** button from the **Toolbar** and save the file at desired location.

PRACTICAL - 2

In this practical, we will create the sketch as shown in Figure-145.

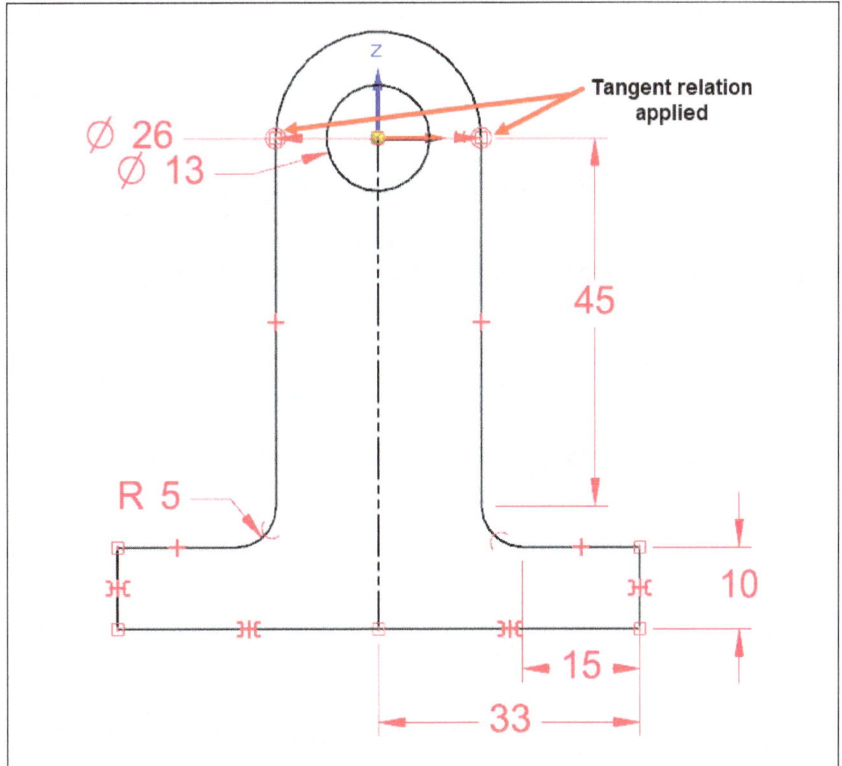

Figure-145. Practical 2

Steps to be performed:

- Start a new part file.
- Select a sketching plane and activate sketching mode.
- Create the sketch using **Line** tool.

- Apply the dimensions using dimension tools.
- Save the file.

Starting a New Sketch

- Start FreeCAD if not started yet.
- Select the **Sketcher** option from the **Switch between workbenches** drop-down in the **Toolbar**. The sketcher environment will be displayed.
- Click on the **Create a new sketch** tool from the **Toolbar** or **Sketch** menu. The **Choose orientation** dialog box will be displayed.
- Select the **XY-Plane** tool from the **Sketch orientation** area of dialog box to create sketch on XY plane.
- Click on the **OK** button from the dialog box. The tools to create sketch will become active.

Creating Sketch

- Click on the **Create Line** tool from the **Toolbar** or **Sketcher geometries** cascading menu of **Sketch** menu to activate line tool. You will be asked to specify start point of line.
- Create a horizontal line of length approximately **66** mm starting from origin.
- Select the **Toggle construction geometry** tool from the **Toolbar** or **Sketcher geometries** cascading menu of **Sketch** menu to activate construction mode.
- Create a vertical line of approximately **60** mm length; refer to Figure-146.

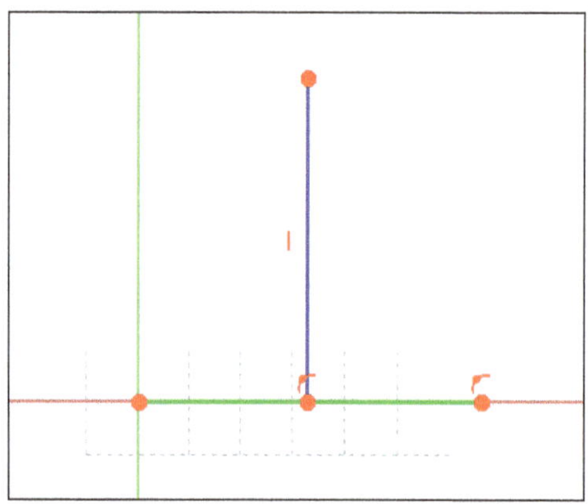

Figure-146. Lines created

- Exit the construction mode by selecting the **Toggle construction geometry** tool again from **Toolbar**.
- Click on the **Create Circle** tool from the **Sketcher geometries** cascading menu of the **Sketch** menu or **Toolbar**. You will be asked to specify location of center of circle.
- Click on the open end point of construction line and create a circle of diameter approximately **13**; refer to Figure-147.

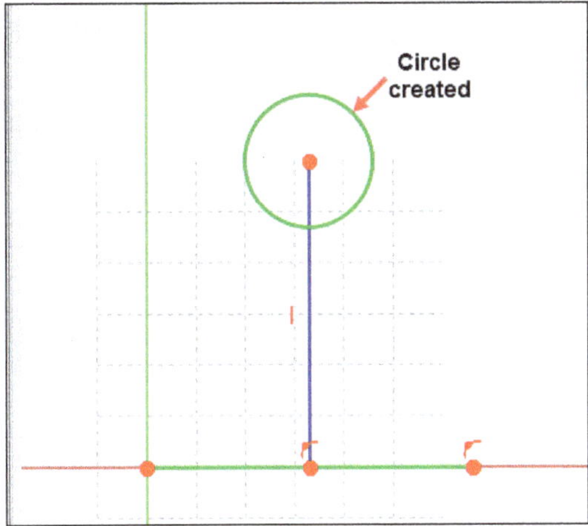

Figure-147. Circle created1

- Create a circle of diameter approximately **26** using the same center as previous circle; refer to Figure-148.

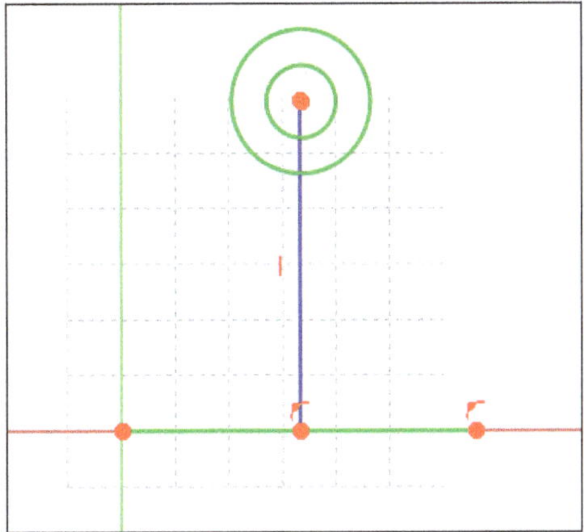

Figure-148. Two circles created

- Click on the **Create Line** tool from the **Toolbar** and create line sketch using approximate dimension as per the drawing; refer to Figure-149.

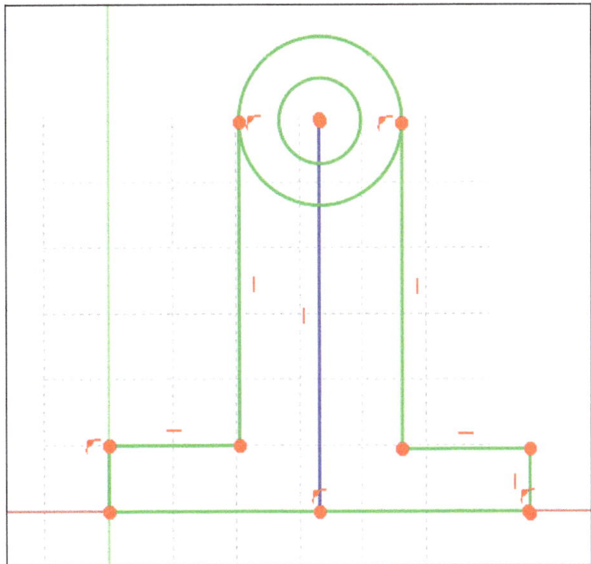

Figure-149. Line sketch created

Trimming and Apply Fillets

- Click on the **Trim edge** tool from the **Sketcher geometries** cascading menu of the **Sketch** menu and click on the inner section of larger circle to trim extra portion; refer to Figure-150.

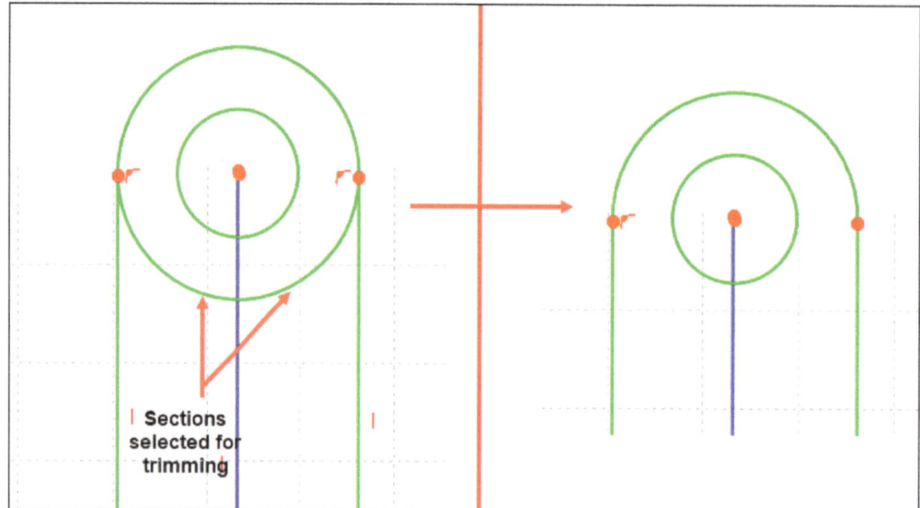

Figure-150. Trimming circle

- Press **ESC** once to exit the **Trim edge** tool and click on the **Create fillet** tool from the **Sketcher geometries** cascading menu of **Sketch** menu. You will be asked to select two entities between which the fillet will be applied.
- Select the lines in pair near their connection points as shown in Figure-151.

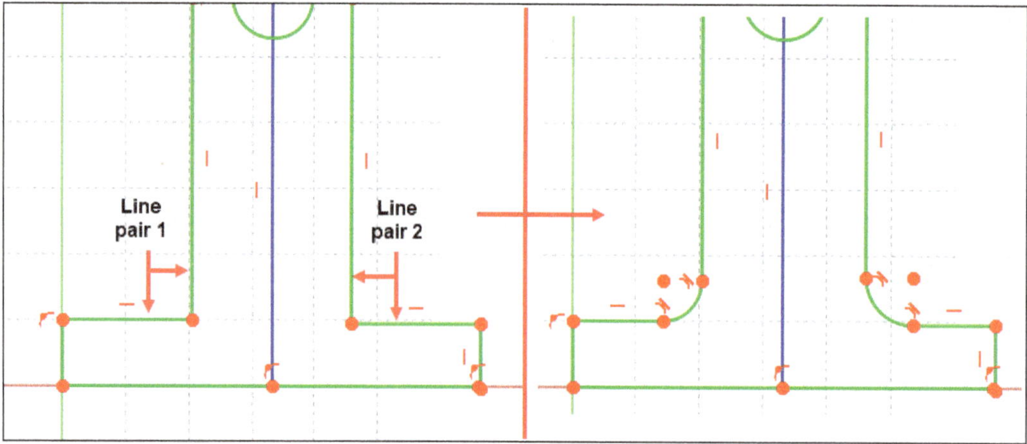

Figure-151. Applying fillets

Applying Constraints and Dimensions

- Select two bottom corner points and centerline passing through the sketch and then select the **Constraint symmetrical** tool from **Sketcher constraints** cascading menu of the **Sketch** menu to keep centerline at the middle of sketch; refer to Figure-152.

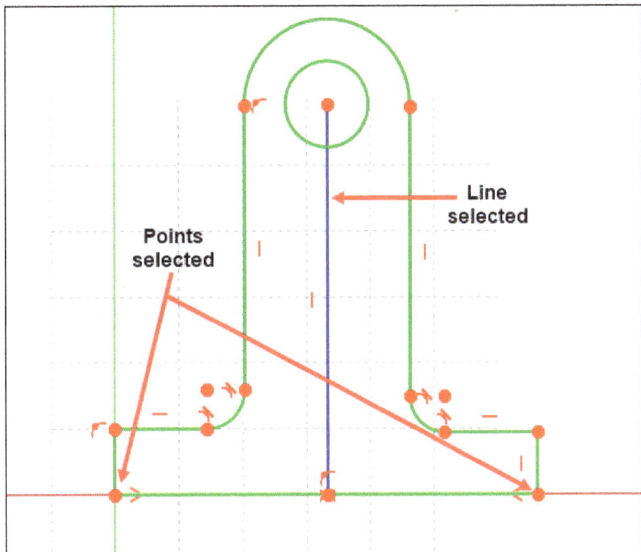

Figure-152. Selection for symmetrical constraint

- If there are any disconnection between points that should be coincident then apply the coincident constraint on those points. Note that you may need to zoom on the intersection points to check whether they are actually coincident or not.
- Click on the **Constrain Distance** tool from **Toolbar** or **Sketcher constraints** cascading menu in the **Sketch** menu and apply the dimensions as shown in Figure-153.

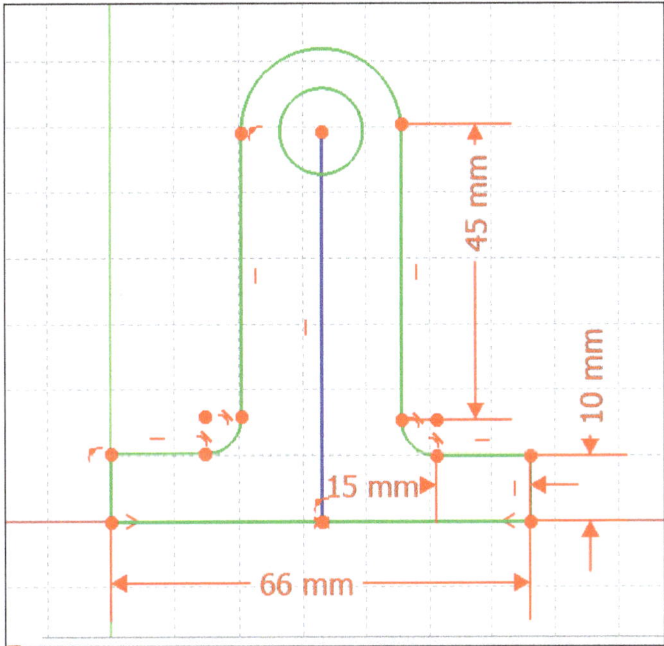

Figure-153. Distance dimensions applied

- Click on the **Constraint equal** tool from **Sketcher constraints** cascading menu of **Sketch** menu and one by one make the entities on both sides of centerline equal; refer to Figure-154.

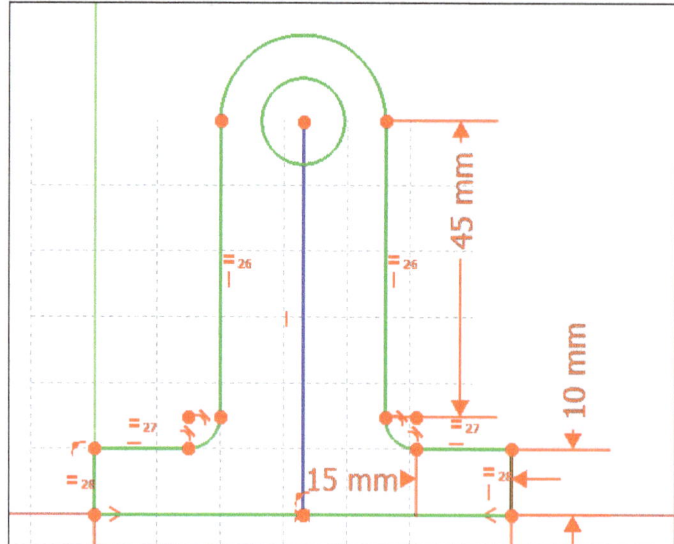

Figure-154. After applying equal constraint

- Select the **Constrain diameter** tool from the **Sketcher constraints** cascading menu of the **Sketch** menu and apply diameter dimensions to arc and circles. The fully defined sketch should display as shown in Figure-155.

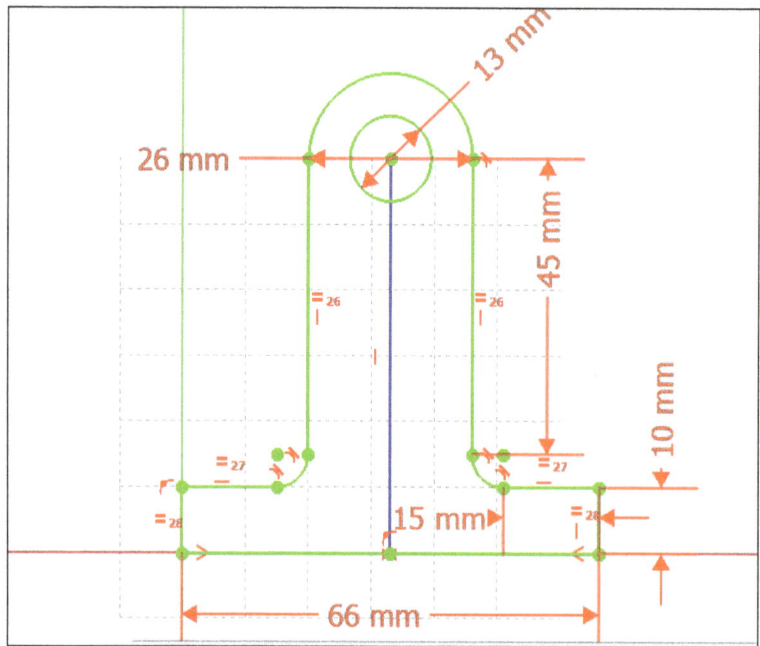

Figure-155. Sketch for Practical 2

Close the sketcher and save the file as discussed earlier.

PRACTICE 1

In this practice session, you will create a sketch for the drawing shown in Figure-156.

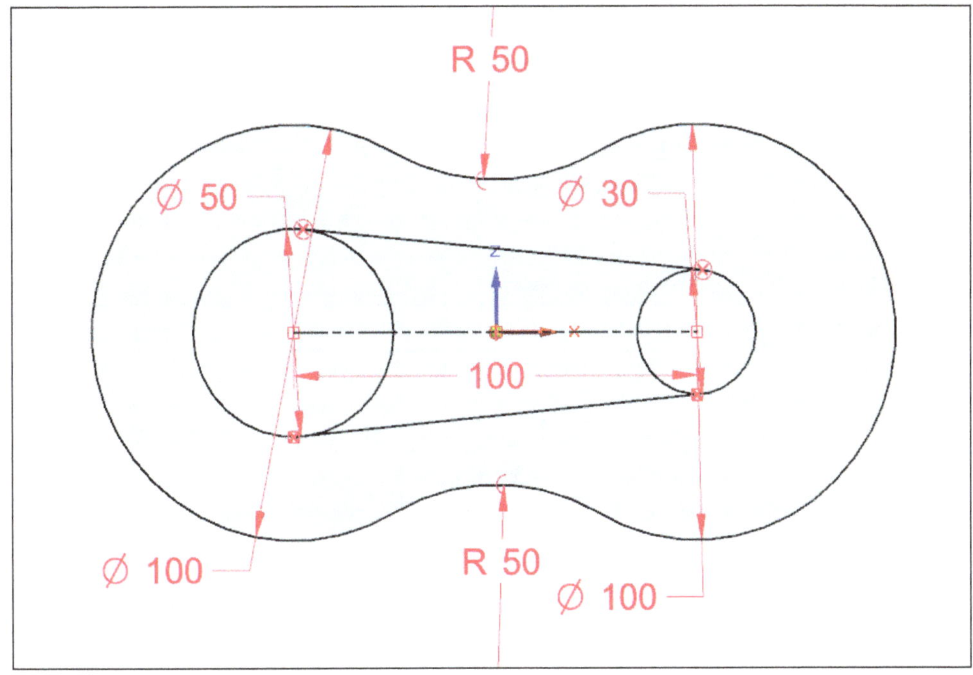

Figure-156. Practice 1

PRACTICE 2

In this practice session, you will create a sketch for drawing shown in Figure-157.

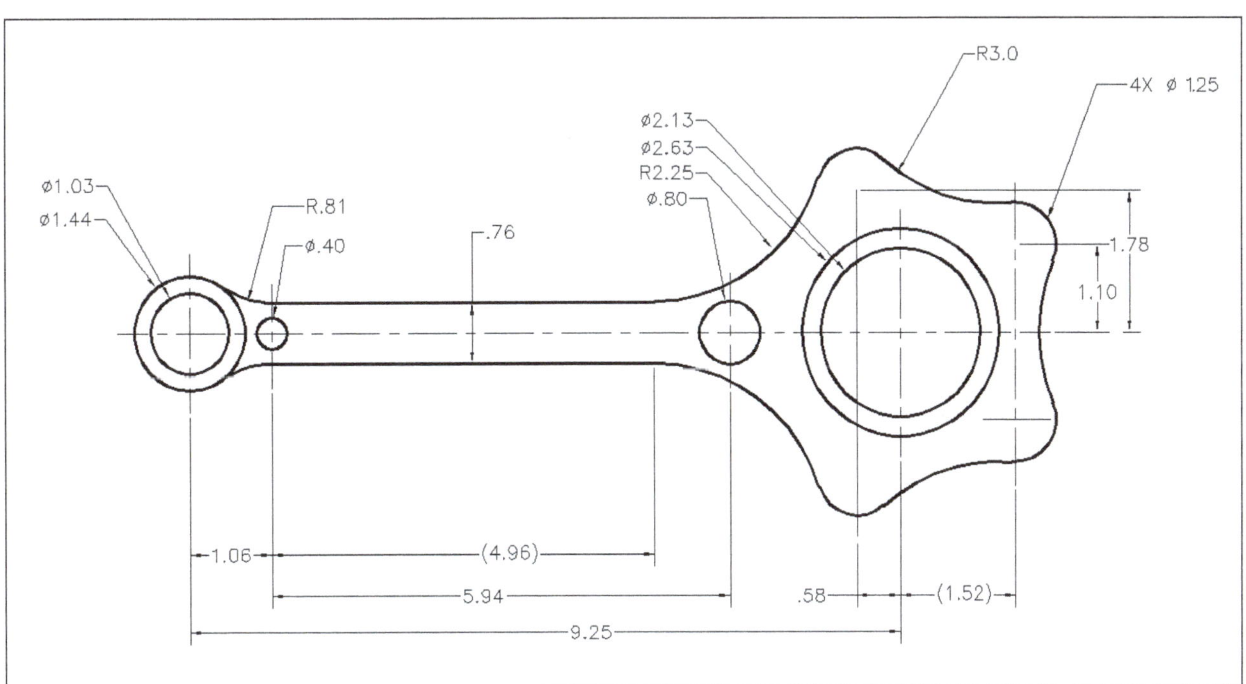

Figure-157. Practice 2

PRACTICE 3

In this practice session, you will create a sketch for the drawing given in Figure-158.

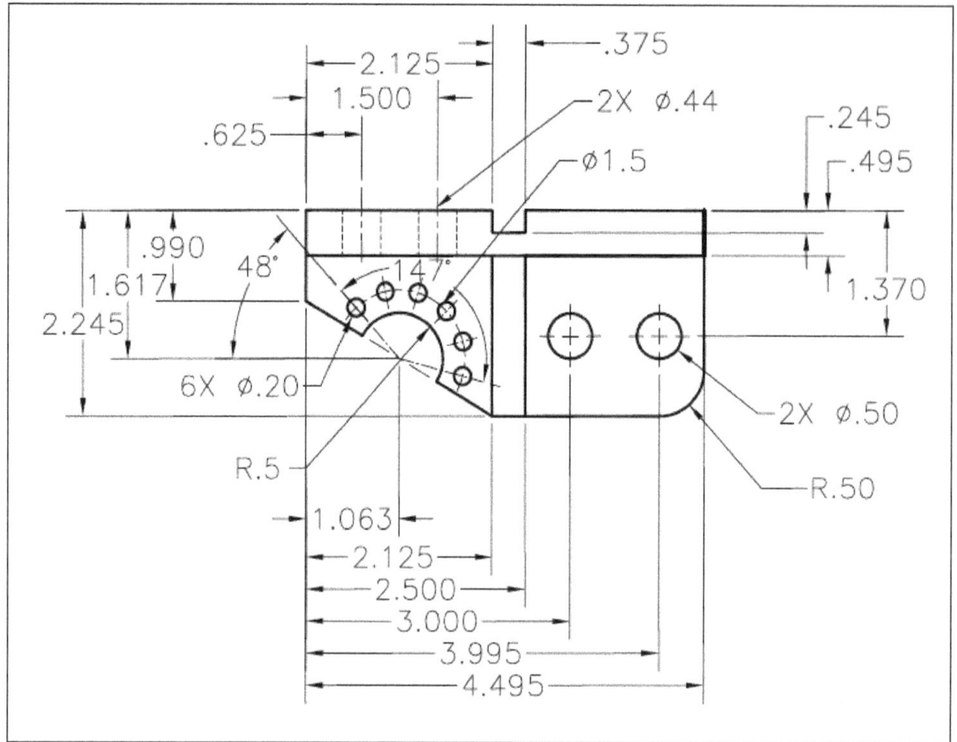

Figure-158. Practice 3

PRACTICE 4

In this practice session, you will create sketches of different views for the drawing given in Figure-159.

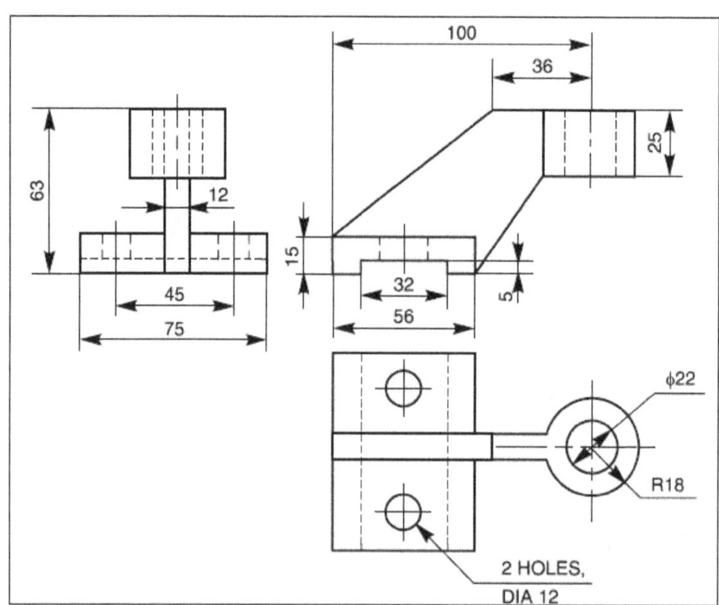

Figure-159. Practice 4

Chapter 3

Solid Modeling

Topics Covered

The major topics covered in this chapter are:

- ***Introduction to Solid Modeling***
- ***Starting Part Design***
- ***Part Design Helper Tools***
- ***Part Design Modeling Tools***
- ***Datum Tools***
- ***Additive Tools***
- ***Subtractive Tools***
- ***Transformation Tools***
- ***Dress-up Tools***
- ***Boolean Operation***
- ***Involute Gear Tool***
- ***Shaft Design Wizard***

INTRODUCTION

Solid Modeling is the most advanced method of geometric modeling in three dimensions. Solid modeling is the representation of the solid objects on your computer. Now, we will learn about various tools and commands used to create solid models.

In FreeCAD, the **Part Design Workbench** provides advanced tools for modeling complex solid parts. It is mostly focused on creating mechanical parts that can be manufactured and assembled into a finished product. Nevertheless, the created solids can be used in general for any other purpose such as architectural design, finite element analysis, or machining and 3D printing.

STARTING PART DESIGN

The **Part Design Workbench** is intrinsically related to the **Sketcher Workbench**. The user normally creates a sketch then uses the **Part Design Pad** tool to extrude it and create a basic solid and then this solid is further modified.

- To start a new part design file, click on the **New** button from **File** menu and select **Part Design** workbench from **Switch between workbenches** drop-down in the **Toolbar**; refer to Figure-1. The tools to create 3D model will be displayed in the **Toolbar**; refer to Figure-2.

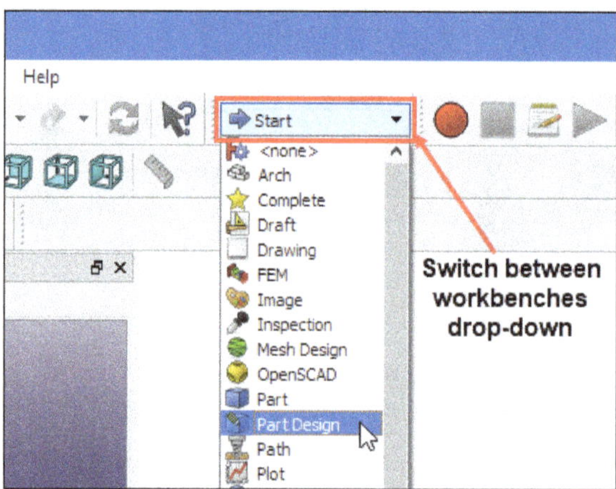

Figure-1. Part Design workbench

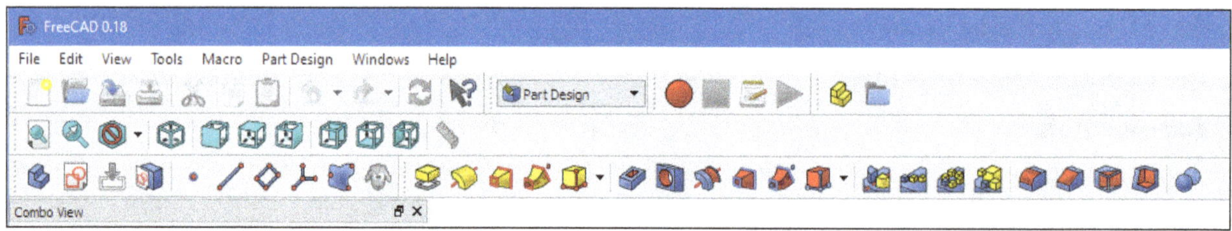

Figure-2. Part Design tools

Part Design helper tools

The **Part Design** helper tools help in creating a body and make it active for creating and editing the sketch. You can also map a sketch to desired face through the helper tools. These tools are available in the **Toolbar** of **Part Design** workbench; refer to Figure-3. The procedures to use these tools are discussed next.

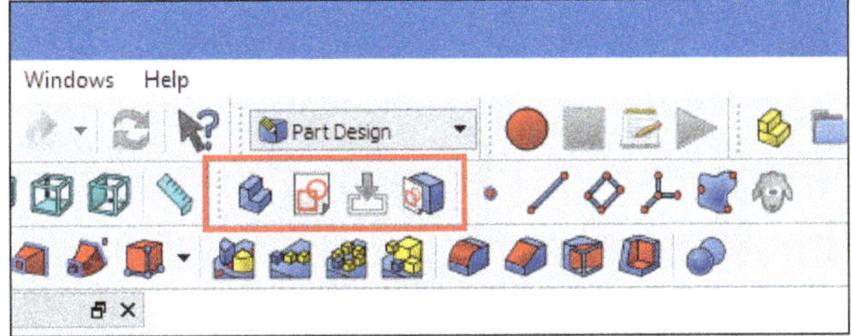

Figure-3. Part design helper tools

Creating a body

A **Part Design** body is the base element to create solid shapes with the **Part Design** workbench. It can contain sketches, datum objects, and part design features that help in building a single contiguous solid. The procedure to use this tool is discussed next.

- Click on the **Create body** tool from **Toolbar** in the **Part Design** workbench; refer to Figure-4. An empty body is created and automatically becomes active; refer to Figure-5.

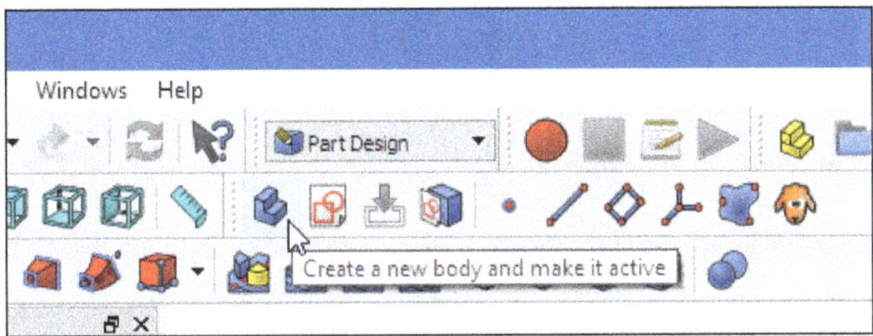

Figure-4. Create body tool

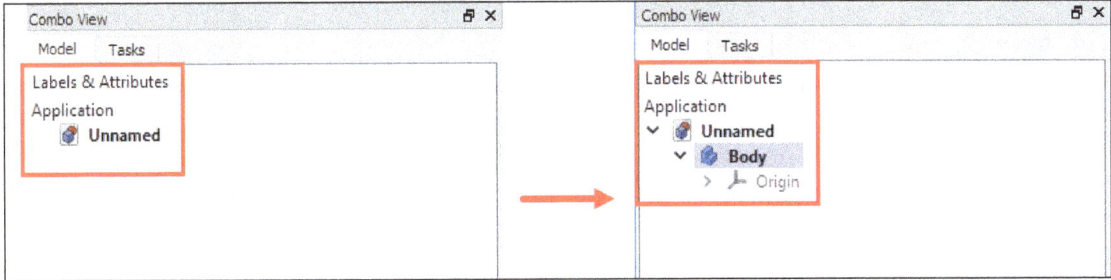

Figure-5. A new empty body created

- Now, you can click **Create sketch** tool to create a sketch in the body that can be used with **Pad** tool.

Creating a sketch

The **Create sketch** tool creates a new sketch on a selected face or plane. If no face is selected while this tool is executed, the user is prompted to select a plane from the **Tasks** panel or from the 3D view area. The procedure to use this tool is discussed next.

- Click on the **Create sketch** tool from **Toolbar** in the **Part Design** workbench; refer to Figure-6. The **Select feature** dialog will be displayed in the **Tasks** panel of **Combo View** to select desired plane; refer to Figure-7. You can also select the plane from 3D view area.

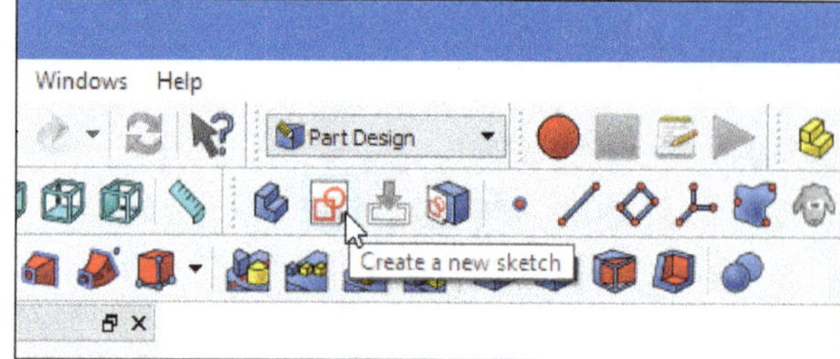

Figure-6. Create sketch tool

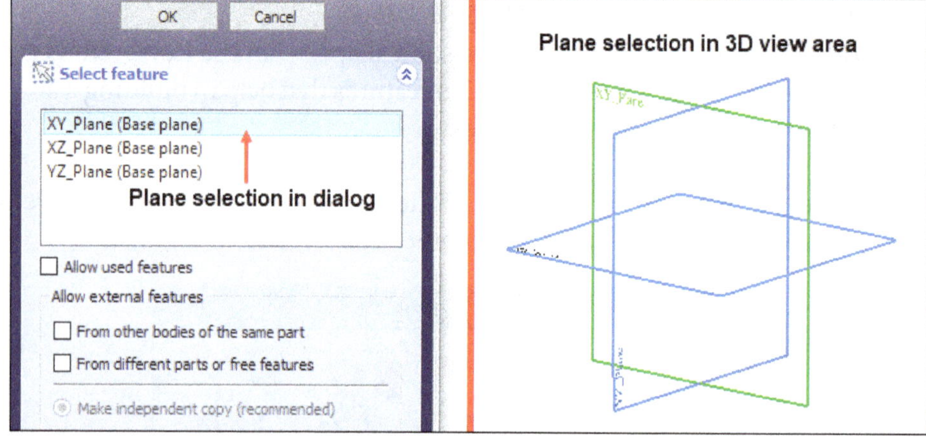

Figure-7. Selecting the plane to create sketch

- After selecting desired plane, click on **OK** button from the dialog. The interface switches to **Sketcher** workbench to create or edit the sketch.

Edit sketch

The **Edit sketch** tool is used to edit an existing sketch. The procedure to use this tool is discussed next.

- Select the sketch you want to edit from the Model tree view or from the 3D view area; refer to Figure-8.

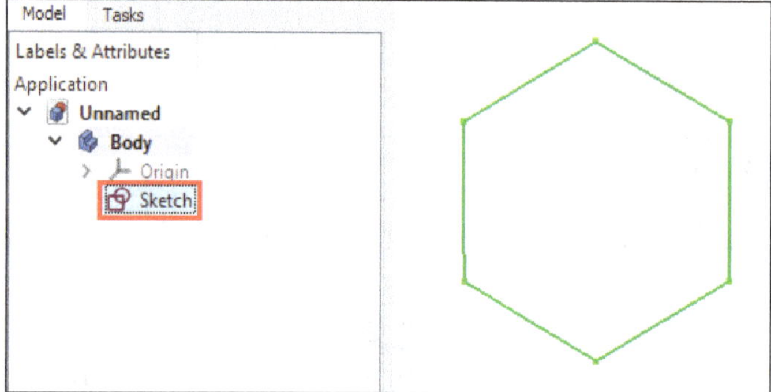

Figure-8. Selecting the sketch to edit

- Click on the **Edit sketch** tool from **Toolbar** in the **Part Design** workbench; refer to Figure-9. The **Sketcher** workbench will be displayed with options to edit the existing sketch.

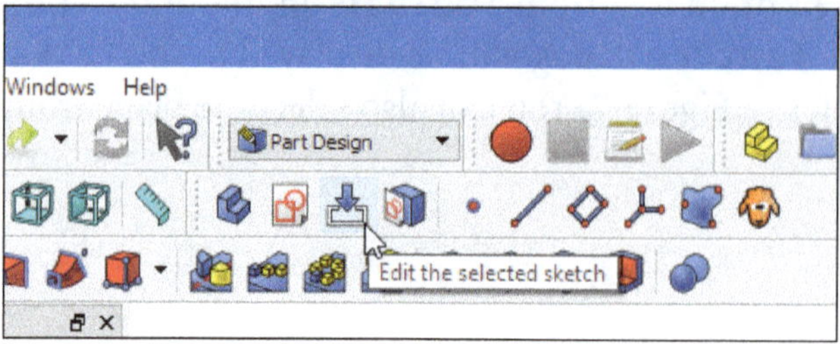

Figure-9. Edit sketch tool

Map sketch to face

The **Map sketch to face** tool maps an existing sketch on the face of a selected body. Note that this tool is not used to create new sketches. It only maps or remaps existing sketch to the face of a solid or a part design feature. The procedure to use this tool is discussed next.

- Select the face of a part design or solid feature on which you want to map the sketch.
- Click on the **Map sketch to face** tool from **Toolbar** in the **Part Design** workbench; refer to Figure-10. The **Select sketch** dialog box will be displayed; refer to Figure-11, asking you to select the sketch entity to be mapped from the list.

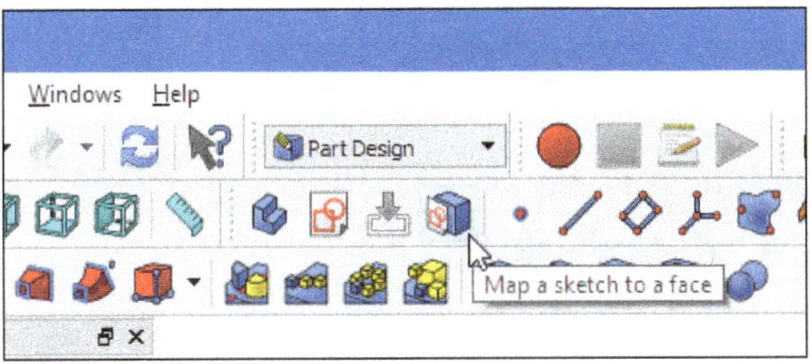

Figure-10. Map sketch to face tool

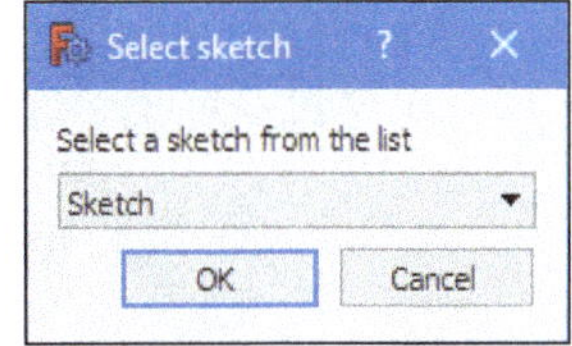

Figure-11. Select sketch dialog box

- Select desired sketch entity from the list which you want to map and click on **OK** button from the dialog box. The **Sketch attachment** dialog box will be displayed; refer to Figure-12, asking you to select method from the list to be used to attach sketch with selected face.

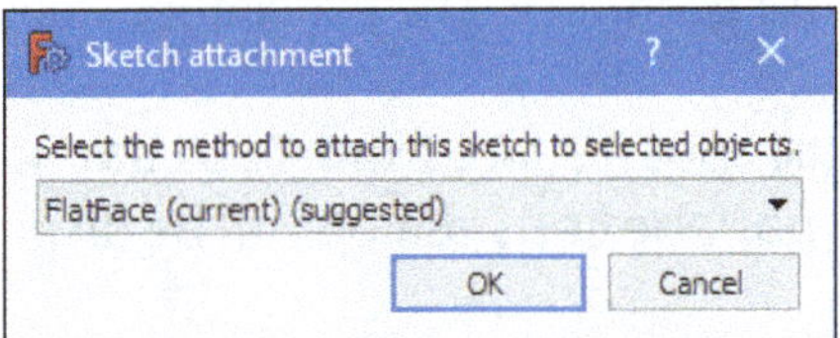

Figure-12. Sketch attachment dialog box

- Select desired method from the list to map the sketch on selected object and click on **OK** button from the dialog box. Generally, it is good to select the suggested method. The selected sketch entity will be mapped to the part design feature; refer to Figure-13.

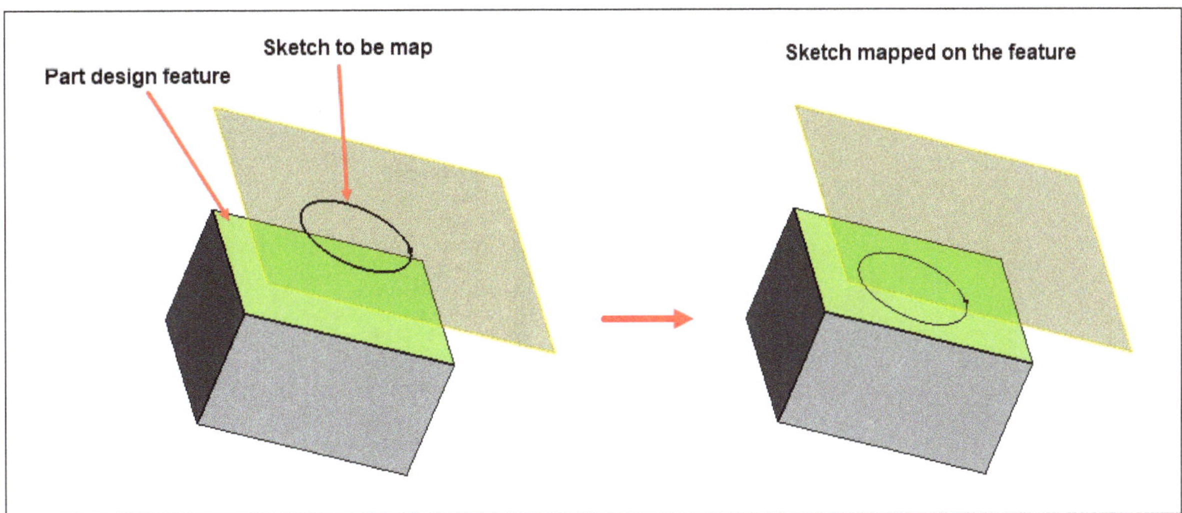

Figure-13. Sketch mapped on the solid feature

PART DESIGN MODELING TOOLS

With the help of Part design modeling tools, you can create a part from a conceptual sketch through solid feature-based modeling, as well as build and modify parts through direct intuitive graphical manipulation.

Datum tools

The **Datum tools** in FreeCAD are referred to as auxiliary geometrical elements which will not form part of the final shape of the body but can be used as references and supports for sketches and other types of features. The **Datum tools** are available in the **Toolbar** of **Part Design** workbench; refer to Figure-14. The procedures to use these tools are discussed next.

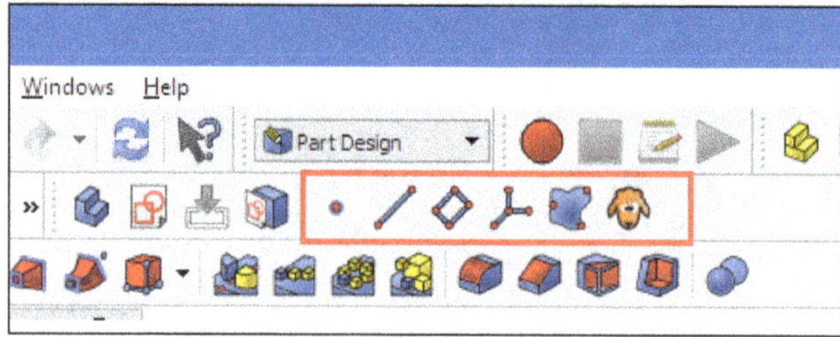

Figure-14. Datum tools

Create a datum point

The **Create a datum point** tool creates a datum point in the active body which can be used as reference for sketches or other datum geometry. The procedure to use this tool is discussed next.

- Click on the **Create a datum point** tool from **Toolbar** in the **Part Design** workbench; refer to Figure-15. The **Point parameters** dialog will be displayed in the **Tasks** panel of **Combo View**; refer to Figure-16.

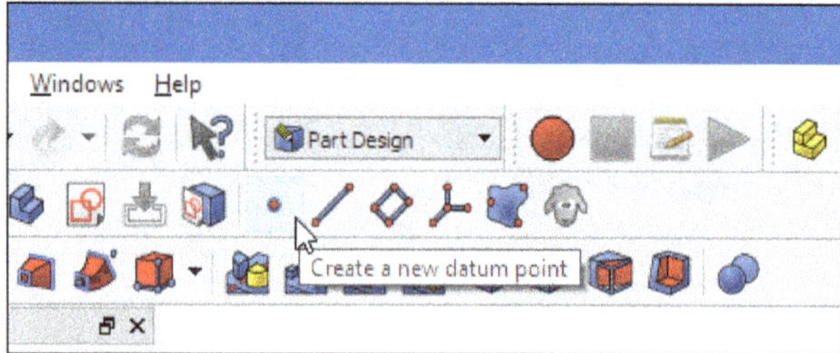

Figure-15. Create a datum point tool

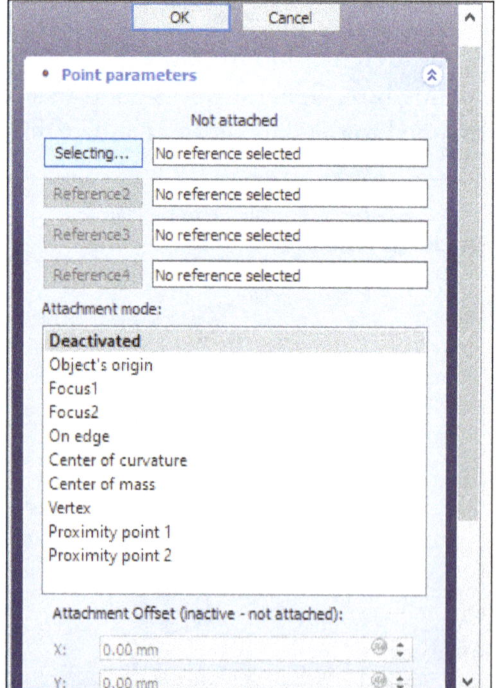

Figure-16. Point parameters dialog

- Select the first reference for attachment mode from the 3D view area or from the attachment modes list available in **Attachment mode** area of the dialog.
- To add an additional reference, click on the **Reference2** button.
- Specify desired offset values in **Attachment Offset** area of the dialog to move point by specified value.
- Click on **OK** button from the dialog. The datum point will be attached to the feature at a specified offset distance; refer to Figure-17.

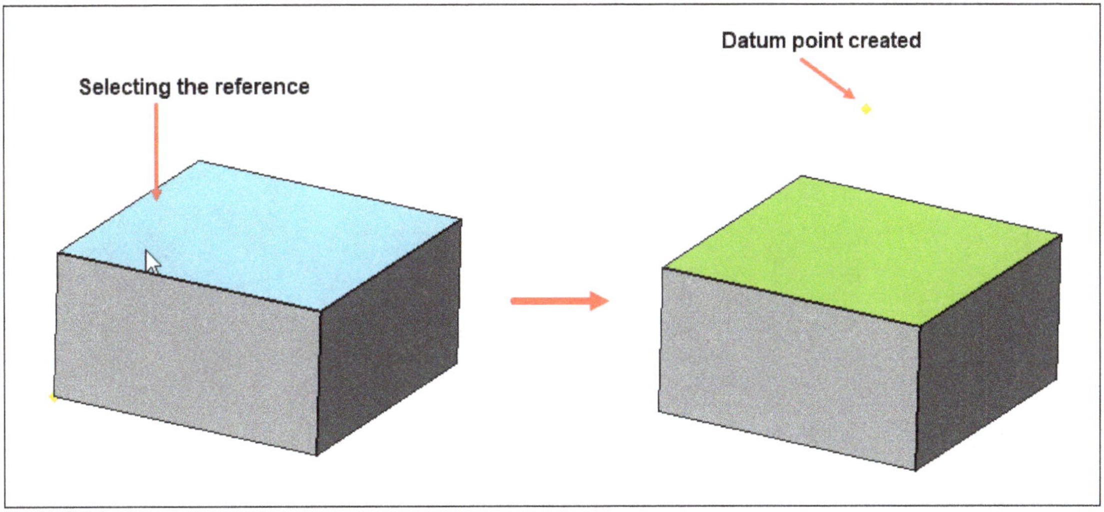

Figure-17. Creation of datum point

Create a datum line

The **Create a datum line** tool creates a datum line which can be used as reference for sketches, other datum geometries, or features. For example, it can be used as revolution axis for **Revolution** and **Groove** features. The procedure to use this tool is discussed next.

- Click on the **Create a datum line** tool from **Toolbar** in the **Part Design** workbench; refer to Figure-18. The **Line parameters** dialog will be displayed in the **Tasks** panel of **Combo View**; refer to Figure-19.

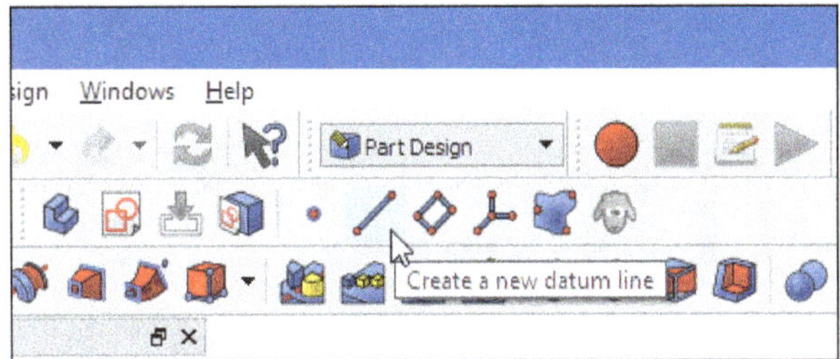

Figure-18. Create a datum line tool

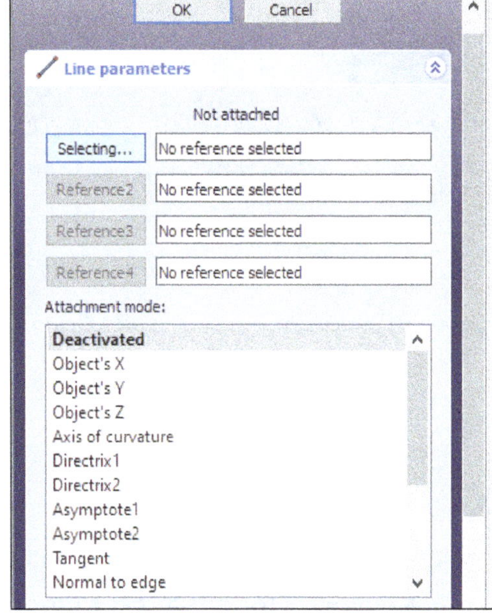

Figure-19. Line parameters dialog

- Select first reference for attachment mode from the 3D view area or from the attachment modes list available in **Attachment mode** area of the dialog.

- To add an additional reference, click on the **Reference2** button and select desired reference.
- Specify desired offset values in the **Attachment Offset** area of the dialog.
- Click on **OK** button from the dialog. The datum line will be created at specified offset distance; refer to Figure-20.

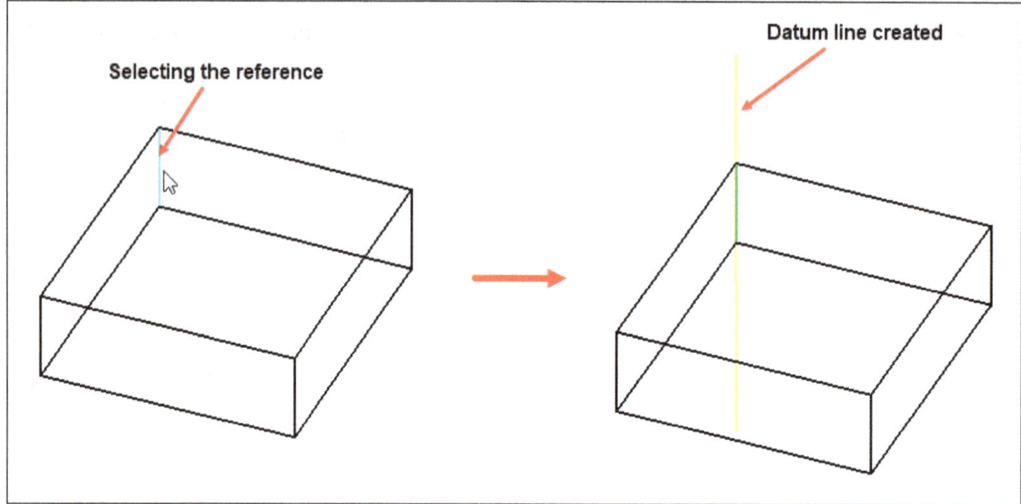

Figure-20. Creation of datum line

Create a datum plane

The **Create a datum plane** tool creates a datum plane which can be used as reference for sketches or other datum geometries. Sketches can also be attached to datum planes. The procedure to use this tool is discussed next.

- Click on the **Create a datum plane** tool from **Toolbar** in the **Part Design** workbench; refer to Figure-21. The **Plane parameters** dialog will be displayed in the **Tasks** panel of **Combo View**; refer to Figure-22.

Figure-21. Create a datum plane tool

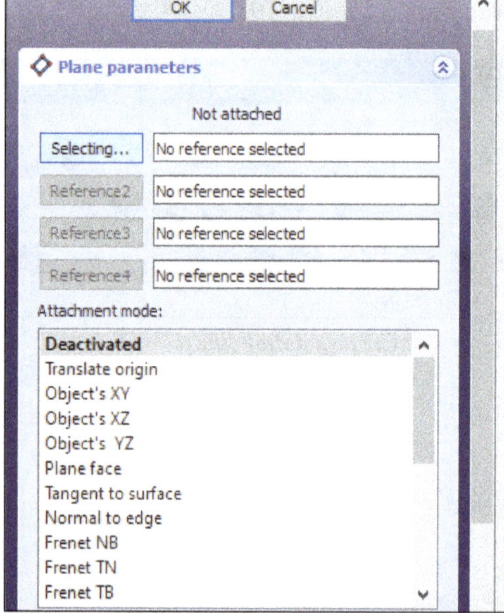

Figure-22. Plane parameters dialog

- Select first reference for attachment mode from the 3D view area or from the attachment modes list available in **Attachment mode** area of the dialog. You can select a point, line, plane, or face as reference.
- To add an additional reference, click on the **Reference2** button.

- Specify desired offset values in the **Attachment Offset** area of the dialog to move plane by specified value in respective directions.
- Click on **OK** button from the dialog. The datum plane will be created at specified offset distance; refer to Figure-23.

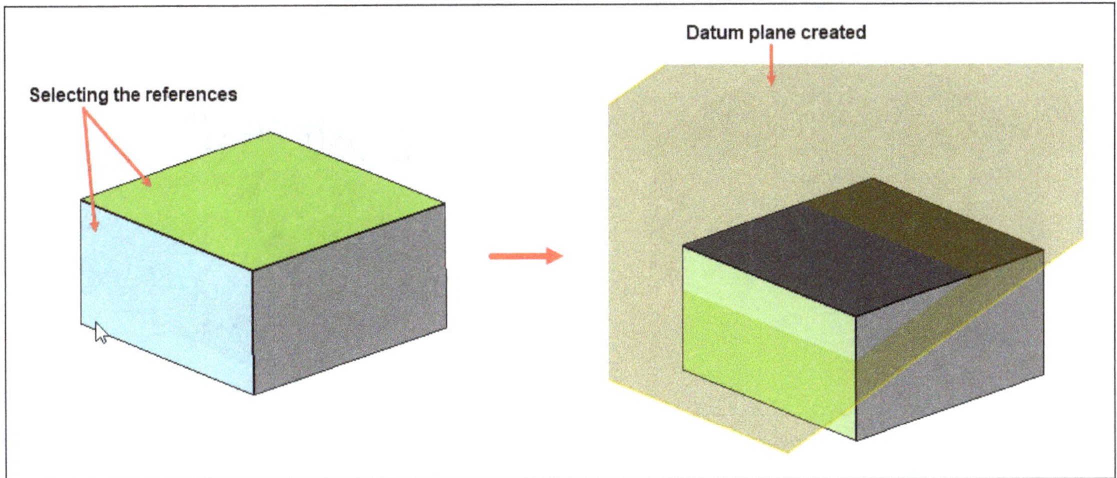

Figure-23. Creation of datum plane

Create a local coordinate system

The **Create a local coordinate system** tool creates a local coordinate system which can be used as reference for other datum geometry. It also helps identify the orientation of the referenced datum geometry in 3D space. The procedure to use this tool is discussed next.

- Click on **Create a local coordinate system** tool from **Toolbar** in the **Part Design** workbench; refer to Figure-24. The **Coordinate System parameters** dialog will be displayed in the **Tasks** panel of **Combo View**; refer to Figure-25.

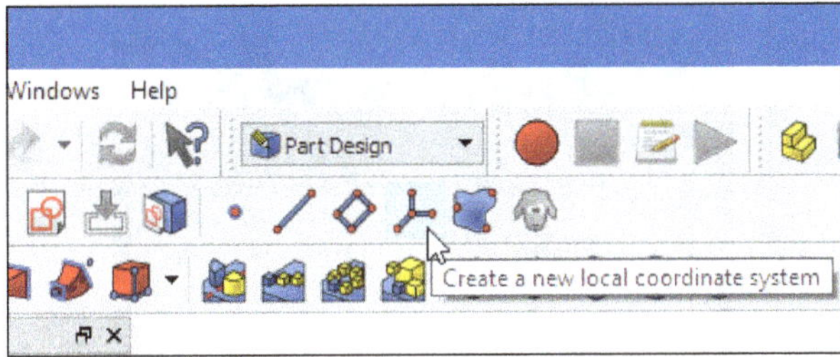

Figure-24. Create a local coordinate system tool

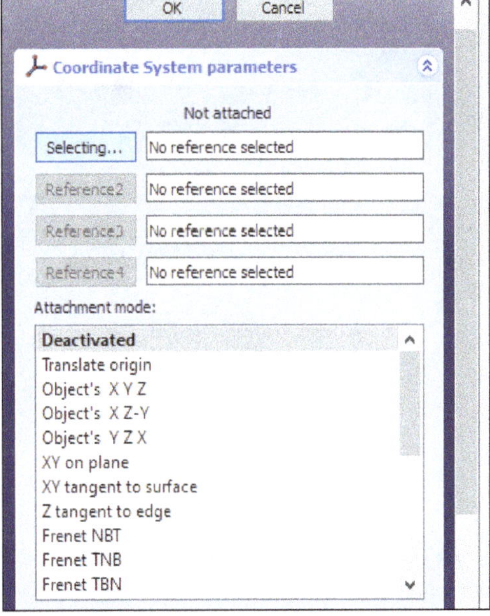

Figure-25. Coordinate System parameters dialog

- Select a point, line, face, or plane as first reference for attachment mode from the 3D view area or from the attachment modes list available in **Attachment mode** area of the dialog.
- To add an additional reference, click on the **Reference2** button and select desired entity.
- Specify desired offset values in the **Attachment Offset** area of the dialog.

- Click on **OK** button from the dialog. The local coordinate system will be created at specified offset distance; refer to Figure-26.

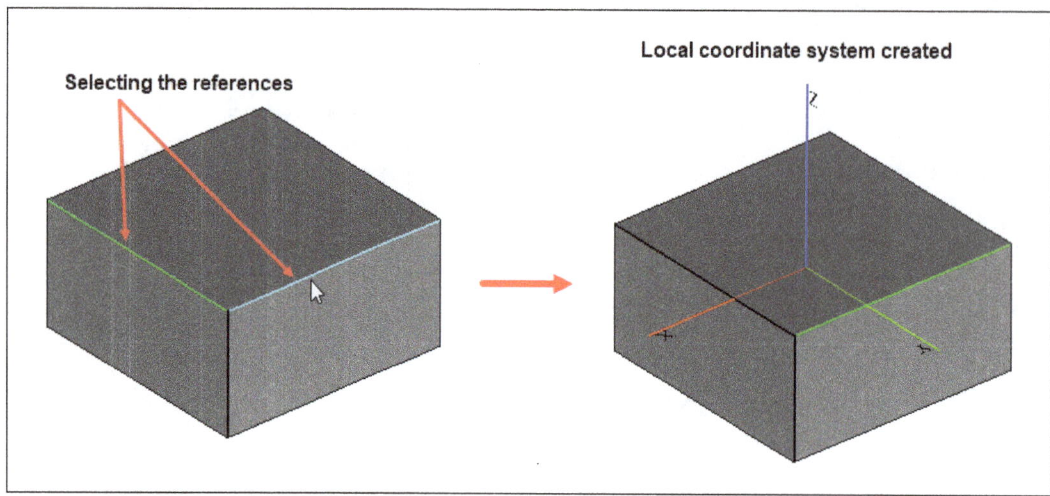

Figure-26. Creation of local coordinate system

Create a shape binder

The **Create a shape binder** tool creates a datum shape binder from selected body inside the active body. A shape binder is a reference object that links edges or faces of another body to active body. It can also be used to link a sketch from one body to another body. The datum shape binder object displays as translucent yellow in the 3D view. The procedure to use this tool is discussed next.

- Click on **Create a shape binder** tool from **Toolbar** in the **Part Design** workbench; refer to Figure-27. The **Datum shape parameters** dialog will be displayed in the **Tasks** panel of **Combo View**; refer to Figure-28.

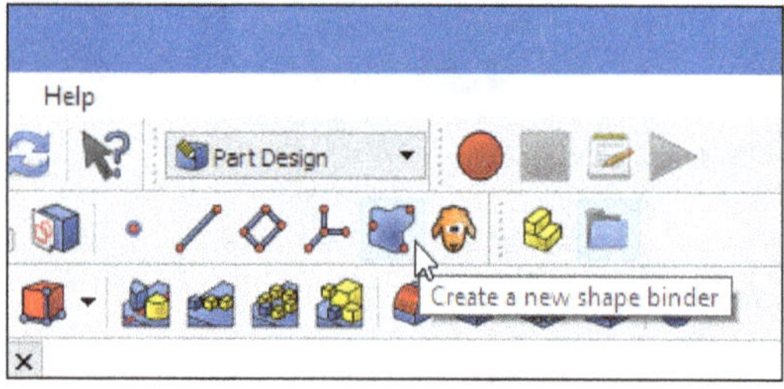

Figure-27. Create a shape binder tool

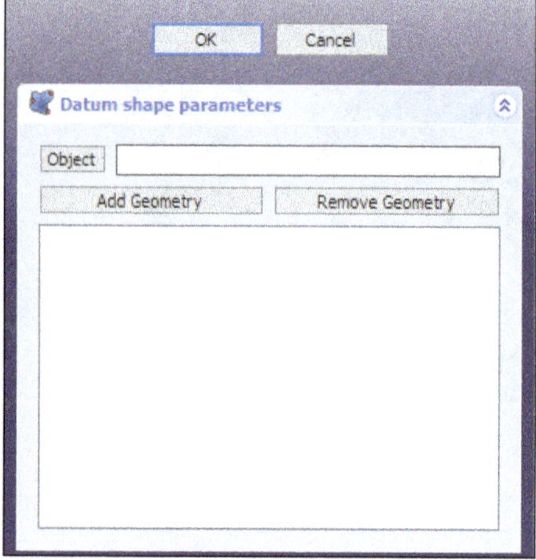

Figure-28. Datum shape parameters dialog

- The **Object** button in the dialog will select the whole body and the **Add Geometry** button will select elements of the object (vertex, edge, face). Click on **Object** button and select desired object in the 3D view area or click on **Add Geometry** button and select elements of the object (vertex, edge, face) in the 3D view area. The datum shape binder object will be created; refer to Figure-29.

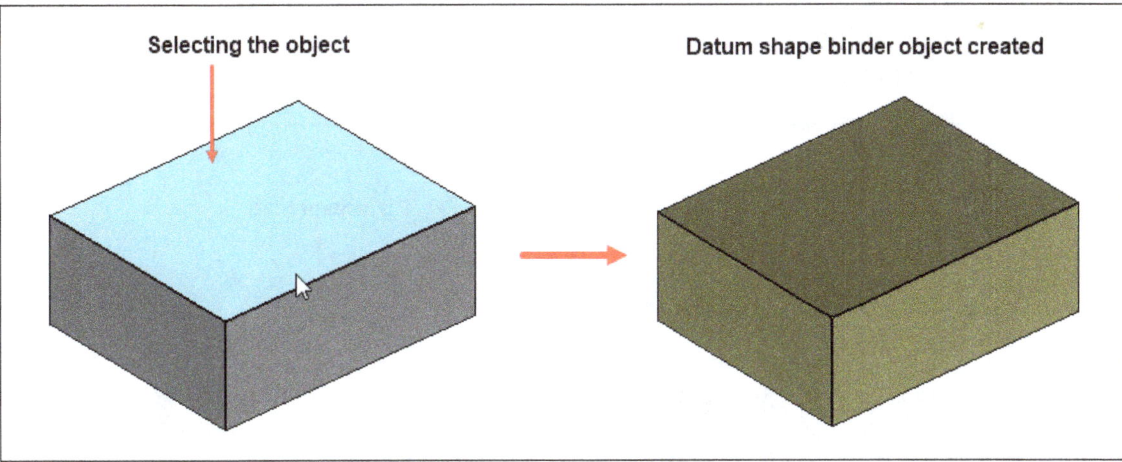

Figure-29. Datum shape binder object created

- To remove selected geometry, click on **Remove Geometry** button from the dialog and select the geometry in the 3D view area.
- To display the datum shape binder object, toggle the visibility of **Pad** feature from **True** to **False** in the **Property editor** of **Combo View**. The datum shape binder object will be displayed; refer to Figure-30.

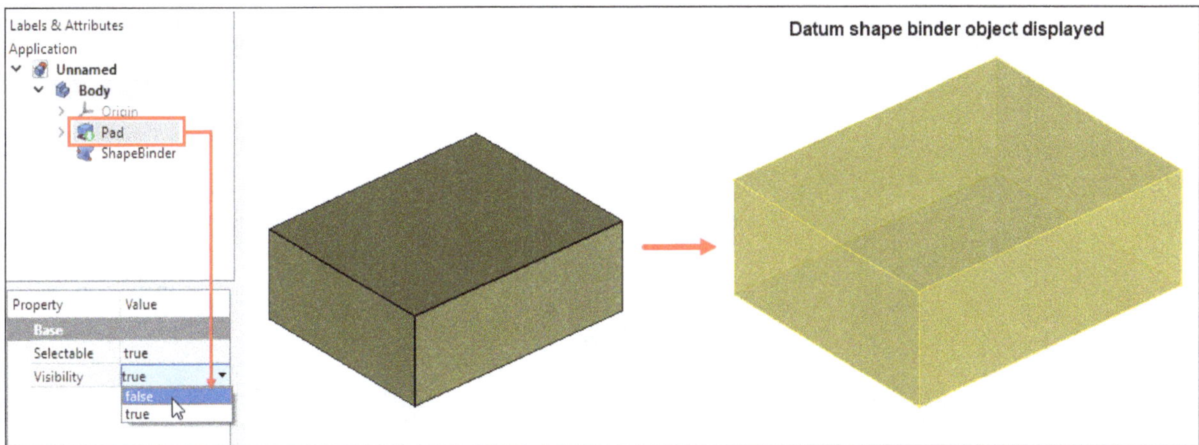

Figure-30. Displaying datum shape binder object

Create a clone

The **Create a clone** tool creates a linked copy of selected object which will follow any future edits to the original object. The procedure to use this tool is discussed next.

- Select the object to be cloned from Model tree view or from the 3D view area.
- Click on the **Create a clone** tool from **Toolbar** in the **Part Design** workbench; refer to Figure-31. The selected object will be cloned.

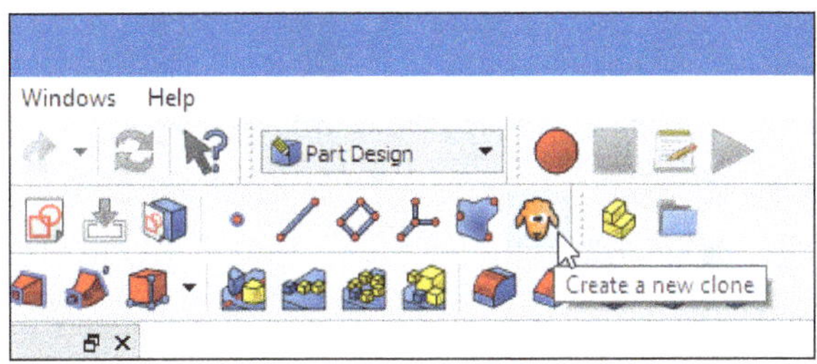

Figure-31. Create a clone tool

- Double-click on cloned object from Model tree view. The **Increments** dialog will be displayed in the **Tasks** panel of **Combo View** and a **Placement triad** will be displayed on the object in the 3D view area; refer to Figure-32.

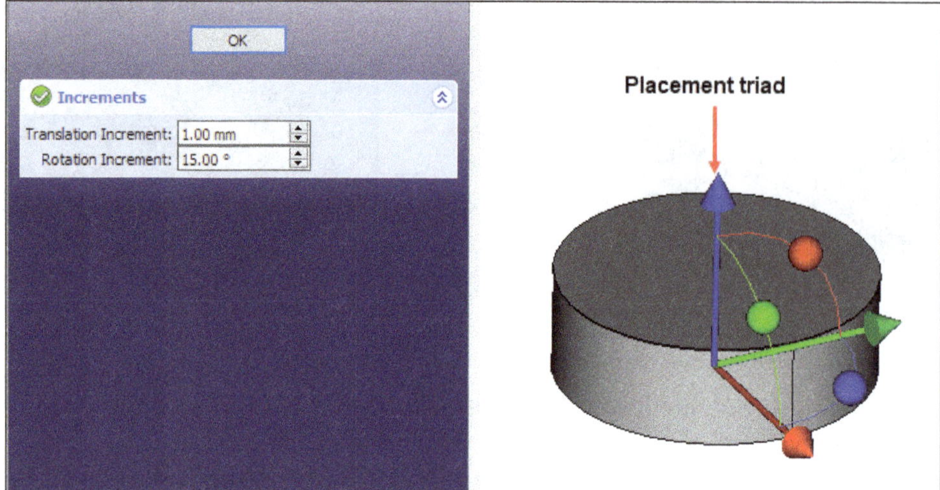

Figure-32. Increments dialog

- Drag the **Placement triad** to place the cloned object in desired location. The object will be placed; refer to Figure-33.

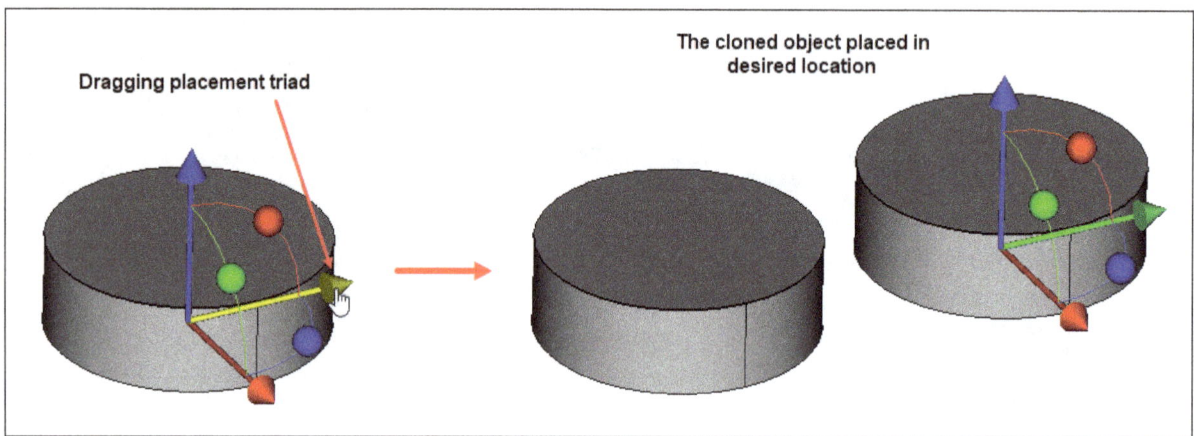

Figure-33. Placement of cloned object

- Click on **OK** button from the dialog.

Additive tools

The **Additive tools** are used for creating base features or adding material to an existing solid body. The **Additive tools** are available in the **Toolbar** of **Part Design** workbench; refer to Figure-34. The procedures to use these tools are discussed next.

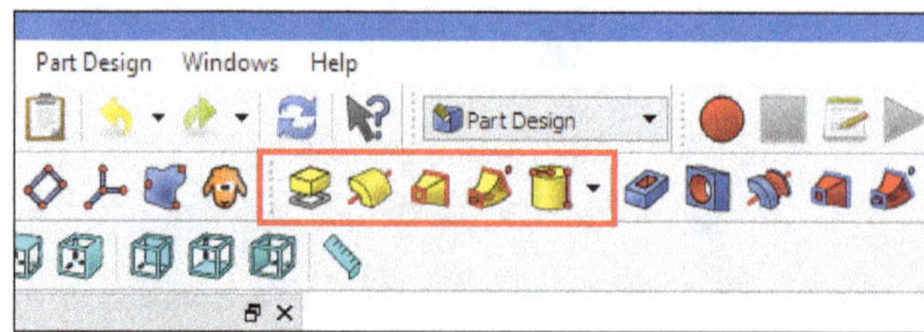

Figure-34. Additive tools

Pad

The **Pad** tool extrudes a sketch to form solid in a direction normal to the sketch plane. The procedure to use this tool is discussed next.

- Select desired sketch from the Model tree view to create the pad feature; refer to Figure-35.

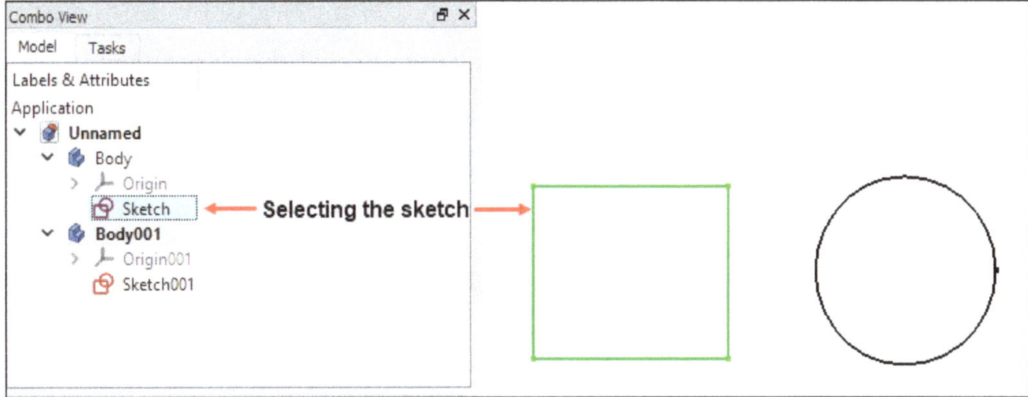

Figure-35. Selecting the sketch to create pad feature

- Click on the **Pad** tool from **Toolbar** in the **Part Design** workbench; refer to Figure-36. The **Pad parameters** dialog will be displayed in the **Tasks** panel of **Combo View** along with the preview of extrude feature; refer to Figure-37.

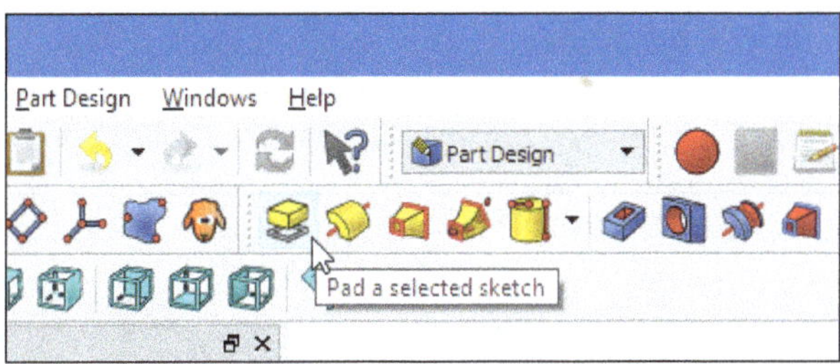

Figure-36. Pad tool

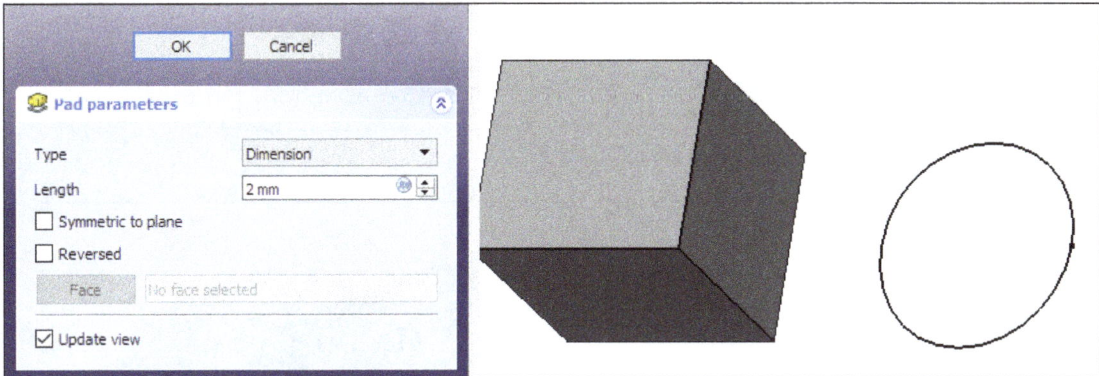

Figure-37. Pad parameters dialog

- Select desired type of extrusion from **Type** drop-down. Select **Dimension** option to extrude the selected sketch on either side of the sketch plane at specified length. Select **To last** option to extrude the sketch upto the last face of the base feature. Select **To first** option to extrude the sketch upto the first face of the base feature. Select **Up to face** option to extrude the sketch upto the selected face of base feature. Select **Two dimensions** option to extrude the sketch by specifying length for both the sides.
- Specify the length of extrusion in the **Length** edit box.

- Select **Symmetric to plane** check box to extrude the sketch on both the sides by same specified length if the **Dimension** option is selected in the **Length** drop-down of dialog.
- Select **Reversed** check box to reverse the direction of extrusion if the **Dimension** option is selected in the **Length** drop-down of dialog.
- Click on **OK** button from the dialog. The selected sketch will be extruded.

Revolution

The **Revolution** tool is used to create a solid by revolving selected sketch or 2D object about given axis. The procedure to use this tool is discussed next.

- Select the sketch to be revolved from **Model** panel in the **Combo View**; refer to Figure-38.
- Click on the **Revolution** tool from **Toolbar** in the **Part Design** workbench; refer to Figure-39. The **Revolution parameters** dialog will be displayed in the **Tasks** panel of **Combo View** along with the preview of revolution feature; refer to Figure-40.

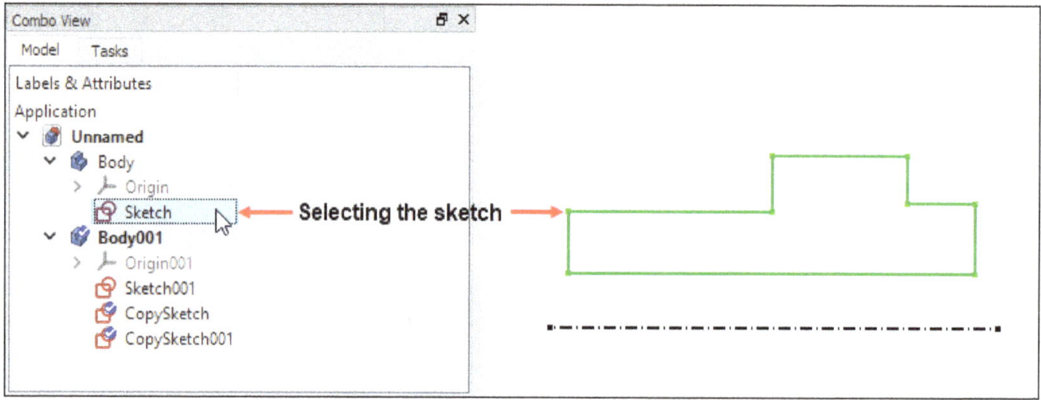

Figure-38. Selecting the sketch to be revolve

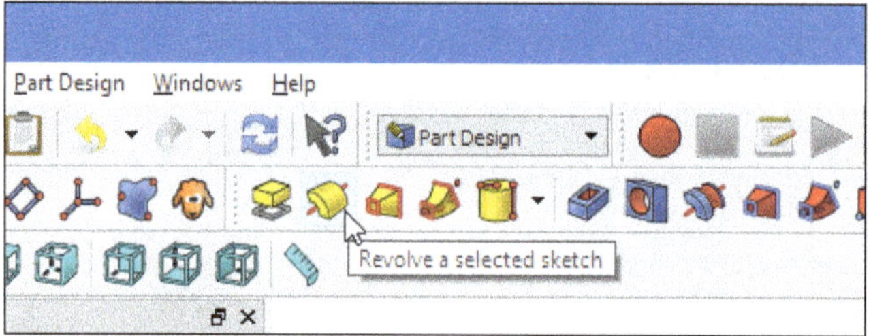

Figure-39. Revolution tool

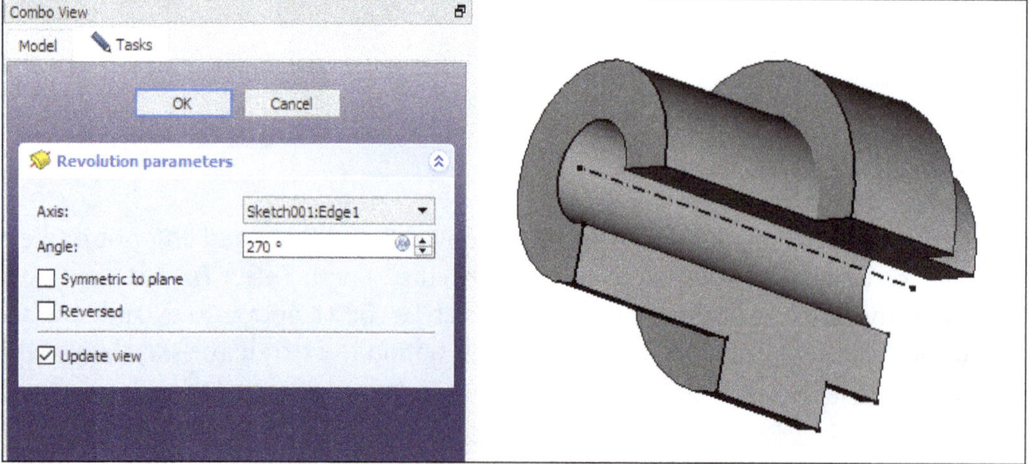

Figure-40. Revolution parameters dialog

- Select desired option from **Axis** drop-down in the dialog about which the sketch is to be revolved. Select **Vertical sketch axis** option to create revolve feature about a vertical sketch axis. Select **Horizontal sketch axis** option to create revolve feature about a horizontal sketch axis. Select **Base X axis**, **Base Y axis**, or **Base Z axis** option to create the revolve feature about **X**, **Y**, or **Z** axis of body's origin, respectively. Select the **Select reference** option to create revolve feature by selecting an edge on the body or a datum line as an axis of revolution.
- Specify the angle of revolution by which the revolution is to be formed in the **Angle** edit box.
- Specify the other parameters as desired and click on **OK** button from the dialog. The revolution feature will be created.

Additive loft

The **Additive loft** tool creates a solid active body by making a transition between two or more sketches. The procedure to use this tool is discussed next.

- Click on **Additive loft** tool from **Toolbar** in the **Part Design** workbench; refer to Figure-41. The **Select feature** dialog will be displayed in the **Tasks** panel of **Combo View** asking you to select the base profile; refer to Figure-42.

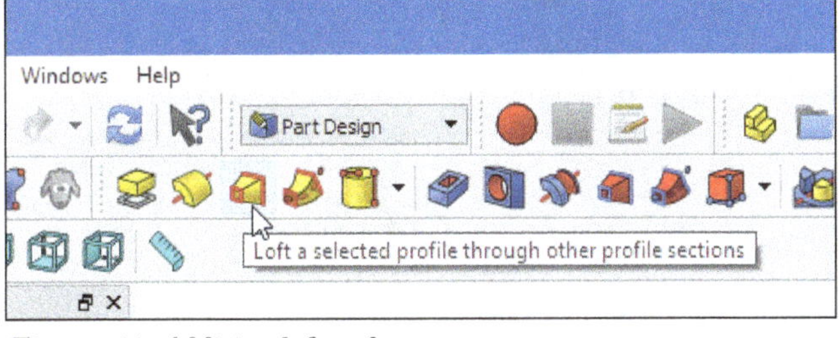

Figure-41. Additive loft tool

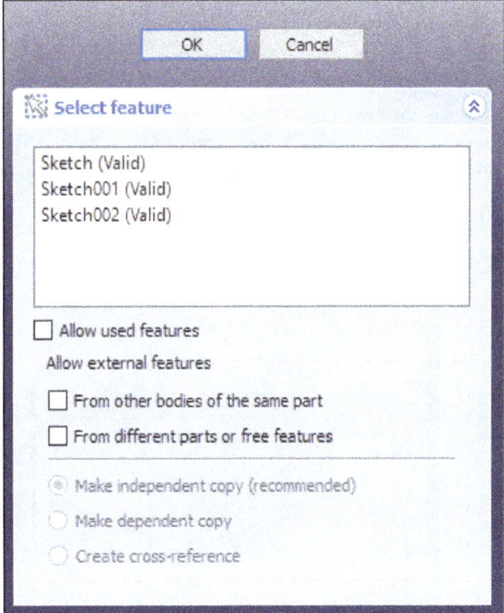

Figure-42. Select feature dialog

- Select a sketch from the dialog to be used as base profile for additive loft feature and click on **OK** button; refer to Figure-43. The **Loft parameters** dialog will be displayed; refer to Figure-44.

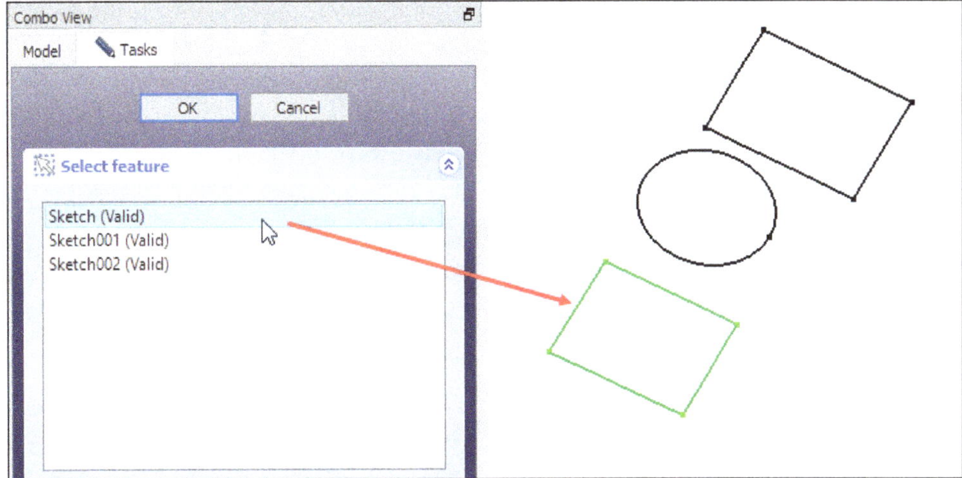

Figure-43. Selecting the sketch for base profile

- The base profile is preselected in the **Object** selection box of **Profile** area. Click on **Add Section** button and select the second sketch in the 3D view area for the first loft section and select the third sketch for the second loft section. The preview of additive loft feature will be displayed; refer to Figure-45.

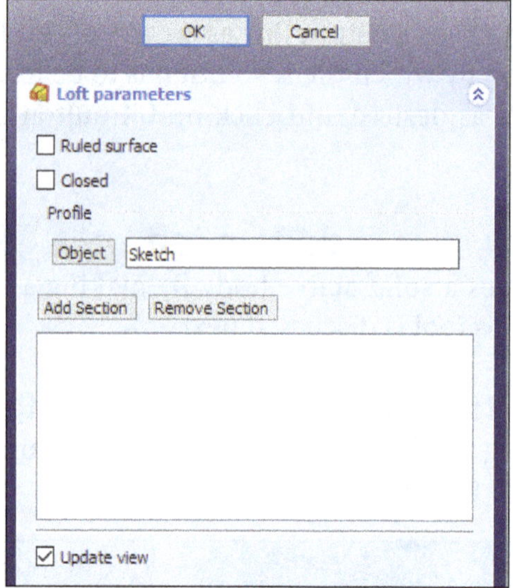

Figure-44. Loft parameters dialog

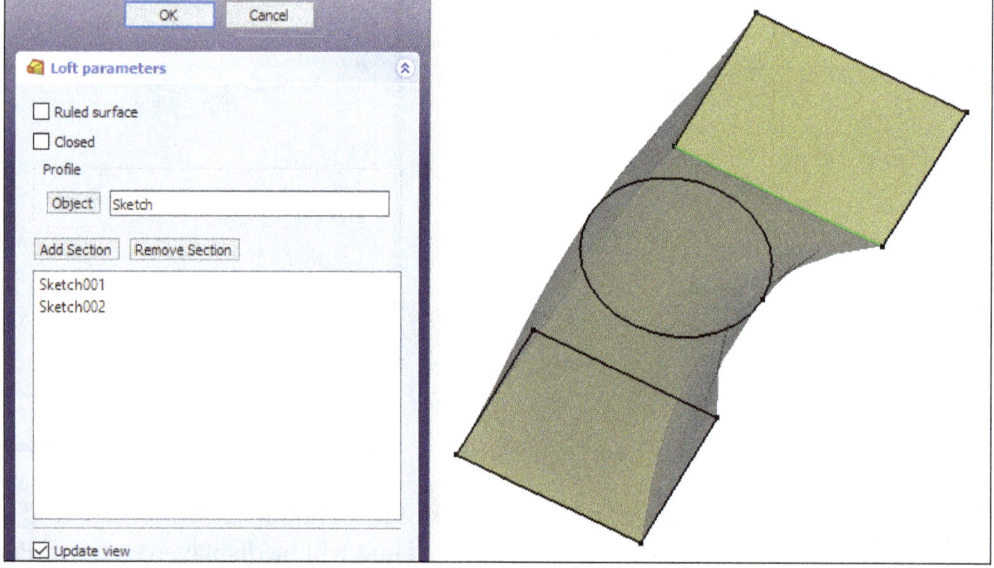

Figure-45. Preview of loft feature

- Select **Ruled surface** check box to make straight transitions between cross-sections.
- Select **Closed** check box to make a transition from the last cross-section to the first to create a loop.
- Specify the parameters as desired and click on **OK** button from the dialog to create the feature.

Additive pipe

The **Additive pipe** tool creates a solid as active body by sweeping one or more sketches along an open or closed path. The procedure to use this tool is discussed next.

- Click on **Additive pipe** tool from **Toolbar** in the **Part Design** workbench; refer to Figure-46. The **Select feature** dialog will be displayed in the **Tasks** panel of **Combo View** as discussed earlier asking you to select the base profile.

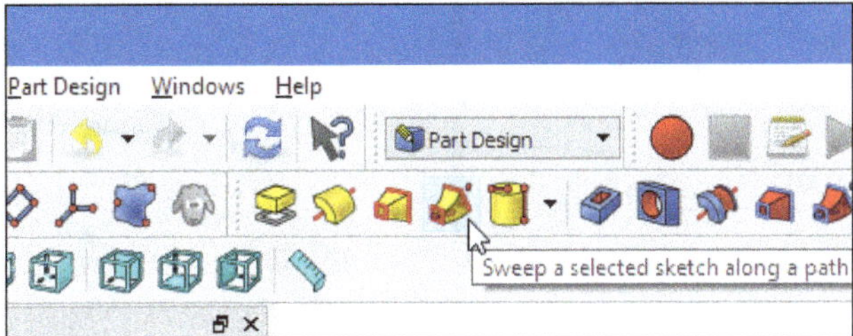

Figure-46. Additive pipe tool

- Select desired sketch for base profile of sweep feature from the dialog; refer to Figure-47 and click on **OK** button. The **Pipe parameters** dialog will be displayed in the **Tasks** panel of **Combo View**; refer to Figure-48.

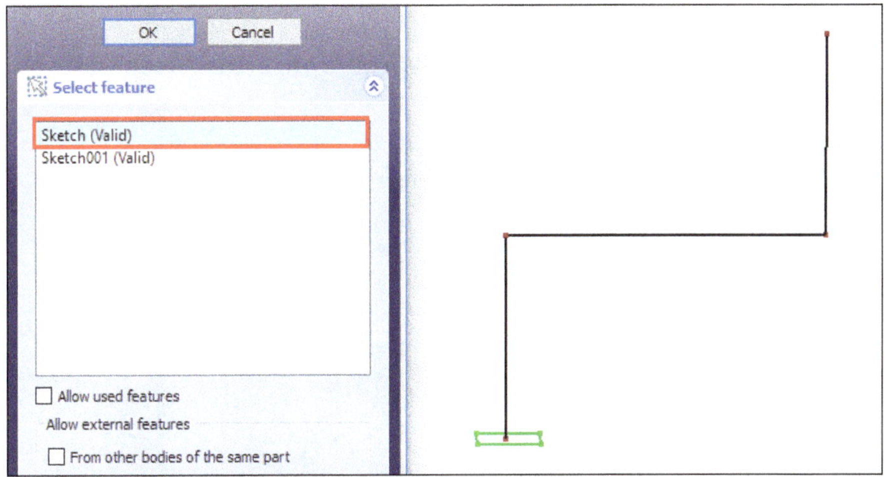

Figure-47. Selecting base profile for sweep feature

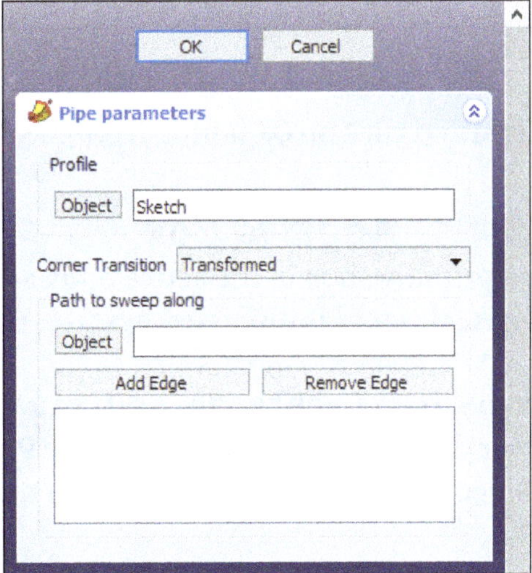

Figure-48. Pipe parameters dialog

- The base profile is preselected in the **Object** selection box of the **Profile** area.
- Click on the **Object** button from **Path to sweep along** area and select an edge of the sketch in 3D view area as a path for sweep feature. The whole sketch will be selected as sweep feature path and the preview of sweep feature will be displayed; refer to Figure-49.

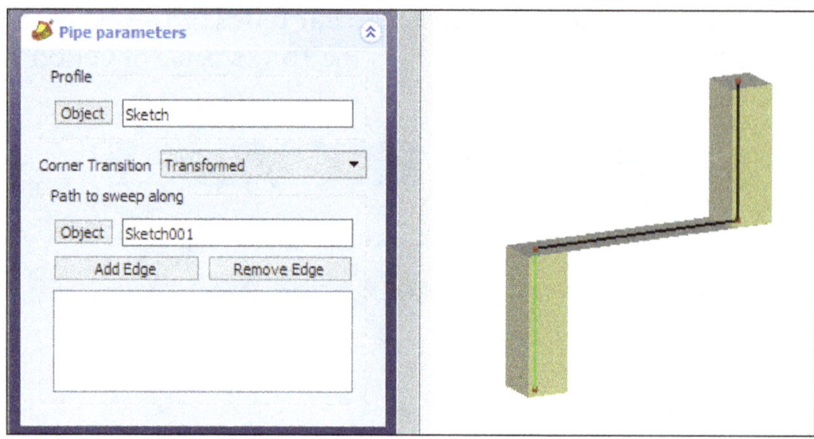

Figure-49. Preview of sweep feature

- Click on **Add Edge** button to select single edge of the sketch from the 3D view area as a path for sweep feature.
- To remove an edge of the sketch as a sweep feature path, select an edge from 3D view area and click on **Remove Edge** button from the dialog.
- Select desired option from **Corner Transition** drop-down. The sweep feature will be created as the selected option; refer to Figure-50.

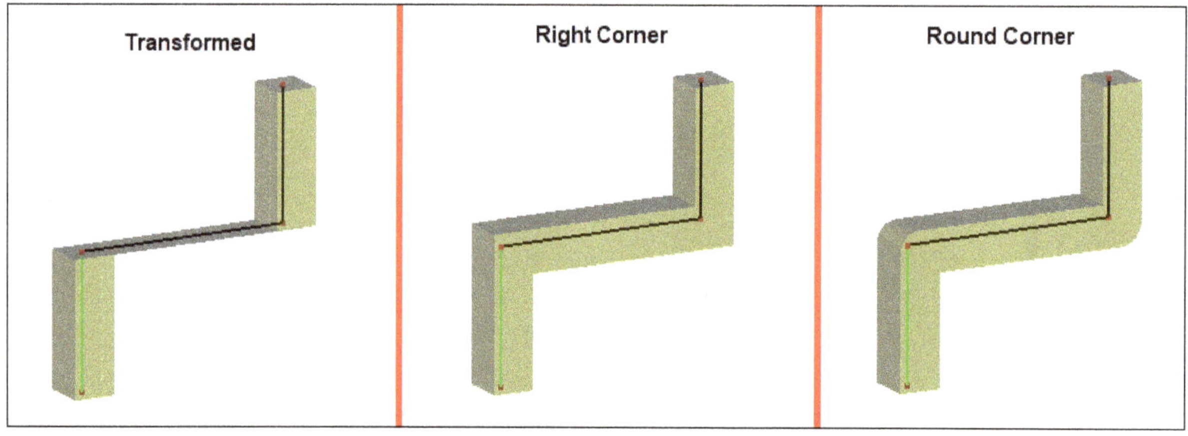

Figure-50. Corner Transition options

- Specify the parameters as desired and click on **OK** button from the dialog to create the sweep feature.

Additive Box

The **Additive Box** tool inserts a primitive box in the active body as the first feature or fuses it to the existing feature. The procedure to use this tool is discussed next.

- Click on the **Additive Box** tool from **Create an additive primitive** drop-down in the **Toolbar** of **Part Design** workbench; refer to Figure-51. The **Primitive parameters** dialog will be displayed in the **Tasks** panel of **Combo View** along with the preview of additive box feature; refer to Figure-52.

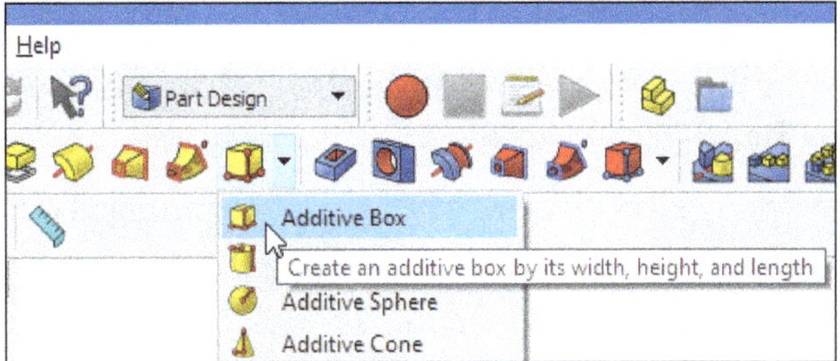

Figure-51. Additive Box tool

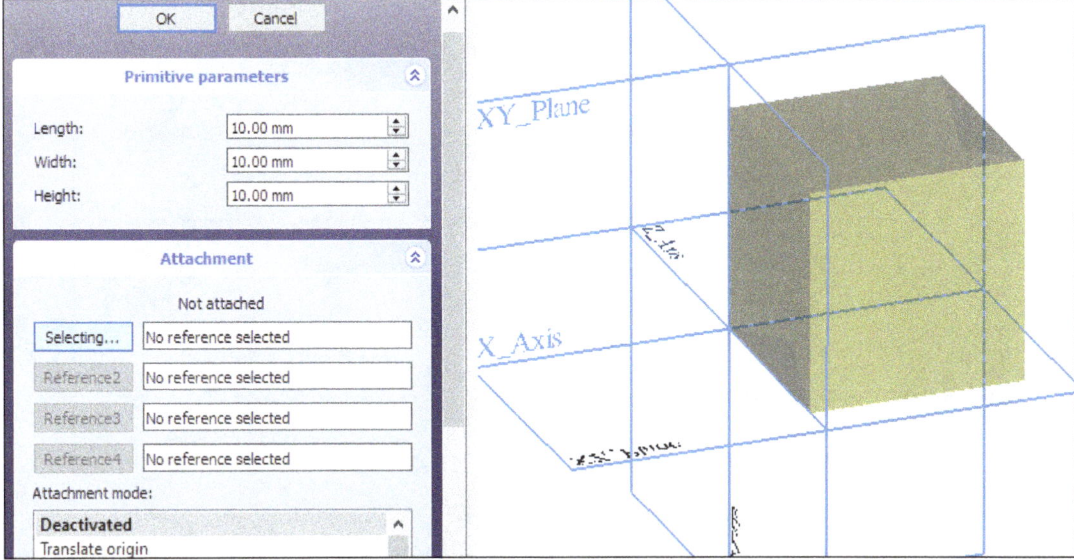

Figure-52. Primitive parameters dialog with preview of additive box feature

- Specify the length, width, and height of the feature in the **Length**, **Width**, and **Height** edit boxes of the dialog, respectively.
- Specify the other parameters in the dialog as discussed earlier.
- Click on **OK** button from the dialog. The additive box feature will be created; refer to Figure-53.

Figure-53. Additive box feature created

Additive Cylinder

The **Additive Cylinder** tool inserts a primitive cylinder in the active body as the first feature or fuses it to the existing feature. The procedure to use this tool is discussed next.

- Click on the **Additive Cylinder** tool from **Create an additive primitive** drop-down in the **Toolbar** of **Part Design** workbench; refer to Figure-54. The **Primitive parameters** dialog will be displayed in the **Tasks** panel of **Combo View** along with the preview of additive cylinder feature; refer to Figure-55.

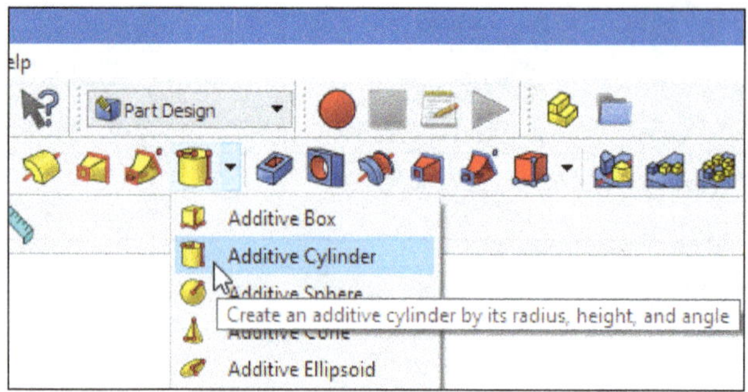

Figure-54. Additive Cylinder tool

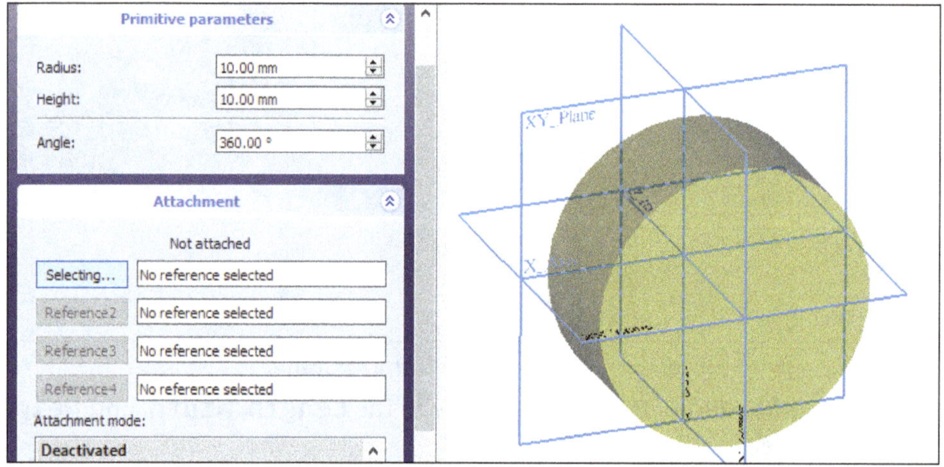

Figure-55. Primitive parameters dialog with preview of additive cylinder feature

- Specify the radius and height of the feature in the **Radius** and **Height** edit boxes, respectively.
- Specify the angle of revolution of the cross-section in the **Angle** edit box of the dialog.
- Specify the other parameters in the dialog as discussed earlier.
- Click on **OK** button from the dialog. The additive cylinder feature will be created; refer to Figure-56.

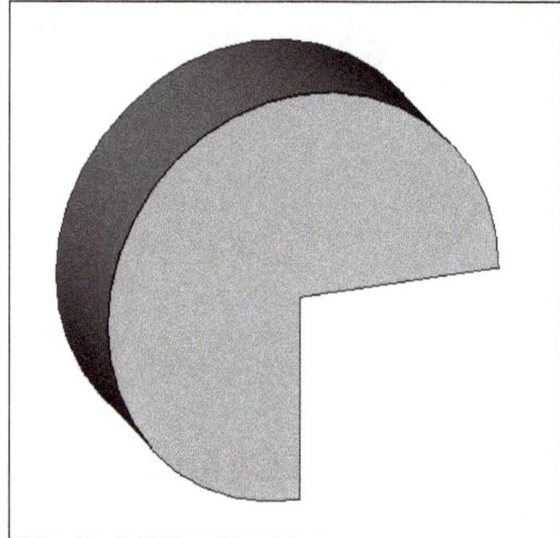

Figure-56. Additive cylinder feature created

Additive Sphere

The **Additive Sphere** tool inserts a primitive sphere in the active body as the first feature or fuses it to the existing feature. The procedure to use this tool is discussed next.

- Click on the **Additive Sphere** tool from **Create an additive primitive** drop-down in the **Toolbar** of **Part Design** workbench; refer to Figure-57. The **Primitive parameters** dialog will be displayed in the **Tasks** panel of **Combo View** along with the preview of additive sphere feature; refer to Figure-58.

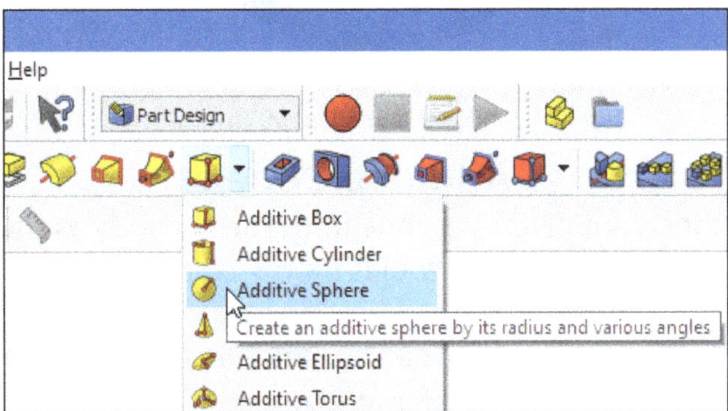

Figure-57. Additive Sphere tool

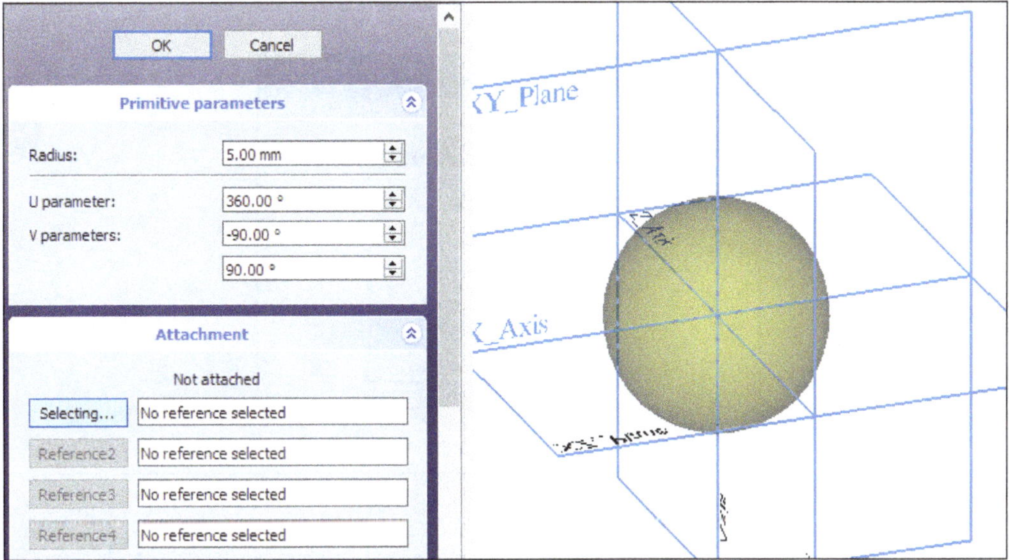

Figure-58. Primitive parameters dialog with preview of additive sphere feature

- Specify desired radius of the sphere in the **Radius** edit box of the dialog.
- Specify desired angle of revolution of the cross section in the **U parameter** edit box of the dialog.
- Specify lower truncation of the sphere parallel to the circular cross section in the **V parameters** edit box and specify upper truncation of the ellipsoid parallel to the circular cross section in the edit box just below to the **V parameters** edit box of the dialog.
- Specify the other parameters in the dialog as discussed earlier.
- Click on **OK** button from the dialog. The additive sphere feature will be created; refer to Figure-59.

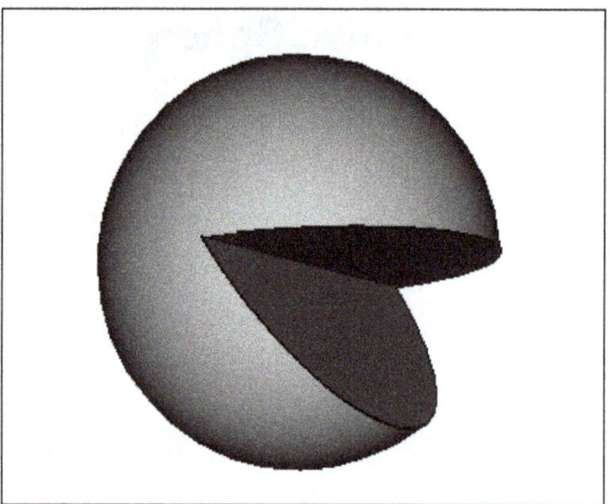

Figure-59. Additive sphere feature created

Additive Cone

The **Additive Cone** tool inserts a primitive cone in the active body as the first feature or fuses it to the existing feature. The procedure to use this tool is discussed next.

- Click on the **Additive Cone** tool from **Create an additive primitive** drop-down in the **Toolbar** of **Part Design** workbench; refer to Figure-60. The **Primitive parameters** dialog will be displayed in the **Tasks** panel of **Combo View** along with the preview of additive cone feature; refer to Figure-61.

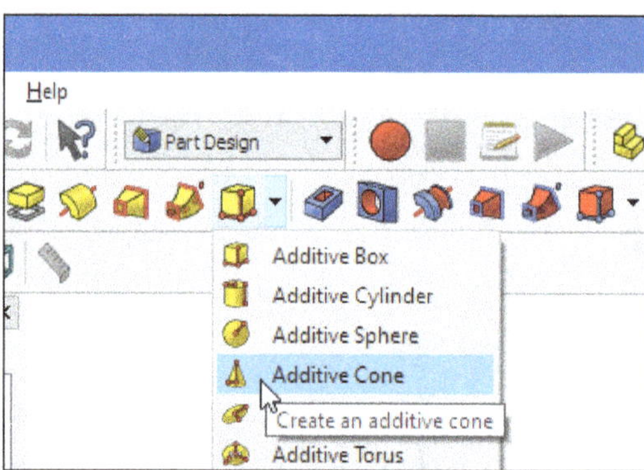

Figure-60. Additive Cone tool

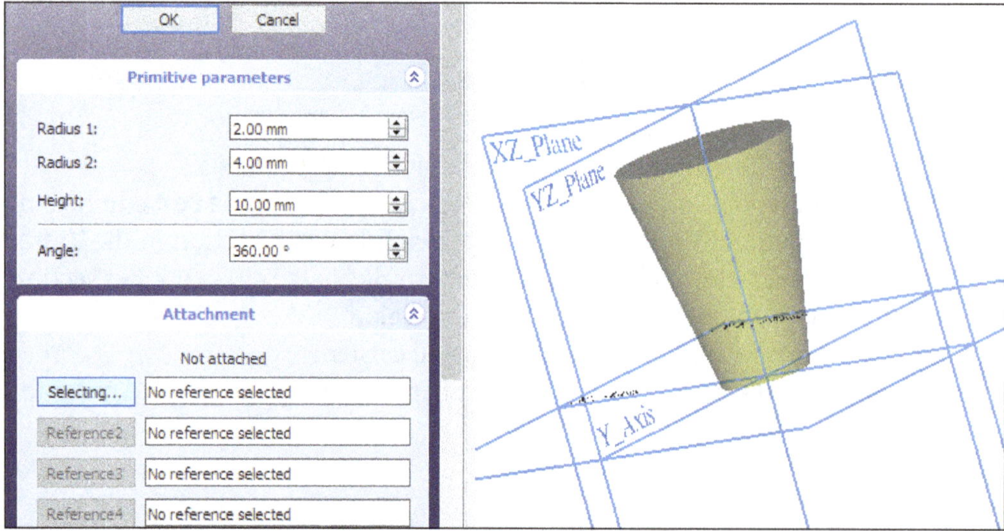

Figure-61. Primitive parameters dialog with preview of additive cone feature

- Specify desired radius value for base of the cone in the **Radius 1** edit box and specify desired radius value for top of the cone in the **Radius 2** edit box of the dialog.
- Specify the height of the cone along its axis in the **Height** edit box of the dialog.
- Specify the angle of revolution of the cross section in the **Angle** edit box of the dialog.
- Specify the other parameters in the dialog as discussed earlier.
- Click on **OK** button from the dialog. The additive cone feature will be created; refer to Figure-62.

Figure-62. Additive cone feature created

Additive Ellipsoid

The **Additive Ellipsoid** tool inserts a primitive ellipsoid in the active body as the first feature or fuses it to the existing feature. The procedure to use this tool is discussed next.

- Click on the **Additive Ellipsoid** tool from **Create an additive primitive** drop-down in the **Toolbar** of **Part Design** workbench; refer to Figure-63. The **Primitive parameters** dialog will be displayed in the **Tasks** panel of **Combo View** along with the preview of additive ellipsoid feature; refer to Figure-64.

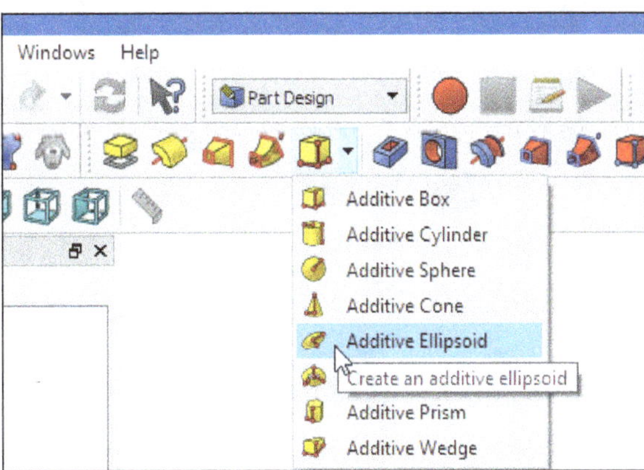

Figure-63. Additive Ellipsoid tool

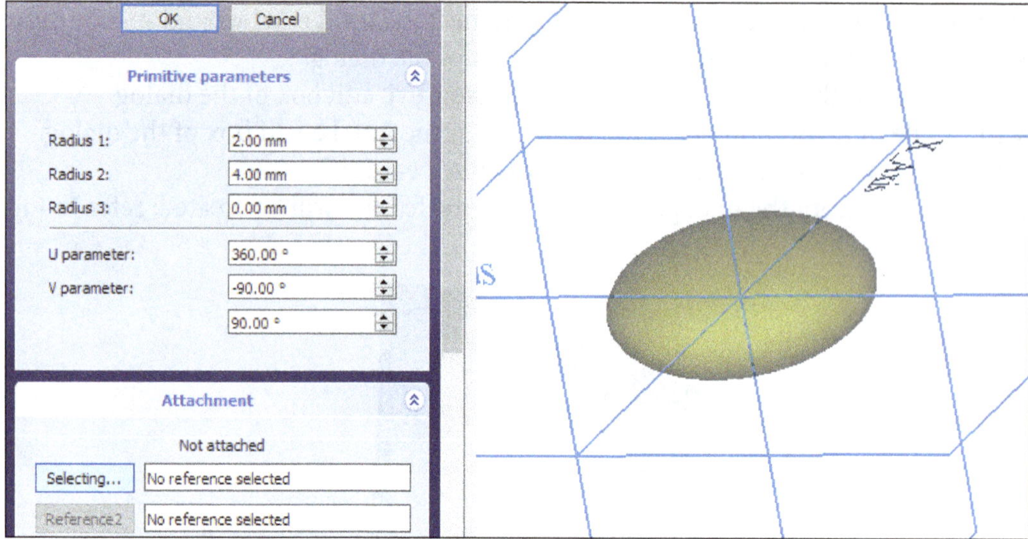

Figure-64. Primitive parameters dialog with preview of additive ellipsoid feature

- Specify desired radius value along the ellipsoid's vertical axis, along the ellipsoid's length, and along the ellipsoid's width in the **Radius 1**, **Radius 2**, and **Radius 3** edit boxes, respectively.
- Specify desired angle of revolution of the elliptical cross section in the **U parameter** edit box of the dialog.
- Specify the lower truncation of the ellipsoid parallel to the circular cross section in the **V parameter** edit box and specify upper truncation of the ellipsoid parallel to the circular cross section in the edit box just below to the **V parameter** edit box.
- Specify other parameters in the dialog as discussed earlier.
- Click on **OK** button from the dialog. The additive ellipsoid feature will be created; refer to Figure-65.

Figure-65. Additive ellipsoid feature created

Additive Torus

The **Additive Torus** tool inserts a primitive torus in the active body as the first feature or fuses it to the existing feature. The procedure to use this tool is discussed next.

- Click on the **Additive Torus** tool from **Create an additive primitive** drop-down in the **Toolbar** of **Part Design** workbench; refer to Figure-66. The **Primitive parameters** dialog will be displayed in the **Tasks** panel of **Combo View** along with the preview of additive ellipsoid feature; refer to Figure-67.

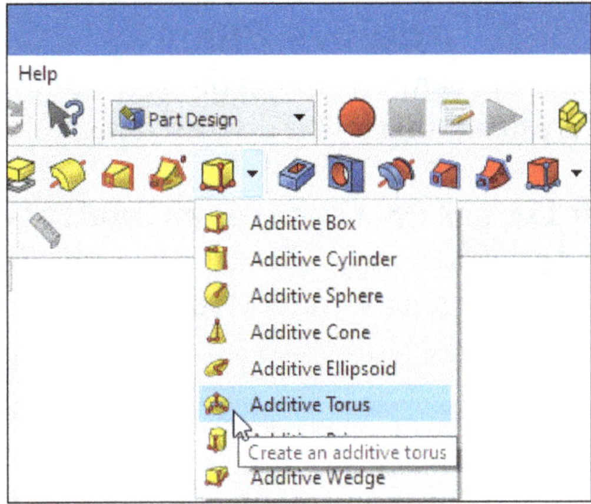

Figure-66. Additive Torus tool

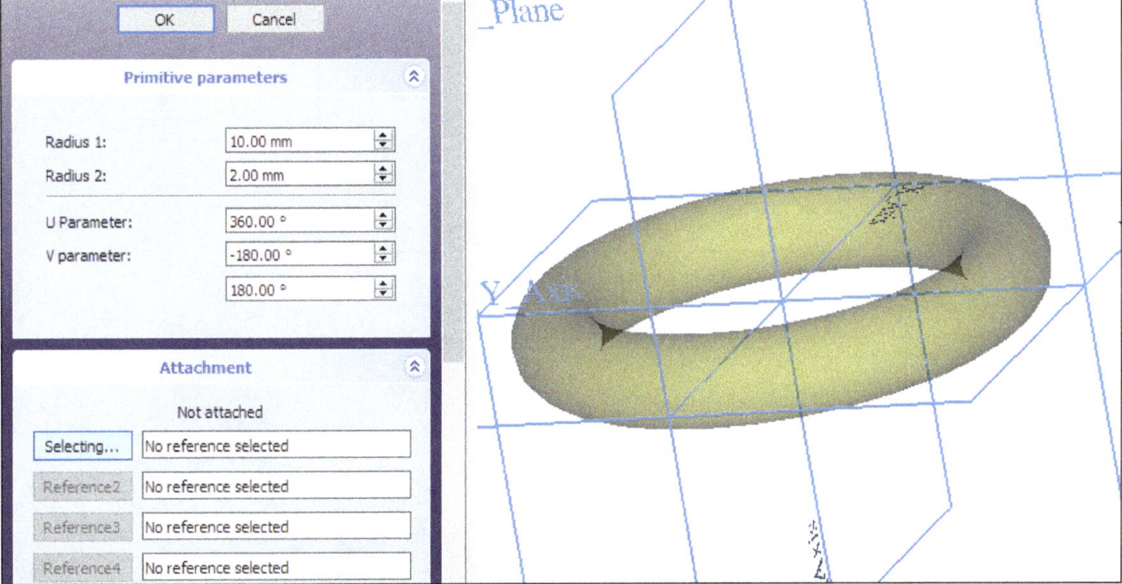

Figure-67. Primitive parameters dialog with preview of additive torus feature

- Specify desired radius of the imaginary orbit around which the circular cross section revolves in the **Radius 1** edit box and specify desired radius of circular cross section defining the form of the torus in **Radius 2** edit box of the dialog.
- Specify other parameters in the dialog as discussed earlier.
- Click on **OK** button from the dialog. The additive torus feature will be created; refer to Figure-68.

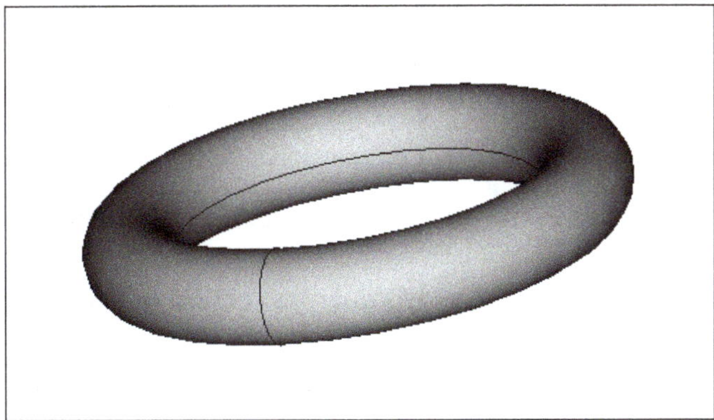

Figure-68. Additive torus feature created

Additive Prism

The **Additive Prism** tool inserts a primitive prism in the active body as the first feature or fuses it to the existing feature. The procedure to use this tool is discussed next.

- Click on the **Additive Prism** tool from **Create an additive primitive** drop-down in the **Toolbar** of **Part Design** workbench; refer to Figure-69. The **Primitive parameters** dialog will be displayed in the **Tasks** panel of **Combo View** along with the preview of additive prism feature; refer to Figure-70.

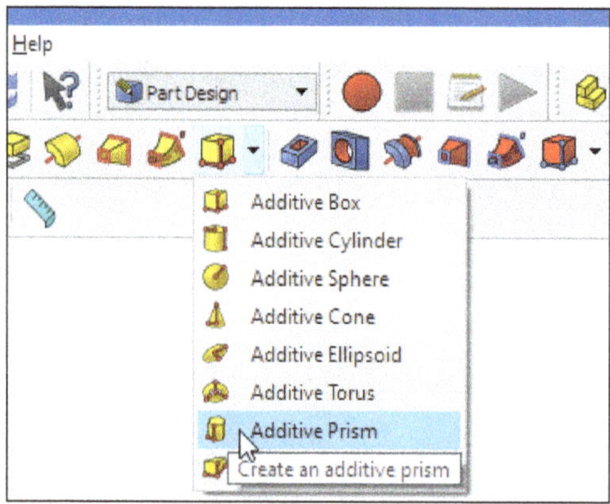

Figure-69. Additive Prism tool

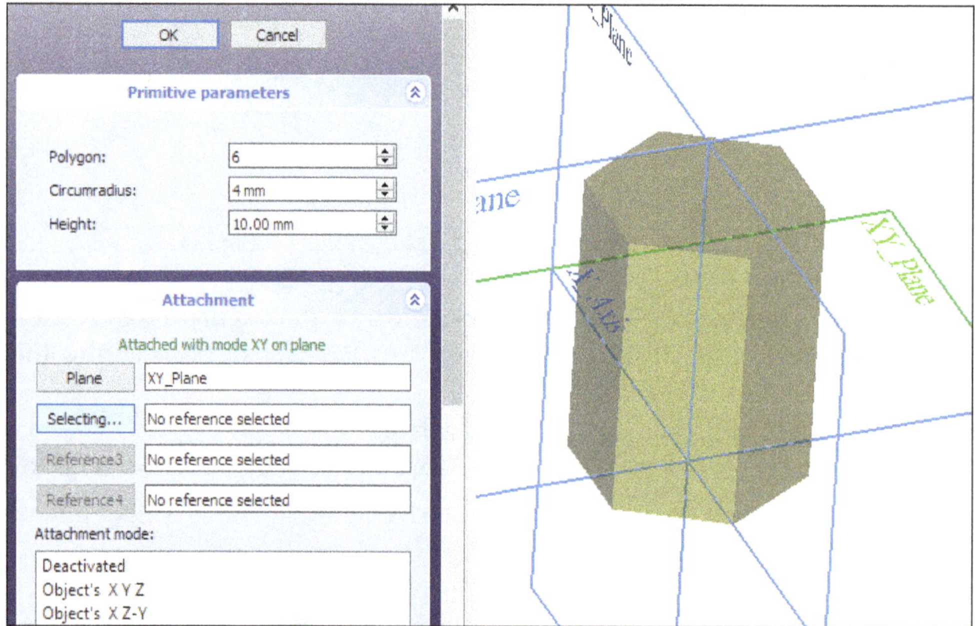

Figure-70. Primitive parameters dialog with preview of additive prism feature

- Specify desired number of sides in the polygon cross section of the prism in the **Polygon** edit box of the dialog.
- Specify the circumscribed radius of the polygon cross section of the prism in the **Circumradius** edit box of the dialog.
- Specify desired height of the prism in the **Height** edit box of the dialog.
- Specify other parameters in the dialog as discussed earlier.
- Click on **OK** button from the dialog. The additive prism feature will be created; refer to Figure-71.

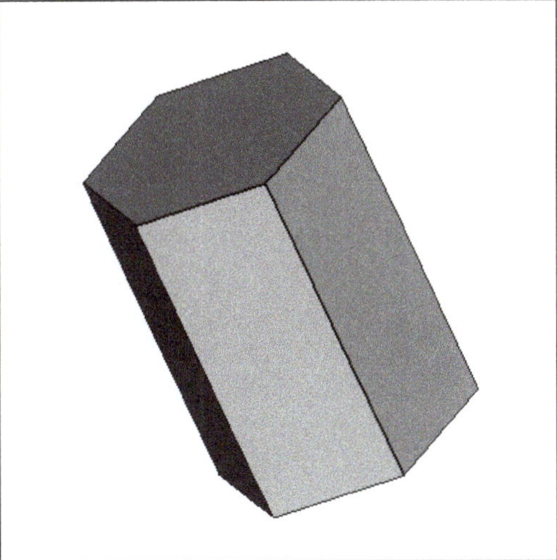

Figure-71. Additive prism feature created

Additive Wedge

The **Additive Wedge** tool inserts a primitive wedge in the active body as the first feature or fuses it to the existing feature. The procedure to use this tool is discussed next.

- Click on the **Additive Wedge** tool from **Create an additive primitive** drop-down in the **Toolbar** of **Part Design** workbench; refer to Figure-72. The **Primitive parameters** dialog will be displayed in the **Tasks** panel of **Combo View** along with the preview of additive wedge feature; refer to Figure-73.

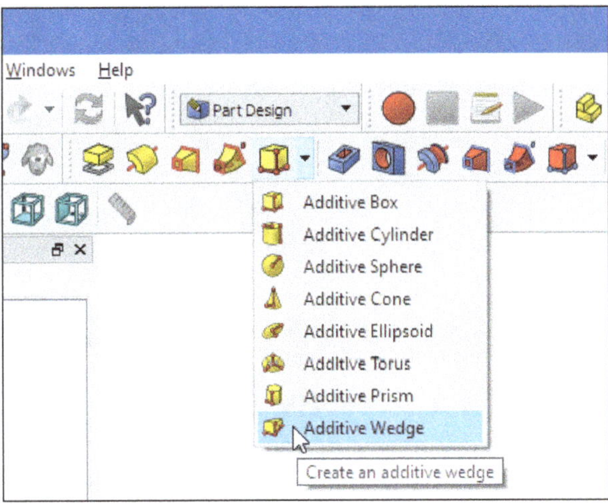

Figure-72. Additive Wedge tool

- Specify the minimum/maximum value for base face of the feature along X axis in the **X min/max** edit boxes, respectively.
- Specify desired height of the wedge in the **Y min/max** edit boxes, respectively.
- Specify the minimum/maximum value for base face of the feature along Z axis in the **Z min/max** edit boxes, respectively.
- Specify the minimum/maximum value for top face of the feature along X axis in the **X2 min/max** edit boxes, respectively.
- Specify the minimum/maximum value for top face of the feature along Z axis in the **Z2 min/max** edit boxes, respectively.

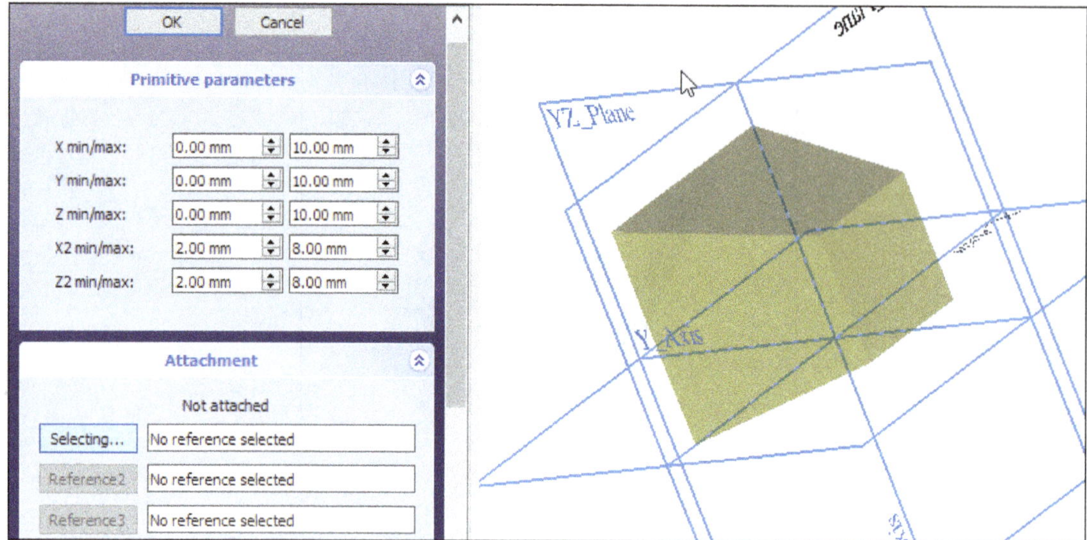

Figure-73. Primitive parameters dialog with preview of additive wedge feature

- Specify the other parameters in the dialog as discussed earlier.
- Click on **OK** button from the dialog. The additive wedge feature will be created; refer to Figure-74.

Figure-74. Additive wedge feature created

Subtractive tools

The **Subtractive tools** are used for subtracting materials from an existing body. The **Subtractive tools** are available in the **Toolbar** of **Part Design** workbench; refer to Figure-75. The procedures to use these tools are discussed next.

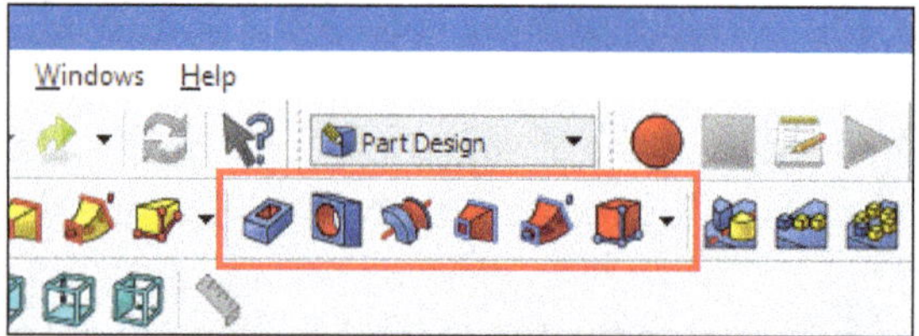

Figure-75. Subtractive tools in Part Design workbench

Pocket

The **Pocket** tool cuts out a solid by extruding a sketch (or a face of the solid) in a straight path and subtracting it from the solid. The procedure to use this tool is discussed next.

- Click on the **Pocket** tool from **Toolbar** in the **Part Design** workbench; refer to Figure-76. The **Pocket parameters** dialog will be displayed in the **Tasks** panel of **Combo View** along with the preview of pocket feature; refer to Figure-77.

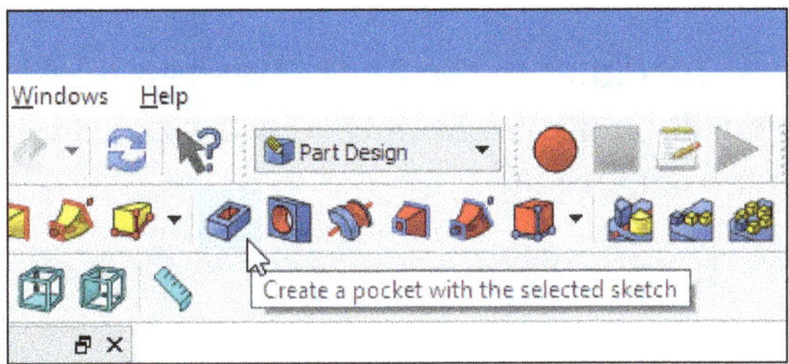

Figure-76. Pocket tool

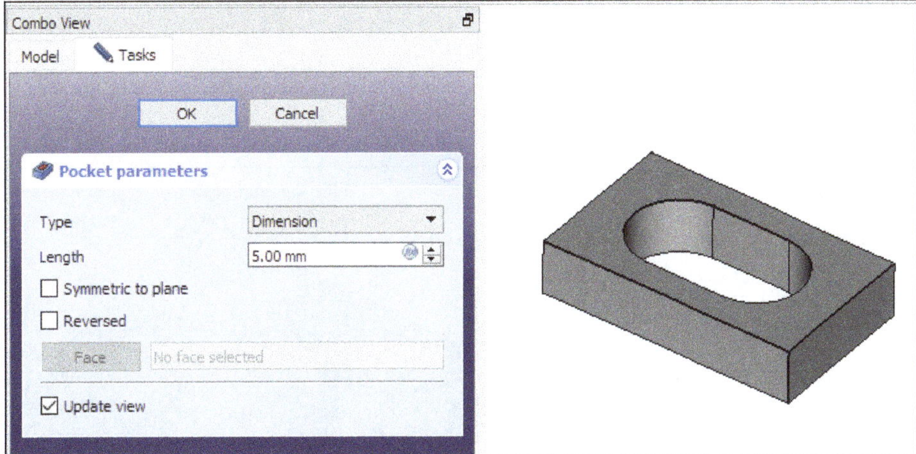

Figure-77. Pocket parameters dialog with preview of pocket feature

- Select desired option from **Type** drop-down in the dialog. Select the **Dimension** option to extrude cut the feature to the defined sketch plane at a specified length. Select the **Through all** option to extrude cut the feature through all material in the extrusion direction. Select the **To first** option to extrude cut the feature upto the first face of the solid entity in the extrusion direction. Select the **Up to face** option to extrude cut the feature upto the selected face. Select the **Two dimensions** option to extrude cut the feature in opposite direction by specifying second length of the extrusion.
- Select **Symmetric to plane** check box to extrude cut the feature in both directions with respect to sketch plane.
- Select **Reversed** check box to reverse the direction of extrusion.
- Click on **OK** button from the dialog. The pocket feature will be created; refer to Figure-78.

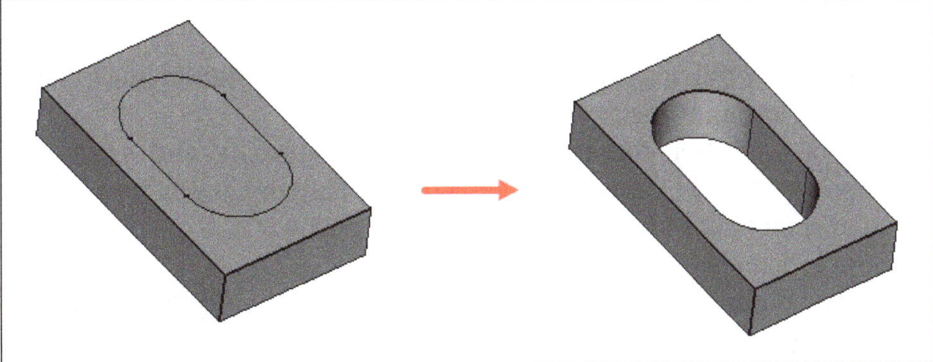

Figure-78. Pocket feature created

Hole

The **Hole** tool creates one or more holes from selected sketch. The sketch must contain one or multiple circles. The procedure to use this tool is discussed next.

- Click on the **Hole** tool from **Toolbar** in the **Part Design** workbench; refer to Figure-79. The **Hole parameters** dialog will be displayed in the **Tasks** panel of **Combo View** along with the preview of hole feature; refer to Figure-80.

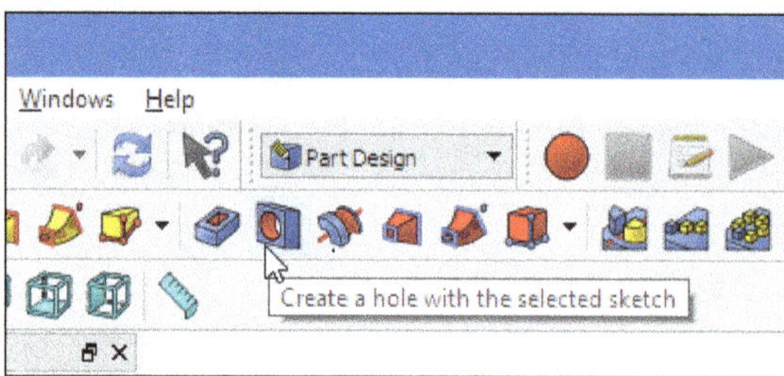

Figure-79. Hole tool

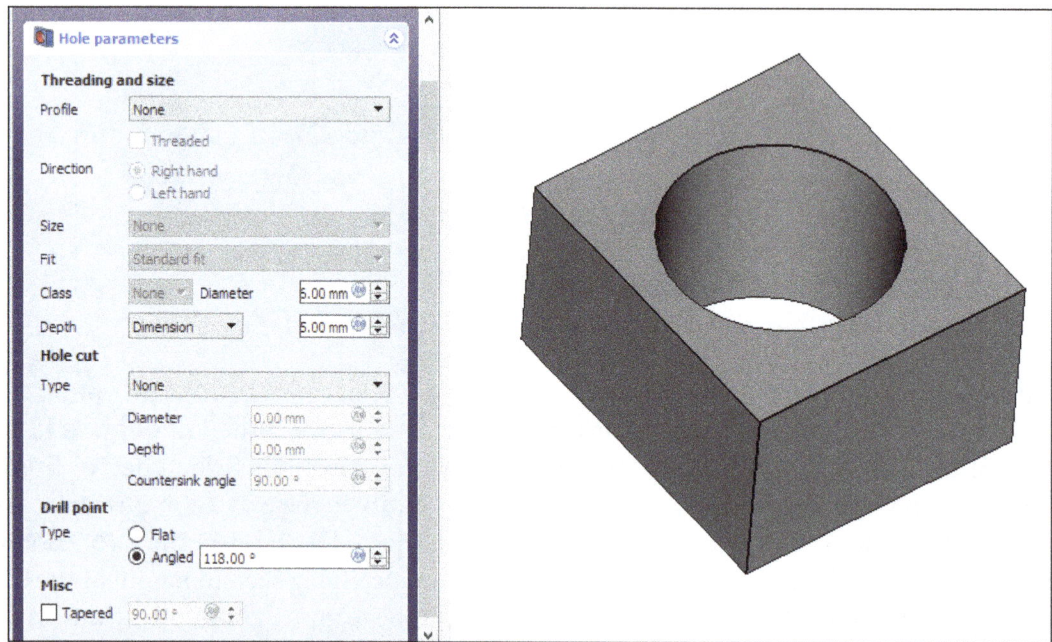

Figure-80. Hole parameters dialog with preview of hole feature

- Select desired option from **Profile** drop-down in the **Threading and size** area of the dialog to specify the type of thread.
- Select the **Threaded** check box to add threading data to the hole feature and the hole minor diameter is used.
- The radio buttons in **Direction** area will be activate when the **Threaded** check box will be selected. Select desired radio button to specify the thread direction.
- Select desired size of thread from **Size** drop down in the **Threading and size** area of the dialog.
- Select desired option from **Fit** drop-down to define standard or close fit for threaded profiles.
- Specify desired tolerance class for hole in the **Class** drop-down.
- Specify desired hole diameter in the **Diameter** edit box.
- Select **Dimension** option from **Depth** drop-down to specify the depth value in the edit box and select **Through All** option from drop-down to cut the hole through the whole body.
- Specify desired type of hole from **Type** drop-down in the **Hole cut** area of the dialog.
- Specify the major diameter for all hole cut types in the **Diameter** edit box from **Hole cut** area of the dialog.

- Specify the depth of hole cut type from the sketch plane in the **Depth** edit box.
- Specify the angle of conical hole cut in the **Countersink angle** edit box.
- Select **Flat** radio button in the **Type** option from **Drill point** area to create the flat bottom of the hole and select **Angled** radio button to create the conical point of the hole.
- Select the **Tapered** check box from **Misc** area of the dialog to specify taper angle of the hole in the edit box.
- Click on **OK** button from the dialog. The hole feature will be created; refer to Figure-81.

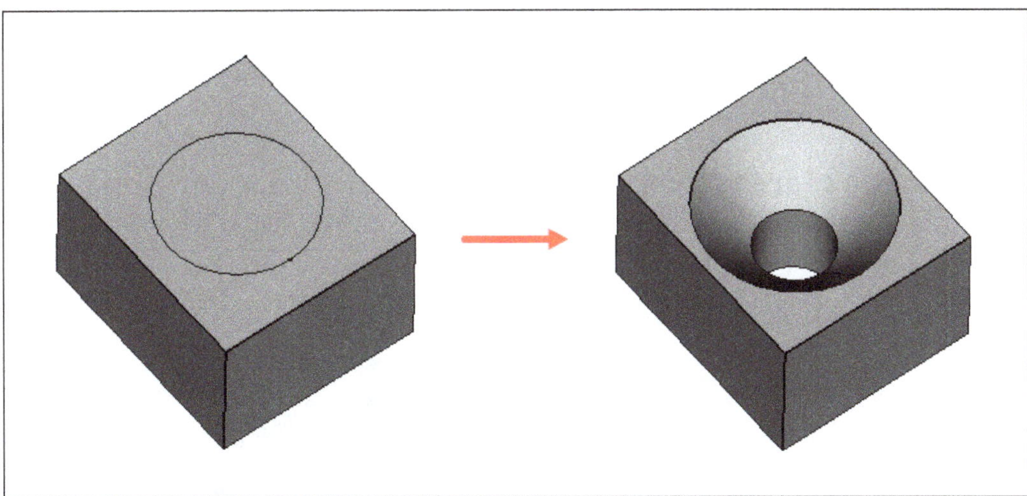

Figure-81. Hole feature created

Groove

The **Groove** tool revolves selected sketch or profile about a given axis, cutting out material from the solid model. The procedure to use this tool is discussed next.

- Select the closed loop sketch to be revolved and click on the **Groove** tool from **Toolbar** in the **Part Design** workbench; refer to Figure-82. The **Revolution parameters** dialog will be displayed in the **Tasks** panel of **Combo View**; refer to Figure-83.

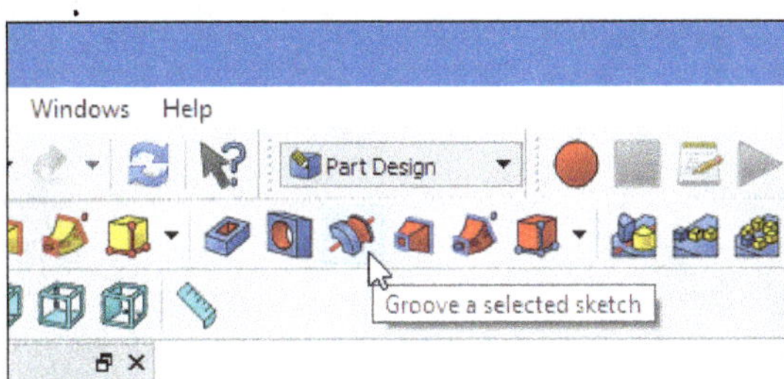

Figure-82. Groove tool

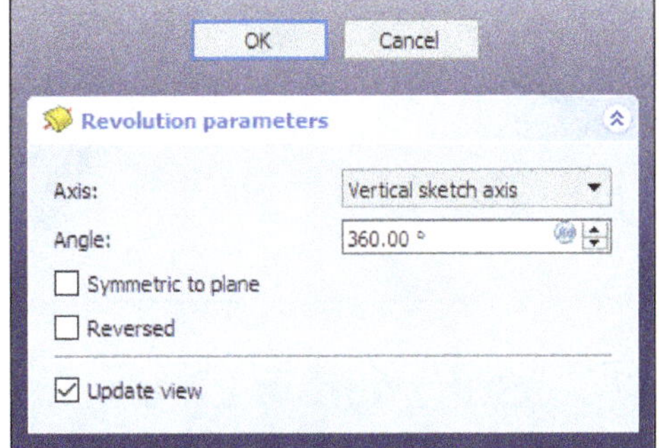

Figure-83. Revolution parameters dialog

- Select the sketch of axis from **Model** panel in the **Combo View** and press **Space** key to toggle the visibility of axis of revolution; refer to Figure-84.

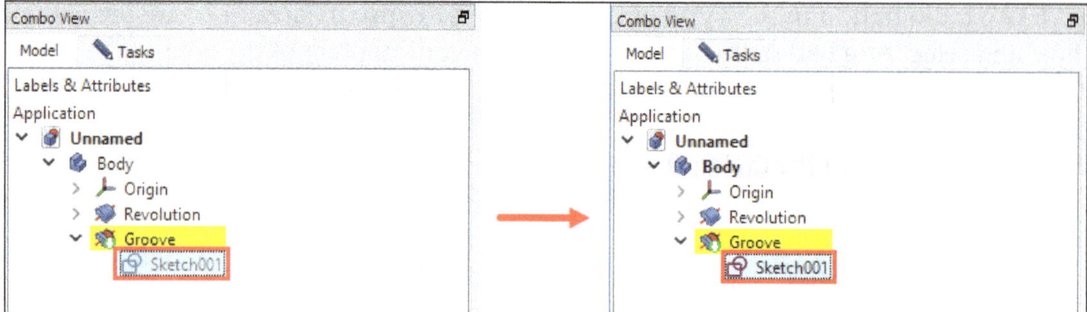

Figure-84. Toggle the visibility of sketch axis

- Select the **Select Reference** option from **Axis** drop-down in the dialog and select the sketch of axis in the 3D view area. The preview of groove feature will be displayed; refer to Figure-85.

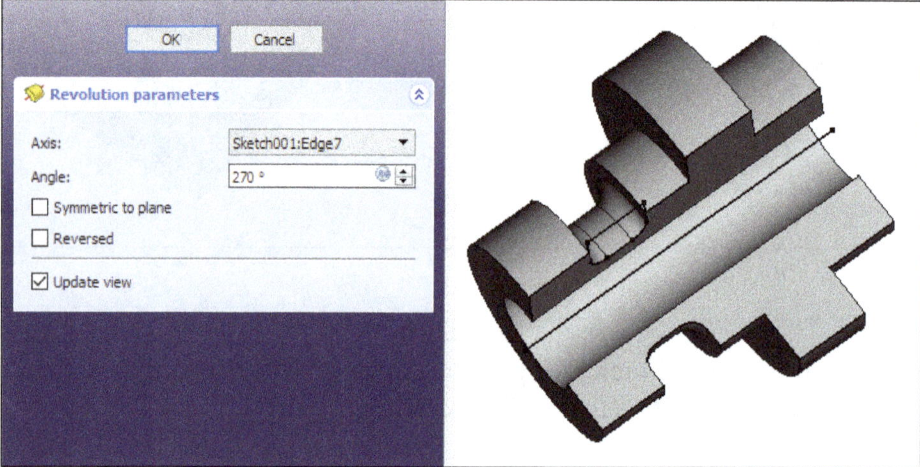

Figure-85. Preview of groove feature

- Specify the angle of revolution in the **Angle** edit box through which the groove is to be formed.
- Specify the other parameters in the dialog as discussed earlier.
- Click on **OK** button from the dialog. The groove feature will be created; refer to Figure-86.

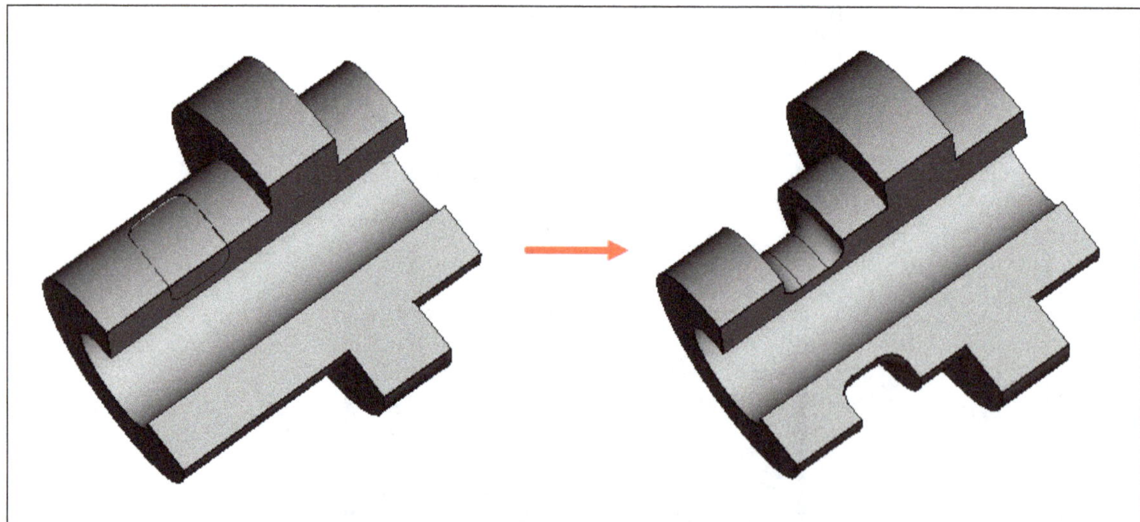

Figure-86. Groove feature created

Subtractive Loft

The **Subtractive Loft** tool creates a subtractive solid in the active body by making a transition between two or more sketches. The procedure to use this tool is discussed next.

- Click on the **Subtractive Loft** tool from **Toolbar** in the **Part Design** workbench; refer to Figure-87. If no object is selected before selecting the tool then the **Select feature** dialog will be displayed in the **Tasks** panel of **Combo View** as discussed earlier asking you to select the base profile.

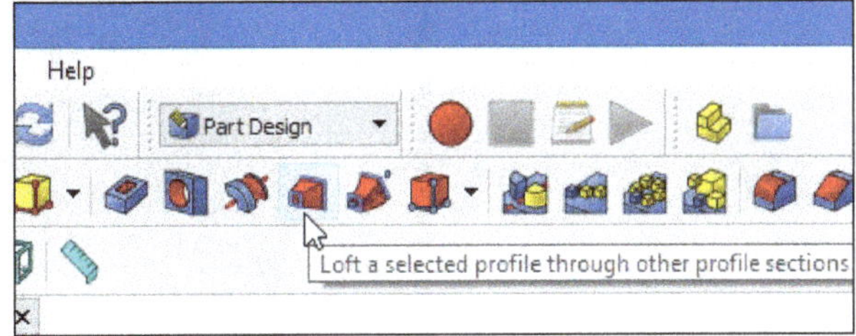

Figure-87. Subtractive loft tool

- Select desired sketch from the dialog as base profile for subtractive loft feature; refer to Figure-88 and click on **OK** button. The **Loft parameters** dialog will be displayed in the **Tasks** panel of **Combo View**; refer to Figure-89.
- The base profile is preselected in the **Object** selection box of **Profile** area. Click on **Add Section** button and select the first, second, and third sketches for loft section. The preview of subtractive loft feature will be displayed; refer to Figure-90.

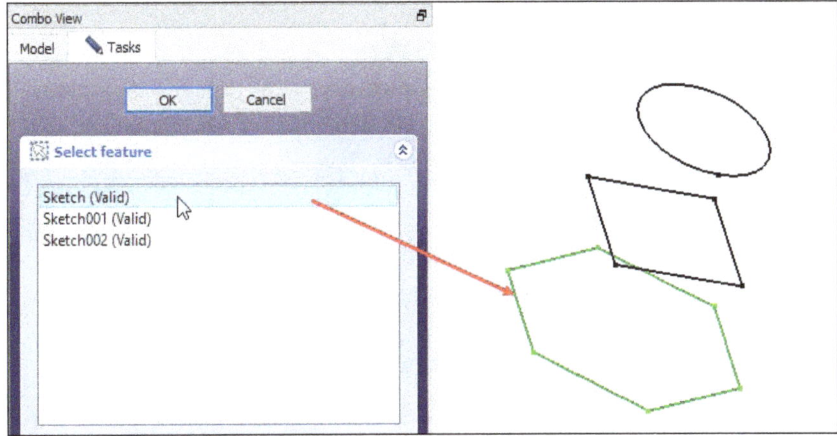

Figure-88. Selecting base profile for subtractive loft feature

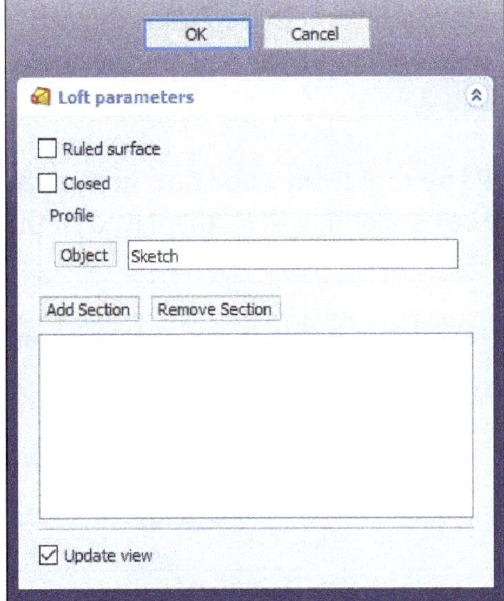

Figure-89. Loft parameters dialog

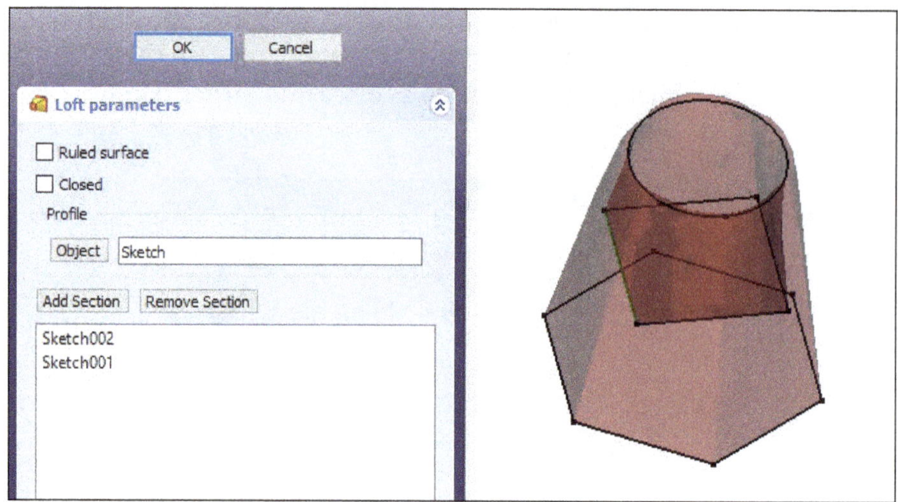

Figure-90. Preview of subtractive loft feature

- Select the **Ruled surface** check box to make straight transitions between cross-sections.
- Select the **Closed** check box to make a transition from the last cross-section to the first by creating a loop.
- Click on **OK** button from the dialog. The subtractive loft feature will be created; refer to Figure-91.

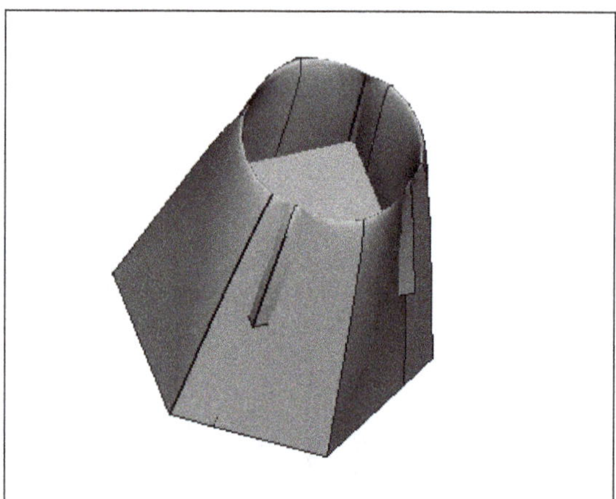

Figure-91. Subtractive loft feature created

Subtractive Pipe

The **Subtractive Pipe** tool creates a subtractive solid in the active body by sweeping one or more sketches along an open or closed path and subtracts it from the active body. The procedure to use this tool is discussed next.

- Click on the **Subtractive Pipe** tool from **Toolbar** in the **Part Design** workbench; refer to Figure-92. The **Pipe parameters** dialog will be displayed in the **Tasks** panel of **Combo View**; refer to Figure-93.

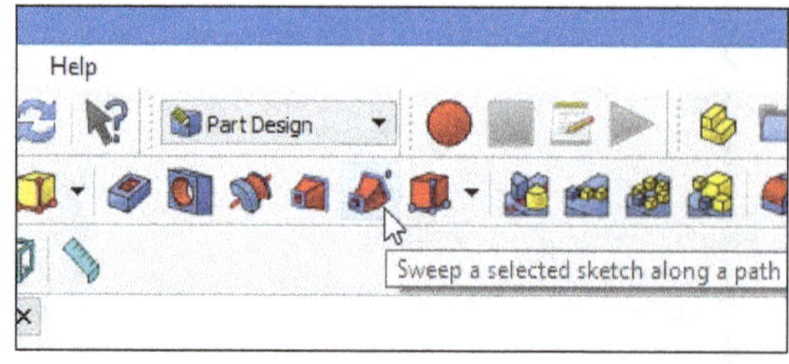

Figure-92. Subtractive pipe tool

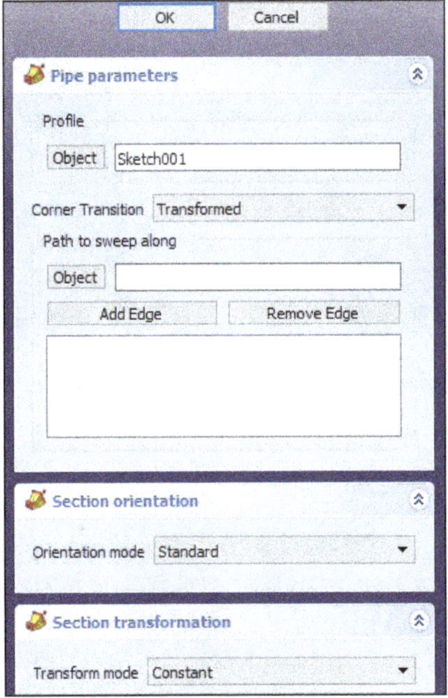

Figure-93. Pipe parameters dialog

- The first cross-section is preselected in the **Object** selection box in the **Profile** area of dialog.
- Click on the **Object** button from **Path to sweep along** area and select an edge of the sketch to be used as path for sweep feature in the 3D view area. The preview of subtractive pipe feature will be displayed; refer to Figure-94.

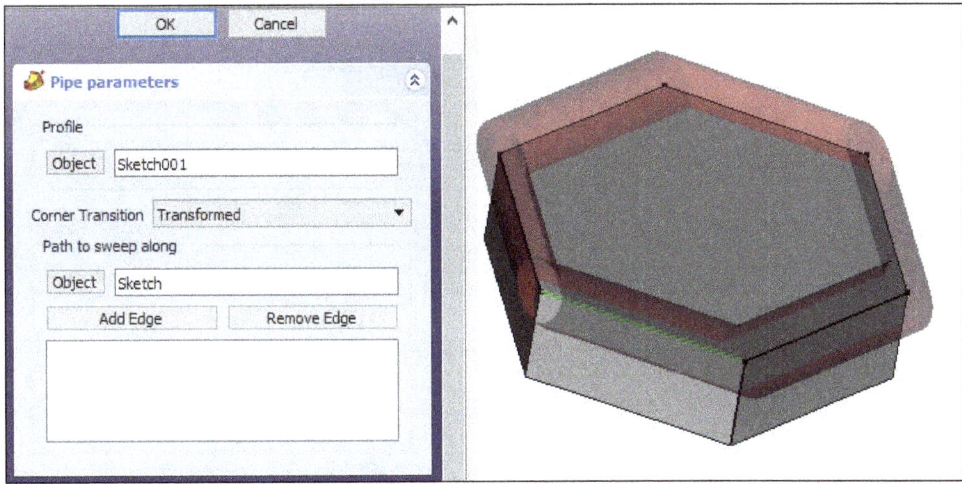

Figure-94. Preview of subtractive pipe feature

- Select desired transition mode from **Corner Transition** drop-down in the dialog.
- Select desired orientation mode from **Orientation mode** drop-down in the **Section orientation** panel of the dialog.
- Select desired transformation mode from **Transform mode** drop-down in the **Section transformation** panel of the dialog.
- Specify the other parameters in the dialog as desired and click on **OK** button. The subtractive pipe feature will be created; refer to Figure-95.

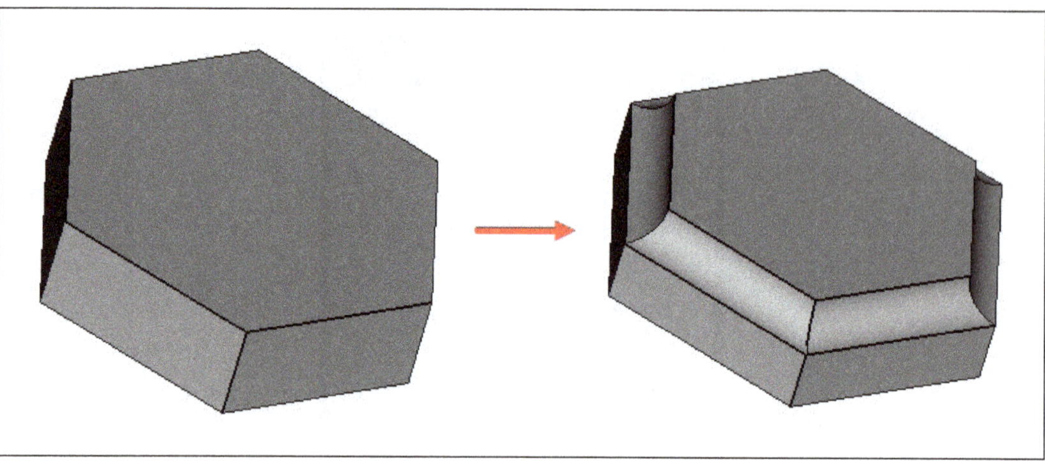

Figure-95. Subtractive pipe feature created

Subtractive Box

The **Subtractive Box** tool inserts a subtractive box in the active body. Its shape is subtracted from the existing solid. The procedure to use this tool is discussed next.
Note that an existing solid body will be required for creating a subtractive box feature.

- Click on the **Subtractive Box** tool from **Create a subtractive primitive** drop-down in the **Toolbar** of **Part Design** workbench; refer to Figure-96. The **Primitive parameters** dialog will be displayed in the **Tasks** panel of **Combo View** along with the preview of subtractive box feature; refer to Figure-97.

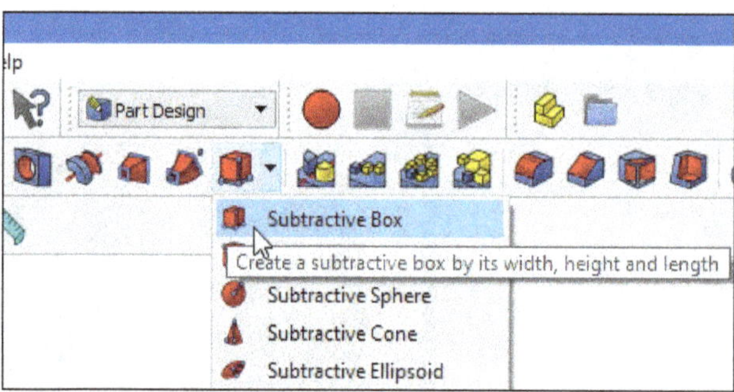

Figure-96. Subtractive box tool

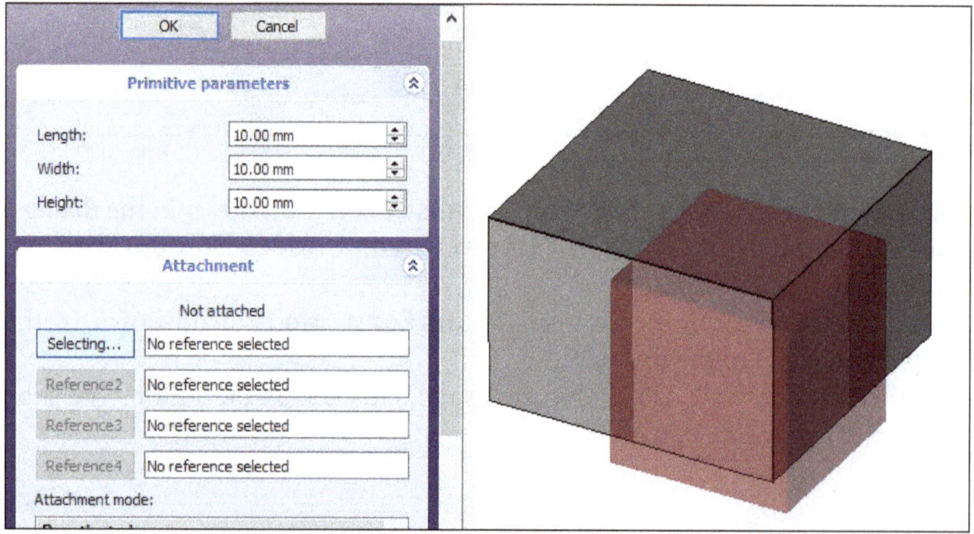

Figure-97. Primitive parameters dialog with preview of subtractive box feature

- Specify the length, width, and height of the feature in the **Length**, **Width**, and **Height** edit boxes, respectively.
- Specify the other parameters in the dialog as discussed earlier.
- Click on **OK** button from the dialog. The subtractive box feature will be created; refer to Figure-98.

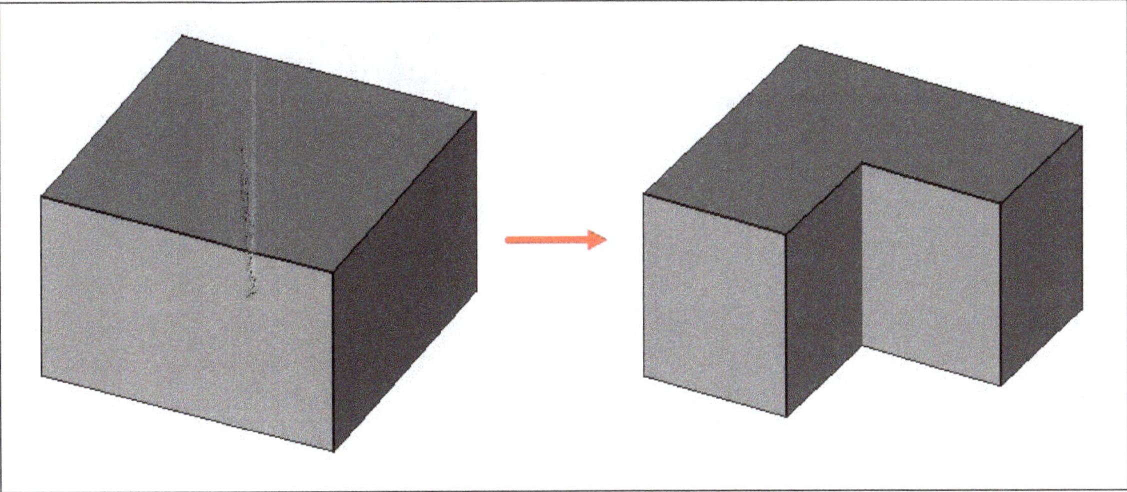

Figure-98. Subtractive box feature created

Subtractive Cylinder

The **Subtractive Cylinder** tool inserts a subtractive cylinder in the active body. Its shape is subtracted from the existing solid. The procedure to use this tool is discussed next.

- Click on the **Subtractive Cylinder** tool from **Create a subtractive primitive** drop-down in the **Toolbar** of **Part Design** workbench; refer to Figure-99. The **Primitive parameters** dialog will be displayed in the **Tasks** panel of **Combo View** along with the preview of subtractive cylinder feature; refer to Figure-100.

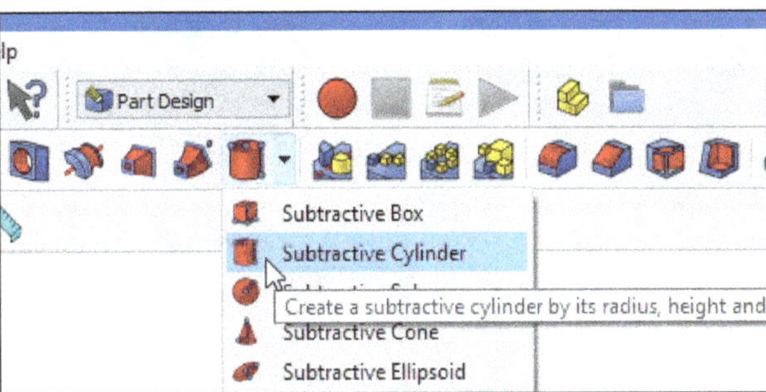

Figure-99. Subtractive cylinder tool

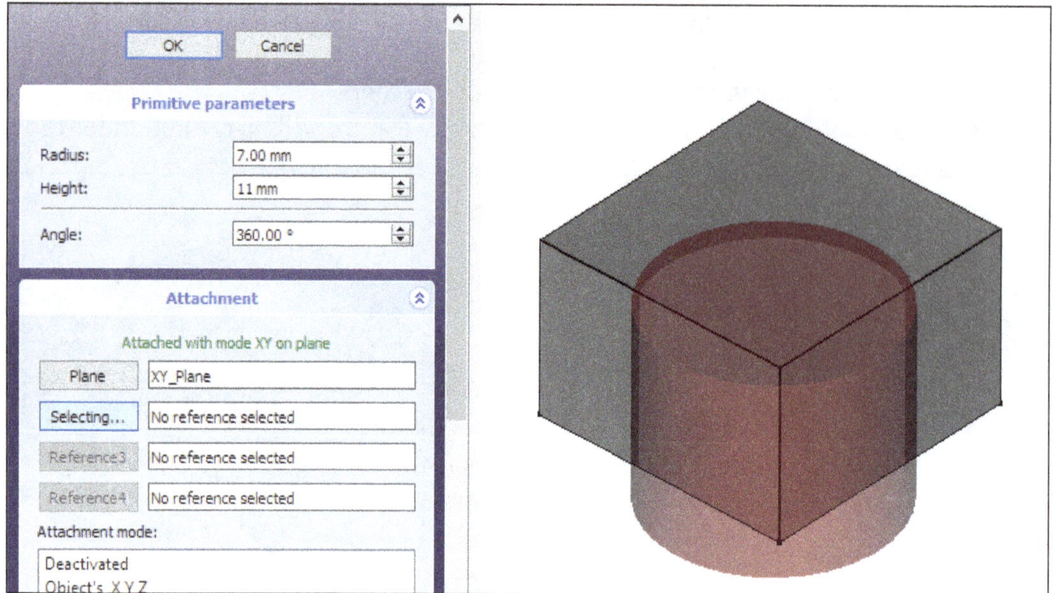

Figure-100. Primitive parameters dialog with preview of subtractive cylinder feature

- Specify desired radius and height of the feature in the **Radius** and **Height** edit boxes, respectively.
- Specify desired value of angle of revolution of the cross-section in the **Angle** edit box.
- Specify the other parameters in the dialog as discussed earlier.
- Click on **OK** button from the dialog. The subtractive cylinder feature will be created; refer to Figure-101.

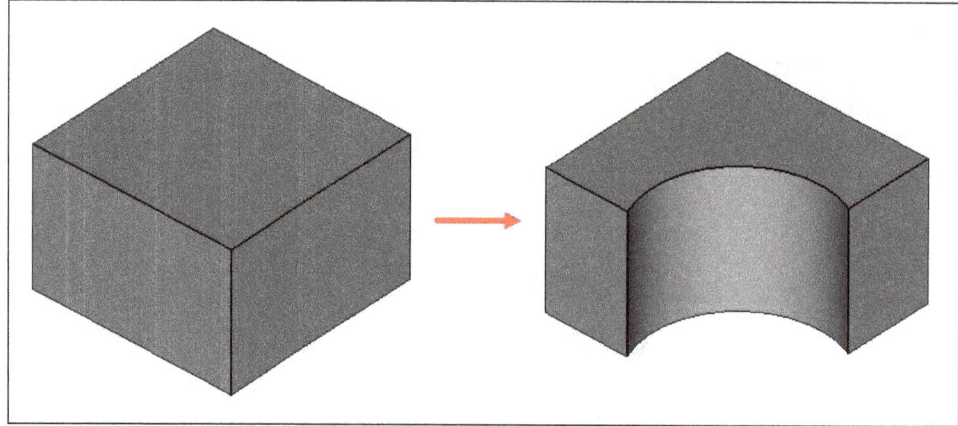

Figure-101. Subtractive cylinder feature created

Subtractive Sphere

The **Subtractive Sphere** tool inserts a subtractive sphere in the active body. Its shape is subtracted from the existing solid. The procedure to use this tool is discussed next.

- Click on the **Subtractive Sphere** tool from **Create a subtractive primitive** drop-down in the **Toolbar** of **Part Design** workbench; refer to Figure-102. The **Primitive parameters** dialog will be displayed in the **Tasks** panel of **Combo View** along with the preview of subtractive sphere feature; refer to Figure-103.

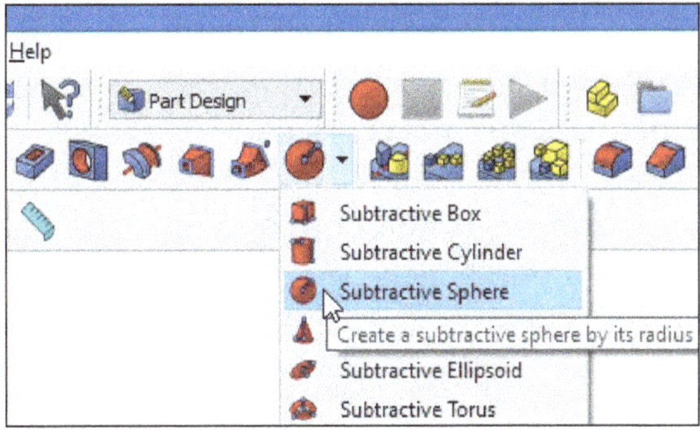

Figure-102. Subtractive sphere tool

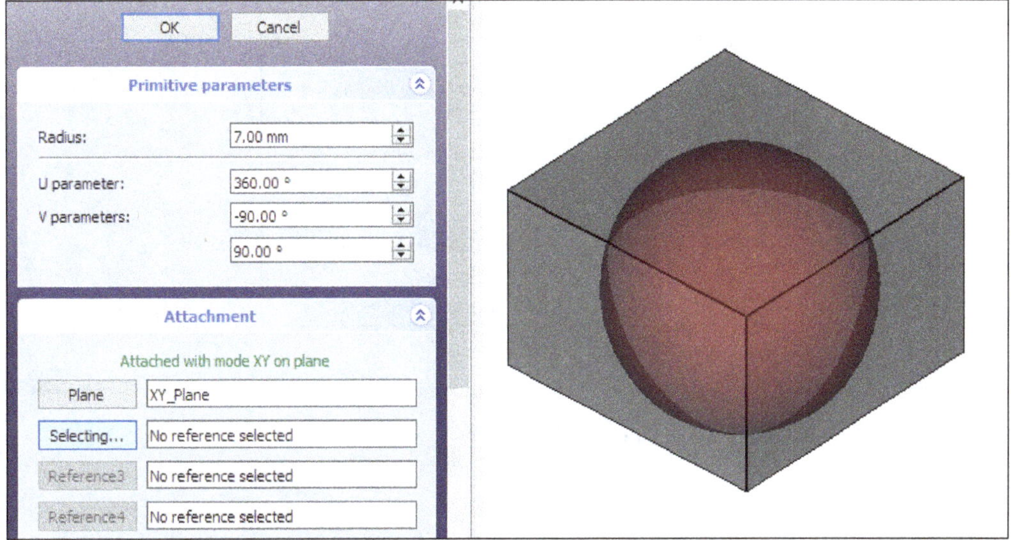

Figure-103. Primitive parameters dialog with preview of subtractive sphere feature

- Specify desired radius of the sphere in the **Radius** edit box.
- Specify desired angle of revolution of the cross section in the **U parameter** edit box.
- Specify lower truncation of the sphere parallel to the circular cross-section in the **V parameters** edit box and specify upper truncation of the ellipsoid parallel to the circular cross-section in the edit box just below to the **V parameters** edit box of the dialog.
- Specify other parameters in the dialog as discussed earlier.
- Click on **OK** button from the dialog. The subtractive sphere feature will be created; refer to Figure-104.

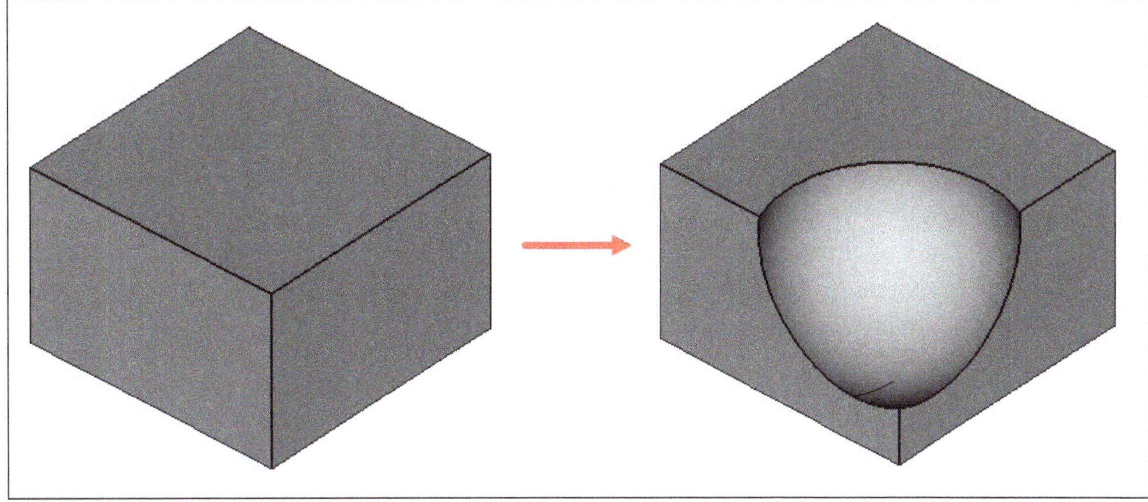

Figure-104. Subtractive sphere feature created

Subtractive Cone

The **Subtractive Cone** tool inserts a subtractive cone in the active body. Its shape is subtracted from the existing solid. The procedure to use this tool is discussed next.

- Click on the **Subtractive Cone** tool from **Create a subtractive primitive** drop-down in the **Toolbar** of **Part Design** workbench; refer to Figure-105. The **Primitive parameters** dialog will be displayed in the **Tasks** panel of **Combo View** along with the preview of subtractive cone feature; refer to Figure-106.

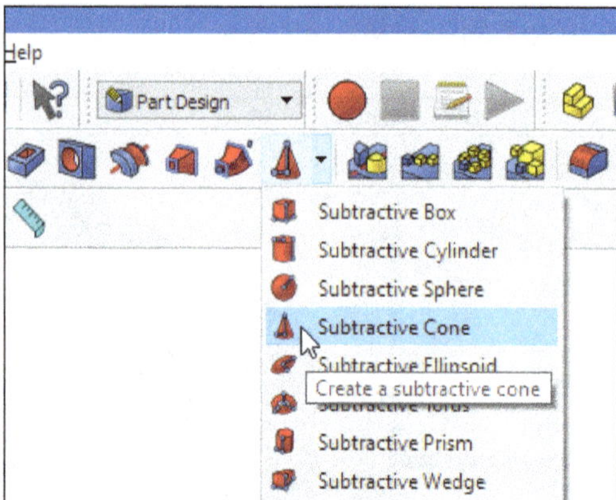

Figure-105. Subtractive cone tool

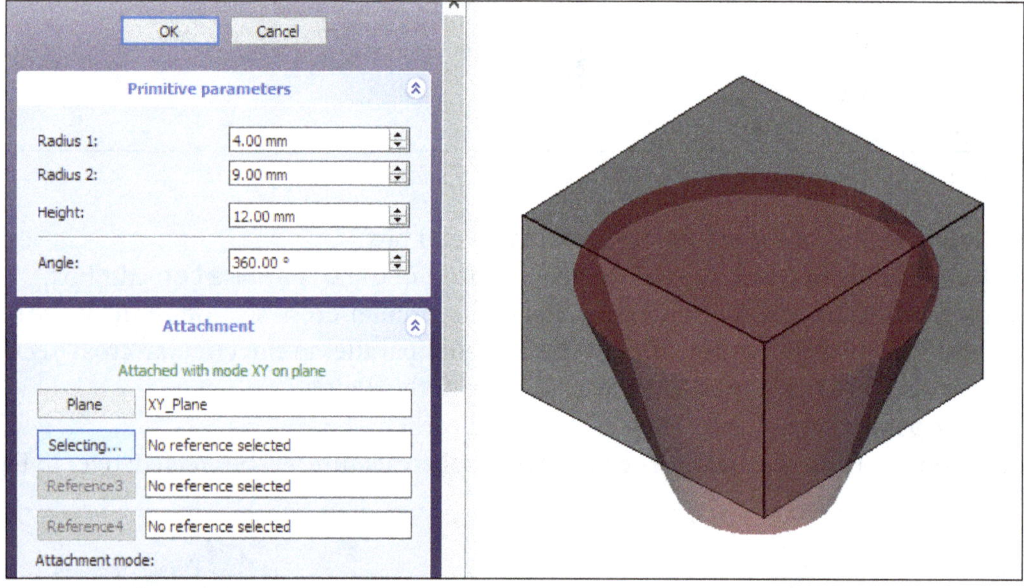

Figure-106. Primitive parameters dialog with preview of subtractive cone feature

- Specify desired radius value for the base of the cone in **Radius 1** edit box and specify desired radius value for the top of the cone in **Radius 2** edit box of the dialog.
- Specify desired height of the cone in the **Height** edit box.
- Specify desired angle of revolution of the cross section in the **Angle** edit box.
- Specify other parameters in the dialog as discussed earlier.
- Click on **OK** button from the dialog. The subtractive cone feature will be created; refer to Figure-107.

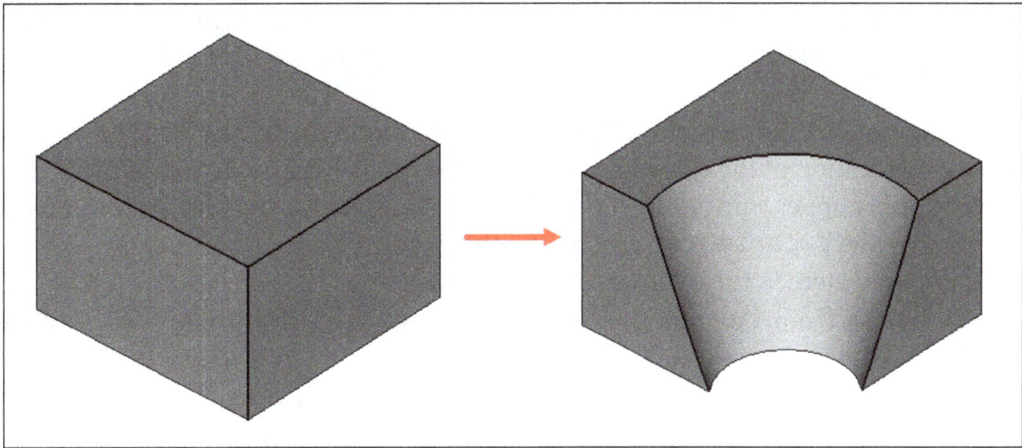

Figure-107. Subtractive cone feature created

Subtractive Ellipsoid

The **Subtractive Ellipsoid** tool inserts a subtractive ellipsoid in the active body. Its shape is subtracted from the existing solid. The procedure to use this tool is discussed next.

- Click on the **Subtractive Ellipsoid** tool from **Create a subtractive primitive** drop-down in the **Toolbar** of **Part Design** workbench; refer to Figure-108. The **Primitive parameters** dialog will be displayed in the **Tasks** panel of **Combo View** along with the preview of subtractive ellipsoid feature; refer to Figure-109.

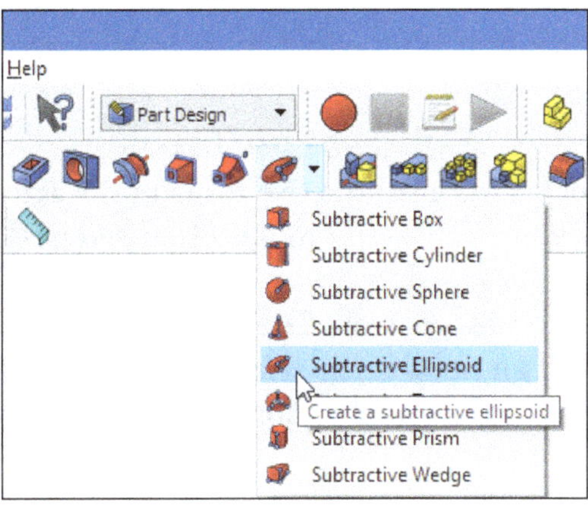

Figure-108. Subtractive ellipsoid tool

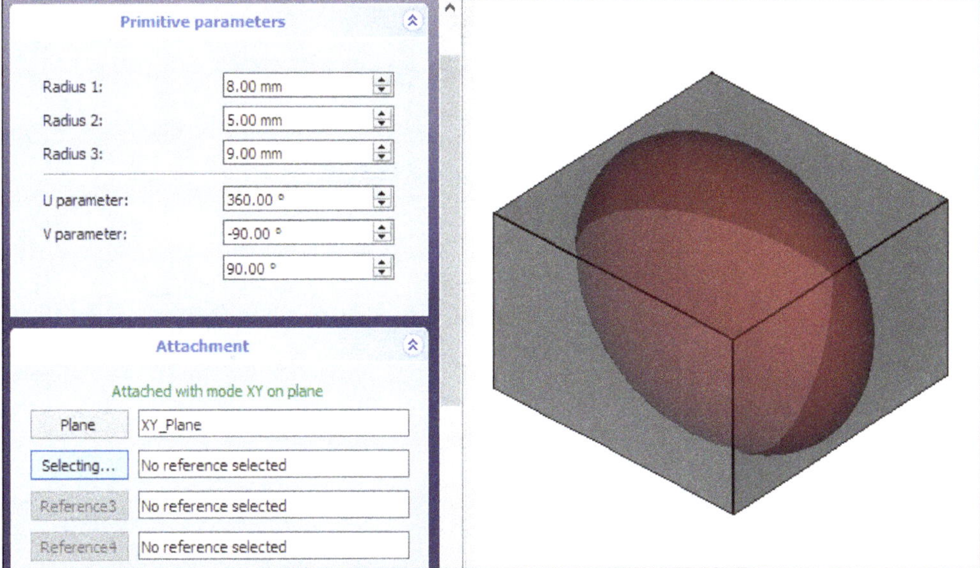

Figure-109. Primitive parameters dialog with preview of subtractive ellipsoid feature

- Specify desired radius value along the ellipsoid's vertical axis in **Radius 1** edit box, along the ellipsoid's length in **Radius 2** edit box, and along the ellipsoid's width in **Radius 3** edit box of the dialog.
- Specify the angle of revolution of the elliptical cross section in the **U parameter** edit box.
- Specify lower truncation of the ellipsoid parallel to the circular cross-section in the **V parameter** edit box and specify upper truncation of the ellipsoid parallel to the circular cross section in the edit box just below to the **V parameter** edit box of the dialog.
- Specify other parameters in the dialog as discussed earlier.
- Click on **OK** button from the dialog. The subtractive ellipsoid feature will be created; refer to Figure-110.

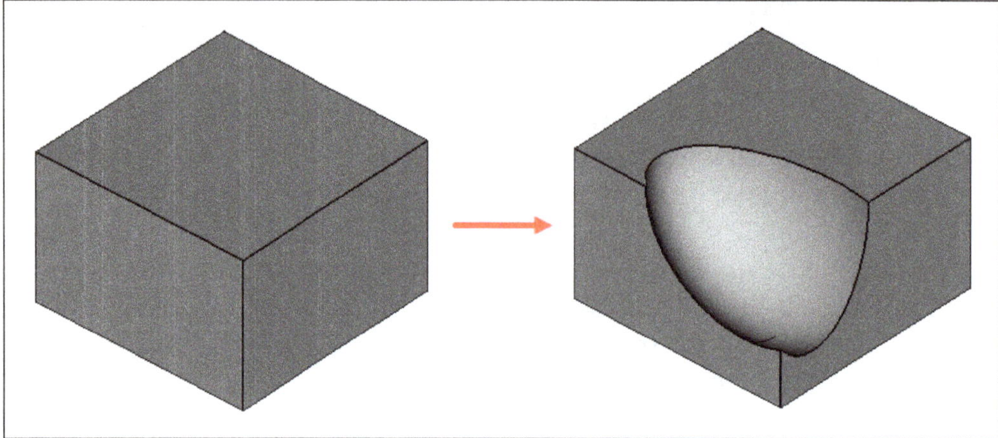

Figure-110. Subtractive ellipsoid feature created

Subtractive Torus

The **Subtractive Torus** tool inserts a subtractive torus in the active body. Its shape is subtracted from the existing solid. The procedure to use this tool is discussed next.

- Click on the **Subtractive Torus** tool from **Create a subtractive primitive** drop-down in the **Toolbar** of **Part Design** workbench; refer to Figure-111. The **Primitive parameters** dialog will be displayed in the **Tasks** panel of **Combo View** along with the preview of subtractive torus feature; refer to Figure-112.

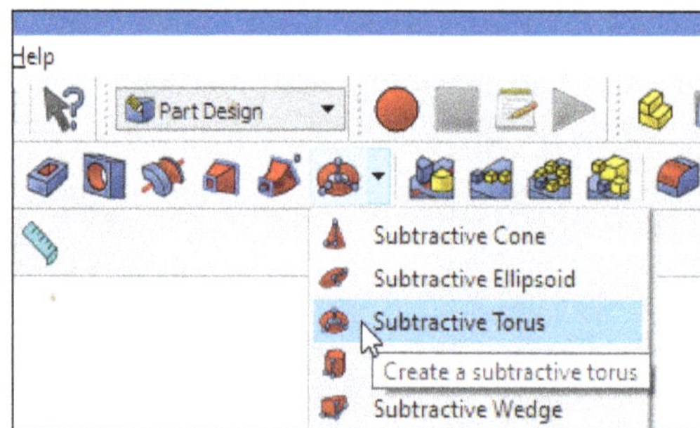

Figure-111. Subtractive torus tool

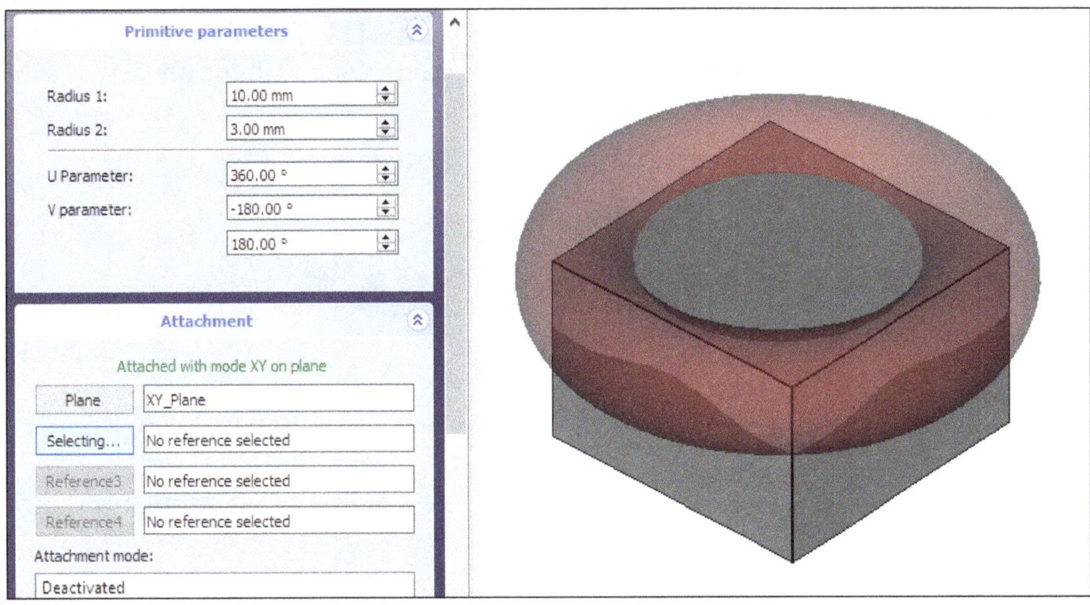

Figure-112. Primitive parameters dialog with preview of subtractive torus feature

- Specify desired radius of the imaginary orbit around which the circular cross-section revolves in **Radius 1** edit box and specify the radius of circular cross-section defining the form of the torus in the **Radius 2** edit box of the dialog.
- Specify the angle of revolution of the circular cross section in the **U parameter** edit box.
- Specify lower truncation of the torus parallel to the circular cross-section in the **V parameter** edit box and specify upper truncation of the ellipsoid parallel to the circular cross section in the edit box just below to the **V parameter** edit box of the dialog.
- Specify other parameters in the dialog as discussed earlier.
- Click on **OK** button from the dialog. The subtractive torus feature will be created; refer to Figure-113.

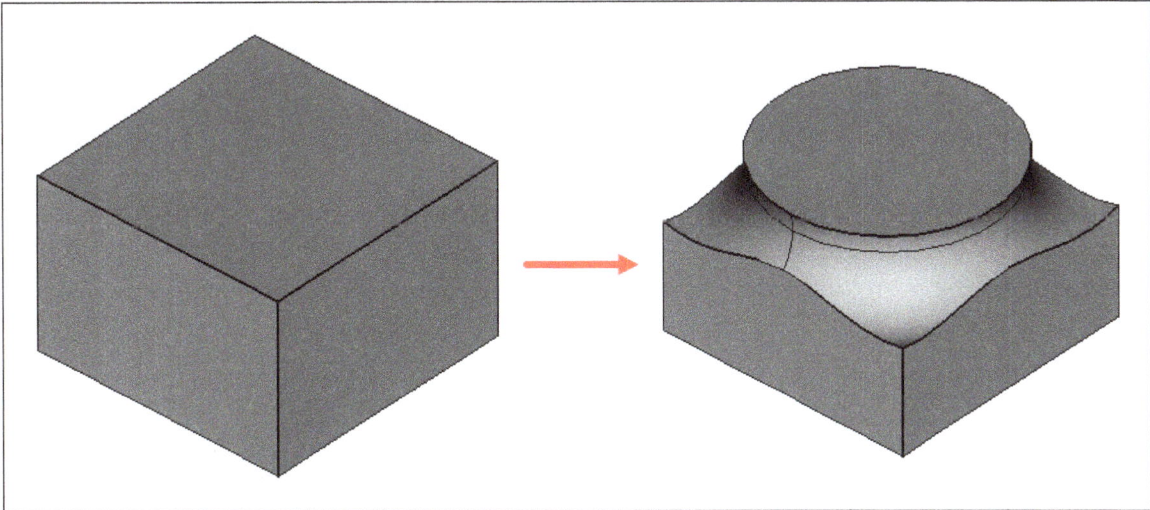

Figure-113. Subtractive torus feature created

Subtractive Prism

The **Subtractive Prism** tool inserts a subtractive prism in the active body. Its shape is subtracted from the existing solid. The procedure to use this tool is discussed next.

- Click on the **Subtractive Prism** tool from **Create a subtractive primitive** drop-down in the **Toolbar** of **Part Design** workbench; refer to Figure-114. The **Primitive parameters** dialog will be displayed in the **Tasks** panel of **Combo View** along with the preview of subtractive prism feature; refer to Figure-115.

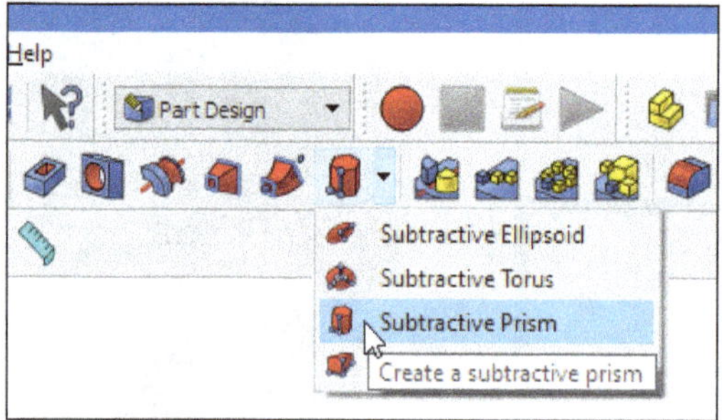

Figure-114. Subtractive prism tool

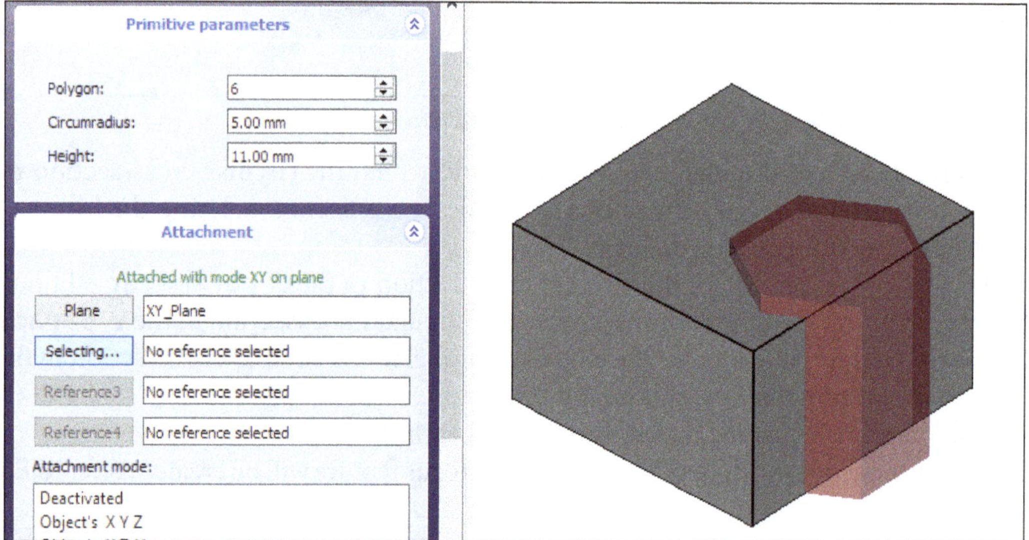

Figure-115. Primitive parameters dialog with preview of subtractive prism feature

- Specify desired number of sides for polygon cross-section of the prism in the **Polygon** edit box.
- Specify circumscribed radius of the polygon cross-section of the prism in the **Circumradius** edit box.
- Specify desired height of the prism in the **Height** edit box of the dialog.
- Specify other parameters in the dialog as discussed earlier.
- Click on **OK** button from the dialog. The subtractive prism feature will be created; refer to Figure-116.

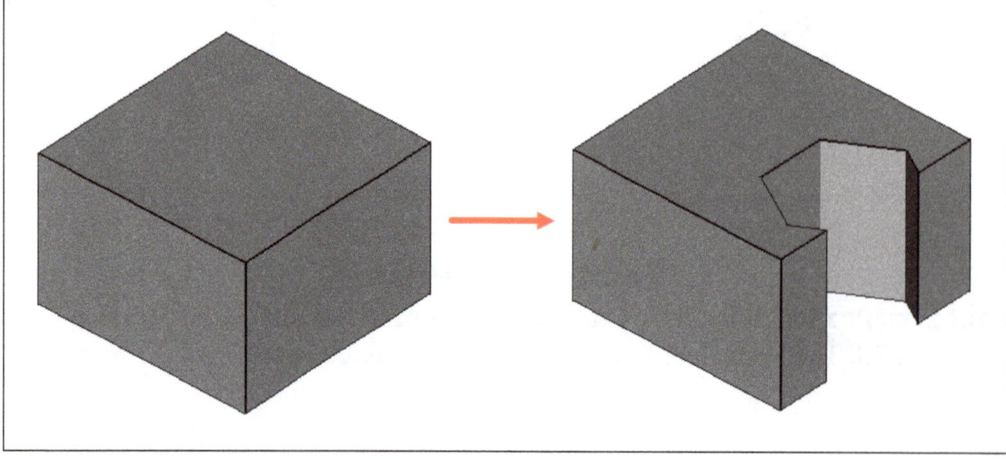

Figure-116. Subtractive prism feature created

Subtractive Wedge

The **Subtractive Wedge** tool inserts a subtractive wedge in the active body. Its shape is subtracted from the existing solid. The procedure to use this tool is discussed next.

- Click on the **Subtractive Wedge** tool from **Create a subtractive primitive** drop-down in the **Toolbar** of **Part Design** workbench; refer to Figure-117. The **Primitive parameters** dialog will be displayed in the **Tasks** panel of **Combo View** along with the preview of subtractive wedge feature; refer to Figure-118.

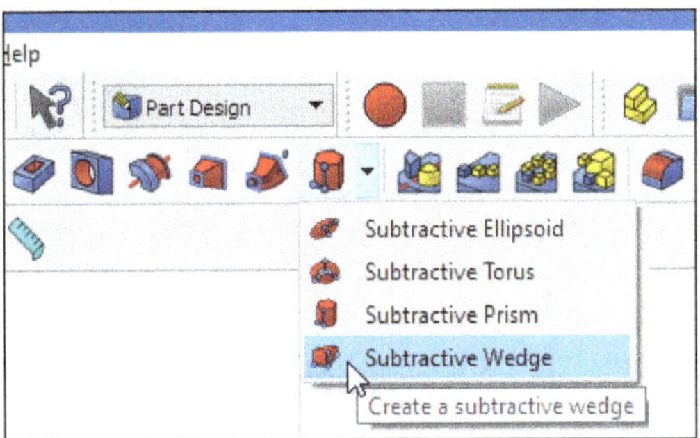

Figure-117. Subtractive wedge tool

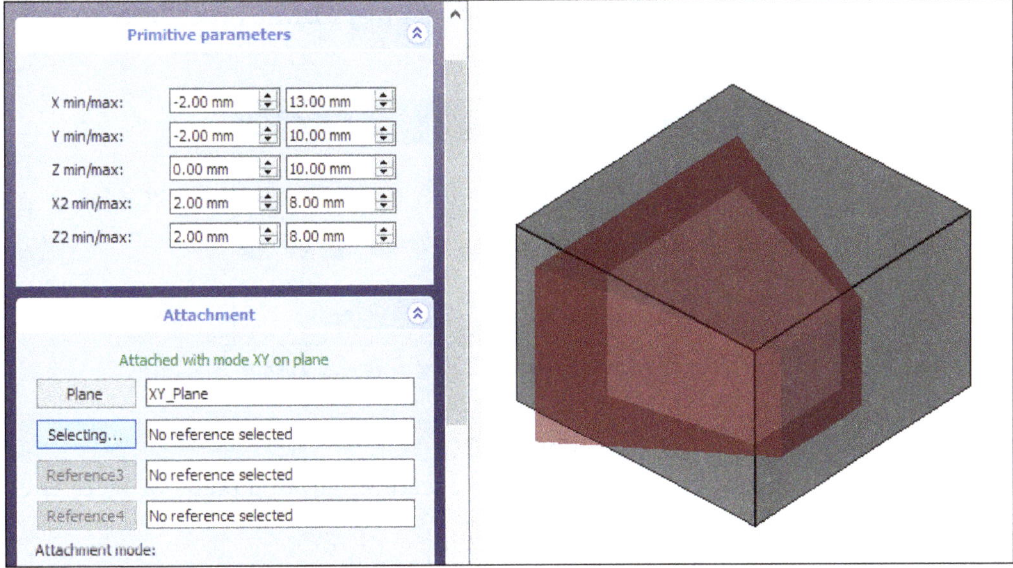

Figure-118. Primitive parameters dialog with preview of subtractive wedge feature

- Specify the minimum/maximum value for base face of the feature along X axis in the **X min/max** edit boxes, respectively.
- Specify desired height of the wedge in the **Y min/max** edit boxes, respectively.
- Specify the minimum/maximum value for base face of the feature along Z axis in the **Z min/max** edit boxes, respectively.
- Specify the minimum/maximum value for top face of the feature along X axis in the **X2 min/max** edit boxes, respectively.
- Specify the minimum/maximum value for top face of the feature along Z axis in the **Z2 min/max** edit boxes, respectively.
- Specify other parameters in the dialog as discussed earlier.
- Click on **OK** button from the dialog. The subtractive wedge feature will be created; refer to Figure-119.

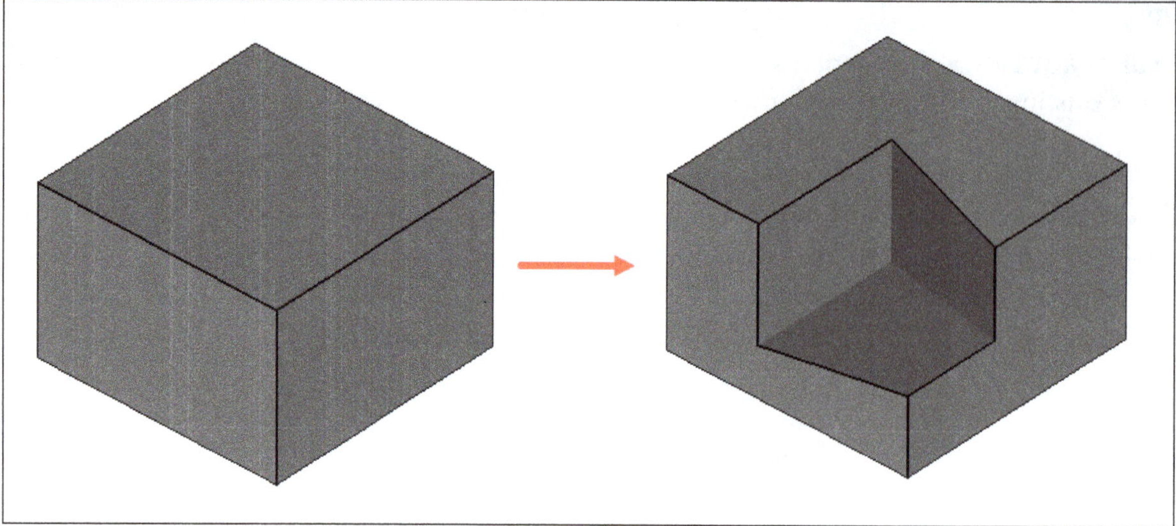

Figure-119. Subtractive wedge feature created

Transformation Tools

The **Transformation** tools are used for transforming existing features into the mirror and pattern. They will allow you to choose which features to transform. The **Transformation** tools are available in the **Toolbar** of **Part Design** workbench; refer to Figure-120. The procedures to use these tools are discussed next.

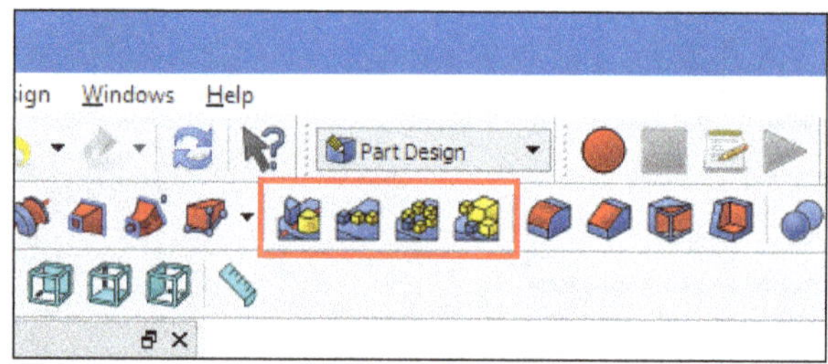

Figure-120. Transformation tools in part design workbench

Mirrored

The **Mirrored** tool creates mirror copy of one or more features about a plane or face. The procedure to use this tool is discussed next.

- Click on the **Mirrored** tool from **Toolbar** in the **Part Design** workbench; refer to Figure-121. The **Select feature** dialog will be displayed in the **Tasks** panel of **Combo View**; refer to Figure-122, asking you to select the feature to be mirrored.

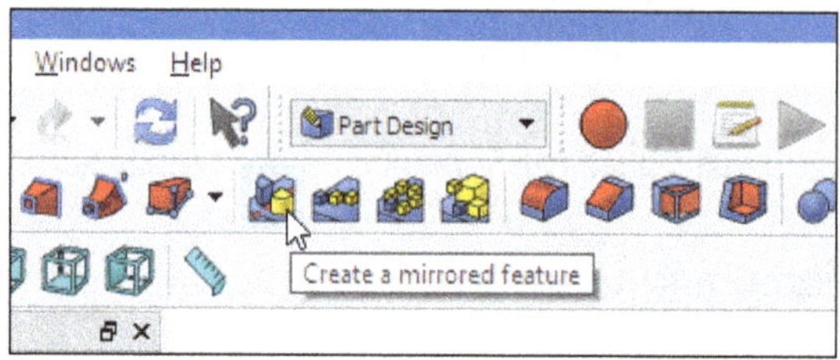

Figure-121. Mirrored tool

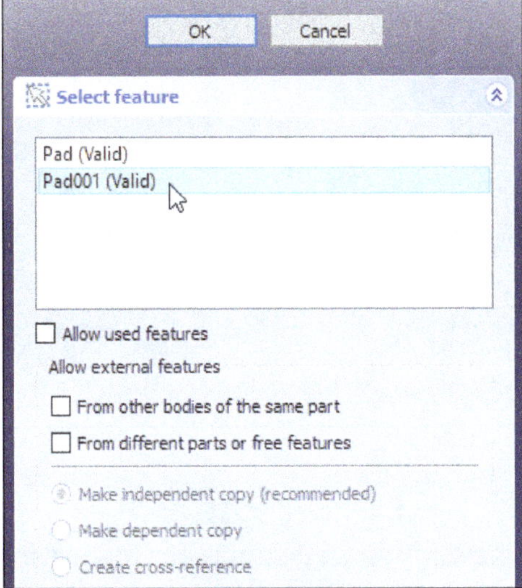

Figure-122. Select feature dialog

- Select the feature from the dialog which you want to mirror and click on **OK** button. The **Mirrored parameters** dialog will be displayed in the **Tasks** panel of **Combo View** along with the preview of mirrored feature; refer to Figure-123.

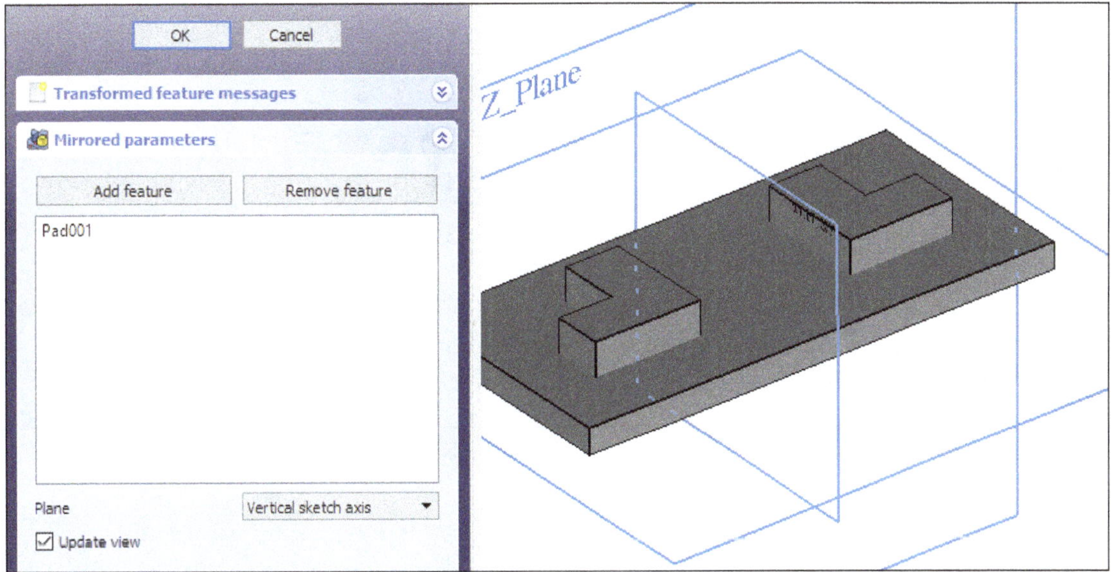

Figure-123. Mirrored parameters dialog with preview of mirrored feature

- To add more features to be mirrored, click on the **Add feature** button from the dialog and select the feature in the 3D view.
- To remove the feature to be mirrored, select the feature from the list and click on **Remove feature** button.
- Select desired plane or face to be used as mirror axis from **Plane** drop-down in the dialog.
- Click on **OK** button from the dialog. The mirrored feature will be created; refer to Figure-124.

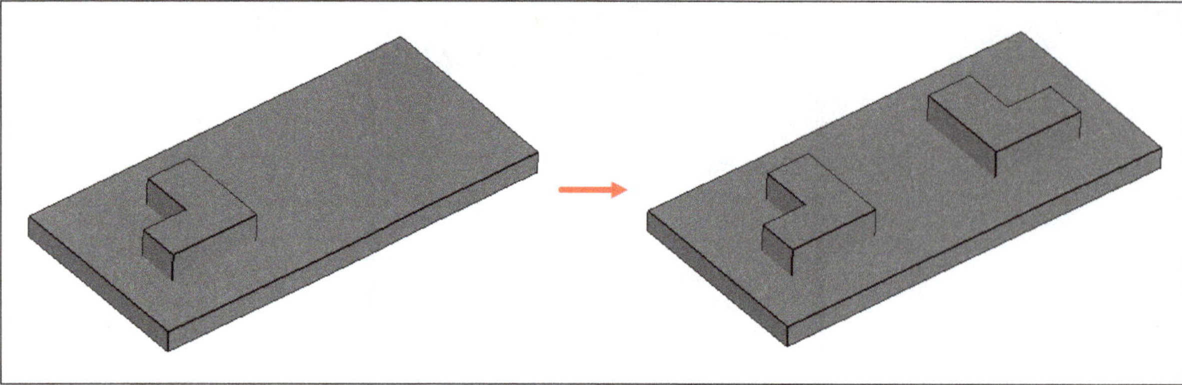

Figure-124. Mirrored feature created

Linear Pattern

The **Linear Pattern** tool creates evenly spaced copies of a feature in a linear direction. The procedure to use this tool is discussed next.

- Click on the **Linear Pattern** tool from **Toolbar** in the **Part Design** workbench; refer to Figure-125. The **Select feature** dialog will be displayed in the **Tasks** panel of **Combo View** as discussed earlier asking you to select the feature to be linear patterned.

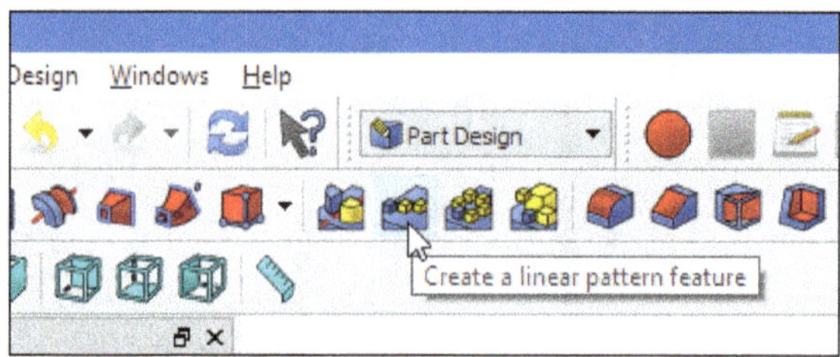

Figure-125. Linear pattern tool

- Select the feature from the dialog to be linear patterned and click on **OK** button. The **LinearPattern parameters** dialog will be displayed in the **Tasks** panel of **Combo View** along with the preview of linear pattern of the feature; refer to Figure-126.

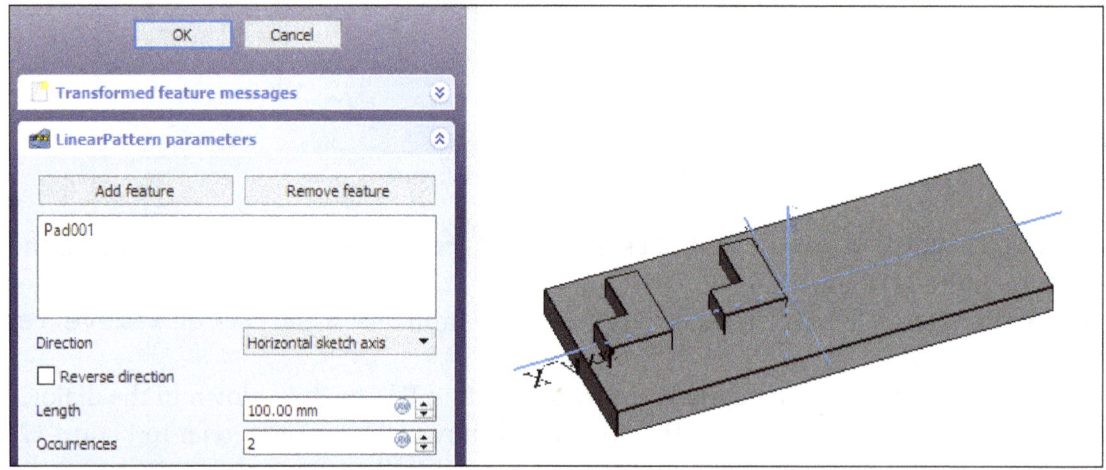

Figure-126. Linear Pattern parameters dialog with preview of linear patterned feature

- To add more features to be patterned, click on **Add feature** button from the dialog and select the feature from Model tree view or from the 3D view.
- To remove a feature from the list of features being patterned, select the feature from the list and click on **Remove feature** button.

- Specify desired direction of pattern from the **Direction** drop-down.
- Select the **Reverse direction** check box to reverse the direction of linear pattern.
- Specify the distance between patterned feature and the original feature in the **Length** edit box.
- Specify desired number of occurrences of linear pattern in the **Occurrences** edit box.
- Click on **OK** button from the dialog. The linear pattern of the feature will be created; refer to Figure-127.

Figure-127. Linear pattern of the feature created

Polar Pattern

The **Polar Pattern** tool creates a set of copies of selected features rotated around a chosen axis. The procedure to use this tool is discussed next.

- Click on the **Polar Pattern** tool from **Toolbar** in the **Part Design** workbench; refer to Figure-128. The **Select feature** dialog will be displayed in the **Tasks** panel of **Combo View** as discussed earlier asking you to select the feature.

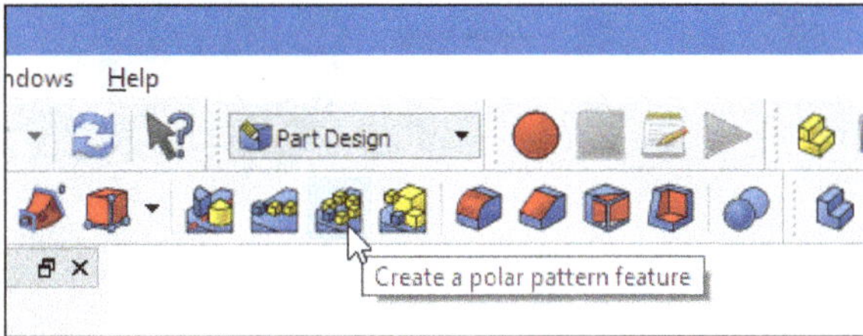

Figure-128. Polar pattern tool

- Select the feature to be polar patterned from the dialog and click on **OK** button. The **PolarPattern parameters** dialog will be displayed in the **Tasks** panel of **Combo View** along with the preview of polar pattern of the feature; refer to Figure-129.

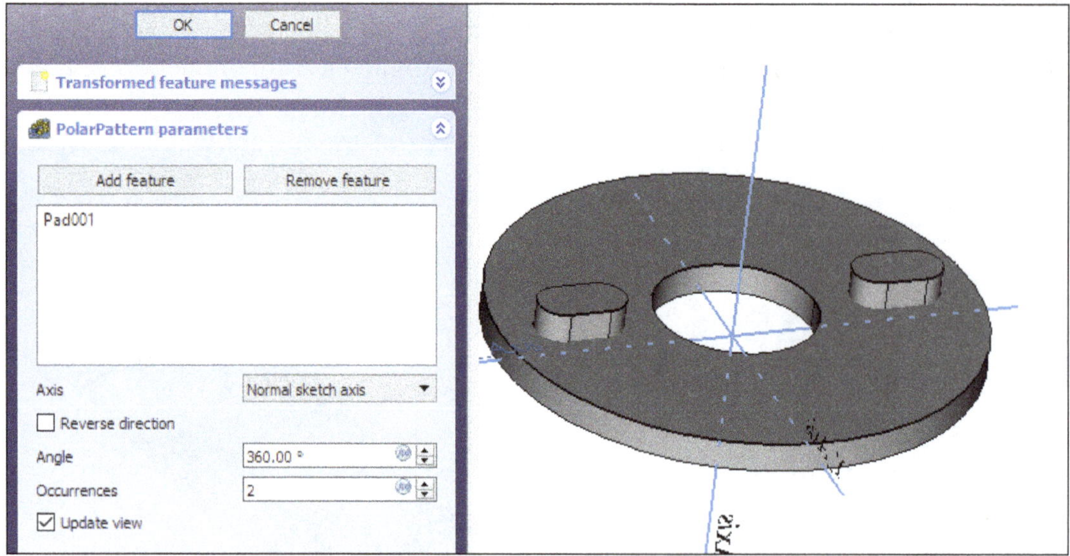

Figure-129. Polar Pattern parameters dialog with preview of polar patterned feature

- Specify desired axis of rotation for polar pattern from **Axis** drop-down in the dialog.
- Select the **Reverse direction** check box to reverse the direction of polar pattern.
- Specify the angle between patterned feature and the original feature in the **Angle** edit box.
- Specify desired number of occurrences of polar pattern in the **Occurrences** edit box.
- Click on **OK** button from the dialog. The polar pattern of the feature will be created; refer to Figure-130.

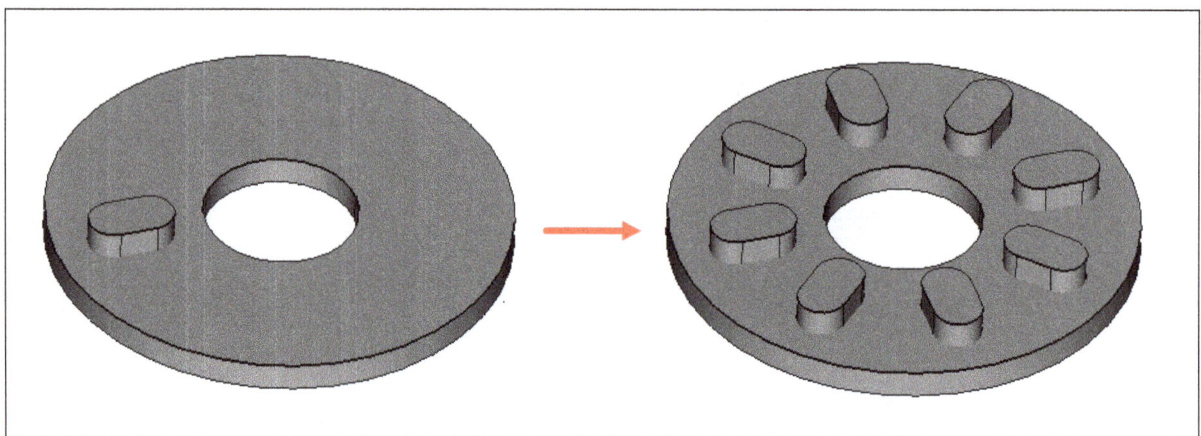

Figure-130. Polar pattern of the feature created

Create Multi Transform

The **Multi Transform** tool takes one or a set of part features as its input and allows the user to apply multiple transformations to that feature or a set of features progressively in sequence - creating a combined or compound transformation. The procedure to use this tool is discussed next.

- Click on the **Multi transform** tool from **Toolbar** in the **Part Design** workbench; refer to Figure-131. The **Select feature** dialog will be displayed in the **Tasks** panel of **Combo View** as discussed earlier asking you to select the feature.

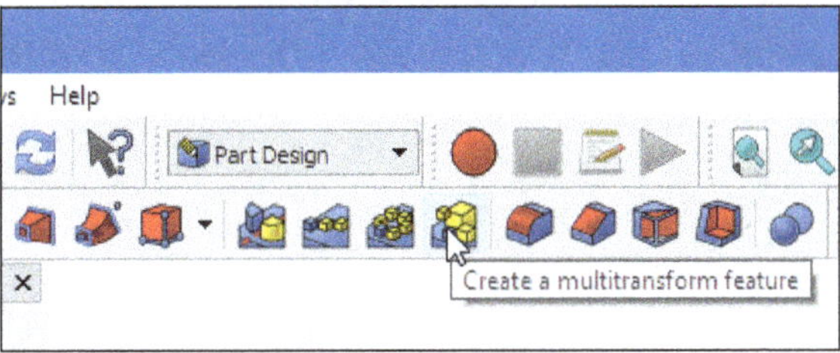

Figure-131. Multi transform tool

- Select the feature you want to be multi transformed and click on **OK** button from the dialog. The **MultiTransform parameters** dialog will be displayed in the **Tasks** panel of **Combo View**; refer to Figure-132.

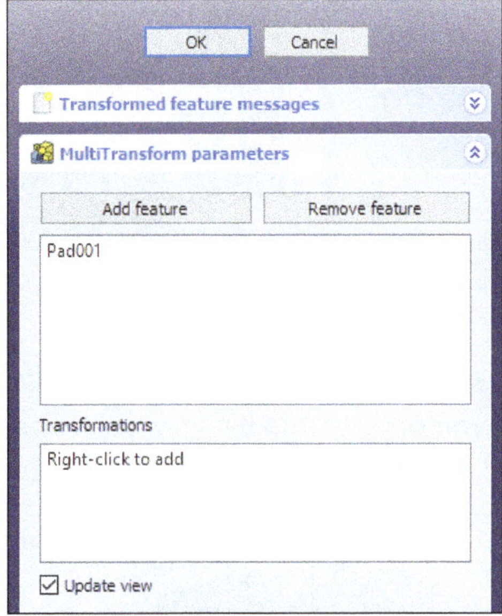

Figure-132. MultiTransform parameters dialog

- If you want to include additional features for the transformations, click on the **Add feature** button from the dialog and select the feature you want to add from the Model tree view or from the 3D view.
- Click **RMB** in the **Transformations** list view in the dialog. A shortcut menu will be displayed; refer to Figure-133.

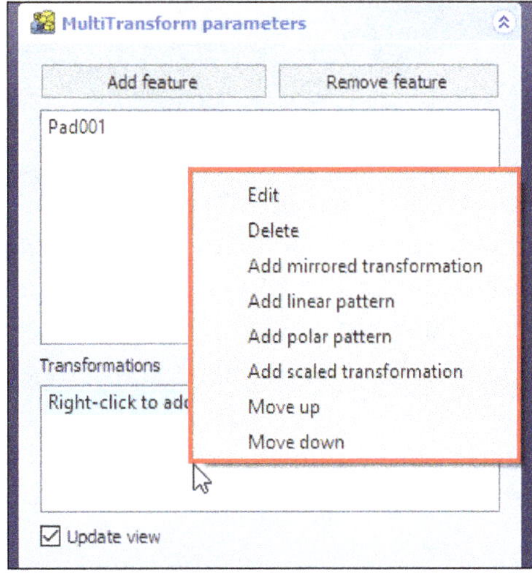

Figure-133. Shortcut menu

- Select desired transformation option from the shortcut menu. The parameters related to the selected transformation will be displayed in the dialog along with the preview of linear pattern of the feature; refer to Figure-134.

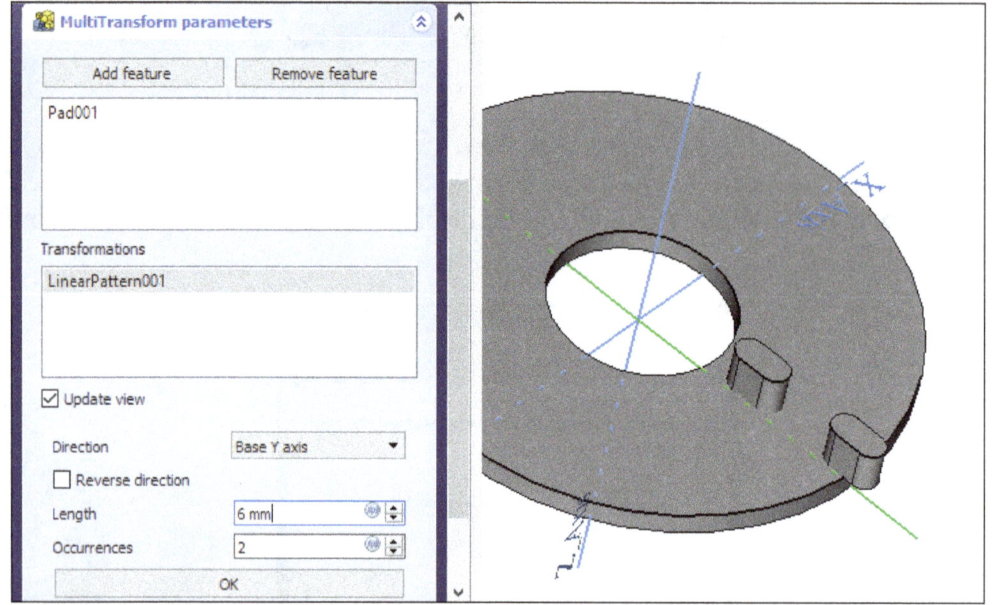

Figure-134. Linear pattern type parameters displayed along with the preview of linear pattern of the feature

- Specify all the parameters related to transformation as discussed earlier.
- Click on **OK** button below **Transformations** list view area. The linear pattern of the feature will be created; refer to Figure-135.

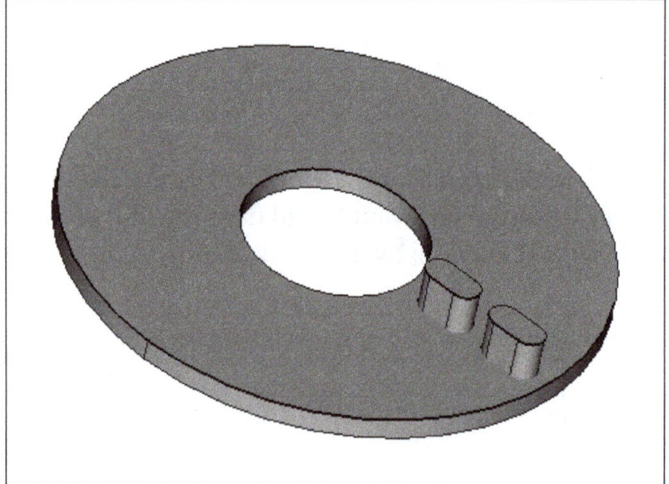

Figure-135. Linear pattern of the feature created

- If you want to add more transformation to the feature then click **RMB** in the **Transformations** list view in the dialog and select desired pattern type from the shortcut menu displayed.
- If you have selected **Add polar pattern** option then the parameters related to selected polar pattern type will be displayed in the dialog along with the preview of polar pattern of the feature; refer to Figure-136.
- Specify all the parameters related to polar pattern type as discussed earlier.
- Click on **OK** button below **Transformations** list box. The polar pattern of the feature will be created; refer to Figure-137.

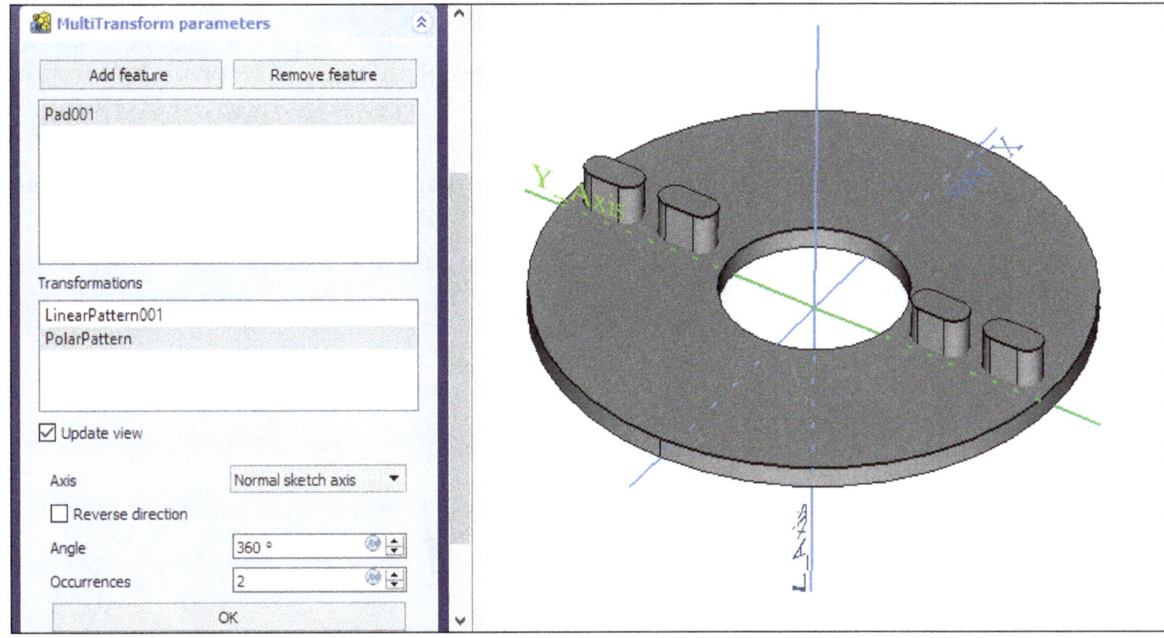

Figure-136. Polar pattern type parameters displayed along with the preview of polar pattern of the feature

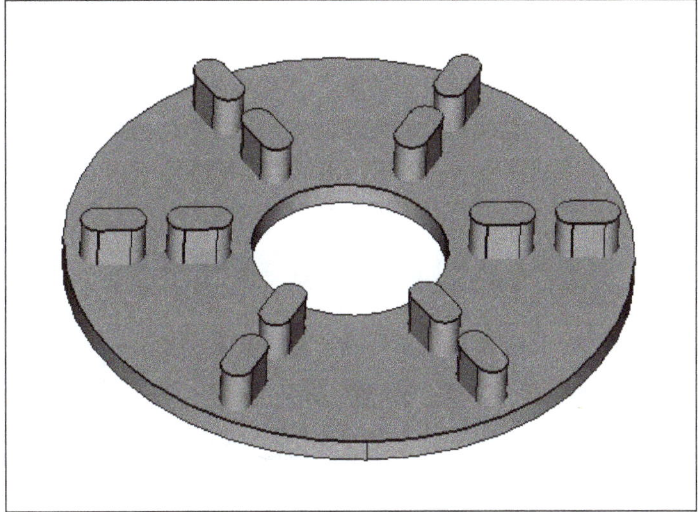

Figure-137. Polar pattern of the feature created

- Click on **OK** button to close the dialog.

Dress-up tools

The **Dress-up** tools are used to apply treatment to the selected edges or faces. These tools are available in the **Toolbar** of **Part Design** workbench; refer to Figure-138. The procedures to use these tools are discussed next.

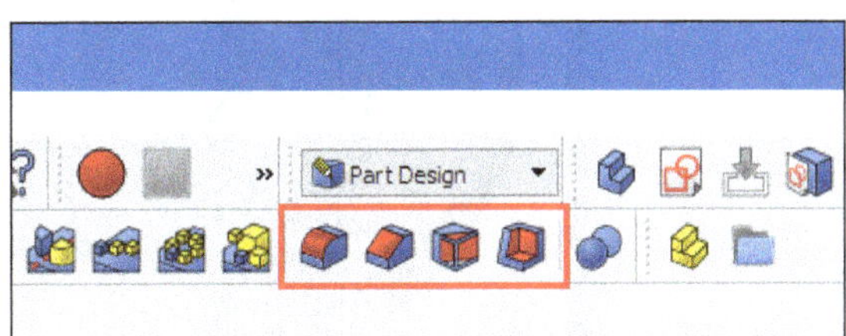

Figure-138. Dress up tools in part design workbench

Fillet

The **Fillet** tool creates fillets (rounds) on the selected edges of an object. The procedure to use this tool is discussed next.

- Select an edge or multiple edges or face of an object on which you want to apply fillet; refer to Figure-139.

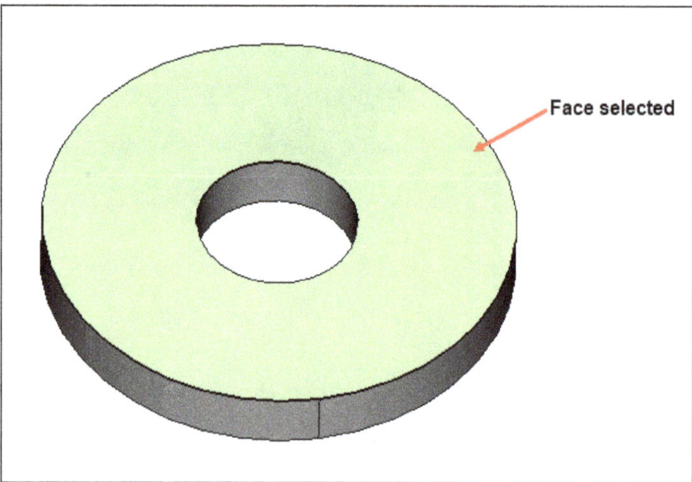

Figure-139. Face selected to create fillet

- Click on the **Fillet** tool from **Toolbar** in the **Part Design** workbench; refer to Figure-140. The **Fillet parameters** dialog will be displayed in the **Tasks** panel of **Combo View** along with the preview of fillets created on the object; refer to Figure-141.

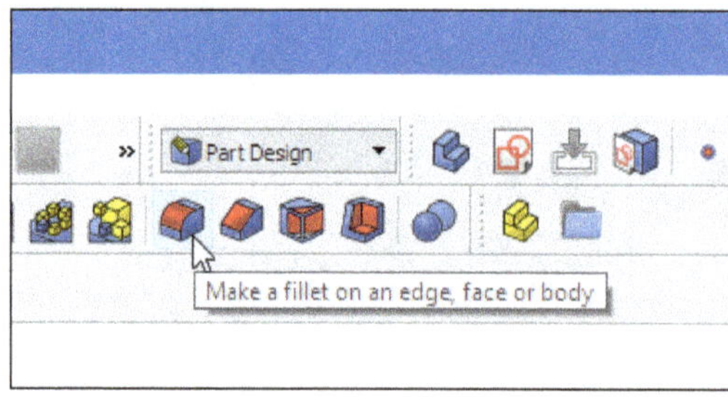

Figure-140. Fillet tool

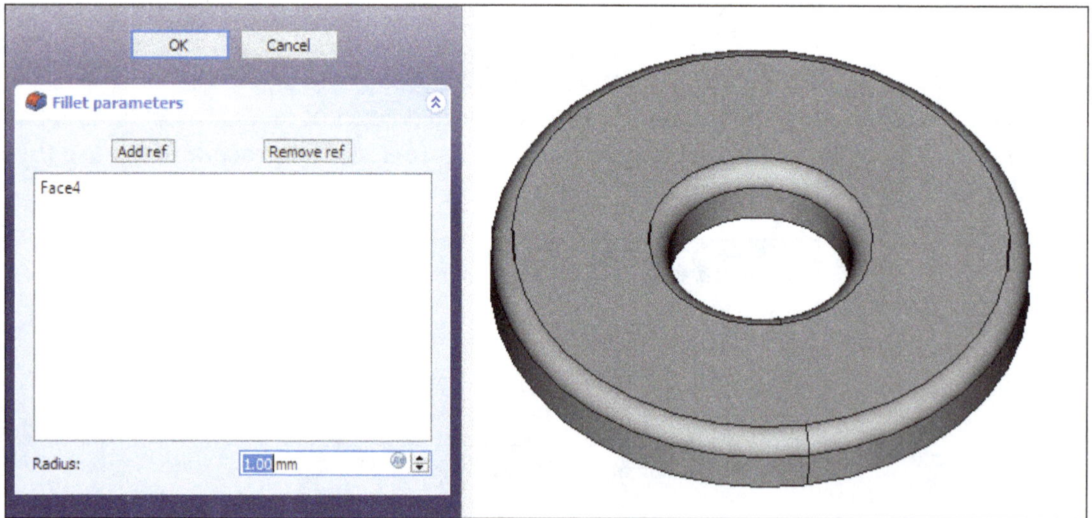

Figure-141. Fillet parameters dialog along with the preview of fillets

- Specify desired radius of the fillet in the **Radius** edit box.
- If you want to add more edges or faces to be fillet then click on the **Add ref** button from the dialog and select the edges or faces from the 3D view area.
- If you want to remove the fillet then select the face or an edge from the list view area in the dialog and click on the **Remove ref** button.
- Click on **OK** button from the dialog. The fillets will be created; refer to Figure-142.

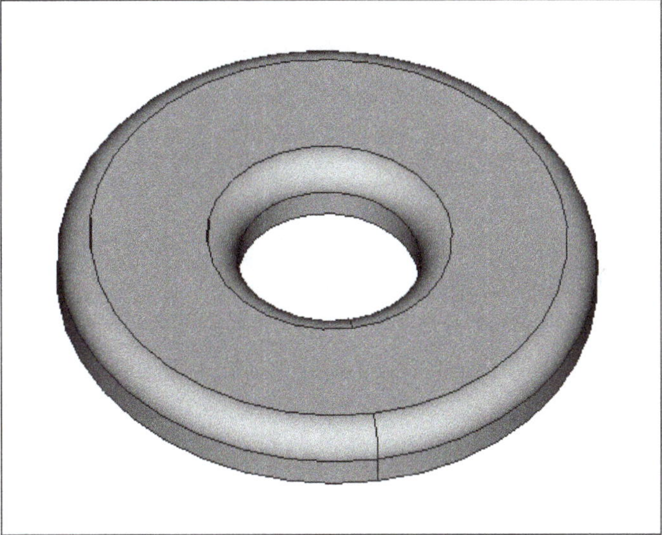

Figure-142. Fillets created on the object

Chamfer

The **Chamfer** tool creates chamfers on the selected edges of an object. The procedure to use this tool is discussed next.

- Select an edge or multiple edges or face of an object on which you want to apply chamfer as discussed for fillet tool.
- Click on the **Chamfer** tool from **Toolbar** in the **Part Design** workbench; refer to Figure-143. The **Chamfer parameters** dialog will be displayed in the **Tasks** panel of **Combo View** along with the preview of chamfers created on the object; refer to Figure-144.

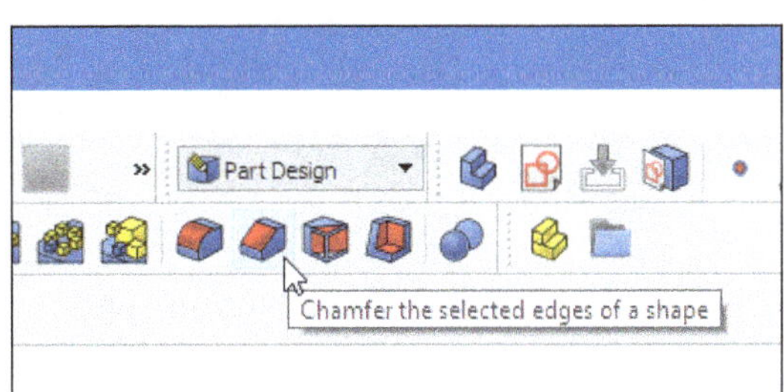

Figure-143. Chamfer tool

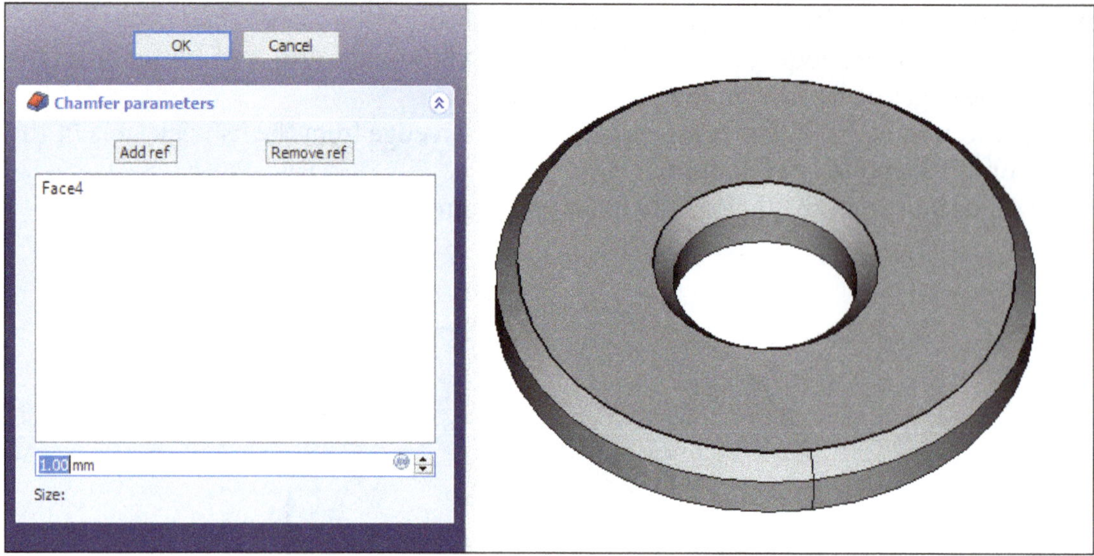

Figure-144. Chamfer parameters dialog along with the preview of chamfers

- Specify desired size of the chamfer in the **Size** edit box.
- The other parameters of the dialog is same as discussed for **Fillet** tool.
- Click on **OK** button from the dialog. The chamfers will be created; refer to Figure-145.

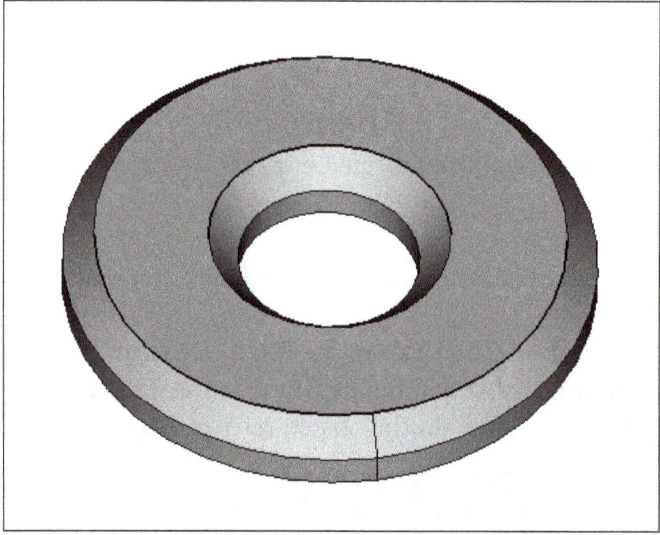

Figure-145. Chamfers created on the object

Draft

The **Draft** tool creates angular draft on the selected faces of an object. The procedure to use this tool is discussed next.

- Select one or multiple faces of an object on which you want to apply an angular draft; refer to Figure-146.
- Click on the **Draft** tool from **Toolbar** in the **Part Design** workbench; refer to Figure-147. The **Draft parameters** dialog will be displayed in the **Tasks** panel of **Combo View** along with the preview of draft applied; refer to Figure-148.

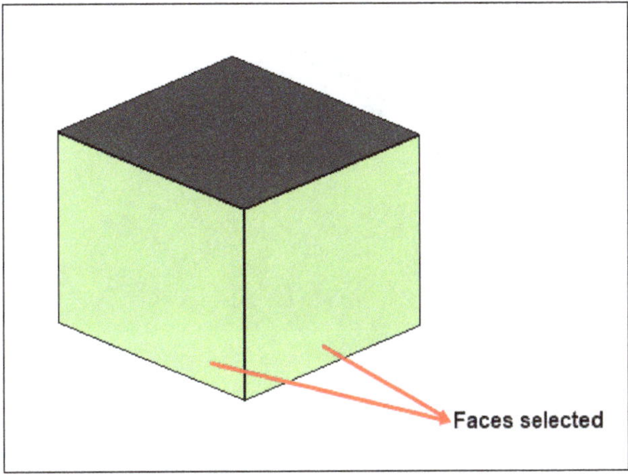

Figure-146. Faces selected for angular draft

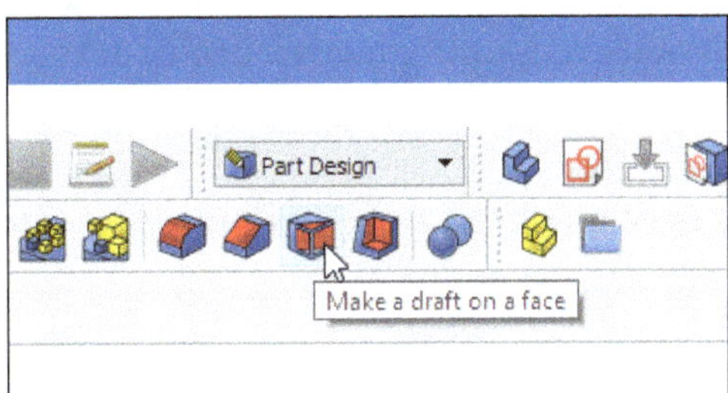

Figure-147. Draft tool

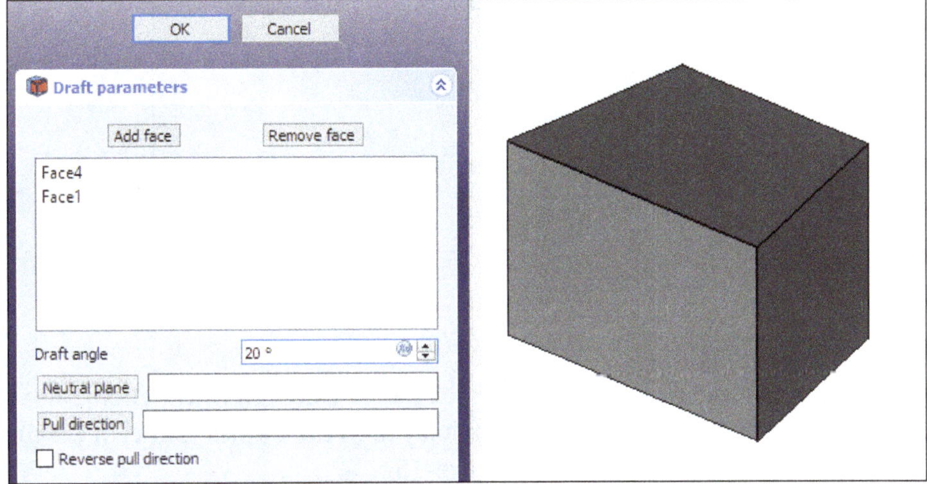

Figure-148. Draft parameters dialog along with the preview of draft applied

- If you want to add more faces to apply the draft then click on the **Add face** button from the dialog and select the faces of an object in the 3D view area.
- If you want to remove the faces then select the faces from the list view area in the dialog and click on **Remove face** button.
- Specify desired angle of draft in the **Draft angle** edit box.
- Click on the **Neutral plane** button and select the face to be used as reference for defining draft angle.
- Click on **Pull direction** button and select desired edge of an object to define direction in which draft will be applied.
- Select the **Reverse pull direction** check box to reverse the direction of draft pulling.
- Click on **OK** button from the dialog. The draft will be created on the object; refer to Figure-149.

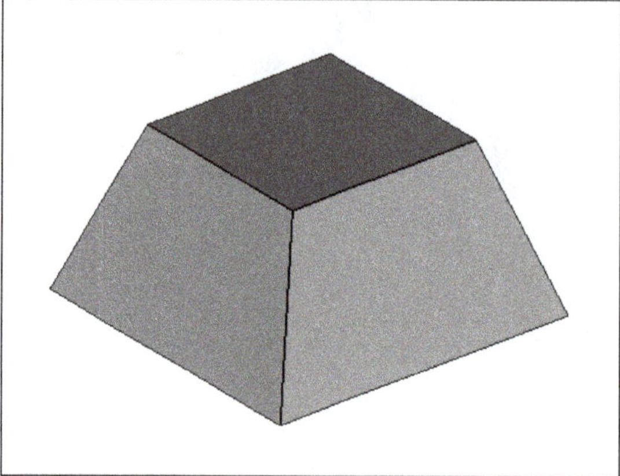

Figure-149. Draft created on the object

Thickness

The **Thickness** tool works on a solid body and transforms it into a thick-walled hollow object with at least one open face giving to each of its remaining faces a uniform thickness. The procedure to use this tool is discussed next.

- Select the face or multiple faces of an object to be hollowed; refer to Figure-150.

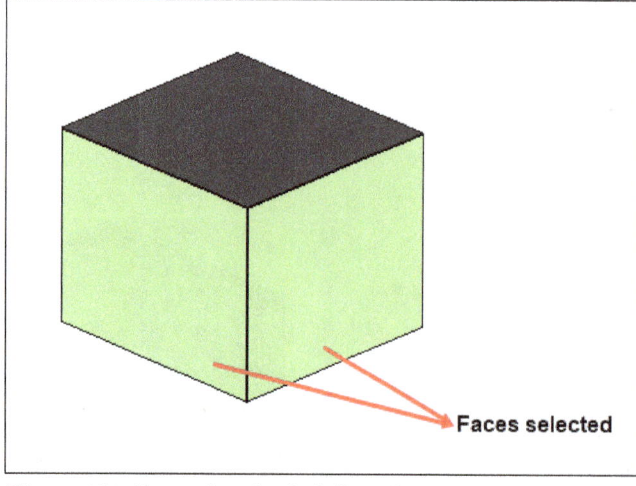

Figure-150. Faces selected to be hollowed

- Click on the **Thickness** tool from **Toolbar** in the **Part Design** workbench; refer to Figure-151. The **Thickness parameters** dialog will be displayed in the **Tasks** panel of **Combo View** along with the preview of hollowed faces; refer to Figure-152.

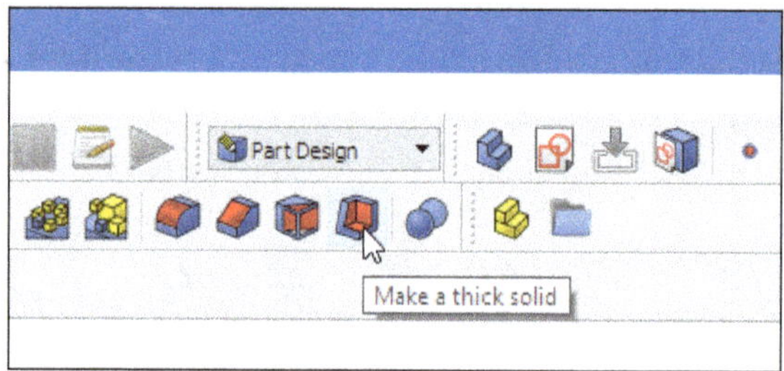

Figure-151. Thickness tool

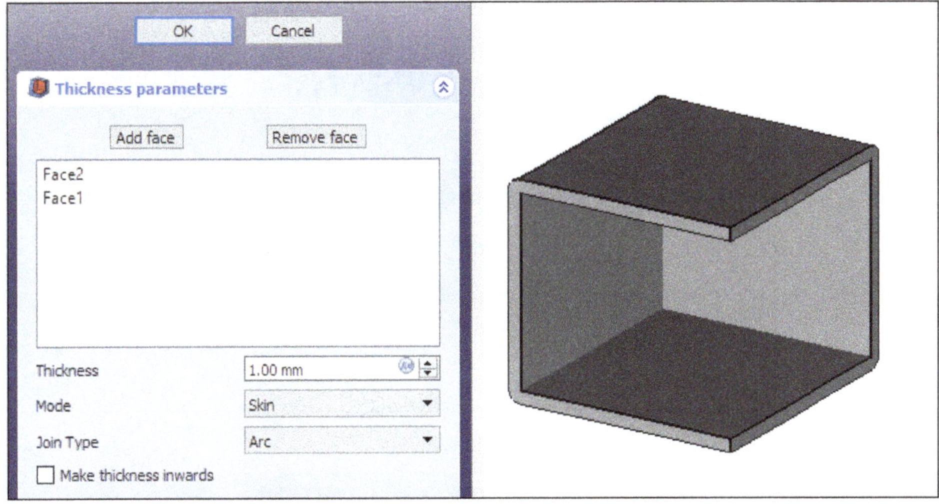

Figure-152. Thickness parameters dialog along with the preview of opened faces

- If you want to add more faces to be hollowed then click on the **Add face** button from the dialog and select the faces of the object from the 3D view area.
- If you want to remove the faces then select the faces from the list of the dialog and click on the **Remove face** button.
- Specify desired wall thickness value in the **Thickness** edit box.
- Specify desired mode of thickness from the **Mode** drop-down. Select the **Skin** option if you want to get an item like a vase, headless but with the bottom. Select the **Pipe** option if you want to get an object like a pipe, headless, and bottomless.
- Select desired option from the **Join Type** drop-down. Select the **Arc** option if you want to remove the outer edges and create a fillet with radius equal to the defined thickness. Select the **Intersection** option when faces are offset outward, sharp edges are kept between faces.
- Select the **Make thickness inwards** check box if you want to offset the faces inwards.
- Click on **OK** button from the dialog. The thick-walled hollow object with opened faces will be created; refer to Figure-153.

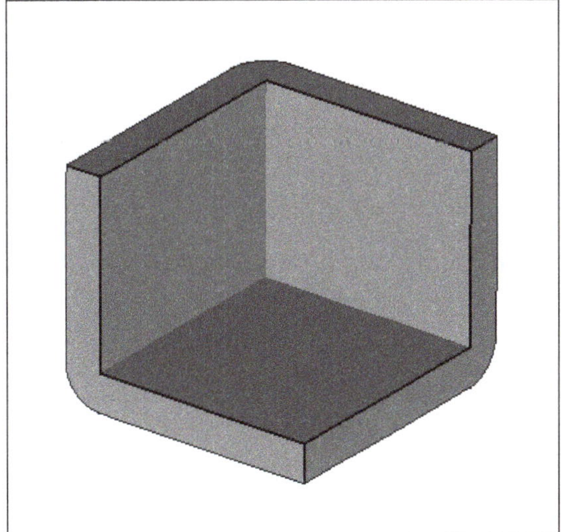

Figure-153. Thick walled hollow object with opened faces created

Boolean operation

The **Boolean** tool applies a boolean operation (fuse, cut, or common) to the Part Design bodies. The procedure to use this tool is discussed next.

- Activate the body on which you want to apply the boolean feature.
- Click on the **Boolean** tool from **Toolbar** in the **Part Design** workbench; refer to Figure-154. The **Boolean** parameters dialog will be displayed in the **Tasks** panel of **Combo View**; refer to Figure-155.

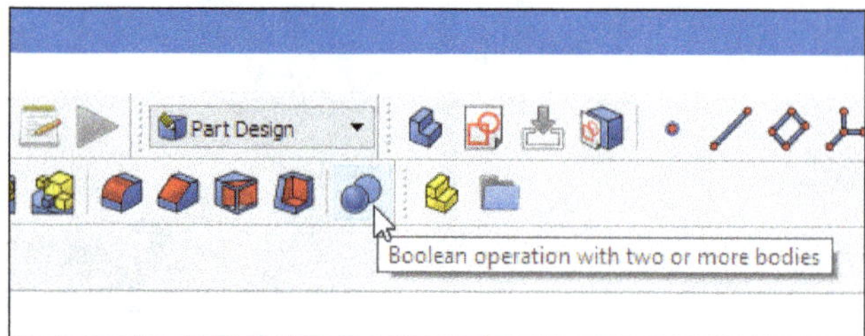

Figure-154. Boolean tool

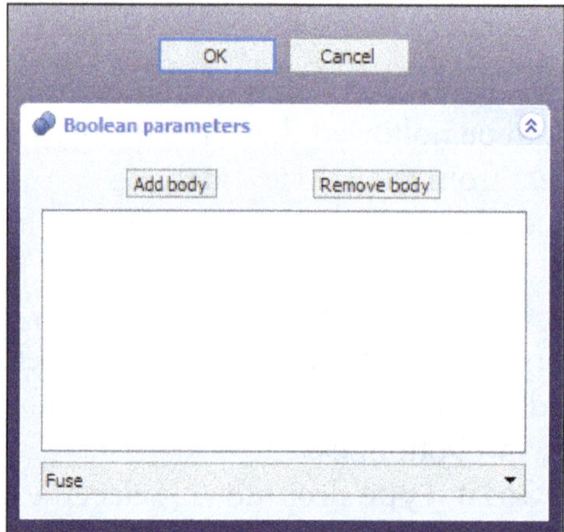

Figure-155. Boolean parameters dialog

- Click on the **Add body** button from the dialog, the active body temporarily disappears from the 3D view; refer to Figure-156.

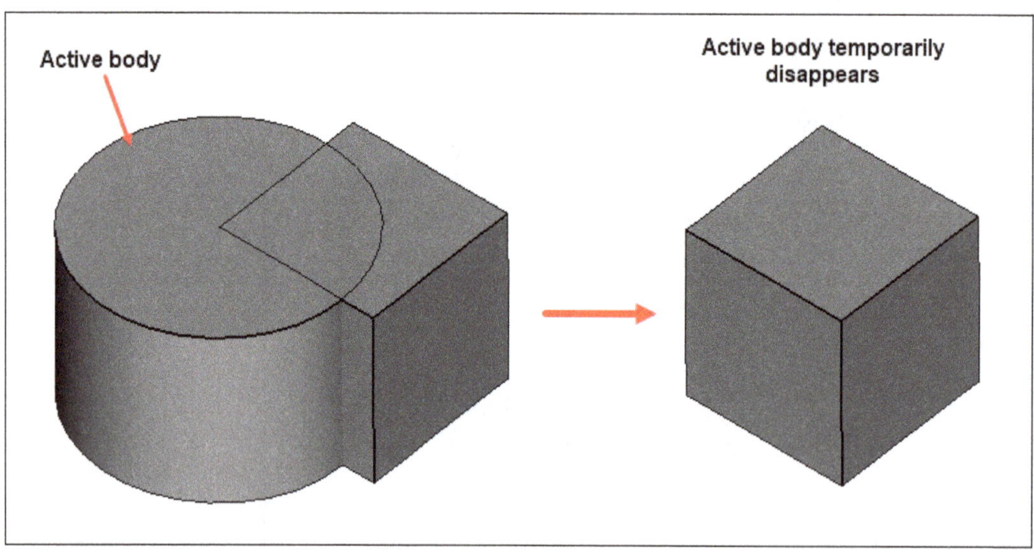

Figure-156. Active body temporarily disappears on clicking Add body button

- Select the body or multiple bodies which you want to use for creating the boolean feature from the 3D view area.
- Select desired type of boolean operation from the drop-down in the dialog. The preview of cut boolean feature will be displayed; refer to Figure-157.

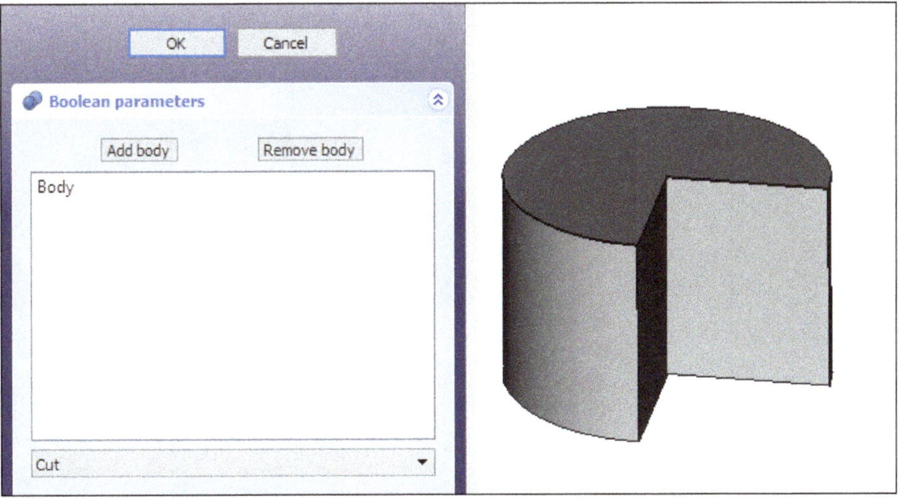

Figure-157. Preview of cut boolean feature

- Select the **Fuse** option from drop-down to merge the tool body or bodies to the active body. Select the **Cut** option to subtract the tool body or bodies from the active body. Select the **Common** option to extract the intersection from the selected body or bodies with the active body.
- If you want to remove the body to be boolean then select the body from the 3D view area and click on **Remove body** button from the dialog.
- Click on **OK** button from the dialog. The cut boolean feature will be created; refer to Figure-158.

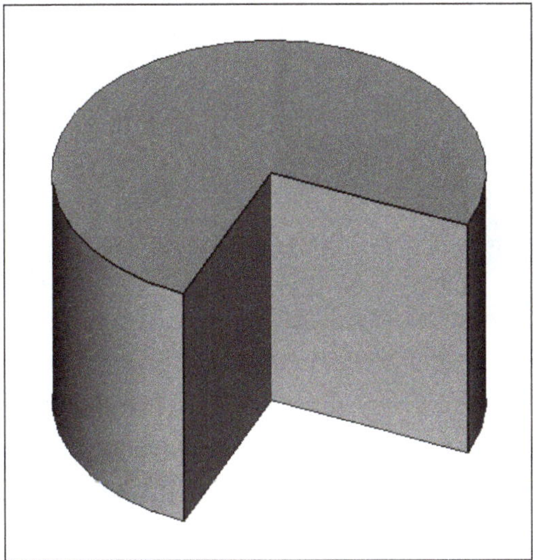

Figure-158. Cut boolean feature created

Creating Involute Gear

The **Involute gear** tool is used to generate profile for creating involute gear. You can later use this profile for **Pad** tool to create solid gear feature. The procedure to use this tool is given next.

- Click on the **Involute gear** tool from the **Part Design** menu; refer to Figure-159. The profile of involute gear will be displayed with **Involute parameter** dialog in **Tasks** panel of **Combo View**; refer to Figure-160.

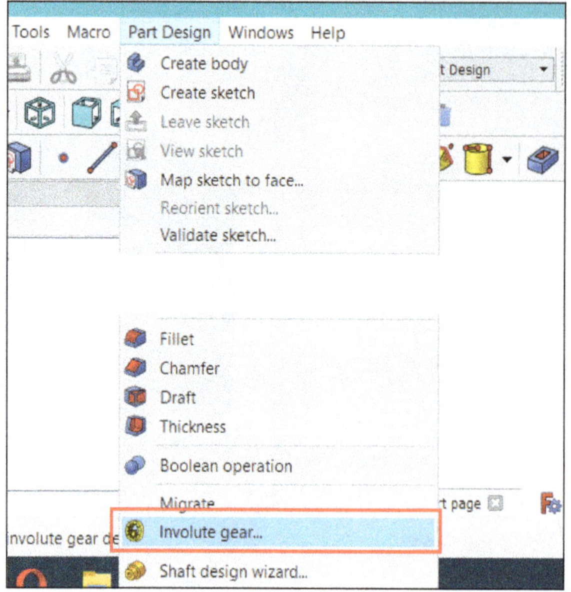

Figure-159. Involute gear tool

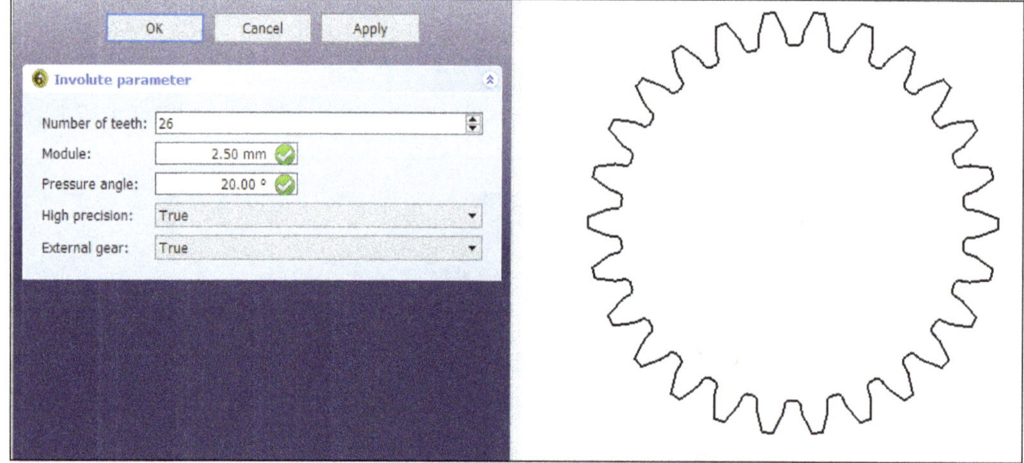

Figure-160. Involute parameters dialog with preview

- Specify the number of teethes to be created in profile in the **Number of teeth** edit box of dialog.
- Specify the other parameters like module of gear, pressure angle, and so on in respective edit boxes of the dialog and then click on the **OK** button to create the profile.

Select the profile created and click on the **Pad** tool from **Toolbar** to create solid gear.

Shaft Design Wizard

The **Shaft design wizard** tool is used to design shaft based on specified parameters. The procedure to use this tool is given next.

- Click on the **Shaft design wizard** tool from **Part Design** menu. A dialog will be displayed in the **Tasks** panel of **Combo View** to specify parameters for creating shaft; refer to Figure-161.

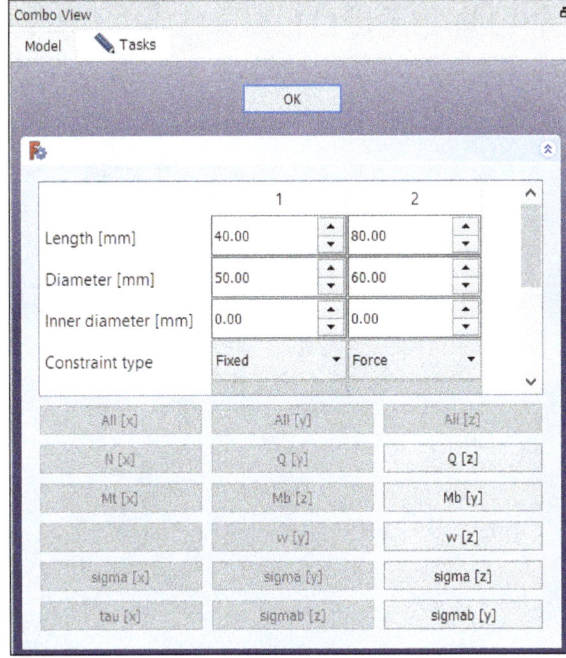

Figure-161. Dialog for shaft design

- Specify length, diameter, inner diameter, and constraint type for shaft sections in respective fields of dialog.
- If you want to add more sections to the shaft then right-click in the dialog and select **Add column** option; refer to Figure-162.

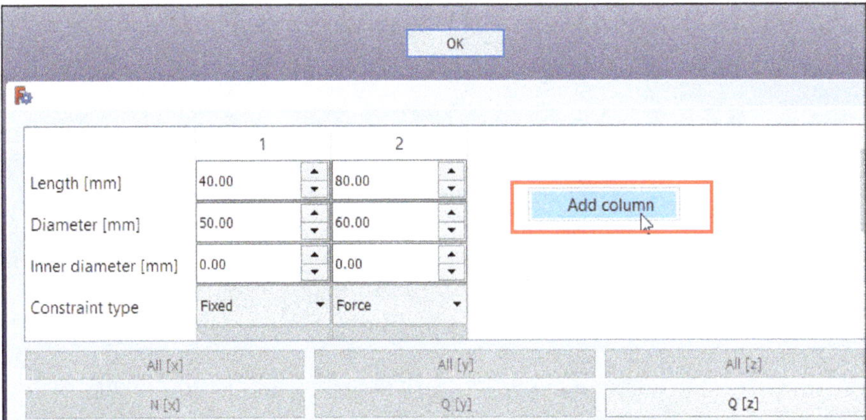

Figure-162. Add column option

- You can check various parameters of shaft using the buttons at the bottom in the dialog.

For Student Notes

Chapter 4

Solid Modeling

Topics Covered

The major topics covered in this chapter are:

- ***Introduction to Solid Modeling***
- ***Starting Part Workbench***
- ***Primitive Tools***
- ***Modifying Object Tools***
- ***Offset Tools***
- ***Compound Tools***
- ***Join Features Tools***
- ***Splitting Tools***
- ***Measure Tools***

INTRODUCTION

The solid modeling capabilities of FreeCAD are based on the **OpenCASCADE Technology** (OCCT) kernel, a professional-grade CAD system that features advanced 3D geometry creation and manipulation. The Part Workbench is a layer sitting on top of the OCCT libraries that gives the user access to OCCT geometric primitives and functions. Essentially, all 2D and 3D drawing functions in every workbench (Draft, Sketcher, PartDesign, etc.), are based on these functions exposed by the Part Workbench. Therefore, the **Part Workbench** is considered the core component of the modeling capabilities of FreeCAD.

Part objects are more complex than mesh objects created with the **Mesh Workbench** as they permit more advanced operations like coherent boolean operations, modifications history, and parametric behaviour.

The **Part Workbench** is the basic layer that exposes the OCCT drawing functions to all workbenches in FreeCAD; refer to Figure-1.

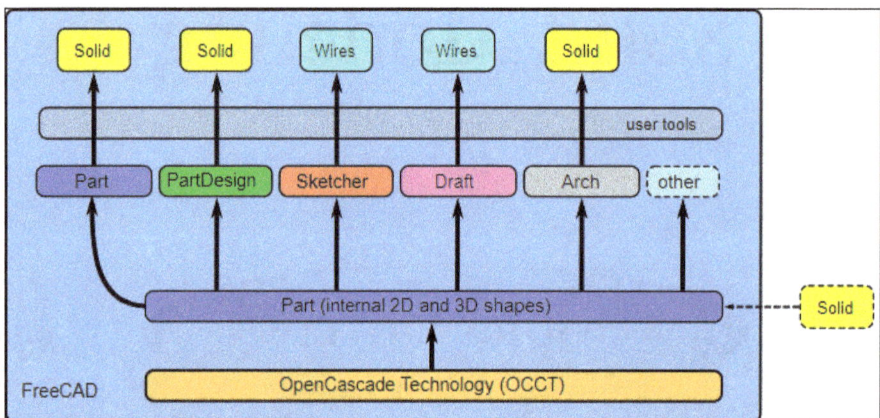

Figure-1. OpenCascade Technology

STARTING PART WORKBENCH

The objects created with the **Part Workbench** are relatively simple; they are intended to be used with boolean operations (unions and cuts) in order to build more complex shapes. This modeling paradigm is known as the constructive solid geometry (CSG) workflow and it was the traditional methodology used in early CAD systems.

- To start a new part file, click on the **New** button from **File** menu and select **Part** workbench from **Switch between workbenches** drop-down in the **Toolbar**; refer to Figure-2. The tools to create 3D models will be displayed in the **Toolbar**; refer to Figure-3.

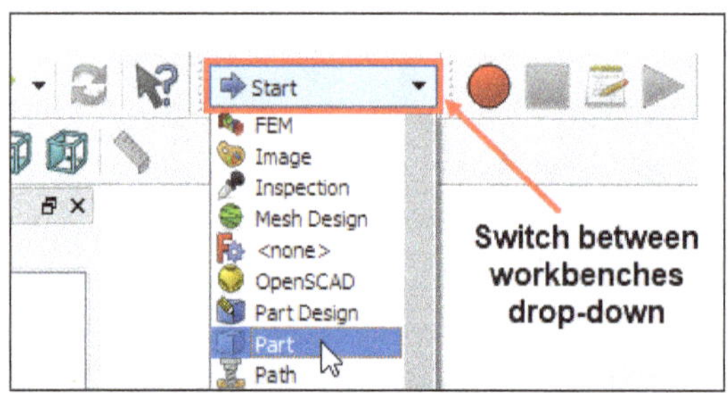

Figure-2. Part workbench

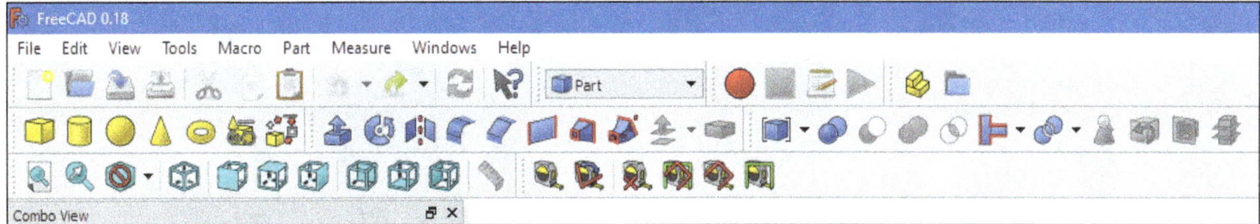

Figure-3. Part workbench tools

Primitive tools

Primitives are the tools for creating primitive objects. These tools are available in the **Toolbar** of **Part** workbench; refer to Figure-4. The procedures to use these tools are discussed next.

Figure-4. Primitive tools

Box

The **Box** tool inserts a parametric, rectangular cuboid, geometric primitive into the active document. The procedure to use this tool is discussed next.

- Click on the **Box** tool from **Toolbar** in the **Part** workbench; refer to Figure-5. A rectangular cuboid will be inserted in the 3D view area; refer to Figure-6.

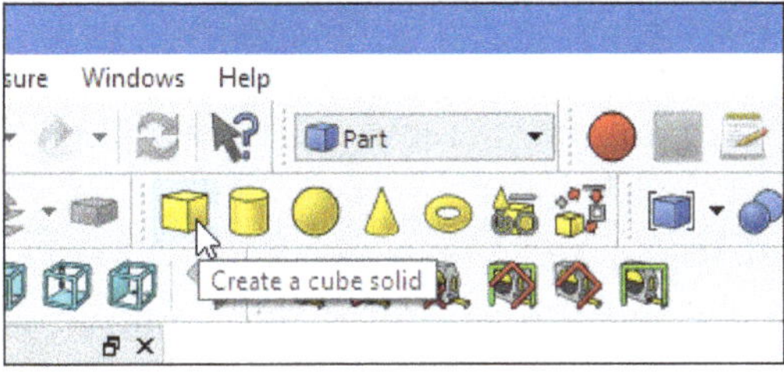

Figure-5. Part box tool

Figure-6. Rectangular cuboid primitive inserted

- If you want to edit the properties of inserted cuboid then select the cube from the Model tree view; refer to Figure-7. The **Property editor** dialog will be displayed in the **Model** tab of **Combo View**; refer to Figure-8.

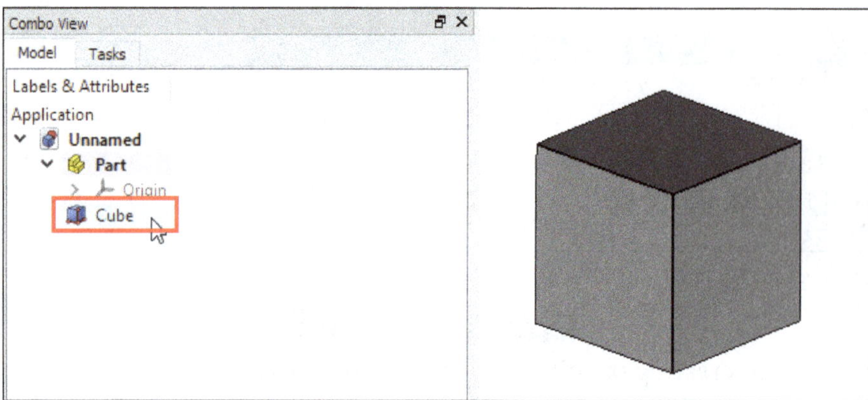

Figure-7. Selecting the object

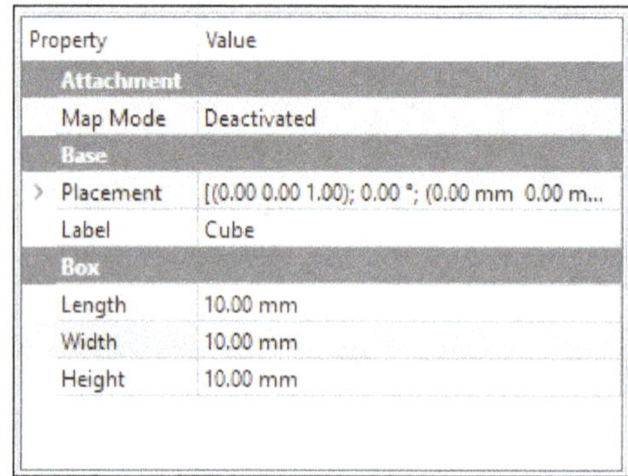

Figure-8. Property editor dialog

- Click on the **Placement** node from **Base** section of the dialog. The parameters related to placement will be displayed to specify the orientation and position of the cube in the 3D view area.
- Specify desired label for the cube in the **Label** edit box.
- Specify desired length, width, and height of the cube in the **Length**, **Width**, and **Height** edit boxes, respectively from **Box** section of the dialog. The properties of the rectangular cuboid will be edited.

Cylinder

The **Cylinder** tool creates a simple parametric cylinder with position, angle, radius, and height parameters. The procedure to use this tool is discussed next.

- Click on the **Cylinder** tool from **Toolbar** in the **Part** workbench; refer to Figure-9. A cylinder will be inserted in the 3D view area; refer to Figure-10.

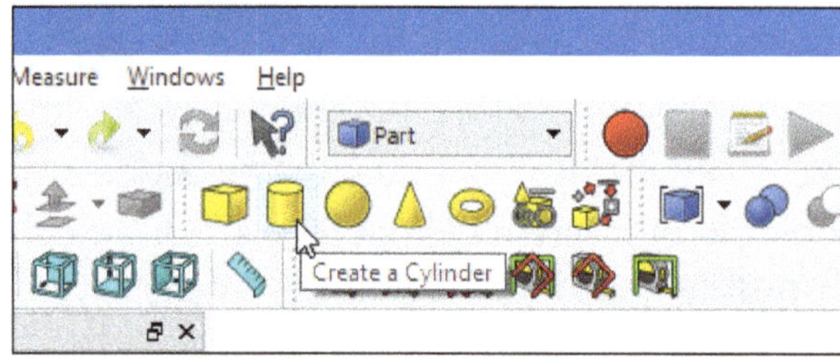

Figure-9. Part Cylinder tool

- The procedure to edit the properties of cylinder is same as discussed for **Box** tool.

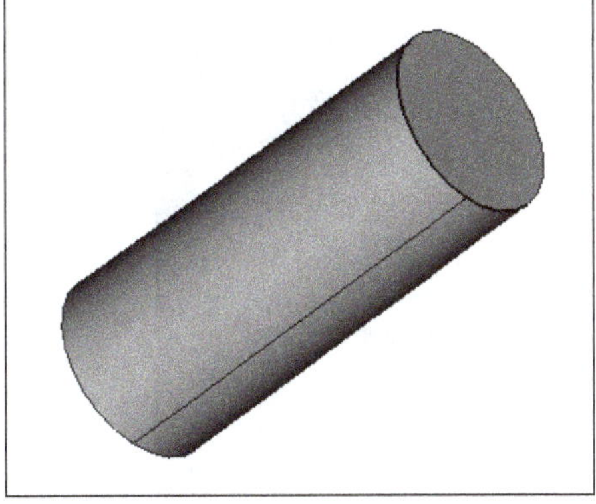

Figure-10. Cylinder primitive inserted

Sphere

The **Sphere** tool creates a simple parametric sphere with position, angle1, angle2, angle3, and radius parameters. The procedure to use this tool is discussed next.

- Click on the **Sphere** tool from **Toolbar** in the **Part** workbench; refer to Figure-11. The sphere will be inserted in the 3D view area; refer to Figure-12.

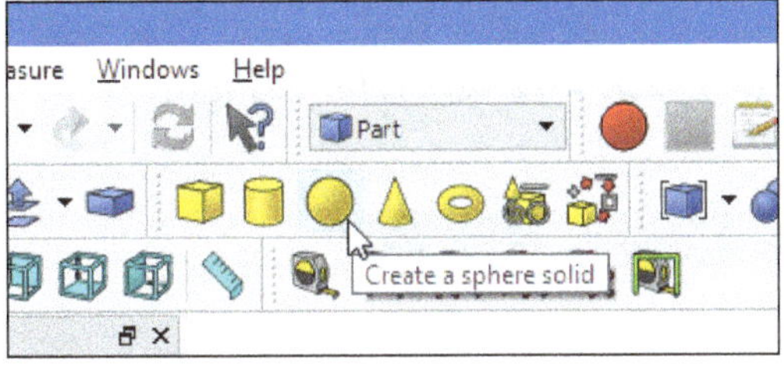

Figure-11. Part sphere tool

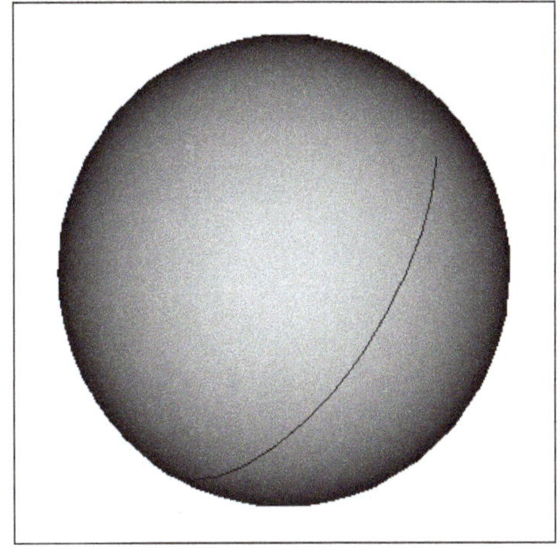

Figure-12. Sphere primitive inserted

- The procedure to edit the properties of sphere is same as discussed for earlier tools.

Cone

The **Cone** tool creates a simple parametric truncated cone. The procedure to use this tool is discussed next.

- Click on the **Cone** tool from **Toolbar** in the **Part** workbench; refer to Figure-13. The cone will be inserted in the 3D view area; refer to Figure-14.

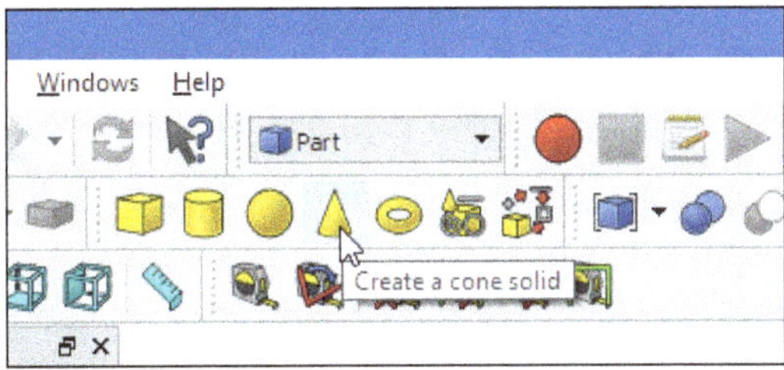

Figure-13. Part Cone tool

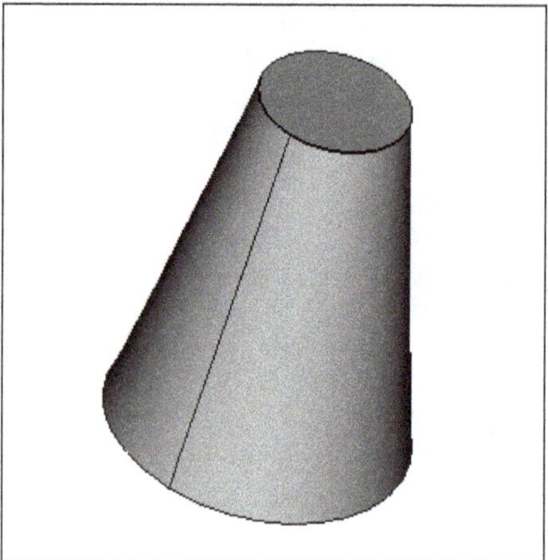

Figure-14. Cone primitive inserted

- The procedure to edit the properties of cone is same as discussed for earlier tools.

Torus

The **Torus** tool creates a simple parametric torus with position, angle1, angle2, angle3, radius1, and radius2 as parameters. The procedure to use this tool is discussed next.

- Click on the **Torus** tool from **Toolbar** in the **Part workbench**; refer to Figure-15. The torus will be inserted in the 3D view area; refer to Figure-16.

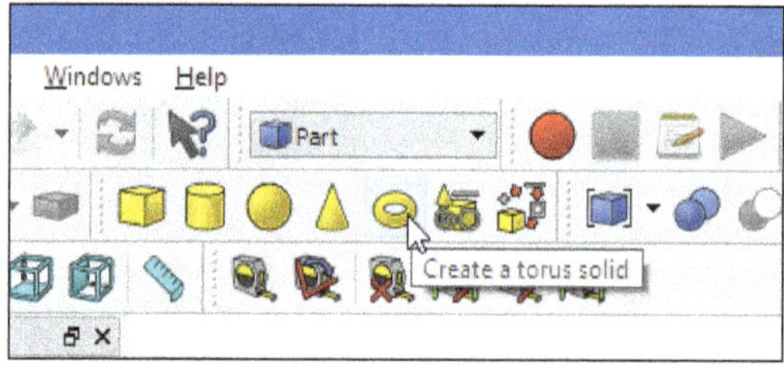

Figure-15. Part Torus tool

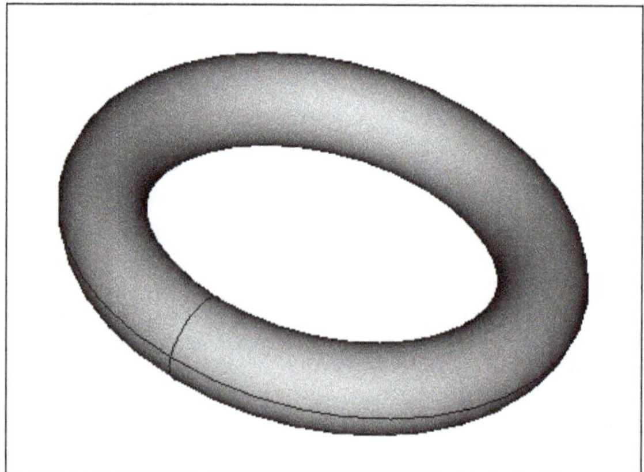

Figure-16. Torus primitive inserted

- The procedure to edit the properties of torus is same as discussed for earlier tools.

Primitives

The **Primitives** tool is used to create any of the parametric geometric primitives defined in the part workbench. The procedure to use this tool is discussed next.

- Click on the **Primitives** tool from **Toolbar** in the **Part** workbench; refer to Figure-17. The **Geometric Primitives** dialog will be displayed in the **Tasks** panel of **Combo View**; refer to Figure-18.

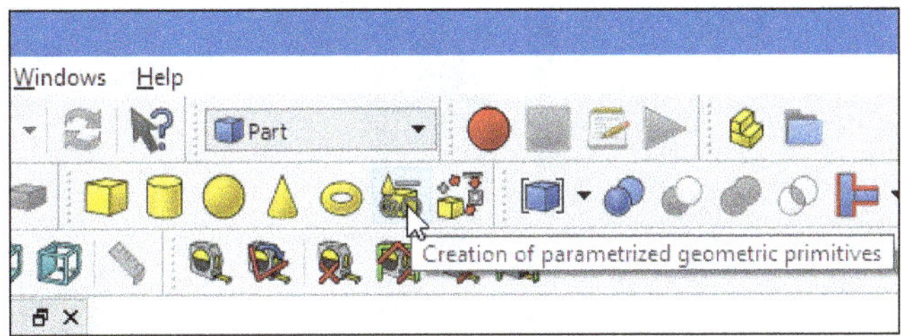

Figure-17. Part Primitives tool

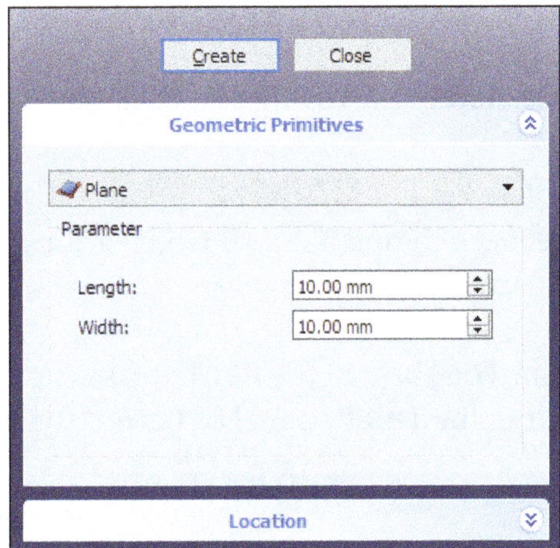

Figure-18. Geometric Primitives dialog

- Click on the **Primitive type** drop-down in the dialog. The list of available parametric geometric primitives will be displayed; refer to Figure-19.

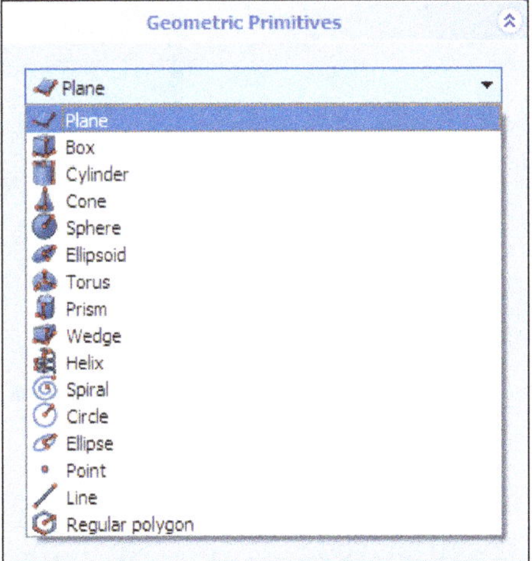

Figure-19. List of available geometric primitives

- Select desired primitive type from the drop-down to be created. The parameters related to the selected primitive type will be displayed in the dialog.
- Specify the parameters as desired and click on **Create** button from the dialog. The selected geometric primitive will be created; refer to Figure-20.

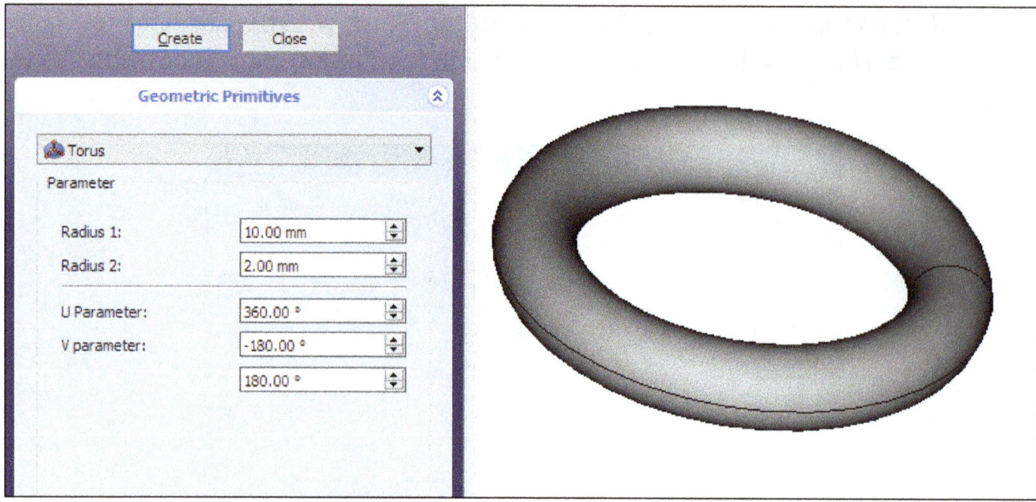

Figure-20. Geometric primitive created

- Specify all the parameters as desired in the dialog and click on **Close** button to close the dialog.

Builder

The **Builder** tool is used to create more complex shapes from various parametric geometric primitives. The procedure to use this tool is discussed next.

- Click on the **Builder** tool from **Toolbar** in the **Part** workbench; refer to Figure-21. The **Create shape** dialog will be displayed in the **Tasks** panel of **Combo View**; refer to Figure-22.

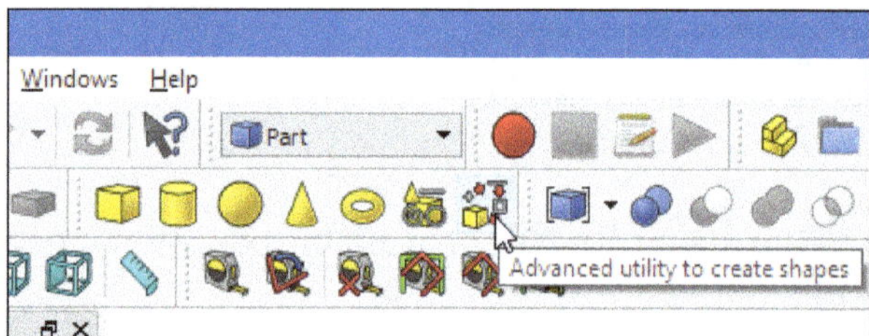

Figure-21. Part Builder tool

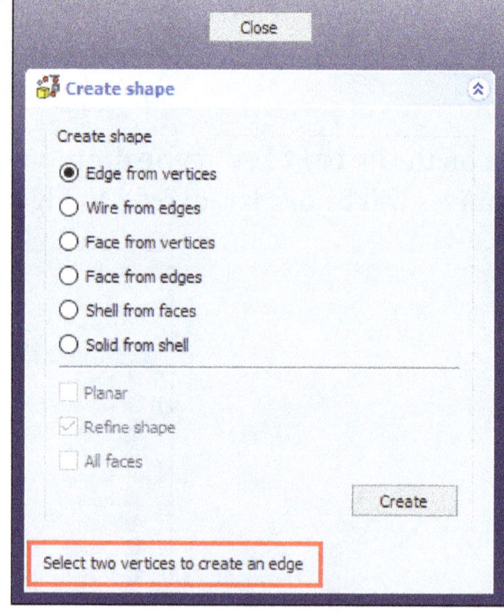

Figure-22. Create shape dialog

- Select desired radio button from **Create shape** section for the shape to be created. The conditions to create the selected shape will be displayed at the bottom of the dialog; refer to Figure-22.
- Select the objects as required for the shape and click on **Create** button from the dialog. The shape will be created; refer to Figure-23.

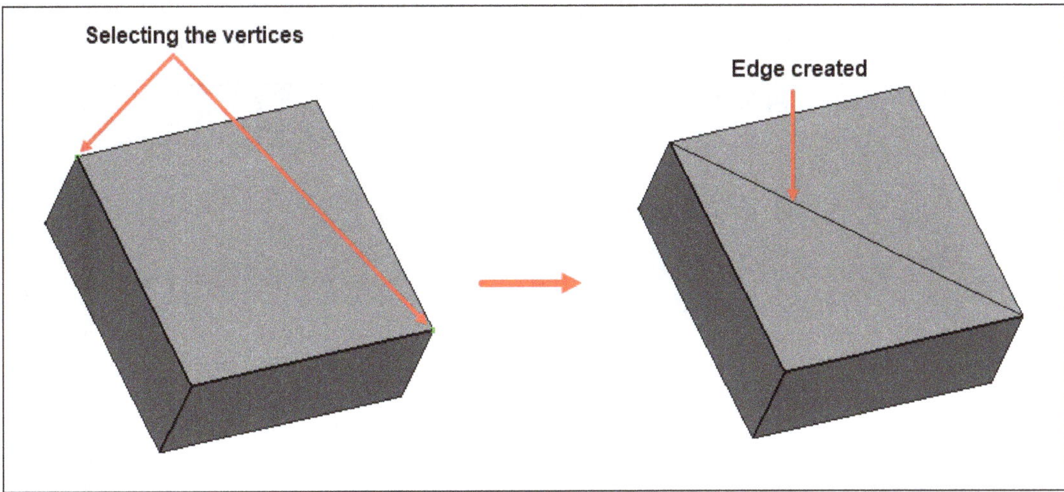

Figure-23. Edge shape created

- Click on **Close** button to close the dialog.

Modifying Object tools

Modifying object tools are used for modifying existing objects. They will allow you to choose which object to modify. These tools are available in the **Toolbar** of **Part** workbench; refer to Figure-24. The procedures to use these tools are discussed next.

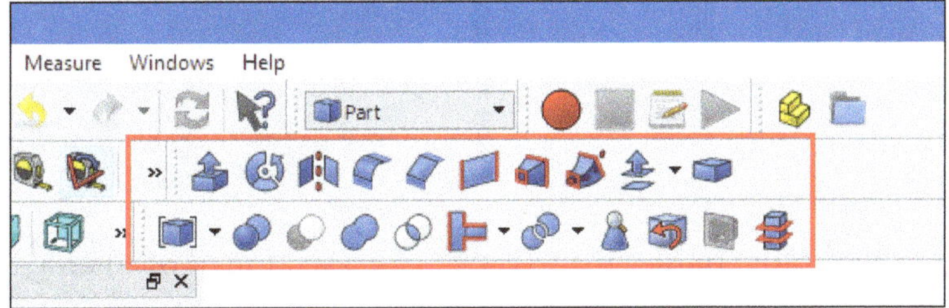

Figure-24. Modifying object tools

Extrude

The **Extrude** tool extends a shape by a specified distance in a specified direction. The output shape type will vary depending on the input shape type and the options selected. The procedure to use this tool is discussed next.

- Select desired shape from the 3D view area or from the Model tree view which you want to extrude; refer to Figure-25.

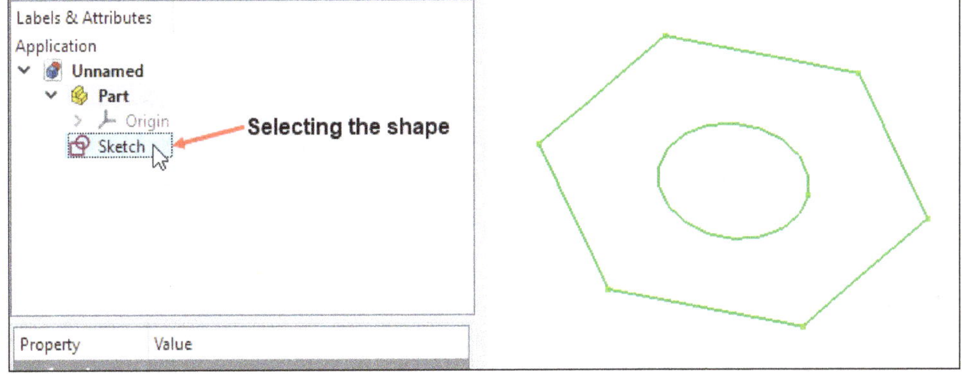

Figure-25. Selecting the sketch for extrusion

- Click on the **Extrude** tool from **Toolbar** in the **Part** workbench; refer to Figure-26. The **Extrude** dialog will be displayed in the **Tasks** panel of **Combo View**; refer to Figure-27.

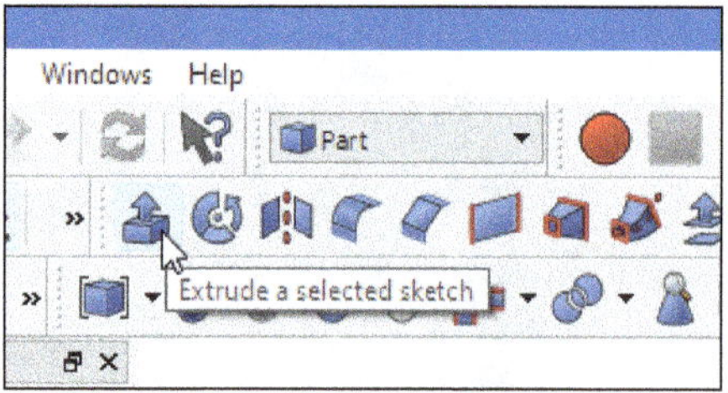

Figure-26. Extrude tool

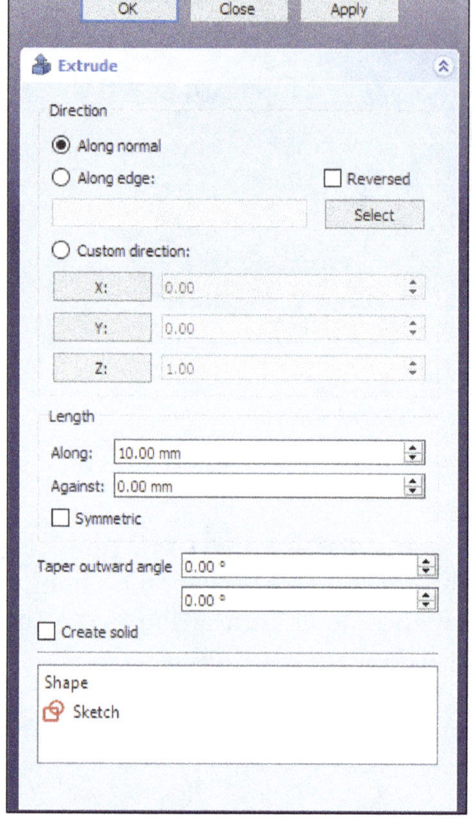

Figure-27. Extrude dialog

- Select the **Along normal** radio button from **Direction** section of the dialog to create the extrusion in a normal direction.
- Select the **Along edge** radio button and click on **Select** button below the radio button to select the edge along the direction of which the extrusion will be created.
- Select the **Reversed** check box next to the **Along edge** radio button to reverse the direction of extrusion along an edge.
- Select the **Custom direction** radio button and specify desired value in **X**, **Y**, and **Z** edit boxes to specify the extrusion direction along x, y, and z axis, respectively.
- Specify desired length of extrusion along normal direction in the **Along** edit box from **Length** section of the dialog.
- Specify the length of extrusion along reverse direction in the **Against** edit box.
- Select the **Symmetric** check box to create the extrusion in both the directions.
- Specify desired value in **Taper outward angle** edit boxes to apply the draft to extrusion side faces.
- Select the **Create solid** check box to extrude the closed wire or edge as a solid model.
- The selected shape to be extruded will display in the **Shape list** area of the dialog. If you want to select more shapes to extrude then select desired shape from the **Shape list** area.
- Click on **Apply** button from the dialog. The preview of extrusion will be created; refer to Figure-28.

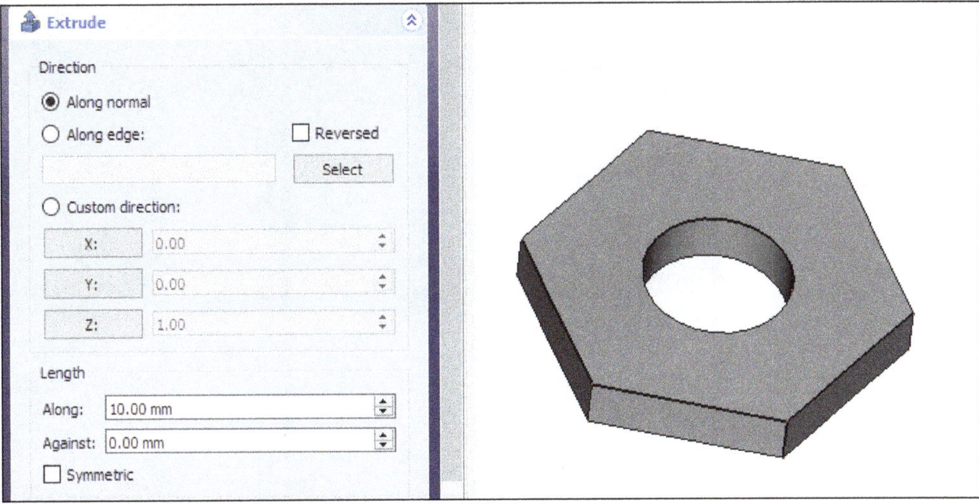

Figure-28. Preview of extrusion

- Click on **OK** button to close the dialog. The extrusion will be created.

Revolve

The **Revolve** tool revolves selected object about an axis. The procedure to use this tool is discussed next.

- Select desired shape from 3D view area or from the Model tree view which you want to revolve; refer to Figure-29.

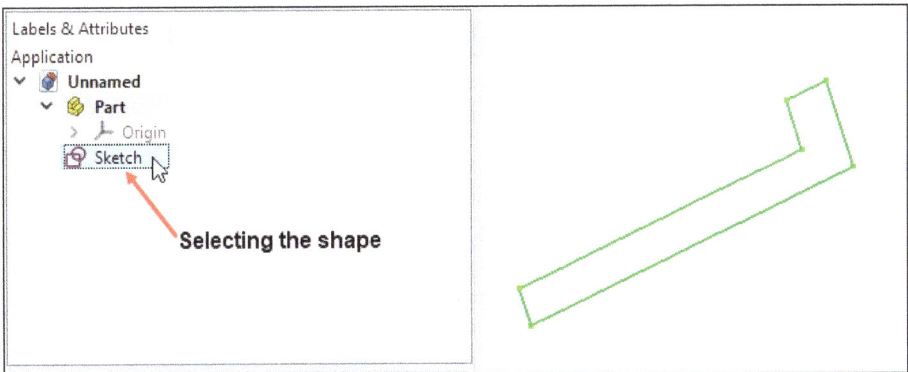

Figure-29. Selecting the shape to revolve

- Click on the **Revolve** tool from **Toolbar** in the **Part** workbench; refer to Figure-30. The **Revolve** dialog will be displayed in the **Tasks** panel of **Combo View**; refer to Figure-31.

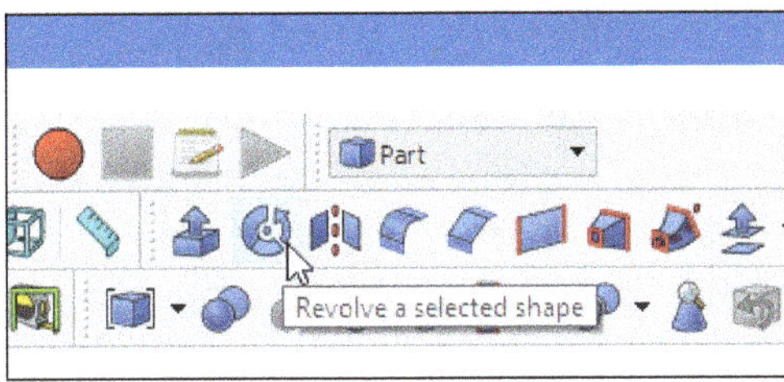

Figure-30. Revolve tool

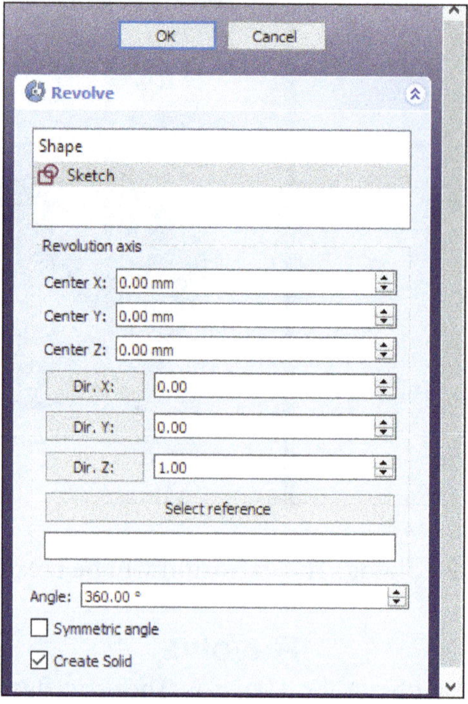

Figure-31. Revolve dialog

- The selected shape will display in the **Shape list** area of the dialog. If you want to select more shapes then select the shape from shape list area.
- Specify the parameters as desired in the **Revolution axis** section of the dialog.
- Click on the **Select reference** button from **Revolution axis** section. You are asked to select the line or an arc.
- Select the line or an arc from 3D view area which you want to make reference for the revolution of shape; refer to Figure-32.

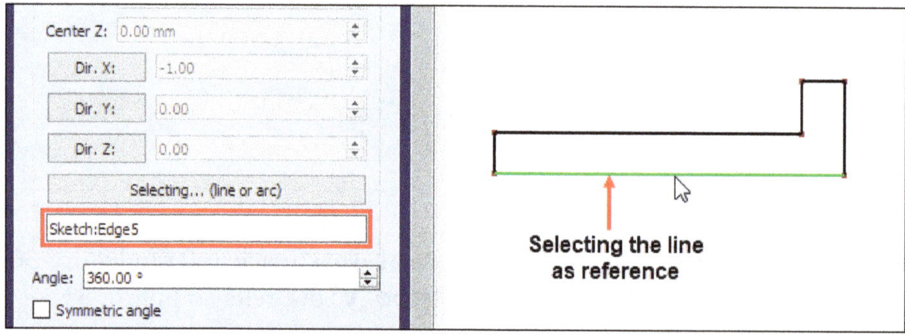

Figure-32. Selecting the line as reference for revolution

- Specify desired angle of revolution in the **Angle** edit box.
- Select the **Symmetric angle** check box to revolve the shape in both the direction at the given angle.
- Select the **Create solid** check box to revolve the shape as a solid model.
- Click on **OK** button from the dialog. The revolved model of the shape will be created; refer to Figure-33.

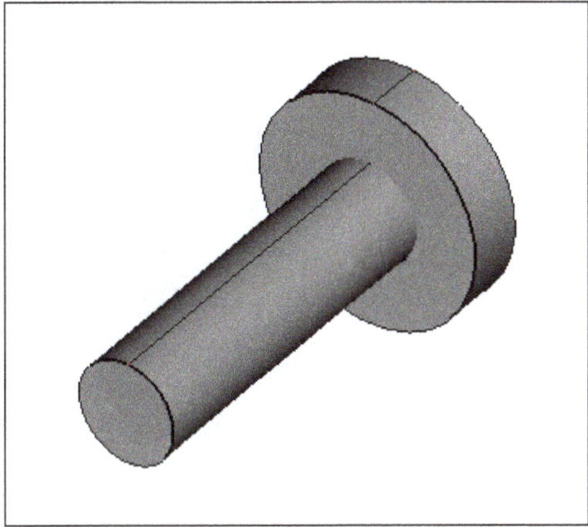

Figure-33. Revolved model of the shape created

Mirror

The **Mirror** tool creates a new object (image) which is a reflection of the original object (source). The image object is created on the other side of mirror plane. The procedure to use this tool is discussed next.

- Click on the **Mirror** tool from **Toolbar** in the **Part** workbench; refer to Figure-34. The **Mirroring** dialog will be displayed in the **Tasks** panel of **Combo View**; refer to Figure-35.

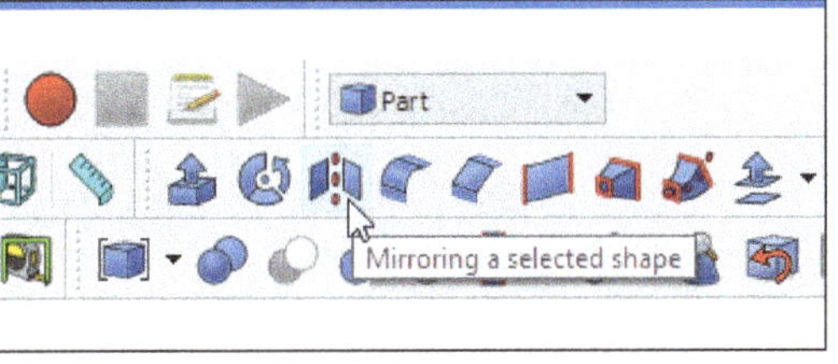

Figure-34. Mirror tool

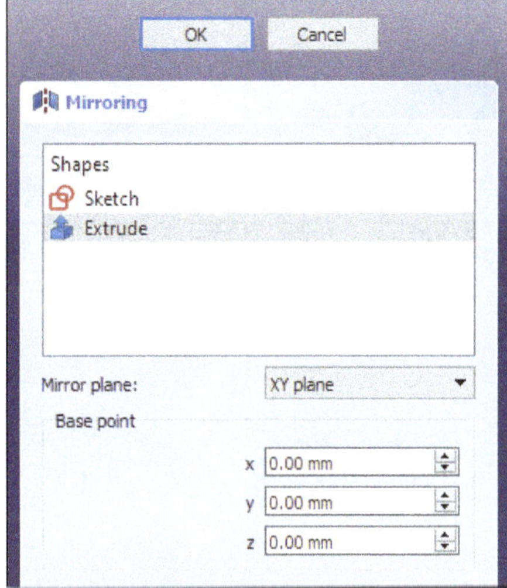

Figure-35. Mirroring dialog

- Select desired shape which you want to be mirror from **Shapes list** area of the dialog.
- Select desired mirror plane from the **Mirror plane** drop-down.
- Specify desired value in **x**, **y**, and **z** edit boxes from the **Base point** section of the dialog to move the mirror plane parallel to the selected standard mirror plane.
- Click on **OK** button from the dialog. The mirrored entity of the shape will be created; refer to Figure-36.

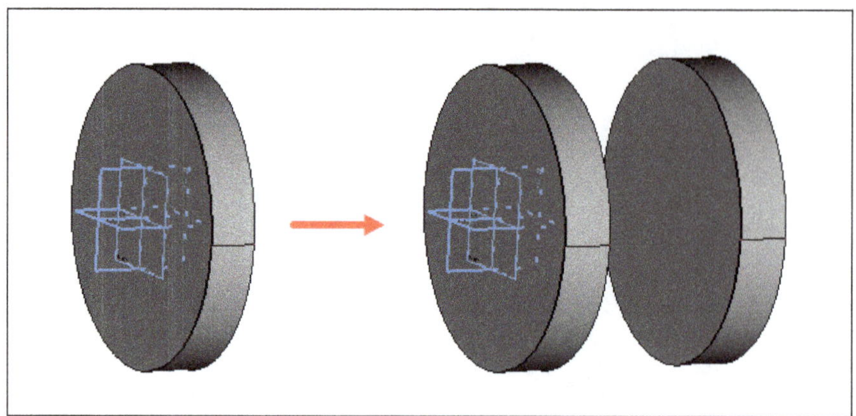

Figure-36. The mirrored entity of the shape

Fillet

The **Fillet** tool creates a fillet (round) on the selected edges of an object. The procedure to use this tool is discussed next.

- Click on the **Fillet** tool from **Toolbar** in the **Part** workbench; refer to Figure-37. The **Fillet Edges** dialog will be displayed in the **Tasks** panel of **Combo View**; refer to Figure-38.

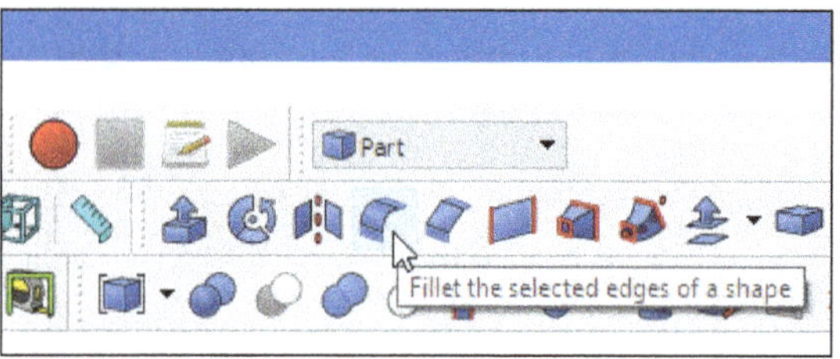

Figure-37. Fillet tool

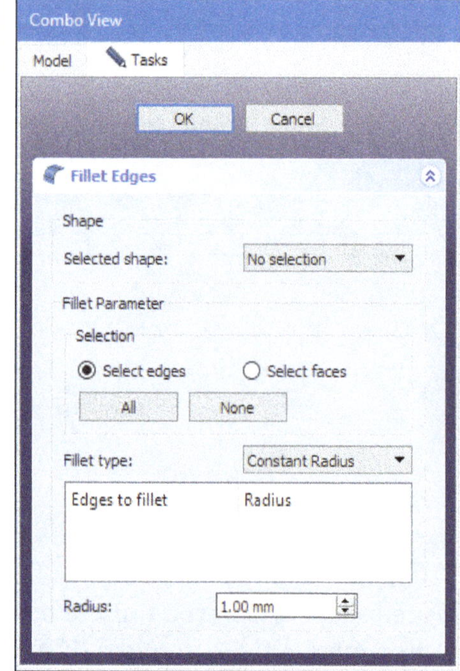

Figure-38. Fillet Edges dialog

- Click on the **Selected shape** drop-down from **Shape** section of the dialog and select the shape from the drop-down list you want to be fillet. All the edges of the selected shape will be displayed in the **Shapes list** area of the dialog.
- If you want to select the edges of the shape then select the **Select edges** radio button from **Selection** area in the **Fillet Parameter** section of the dialog and if you want to select the faces of the shape then select the **Select faces** radio button from the **Selection** area.
- Click on the **All** button from the **Selection** area to select all the edges available in the **Shapes list** area of the dialog and click on **None** button to deselect all the edges from the **Shapes** list area; refer to Figure-39.

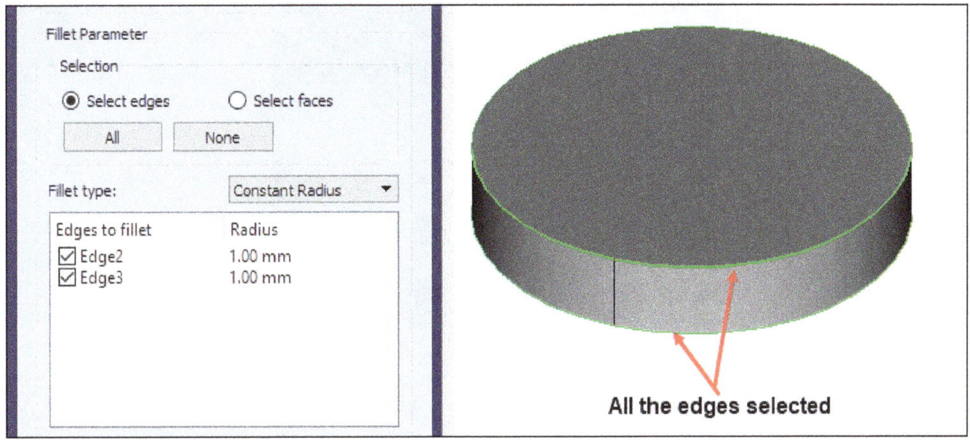

Figure-39. All the edges selected to be fillet

- Select desired type of fillet from **Fillet type** drop-down in the **Fillet Parameter** section. Select **Constant Radius** option to specify same fillet radius on both ends of the edges. Select **Variable Radius** option to specify different radius values at start and end points of the edges.
- Click on **OK** button from the dialog. The fillets will be created on the edges; refer to Figure-40.

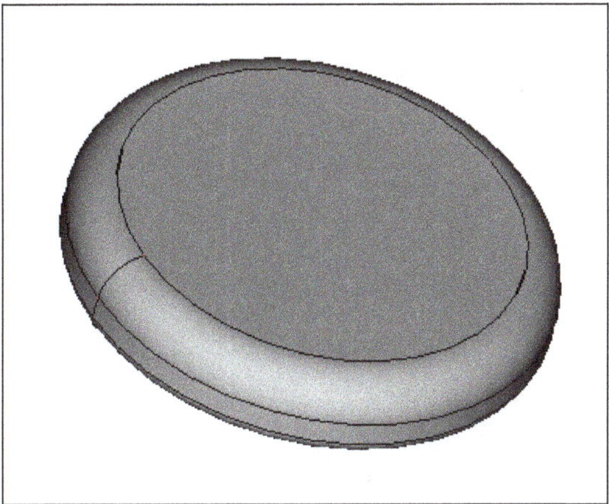

Figure-40. Fillets created

Chamfer

The **Chamfer** tool chamfers selected edges of an object. The procedure to use this tool is discussed next.

- Click on the **Chamfer** tool from **Toolbar** in the **Part** workbench; refer to Figure-41. The **Chamfer Edges** dialog will be displayed in the **Tasks** panel of **Combo View**; refer to Figure-42.

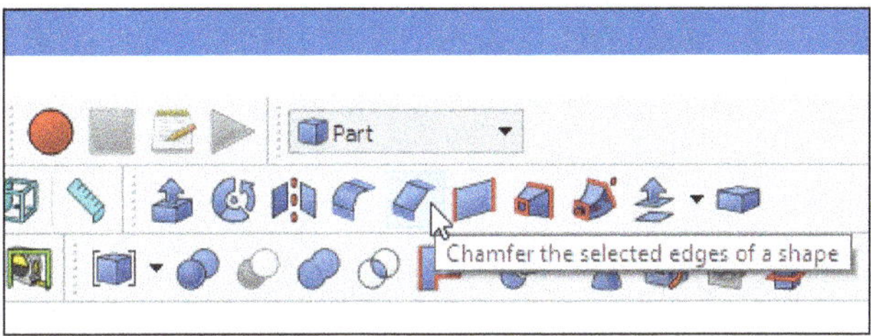

Figure-41. Chamfer tool

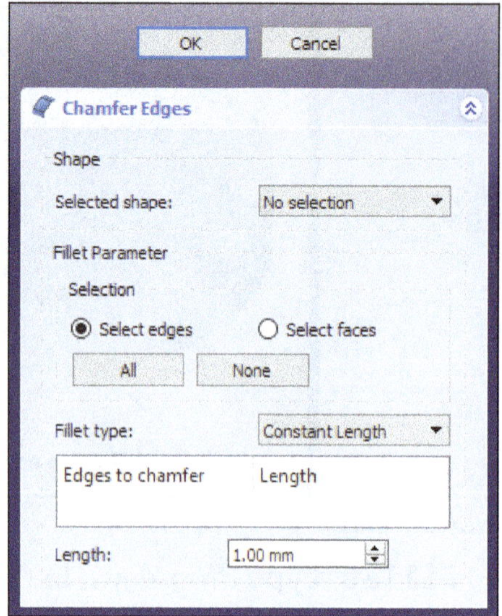

Figure-42. Chamfer Edges dialog

- The procedure to create the chamfer is same as discussed for **Fillet** tool.
- Specify the parameters and click on **OK** button from the dialog. The chamfers will be created on the edges; refer to Figure-43.

Figure-43. Chamfer created

Ruled Surface

The **Ruled Surface** tool is used to create a ruled surface using two edges or sketch curves. The procedure to use this tool is discussed next.

- Select the two edges or two wires from the 3D view area; refer to Figure-44.

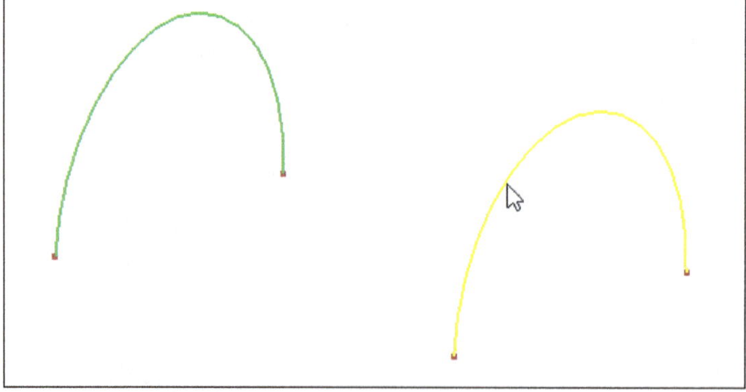

Figure-44. Selecting the wires

- Click on the **Ruled Surface** tool from **Toolbar** in the **Part** workbench; refer to Figure-45. The ruled surface will be created; refer to Figure-46.

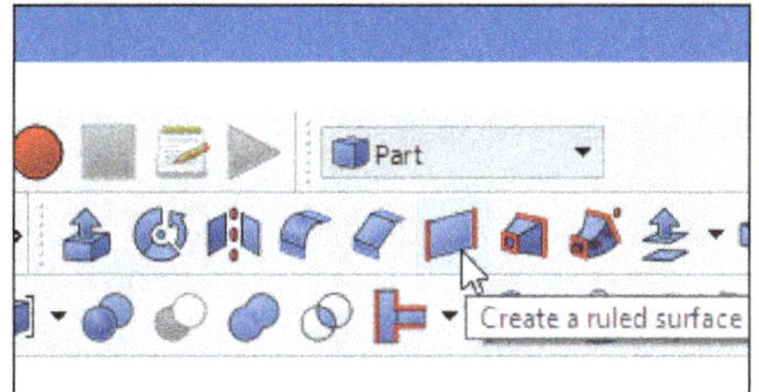

Figure-45. Ruled surface tool

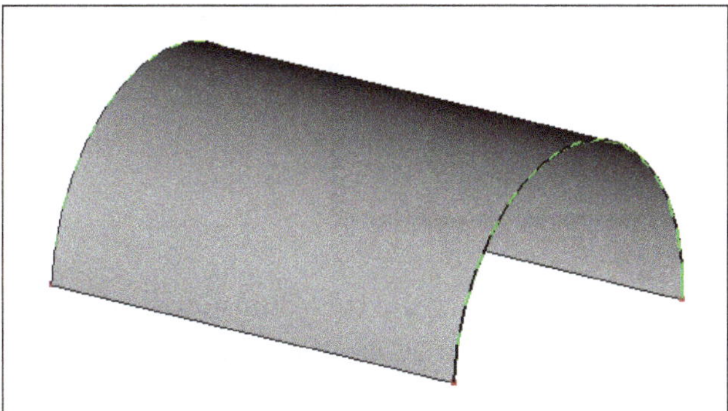

Figure-46. Ruled surface created

Loft

The **Loft** tool is used to create a face, shell, or a solid shape from two or more profiles. The profiles can be a point (vertex), line (Edge), wire, or face. Edges and wires may be either open or closed. The procedure to use this tool is discussed next.

- Click on the **Loft** tool from **Toolbar** in the **Part** workbench; refer to Figure-47. The **Loft** dialog will be displayed in the **Tasks** panel of **Combo View**; refer to Figure-48.

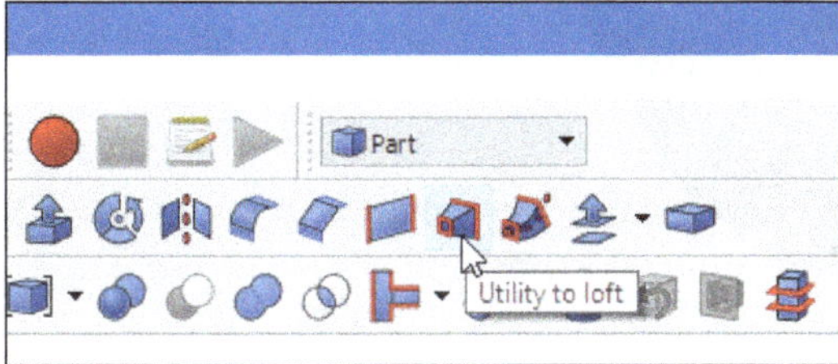

Figure-47. Loft tool

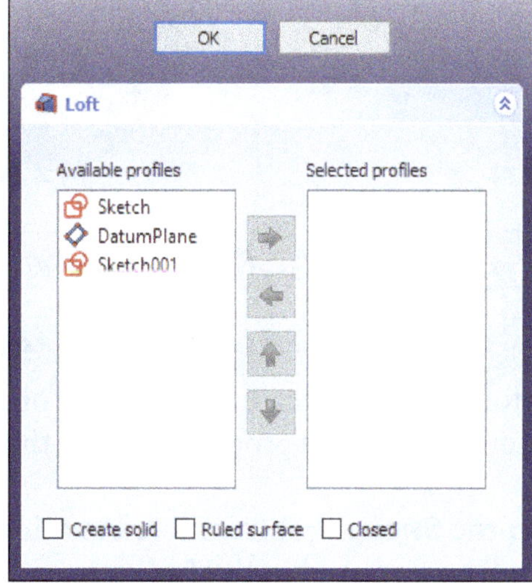

Figure-48. Loft dialog

- The list of profiles which are available to be loft will display in the **Available profiles** list area in the dialog.
- To create the loft feature, add the two or more available profiles one by one in the **Selected profiles** list area by clicking on **Add** button.
- Select desired profile from the **Selected profiles** list area as a base profile for loft; refer to Figure-49.

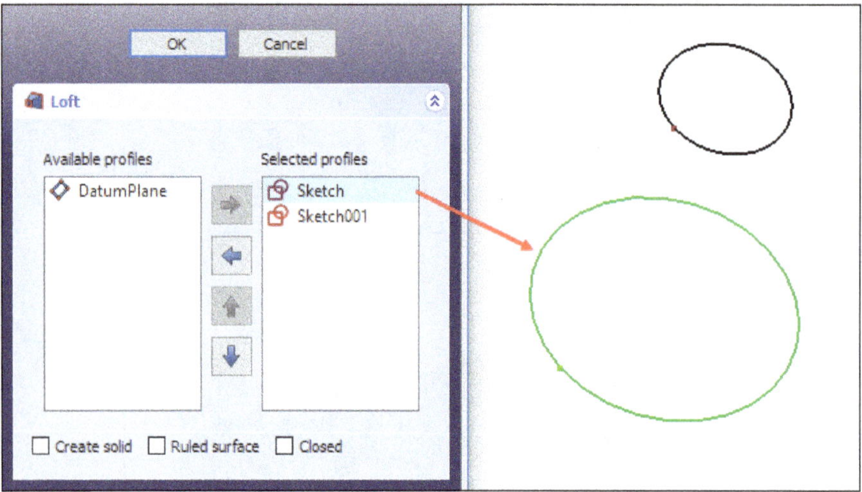

Figure-49. Selecting the base profile for loft

- Select the **Create solid** check box to create the solid model of the profiles, if the profiles are of closed geometry. If the profiles are of open geometry, it creates a face or shell of the profiles.
- Select the **Ruled surface** check box to create a face or ruled surface using the profiles.
- Select the **Closed** check box to loft the last profile to the first profile creating a closed body.
- Click on **OK** button from the dialog. The loft feature will be created; refer to Figure-50.

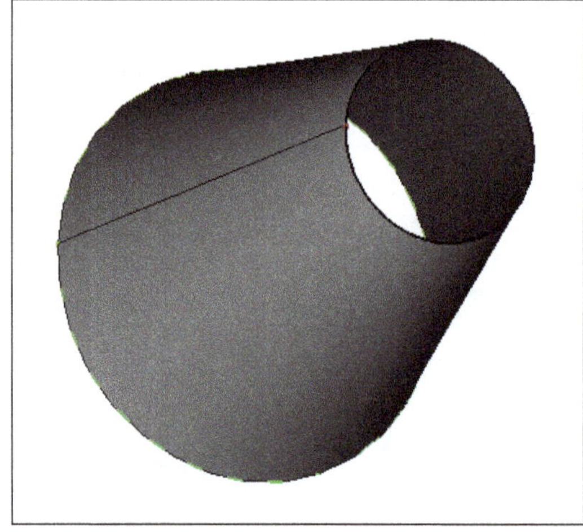

Figure-50. Loft feature created

Sweep

The **Sweep** tool is used to create a face, a shell, or a solid shape from one or more profiles (cross-sections) projected along a path. The procedure to use this tool is discussed next.

- Click on the **Sweep** tool from **Toolbar** in the **Part** workbench; refer to Figure-51. The **Sweep** dialog will be displayed in the **Tasks** panel of **Combo View**; refer to Figure-52.

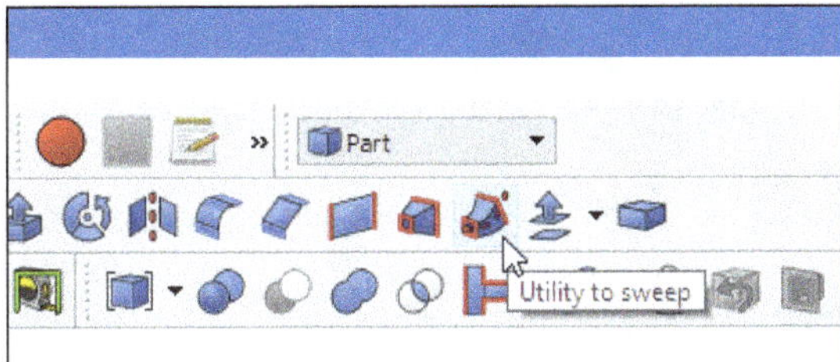

Figure-51. Sweep tool

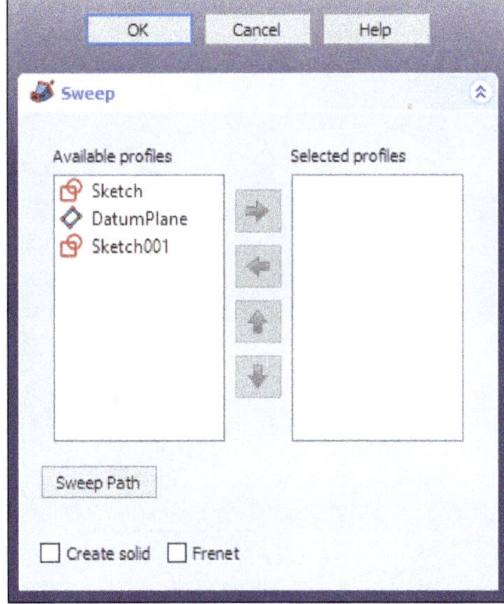

Figure-52. Sweep dialog

- Select the profile to be used as a base profile for sweep feature from the **Available profiles** list area of the dialog and add it in the **Selected profiles** list area by clicking on **Add** button.
- Select the profile from the **Selected profiles** list area and click on **Sweep Path** button. You are asked to select the profile for sweep path from the 3D view area.
- Select the profile from the 3D view area to be used as path for creating a sweep feature; refer to Figure-53.

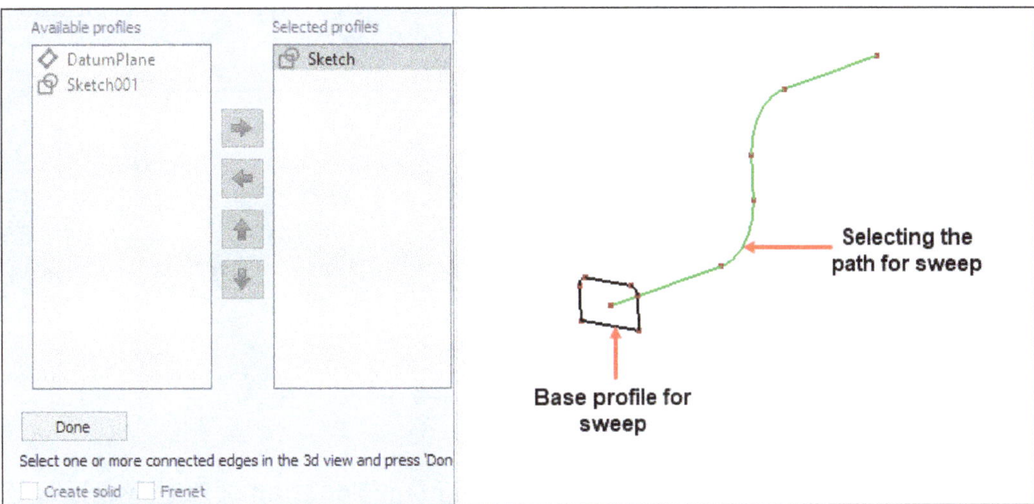

Figure-53. Selecting the path for sweep feature

- Click on **Done** button from the dialog.
- Select the **Create solid** check box to create the solid model of the profiles, if the profiles are of closed geometry. If the profiles are of open geometry, it creates a face or shell of the profiles.
- Select the **Frenet** check box to compute the profile orientation basing on local curvature and tangency vectors of the path.
- Click on **OK** button from the dialog. The sweep feature will be created; refer to Figure-54.

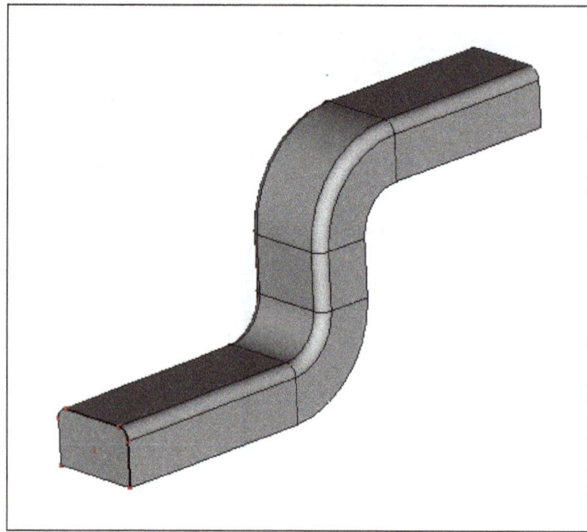

Figure-54. Sweep feature created

Offset tools

The **Offset** tools are used to create 2D as well as 3D copies of selected shapes. The tools available to create 2D and 3D offsets are discussed next.

3D Offset

The **3D Offset** tool creates parallel copies of selected shape at a certain distance from the base shape, giving a new object. The procedure to use this tool is discussed next.

- Select the object from the Model tree view or from the 3D view area of which you want to create the offset; refer to Figure-55.

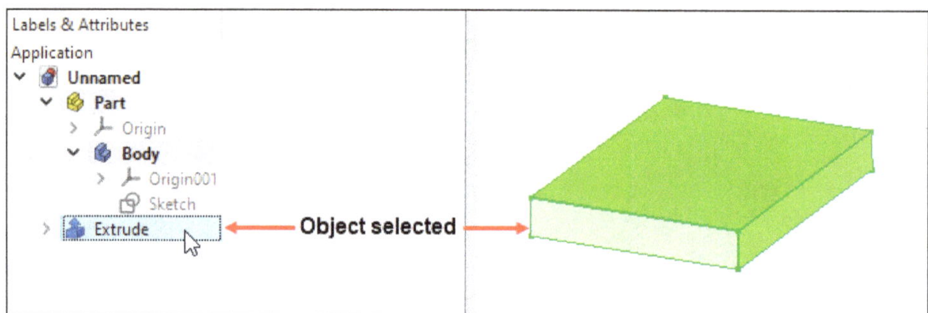

Figure-55. Object selected to create the 3D offset

- Click on the **3D Offset** tool from **Offset** drop-down in the **Toolbar** of **Part** workbench; refer to Figure-56. The **Offset** dialog will be displayed in the **Tasks** panel of **Combo View** along with the preview of offset creation; refer to Figure-57.
- Select the mode of offset creation from the **Mode** drop-down. Select the **Skin** option to create a new shape around the source shape.
- Select desired option from the **Join type** drop-down. Select the **Arc** and **Tangent** option to create the offset of rounded corners. Select the **Intersection** option to create the offset of sharp corners by linear extension of the edges.
- Select the **Intersection** check box to allow the offsets pointing inwards to overflood the gap by intersecting the resulting shape until opposite faces are reached.
- Select the **Fill offset** check box while using the 2 dimensional shape in which the gap in between the two shapes gets filled.
- Click on **OK** button from the dialog. The 3D offset of the original object will be created; refer to Figure-58.

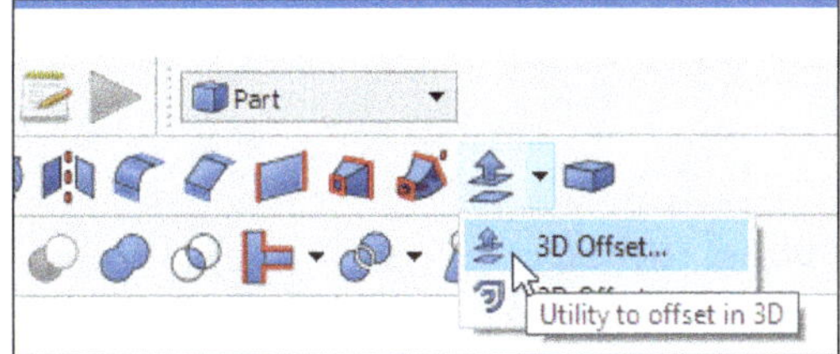

Figure-56. 3D Offset tool

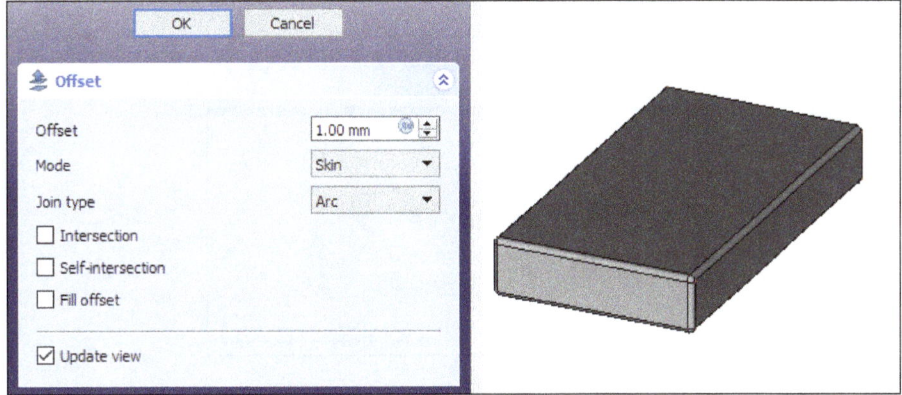

Figure-57. 3D Offset dialog with the preview of offset creation

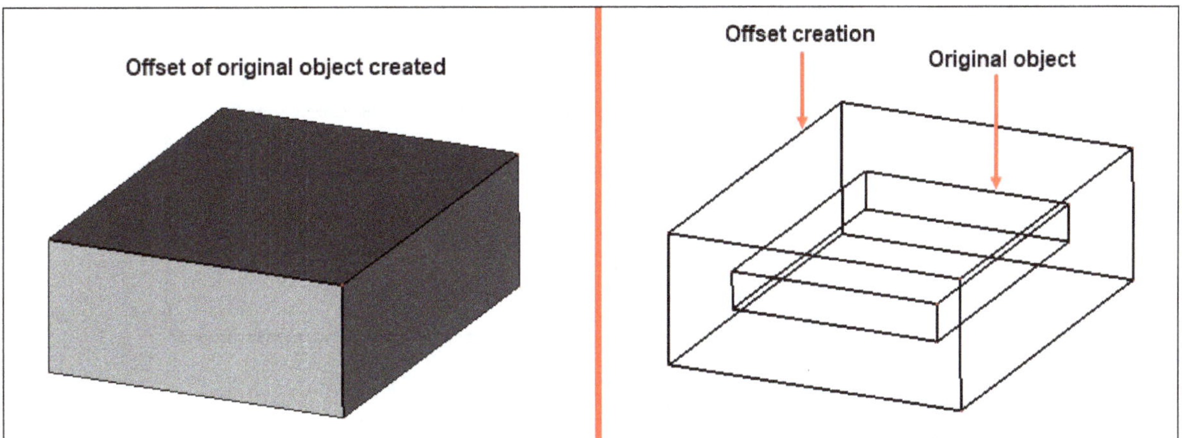

Figure-58. 3D offset of original object created

2D Offset

The **2D Offset** tool constructs a wire parallel to the original wire at a certain distance from it or enlarges/ shrinks a planar face. The procedure to use this tool is discussed next.

- Select the object from the Model tree view or from the 3D view area which you want to offset; refer to Figure-59.
- Click on the **2D Offset** tool from the **Offset** drop-down in the **Toolbar** of **Part** workbench; refer to Figure-60. The **Offset** dialog will be displayed in the **Tasks** panel of **Combo View** along with the preview of offset creation; refer to Figure-61.

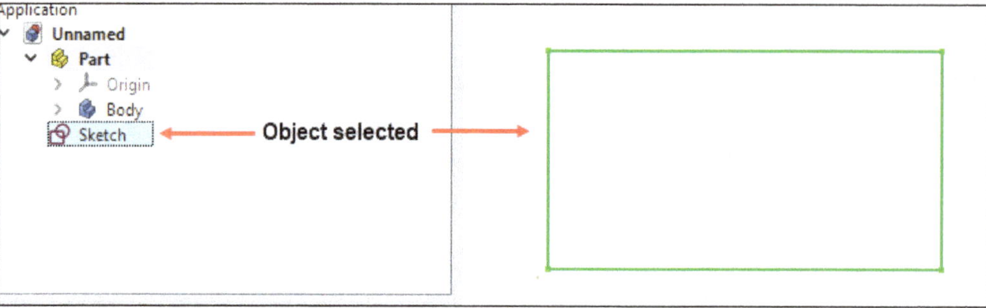

Figure-59. Object selected to create the 2D offset

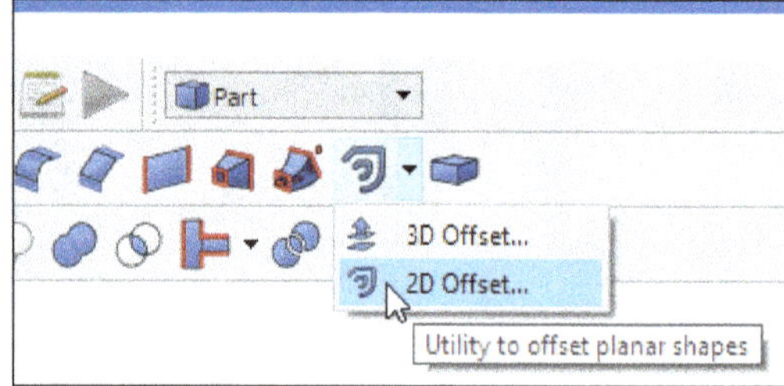

Figure-60. 2D Offset tool

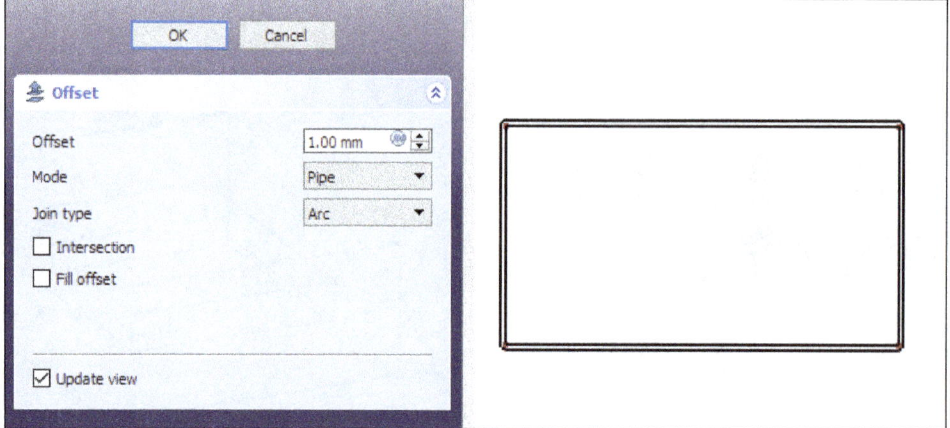

Figure-61. 2D Offset dialog with the preview of offset creation

- The parameters in the 2D **Offset** dialog are same as discussed for **3D Offset** tool.
- Specify the parameters and click on **OK** button from the dialog. The 2D offset of the original object will be created; refer to Figure-62.

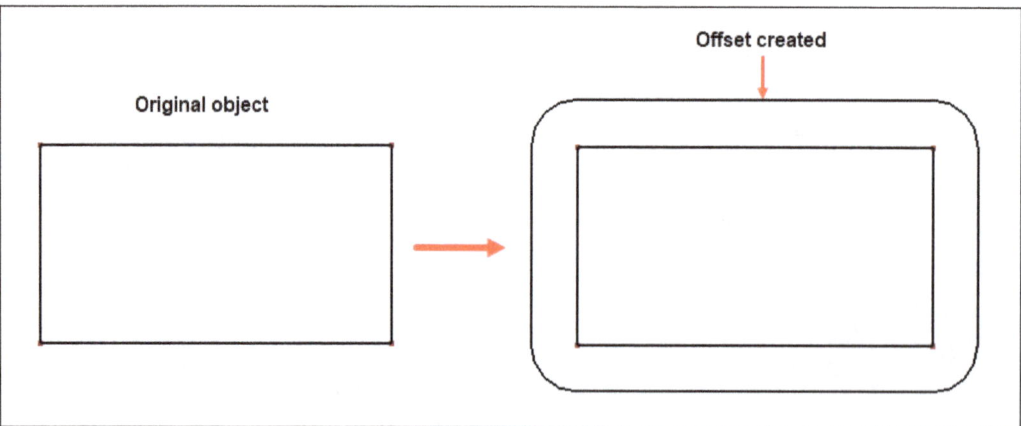

Figure-62. 2D offset of original object created

Thickness

The **Thickness** tool works on a solid shape and transforms it into a hollow object giving each of its faces a defined thickness. The procedure to use this tool is discussed next.

- Select one or more faces of the solid object from the 3D view area which you want to remove after applying thickness; refer to Figure-63.

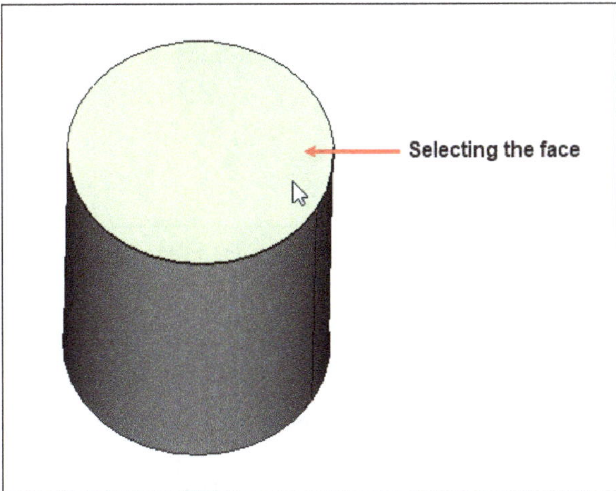

Figure-63. Selecting the face of solid object

- Click on the **Thickness** tool from **Toolbar** in the **Part** workbench; refer to Figure-64. The **Thickness** dialog will be displayed in the **Tasks** panel of **Combo View** along with the preview of thickness defined to the object; refer to Figure-65.

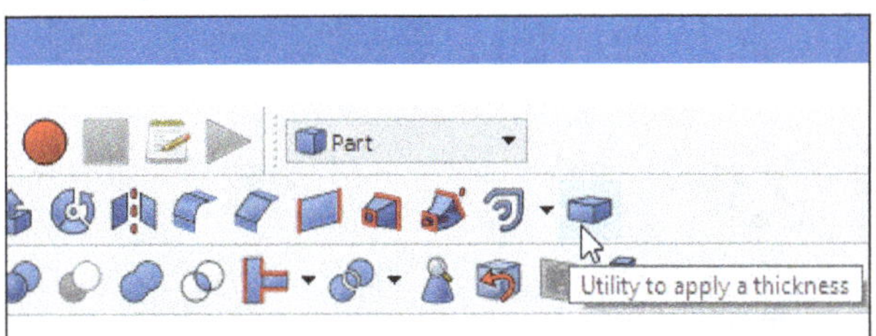

Figure-64. Thickness tool

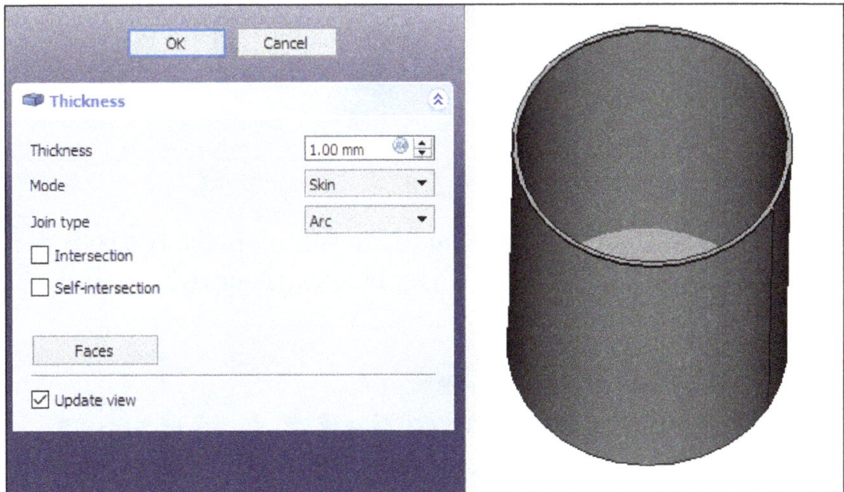

Figure-65. Thickness dialog with the preview of thickness defined

- The parameters of **Thickness** dialog are same as discussed for **Offset** tools.
- Specify the parameters and click on **OK** button from the dialog. The thickness will be defined to the object; refer to Figure-66.

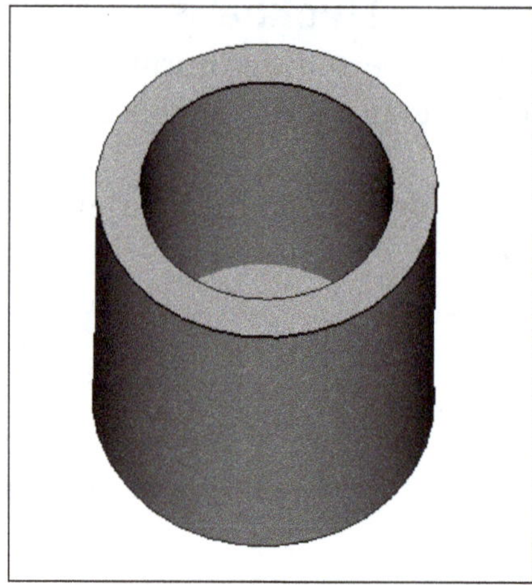

Figure-66. Thickness defined to the object

Compound tools

The **Compound** tools are used to create a set of shapes, to split up the shapes, or to extract individual pieces from split shapes. These tools are discussed next.

Make Compound

The **Make Compound** tool creates a compound of any kind of topological shapes. These can be solids, meshes, or any other kind of topological shapes. A compound is a set of shapes grouped into one object. The procedure to use this tool is discussed next.

- Select the topological shapes from the Model tree view to be added to the compound; refer to Figure-67.

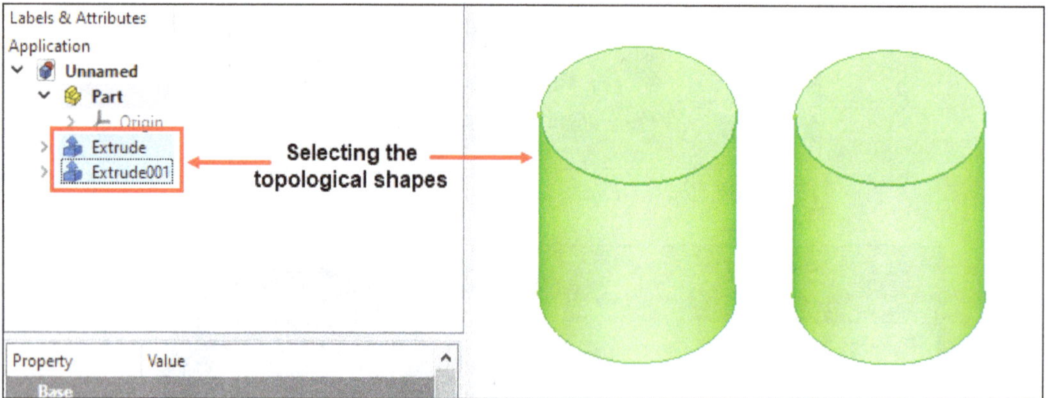

Figure-67. Selecting the topological shapes

- Click on the **Make compound** tool from **Compound** drop-down in the **Toolbar** of **Part** workbench; refer to Figure-68. The compound of the selected topological shapes will be created; refer to Figure-69.

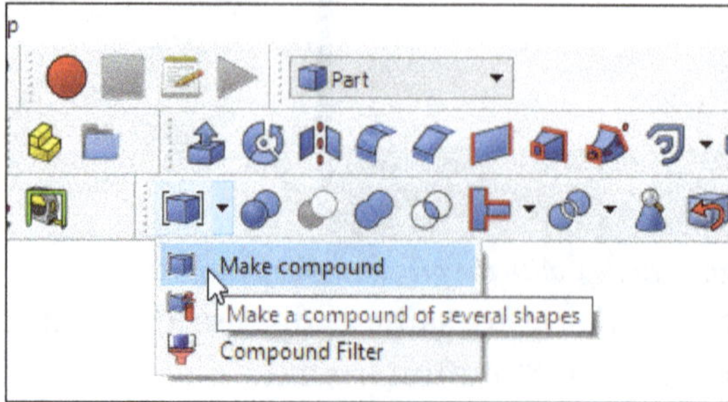

Figure-68. Make compound tool

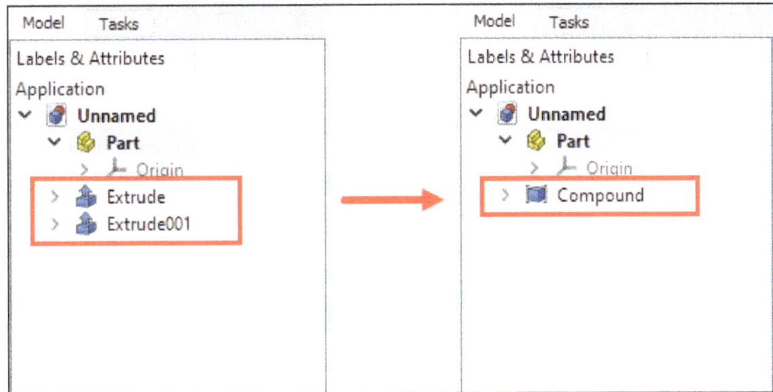

Figure-69. Compound of the shapes created

Explode Compound

The **Explode compound** tool is used to split up compounds of shapes to make each contained shape available as a separate object in Model tree view. The procedure to use this tool is discussed next.

- Select the compound body to be split into individual objects from the Model tree view; refer to Figure-70.

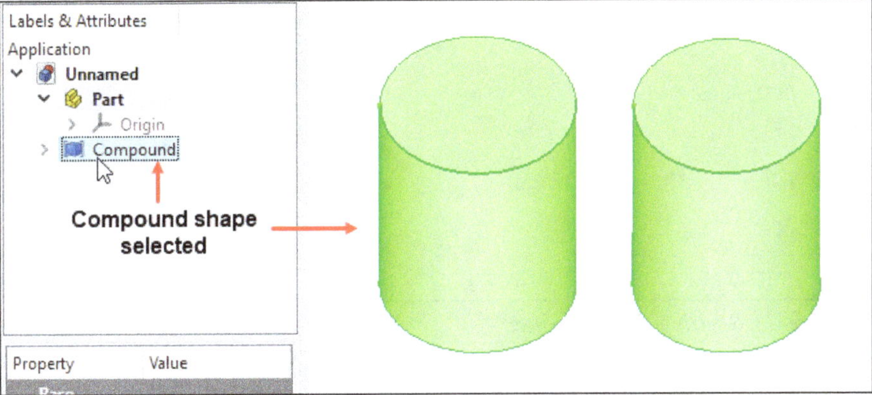

Figure-70. Compound shape selected

- Click on the **Explode compound** tool from **Compound** drop-down in the **Toolbar** of **Part** workbench; refer to Figure-71. The Compound shape will explode to generate separate objects; refer to Figure-72.

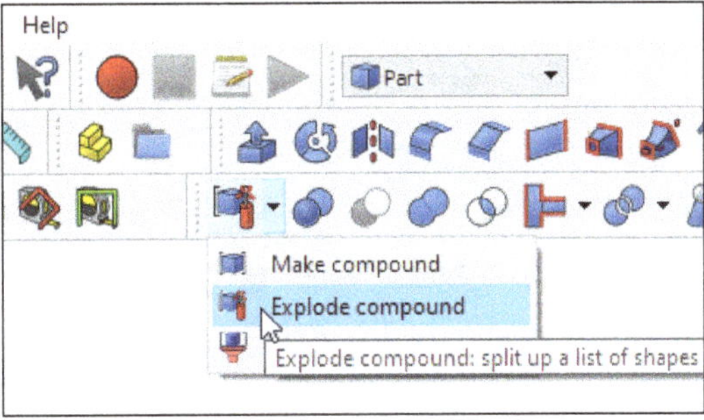

Figure-71. Explode compound tool

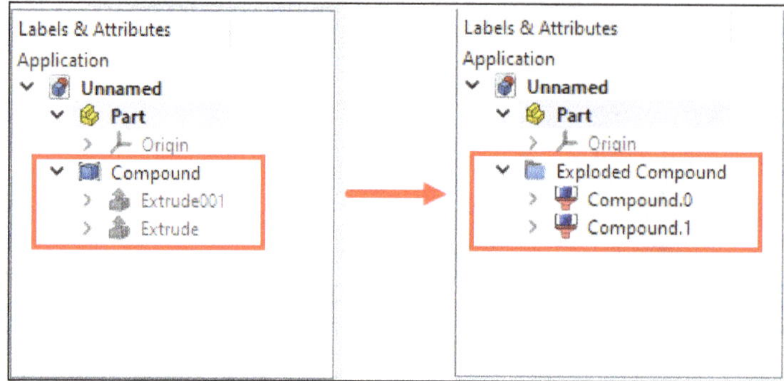

Figure-72. Compound shape exploded

Compound Filter

The **Compound Filter** tool can be used to extract individual pieces of the result of a **Part Slice** operation with which you have split an object. The procedure to use this tool is discussed next.

- Select the sliced object from Model tree view or from the 3D view area to extract the individual pieces; refer to Figure-73.

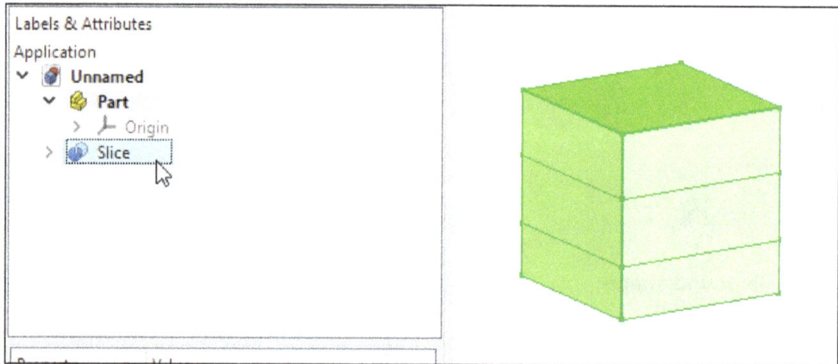

Figure-73. Selecting the sliced object

- Click on the **Compound Filter** tool from **Compound** drop-down in the **Toolbar** of **Part** workbench; refer to Figure-74. The compound filter object will be created; refer to Figure-75.

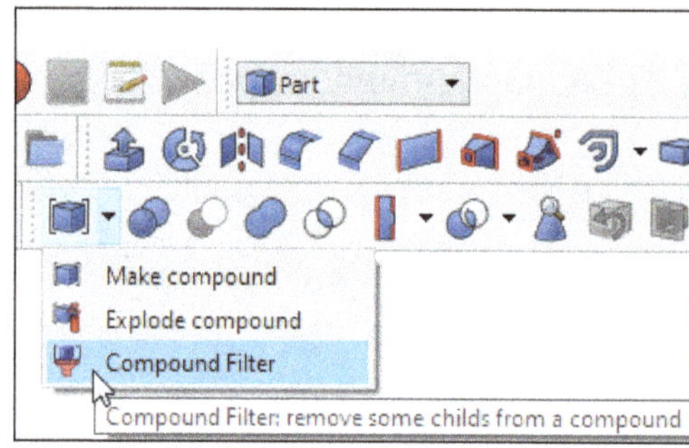

Figure-74. Compound Filter tool

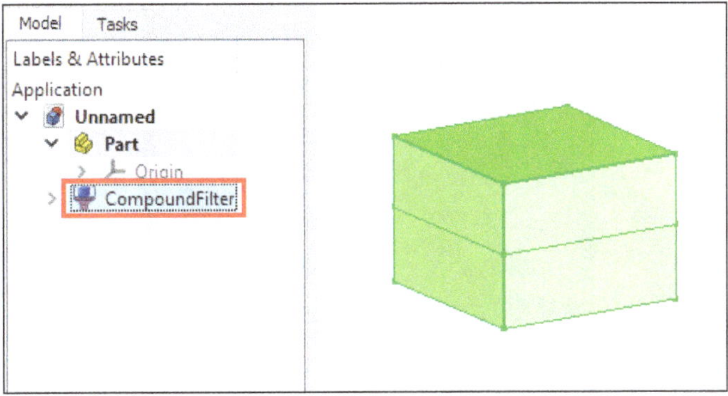

Figure-75. Compound filter object created

- Select the compound filter object from the Model tree view. The **Property editor** dialog will be displayed in the **Model** tab of **Combo View** as discussed earlier.
- Select the **specific items** option from **Filter Type** drop-down in the **Compound Filter** section of the dialog.
- Specify the number of elements you want to extract in the **items** edit box of the dialog. For a single piece, this is a number starting with 0, i.e. if you want to extract the first element enter 0 in this edit box, 1 for the next element.
- After specifying the parameters, click **LMB** in the 3D view area. The element will be extracted; refer to Figure-76.

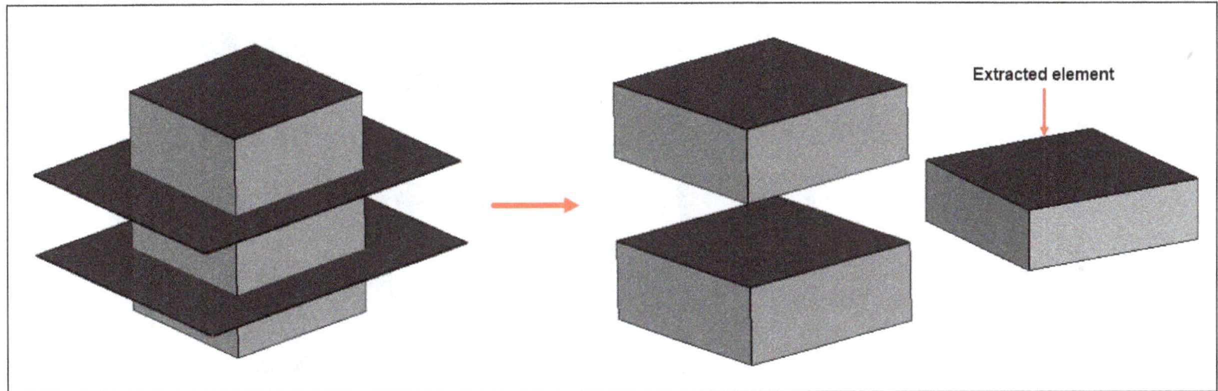

Figure-76. Element extracted

Boolean

The **Boolean** tool is a generic all-in-one boolean tool. It allows you to specify the objects and operation to perform via a single dialog. The procedure to use this tool is discussed next.

- Click on the **Boolean** tool from **Toolbar** in the **Part** workbench; refer to Figure-77. The **Boolean Operation** dialog will be displayed in the **Tasks** panel of **Combo View**; refer to Figure-78.

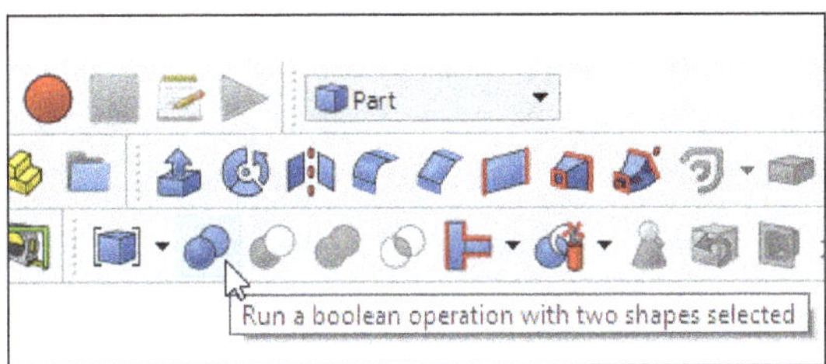

Figure-77. Boolean tool

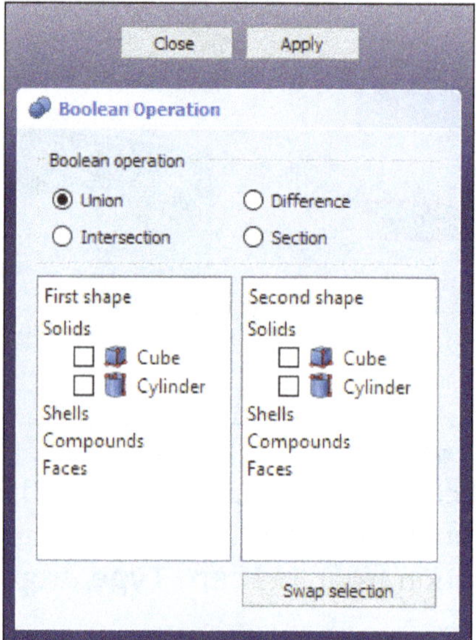

Figure-78. Boolean Operation dialog

- The objects available to perform boolean operation will display in the both **First shape** list and **Second shape** list area of the dialog; refer to Figure-79.

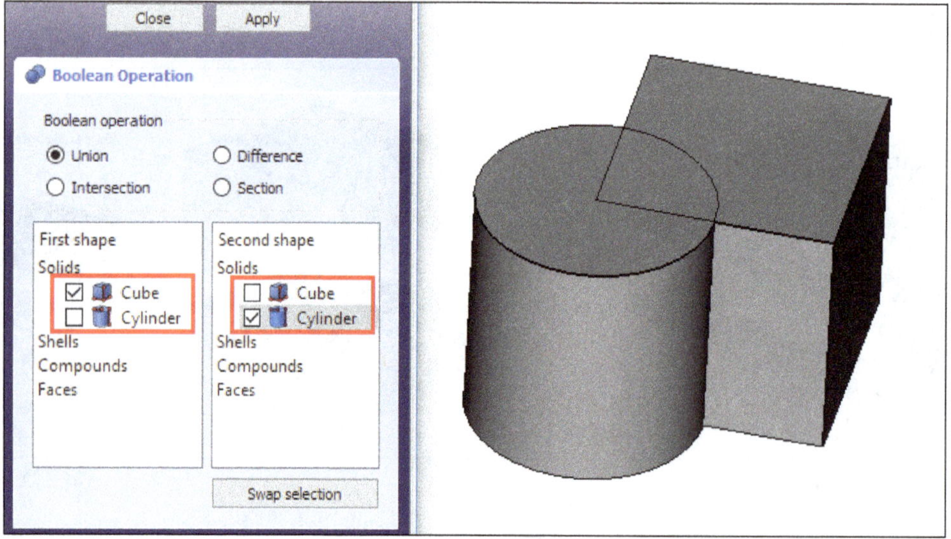

Figure-79. Objects to perform boolean operation

- Select the first object to be used in boolean operation from the **First shape** list area and select the second object from the **Second shape** list area.
- If you want to swap the selection of objects in the shape lists area then click on the **Swap selection** button at the bottom of the dialog.
- Select the type of boolean operation you want to perform from the **Boolean operation** area of the dialog. Select the **Union** radio button if you want to unite the selected part objects into one. Select the **Difference** radio button if you want to subtract second object from first object. Select the **Intersection** radio button if you want to extract the common part between selected part objects. Select the **Section** radio button if you want to extract a section from the intersection of two selected shapes.
- Click on **Apply** button from the dialog. The boolean operation will be performed; refer to Figure-80.
- Click on **Close** button to close the dialog.

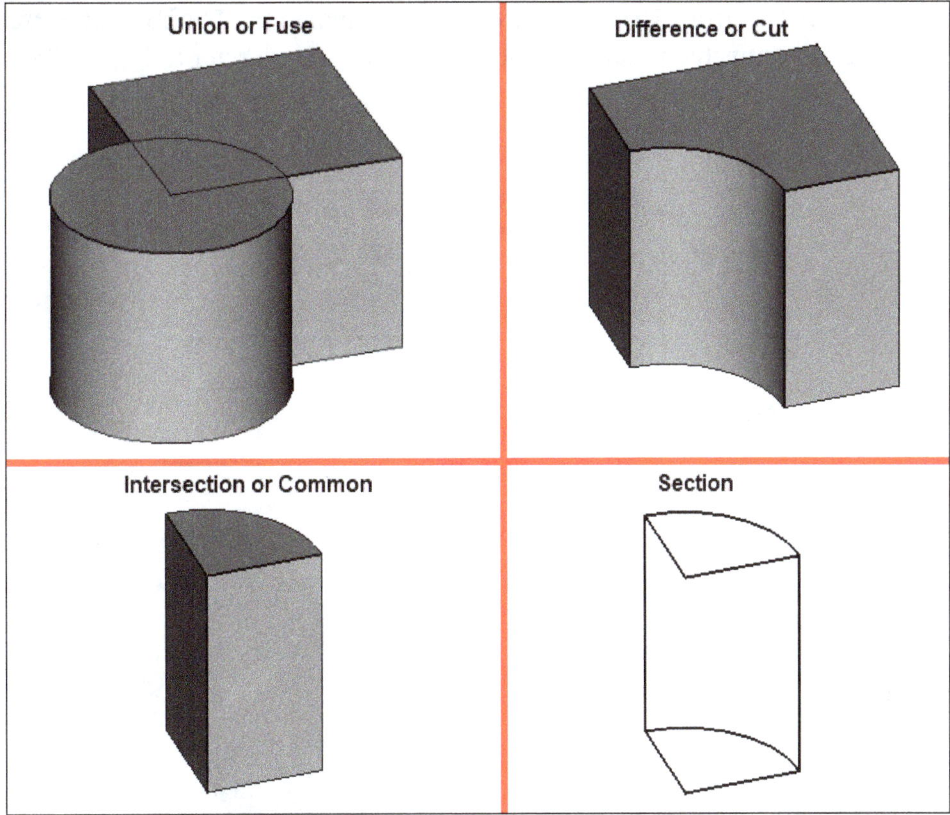

Figure-80. Types of boolean operation

If you want the quicker access to these operations, use the **Fuse** tool, **Cut** tool, **Common** tool, and **Section** tool from **Toolbar** in the **Part** workbench.

Join Features tools

The **Join Features** tools are used to connect, embed, and cutout the walled objects. These tools are discussed next.

Connect

The **Connect** tool connects the interiors of two walled objects (e.g., pipes). It can also join shells and wires. The procedure to use this tool is discussed next.

- Select two or more objects from the Model tree view or from the 3D view area which you want to connect; refer to Figure-81.

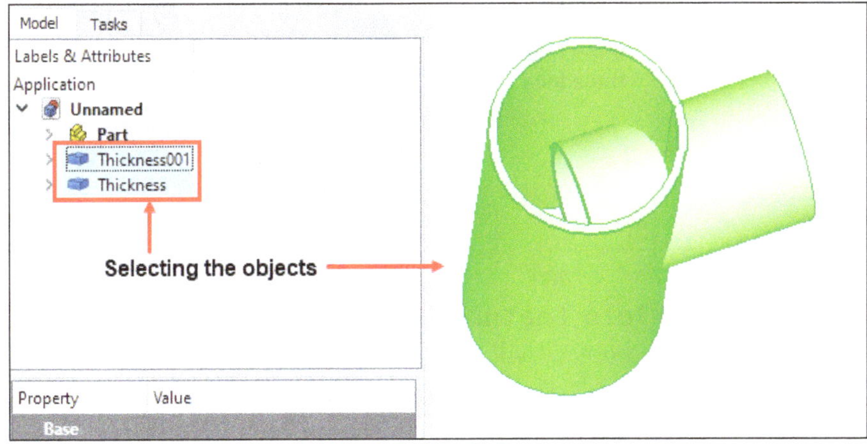

Figure-81. Selecting the objects to connect

- Click on the **Connect** tool from **Join Features** drop-down in the **Toolbar** of **Part** workbench; refer to Figure-82. A connected parametric object will be created; refer to Figure-83.

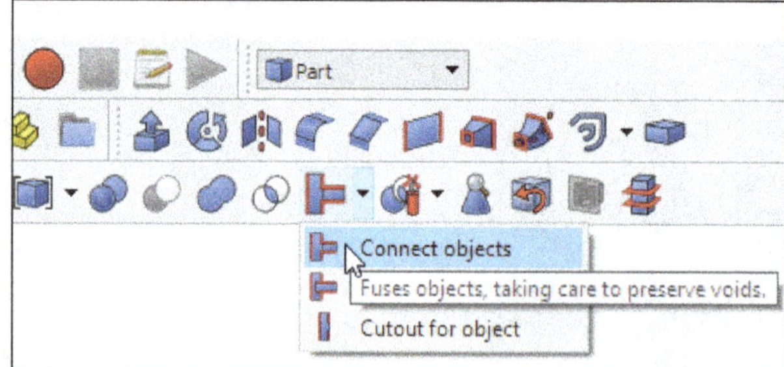

Figure-82. Connect tool

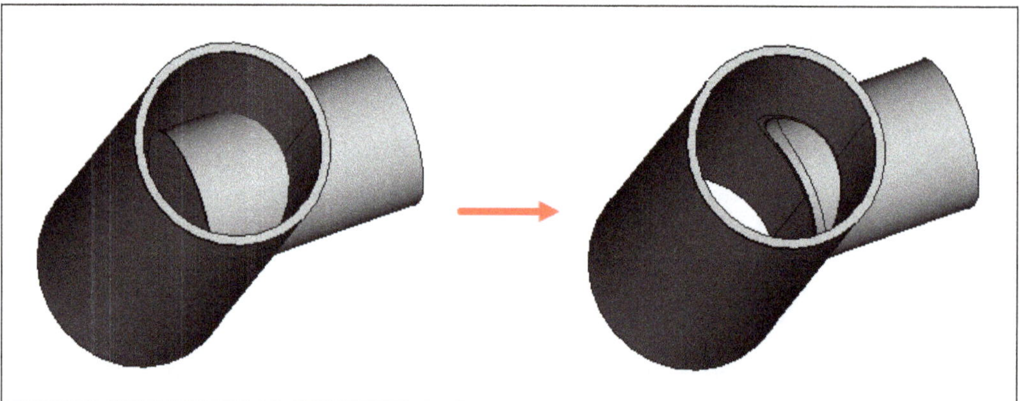

Figure-83. Connect objects created

Embed

The **Embed** tool embeds a walled object (e.g., a pipe) into another walled object. The procedure to use this tool is discussed next.

- Select the base object first and then select the object to be embedded from the Model tree view or from the 3D view area; refer to Figure-84.

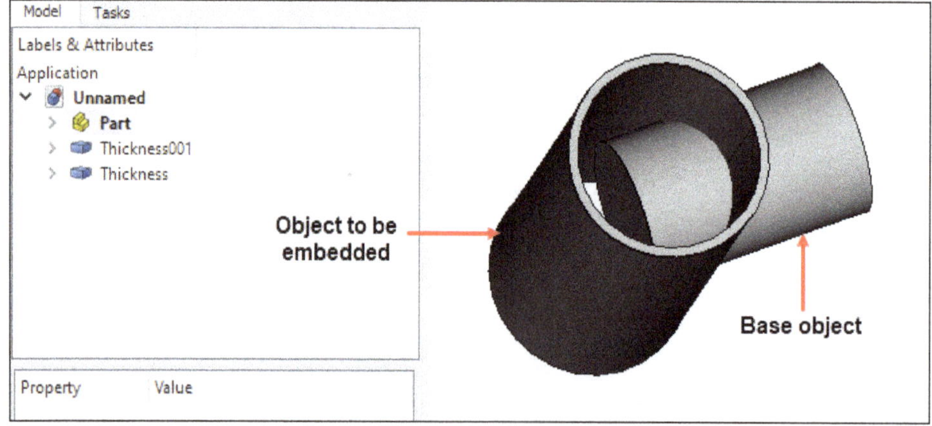

Figure-84. Selection of objects to embed

- Click on the **Embed** tool from **Join Features** drop-down in the **Toolbar** of **Part** workbench; refer to Figure-85. The embeded object will be created; refer to Figure-86.

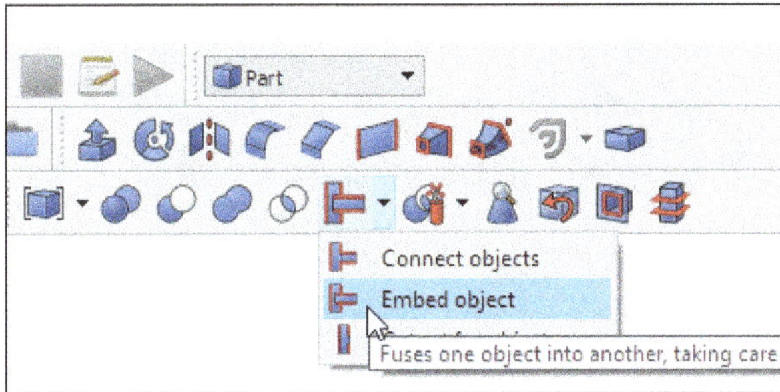

Figure-85. Embed tool

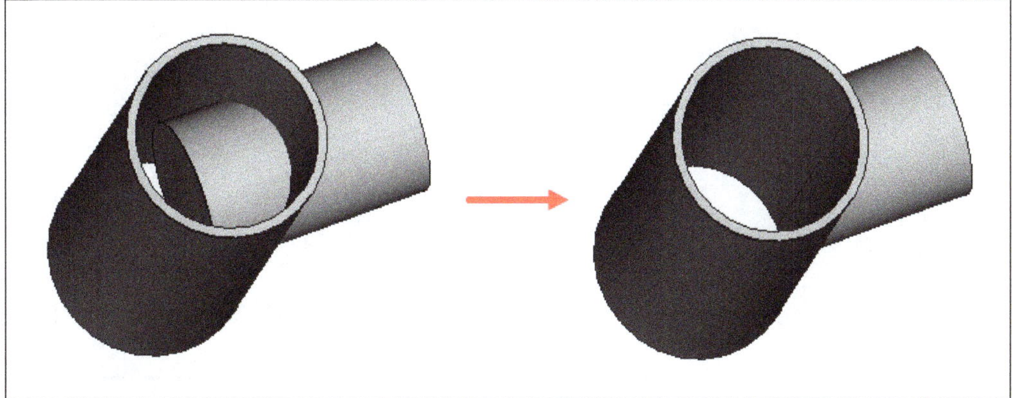

Figure-86. Embed object created

Cutout

The **Cutout** tool creates a cutout in a walled object (e.g., a pipe) to fit another walled object. The procedure to use this tool is discussed next.

- Select the base object first, and then select the object to define the cutout from the Model tree view or from the 3D view area; refer to Figure-87.

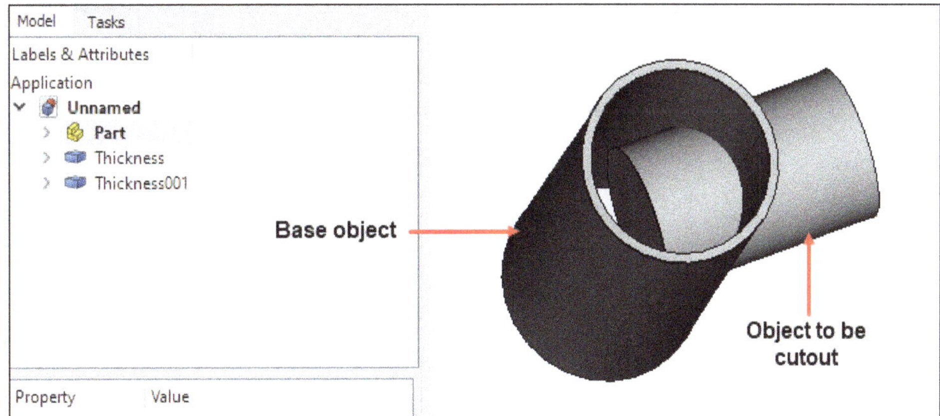

Figure-87. Selection of objects to be cutout

- Click on the **Cutout** tool from **Join Features** drop-down in the **Toolbar** of **Part** workbench; refer to Figure-88. The cutout object will be created; refer to Figure-89.

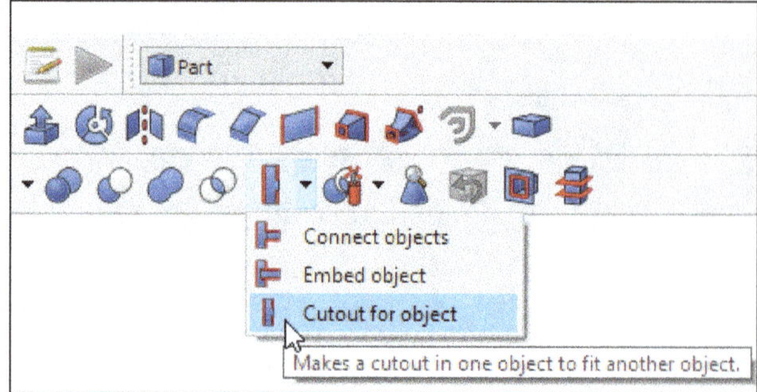

Figure-88. Cutout tool

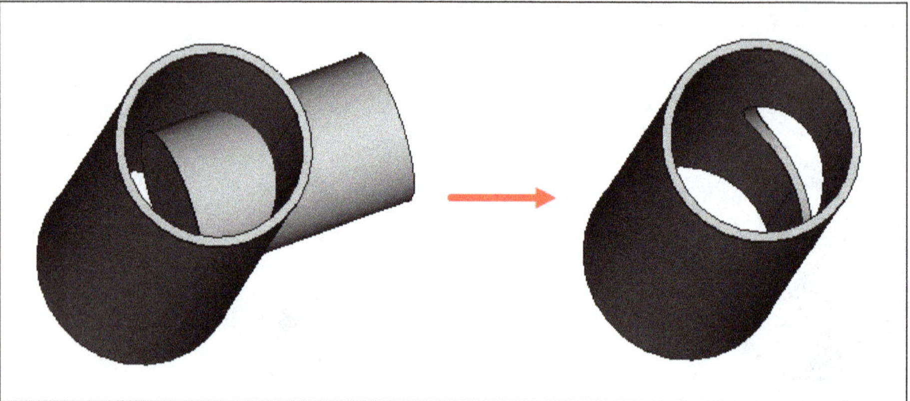

Figure-89. Cutout object created

Splitting tools

The **Splitting** tools are used to compute boolean fragments and split with other shapes. These tools are discussed next.

Boolean Fragments

The **Boolean fragments** tool is used to compute all fragments that can result from applying boolean operations between input shapes. The procedure to use this tool is discussed next.

- Select the objects to create boolean fragments from the Model tree view or from the 3D view area; refer to Figure-90.

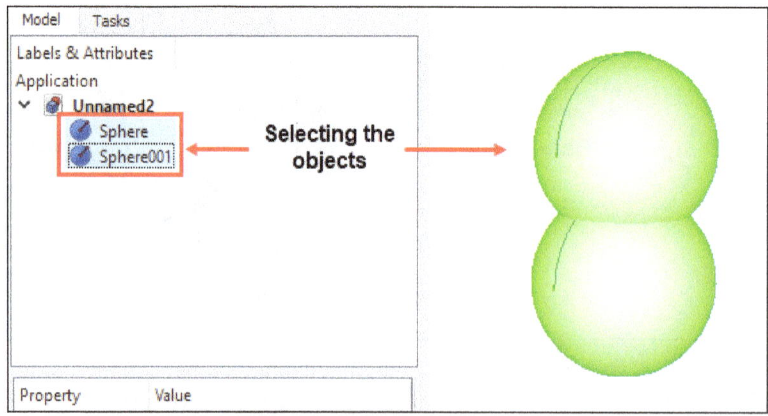

Figure-90. Selecting the objects to create boolean fragments

- Click on the **Boolean fragments** tool from **Splitting tools** drop-down in the **Toolbar** of **Part** workbench; refer to Figure-91. The boolean fragments of the objects will be created; refer to Figure-92.

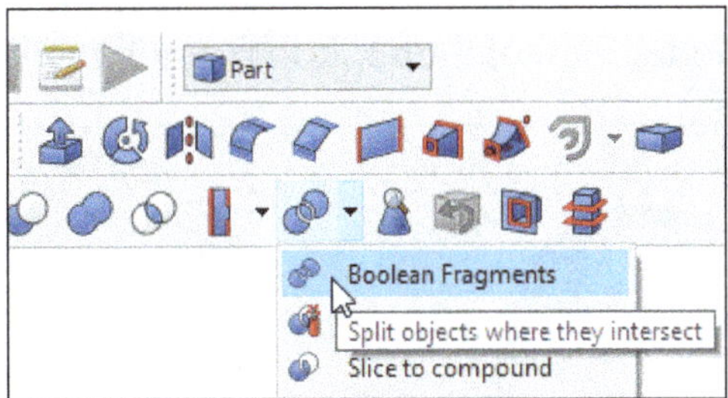

Figure-91. Boolean fragments tool

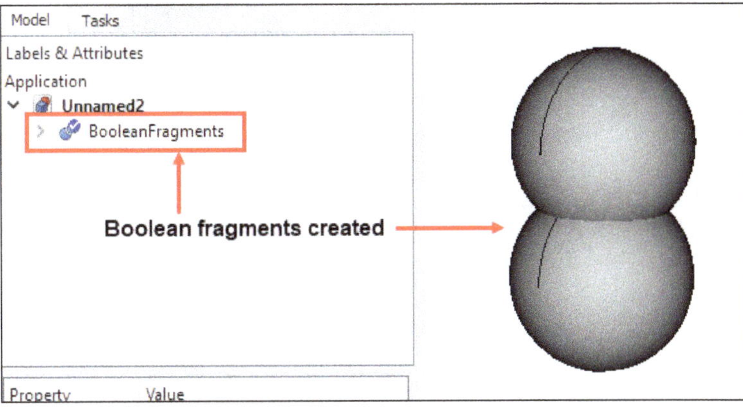

Figure-92. Boolean fragments created

- The individual pieces of the boolean fragments will not be displayed immediately as the pieces remain grouped together.
- If you want to extract the individual pieces of the object then explode the object using **Downgrade** tool.
- Select both the objects from 3D view area and click on the **Downgrade** tool from **Toolbar** in the **Draft** workbench; refer to Figure-93. The exploded pieces of the boolean fragments will be displayed in the Model tree view; refer to Figure-94.

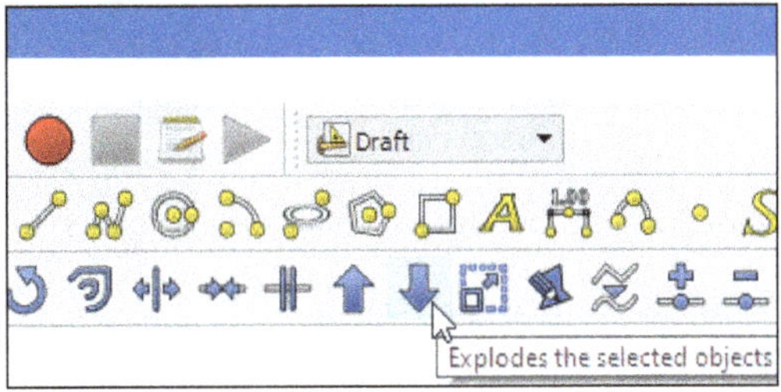

Figure-93. Downgrade tool

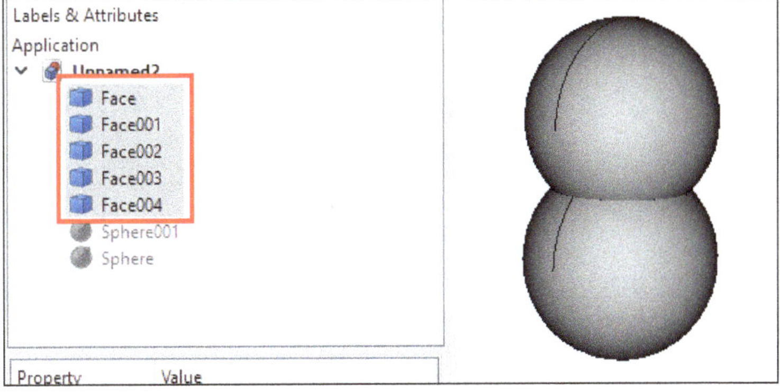

Figure-94. Exploded pieces of the boolean fragments

- Select the pieces of the object one by one from 3D view area and move apart manually using the **Property editor**. The exploded view of the boolean fragments will be displayed; refer to Figure-95.

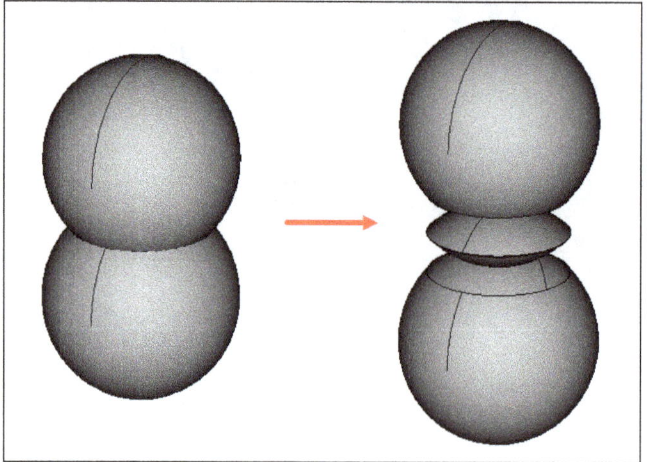

Figure-95. Exploded view of the boolean fragments

Slice Apart

The **Slice apart** tool is used to split shapes by intersection with other shapes. The procedure to use this tool is discussed next.

- Select the base object first which is to be sliced and then select the other object to slice with from the Model tree view or from the 3D view area; refer to Figure-96.

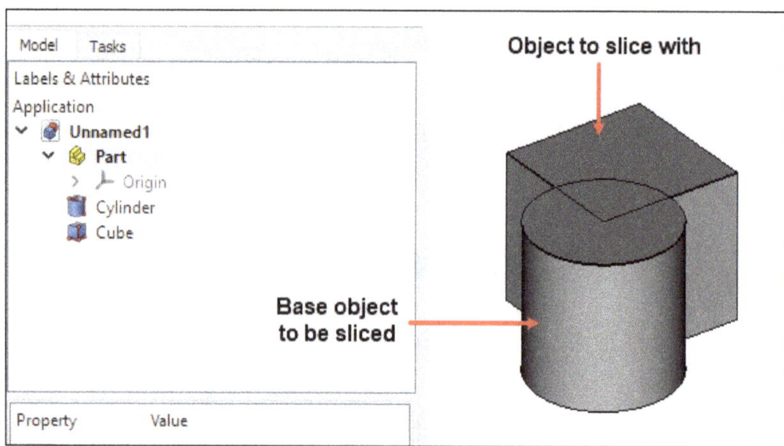

Figure-96. Selection of objects to slice apart

- Click on the **Slice apart** tool from **Splitting tools** drop-down in the **Toolbar** of **Part** workbench; refer to Figure-97. The sliced part of base object will be created; refer to Figure-98.

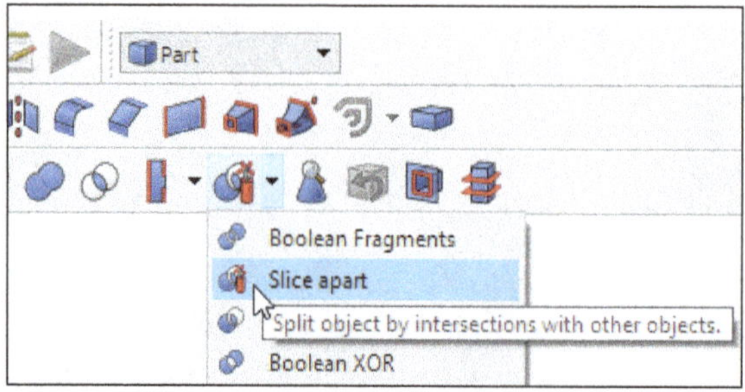

Figure-97. Slice apart tool

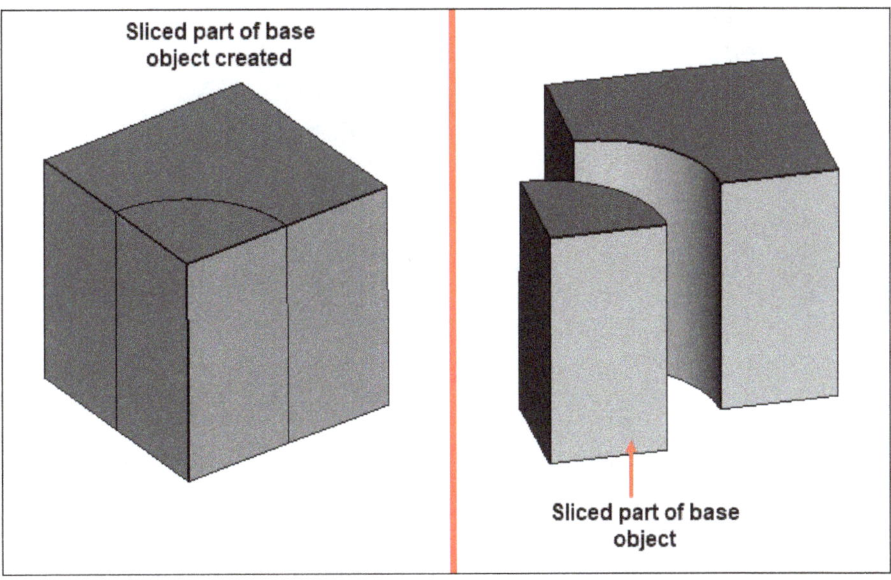

Figure-98. Sliced part of base object created

Slice

The **Slice** tool also known as **Slice to compound** tool is used to split shapes by intersection with other shapes. The procedure to use this tool is discussed next.

- First select the base object to be sliced and then select the other object to slice with from the Model tree view or from the 3D view area; refer to Figure-99.

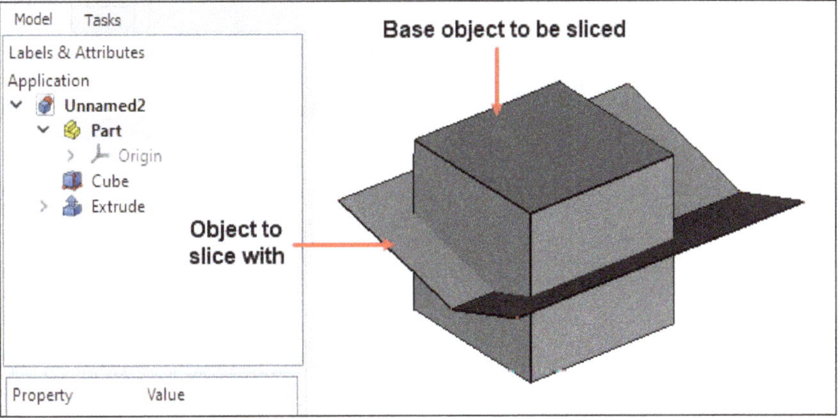

Figure-99. Selection of objects to slice

- Click on the **Slice** tool from **Splitting tools** drop-down in the **Toolbar** of **Part** workbench; refer to Figure-100. The sliced object will be created; refer to Figure-101.

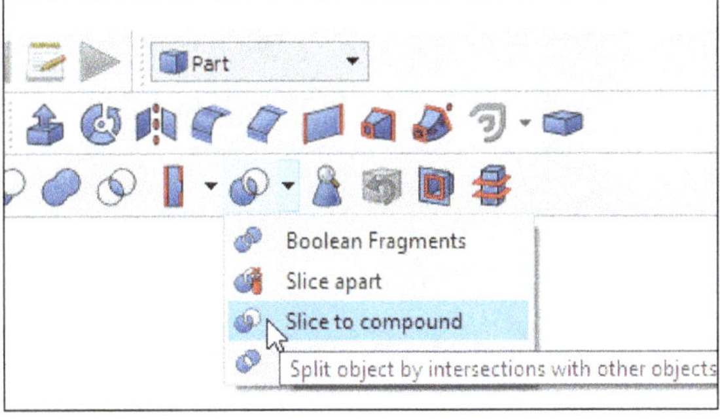

Figure-100. Slice tool

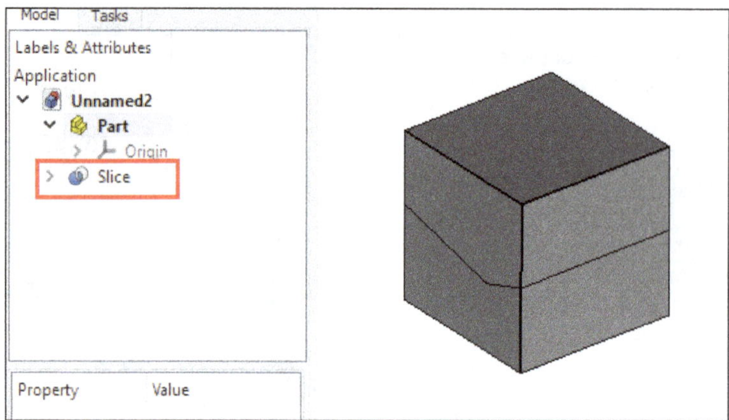

Figure-101. Sliced object created

- If you want to break the sliced object into individual pieces then select the object from the Model tree view and explode the object using **Explode compound** tool which we have discussed earlier. The sliced object will be exploded.
- If you want to view the exploded slice object individually then select the object from the Model tree view or from the 3D view area and move apart manually using the **Property editor**. The exploded view of sliced object will be displayed; refer to Figure-102.

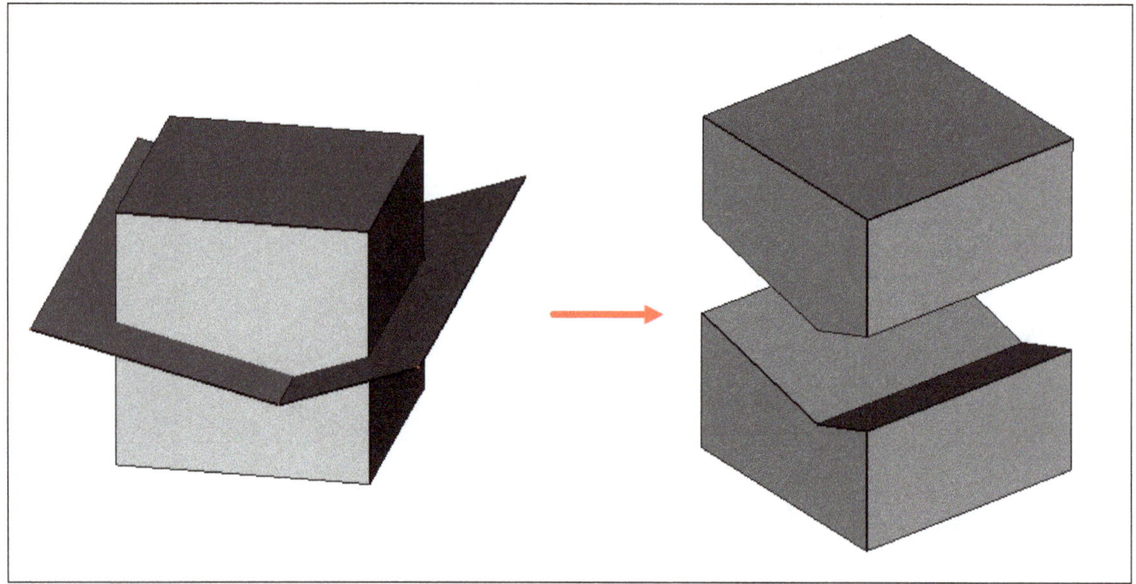

Figure-102. Exploded view of sliced object

Check Geometry

The **Check Geometry** tool runs a verification and reports if geometry is a valid solid. The procedure to use this tool is discussed next.

- Select the geometry from the Model tree view which you want to check for valid solid; refer to Figure-103.

Figure-103. Selecting the geometry to check

- Click on the **Check Geometry** tool from **Toolbar** in the **Part** workbench; refer to Figure-104. The **Check Geometry** dialog will be displayed in the **Tasks** panel of **Combo View** showing the result of the scan; refer to Figure-105.

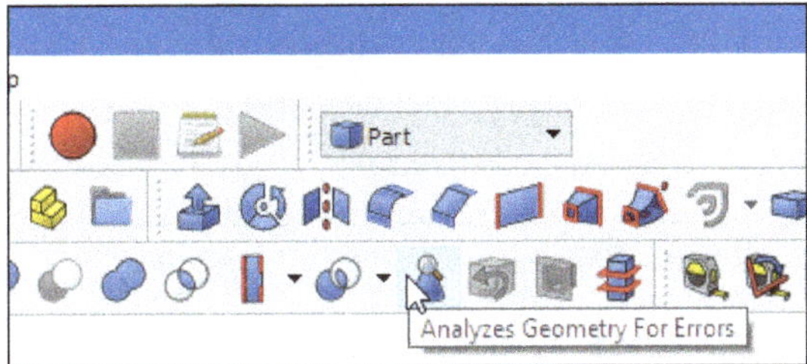

Figure-104. Check geometry tool

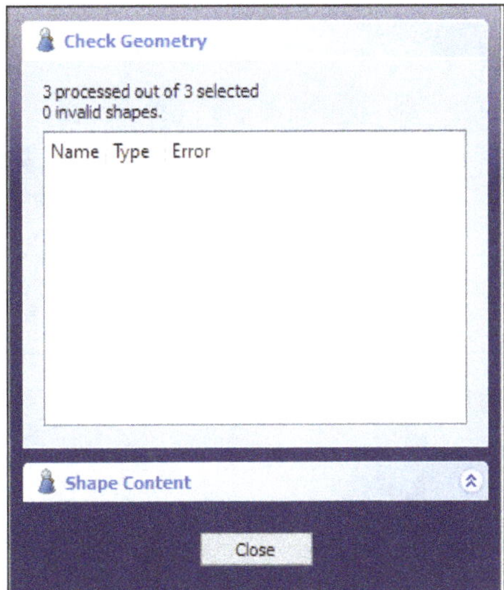

Figure-105. Check Geometry dialog

- If the result of the scan showing errors, click on a specific error message in the dialog. The corresponding geometric object will be highlighted in the 3D view area.

Note that till the FreeCAD vesion 0.18, it has no automatic repair methods for solids, so you need to look at the steps used in FreeCAD and try to fix the error yourself.

- Click on **Close** button to close the dialog.

Defeaturing

The **Defeaturing** tool is intended for removal of selected features from the model. In this context, features are meant as holes, protrusions, gaps, chamfers, fillets, etc. found on the model. The procedure to use this tool is discussed next.

- Select the faces on the model from the 3D view area which you want to defeature; refer to Figure-106.

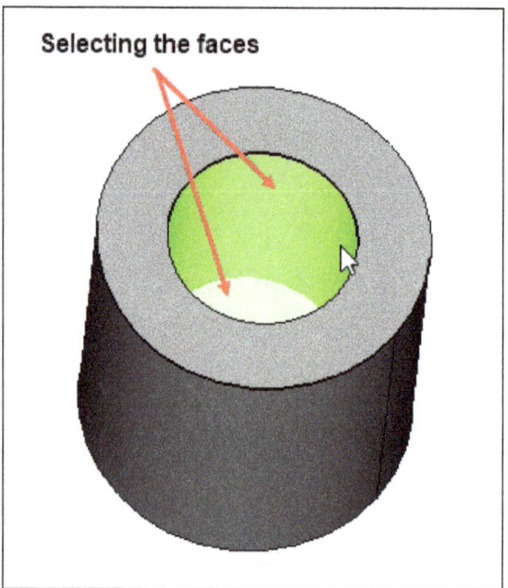

Figure-106. Selecting the faces to defeature

- Click on the **Defeaturing** tool from **Toolbar** in the **Part** workbench; refer to Figure-107. The selected feature will be removed from the model; refer to Figure-108.

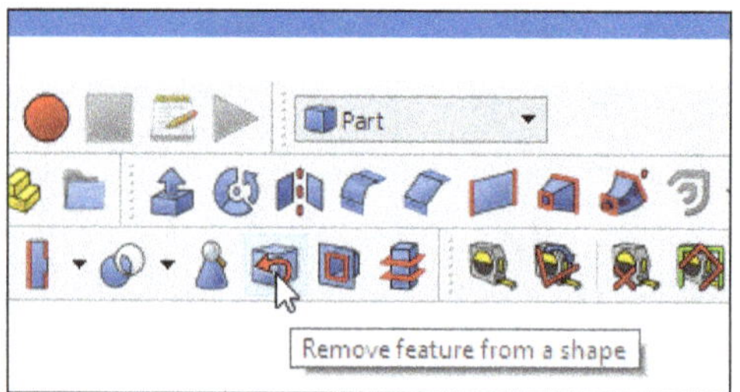

Figure-107. Defeaturing tool

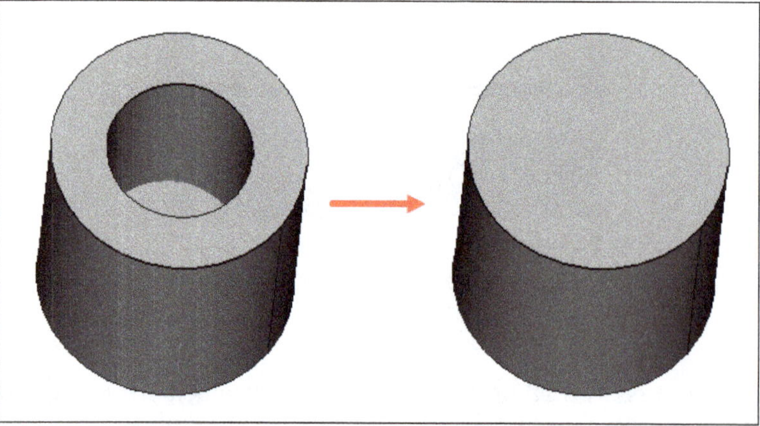

Figure-108. Defeatured model created

We have already discussed the **Section** tool in this chapter.

Cross-Sections

The **Cross-sections** tool creates one or more cross-sections using the selected shape. The procedure to use this tool is discussed next.

- Select the shape from which you want to create the cross-sections from the Model tree view or from the 3D view area; refer to Figure-109.

Figure-109. Selecting the shape to cross-section

- Click on the **Cross-sections** tool from **Toolbar** in the **Part** workbench; refer to Figure-110. The **Cross sections** dialog will be displayed in the **Tasks** panel of **Combo View** along with the preview of cross section; refer to Figure-111.

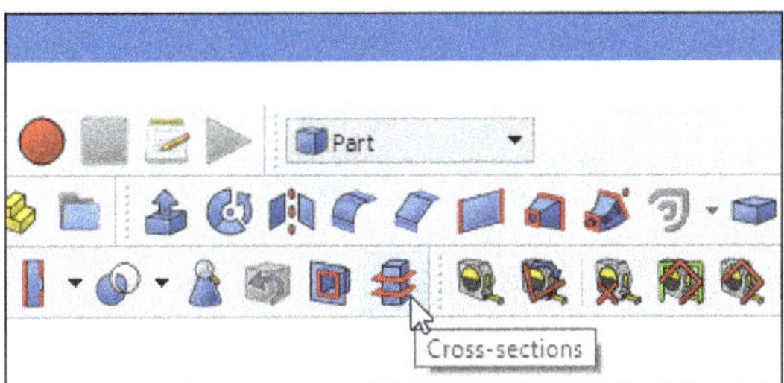

Figure-110. Cross sections tool

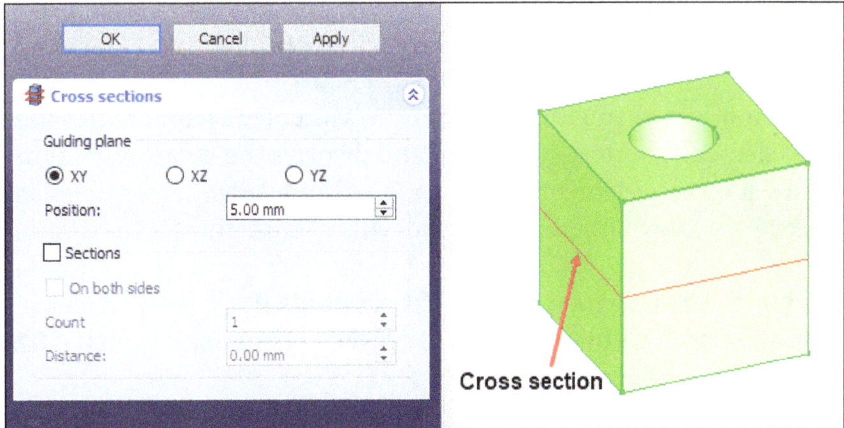

Figure-111. Cross sections dialog with preview of cross section

- Select desired guiding plane of creating the cross-section from the **Guiding plane** section of the dialog. Select the **XY**, **XZ**, or **YZ** radio button to create the cross-section along xy, xz, or yz plane, respectively.
- Specify the position of cross section in the **Position** edit box.
- Select the **Sections** check box to edit the parameters of cross-section.

- Select **On both sides** check box to create the cross-section on both sides of the selected guiding plane.
- Specify the number of cross sections to be created in the **Count** edit box.
- Specify distance between the cross sections in the **Distance** edit box of the dialog.
- Click on the **Apply** button. The cross sections of the shape will be created; refer to Figure-112.

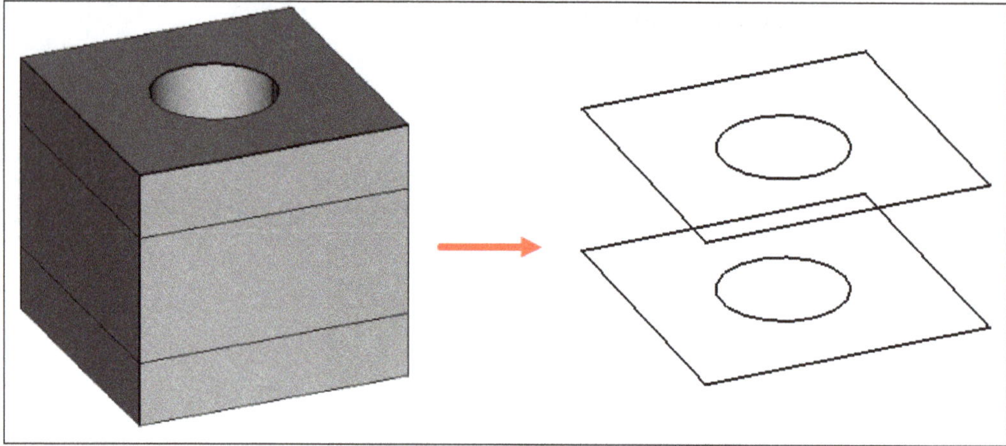

Figure-112. Cross sections of the shape created

- Click on **OK** button to close the dialog.

Measure Tools

The **Measure** tools allow linear and angular measurement between points, edges, and faces. These tools are available in the **Toolbar** of **Part** workbench; refer to Figure-113. The procedures to use these tools are discussed next.

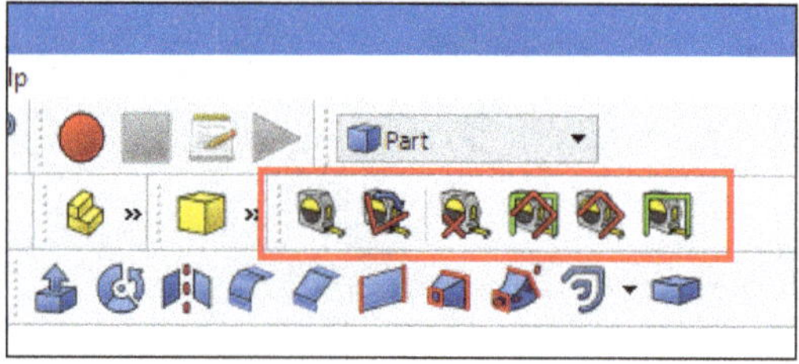

Figure-113. Measure tools

Measure Linear

The **Measure Linear** tool allows you to make linear measurements. It measures the distance between two selected topology elements (vertex, edge, face) and displays the measurement in the 3D view. The shortest distance between the two elements is shown in red and delta measurements (distances parallel to standard X, Y, Z axes) are shown in green. The procedure to use this tool is discussed next.

- Click on the **Measure Linear** tool from **Toolbar** in the **Part** workbench; refer to Figure-114. The **Selections** and **Control** dialog will be displayed in the **Tasks** panel of **Combo View** asking you to select the elements; refer to Figure-115.

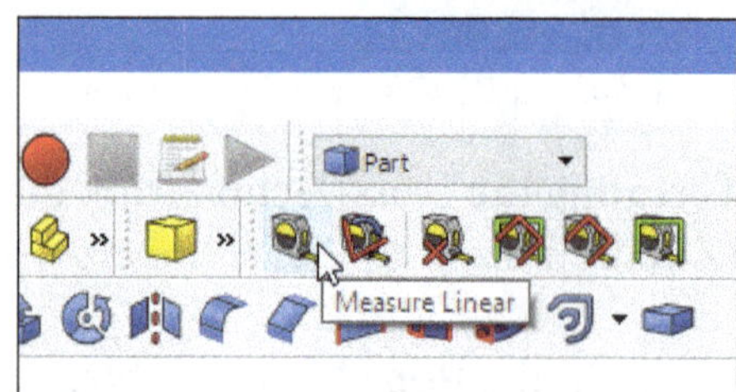

Figure-114. Measure linear tool

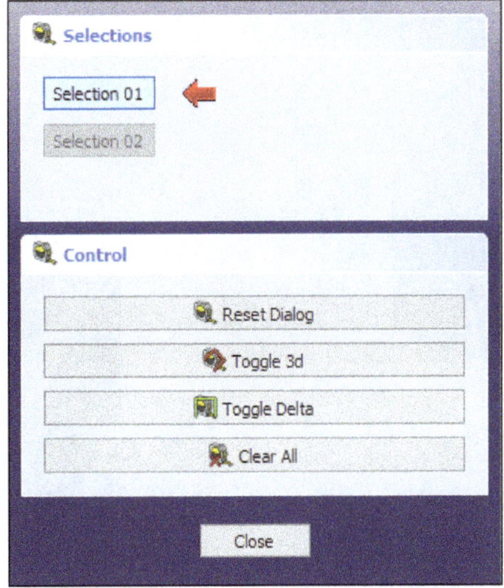

Figure-115. Selections and Control dialog

- Select the vertex, edge, or face from both the shapes in the 3D view area between which you want to measure the linear distance. The right tick mark will be displayed in front of both **Selection 01** and **Selection 02** buttons of the **Selections** dialog; refer to Figure-116.

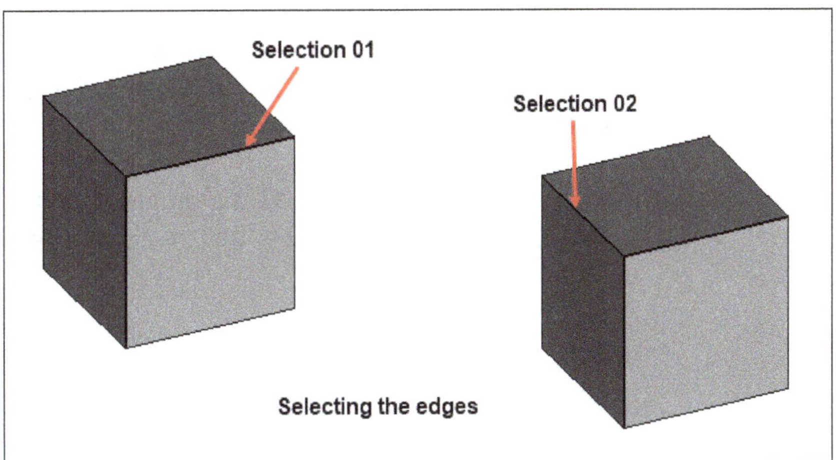

Figure-116. Selecting the edges to measure distance

- In **Control** dialog, click on the **Reset Dialog** button to reset all the selections.
- Click on the **Toggle 3d** button to measure the shortest distance between the two elements; refer to Figure-117.

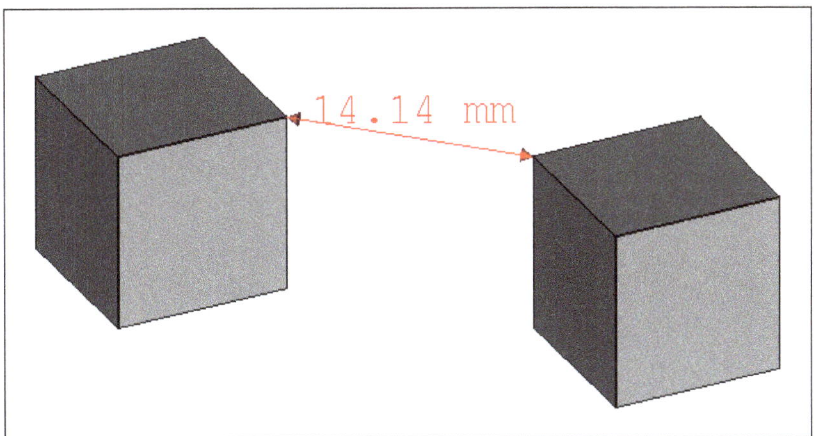

Figure-117. Shortest distance between elements

- Click on the **Toggle Delta** button to create delta measurement between the elements; refer to Figure-118.

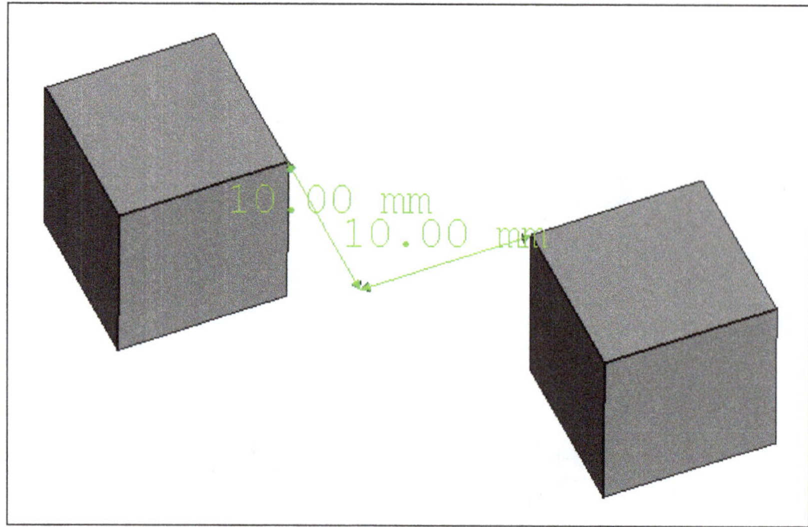

Figure-118. Delta measurement between elements

- Click on the **Clear All** button to clear all the measurements created between the elements.
- Click on **Close** button to close the dialog.

Measure Angular

The **Measure Angular** tool measures the angle between two straight edges, two planar faces, or one straight edge and a planar face and displays the measurement in the 3D view. The procedure to create angular measurement is same as discussed for linear measurement.

If you want the quicker access to the options used in **Control** dialog, use **Clear All** button, **Toggle All** button, **Toggle 3d** button, and **Toggle Delta** button from **Toolbar** in the **Part** workbench.

Chapter 5

Solid Modeling Practice

Topics Covered

The major topics covered in this chapter are:

- ***Practical 1***
- ***Practical 2***
- ***Practical 3***
- ***Practices***

PRACTICAL 1

Create the model (isometric view) as shown in Figure-1. The views of the model with dimensions are given in Figure-2.

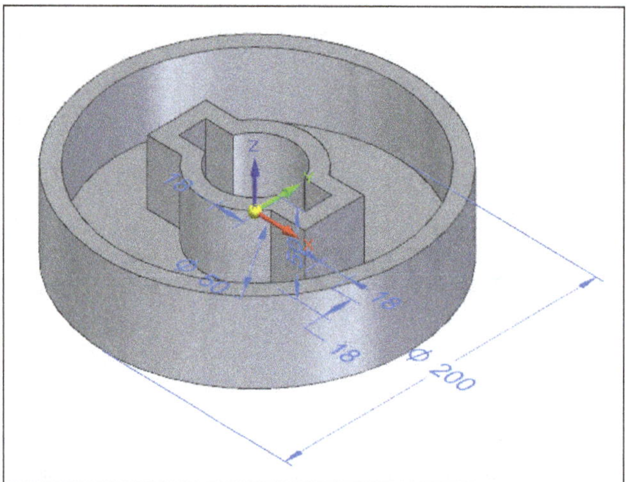

Figure-1. Practical Model 1

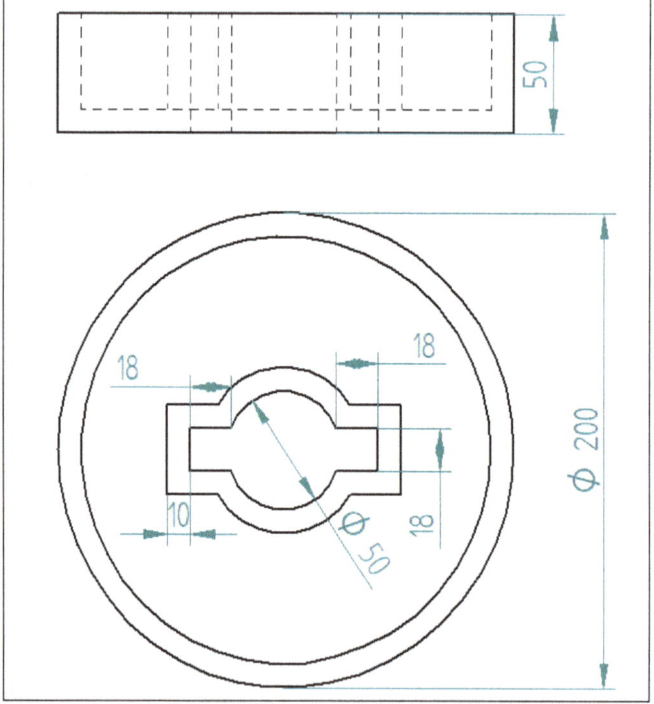

Figure-2. Views for practical 1

Before we start working on the practical, it is important to understand two terms; first angle projection and third angle projection. These are the standards of placing views in the engineering drawing. The views placed in the above figure are using third angle projection. In first angle projection, the top view of model is placed below the front view and right side view is placed at left of the front view. You will learn more about projection in chapter related to drafting.

Starting Part Design Environment and Creating Sketch

- Double-click on the FreeCAD icon from desktop if you have not started FreeCAD yet.
- Start a new document by clicking on the **New** button from **Toolbar** or press **CTRL+N** shortcut key.
- Select the **Part Design** option from **Switch between workbenches** drop-down in the **Toolbar**. The tools to create solid model will be displayed.
- Click on the **Create sketch** tool from the **Toolbar** or **Part Design** menu. The **Select feature** dialog will be displayed in **Combo View** and you will be asked to select a plane for creating sketch.
- Select the XY plane from the list in the dialog and click on the **OK** button. The sketching environment will be displayed.
- Create a circle of diameter **200** with its center at origin; refer to Figure-3. Note that as we create a circle, we have also applied the diameter dimension to it. In FreeCAD, it is better to apply dimension as soon as you create the entity if the entity is independent.

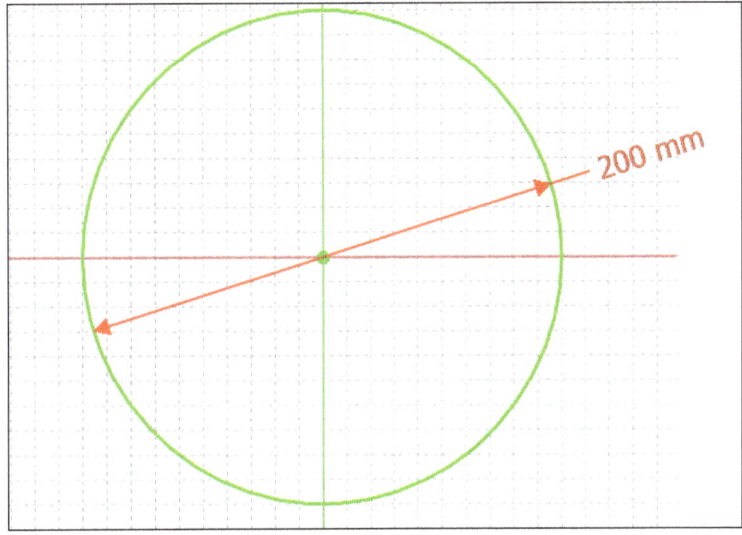

Figure-3. Circle to be created

- Create a circle of diameter **50** at the center of earlier created circle and then draw a rectangle as shown in Figure-4.
- Trim the inner portion of circle and rectangle, and then apply the dimensions as shown in Figure-5.

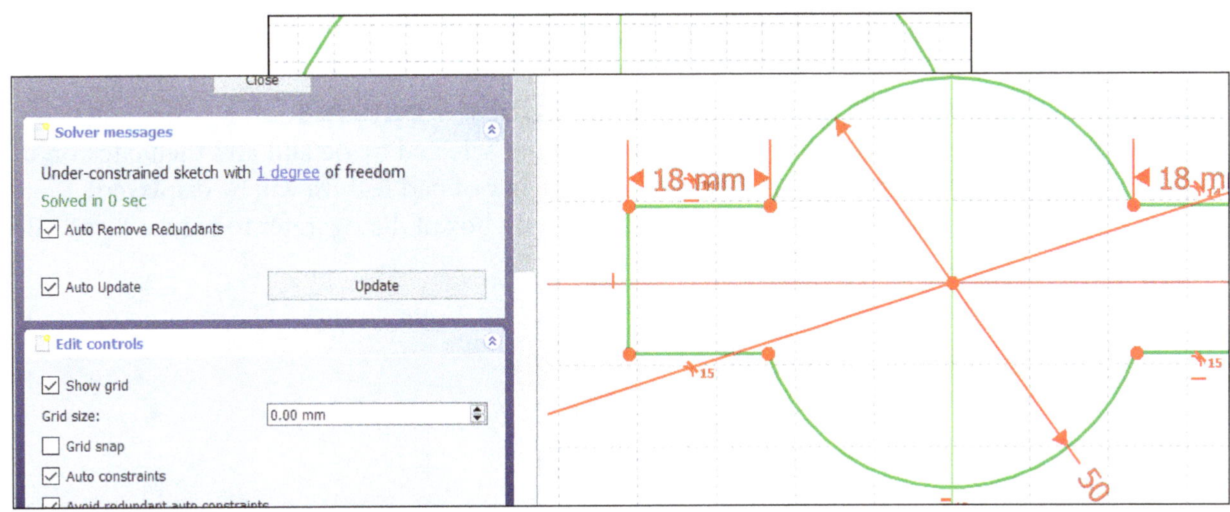

Figure-5. Sketch after applying dimensions

- Note that there is 1 degree of freedom left in the sketch. If you drag the vertical line of rectangle then you will find that it is free to move up and down. To constraint this movement, select the points of rectangle as shown in Figure-6 and click on the **Constrain vertically** tool from the **Toolbar**.

The sketch will become fully constraint; refer to Figure-7.

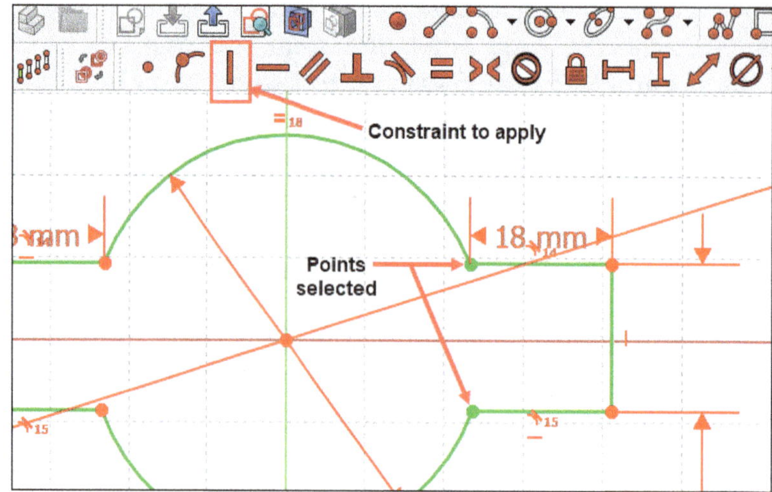

Figure-6. Applying vertical constrain

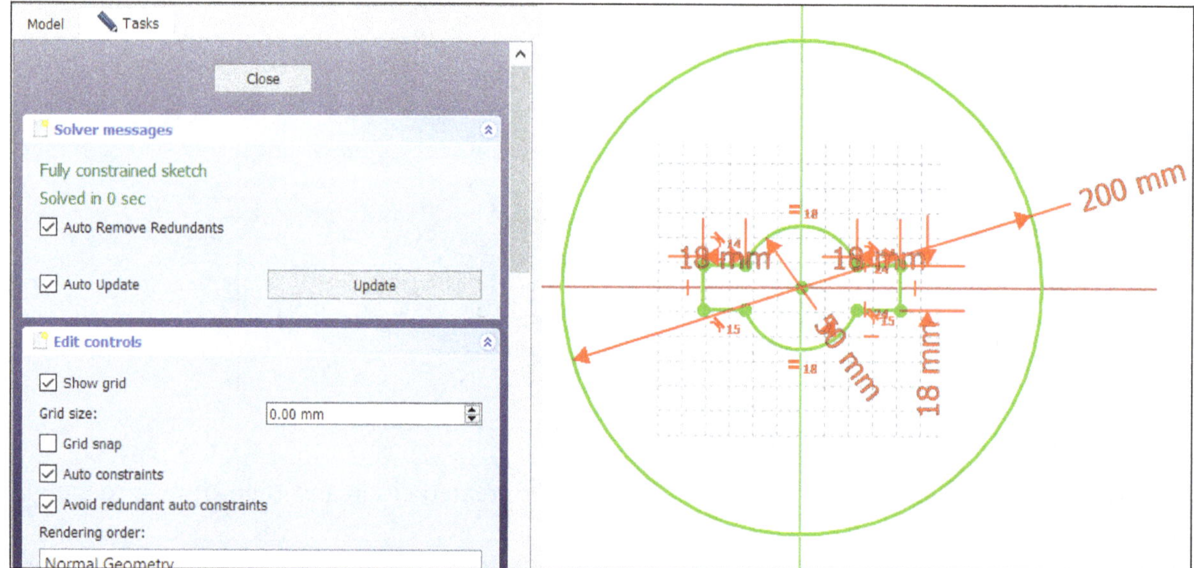

Figure-7. After applying constraints

- Click on the **Close** button to exit the sketching environment.

Creating Pad and Thick Solid Features

- Select the newly created sketch from **Model Tree** if not selected by default and then click on the **Pad** tool from **Toolbar** or **Part Design** menu. Preview of pad feature will be displayed.
- Specify the length of pad feature as **50** in the **Length** edit box of dialog; refer to Figure-8 and click on the **OK** button. The feature will be created.

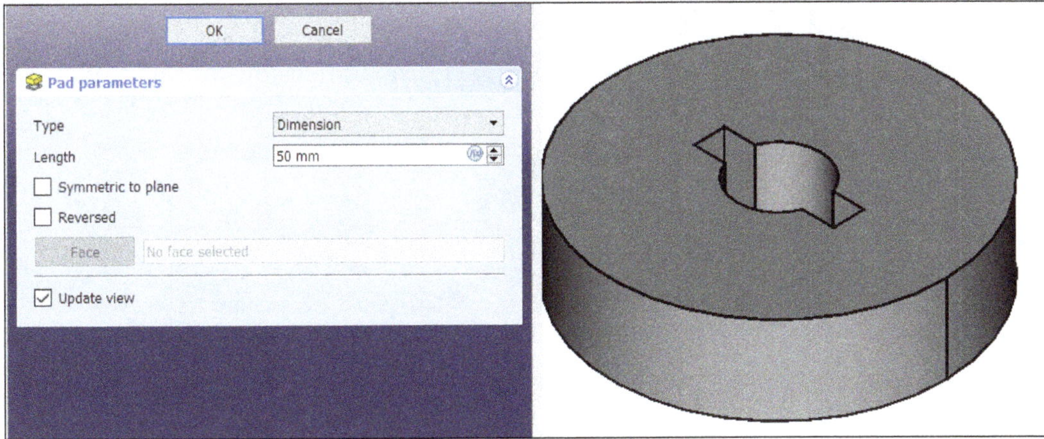

Figure-8. Creating pad feature

- Select the top flat face of the model from drawing area and click on the **Thickness** tool from the **Toolbar** or **Part Design** menu; refer to Figure-9. Preview of thickness feature will be displayed. (Do not worry if everything goes blank!!)

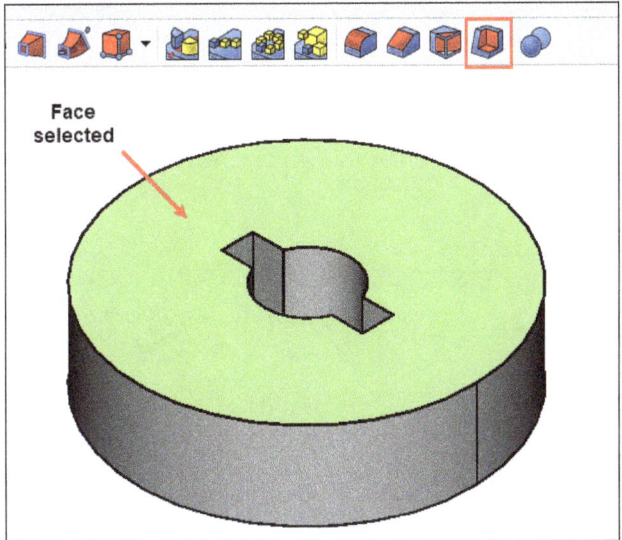

Figure-9. Face selected for thickness feature

- Specify the parameters shown in Figure-10 and click on the **OK** button to create the final model.

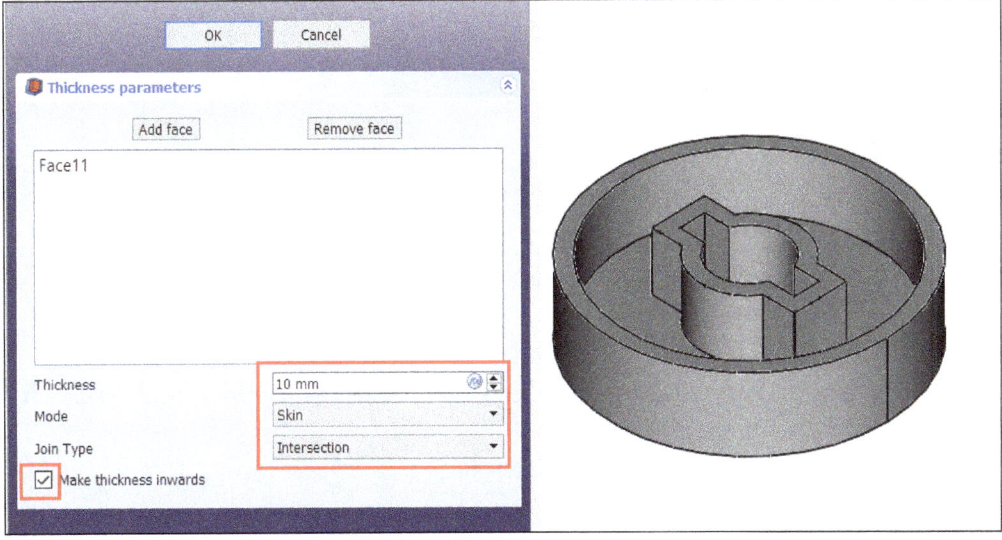

Figure-10. Preview of thickness feature

PRACTICAL 2

Create the model (isometric view) as shown in Figure-11. The dimensions of the model are given in Figure-12.

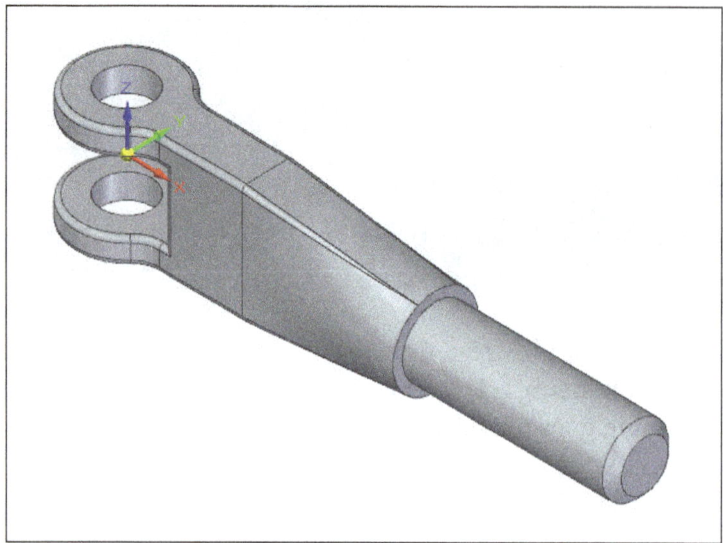

Figure-11. Model for Practical 2

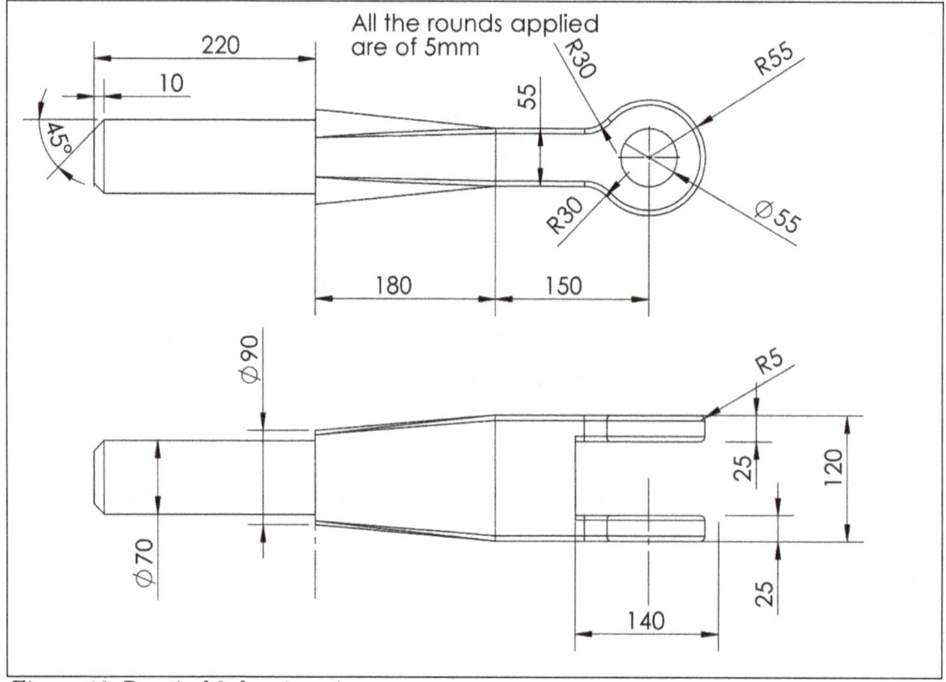

Figure-12. Practical 2 drawing views

Creating first pad feature

This model is a combination of pad and loft features. We will create the model in multiple steps with first feature acting as base for all the other features.

- Start FreeCAD if not started yet and create a new document.
- Switch to the **Part Design** workbench using **Part Design** option from **Switch between workbenches** drop-down in the **Toolbar**.

- Click on the **Create sketch** tool from the **Toolbar** or **Part Design** menu. The **Select feature** dialog box will be displayed and you will be asked to specify a plane for creating sketch.
- Select the **XY** plane from dialog or drawing area and click on the **OK** button. The sketching environment will be displayed.
- Create the sketch as shown in Figure-13.

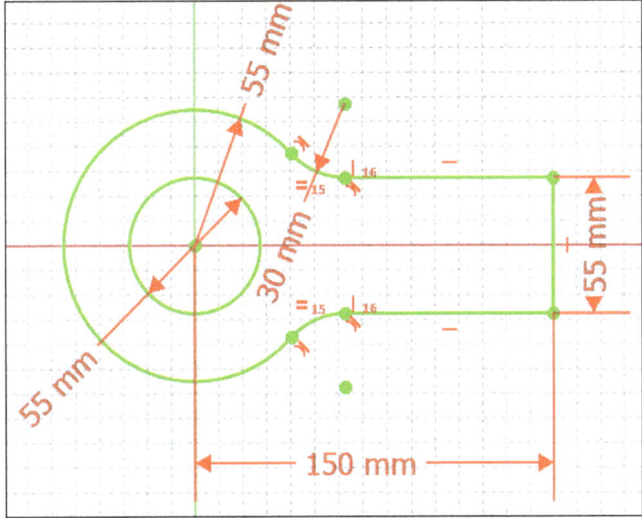

Figure-13. Sketch created on XY plane

- Click on the **Close** button from dialog after creating sketch to exit sketching mode.
- Click on the **Pad** tool from **Toolbar** after selecting newly created sketch. The **Pad parameters** dialog will be displayed with preview of feature.
- Select the **Symmetric to plane** check box from dialog and specify length of extrude feature as **120** mm; refer to Figure-14.

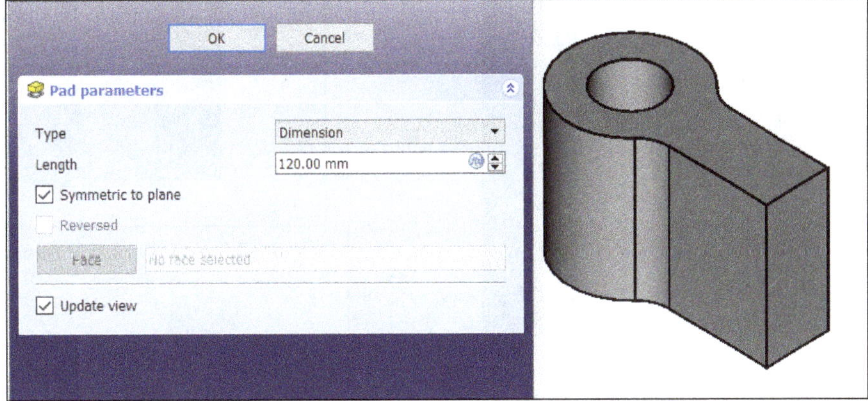

Figure-14. Pad parameters specified

- Click on the **OK** button from the dialog to create the feature.

Creating Loft feature

- Select the flat side face of model as shown in Figure-15 and click on the **Create sketch** tool from **Toolbar**. The sketching environment will be displayed.

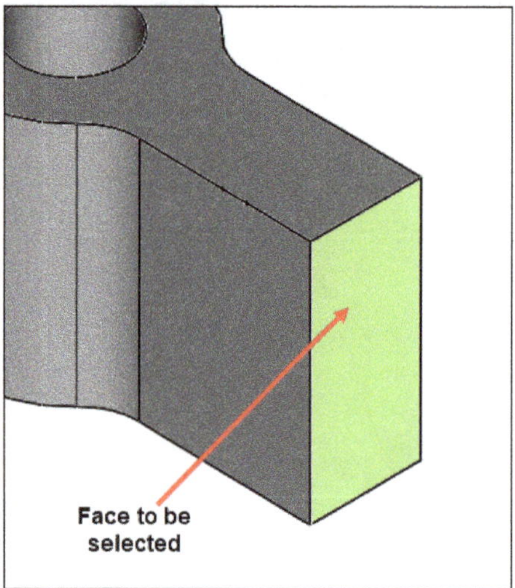

Figure-15. Face selected for sketching

- Click on the **External geometry** tool from the **Sketcher geometries** cascading menu or **Toolbar**. You will be asked to select edges to be projected in the sketch.
- Select the boundary edges of flat face selected earlier; refer to Figure-16. Note the projected edges are not part of sketch, they are for reference only.
- Click on the **Create rectangle** tool from **Toolbar** or **Sketcher geometries** cascading menu of the **Sketch** menu and create a rectangle overlapping the projected edges; refer to Figure-17.

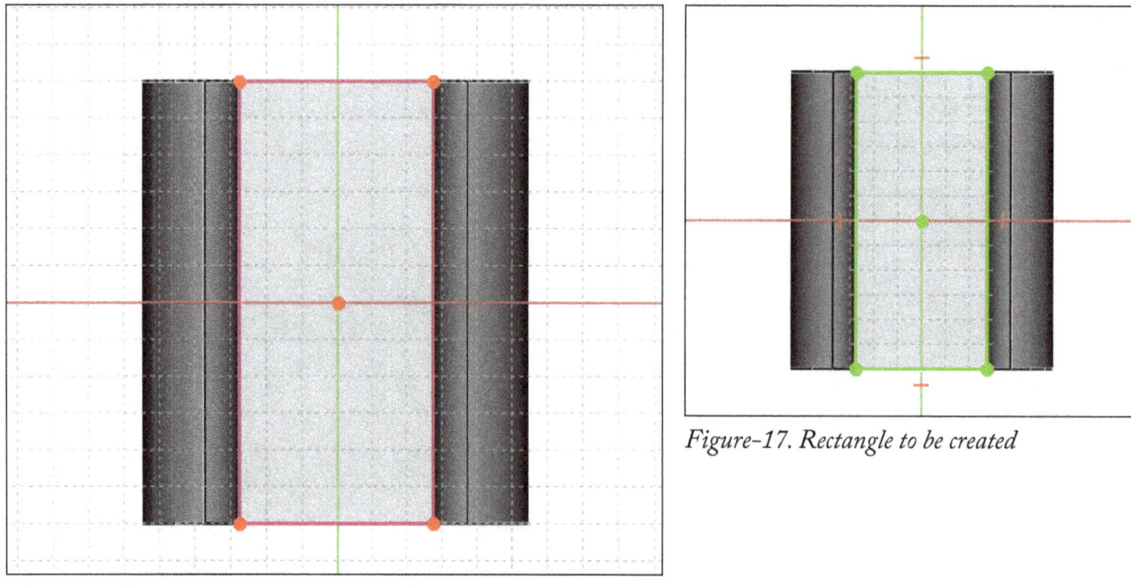

Figure-16. Edges of face projected in sketch

Figure-17. Rectangle to be created

- Click on the **Close** button to exit sketching environment. Click in the empty area to exit any entity selected.
- Click on the **Create a datum plane** tool from the **Part Design** menu or **Toolbar**. The **Plane parameters** dialog will be displayed in **Combo View**.
- Select flat side face of model as reference and specify **Z** offset distance as 180; refer to Figure-18.

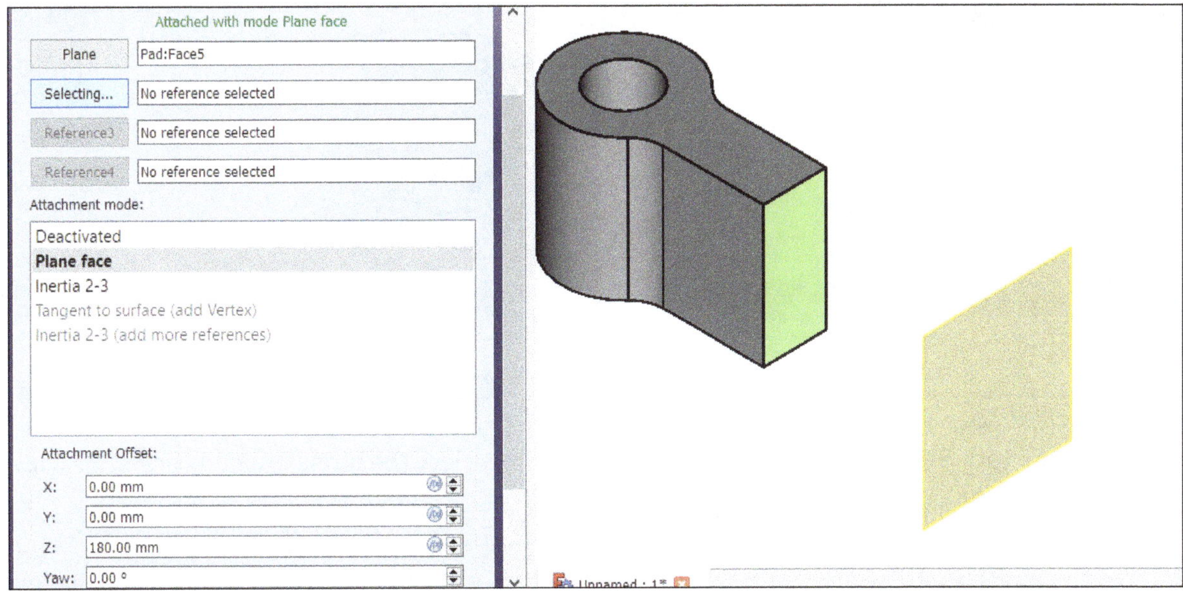

Figure-18. Plane to be created

- Click on the **OK** button from the dialog to create plane.
- Select the newly created plane and click on the **Create sketch** tool from **Part Design** menu or **Toolbar**. The sketching environment will be displayed.
- Create a circle of diameter **90** mm at the origin; refer to Figure-19.

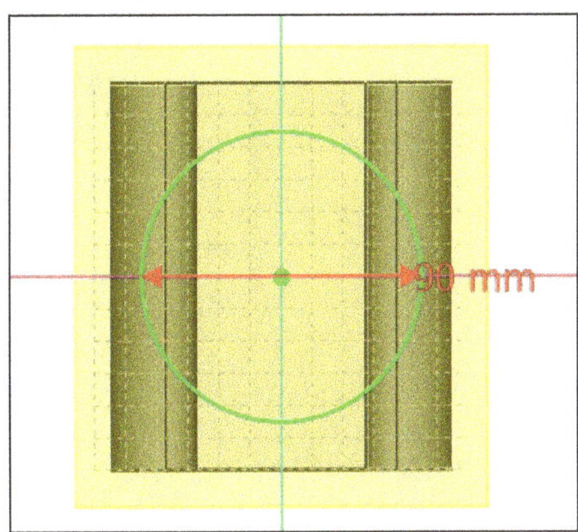

Figure-19. Circle to create

- Click on the **Close** button from the dialog to exit sketching mode.
- Click on the **Additive loft** tool from **Part Design** menu or **Toolbar**. The **Select feature** dialog will be displayed in **Combo View**.
- Select the sketch earlier created on flat face of model from the list; refer to Figure-20 and click on the **OK** button from the dialog. The **Loft parameters** dialog will be displayed and you will be asked to add section sketches for loft feature.

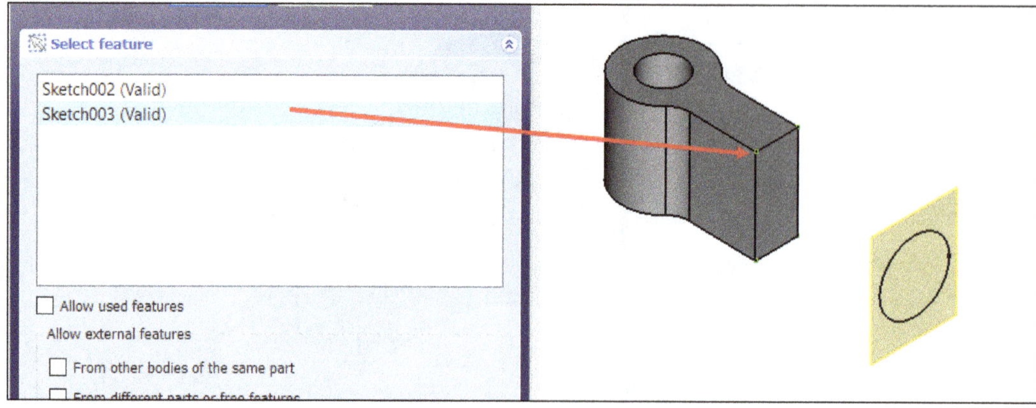

Figure-20. Sketch selected for base

- Click on the **Add Section** button from the dialog and select the circle created earlier. Preview of loft feature will be displayed; refer to Figure-21.

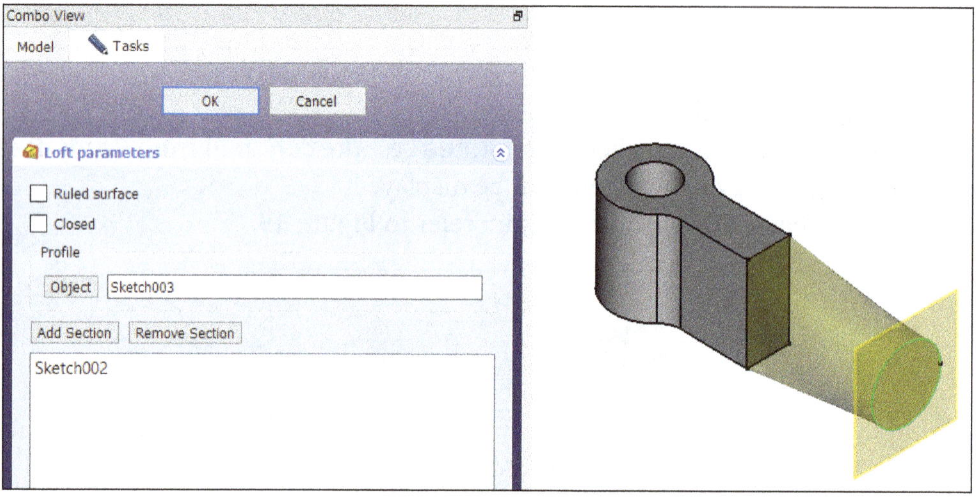

Figure-21. Preview of loft

- Click on the **OK** button from the dialog to create the feature.

Creating Second Pad Feature

- Select the datum plane earlier created and press **SPACEBAR** to hide it.
- Select the flat face of recently created loft feature and click on the **Create sketch** tool from **Part Design** menu. The sketching environment will be activated.
- Create a circle of diameter **70** at the origin; refer to Figure-22.

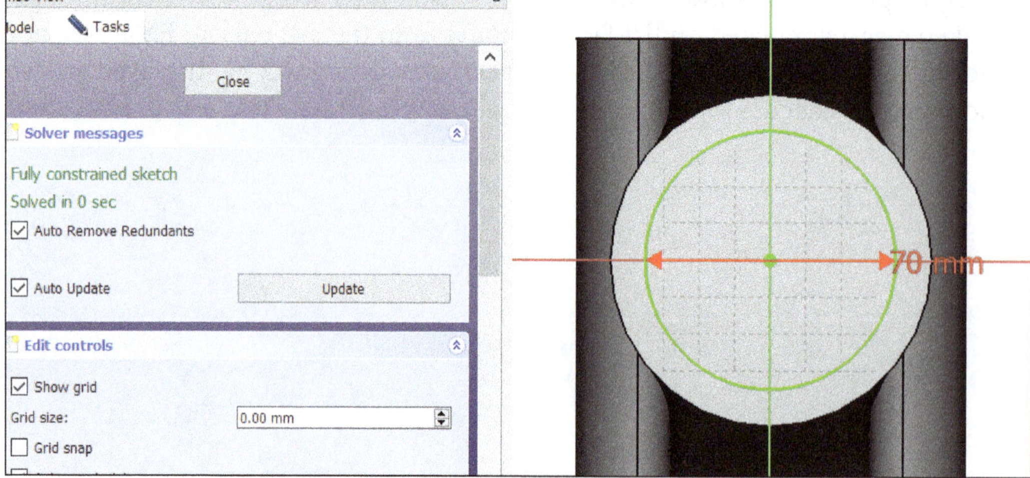

Figure-22. Circle created on loft face

- Click on the **Close** button from the dialog to exit sketching environment.
- Click on the **Pad** tool from the **Toolbar** or **Part Design** menu. Preview of pad feature will be displayed using the recently created sketch.
- Specify the length of feature as **220** in the **Length** edit box; refer to Figure-23 and click on the **OK** button to create the feature.

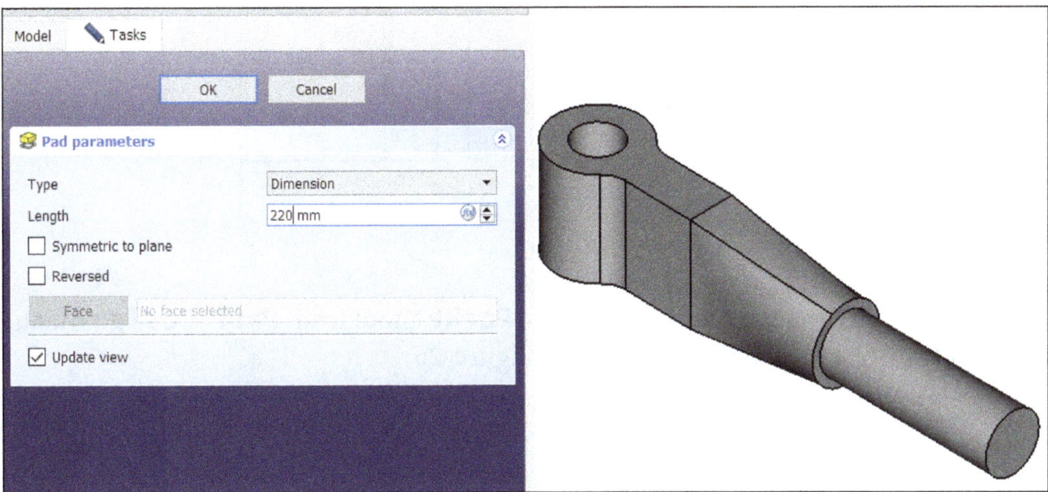

Figure-23. Pad feature after applying length value

Creating Pocket Feature

- Make sure nothing is selected from the model and then click on the **Create sketch** tool from **Part Design** menu. The **Select feature** dialog will be displayed with list of planes available.
- Select the **XZ Plane** from the dialog and click on the **OK** button. The sketching environment will be displayed; refer to Figure-24.

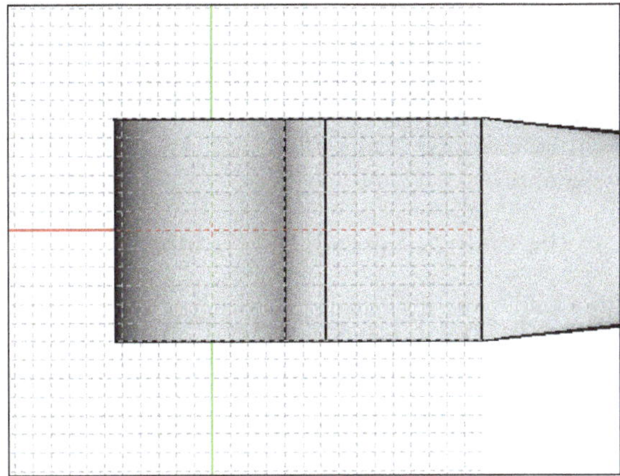

Figure-24. Sketching environment

- Create a rectangle as shown in Figure-25. You may need to switch draw style to Wireframe for creating rectangle.

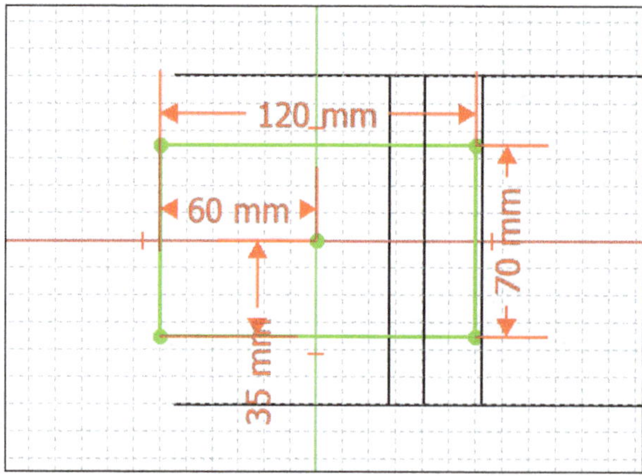

Figure-25. Rectangle to create

- After creating sketch, click on the **Close** button.
- Make sure the sketch is selected and click on the **Pocket** tool from **Part Design** menu. Preview of the pocket feature will be displayed; refer to Figure-26.

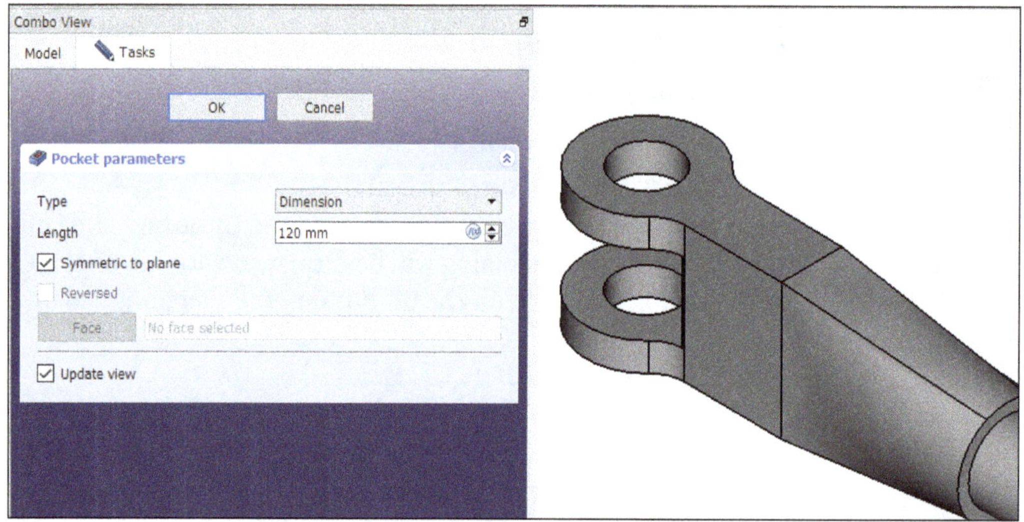

Figure-26. Preview of pocket feature

- Click on the **OK** button from the dialog to create the feature.

Applying Fillets and Chamfers

- Select the edges of model and click on the **Fillet** tool from the **Part Design** menu. Preview of the fillet will be displayed; refer to Figure-27.

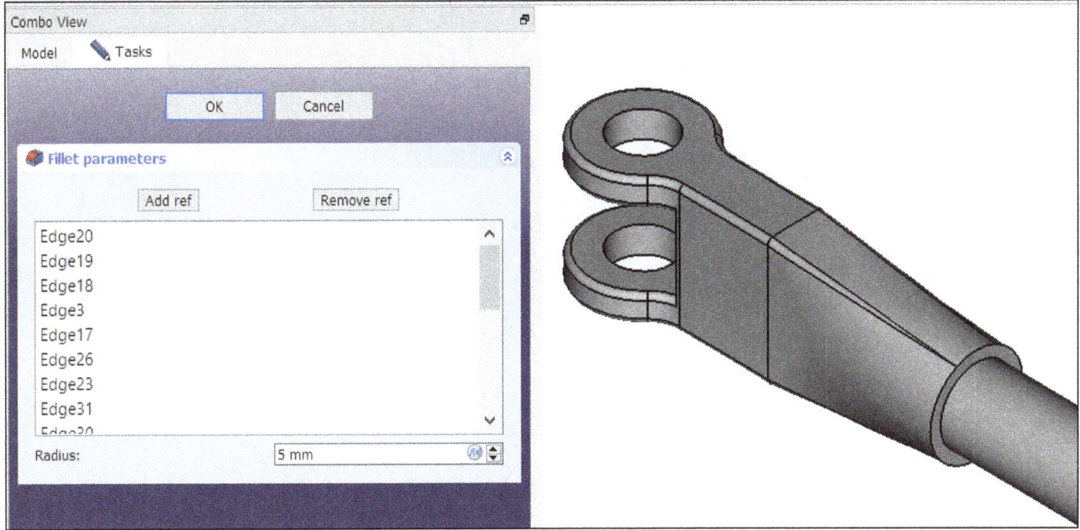

Figure-27. Preview of model after fillet

- Click on the **OK** button from the dialog to apply fillets.
- Select the round edge of pad feature created earlier and click on the **Chamfer** tool from **Part Design** menu.
- Specify the parameters as shown in Figure-28 and click on the **OK** button to apply chamfer.

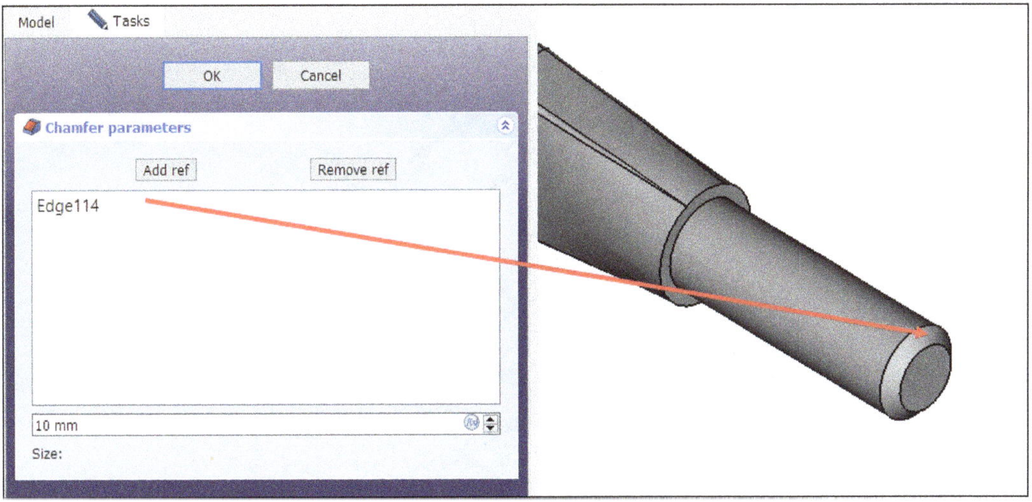

Figure-28. Preview of chamfer applied

- Save the file at desired location.

PRACTICE 1 TO 4

Create 3D models for drawings shown in Figure-29, Figure-30, Figure-31, and Figure-32.

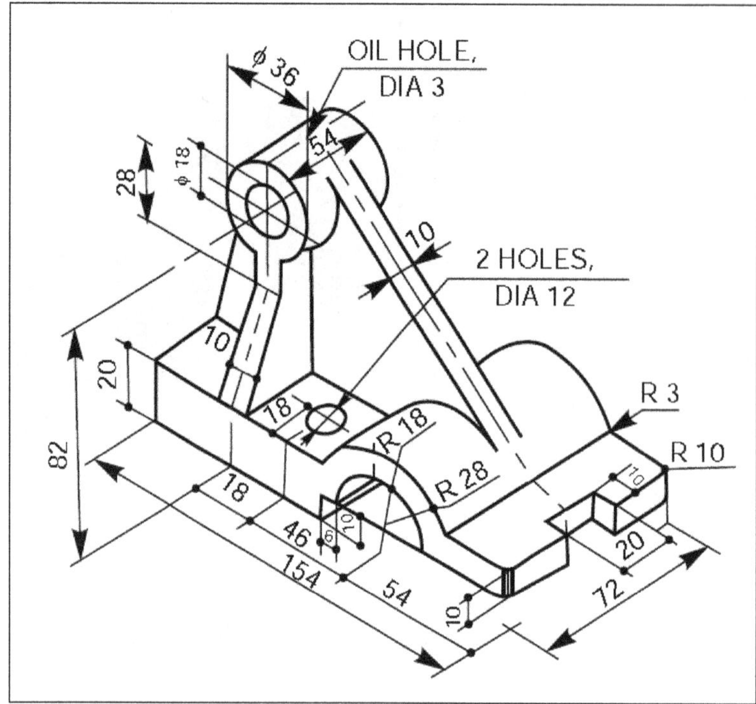

Figure-29. Practice 1

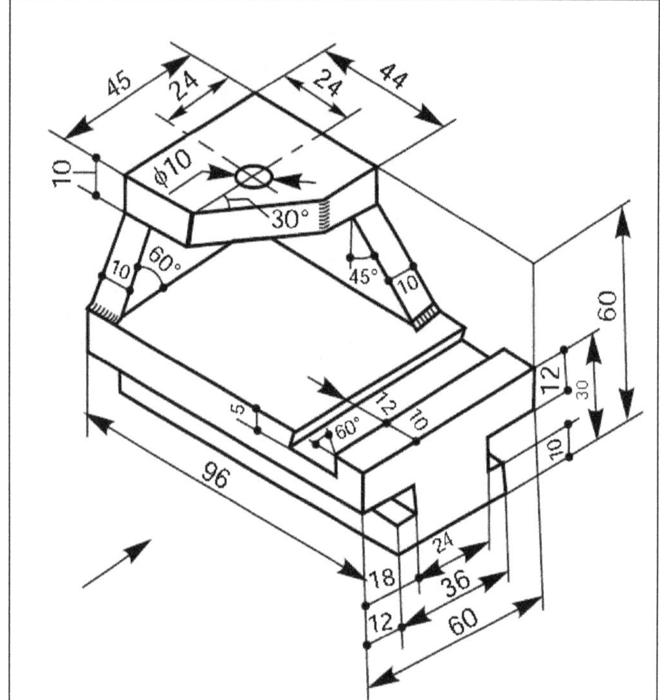

Figure-30. Practice 2

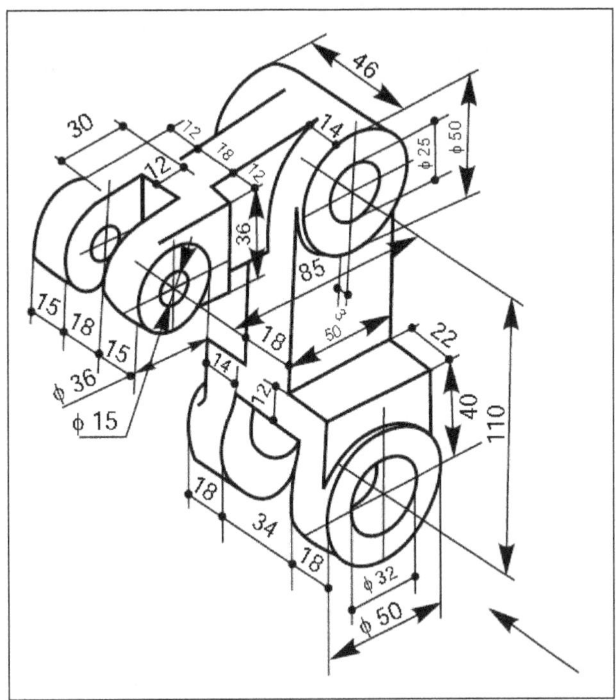

Figure-31. Practice 3

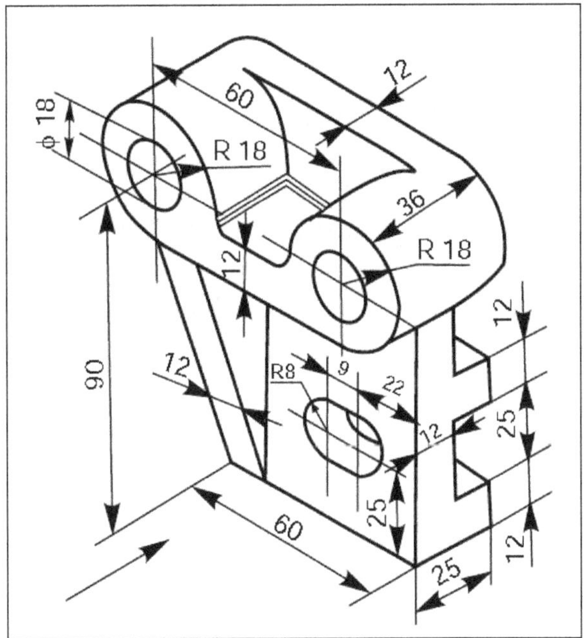

Figure-32. Practice 4

PRACTICAL - 3

Create the model (isometric view) as shown in Figure-33. The dimensions and view are given in Figure-34.

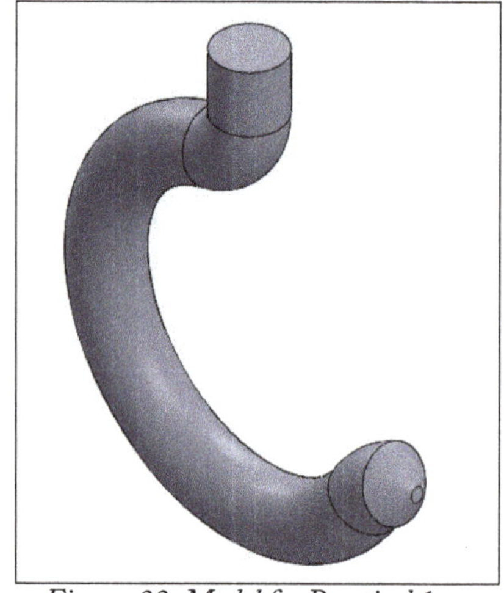

Figure-33. Model for Practical 1

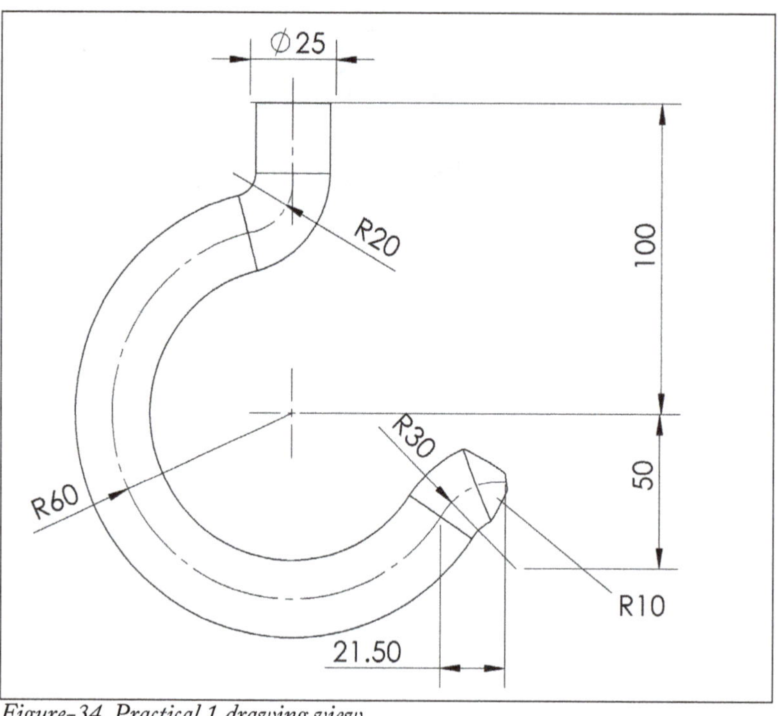

Figure-34. Practical 1 drawing view

Creating Sketches for Hook

- Start FreeCAD application using desktop icon if not started yet.
- Create a new document and switch to **Part Design** workbench.
- Click on the **Create sketch** tool from **Toolbar**. The **Select feature** dialog will be displayed in **Combo View** and you will be asked to select sketching plane.
- Select the **XY Plane** from the list and click on the **OK** button. The sketching environment will be displayed.
- Create the sketch as shown in Figure-35.

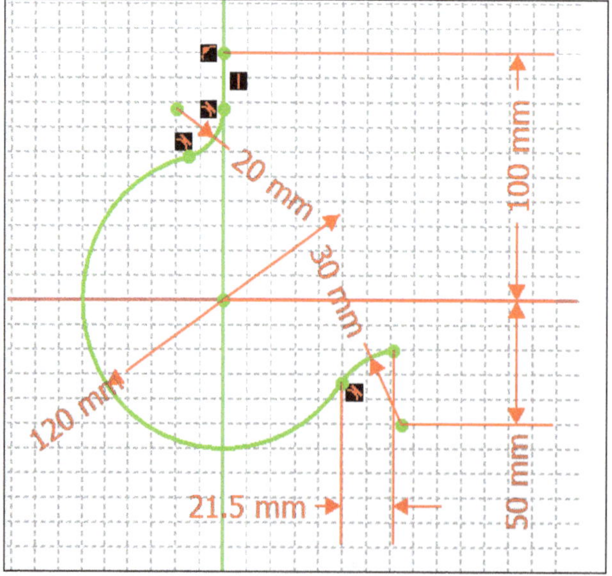

Figure-35. Sketch to be created

- Click on **Close** button from the dialog to exit sketching environment.
- Click in the empty area to make sure nothing is selected.
- Select straight line of sketch as shown in Figure-36 and click on the **Create a datum plane** tool from **Toolbar** or **Part Design** menu. Preview of plane will be displayed normal to selected line segment; refer to Figure-37.

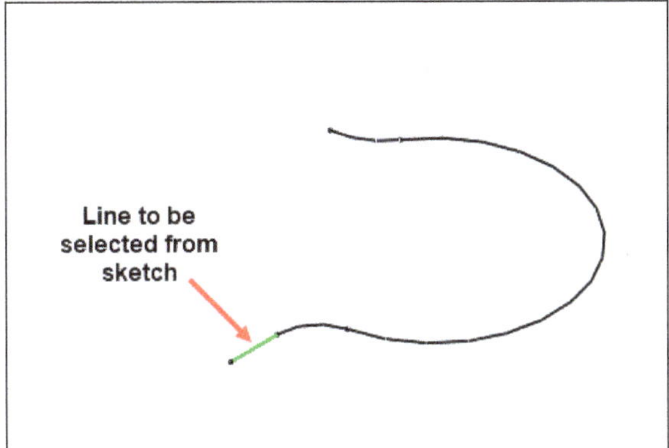

Figure-36. Line selected for plane creation

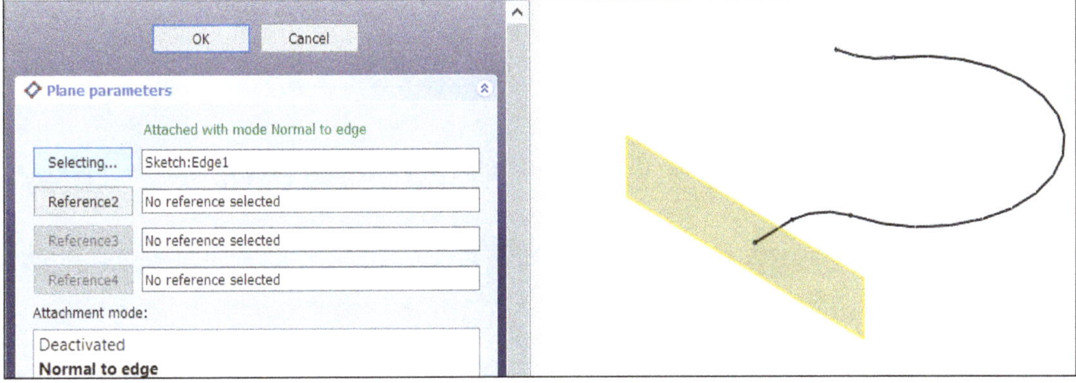

Figure-37. Preview of plane

- Click on the **OK** button from the dialog to create plane.
- Select the newly created plane from drawing area and click on the **Create sketch** tool from **Part Design** menu or **Toolbar**. The sketching environment will become active.
- Create a circle of diameter **25** mm at the origin; refer to Figure-38.

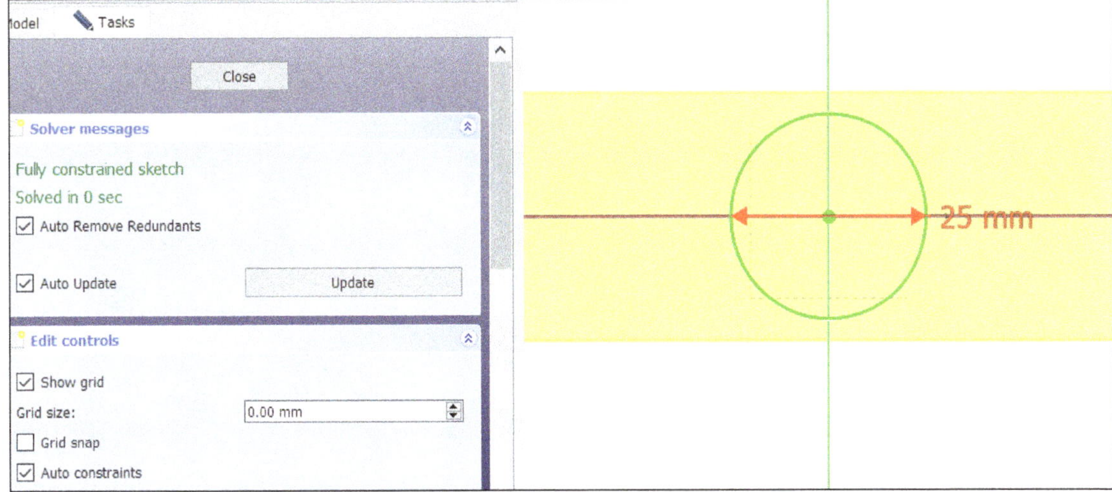

Figure-38. Second sketch to be created

- Click on the **Close** button from the dialog to exit sketching environment. Click in the empty area of drawing to make sure nothing is selected.

Creating Sweep Feature

- Select the newly created circle sketch and click on the **Additive pipe** tool from the **Part Design** menu. The **Pipe parameters** dialog will be displayed with earlier selected sketch defined as section profile; refer to Figure-39.

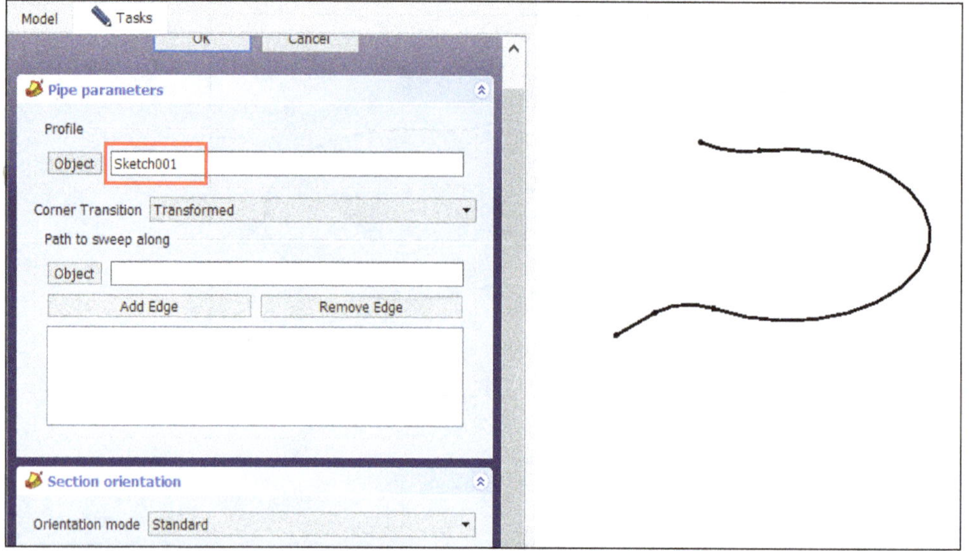

Figure-39. Pipe parameters dialog2

- Click on the **Add Edge** button from dialog and select the line in profile sketch. Preview of pipe feature will be displayed; refer to Figure-40.

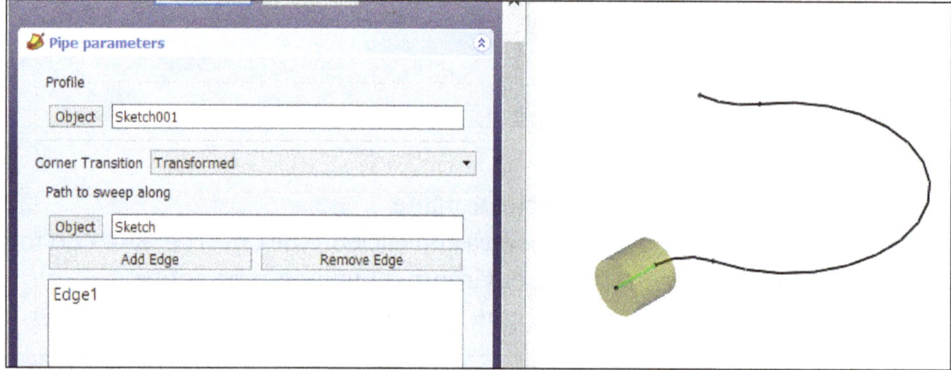

Figure-40. Preview of pipe feature

- Click on the **Add Edge** button again and select next curve in the sketch. Repeat the steps until all the curves of sketch are added as path; refer to Figure-41.
- Click on the **OK** button from dialog to create the feature.

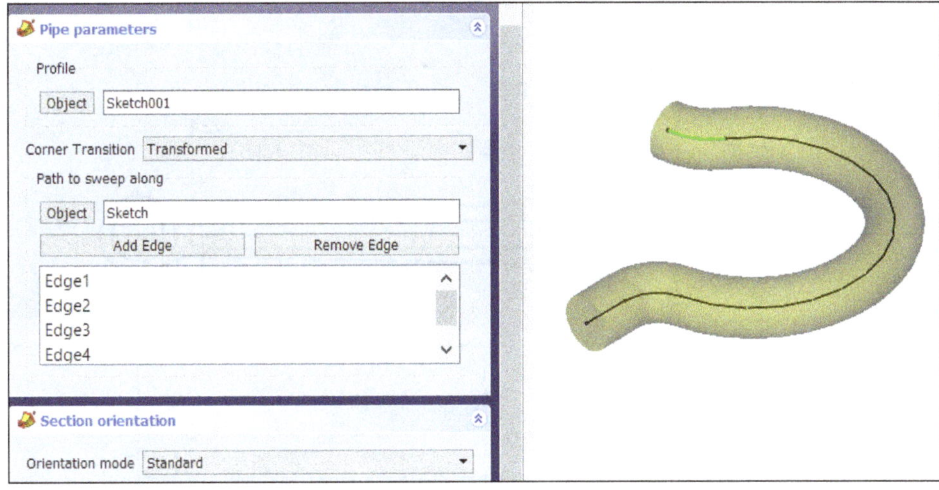

Figure-41. Preview of feature

- Apply a fillet of radius **10** mm at the end edge of model; refer to Figure-42.

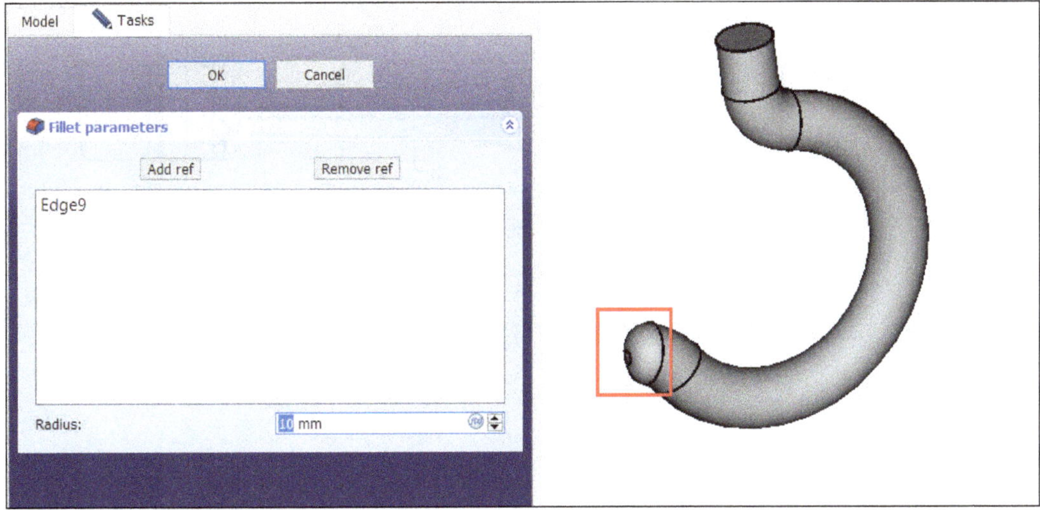

Figure-42. Applying fillet

Practice 5

Create the model as shown in Figure-43. The dimensions are given in Figure-44.

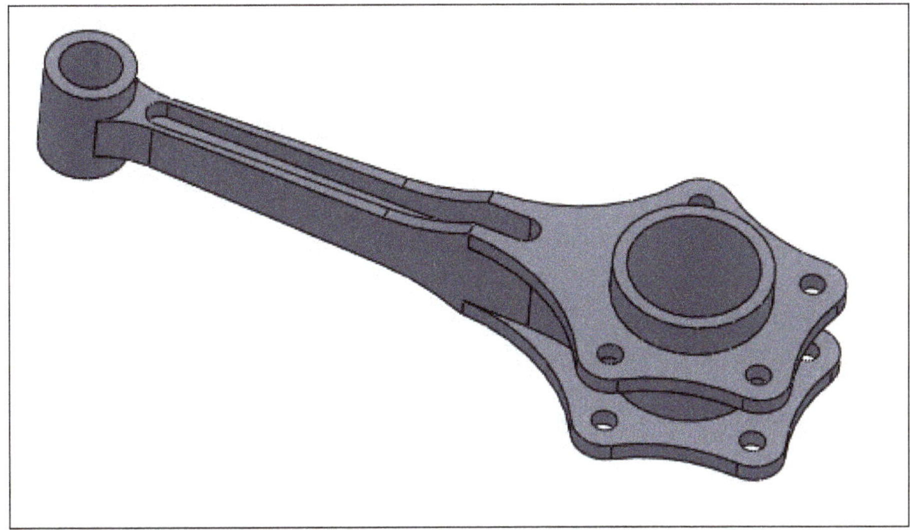

Figure-43. Practice 5

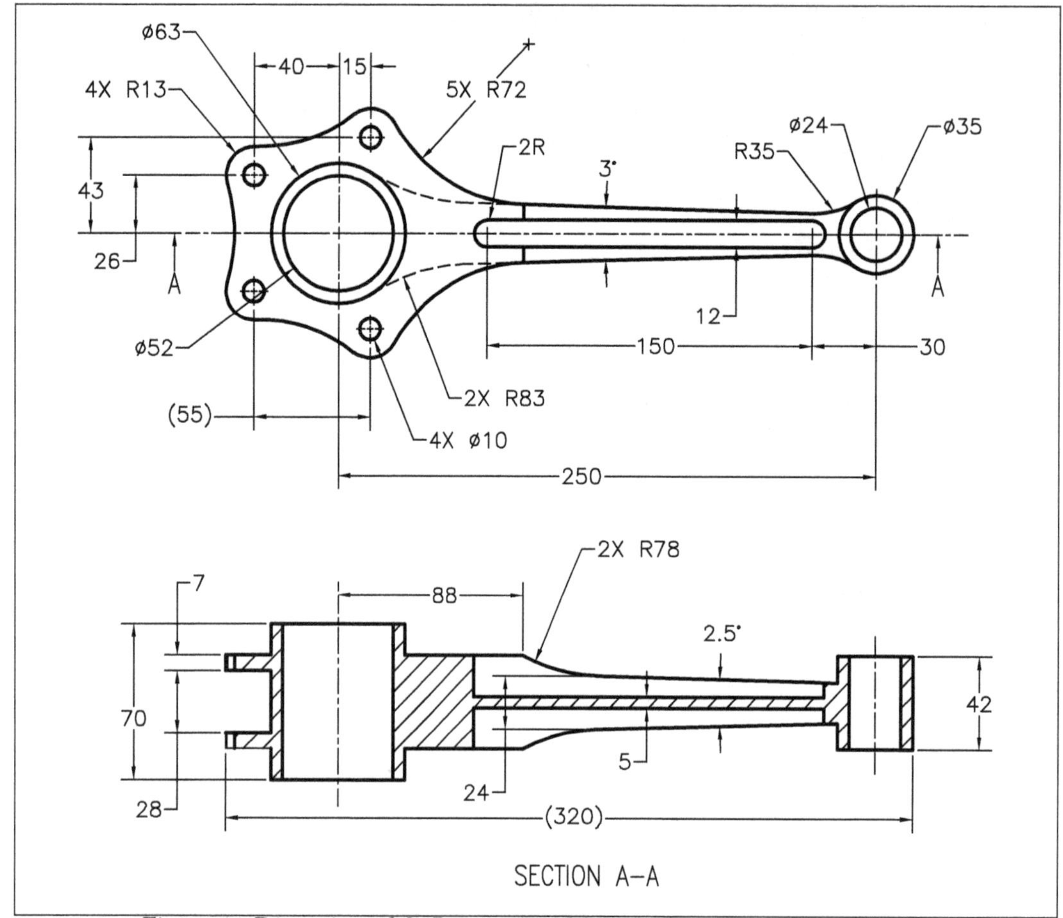

Figure-44. Dimensions of the Practice 5 model

PRACTICE 6

Create the model using the drawings shown in Figure-45.

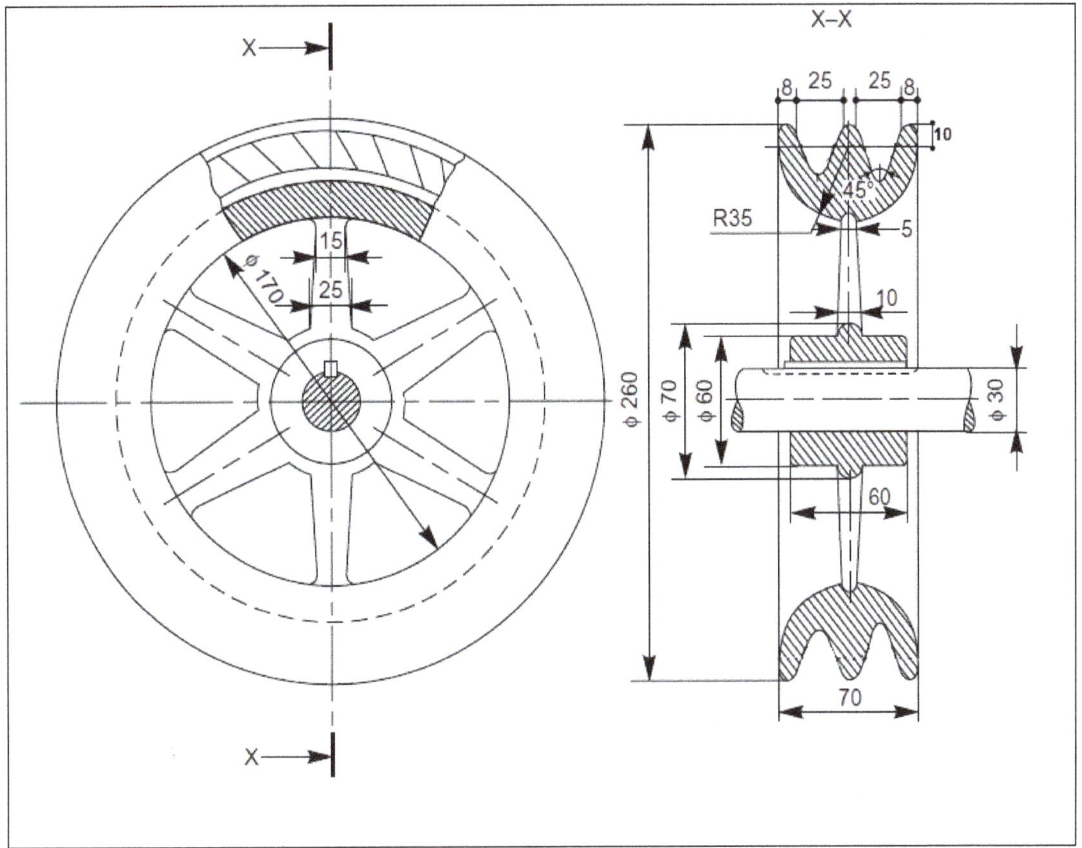

Figure-45. Rope Pulley

PRACTICE 7

Create the model as shown in Figure-46. Dimensions are given in Figure-47. Assume the missing dimensions.

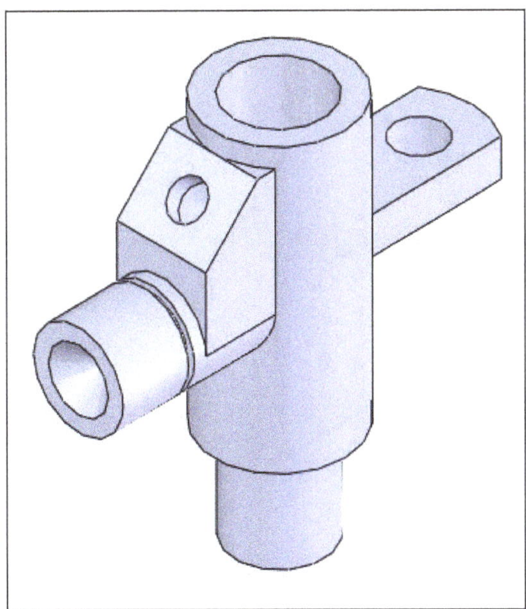

Figure-46. Practice 7 model

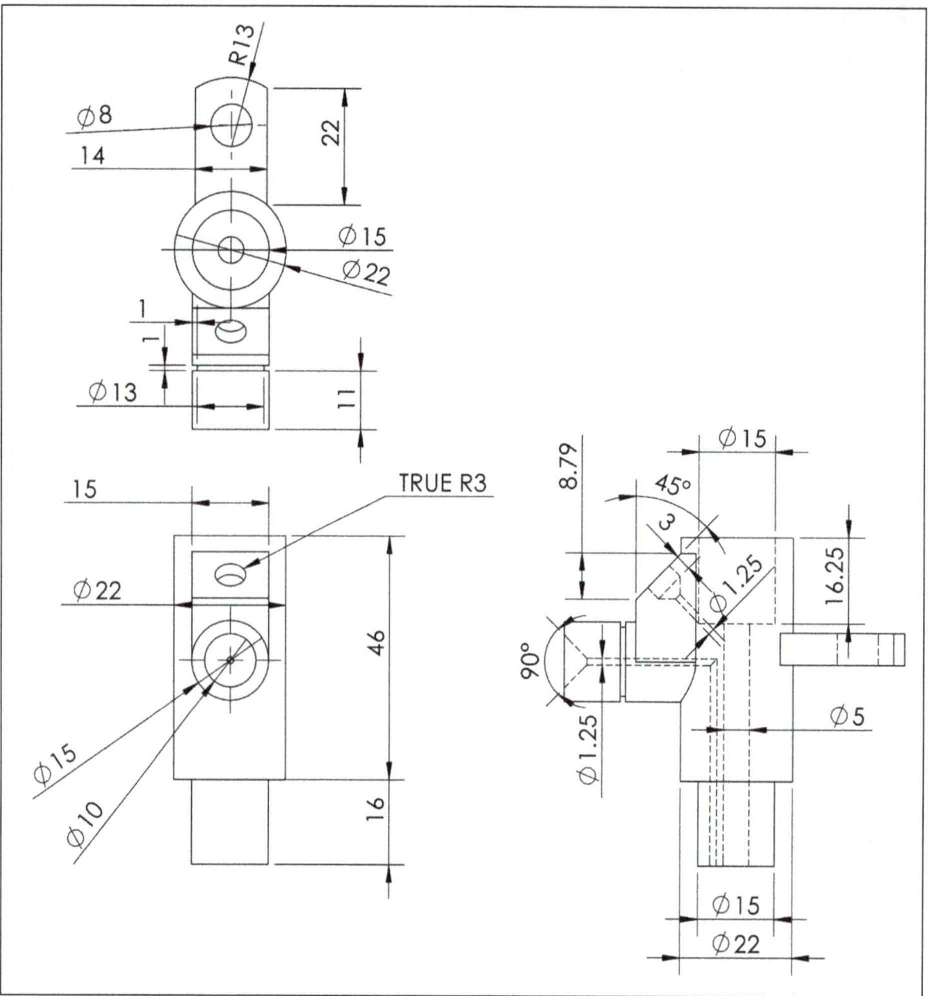

Figure-47. Practice 7

Practice 8

Create the model by using the dimensions given in Figure-48.

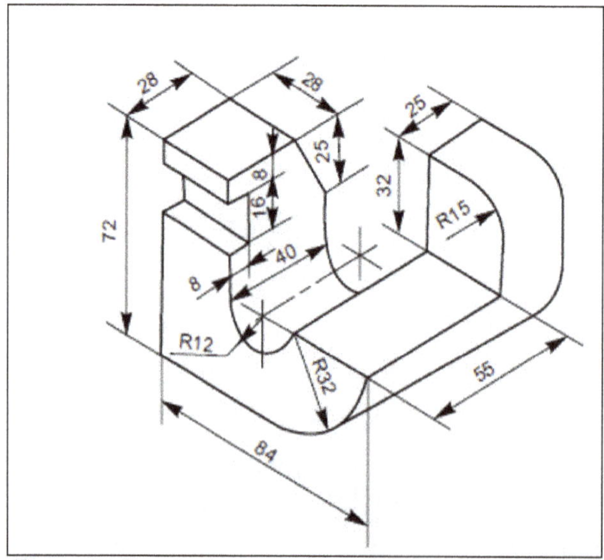

Figure-48. Practice 8

Practice 9

Create the model by using the dimensions given in Figure-49.

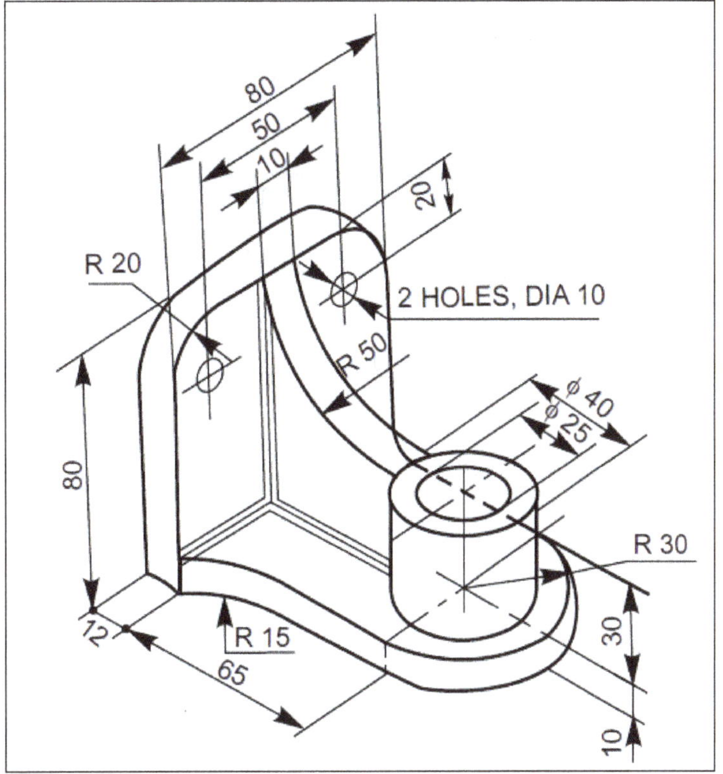

Figure-49. Practice 9

Practice 10

Create the model by using the dimensions given in Figure-50.

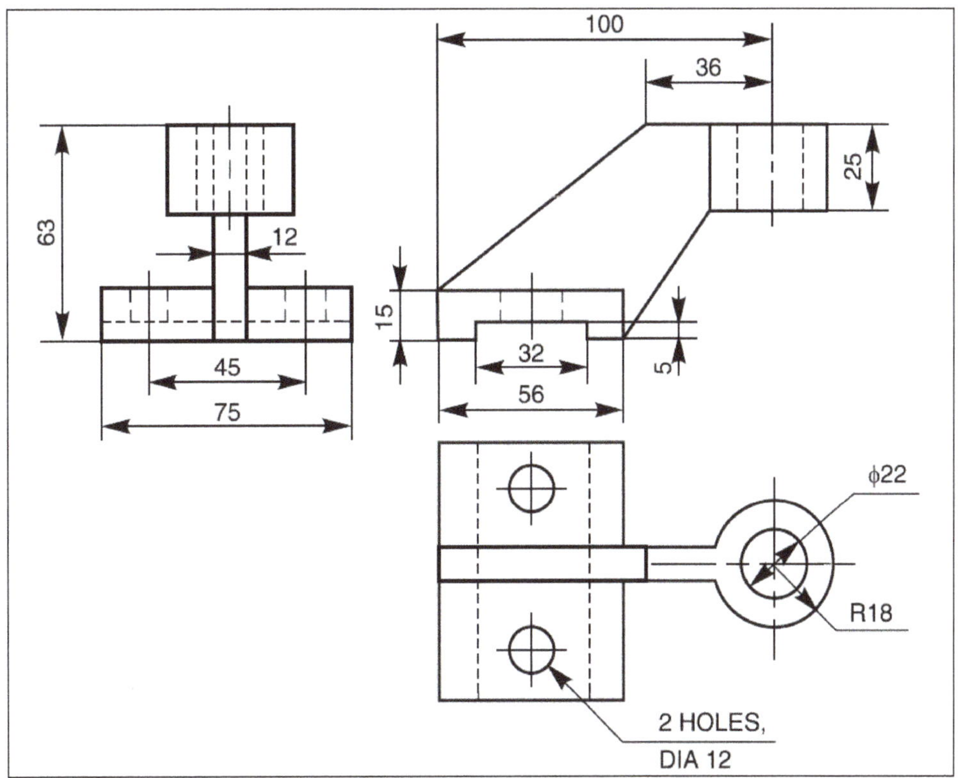

Figure-50. Practice 10

PRACTICE 11

Create a ring nut with value of **D** as **5**, **6**, **8**, and **10** using equation and design table. Dimensions are given in Figure-51.

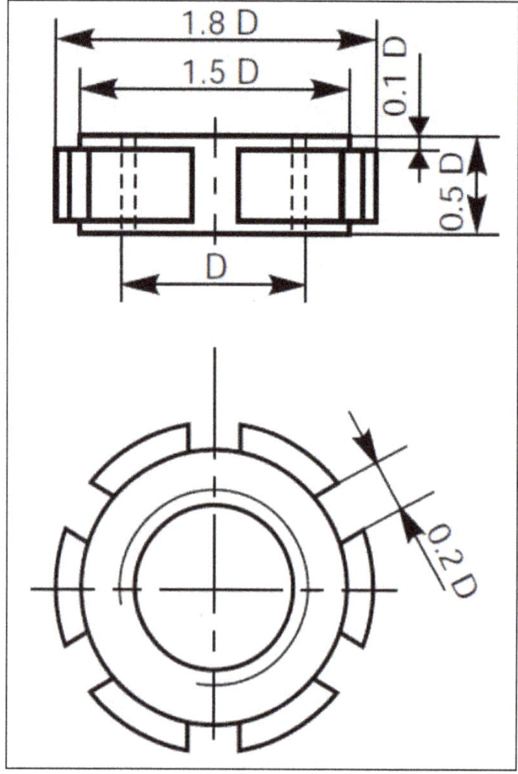

Figure-51. Ring Nut

Chapter 6

Drafting

Topics Covered

The major topics covered in this chapter are:

- ***Introduction to Drafting***
- ***Drawing Objects Tools***
- ***Annotation Objects Tools***
- ***Modifying Objects Tools***
- ***Draft Tray Toolbar***
- ***Draft Snap Toolbar***

INTRODUCTION

The tools in **Draft Workbench** are used to draw simple 2D objects and modify them afterwards. It also provides tools to define a working plane, a grid, and a snapping system to precisely control the position of your geometry. Example of the functionality of the **Draft Workbench** with many elements like lines, curves, arcs, and polygons; refer to Figure-1.

Figure-1. Draft workbench example

The created 2D objects can be used for general drafting in a way similar to Inkscape or Autocad. These 2D shapes can also be used as base features of 3D objects created with other workbenches, for example, the **Part** and **Arch** workbenches. Conversion of draft objects to sketches is also possible which means that the shapes can also be used with the **PartDesign** workbench for the creation of solid bodies.

STARTING DRAFT WORKBENCH

FreeCAD is primarily a 3D modeling application and thus its 2D tools aren't as advanced as in other drawing programs. If your primary goal is the production of complex 2D drawings and DXF files and you don't need 3D modeling, you may wish to consider a dedicated software program for technical drafting such as LibreCAD, QCad, or others.

- To start a new draft file, click on the **New** button from **File** menu and select **Draft** workbench from **Switch between workbenches** drop-down in the **Toolbar**; refer to Figure-2. The tools to perform drafting will be displayed in the **Toolbar**; refer to Figure-3.

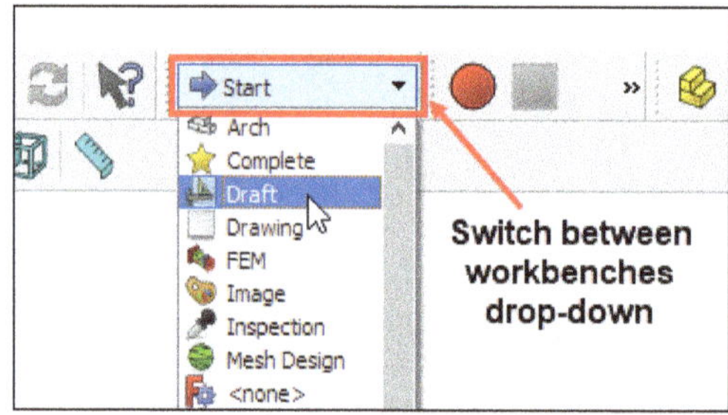

Figure-2. Draft workbench

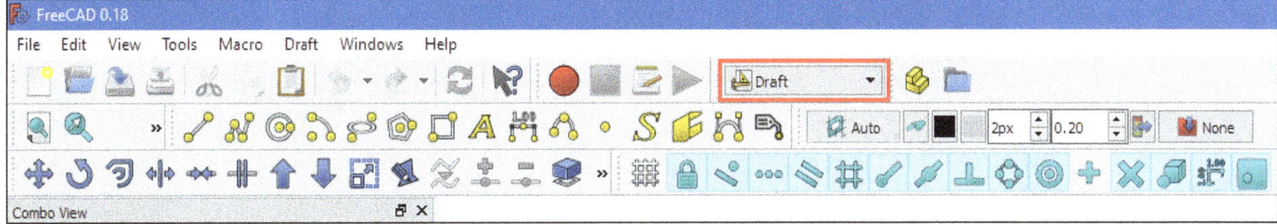
Figure-3. Draft workbench tools

Drawing objects tools

The Drawing objects tools are used for creating 2D objects. These tools are available in the **Toolbar** of **Draft** workbench; refer to Figure-4. The procedures to use these tools are discussed next.

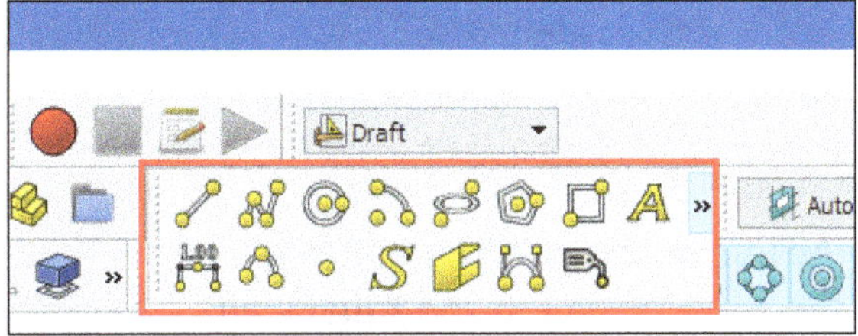
Figure-4. Drawing objects tools

Creating Line

The **Line** tool creates a straight line defined by two points. The procedure to use this tool is discussed next.

- Click on the **Line** tool from **Toolbar** in the **Draft** workbench; refer to Figure-5. The **Line** dialog will be displayed in the **Tasks** panel of **Combo View** along with the plus sign in place of original cursor; refer to Figure-6. You will be asked to specify the first point.

Figure-5. Line tool

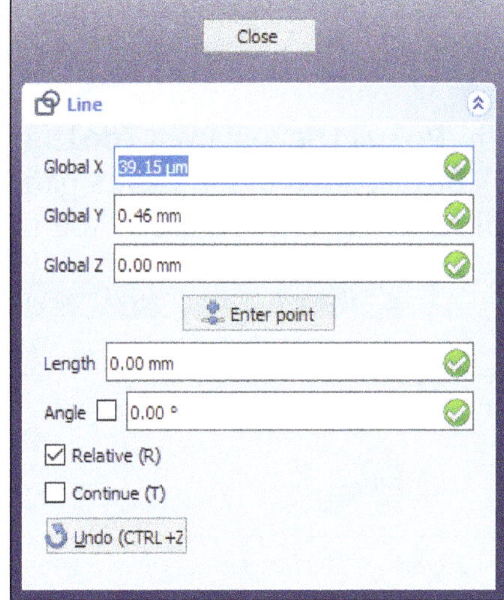

Figure-6. Line dialog

- Click on the 3D view area to specify the first point or enter desired values for x, y, and z coordinates in the **Global X**, **Global Y**, and **Global Z** edit boxes of the dialog, respectively.
- After specifying coordinates for the first point in dialog, click on the **Enter point** button from the dialog. You will be asked to specify the second point.
- Move the cursor away and click at desired location to specify the second point or enter desired values for the coordinates in their respective edit boxes.
- After specifying coordinates for the second point in dialog, click on the **Enter point** button again. The line will be created; refer to Figure-7.

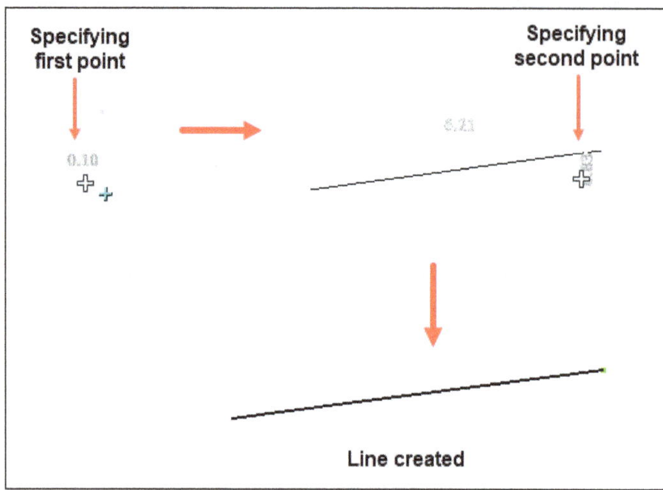

Figure-7. Line created

- Specify desired length of the line in the **Length** edit box.
- Specify desired value for the angle of line in the **Angle** edit box. Select the check box next to the **Angle** to constrain the pointer to the specified angle.
- On selecting the **Relative (R)** check box, the coordinates of second point are relative to the first one. If not selected, they are absolute taken from the origin.
- On selecting the **Continue (T)** check box, the **Line** tool will restart after you give the second point allowing you to draw another line segment without clicking on the tool button again.
- Click on **Undo (CTRL+Z)** button to undo the last point.
- Click on **Close** button to close the dialog.

Creating Wire

The **Wire** or **Polyline** tool creates a sequence of several line segments. This tool allows you to enter more than two points. The procedure to use this tool is discussed next.

- Click on the **Polyline** tool from **Toolbar** in the **Draft** workbench; refer to Figure-8. The **DWire** dialog will be displayed in the **Tasks** panel of **Combo View** along with the plus sign in place of original cursor; refer to Figure-9. You will be asked to specify first point of the line.

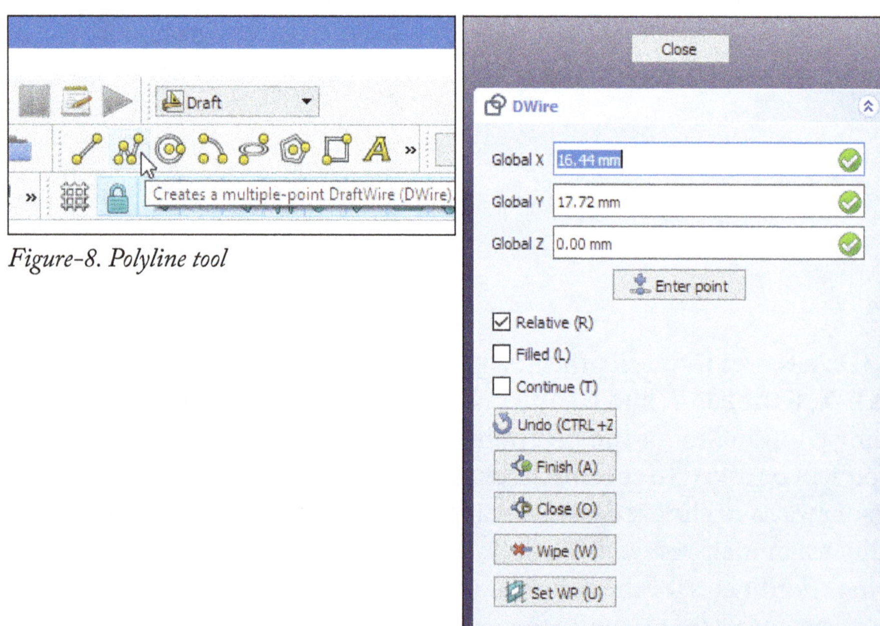

Figure-8. Polyline tool

Figure-9. DWire dialog

- Click in the 3D view area to specify the first point or enter desired values for x, y, and z coordinates in the **Global X**, **Global Y**, and **Global Z** edit boxes of the dialog, respectively.
- After specifying coordinates for the first point in dialog, click on the **Enter point** button from the dialog. You will be asked to specify the additional points.
- Move the cursor away and click at desired locations to specify the additional points for the polyline or enter desired values for the coordinates in their respective edit boxes.
- After specifying coordinates for additional points, click on the **Enter point** button again. The polyline will be created; refer to Figure-10.

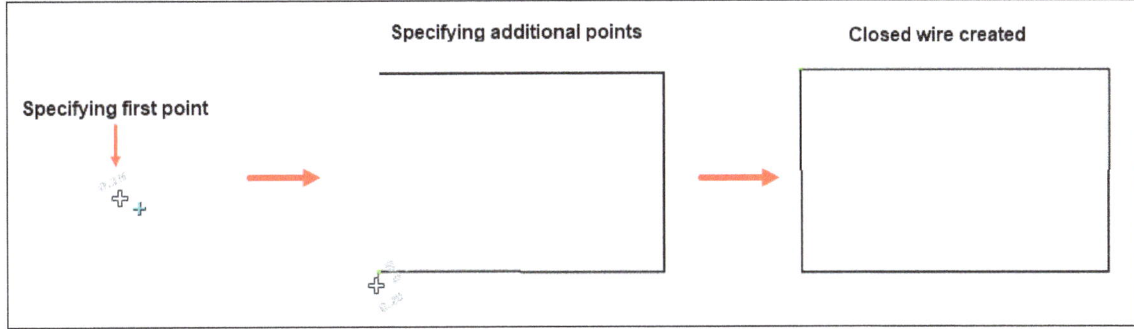

Figure-10. Polyline created

- On selecting the **Filled** check box, a closed wire or a polyline will create a face.
- Click on the **Finish (A)** button to finish the polyline, leaving it open.
- Click on the **Close (O)** button to close the wire, that is, a line segment will be added from the last point to the first one to form a face.
- Click on the **Wipe (W)** button to remove the line segments already placed but keep editing the wire from the last point.
- Click on the **Set WP (U)** button to adjust the current working plane in the orientation of the last point.
- Other parameters of the **DWire** dialog have been discussed in the **Line** tool.
- Click on **Close** button to close the dialog.

Creating Circle

The **Circle** tool creates a circle in the current work plane by entering two points, the center and the radius, or by picking tangents, or any combination of these. The procedure to use this tool is discussed next.

- Click on the **Circle** tool from **Toolbar** in the **Draft** workbench; refer to Figure-11. The **Circle** dialog will be displayed in the **Tasks** panel of **Combo View** along with the plus sign in place of original cursor; refer to Figure-12. You will be asked to specify the first point.

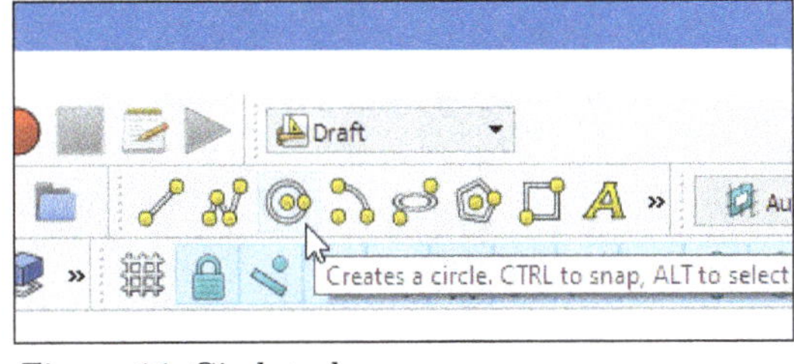

Figure-11. Circle tool

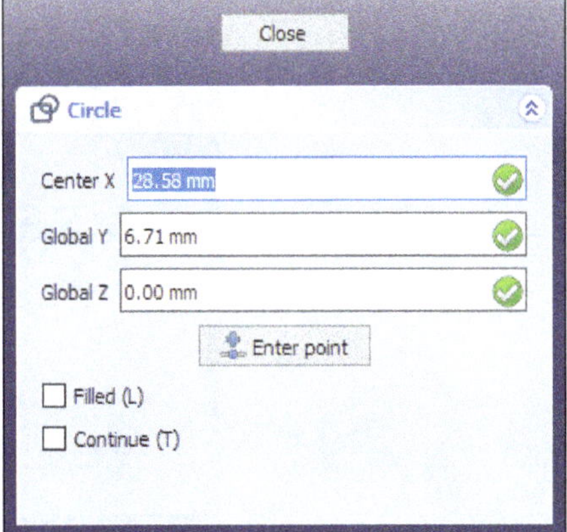

Figure-12. Circle dialog

- Click in the 3D view area to specify the first point or enter desired values for x, y, and z coordinates in the **Center X**, **Global Y**, and **Global Z** edit boxes of the dialog, respectively.
- After specifying coordinates for the first point in dialog, click on the **Enter point** button. You will be asked to specify the radius of the circle.
- Move the cursor away and click at desired location to specify the second point for the circle or enter desired radius value in the **Radius** edit box of the dialog. The circle will be created; refer to Figure-13.

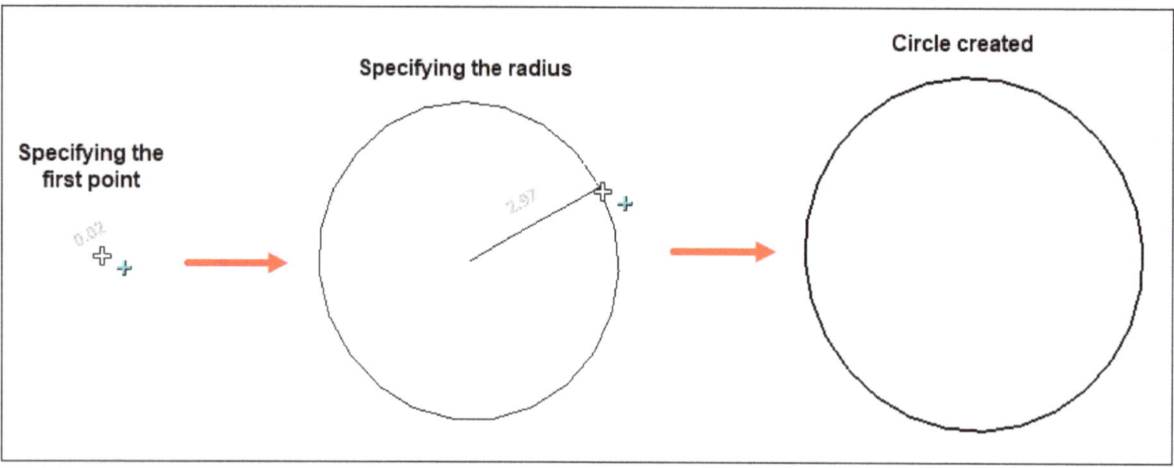

Figure-13. Circle created

- Other parameters of the **Circle** dialog have been discussed in the previous tool.
- Click on **Close** button to close the dialog.

Creating Arc

The **Arc** tool creates a circular arc in the current work plane by entering four points, the center, a point that defines the radius, the first end point and the second end point, or by picking tangents, or any combination of these. The procedure to use this tool is discussed next.

- Click on the **Arc** tool from **Toolbar** in the **Draft** workbench; refer to Figure-14. The **Arc** dialog will be displayed in the **Tasks** panel of **Combo View** along with the plus sign in place of original cursor; refer to Figure-15. You will be asked to specify the first point.

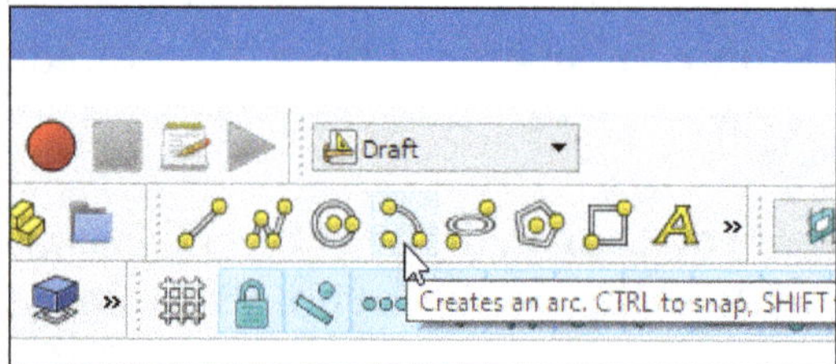

Figure-14. Arc tool

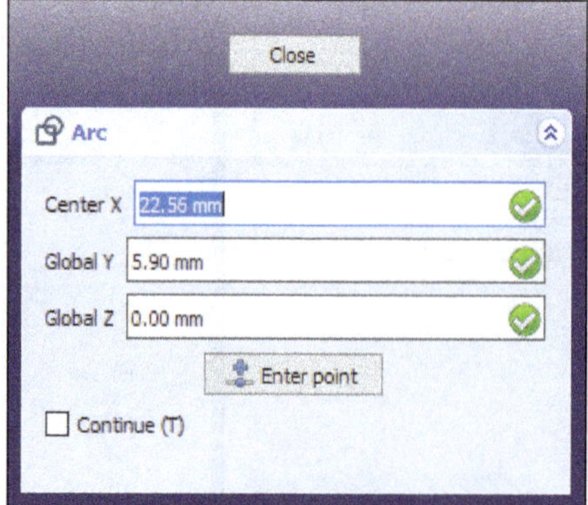

Figure-15. Arc dialog

- Click in the 3D view area to specify the first point or enter desired values for x, y, and z coordinates in the **Center X**, **Global Y**, and **Global Z** edit boxes of the dialog, respectively.
- After specifying coordinates in the dialog, click on the **Enter point** button. You will be asked to specify the radius of arc.
- Move the cursor away and click at desired location to specify the second point or enter desired radius value for the arc in the **Radius** edit box. You will be asked to specify the start angle.
- Click in the 3D view area to specify the start angle for the arc or enter desired angle value in the **Start angle** edit box of the dialog. You will be asked to specify the aperture angle.
- Click in the 3D view area to specify the aperture angle of an arc or enter desired angle value in the **Aperture** edit box of the dialog. The arc will be created; refer to Figure-16.

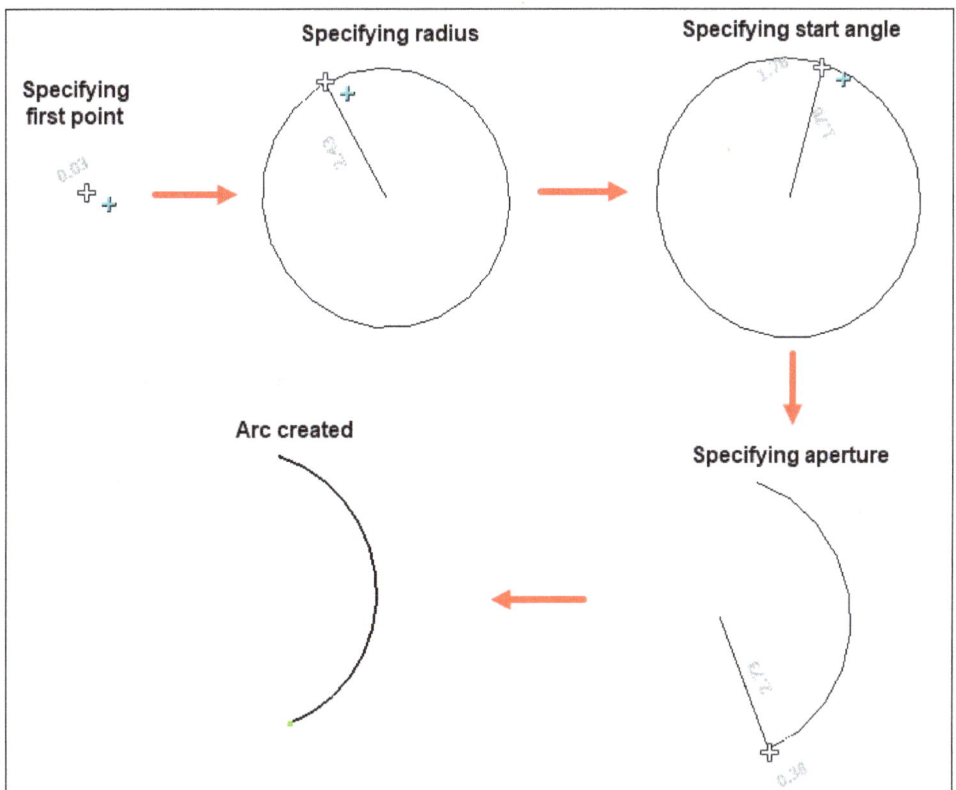

Figure-16. Arc created

- Other parameters of the **Arc** dialog have been discussed earlier.
- Click on **Close** button to close the dialog.

Creating Ellipse

The **Ellipse** tool creates an ellipse in the current work plane by entering two points, defining the corners of a rectangular box in which the ellipse will fit. This tool can also be used to create elliptical arcs by specifying the start and end angles. The procedure to use this tool is discussed next.

- Click on the **Ellipse** tool from **Toolbar** in the **Draft** workbench; refer to Figure-17. The **Ellipse** dialog will be displayed in the **Tasks** panel of **Combo View** along with the plus sign in place of original cursor; refer to Figure-18. You will be asked to specify the first point.

Figure-17. Ellipse tool

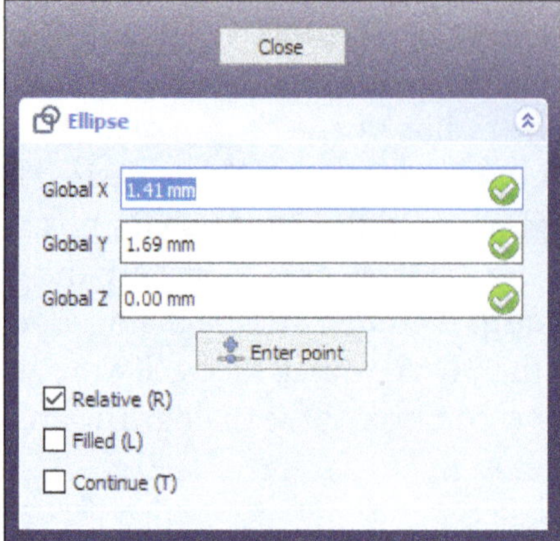

Figure-18. Ellipse dialog

- Click in the 3D view area to specify the first point or enter desired values for the x, y, and z coordinates in the **Global X**, **Global Y**, and **Global Z** edit boxes of the dialog, respectively.
- After specifying coordinates for the first point in dialog, click on the **Enter point** button from the dialog. You will be asked to specify the second point.
- Move the cursor away and click at desired location to specify the second point or enter desired values for the coordinates in their respective edit boxes.
- After specifying coordinates for the second point in dialog, click on the **Enter point** button again. The ellipse will be created; refer to Figure-19.

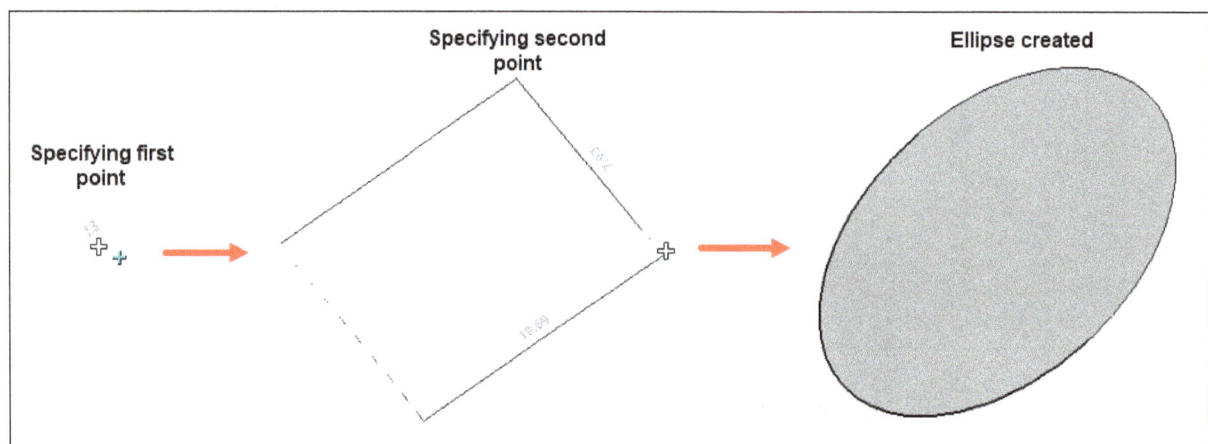

Figure-19. Ellipse created

- Other parameters in the **Ellipse** dialog have been discussed earlier.
- Click on **Close** button to close the dialog.

Creating Polygon

The **Polygon** tool creates a regular polygon inscribed in a circumference by picking two points, the center and the radius. The procedure to use this tool is discussed next.

- Click on the **Polygon** tool from **Toolbar** in the **Draft** workbench; refer to Figure-20. The **Polygon** dialog will be displayed in the **Tasks** panel of **Combo View** along with the plus sign in place of original cursor; refer to Figure-21. You will be asked to specify the first point.

Figure-20. Polygon tool

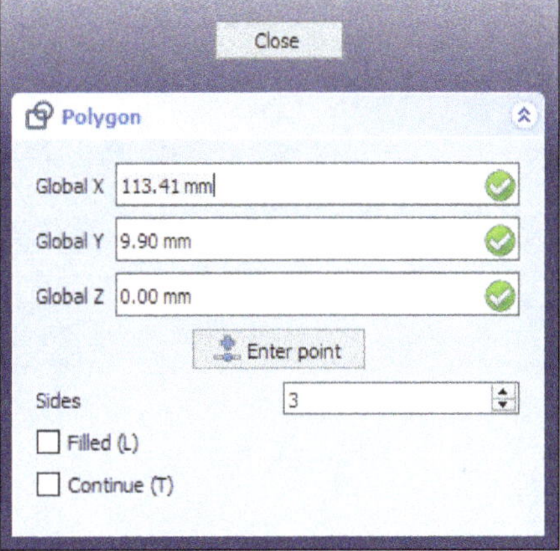

Figure-21. Polygon dialog

- Click in the 3D view area to specify the first point or enter desired values for the x, y, and z coordinates in the **Global X**, **Global Y**, and **Global Z** edit boxes of the dialog, respectively.
- After specifying coordinates in the dialog, click on the **Enter point** button from the dialog. You will be asked to specify the sides and radius.
- Specify desired number of sides for the polygon in the **Sides** edit box of the dialog.
- Move the cursor away and click at desired location to specify the radius of the polygon or enter desired value of radius in the **Radius** edit box of the dialog. The regular polygon will be created; refer to Figure-22.

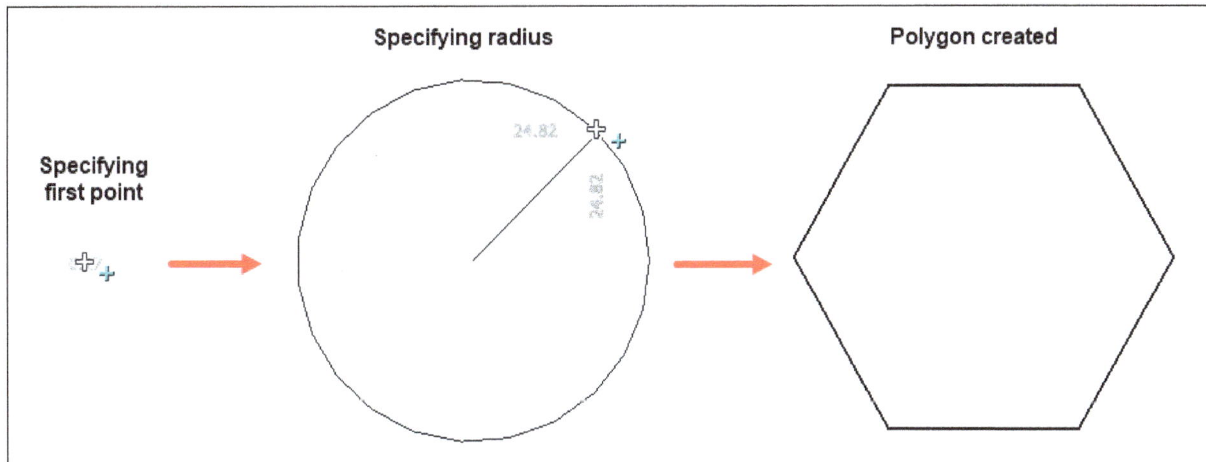

Figure-22. Regular polygon created

- Other parameters in the **Polygon** dialog have been discussed earlier.
- Click on **Close** button to close the dialog.

Creating Rectangle

The **Rectangle** tool creates a rectangle by picking two points. The procedure to use this tool is discussed next.

- Click on the **Rectangle** tool from **Toolbar** in the **Draft** workbench; refer to Figure-23. The **Rectangle** dialog will be displayed in the **Tasks** panel of **Combo View** along with the plus sign in place of original cursor; refer to Figure-24. You will be asked to specify the first point.

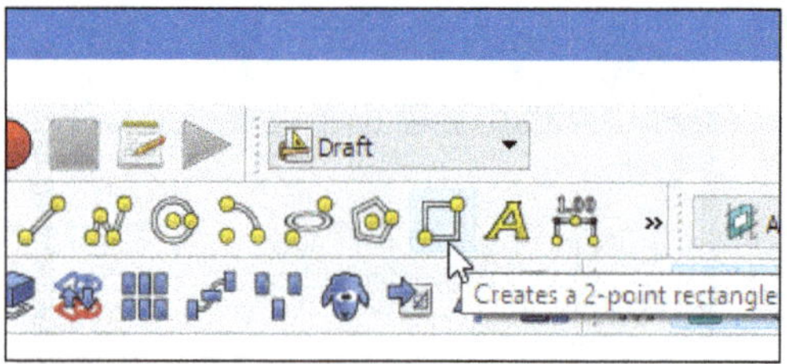

Figure-23. Rectangle tool

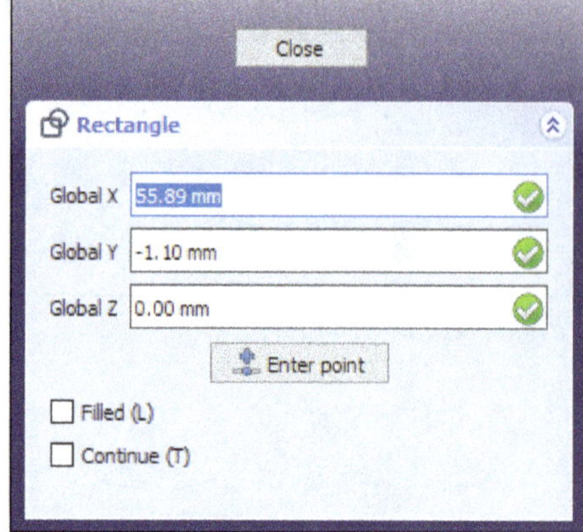

Figure-24. Rectangle dialog

- Click in the 3D view area to specify the first point or enter desired values for the x, y, and z coordinates in the **Global X**, **Global Y**, and **Global Z** edit boxes of the dialog, respectively.
- After specifying coordinates for the first point in dialog, click on the **Enter point** button from the dialog. You will be asked to specify the second point.
- Move the cursor away and click at desired location to specify the second point or enter desired values for the coordinates in their respective edit boxes of the dialog.
- After specifying coordinates for the second point in dialog, click on the **Enter point** button again. The rectangle will be created; refer to Figure-25.

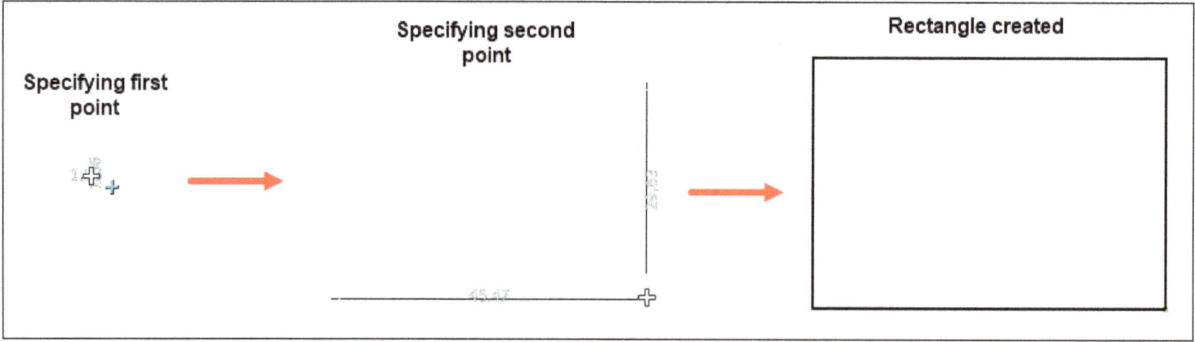

Figure-25. Rectangle created

- Other parameters in the **Rectangle** dialog have been discussed earlier.
- Click on **Close** button to close the dialog.

Creating Text

The **Text** tool inserts a multi-line textbox at a given point. The procedure to use this tool is discussed next.

- Click on the **Text** tool from **Toolbar** in the **Draft** workbench; refer to Figure-26. The **Text** dialog will be displayed in the **Tasks** panel of **Combo View** along with the plus sign in place of original cursor; refer to Figure-27. You will be asked to specify the location of text.

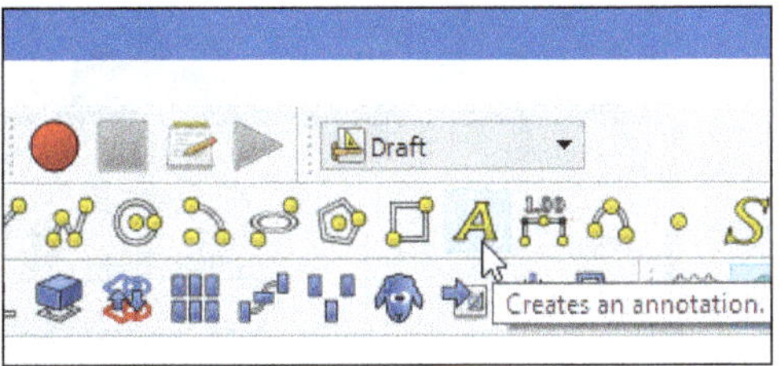

Figure-26. Text tool

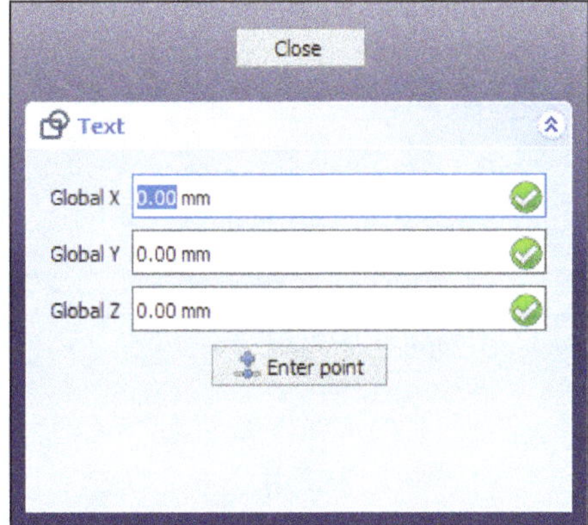

Figure-27. Text dialog

- Click in the 3D view area to specify desired location to place text or enter desired values for the x, y, and z coordinates in the **Global X**, **Global Y**, and **Global Z** edit boxes of the dialog, respectively.
- After specifying coordinates in the dialog, click on the **Enter point** button from the dialog. You will be asked to specify text in the **Text** dialog; refer to Figure-28.

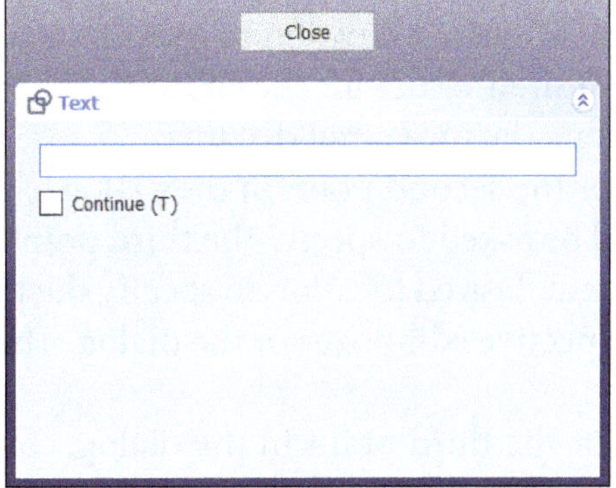

Figure-28. Text dialog asking to specify text

- Specify desired text in the edit box of the dialog and press **ENTER** key twice from the keyboard. The text will be created; refer to Figure-29.

Figure-29. Text created

Creating Dimension

The **Dimension** tool creates an object that measures and displays the distance between two points; a third point specifies the position of the dimension line. The procedure to use this tool is discussed next.

- Click on the **Dimension** tool from **Toolbar** in the **Draft** workbench; refer to Figure-30. The **Dimension** dialog will be displayed in the **Tasks** panel of **Combo View** along with the plus sign in place of original cursor; refer to Figure-31. You will be asked to specify the first point or select the edge.

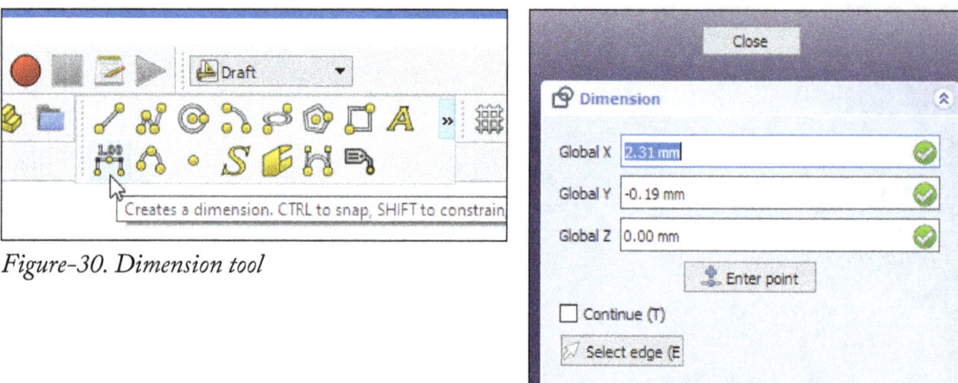

Figure-30. Dimension tool

Figure-31. Dimension dialog

- Click in the 3D view area to specify first point from where you want to measure the distance or enter desired values for the x, y, and z coordinates in the **Global X**, **Global Y**, and **Global Z** edit boxes of the dialog, respectively.
- After specifying coordinates for the first point in the dialog, click on the **Enter point** button from the dialog. You will be asked to specify the second point.
- Move the cursor away and click at desired location to specify second point upto which you want to measure the distance or enter desired values for the coordinates in their respective edit boxes of the dialog. The first two points define the measured distance.
- After specifying coordinates for the second point in the dialog, click on the **Enter point** button again from the dialog. You will be asked to specify the third point.
- Move the cursor away and click at desired location to specify the third point or enter desired values for the coordinates in their respective edit boxes of the dialog. The third point defines the position of the measurement line.
- After specifying coordinates for the third point in the dialog, click on the **Enter point** button again from the dialog. The dimension will be created; refer to Figure-32.

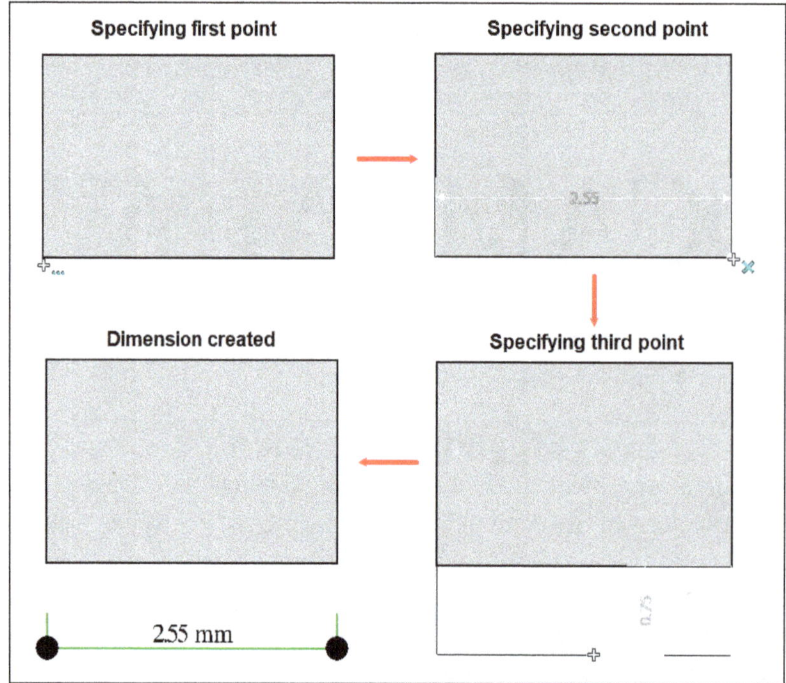

Figure-32. Dimension created

- If you want to create the dimension by selecting the edge then on selecting the tool, the **Dimension** dialog will be displayed as discussed earlier.
- Click on the **Select edge** button from the **Dimension** dialog. You will be asked to select the edge to be measured.
- Select the edge of the model whose distance is to be measured by this dimension; refer to Figure-33. You will be asked to specify the point to place the measurement line.
- Move the cursor away and click at desired location to specify the point to position the measurement line or enter desired values for the x, y, and z coordinates in the **Global X**, **Global Y**, and **Global Z** edit boxes of the dialog, respectively.
- After specifying coordinates for the point in the dialog, click on the **Enter point** button from the dialog. The dimension will be created; refer to Figure-34.

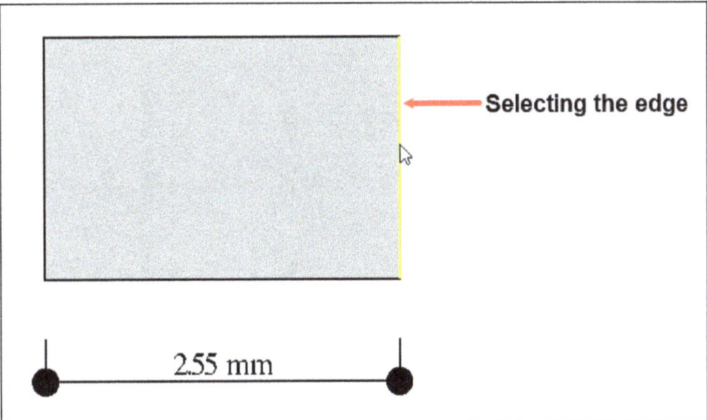

Figure-33. Selecting the edge to be measured

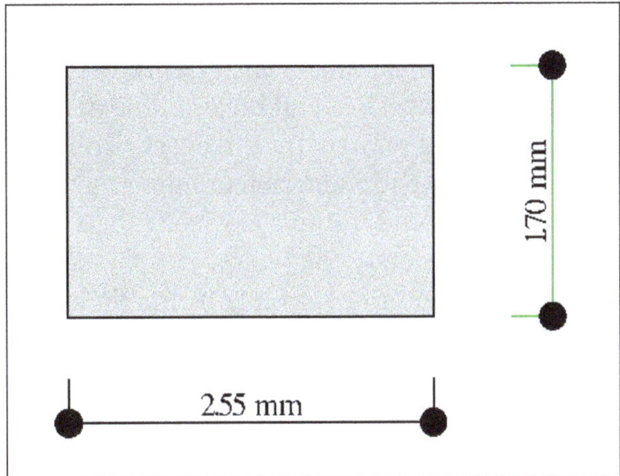

Figure-34. Dimension created

- Other parameters in the **Dimension** dialog have been discussed earlier.
- Click on the **Close** button to close the dialog.

Creating B-Spline

The **B-Spline** tool creates a B-spline curve from several points. The procedure to use this tool is discussed next.

- Click on the **B-spline** tool from **Toolbar** in the **Draft** workbench; refer to Figure-35. The **BSpline** dialog will be displayed in the **Tasks** panel of **Combo View** along with the plus sign in place of original cursor; refer to Figure-36. You will be asked to specify the first point.

Figure-35. B-Spline tool

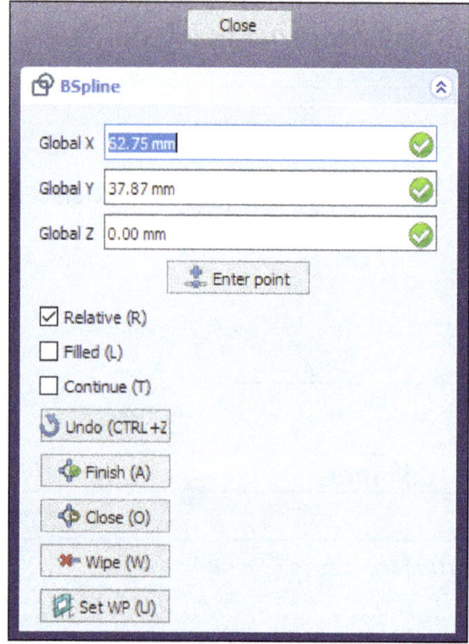

Figure-36. BSpline dialog

- Click in the 3D view area to specify the first point or enter desired values for the x, y, and z coordinates in the **Global X**, **Global Y**, and **Global Z** edit boxes of the dialog, respectively.
- After specifying coordinates for the first point in the dialog, click on the **Enter point** button from the dialog. You will be asked to specify the additional points.
- Move the cursor away and click at desired locations to specify the additional points or enter desired values for the coordinates in their respective edit boxes.
- After specifying coordinates for the additional points in the dialog, click on the **Enter point** button again or press **Esc** key. The B-spline will be created; refer to Figure-37.

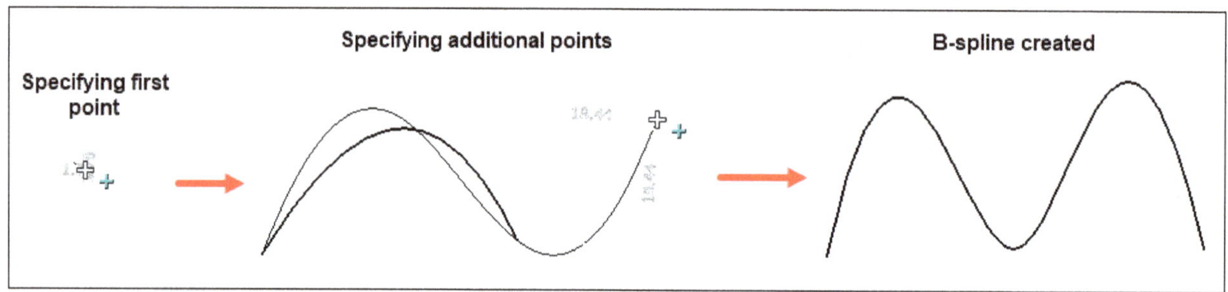

Figure-37. B-spline created

- Other parameters in the **BSpline** dialog have been discussed earlier.
- Click on **Close** button to close the dialog.

Creating Point

The **Point** tool creates a simple point in the current work plane, handy to serve as reference for placing lines, wires, or other objects later. The procedure to use this tool is discussed next.

- Click on the **Point** tool from **Toolbar** in the **Draft** workbench; refer to Figure-38. The **Point** dialog will be displayed in the **Tasks** panel of **Combo View** along with the plus sign in place of original cursor; refer to Figure-39. You will be asked to specify the location of point.

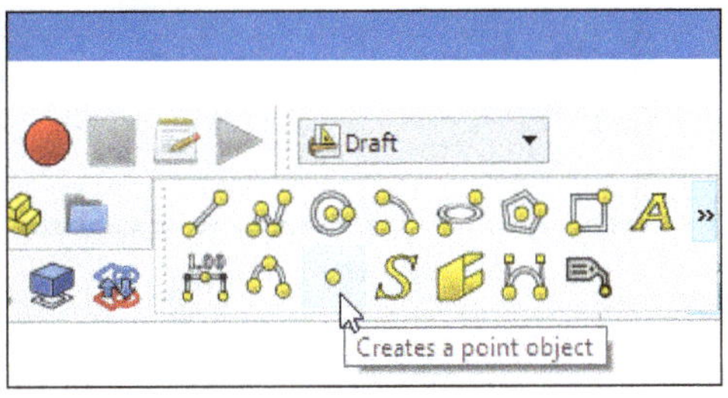

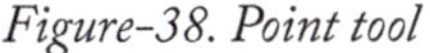

Figure-38. Point tool

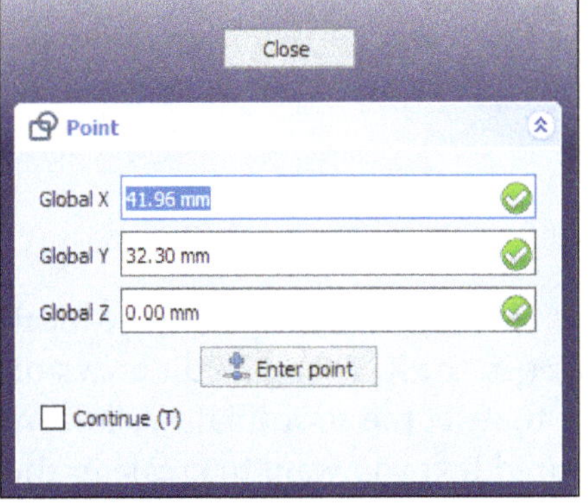

Figure-39. Point dialog

- Click in the 3D view area to specify the location of point or enter desired values for x, y, and z coordinates in the **Global X**, **Global Y**, and **Global Z** edit boxes of the dialog, respectively.
- After specifying coordinates in the dialog, click on the **Enter point** button from the dialog. The point will be created; refer to Figure-40.

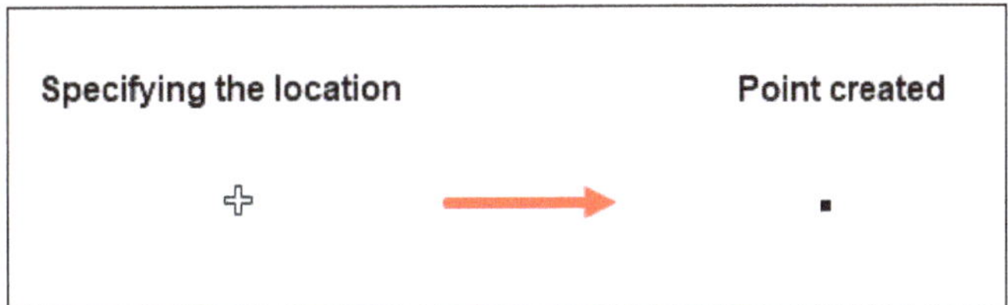

Figure-40. Point created

- Other parameters in the **Point** dialog have been discussed earlier.
- Click on **Close** button to close the dialog.

Creating Shape String

The **Shape String** tool inserts a compound shape that represents a text string. Text height, tracking, and font can be specified. The resulting shape can be used with the **Extrude** tool to create 3D letters. The procedure to use this tool is discussed next.

- Click on the **Shape String** tool from **Toolbar** in the **Draft** workbench; refer to Figure-41. The **ShapeString** dialog will be displayed in the **Tasks** panel of **Combo View** along with the plus sign in place of original cursor; refer to Figure-42. You will be asked to specify the location of shape string.

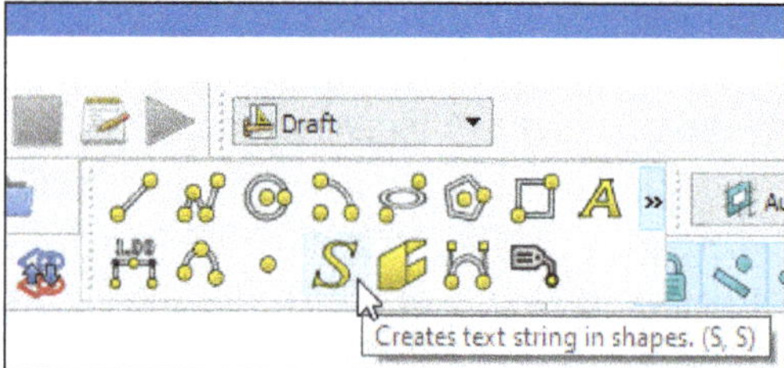

Figure-41. Shape String tool

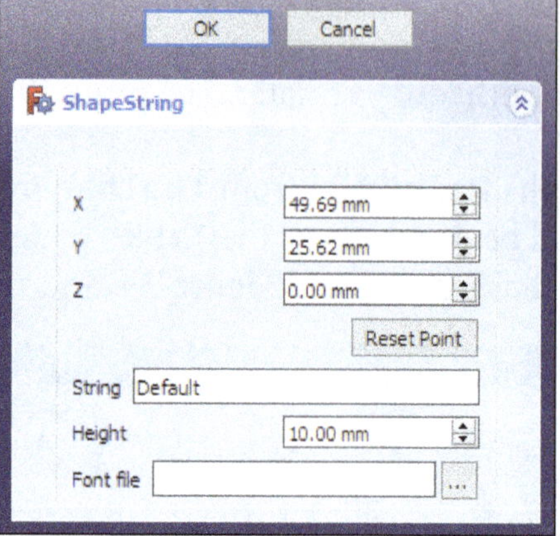

Figure-42. ShapeString dialog

- Click in the 3D view area to specify the location of shape string or enter desired values for x, y, and z coordinates in the **X**, **Y**, and **Z** edit boxes of the dialog, respectively.
- If you want to reset the coordinates, click on the **Reset Point** button from the dialog.
- Specify desired text you want to create in the **String** edit box.
- Specify the height of the text in the **Height** edit box.
- Click on the button ... available next to the **Font file** edit box. The **Select a file** dialog box will be displayed; refer to Figure-43.

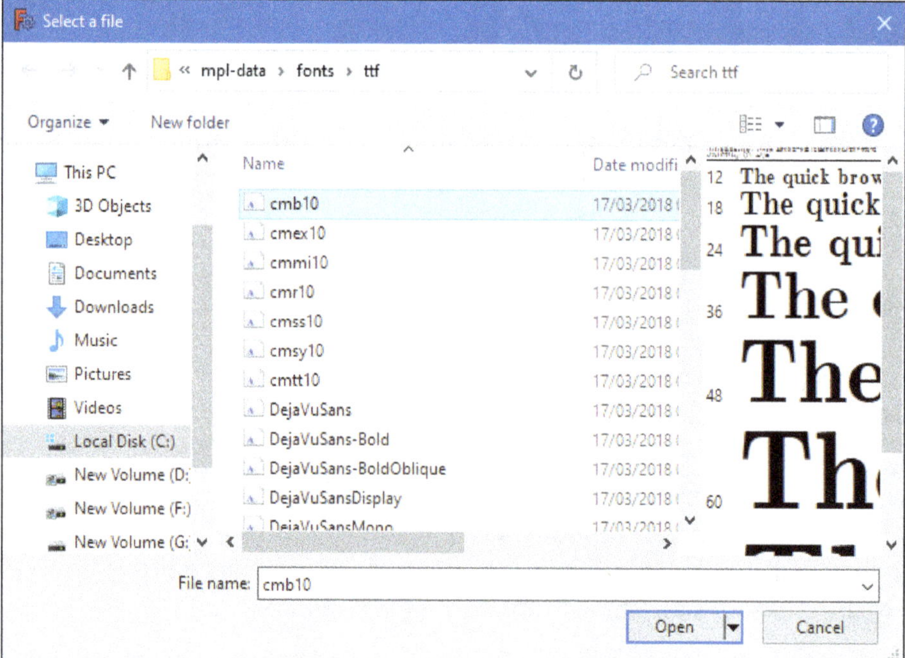

Figure-43. Select a file dialog box

- Select desired font file from the dialog box and click on **Open** button. The selected font file will be displayed in the **Font file** edit box of the dialog; refer to Figure-44. The font file is available in the installation folder of FreeCAD software.

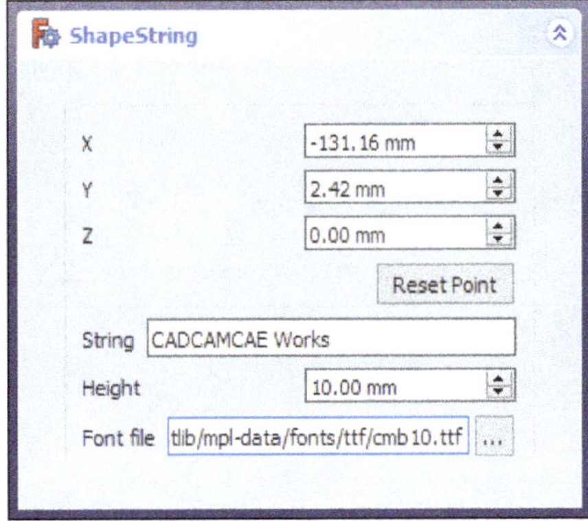

Figure-44. Font file selected

- Click on **OK** button from the dialog. The Shape string will be created; refer to Figure-45.

Figure-45. Shape string created

Creating Facebinder

The **Facebinder** tool creates a surface object from the selected faces of a solid object. It is parametric, meaning that if you modify the original object, the facebinder updates accordingly. It can be used to create an extrusion from a collection of faces from other objects. The procedure to use this tool is discussed next.

- Select desired faces from the solid object which you want to create the facebinder; refer to Figure-46.

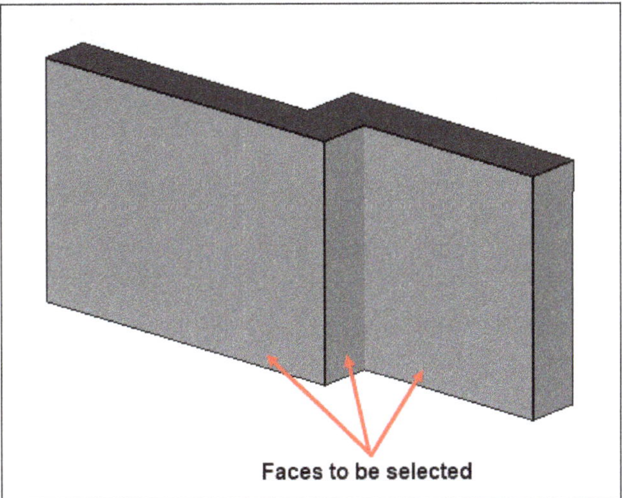

Figure-46. Selecting the faces to create facebinder

- Click on the **Facebinder** tool from **Toolbar** in the **Draft** workbench; refer to Figure-47. The facebinder object will be created; refer to Figure-48.
- If you want to see the facebinder object, select the original solid object from the Model tree view and hide it using **Space** key from the keyboard. The facebinder object will be displayed in the 3D view area; refer to Figure-49.

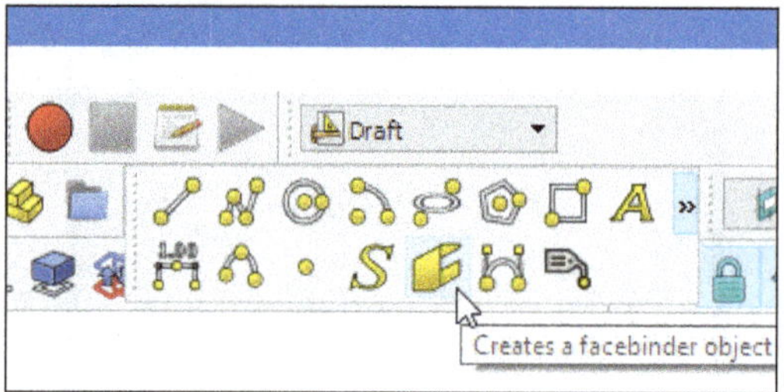

Figure-47. Facebinder tool

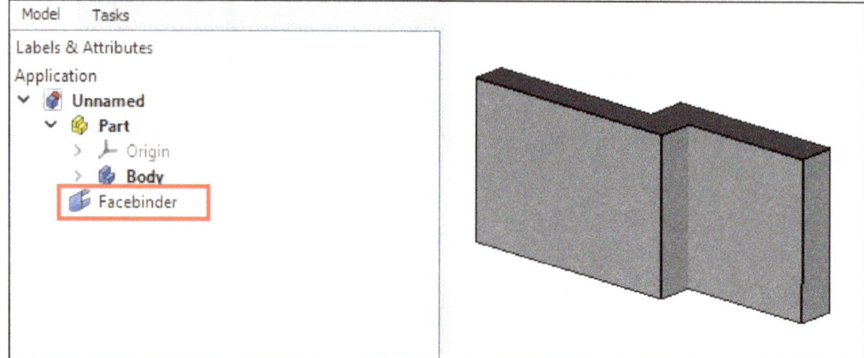

Figure-48. Facebinder object created

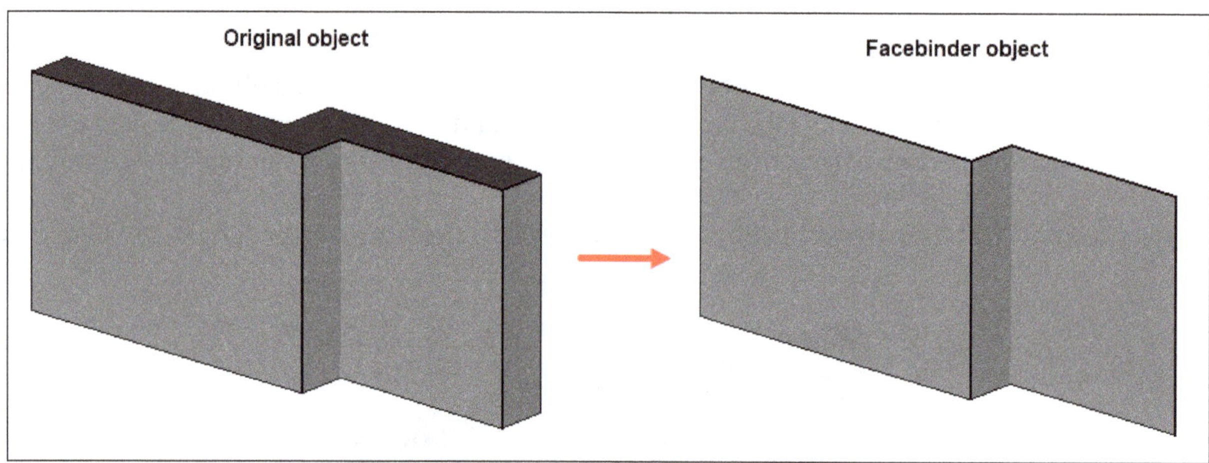

Figure-49. Facebinder object displayed

- You can edit the facebinder object using **Faces** dialog.
- To add the face to the facebinder object, press and hold the **Ctrl** key from the keyboard and select the face or multiple faces from the solid object in the 3D view area; refer to Figure-50.

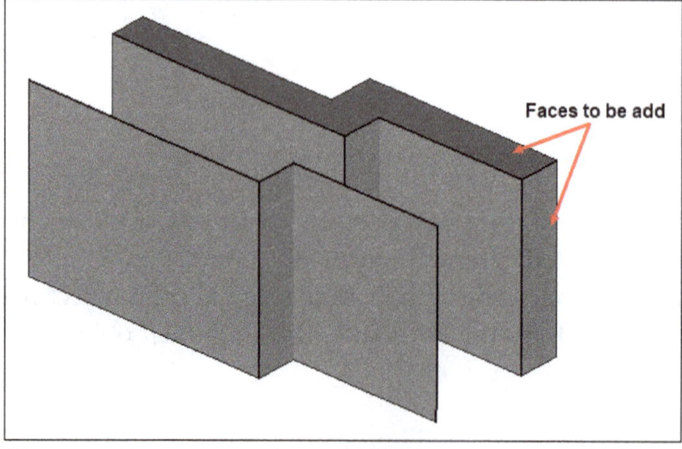

Figure-50. Faces to be add from solid object

- Holding the **Ctrl** key, double-click on the facebinder object from the Model tree view. The **Faces** dialog will be displayed in the **Tasks** panel of **Combo View**; refer to Figure-51.

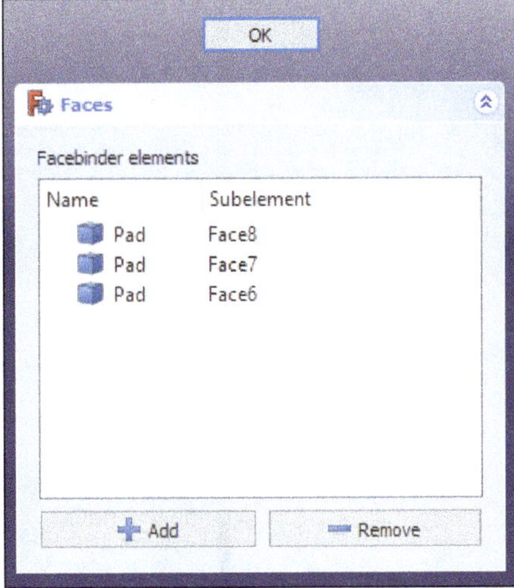

Figure-51. Faces dialog

- Click on **Add** button from the dialog, the faces added in the dialog will be added in the facebinder object; refer to Figure-52.

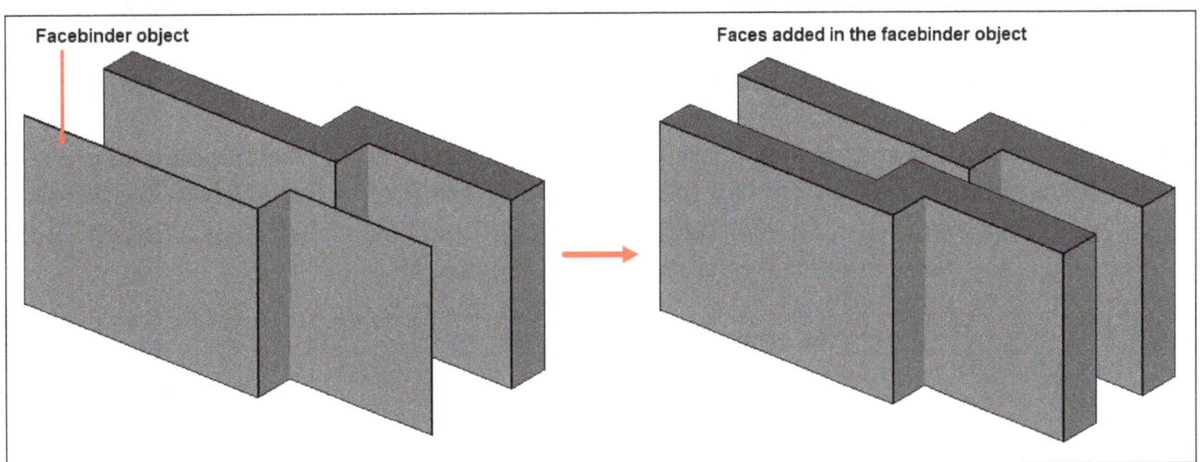

Figure-52. Faces added in the facebinder object

- If you want to remove the face from the facebinder object then select the face from the **Facebinder elements** list view area of the dialog and click on **Remove** button. The face will be removed.
- Click on **OK** button from the dialog to complete the procedure.

Creating Bezier Curve

The **Bezier Curve** tool creates a bezier curve or a piecewise bezier curve from several points. This tool uses control points to define the direction of the curve.The procedure to use this tool is discussed next.

- Click on the **Bezier Curve** tool from **Toolbar** in the **Draft** workbench; refer to Figure-53. The **BezCurve** dialog will be displayed in the **Tasks** panel of **Combo View** along with the plus sign in place of original cursor; refer to Figure-54. You will be asked to specify the first point.

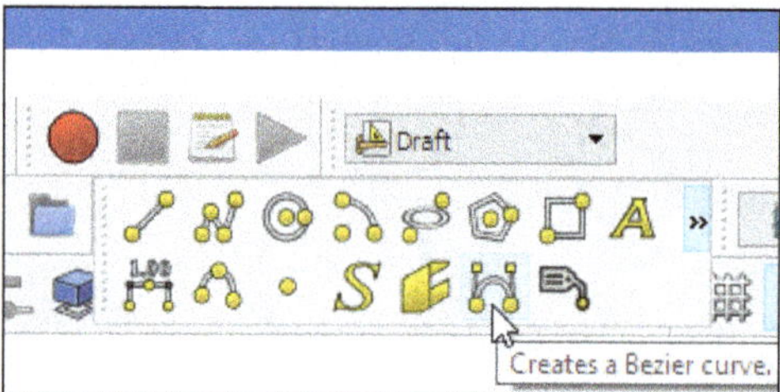

Figure-53. Bezier curve tool

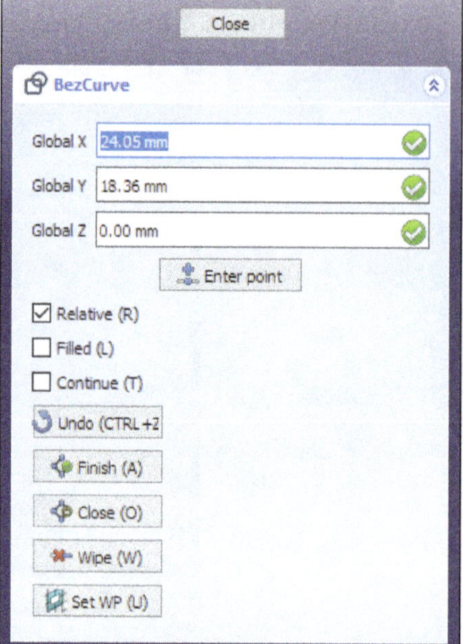

Figure-54. BezCurve dialog

- Click in the 3D view area to specify the first point or enter desired values for the x, y, and z coordinates in the **Global X**, **Global Y**, and **Global Z** edit boxes of the dialog, respectively.
- After specifying coordinates for the first point in the dialog, click on the **Enter point** button from the dialog.
- Move the cursor away and click at desired locations to specify the additional points to create the bezier curve or enter desired values for the coordinates in their respective edit boxes of the dialog.
- After specifying coordinates for the additional points in the dialog, click on the **Enter point** button again. The bezier curve will be created; refer to Figure-55.

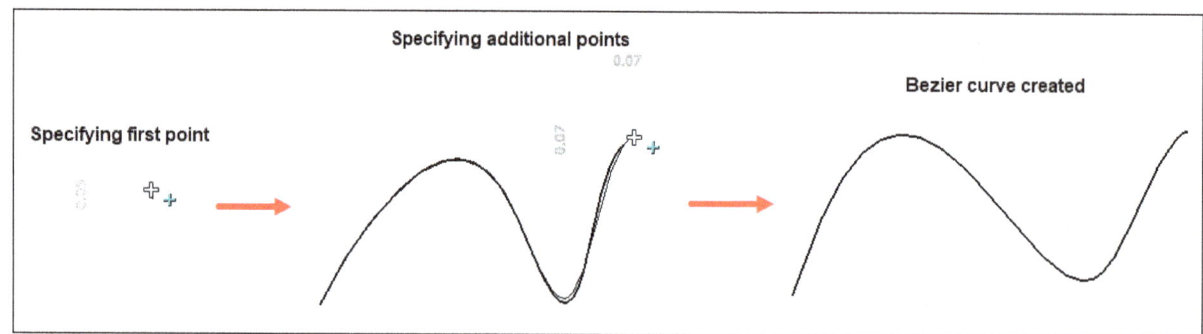

Figure-55. Bezier curve created

- Other parameters in the **BezCurve** dialog have been discussed earlier.
- Click on **Close** button to close the dialog.

Creating Label

The **Label** tool inserts a multi-line textbox with a 2-segment leader line and an arrow. If an object or a sub-element (face, edge, or vertex) is selected when starting the command, the Label can be made to display a certain attribute of the selected element including position, length, area, volume, or material. The procedure to use this tool is discussed next.

- Click on the **Label** tool from **Toolbar** in the **Draft** workbench; refer to Figure-56. The **Label** and **Label type** dialogs will be displayed in the **Tasks** panel of **Combo View** along with the plus sign in place of original cursor; refer to Figure-57. You will be asked to specify the first point.

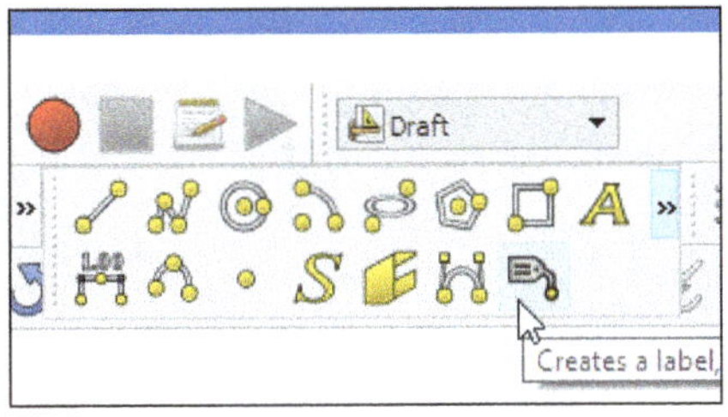

Figure-56. Label tool

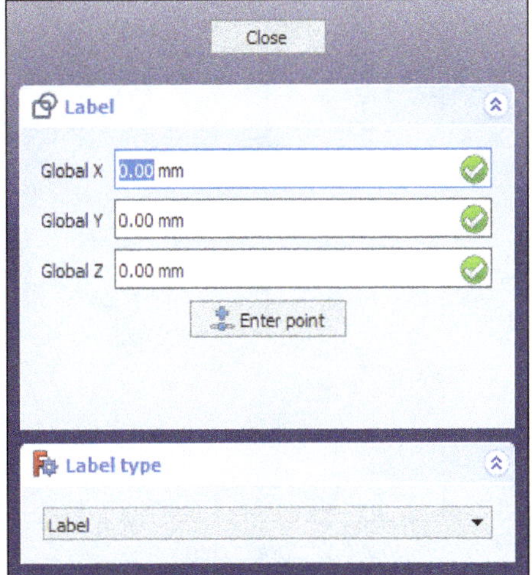

Figure-57. Label dialog

- Select desired label type from the drop-down available in the **Label type** dialog. Select **Custom** option to display the contents of custom text. Select **Name** option to display the internal name of target object. Select **Label** option to display the label of target object. Select **Position** option to display the coordinates of the base point of the target object. Select **Length** option to display the length of the target sub-element. Select **Area** option to display the area of the target sub-element. Select **Volume** option to display the volume of target object. Select **Tag** option to display the tag attribute of the target object. Select **Material** option to display the label of the material of the target object.
- Click in the 3D view area to specify the first point of label or enter desired values for the x, y, and z coordinates in the **Global X**, **Global Y**, and **Global Z** edit boxes of the dialog, respectively.
- After specifying coordinates for the first point in the dialog, click on the **Enter point** button from the dialog. You will be asked to specify the second point of label or the start point of a horizontal or vertical leader.
- Move the cursor away and click at desired location to specify the second point or enter desired value for the coordinates in their respective edit boxes.
- After specifying coordinates for the second point, click on the **Enter point** button from the dialog. You will be asked to specify the third point of label or the base point of the text.
- Move the cursor away and click at desired location to specify the third point or enter desired values for the coordinates in their respective edit boxes.
- After specifying coordinates for the third point in the dialog, click on the **Enter point** button again. The label will be created; refer to Figure-58.

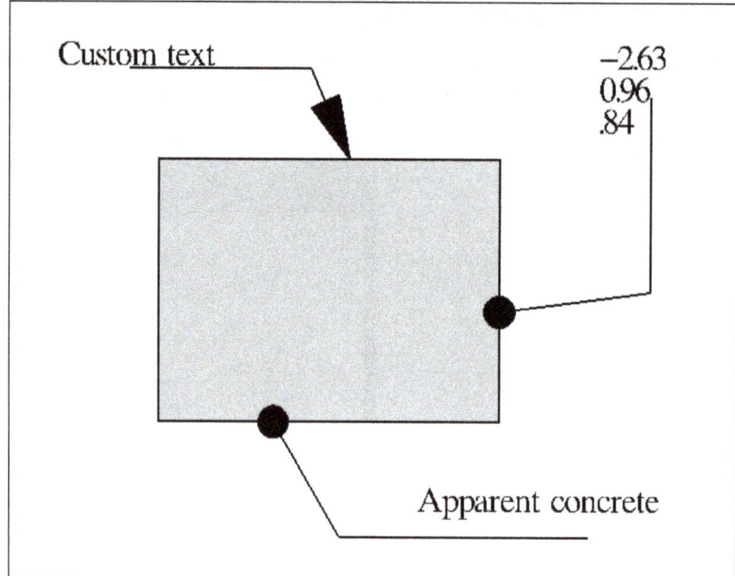

Figure-58. Labels created

Modifying objects tools

The **Modifying objects** tools are used for modifying existing objects. They work on selected objects but if no object is selected, you will be asked to select one. These tools are available in the **Toolbar** of **Draft** workbench; refer to Figure-59. These tools are discussed next.

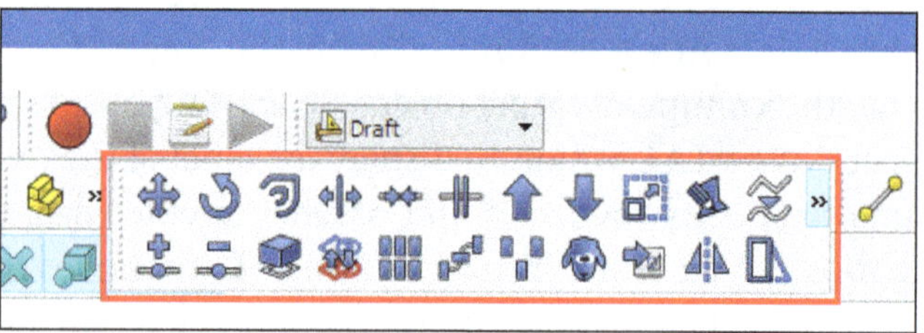

Figure-59. Modifying object tools

Moving object

The **Move** tool moves or copies the selected objects from one point to another. The procedure to use this tool is discussed next.

- Select the object from the Model tree view which you want to move; refer to Figure-60.

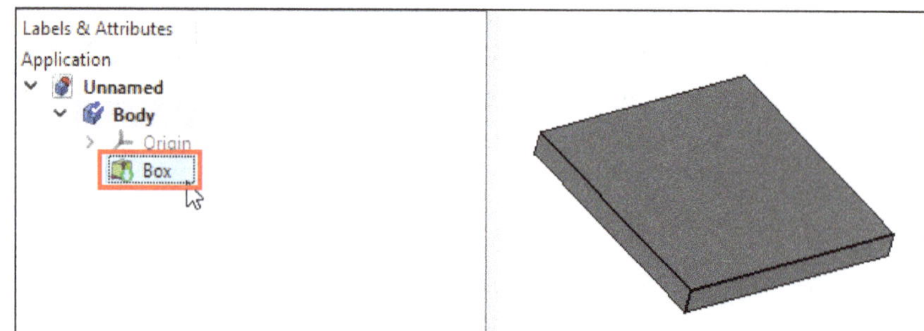

Figure-60. Selecting the object to move

- Click on the **Move** tool from **Toolbar** in the **Draft** workbench; refer to Figure-61. The **Move** dialog will be displayed in the **Tasks** panel of **Combo View** along with the plus sign in place of original cursor; refer to Figure-62. You will be asked to specify the base point to move or copy an object.

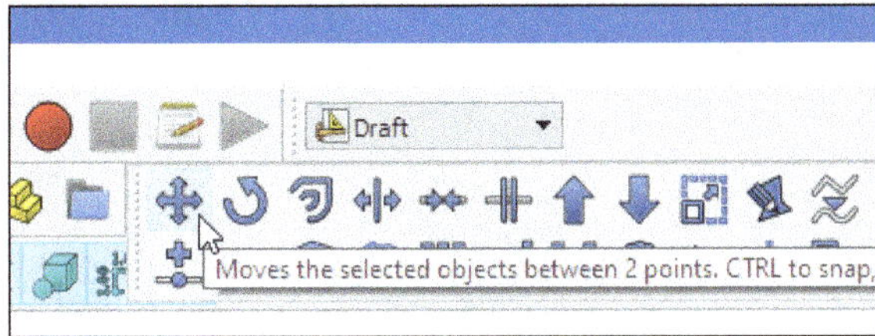

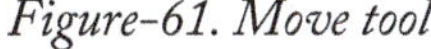

Figure-61. Move tool

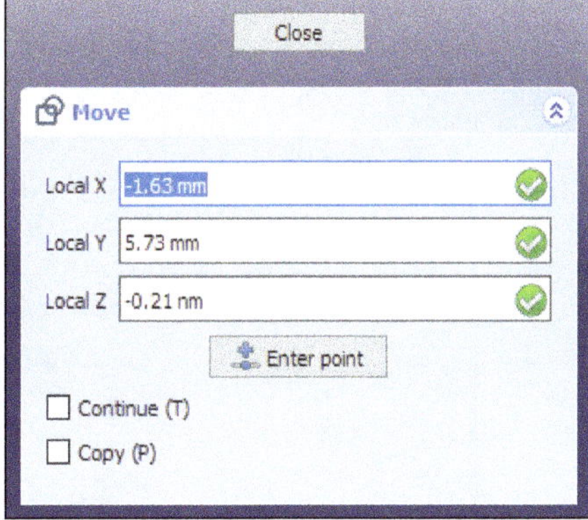

Figure-62. Move dialog

- Click in the 3D view area to specify the base point of moving the object or enter desired values for x, y, and z coordinates in the **Local X**, **Local Y**, and **Local Z** edit boxes of the dialog, respectively.
- After specifying coordinates in the dialog, click on the **Enter point** button from the dialog. You will be asked to specify the new position for the base point.
- Move the cursor away and click at desired location to specify the new position of the base point or enter desired values for the coordinates in their respective edit boxes of the dialog.
- After specifying coordinates in the dialog, click on the **Enter point** button again. The object will be moved to the new position; refer to Figure-63.

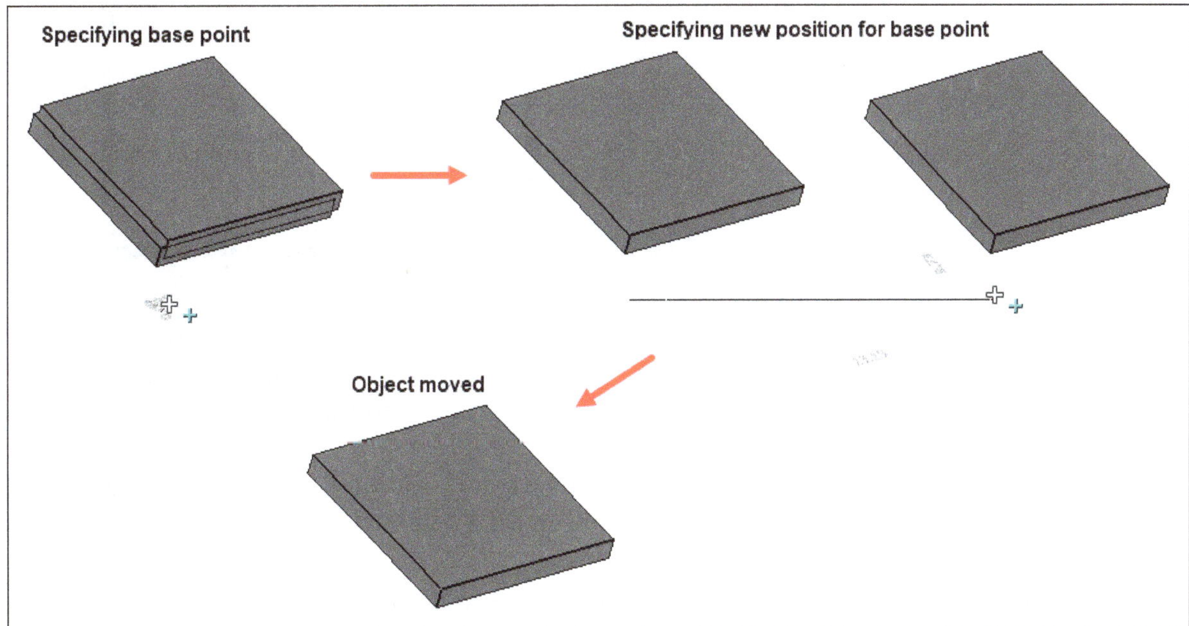

Figure-63. Object moved

- Select **Copy** check box from the dialog to keep the original object in its place and make a copy at the second point.
- Other parameters in the **Move** dialog have been discussed earlier.
- Click on **Close** button to close the dialog.

Rotating object

The **Rotate** tool rotates or copies the selected objects by a given angle around a reference point. The procedure to use this tool is discussed next.

- Select the object from the Model tree view which you want to rotate; refer to Figure-64.

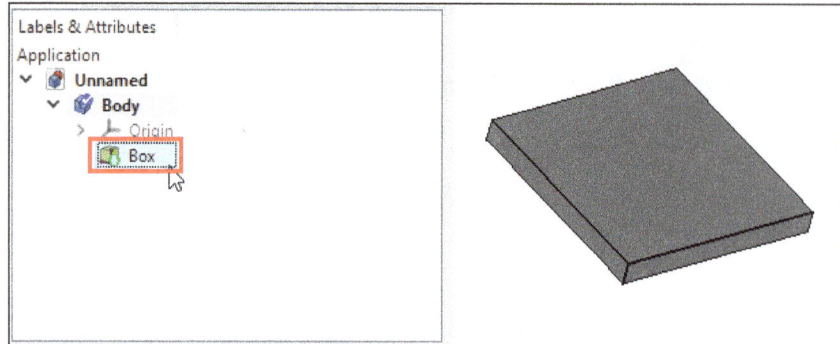

Figure-64. Selecting the object to rotate

- Click on the **Rotate** tool from **Toolbar** in the **Draft** workbench; refer to Figure-65. The **Rotate** dialog will be displayed in the **Tasks** panel of **Combo View** along with the plus sign in place of original cursor; refer to Figure-66. You will be asked to specify the base point through which the axis of rotation will pass.

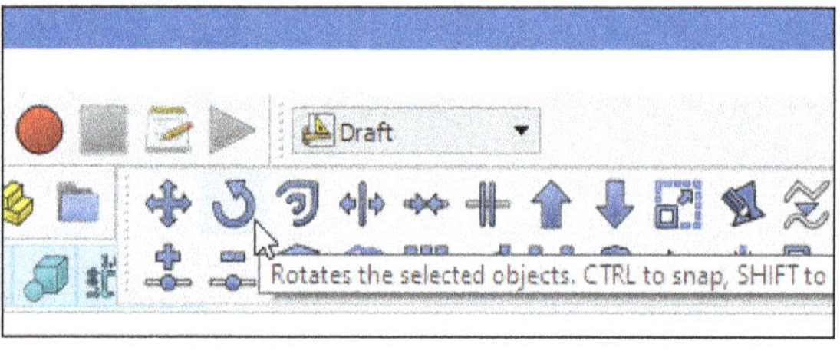

Figure-65. Rotate tool

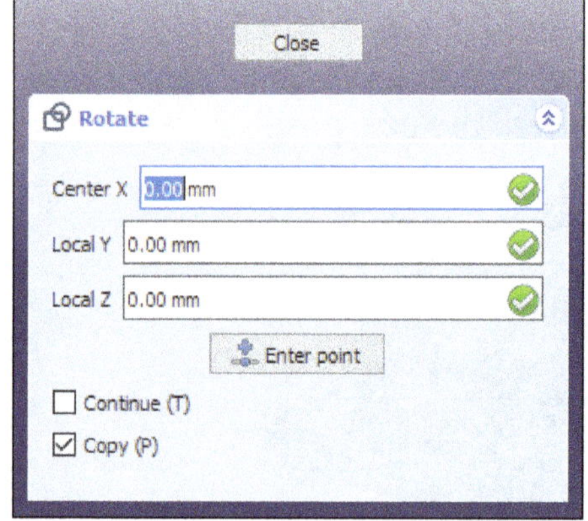

Figure-66. Rotate dialog

- Click in the 3D view area to specify the base point of rotating the object or enter desired values for x, y, and z coordinates in the **Center X**, **Local Y**, and **Local Z** edit boxes of the dialog, respectively.
- After specifying coordinates in the dialog, click on the **Enter point** button from the dialog. You will be asked to specify the base angle.
- Move the cursor away and click at desired location to specify the base angle for the object that will rotate around the base point or enter desired value in the **Base angle** edit box of the dialog. You will be asked to specify the rotation angle.
- Rotate the cursor away and click at desired location to specify the rotation angle for the object or specify desired value in the **Rotation** edit box of the dialog. The object will be rotated; refer to Figure-67.

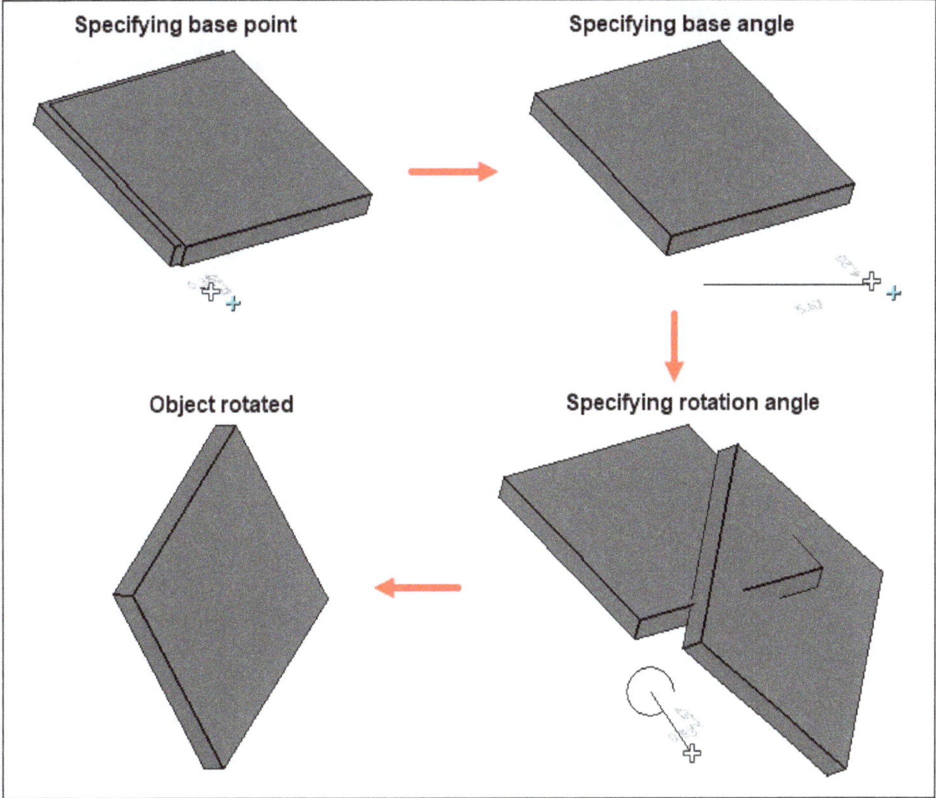

Figure-67. Object rotated

- Other parameters in the **Rotate** dialog have been discussed earlier.
- Click on **Close** button to close the dialog.

Offsetting object

The **Offset** tool moves the selected object by a given distance (offset) perpendicular to itself. The procedure to use this tool is discussed next.

- Select the object from the Model tree view which you want to offset; refer to Figure-68.

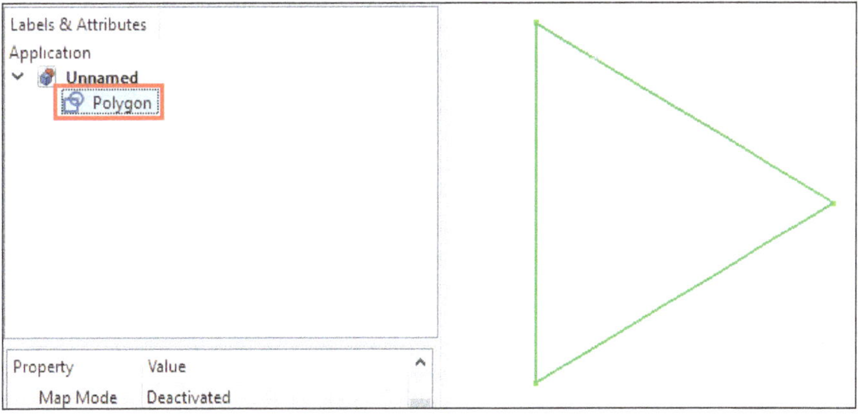

Figure-68. Selecting the object to offset

- Click on the **Offset** tool from **Toolbar** in the **Draft** workbench; refer to Figure-69.The **Offset** dialog will be displayed in the **Tasks** panel of **Combo View** along with the plus sign in place of original cursor; refer to Figure-70. You will be asked to specify the offset distance.

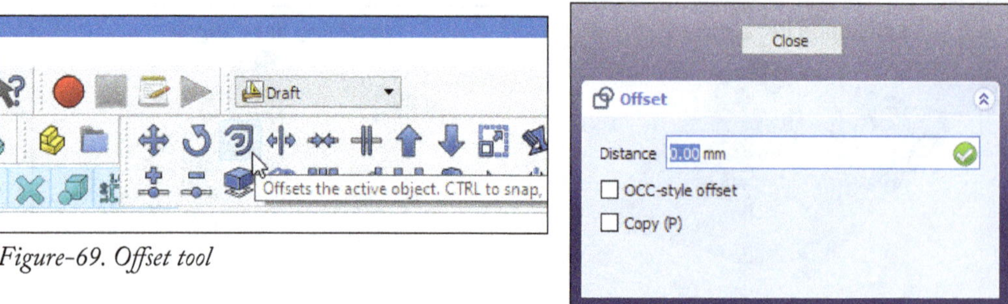

Figure-69. Offset tool

Figure-70. Offset dialog

- Move the cursor away and click in the 3D view area to specify the offset distance or enter desired value in the **Distance** edit box of the dialog and press **ENTER** key from keyboard. The offset of the object will be created; refer to Figure-71.

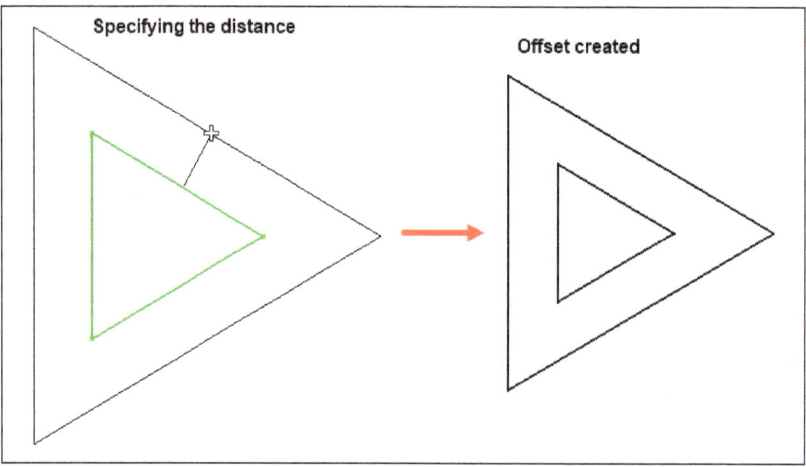

Figure-71. Offset created

- Click on **Close** button to close the dialog.

Trimming/Extending object

The **Trimex** tool trims or extends lines and wires so that they end at an intersection with another curve or edge. This tool also extrudes faces created from closed wires. The procedure to use this tool is discussed next.

- Select the line from the 3D view area which you want to extend; refer to Figure-72.

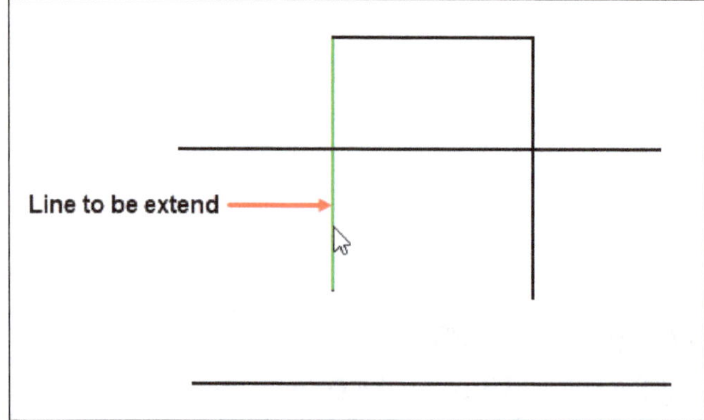

Figure-72. Selecting the line to be extend

- Click on the **Trimex** tool from **Toolbar** in the **Draft** workbench; refer to Figure-73. The **Trim** dialog will be displayed in the **Tasks** panel of **Combo View** and the cursor will be attached to the selected line; refer to Figure-74. You will be asked to specify the distance.

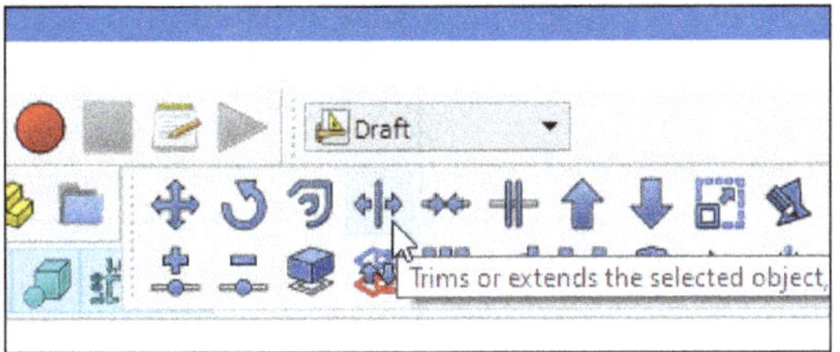

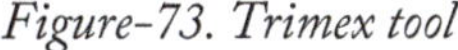

Figure-73. Trimex tool

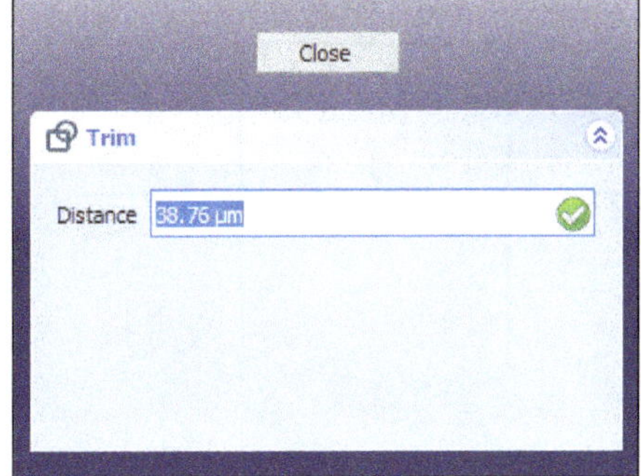

Figure-74. Trim dialog

- Move the cursor away to the distance upto which you want to extend the line and click **LMB** or enter desired value in the **Distance** edit box of the dialog and press **Enter** key from keyboard. The line will be extended; refer to Figure-75.

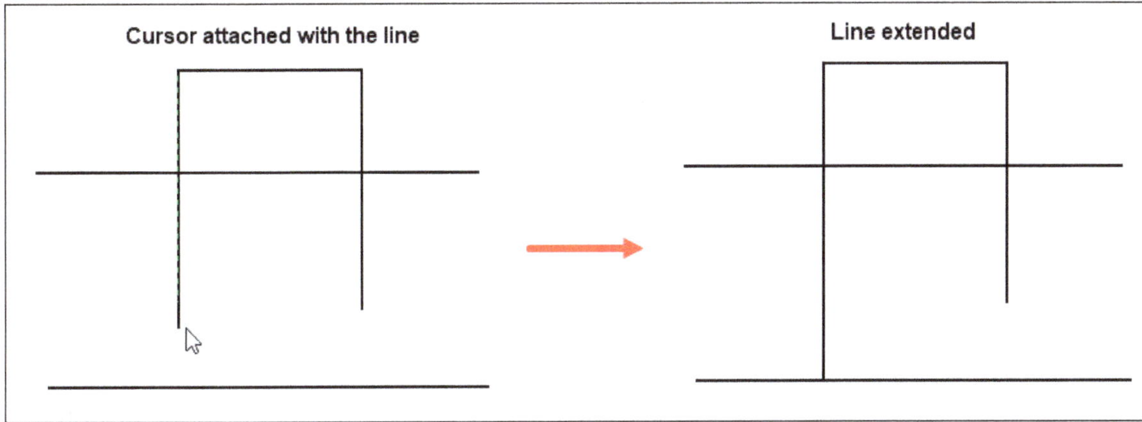

Figure-75. Line extended

- If you want to trim the line then select the line from the 3D view area which you want to trim; refer to Figure-76.

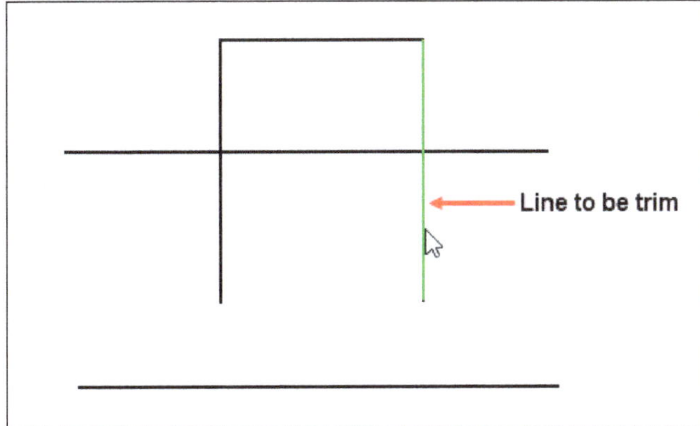

Figure-76. Selecting the line to be trim

- After selecting the line, click on the **Trimex** tool. The **Trim** dialog will be displayed and the cursor will be attached to the selected line as discussed earlier. You will be asked to specify the distance.
- Move the cursor away to the distance upto which you want to trim the line and click **LMB** or enter desired value in the **Distance** edit box of the dialog and press **Enter** key from keyboard. The line will be trimmed; refer to Figure-77.

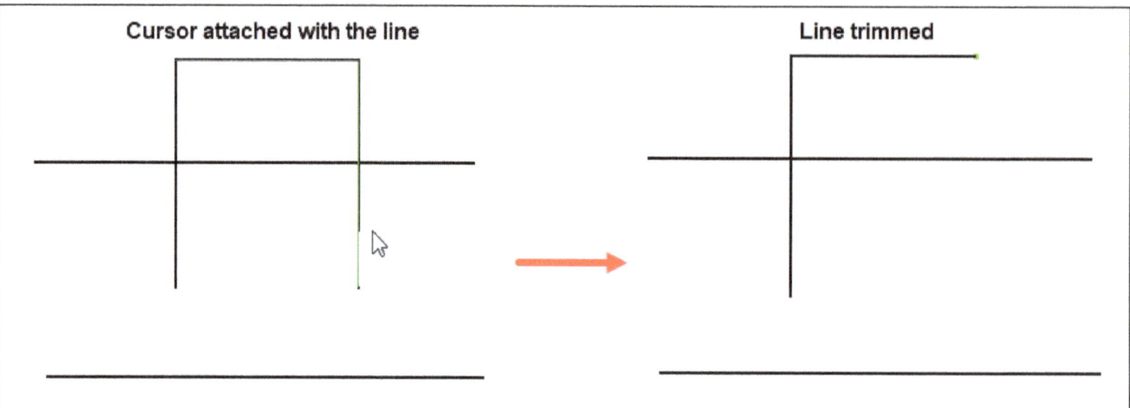

Figure-77. Line trimmed

- If you want to extrude the face created from closed wire then select the face from the Model tree view or from the 3D view area; refer to Figure-78.

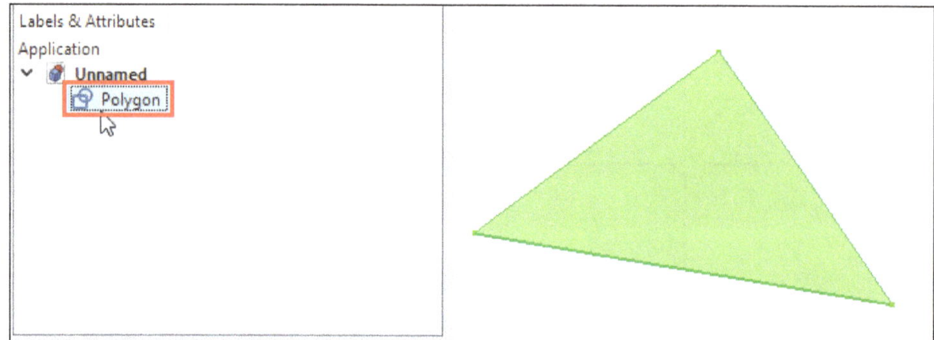

Figure-78. Selecting the face to extrude

- After selecting the face, click on the **Trimex** tool. The **Trim** dialog will be displayed and the cursor will be attached to the extrusion of face. You will be asked to specify the distance.
- Move the cursor away to the distance upto which you want to extrude the face and click **LMB** or enter desired value in the **Distance** edit box of the dialog and press **Enter** key from keyboard. The face will be extruded; refer to Figure-79.

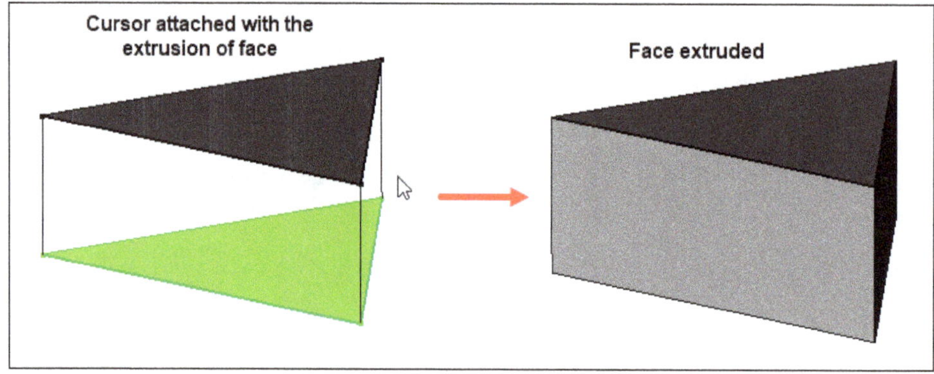

Figure-79. Face extruded

- Click on **Close** button to close the dialog.

Joining objects

The **Join** tool attempts to join all wires currently in the selection into a single wire. The procedure to use this tool is discussed next.

- Select two or more lines or curves from the Model tree view or from the 3D view area to make a single wire; refer to Figure-80.

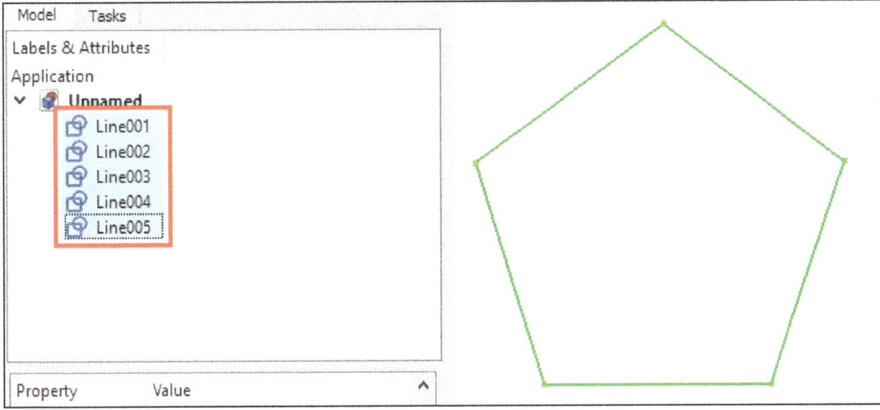

Figure-80. Selecting the lines to create single wire

- Click on the **Join** tool from **Toolbar** in the **Draft** workbench; refer to Figure-81. The single wire will be created; refer to Figure-82.

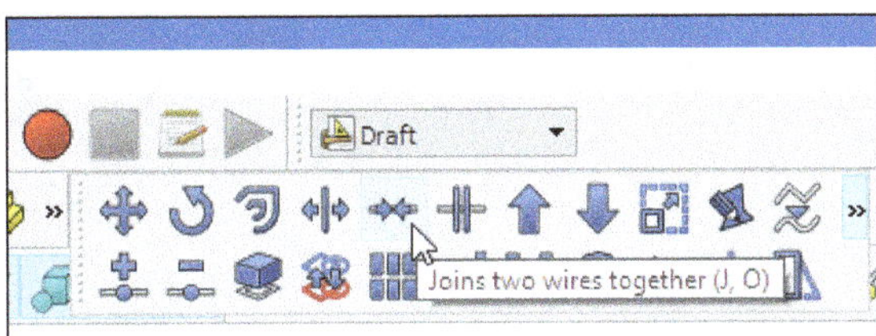

Figure-81. Join tool

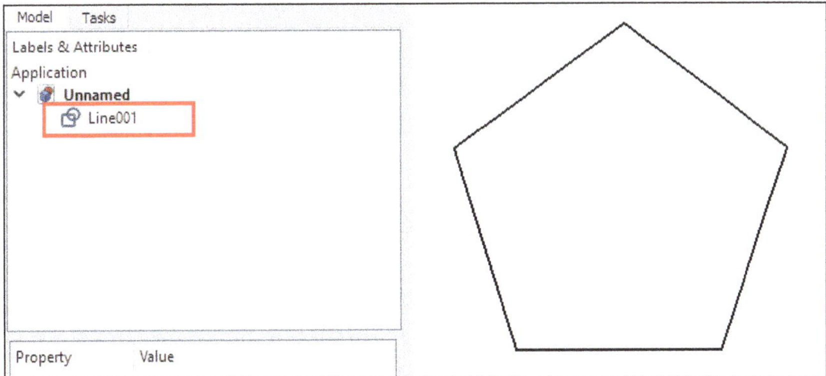

Figure-82. Single wire created

Splitting object

The **Split** tool attempts to split an existing wire at a specified edge or point. The procedure to use this tool is discussed next.

- Click on the **Split** tool from **Toolbar** in the **Draft** workbench; refer to Figure-83.

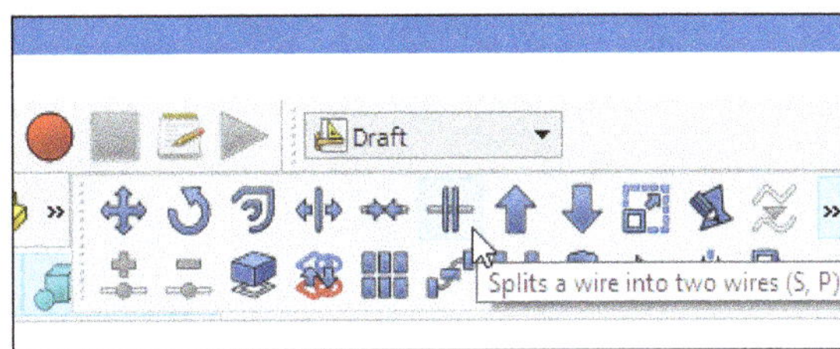

Figure-83. Split tool

- Click on the wire at the location from where you want to split the wire. The wire will be splitted from that location; refer to Figure-84.

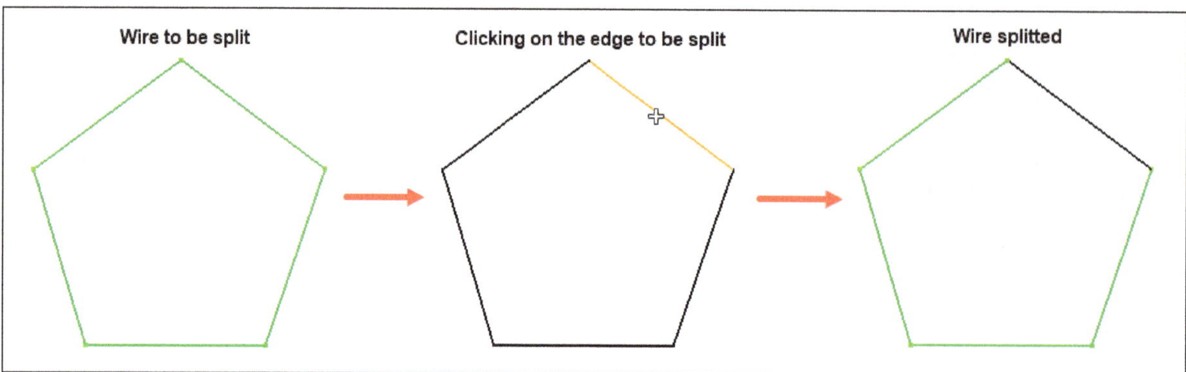

Figure-84. Wire splitted

Upgrading object

The **Upgrade** tool performs things such as creating faces and fusing different elements. The procedure to use this tool is discussed next.

- Select the object from the Model tree view or from the 3D view area which you want to upgrade; refer to Figure-85.

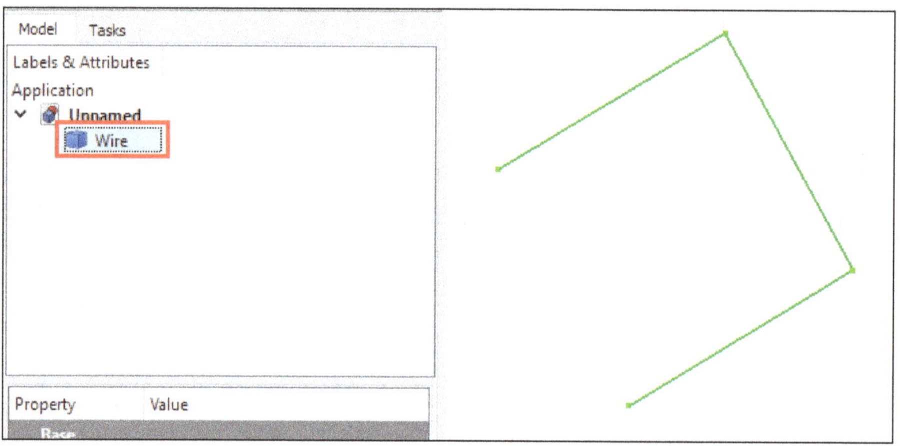

Figure-85. Selecting the object to upgrade

- Click on the **Upgrade** tool from **Toolbar** in the **Draft** workbench; refer to Figure-86. The opened wire will be upgraded to a closed wire; refer to Figure-87.

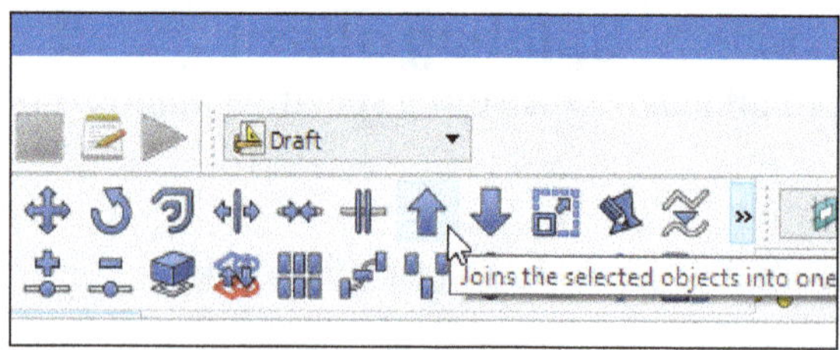

Figure-86. Upgrade tool

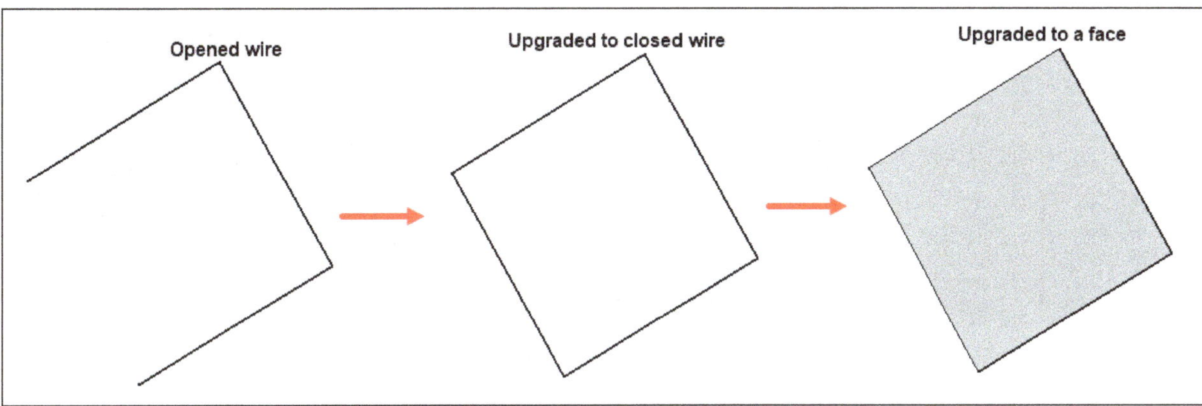

Figure-87. Upgradation of an object

- If you want to upgrade the closed wire then click on the **Upgrade** tool again. The closed wire will be upgraded to a face; refer to Figure-87.

Downgrading object

The **Downgrade** tool performs things such as breaking faces and deconstructing wires into their individual objects. The procedure to use this tool is discussed next.

- Select the object from the Model tree view or from the 3D view area which you want to downgrade; refer to Figure-88.

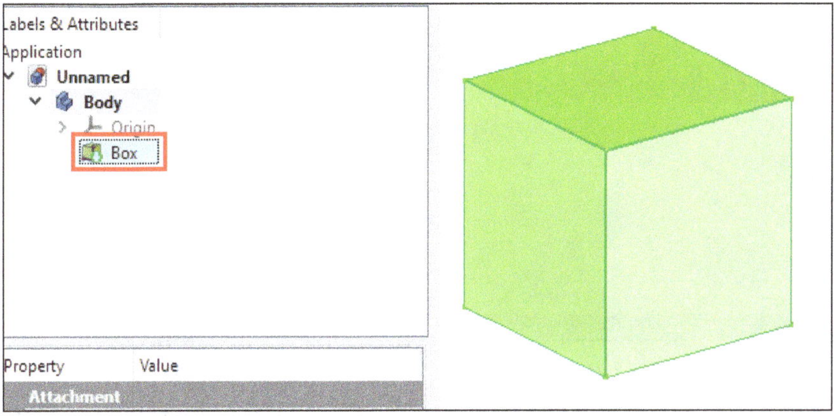

Figure-88. Selecting the object to downgrade

- Click on the **Downgrade** tool from **Toolbar** in the **Draft** workbench; refer to Figure-89. The object will be downgraded into the split faces; refer to Figure-90.

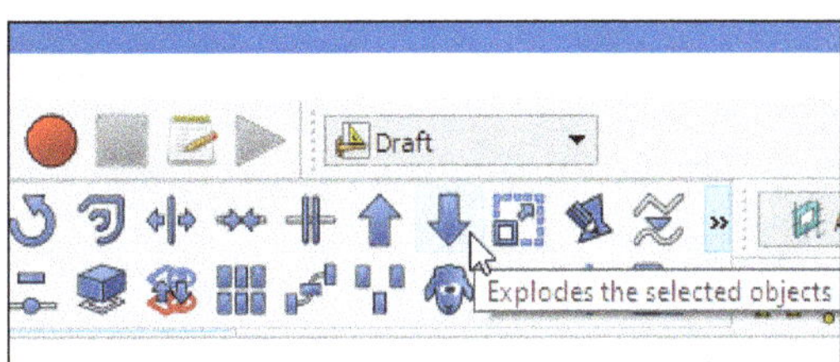

Figure-89. Downgrade tool

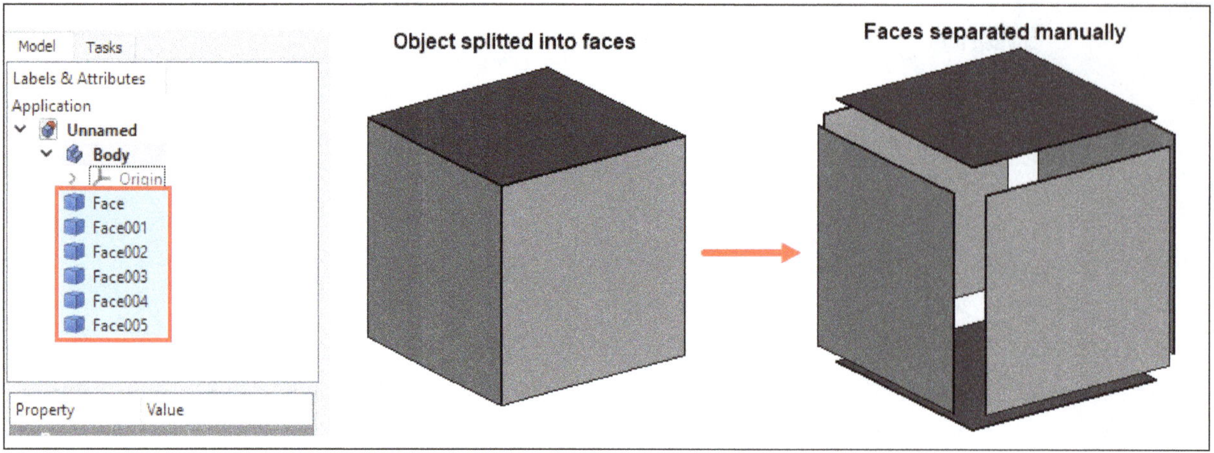

Figure-90. Object downgraded into the split faces

Scaling object

The **Scale** tool scales or copies selected objects around a base point. The procedure to use this tool is discussed next.

- Select the object from the Model tree view or from the 3D view area which you want to scale; refer to Figure-91.

Figure-91. Selecting the object to scale

- Click on the **Scale** tool from **Toolbar** in the **Draft** workbench; refer to Figure-92. The **Scale** dialog will be displayed in the **Tasks** panel of **Combo View** along with the plus sign in place of original cursor; refer to Figure-93. You will be asked to specify the base point.

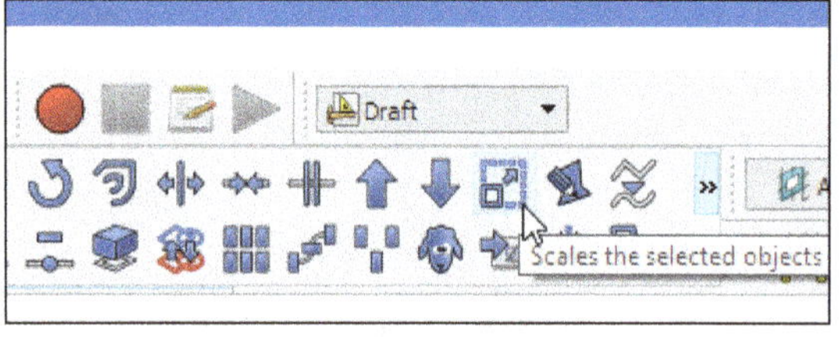

Figure-92. Scale tool

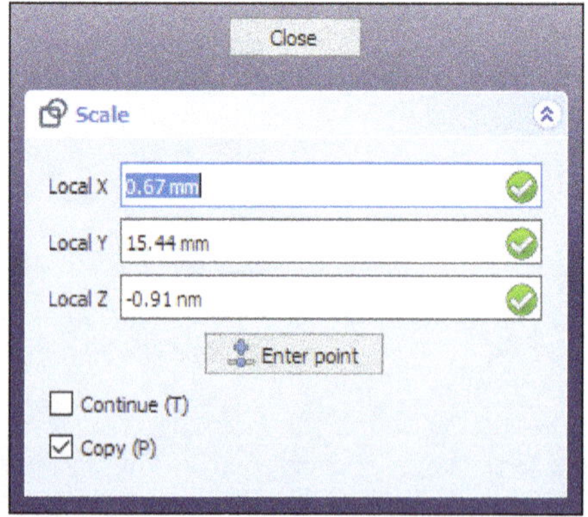

Figure-93. Scale dialog

- Click in the 3D view area to specify the base point for scaling the object or enter desired values for the x, y, and z coordinates in the **Local X**, **Local Y**, and **Local Z** edit boxes of the dialog, respectively.
- Other parameters in the **Scale** dialog have been discussed earlier.
- After specifying coordinates in the dialog, click on **Enter point** button from the dialog. The **Scale** dialog will be modified; refer to Figure-94.

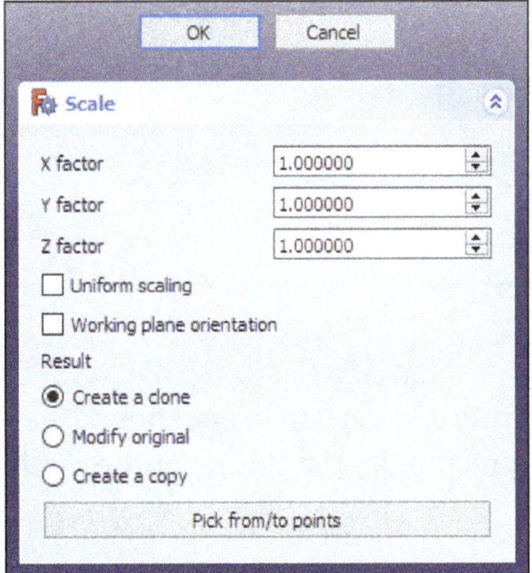

Figure-94. Modified Scale dialog

- Specify desired values for scaling the object along x, y, and z directions in the **X factor**, **Y factor**, and **Z factor** edit boxes of the dialog, respectively. The preview of the scaled object will be displayed; refer to Figure-95.

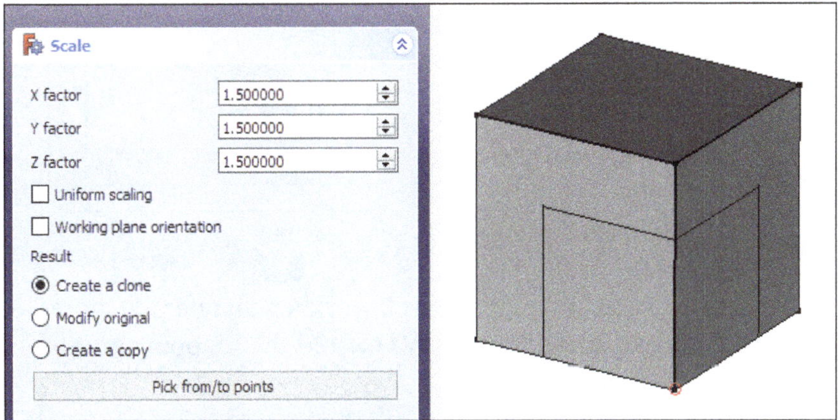

Figure-95. Preview of scaled object

- Select **Uniform scaling** check box to lock the same value for X, Y, and Z factors.
- Select **Working plane orientation** check box to lock the scaling of an object along the current working plane.
- Select **Create a clone** radio button from the **Result** section of the dialog to create the clone of original object.
- Select **Modify original** radio button to modify the size of original object.
- Select **Create a copy** radio button to create the scaled copy of the original object.
- Click on **OK** button from the dialog. The object will be scaled; refer to Figure-96.

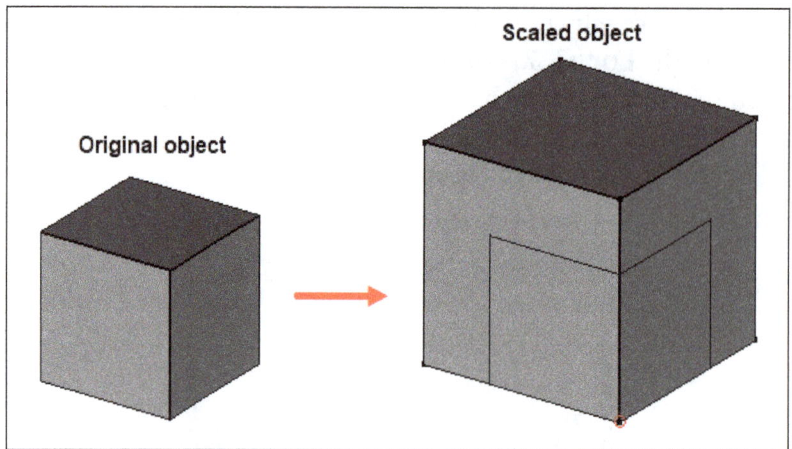

Figure-96. Original object scaled

Editing object

The **Edit** tool allows you to graphically edit certain properties of the selected object such as, the vertices of a wire, the length and width of a rectangle, the radius of the circle, and so on. The procedure to use this tool is discussed next.

- Select the object you want to edit from the Model tree view or from the 3D view area; refer to Figure-97.

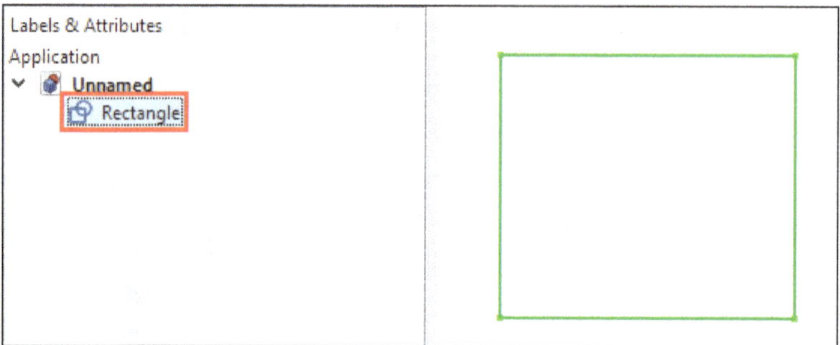

Figure-97. Selecting the object to edit

- Click on the **Edit** tool from **Toolbar** in the **Draft** workbench; refer to Figure-98. The **Edit** dialog will be displayed in the **Tasks** panel of **Combo View** and the snapping points will be displayed on the object; refer to Figure-99.

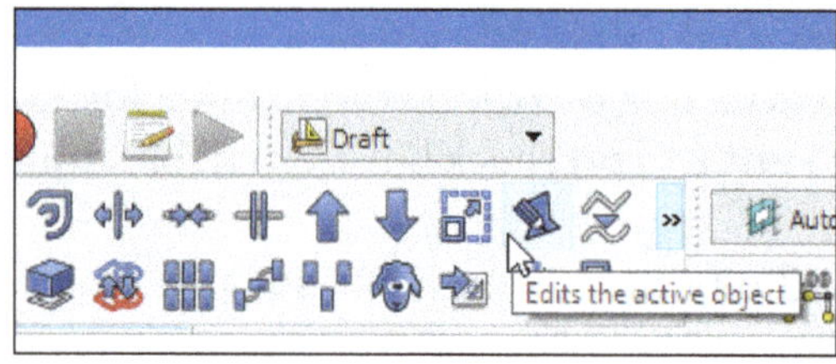

Figure-98. Edit tool

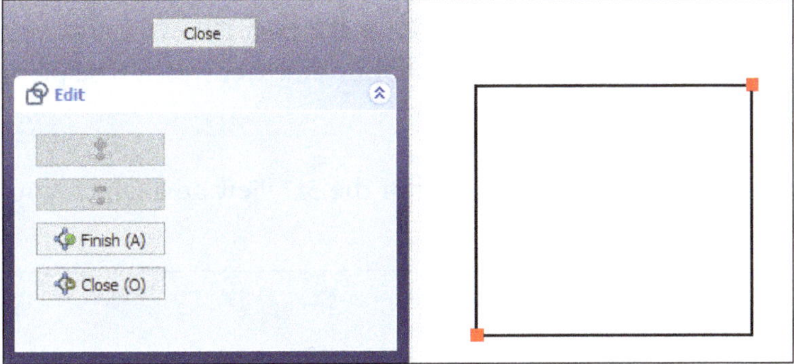

Figure-99. Edit dialog with the snapping points on object

- Click on the snapping point from where you want to edit the object. The **Point** dialog will be displayed in the **Tasks** panel of **Combo View** along with the plus sign in place of original cursor asking you to specify another point; refer to Figure-100.

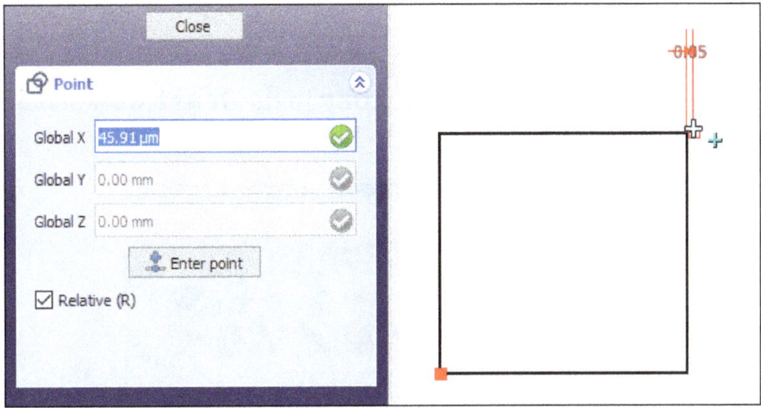

Figure-100. Point dialog

- Move the cursor away and click **LMB** at desired location to specify the another point of object or enter desired values for the x, y, and z coordinates in the **Global X**, **Global Y**, and **Global Z** edit boxes of the dialog, respectively.
- After specifying coordinates for another point in the dialog, click on **Enter point** button from the dialog. The object will be edited and the **Edit** dialog will be displayed again; refer to Figure-101.

Figure-101. Object edited

- Click on **Add point** button from the dialog to add a new point on the wire or spline.
- Click on **Remove point** button to remove the existing point from the wire or spline.
- Click on **Finish** button to finish the editing operation and click on **Close** button to close the dialog.

Converting Wire to B-Spline

The **Wire to B-Spline** tool converts wire to B-Spline and vice-versa. The procedure to use this tool is discussed next.

- Select the wire from the Model tree view or from the 3D view area which you want to convert into B-Spline; refer to Figure-102.

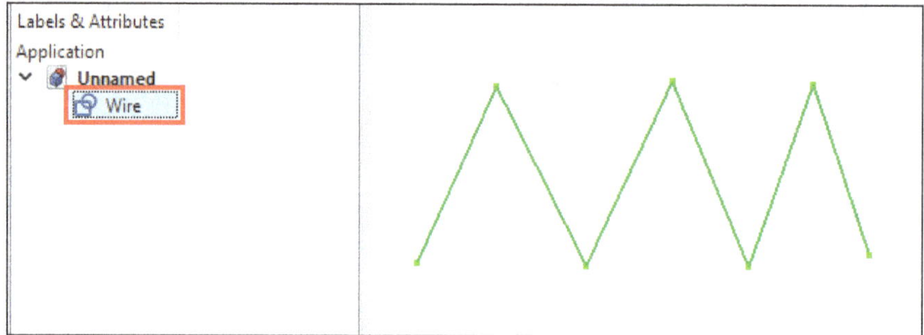

Figure-102. Selecting the wire to create B Spline

- Click on the **Wire to BSpline** tool from **Toolbar** in the **Draft** workbench; refer to Figure-103. The wire will be converted into B-Spline; refer to Figure-104.

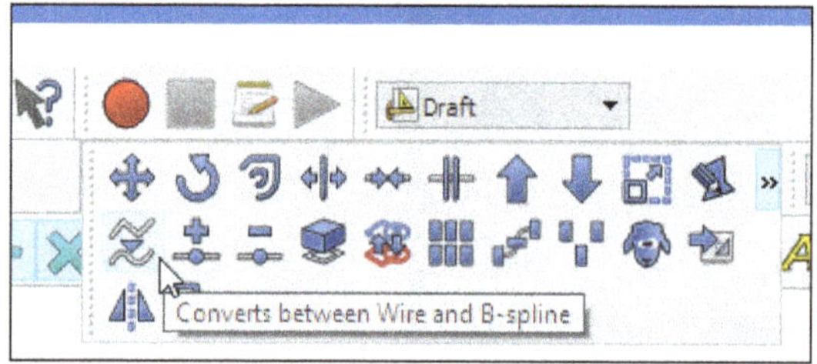

Figure-103. Wire to BSpline tool

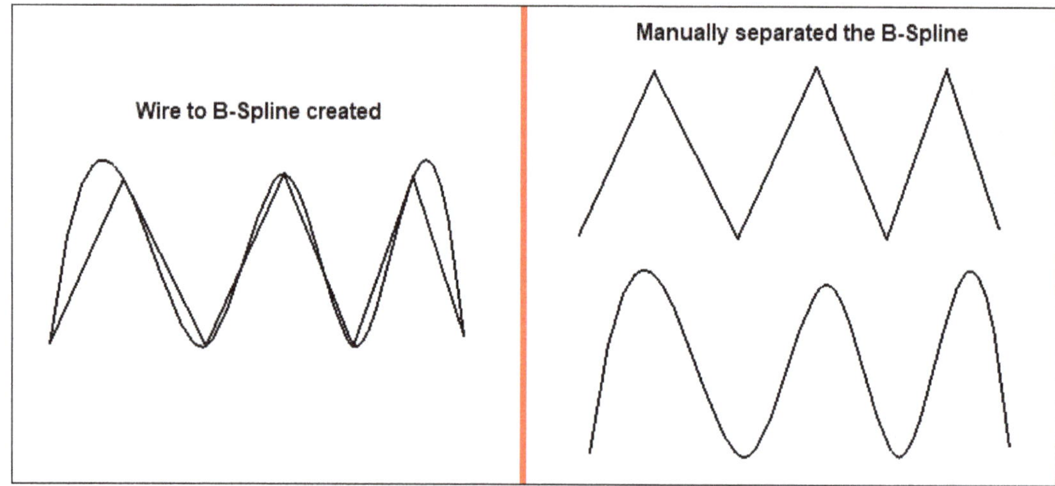

Figure-104. Wire converted to BSpline

Adding points (Obsolete in version 0.19)

The **Add point** tool allows you to add points to already created **Draft Wires** and **Draft BSplines**. The procedure to use this tool have been discussed earlier.

Deleting points (Obsolete in version 0.19)

The **Delete point** tool allows you to remove points from already created **Draft Wires** and **Draft BSplines**. The procedure to use this tool have been discussed earlier.

Creating Shape 2D View

The **Shape 2D View** tool produces a 2D projection from a selected 3D solid object such as those created with the **Part**, **Part Design**, and **Arch** workbenches. The procedure to use this tool is discussed next.

- Rotate the view of 3D object to be projected from the 3D view area, so it reflects the direction of the desired projection. For example, a top view will project the object on the XY plane.
- Select the 3D object from the Model tree view or from the 3D view area; refer to Figure-105.

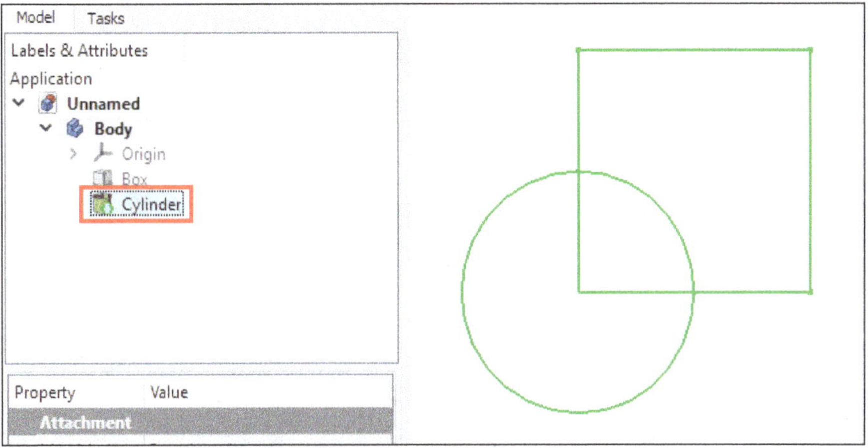

Figure-105. Selecting the 3D object to be project

- Click on the **Shape 2D View** tool from **Toolbar** in the **Draft** workbench; refer to Figure-106. The 2D projection will be created below the selected 3D object, lying on the XY plane; refer to Figure-107.

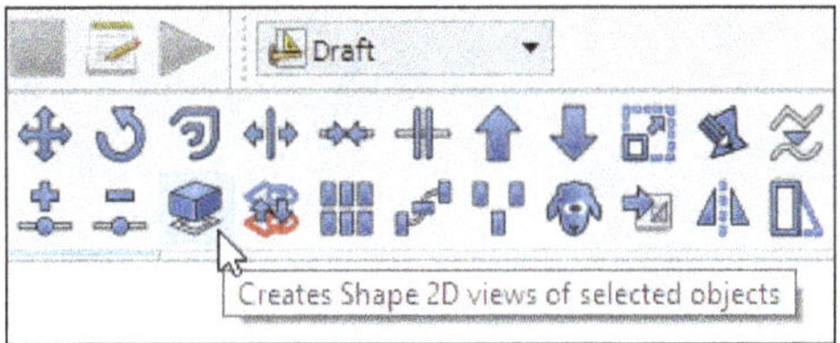

Figure-106. Shape 2D View tool

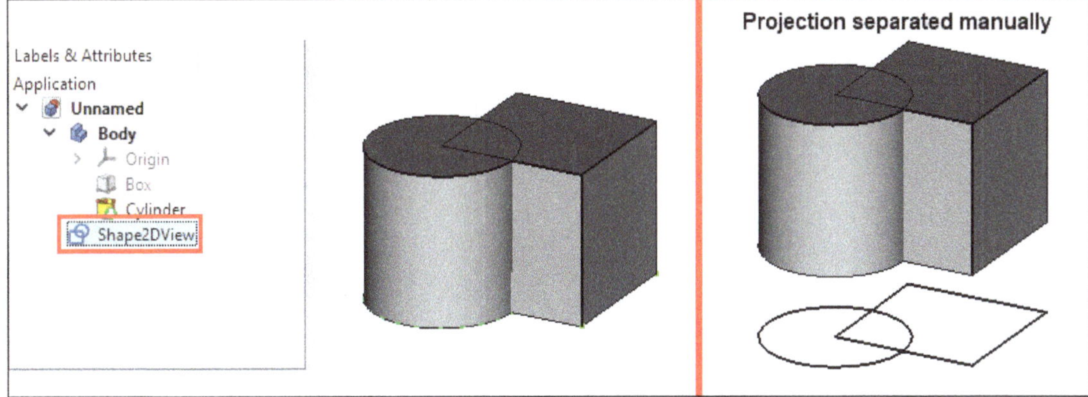

Figure-107. 2D projection created

- You can manually change the position of projection created from the **Property editor** dialog.

Converting Draft to Sketch

The **Draft to Sketch** tool converts objects of **Draft** workbench into the objects of **Sketcher** workbench with constraints and vice-versa. The procedure to use this tool is discussed next.

- Select the object you want to convert from the Model tree view or from the 3D view area; refer to Figure-108.

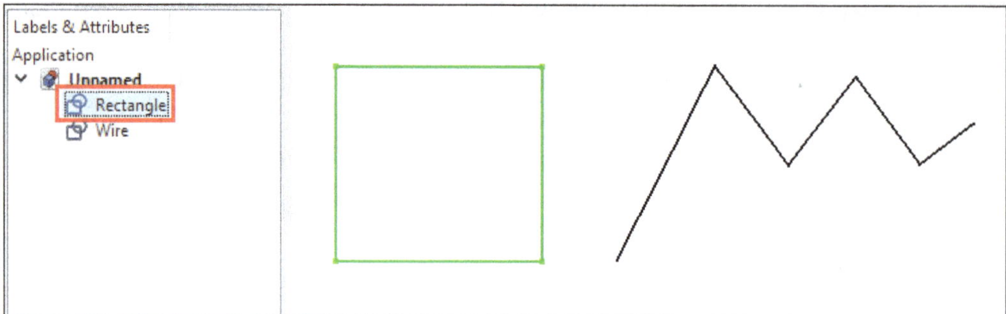

Figure-108. Selecting the object to convert

- Click on the **Draft to Sketch** tool from **Toolbar** in the **Draft** workbench; refer to Figure-109. The objects of **Draft** workbench will be converted into sketches of **Sketcher** workbench with constraints; refer to Figure-110.

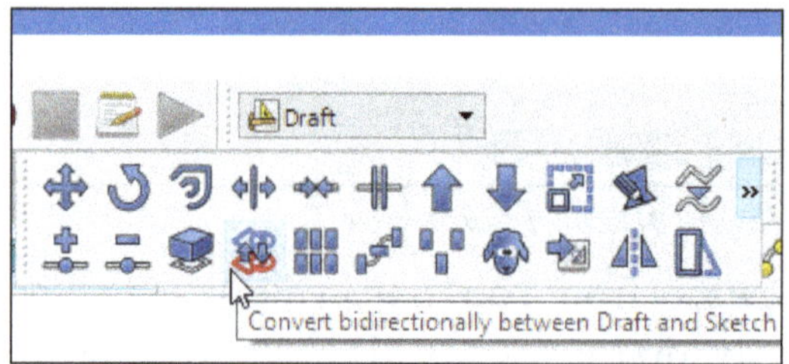

Figure-109. Draft to Sketch tool

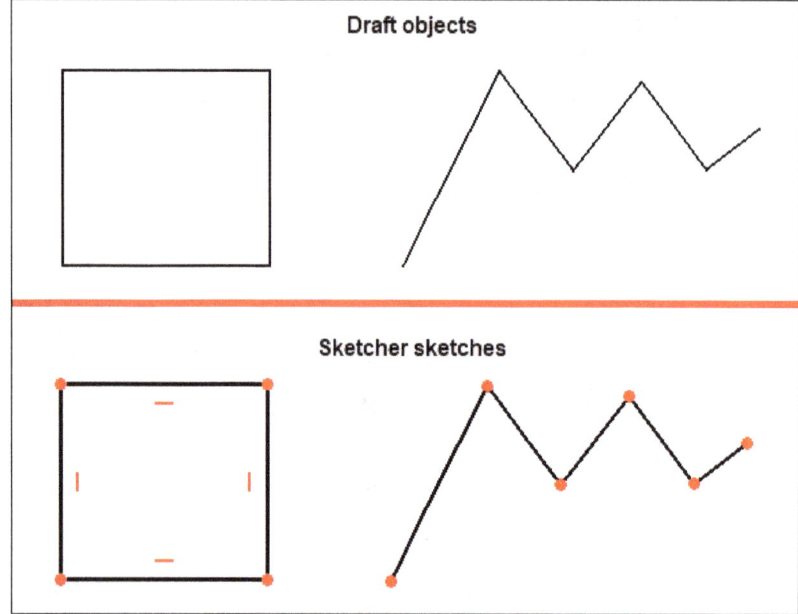

Figure-110. Objects converted into sketches

Creating Array (Obsolete in version 0.19)

The **Array** tool creates an orthogonal (3-axes), polar, or circular array of selected object. The procedure to use this tool is discussed next.

- Select the object from Model tree view or from the 3D view area for which you want to create an array; refer to Figure-111.

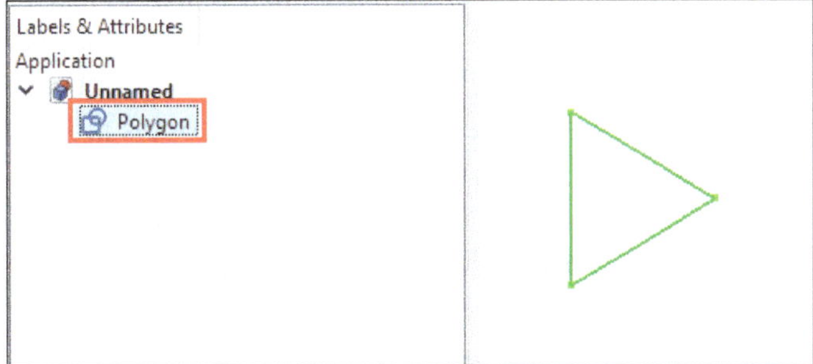

Figure-111. Selecting the object to create array

- Click on the **Array** tool from **Toolbar** in the **Draft** workbench; refer to Figure-112. The array object will be created; refer to Figure-113.

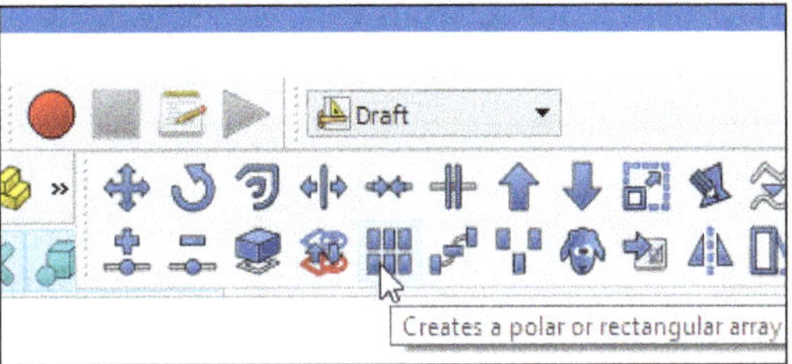

Figure-112. Array tool

Figure-113. Array created

- If you want to change the properties of the array, i.e., to change the number and direction of copies created then select the array object created from the Model tree view. The **Property editor** dialog will be displayed in the **Model** panel of **Combo View**; refer to Figure-114.

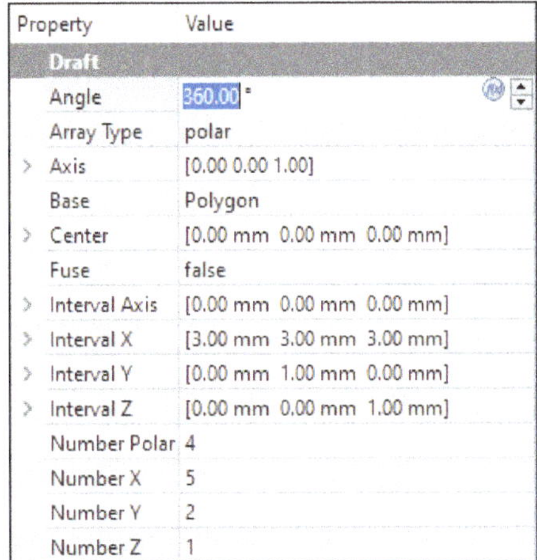

Figure-114. Property editor dialog on selecting array object

- Specify the type of array to be created from **Array Type** drop-down in the dialog. Select **ortho** option from the drop-down to create orthogonal arrays and select **polar** option from the drop-down to create polar arrays.

- For creating the orthogonal arrays, expand the **Interval X** node from **Draft** section of the dialog and specify desired values in the **x** edit box for the interval between each copy on the X axis. Expand the **Interval Y** node and specify desired values in the **y** edit box for the interval between each copy on the Y axis. Expand the **Interval Z** node and specify desired values in the **z** edit box for the interval between each copy on the Z axis. Specify desired number of copies along the X axis in the **Number X** edit box of the dialog. Specify desired number of copies along the Y axis in the **Number Y** edit box. Specify desired number of copies along the Z axis in the **Number Z** edit box of the dialog.
- For creating the polar arrays, specify the aperture of the circular arc to create the array in the **Angle** edit box of the dialog. Expand the **Axis** node and specify desired values in the **x**, **y**, and **z** edit boxes for the normal direction of the array circle. Expand the **Center** node and specify desired values in the **x**, **y**, and **z** edit boxes for the center point of the array circle. Specify desired number of copies to place in the circular arrangement in the **Number Polar** edit box. Expand the **Interval Axis** node and specify desired values in the **x**, **y**, and **z** edit boxes for the interval between each copy on the axis direction.
- After specifying desired parameters, click **LMB** in the 3D view area. The array object will be created; refer to Figure-115.

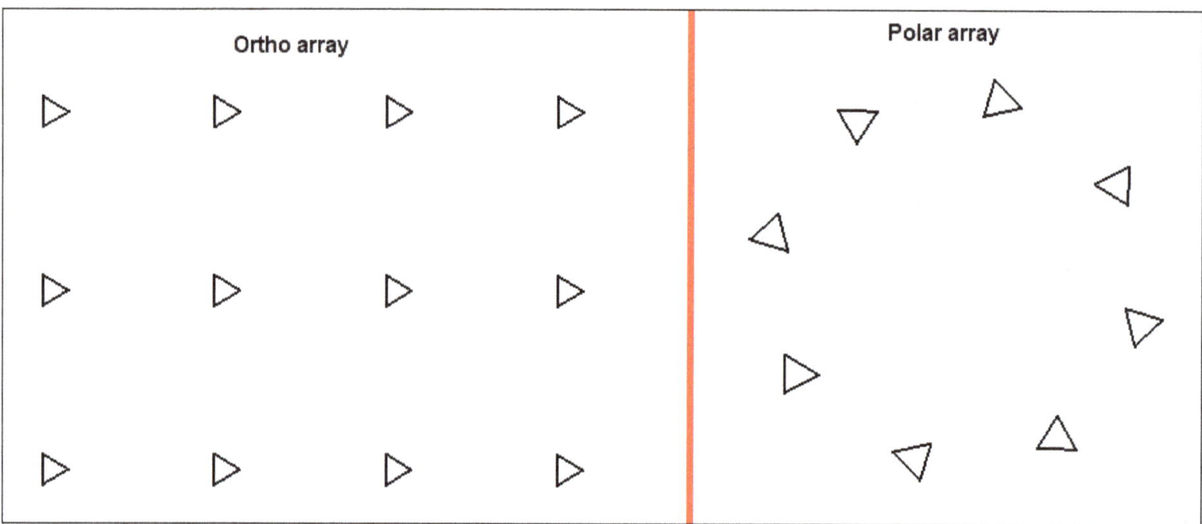

Figure-115. Array created

Creating Path Array

The **Path Array** tool places copies of selected shape along selected path which can be a wire, a B-Spline, and similar edges. The procedure to use this tool is discussed next.

- First, select the object from the Model tree view or from the 3D view area that you want to array and then select the object to be used as path along which the object will be distributed while holding the **CTRL** key; refer to Figure-116.

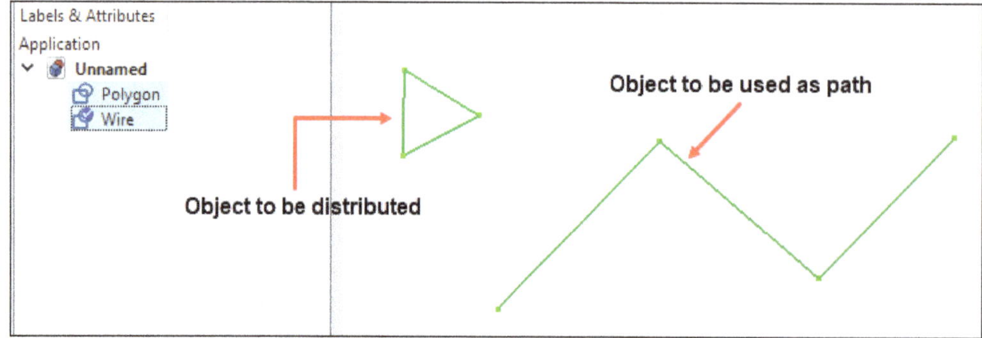

Figure-116. Selecting the object for path array

- Click on the **Path Array** tool from **Toolbar** in the **Draft** workbench; refer to Figure-117. The array objects will be created along path; refer to Figure-118.

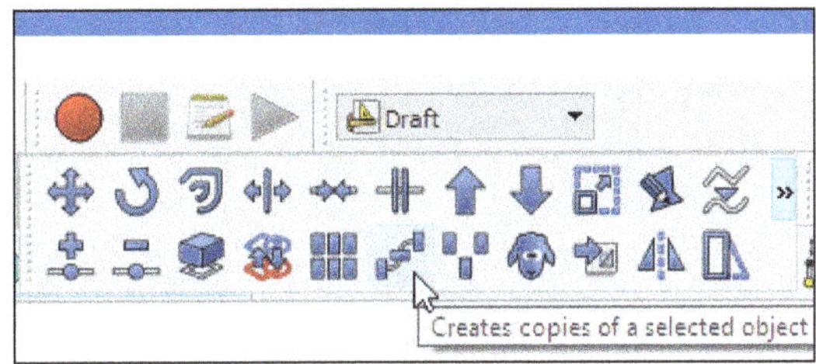

Figure-117. Path Array tool

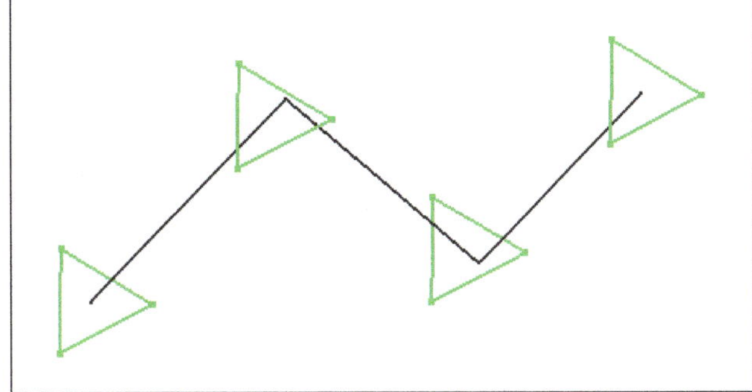

Figure-118. Array object created along path

- You can change the properties of array created from the **Property editor** dialog as discussed earlier.

Creating Point Array

The **Point Array** tool places copies of selected object on selected points. The procedure to use this tool is discussed next.

- First, select the object from the Model tree view or from the 3D view area that you want to distribute and then select the point along which the object will be distributed; refer to Figure-119.

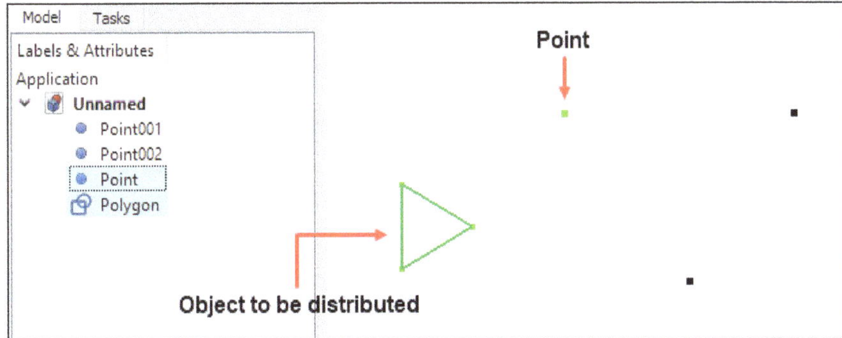

Figure-119. Selecting the object for point array

- Click on the **Point Array** tool from **Toolbar** in the **Draft** workbench; refer to Figure-120. The array object will be created along point; refer to Figure-121.

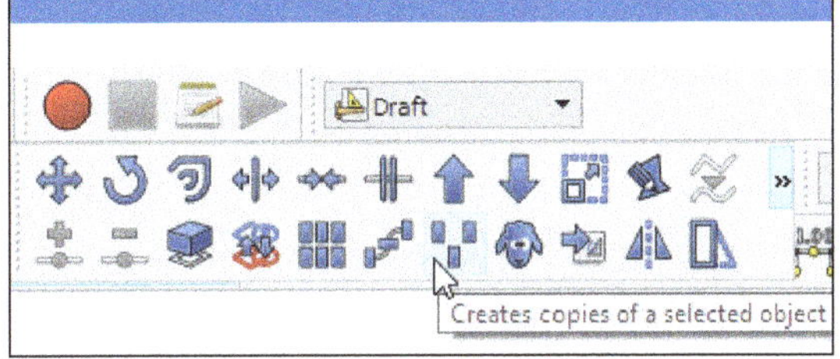

Figure-120. Point Array tool

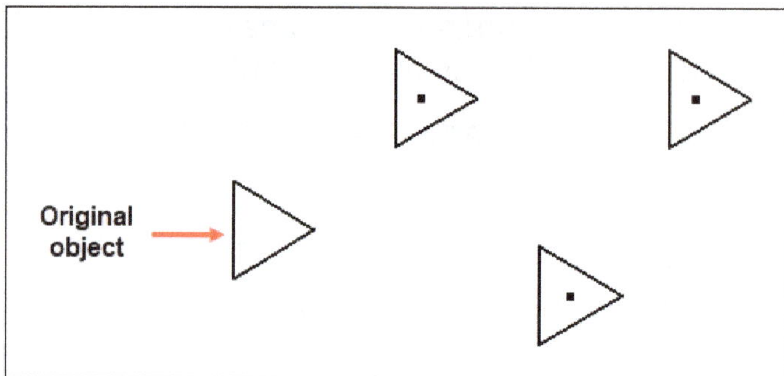

Figure-121. Array object created along point

- You can change the properties of array created from the **Property editor** dialog as discussed earlier.

Creating Clone

The **Clone** tool produces linked copies of selected shape. This means that if the original object changes its shape and properties, all clones change as well. The procedure to use this tool is discussed next.

- Select the object from the Model tree view or from the 3D view area for which you want to create clone; refer to Figure-122.

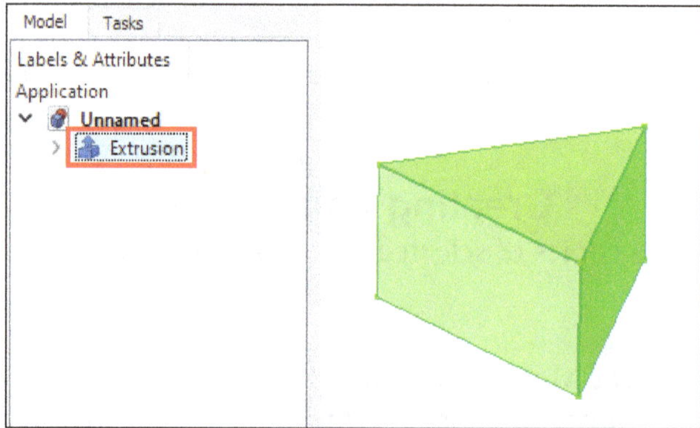

Figure-122. Selecting the object to be cloned

- Click on the **Clone** tool from **Toolbar** in the **Draft** workbench; refer to Figure-123. The cloned object will be created; refer to Figure-124.

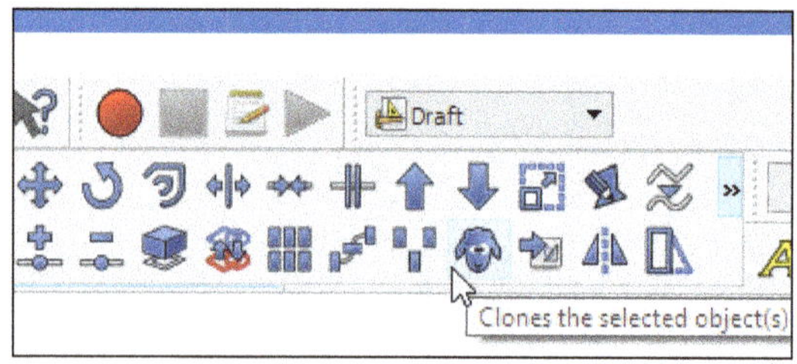

Figure-123. Clone tool

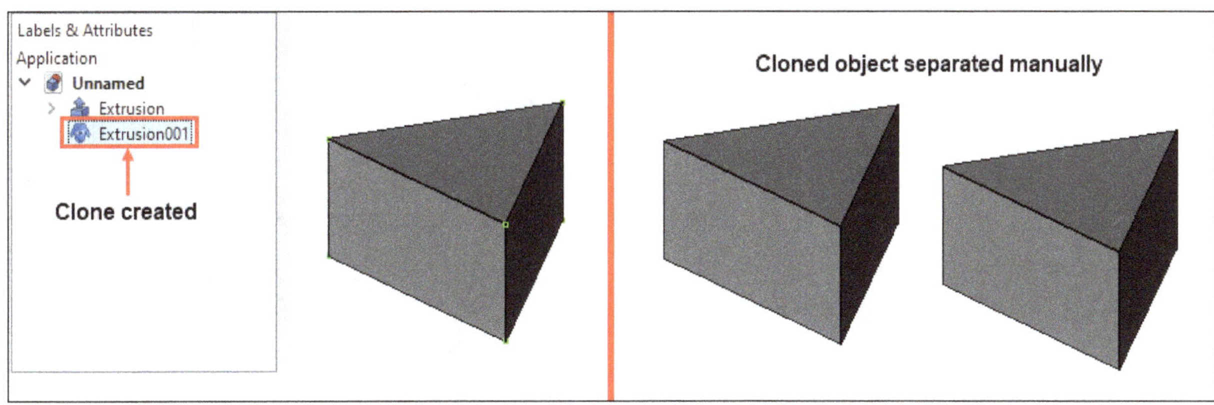

Figure-124. Cloned object created

- You can change the properties of clone created from the **Property editor** dialog as discussed earlier.

Creating Drawing (obsolete in version 0.17)

The **Drawing** tool allows you to put selected objects on a drawing sheet created with the **Drawing** workbench. The procedure to use this tool is discussed next.

- Select the object from the Model tree view or from the 3D view area which you want to put on drawing sheet; refer to Figure-125.

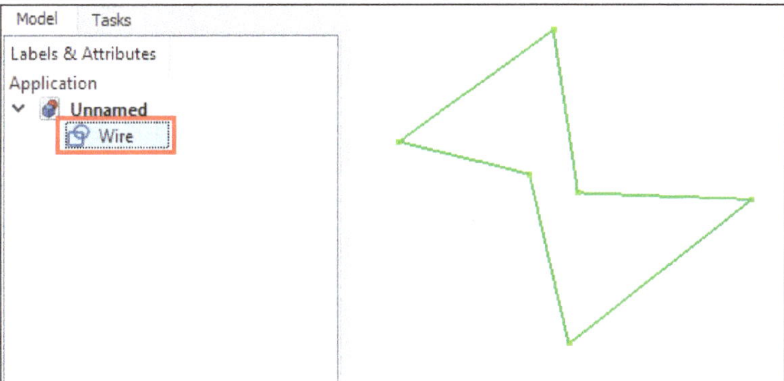

Figure-125. Selecting the object to put on drawing sheet

- Click on the **Drawing** tool from **Toolbar** in the **Draft** workbench; refer to Figure-126. A drawing page of the object will be created in the Model tree view ; refer to Figure-127.

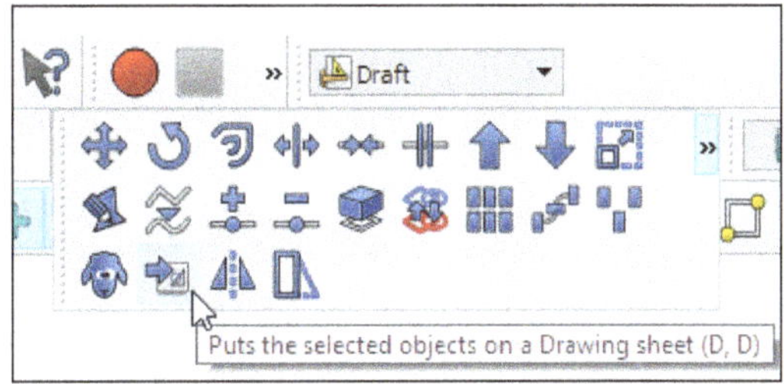

Figure-126. Drawing tool

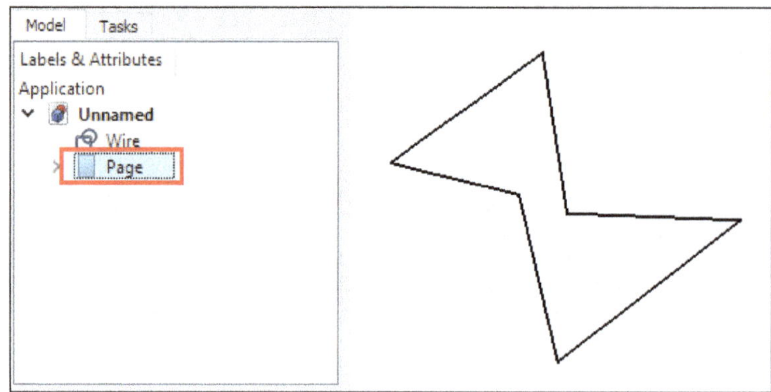

Figure-127. Drawing page created

- Select the page created from the Model tree view and click **RMB**, a shortcut menu will be displayed; refer to Figure-128.

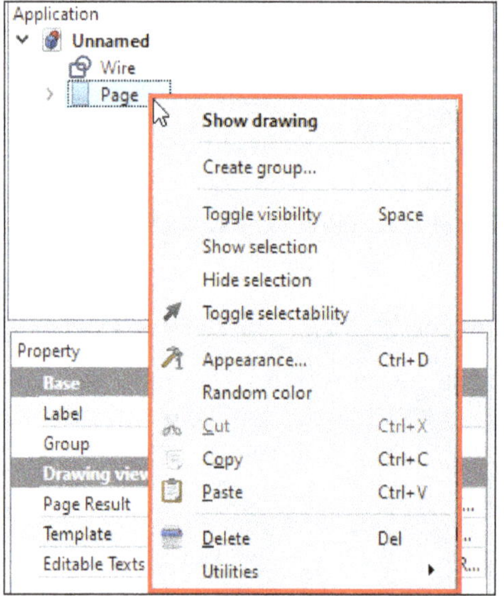

Figure-128. Shortcut menu on selecting page

- Click on the **Show drawing** option from the menu, a new file will be opened with the object put on a drawing sheet; refer to Figure-129.

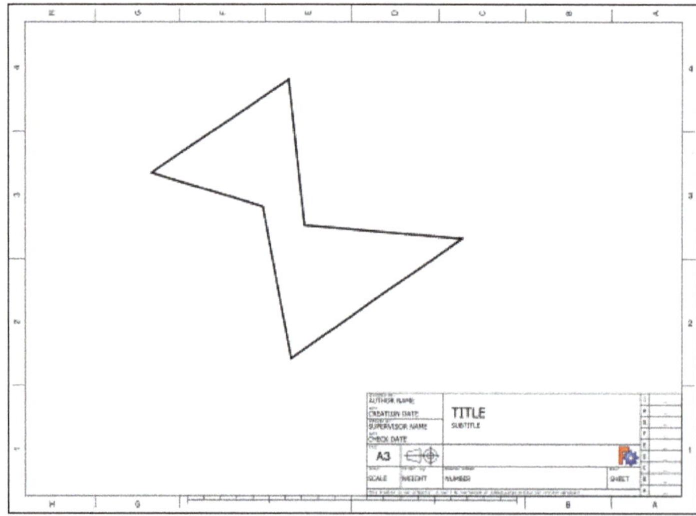

Figure-129. Object put on a drawing sheet

Mirroring object

The **Mirror** tool produces a mirrored copy of a selected object using the mirror operation. The copy just like a **Draft Clone**, is linked to the original object. This means that if the original object changes its shape and properties, the mirrored shape changes as well. The procedure to use this tool is discussed next.

- Select the object from the Model tree view or from the 3D view area which you want to be mirrored; refer to Figure-130.

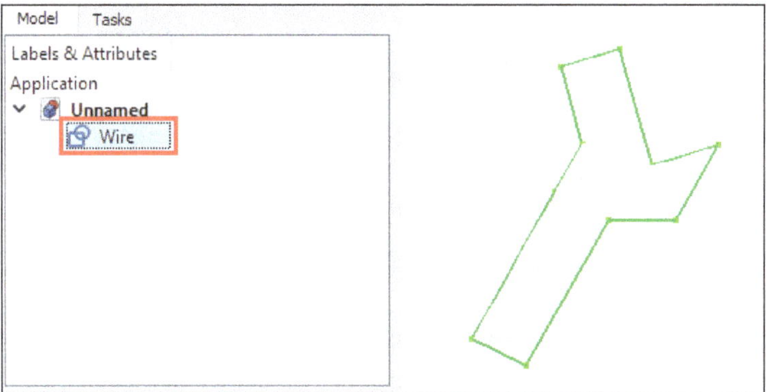

Figure-130. Selecting the object to be mirrored

- Click on the **Mirror** tool from **Toolbar** in the **Draft** workbench; refer to Figure-131. The **Mirror** dialog will be displayed in the **Tasks** panel of **Combo View** along with the plus sign in place of original cursor; refer to Figure-132. You will be asked to specify the first point.

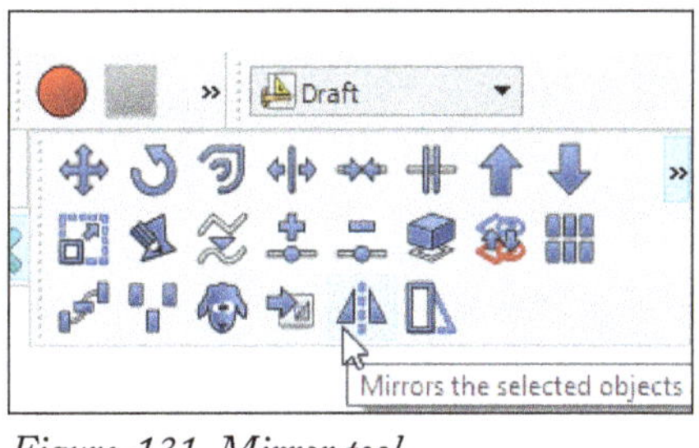

Figure-131. Mirror tool

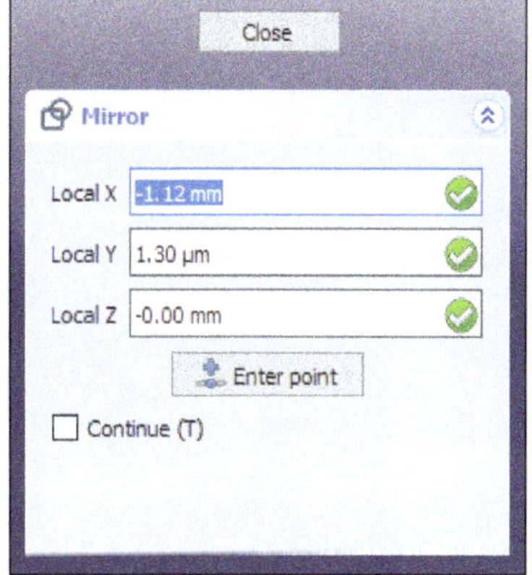

Figure-132. Mirror dialog

- Click in the 3D view area to specify the first point or enter desired values for the x, y, and z coordinates in the **Local X**, **Local Y**, and **Local Z** edit boxes of the dialog, respectively.
- After specifying coordinates for the first point in dialog, click on the **Enter point** button from the dialog. You will be asked to specify the second point.
- Move the cursor away and click at desired location to specify the second point or enter desired values for the coordinates in their respective edit boxes.
- After specifying coordinates for the second point in dialog, click on the **Enter point** button again. The object will be mirrored; refer to Figure-133.

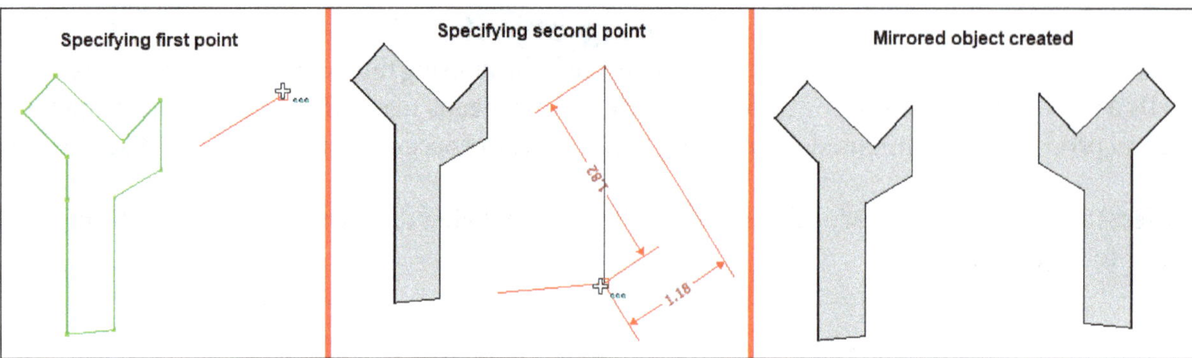

Figure-133. Mirrored object created

- Other parameters in the **Mirror** dialog have been discussed earlier.
- Click on **Close** button to close the dialog.

Stretching object

The **Stretch** tool stretches an object by moving some of its selected vertices. The equivalent action is editing the object and moving the points manually to a new position. The procedure to use this tool is discussed next.

- Select the object from the Model tree view or from the 3D view area which you want to stretch; refer to Figure-134.

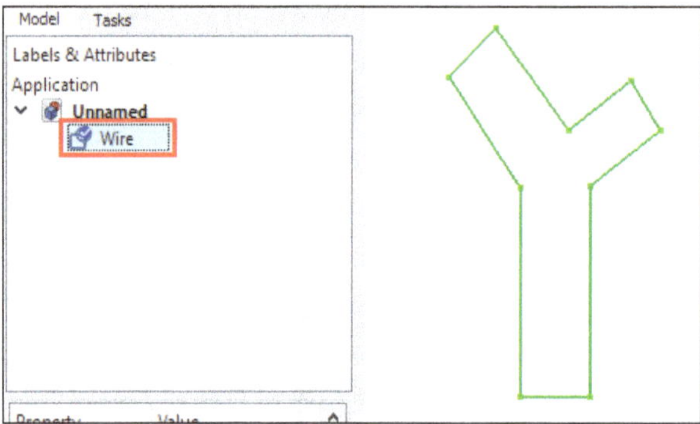

Figure-134. Selecting the object to be stretched

- Click on the **Stretch** tool from **Toolbar** in the **Draft** workbench; refer to Figure-135. The **Stretch** dialog will be displayed in the **Tasks** panel of **Combo View** along with the plus sign in place of original cursor; refer to Figure-136. You will be asked to specify the first point.

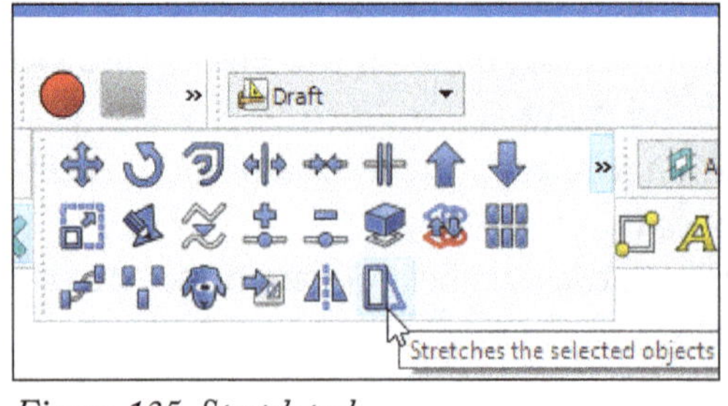

Figure-135. Stretch tool

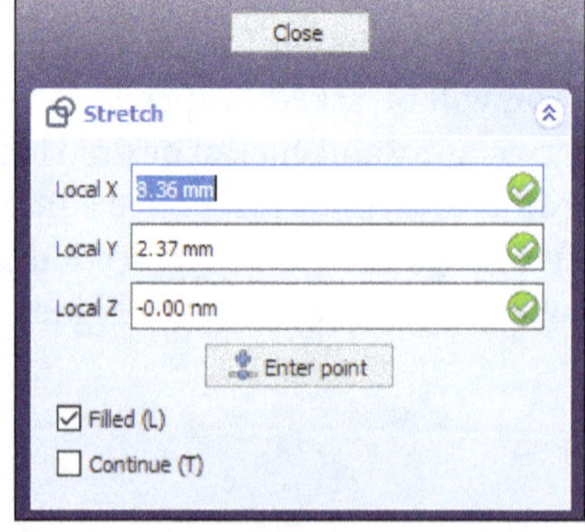

Figure-136. Stretch dialog

- Click in the 3D view area to specify the first point or enter desired values for the x, y, and z coordinates in the **Local X**, **Local Y**, and **Local Z** edit boxes of the dialog, respectively.

- After specifying coordinates for the first point in dialog, click on the **Enter point** button from the dialog. You will be asked to specify the second point.
- Move the cursor away and click at desired location to specify the second point or enter desired values for the coordinates in their respective edit boxes.
- After specifying coordinates for the second point in dialog, click on the **Enter point** button again. You will be asked to specify the third point. The first and second point define a selection rectangle. The vertices of the original object enclosed by this rectangle will become highlighted; refer to Figure-137.

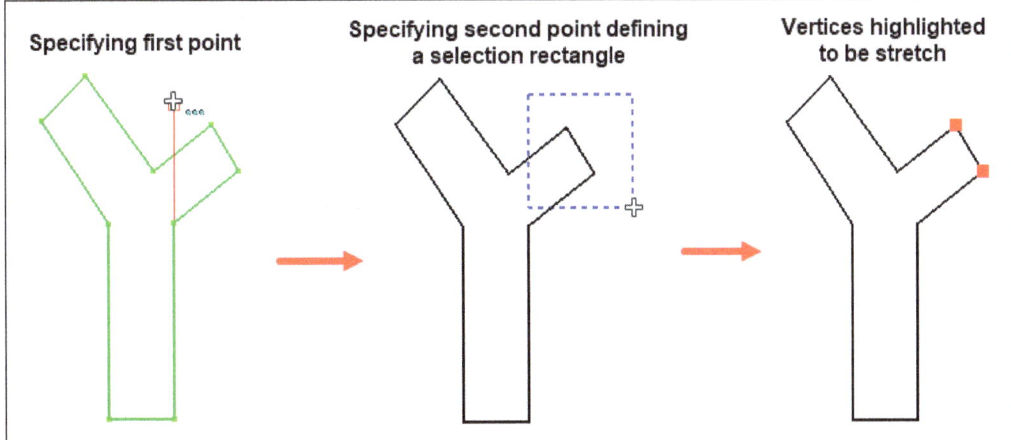

Figure-137. Selecting the vertices to be stretch

- Click in the 3D view area to specify the third point or enter desired values for the coordinates in their respective edit boxes.
- After specifying coordinates for the third point in dialog, click on the **Enter point** button again. You will be asked to specify the fourth point.
- Move the cursor away and click at desired location to specify the fourth point or enter desired values for the coordinates in their respective edit boxes. The third and fourth point define a line whose distance and direction will be used to stretch the figure attached to the highlighted points.
- After specifying coordinates for the fourth point in dialog. Click on the **Enter point** button again. The object will be stretched; refer to Figure-138.

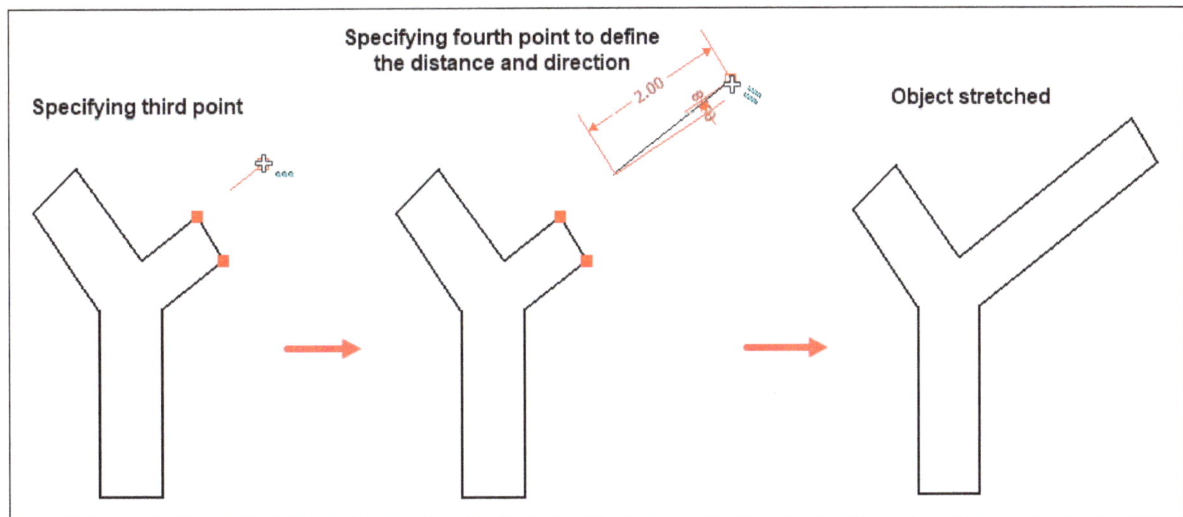

Figure-138. Object stretched

- Other parameters in the **Stretch** dialog have been discussed earlier.
- Click on **Close** button to close the dialog.

Draft Tray Toolbar

The **Draft tray toolbar** allows selecting the working plane together with some visual properties like the line color, shape color, text size, line width, and automatic group. This toolbar is available in the **Draft** workbench; refer to Figure-139. The procedures to use the tools of this toolbar are discussed next.

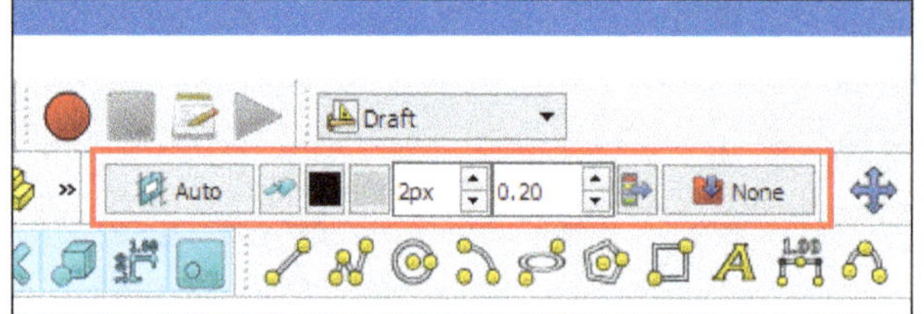

Figure-139. Draft tray toolbar

Creating Working Plane

The **Working Plane** tool sets a working plane from a standard view or selected face. The working plane in the 3D view area indicates where a draft shape will be built. The procedure to use this tool is discussed next.

- Click on the **Select Plane** tool from **Toolbar** in the **Draft** workbench; refer to Figure-140. The **Select Plane** dialog will be displayed in the **Tasks** panel of **Combo View**; refer to Figure-141.

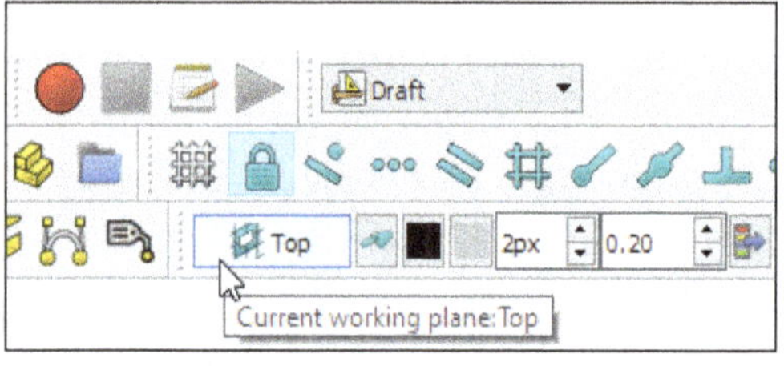

Figure-140. Select Plane tool

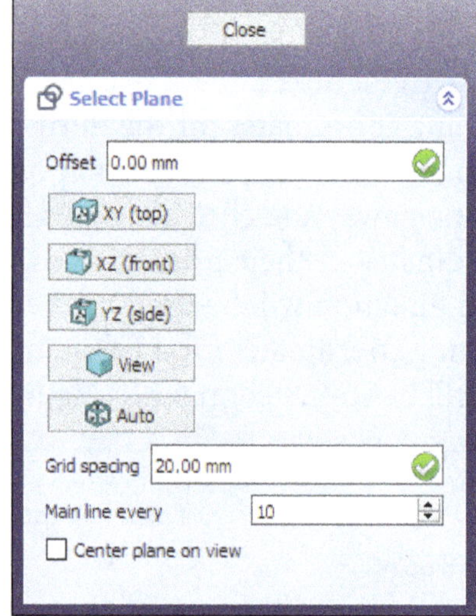

Figure-141. Select Plane dialog

- Specify desired offset value in the **Offset** edit box to set the working plane at specified perpendicular distance from the plane you selected.
- Specify desired value in **Grid spacing** edit box to define the space between two consecutive lines in the grid.
- Specify desired value in **Main line every** edit box to draw a slightly thicker line in the grid at the set value.
- Click on the **XY (top)**, **XZ (front)**, or **YZ (side)** button to set the working plane on XY, XZ, or YZ plane, respectively.
- Click on the **View** button to set the working plane to the current 3D view, perpendicular to the camera axis, and passing through the origin.
- Click on the **Auto** button to unset any current working plane and automatically set a working plane when a tool is used.
- On clicking desired plane, the active plane will be displayed in the toolbar along with the grid view; refer to Figure-142.

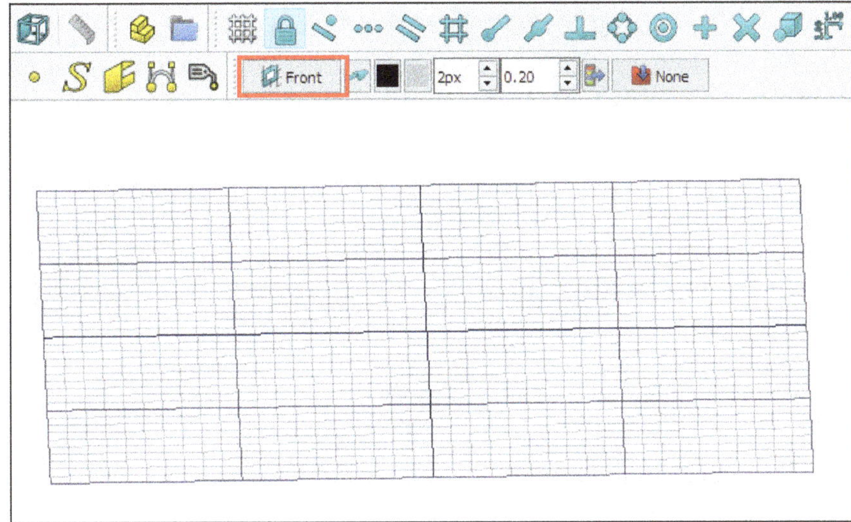

Figure-142. Active plane displayed along with grid view

- Select the **Center plane on view** checkbox to draw the plane and grid closer to the camera view in the 3D view.
- Click on **Close** button to close the dialog.

Toggle Construction Mode

The **Toggle Construction Mode** toggles the draft construction mode on or off. The construction geometry is comprised of lines, points, and other shapes that serve as references or snapping elements which are helpful in building your main geometry. The procedure to use this tool is discussed next.

- Click on the **Toggle construction mode** button from **Toolbar** in the **Draft** workbench; refer to Figure-143.

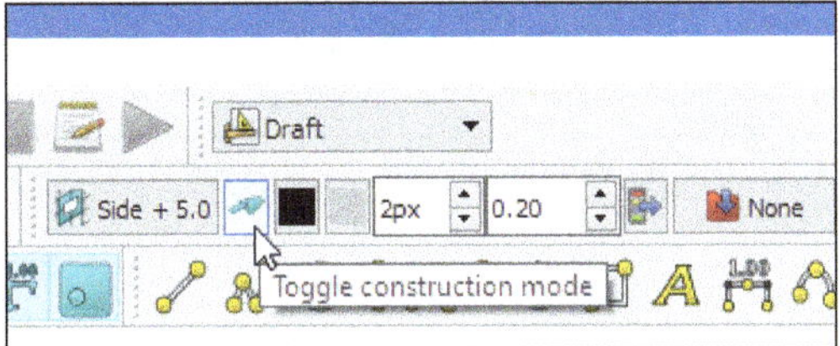

Figure-143. Toggle construction mode button

- Create some objects in the 3D view area as desired. Now, it will be created in construction mode; refer to Figure-144.

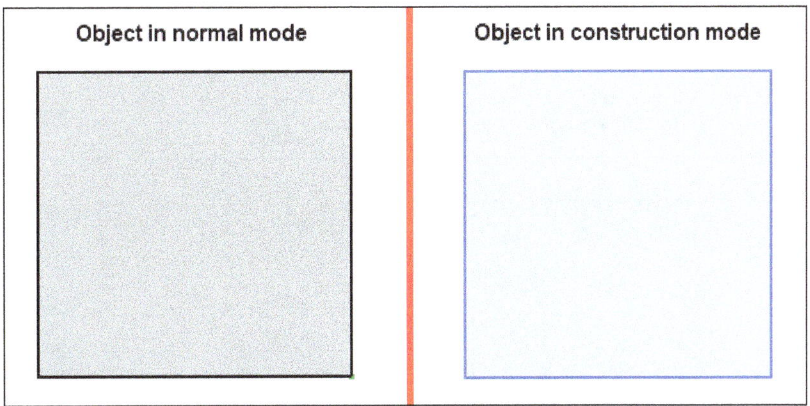

Figure-144. Objects in normal and construction mode

- Click on the **Toggle construction mode** button again to go back to normal mode.

Applying Style settings

The **Apply style** tool allows to set default visual properties like line color or line width to be used by all draft objects created after this tool is used and bulk apply these styles to selected objects. The procedure to use this tool is discussed next.

- Set desired line color and face color by clicking on the **Current line color** and **Current face color** buttons, respectively from the **Toolbar**.
- Specify desired line width and font size in the **Current line width** and **Current font size** edit boxes, respectively from the **Toolbar**; refer to Figure-145.

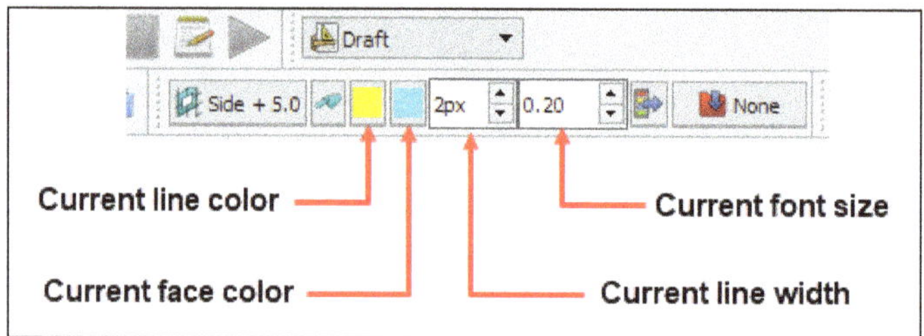

Figure-145. Setting colors and sizes

- After setting desired colors and sizes, select the object or multiple objects on which you want to apply the style; refer to Figure-146.

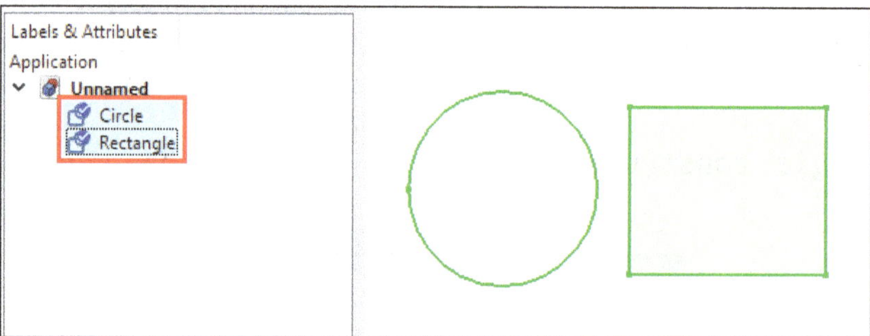

Figure-146. Selecting the objects to apply style

- Click on the **Apply style** tool from **Toolbar** in the **Draft** workbench; refer to Figure-147. The style will be applied; refer to Figure-148.

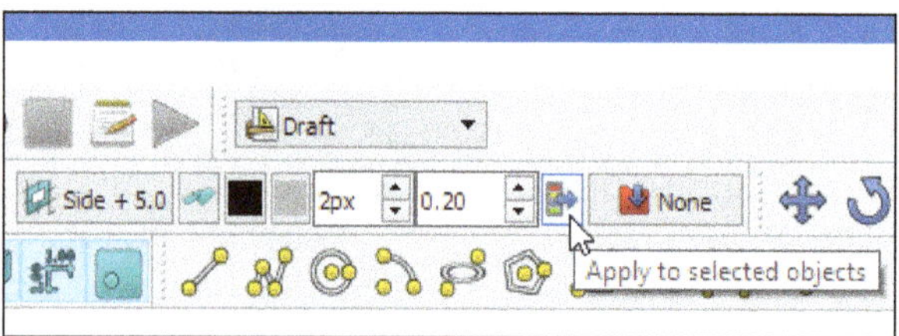

Figure-147. Apply style tool

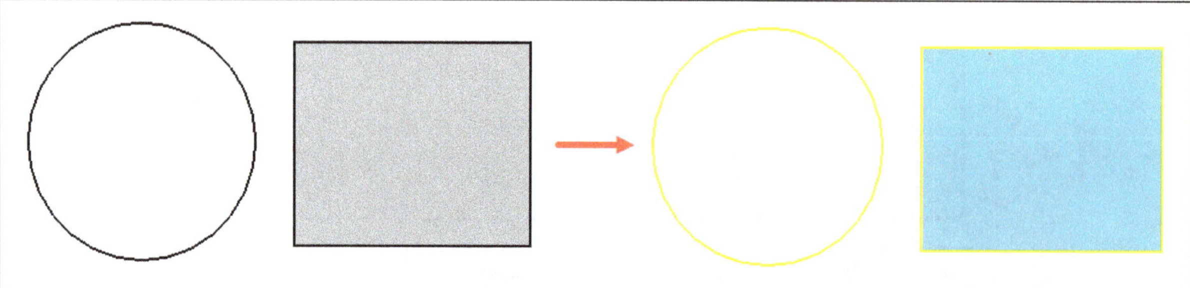
Figure-148. Style applied

Creating Auto Group

The **Auto Group** tool sets a selected **Std Group** or a related element like a **Draft VisGroup**, **Arch Site**, **Arch Building**, or **Arch BuildingPart** as the active auto-group. When an auto-group is set, new objects will be automatically moved to the indicated group upon being created. Auto-grouping works with elements created with the **Draft** and **Arch** workbenches. The procedure to use this tool is discussed next.

- Select the **Std Group**, **Construction group**, or **Draft VisGroup** from the Model tree view; refer to Figure-149.

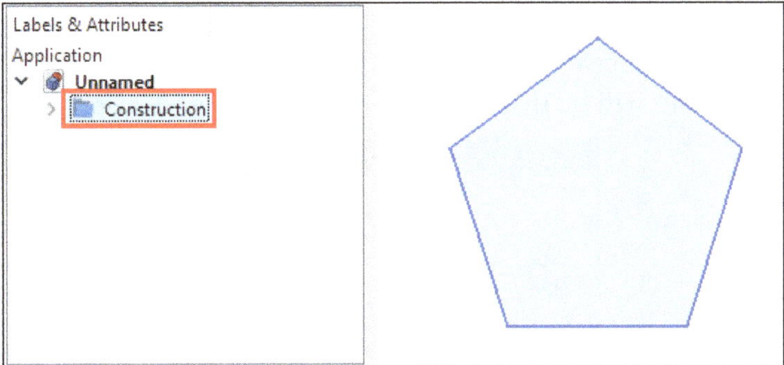

Figure-149. Selecting the construction group

- Click on the **Auto Group** tool button from **Toolbar** in the **Draft** workbench; refer to Figure-150. The **None** button will change with the name of the active auto-group; refer to Figure-151.

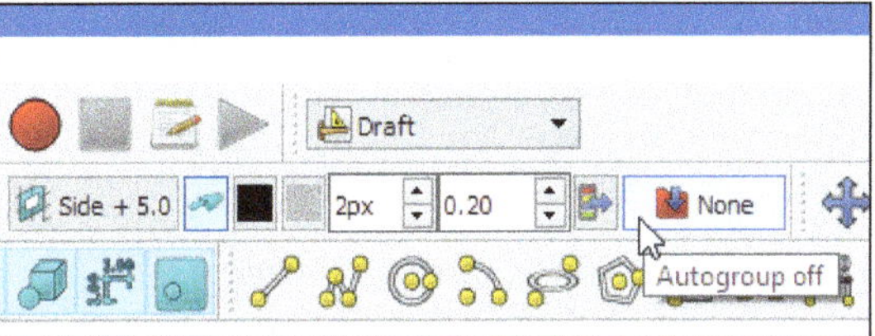

Figure-150. Auto Group tool

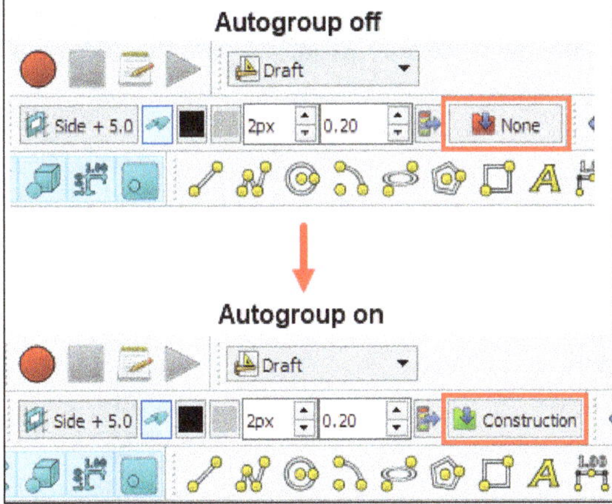

Figure-151. Auto group activated

Draft Snap Toolbar

The **Draft Snap toolbar** allows to select various snapping mode. All the buttons in this toolbar are toggle buttons. This toolbar is available in the **Draft** workbench; refer to Figure-152. The procedures to use the tools of this toolbar are discussed next.

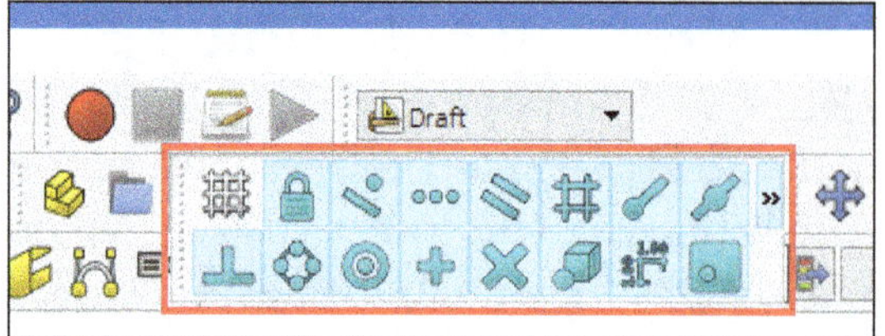

Figure-152. Draft Snap toolbar

Toggle Grid

The **Toggle Grid** tool allows you to show and hide the grid defined in the **Draft Preferences** or with the **Draft Select Plane** tool. The procedure to use this tool is discussed next.

- If you want to set the appearance of the grid then click on the **Edit** menu from the **Toolbar**. The options related to this menu will be displayed; refer to Figure-153.

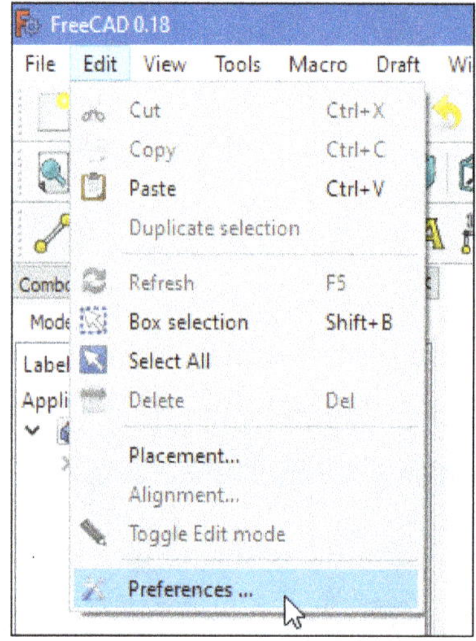

Figure-153. Edit menu options

- Click on the **Preferences** option from the menu. The **Preferences** dialog box will be displayed; refer to Figure-154.

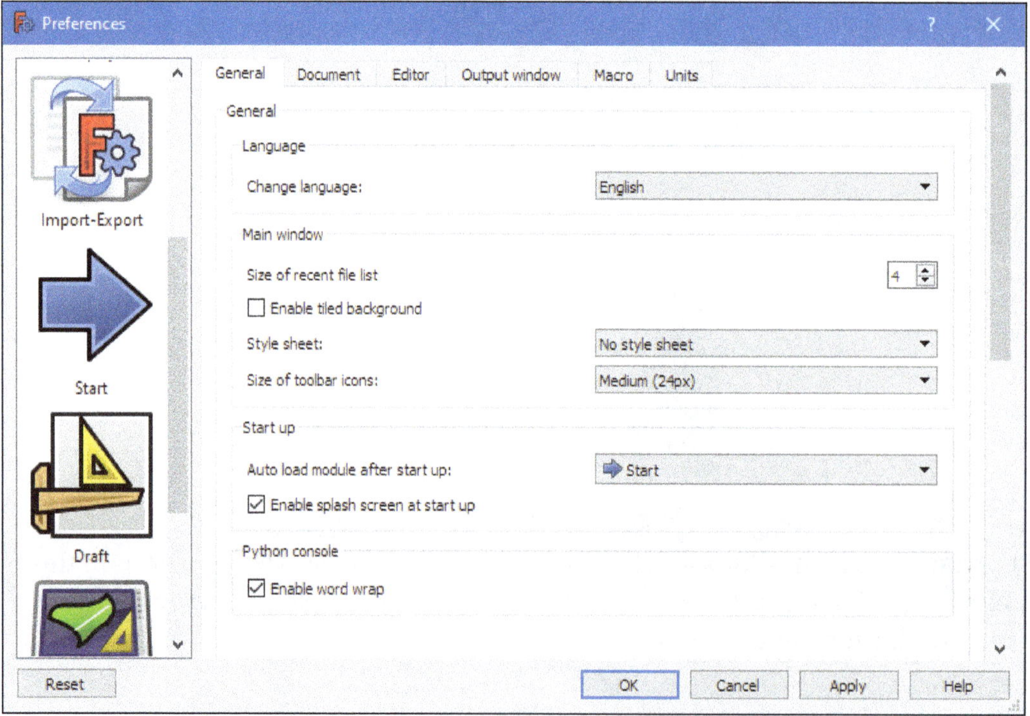

Figure-154. Preferences dialog box

- Click on the **Draft** option from left area of the dialog box. The parameters related to **Draft** will be displayed.
- Click on the **Grid and snapping** tab from **Draft** option of the dialog box. The parameters related to **Grid and snapping** will be displayed.
- Specify desired values in the **Main lines every**, **Grid spacing**, and **Grid size** edit boxes from **Grid** section of the tab; refer to Figure-155.

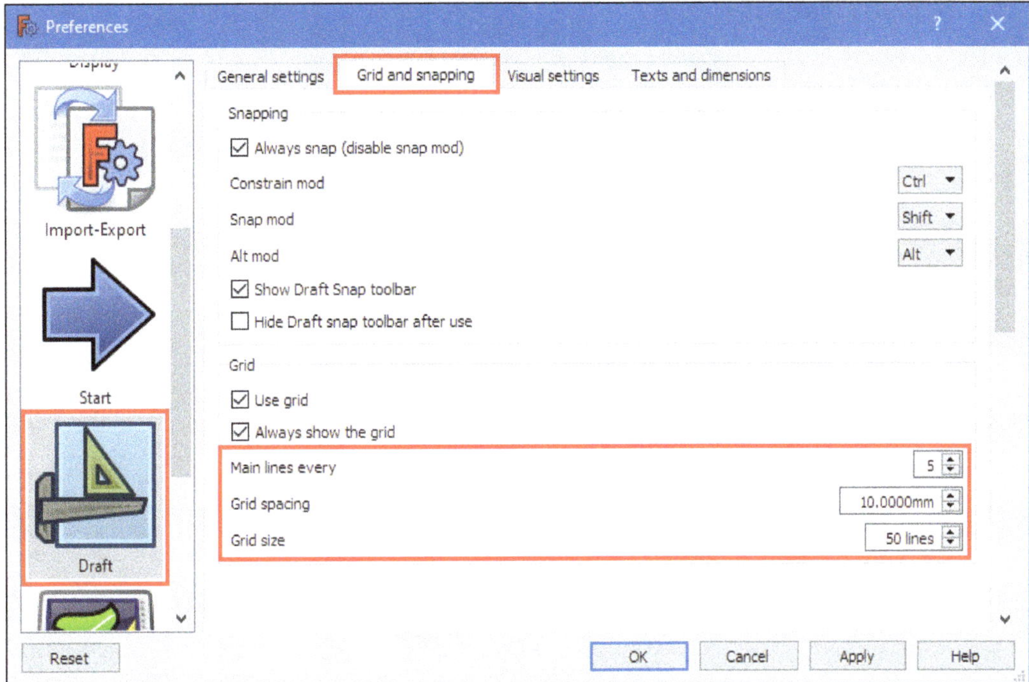

Figure-155. Specifying the grid parameters

- Click on **Apply** and then **OK** button from the dialog box to apply the changes.
- To make the grid visible, click on the **Toggle Grid** button from **Toolbar** in the **Draft** workbench; refer to Figure-156. The grid will be displayed in the 3D view area; refer to Figure-157.

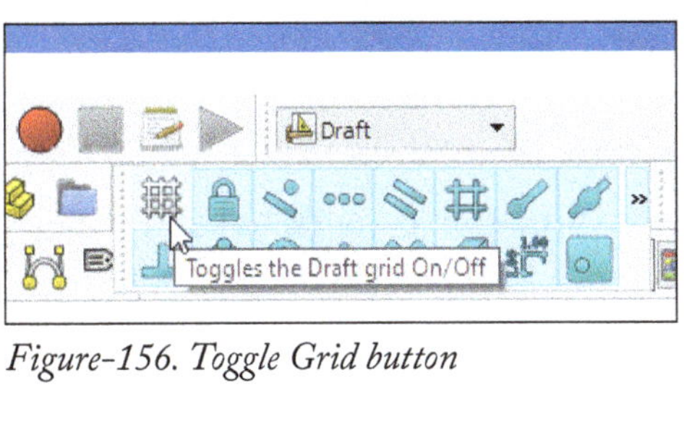

Figure-156. Toggle Grid button

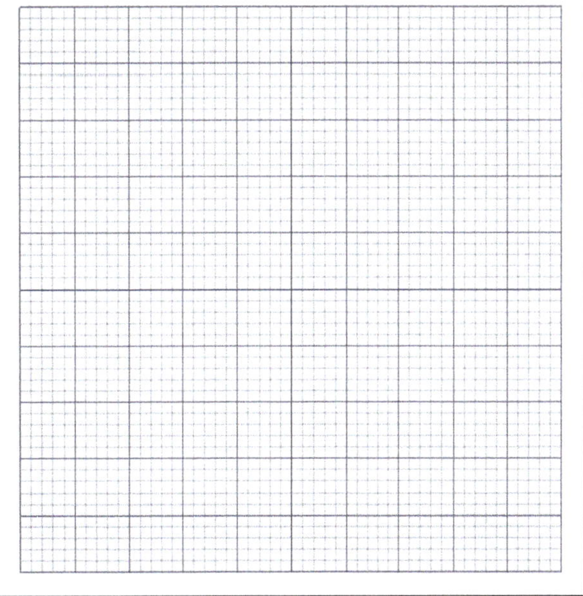

Figure-157. Grid displayed in the 3D view area

- Click on the **Toggle Grid** button again to turn off the draft grid.

Toggle Snap

The **Toggle Snap** tool allows you to activate or deactivate globally the Draft Snap methods. The procedure to use this tool is discussed next.

- If you want to make the snapping methods available then click on the **Toggle Snap** button from **Toolbar** in the **Draft** workbench; refer to Figure-158. The snapping will be displayed while creating the objects; refer to Figure-159.

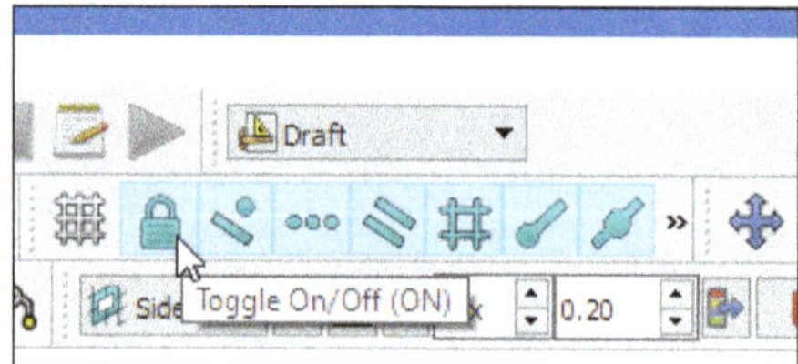

Figure-158. Toggle Snap tool

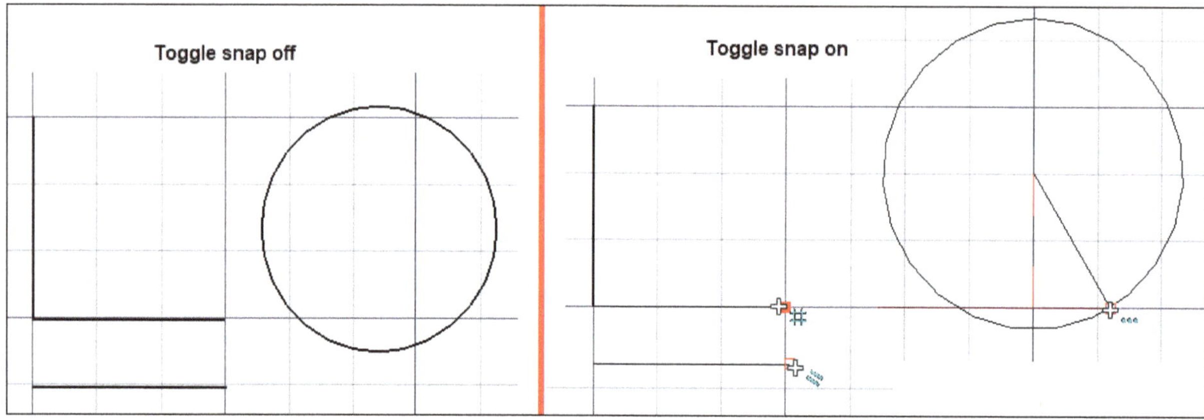

Figure-159. Snapping displayed while creating objects

Near

The **Near** snap button snaps to the closest point or edge on the nearest object.

Extension

The **Extension** snap button snaps to a point on an imaginary line that extends beyond the endpoints of line segments. Hover the mouse over desired object to activate its extension snap.

Parallel

The **Parallel** snap button snaps on an imaginary line parallel to a line segment. Hover the mouse over desired object to activate its parallel snap.

Grid

The **Grid** snap button snaps to the intersection of two grid lines, if the grid is visible.

Endpoint

The **Endpoint** snap button snaps to the endpoints of line, arc, and spline segments.

Midpoint

The **Midpoint** snap button snaps to the middle point of line and arc segments.

Perpendicular 

The **Perpendicular** snap button snaps to a line or edge or to an extension of it to produce a line that is perpendicular to that edge.

Angle

The **Angle** snap button snaps to the points of circles and arcs at specific increments of 30° and 45° on the arc; this includes 0°, 60°, 90°, 180°, 210°, 270°, and other integer multiples.

Center 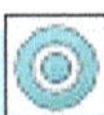

The **Center** snap button snaps to the center point of arcs and circles.

Ortho 

The **Ortho** snap button snaps to a point on an imaginary line that originates from the previous point and extends infinitely at specific increments of 45°; this includes 0°, 90°, 135°, 180°, 225°, 270°, and other integer multiples.

Intersection 

The **Intersection** snap button snaps to the intersection of two line or arc segments. Hover the mouse over the two desired objects to activate their intersection snaps.

Special 

The **Special** snap button snaps to the special location points defined by a particular object such as an Arch Wall.

Dimensions 

The **Dimensions** snap button shows temporary X and Y dimensions between the current point and the last snapped point on screen while snapping.

Working Plane 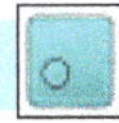

The **Working Plane** button always places a snapped point on the current working plane even if you also use another snapping method and select a point outside of that working plane. In other words, it projects an external snapping point to the current working plane.

FOR STUDENT NOTES

Chapter 7

Arch Modeling

Topics Covered

The major topics covered in this chapter are:

- ***Introduction to Arch Workbench***
- ***Arch Tools***
- ***Axis Tools***
- ***Panel Tools***
- ***Pipe Tools***
- ***Material Tools***

INTRODUCTION

The **Arch** workbench provides a modern building information modeling (BIM) workflow to FreeCAD with support for features like fully parametric architectural entities such as walls, beams, roofs, windows, stairs, pipes, and furniture. It supports industry foundation class (IFC) files and production of 2D floor plans in combination with the **TechDraw** workbench; refer to Figure-1.

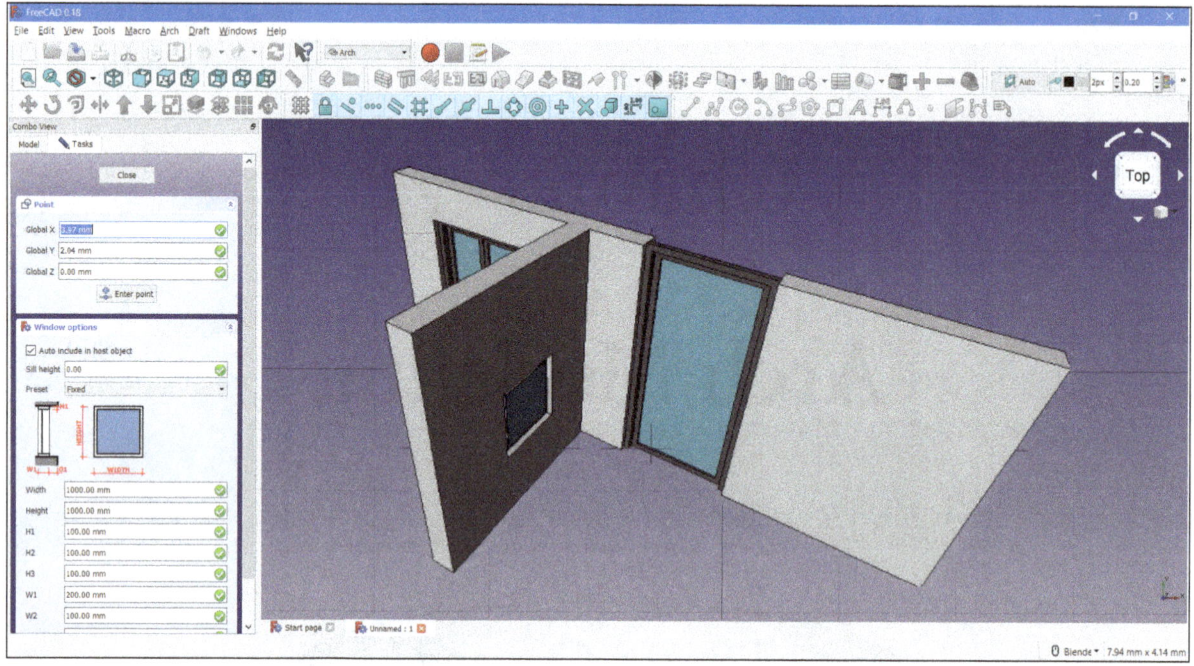

Figure-1. Arch window

The BIM functionality of FreeCAD is now progressively split into this **Arch** workbench which holds basic architectural tools and the **BIM** workbench which is available from the **Addon Manager**. This **BIM** workbench adds a new interface layer on top of the **Arch** tools with the aim of making the BIM workflow more intuitive and user-friendly.

STARTING ARCH WORKBENCH

The **Arch** workbench imports all tools from the **Draft** workbench as it uses its 2D objects to build 3D parametric architectural objects. Nevertheless, Arch can also use solid shapes created with other workbenches like **Part** and **Part Design**.

- To start a new arch file, click on the **New** button from **File** menu and select **Arch** workbench from **Switch between workbenches** drop-down in the **Toolbar**; refer to Figure-2. The tools related to **Arch** workbench will be displayed in the **Toolbar**; refer to Figure-3.

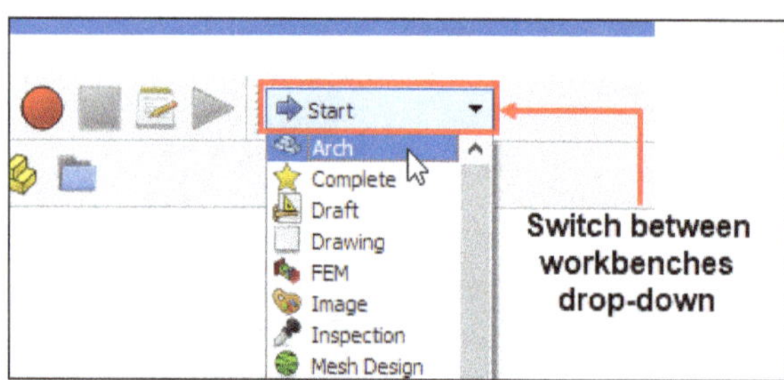

Figure-2. Arch workbench

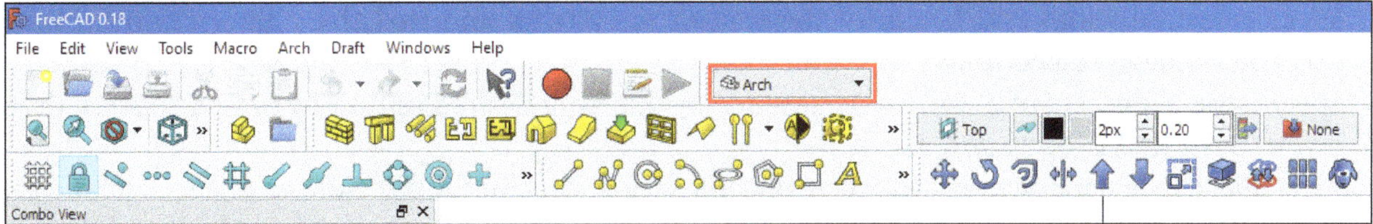

Figure-3. Arch workbench tools

ARCH TOOLS

The **Arch** tools are used for creating architectural objects. These tools are available in the **Toolbar** of **Arch** workbench; refer to Figure-4. These tools are discussed next.

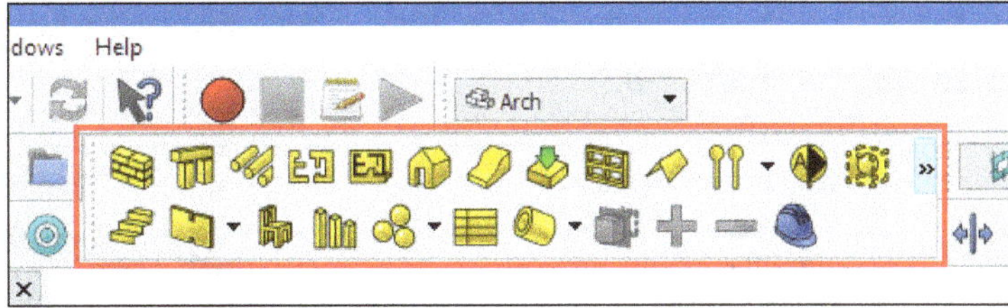

Figure-4. Arch tools

Creating Wall

The **Wall** tool builds a wall object from scratch or on top of any other shape-based or mesh-based object. A wall can be built without any base object in which case it behaves as a cubic volume using length, width, and height properties. When built on top of an existing shape, a wall can be based on a linear 2D object, a flat face, a solid, or a mesh. The procedure to use this tool is discussed next.

- Click on the **Wall** tool from **Toolbar** in the **Arch** workbench; refer to Figure-5. The **Point** and **Wall options** dialogs will be displayed in the **Tasks** panel of **Combo View** along with the plus sign in place of original cursor; refer to Figure-6 and Figure-7. You will be asked to specify the first point.

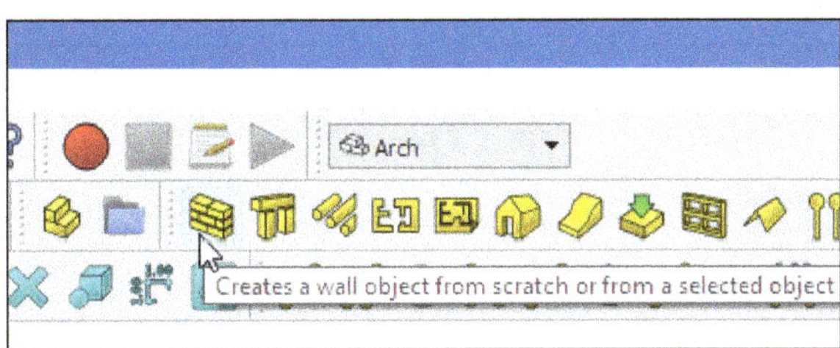

Figure-5. Wall tool

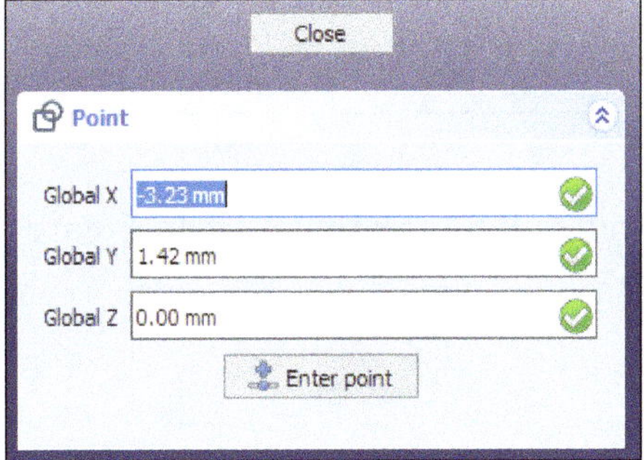

Figure-6. Point dialog

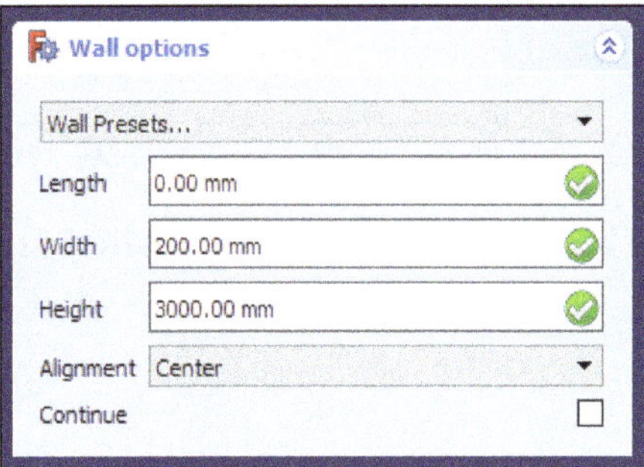

Figure-7. Wall options dialog

- Specify desired width and height of the wall in the **Width** and **Height** edit boxes from **Wall options** dialog, respectively.
- Specify the alignment of a wall on its baseline by selecting desired option from **Alignment** drop-down in the **Wall options** dialog.
- After specifying parameters in **Wall options** dialog, click in the 3D view area to specify the first point or enter desired values for x, y, and z coordinates in the **Global X**, **Global Y**, and **Global Z** edit boxes from **Point** dialog, respectively.
- After specifying coordinates for the first point in dialog, click on the **Enter point** button from **Point** dialog. You will be asked to specify the second point.
- Move the cursor away and click at desired location to specify the second point as well as length of the wall or enter desired values for the coordinates in their respective edit boxes.
- After specifying coordinates for the second point in dialog, click on the **Enter point** button again. The wall will be created; refer to Figure-8.

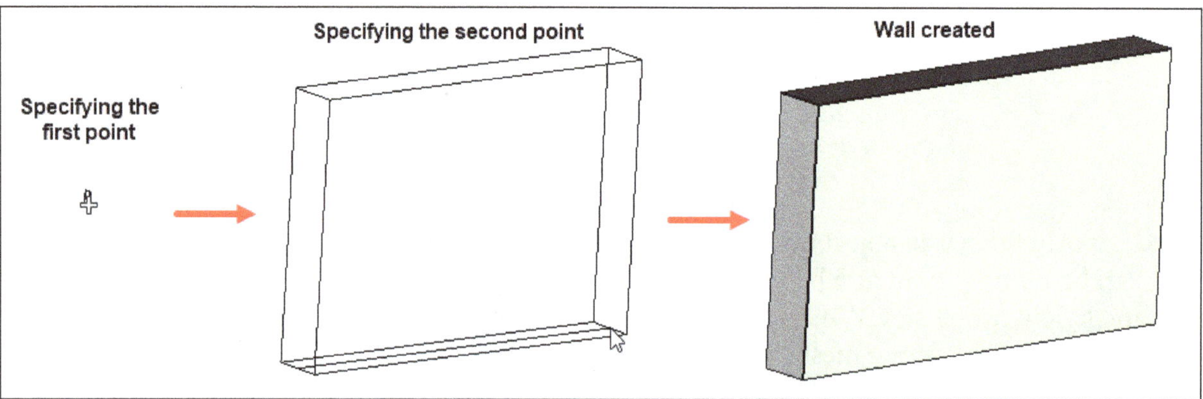

Figure-8. Wall created

- Click on **Close** button to close the dialog.
- If you want to create a wall using base object then select the base geometry objects (Draft object, sketch, etc.) from Model tree view or from the 3D view area to create a wall; refer to Figure-9.

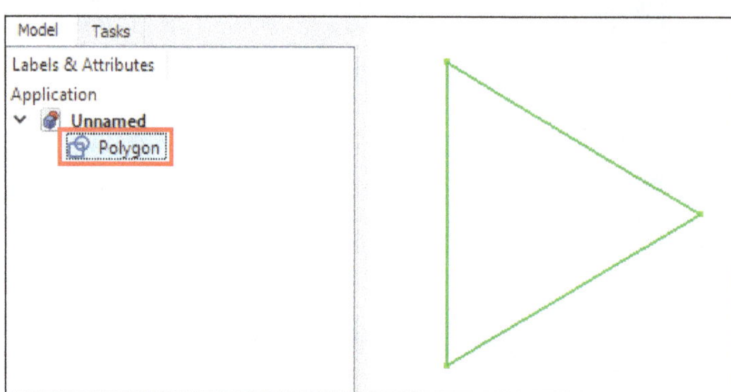

Figure-9. Selecting base object

- Click on the **Wall** tool as discussed earlier. The wall will be created; refer to Figure-10.

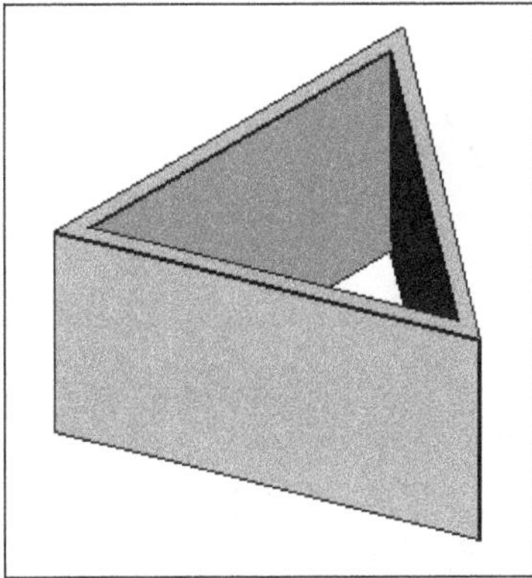

Figure-10. Wall created

- If you want to edit the properties of wall then select the wall created from the Model tree view. The **Property editor** dialog will be displayed in the **Model** panel of **Combo View** with the parameters related to wall; refer to Figure-11.

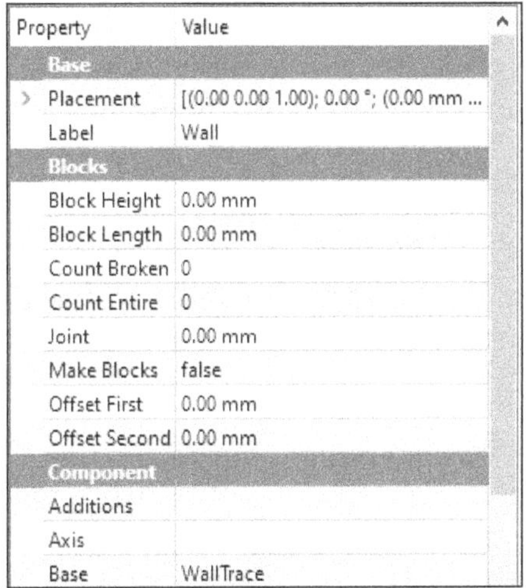

Figure-11. Property editor dialog with wall parameters

- Specify desired height and length of each block in the **Block Height** and **Block Length** edit boxes from **Blocks** section of the dialog, respectively.
- Specify the size of joints between each block in the **Joint** edit box.
- Select desired option from **Make Blocks** drop-down whether to make the wall generate blocks or not.
- Specify the horizontal offset of first line and second line of blocks in the **Offset First** and **Offset Second** edit boxes, respectively.
- Select desired option for the alignment of the wall on its baseline from **Align** drop-down in the **Wall** section of the dialog.
- Enter desired value in the **Face** edit box to specify index of the face from the base object to use. If the value is not set or 0, the whole object is used.
- Specify desired length, width, and height of the wall in the **Length**, **Width**, and **Height** edit boxes, respectively.
- Enter desired value in the **Offset** edit box to specify the distance between the wall and its baseline.
- Enter desired value in the **x**, **y**, and **z** edit boxes from **Normal** cascading menu to specify extrusion for the wall along x, y, and z direction, respectively.
- On specifying parameters in the **Property editor** dialog, the properties of wall will be modified; refer to Figure-12.

Figure-12. Wall edited

Creating Structure

The **Structure** tool allows you to build structural elements such as columns or beams, by specifying their width, length, and height, or by basing them on a 2D profile (face, wire, or sketch). The procedure to use this tool is discussed next.

- Click on the **Structure** tool from **Toolbar** in the **Arch** workbench; refer to Figure-13. Multiple dialogs will be displayed in the **Tasks** panel of **Combo View** along with the plus sign and a structure of column in place of original cursor; refer to Figure-14. You will be asked to specify placement point for structure.

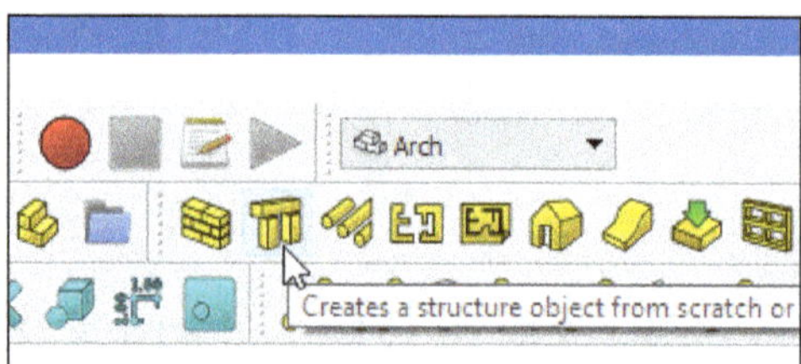

Figure-13. Structure tool

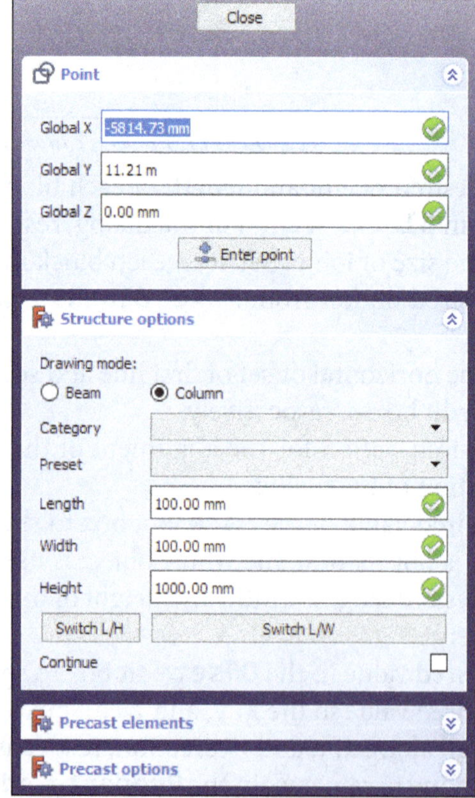

Figure-14. Multiple dialogs displayed

- Select desired drawing mode from **Drawing mode** area of the **Structure options** dialog. In **Beam** mode, you will be asked to specify two points in 3D view area or by entering coordinates. The new structural object will span between these two points. In **Column** mode, you will be asked to specify one point in 3D view area or by entering coordinates. The new structural object will be placed at that point.
- Select desired presets of structure from **Category** drop-down in the **Structure options** dialog.
- Select desired size of the preset from **Preset** drop-down in the **Structure options** dialog.
- On selecting the **Precast concrete** option from **Category** drop-down, the parameters related to the selected structure and preset will be displayed in **Precast elements** dialog and **Precast options** dialog in the **Tasks** panel of **Combo View**; refer to Figure-15 and Figure-16.

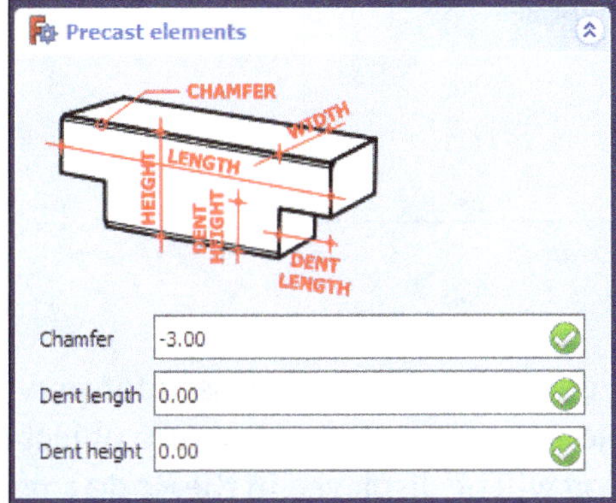

Figure-15. Precast elements dialog

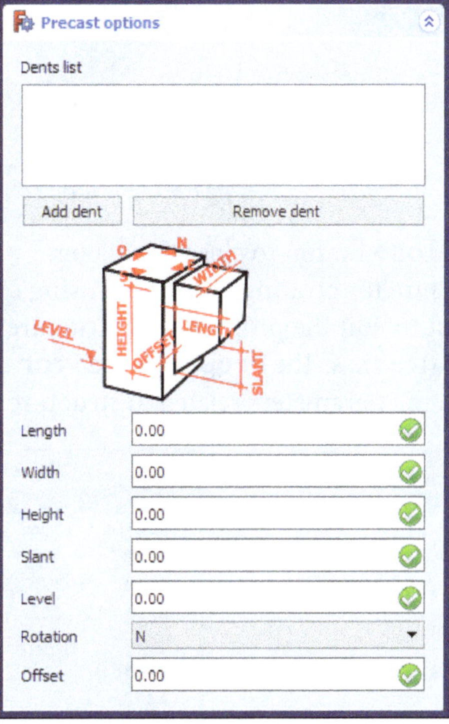

Figure-16. Precast options dialog

- Specify desired parameters in **Precast elements** and **Precast options** dialogs.
- Specify desired length, width, and height of the structure in the **Length**, **Width**, and **Height** edit boxes from **Structure options** dialog, respectively.
- Click on the **Switch L/H** or **Switch L/W** button to switch between length and height values or length and width values, respectively.
- After specifying parameters in all the dialogs, click in the 3D view area to specify the point for creating the structure or enter desired values for the x, y, and z coordinates in the **Global X**, **Global Y**, and **Global Z** edit boxes of the **Point** dialog, respectively.
- After specifying coordinates for the point in dialog, click on the **Enter point** button from **Point** dialog. The structure object will be created; refer to Figure-17.

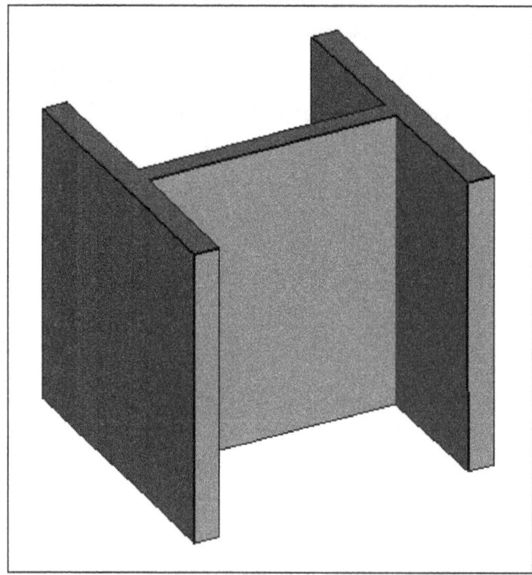

Figure-17. Structure object created

- Click on **Close** button to close the dialog.
- The procedure for creating a structure using base object is same as discussed for previous tool.
- If you want to edit the properties of structure object then select the structure object created from the Model tree view. The **Property editor** dialog will be displayed in the **Model** panel of **Combo View** with the parameters related to structure object; refer to Figure-18.

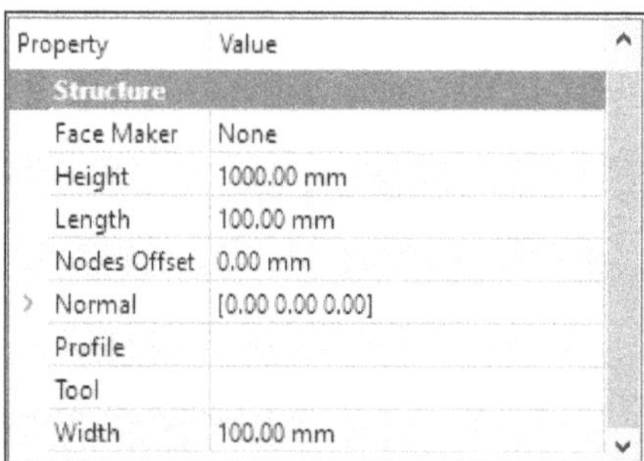

Property	Value
Structure	
Face Maker	None
Height	1000.00 mm
Length	100.00 mm
Nodes Offset	0.00 mm
> Normal	[0.00 0.00 0.00]
Profile	
Tool	
Width	100.00 mm

Figure-18. Property editor dialog with structure parameters

- Select desired option from **Face Maker** drop-down in the **Structure** section of the dialog to specify the type of face generation algorithm to be used for building the profile.
- Specify desired height, length, and width of the structure in the **Height**, **Length**, and **Width** edit boxes, respectively.
- Enter desired value in the **Nodes Offset** edit box to specify the offset between the centerline and the nodes line.
- Enter desired values in **x**, **y**, and **z** edit boxes of **Normal** cascading menu to specify the extrusion of base face of this structure along x, y, and z directions, respectively.
- Click on the **Tool** button to select an optional extrusion path which can be any type of wire.
- On specifying the parameters in the **Property editor** dialog, the properties of structure object will be modified; refer to Figure-19.

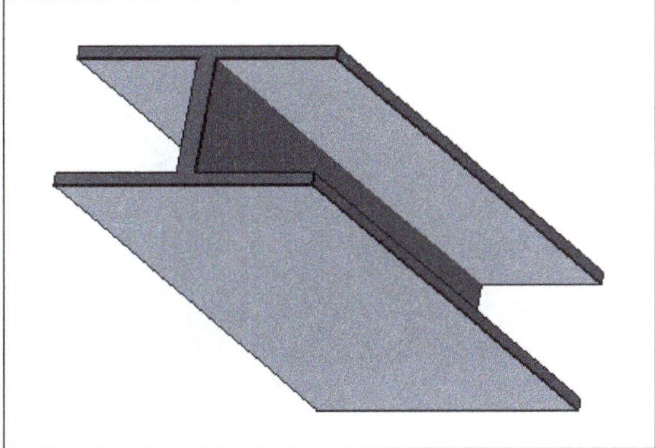

Figure-19. Structure object edited

Creating Rebar

The **Rebar** tool allows you to place reinforcing bars inside structure objects. Rebar objects are based on 2D profiles such as draft objects and sketches that must be drawn on a face of the structural object. The procedure to use this tool is discussed next.

- First, create the structure object using **Structure** tool as discussed earlier; refer to Figure-20.

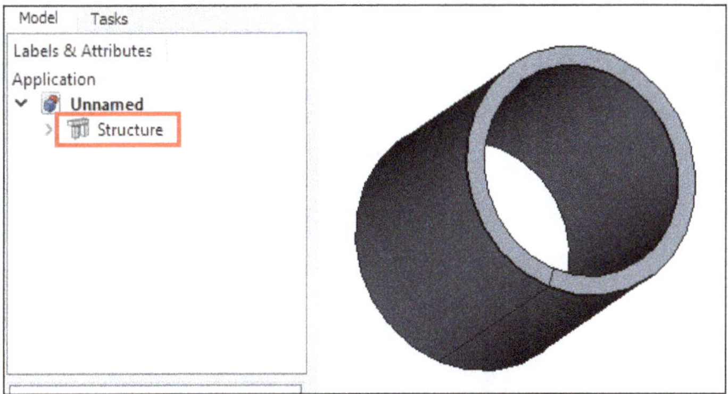

Figure-20. Structure object created

- Switch to **Sketcher** workbench from **Switch between workbenches** drop-down in the **Toolbar**. The tools related to **Sketcher** workbench will be displayed.
- Select desired face of the structural object on which you want to create the sketch; refer to Figure-21.

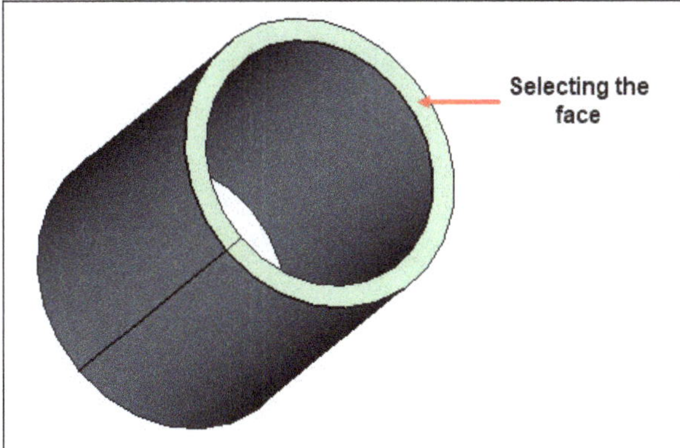

Figure-21. Selecting the face of structural object

- Click on the **New Sketch** tool from **Toolbar** in the **Sketcher** workbench and create desired sketch.
- After creating the sketch, switch to **Arch** workbench from **Switch between workbenches** drop-down.
- Select the newly created sketch from the Model tree view or from the 3D view area; refer to Figure-22.

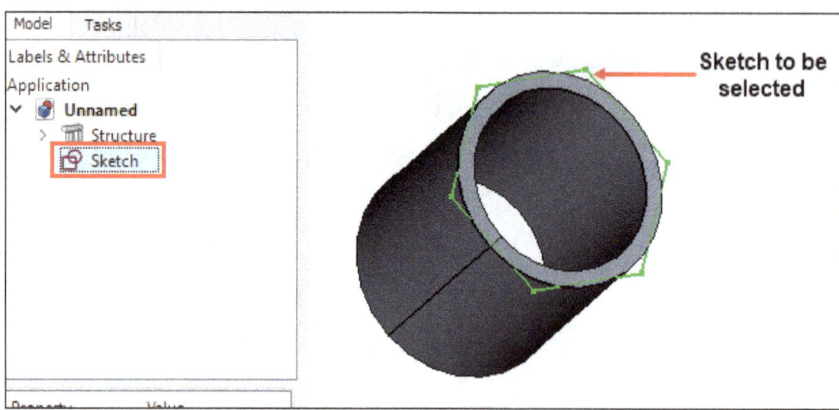

Figure-22. Sketch selected to create rebar

- Click on the **Rebar** tool from **Toolbar** in the **Arch** workbench; refer to Figure-23. The rebar object will be created; refer to Figure-24.

Figure-23. Rebar tool

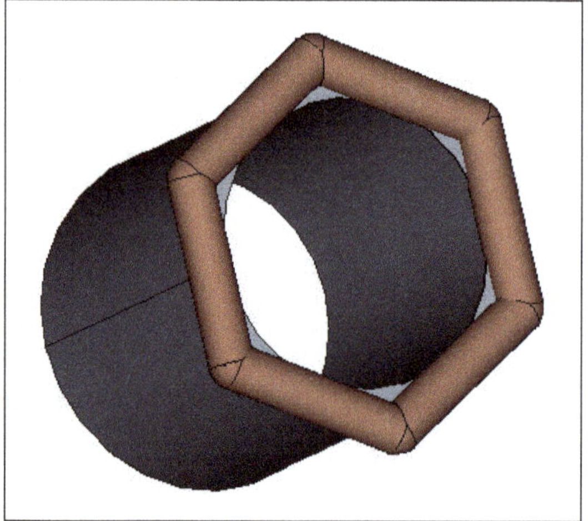

Figure-24. Rebar object created

- If you want to edit the properties of rebar object then select the rebar object created from the Model tree view. The **Property editor** dialog will be displayed in the **Model** panel of **Combo View** with the parameters related to rebar object; refer to Figure-25.

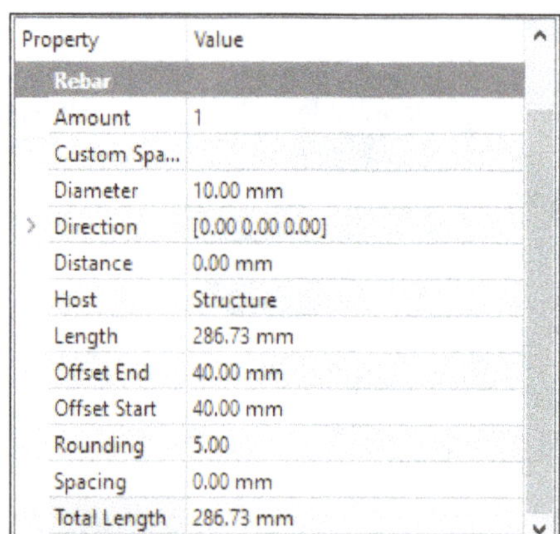

Property	Value
Rebar	
Amount	1
Custom Spa...	
Diameter	10.00 mm
Direction	[0.00 0.00 0.00]
Distance	0.00 mm
Host	Structure
Length	286.73 mm
Offset End	40.00 mm
Offset Start	40.00 mm
Rounding	5.00
Spacing	0.00 mm
Total Length	286.73 mm

Figure-25. Property editor dialog with rebar parameters

- Specify desired amount of bars in the **Amount** edit box from **Rebar** section of the dialog.
- Specify desired diameter of the bars in the **Diameter** edit box.
- Enter desired value in **x**, **y**, and **z** edit boxes from **Direction** cascading menu to specify the spreading of bars along x, y, and z direction, respectively.

- Specify desired offset distance between the border of the structural object and the first bar in the **Offset Start** edit box and between the border of the structural object and the last bar in the **Offset End** edit box of the dialog.
- Specify desired rounding value to be applied to the corner of the bars, expressed in times the diameter in **Rounding** edit box.
- Enter desired value in **Spacing** edit box to specify the distance between the axis of each bar.
- On specifying the parameters in **Property editor** dialog, the properties of rebar object will be modified; refer to Figure-26.

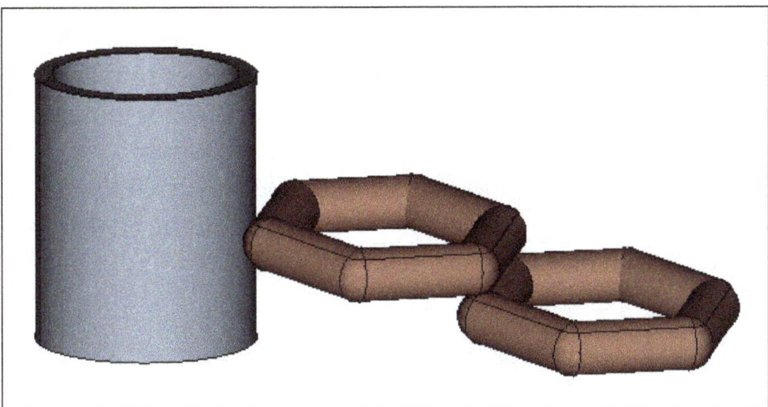

Figure-26. Reber object edited

Creating Building Part

The **BuildingPart** tool is used to create Floor/Storey/Levels as well as all kinds of situations where different Arch/BIM objects need to be grouped and that group might need to be handled as one object or replicated. The procedure to use this tool is discussed next.

- Select one or more objects from Model tree view or from the 3D view area which you want to include in a new building part; refer to Figure-27.

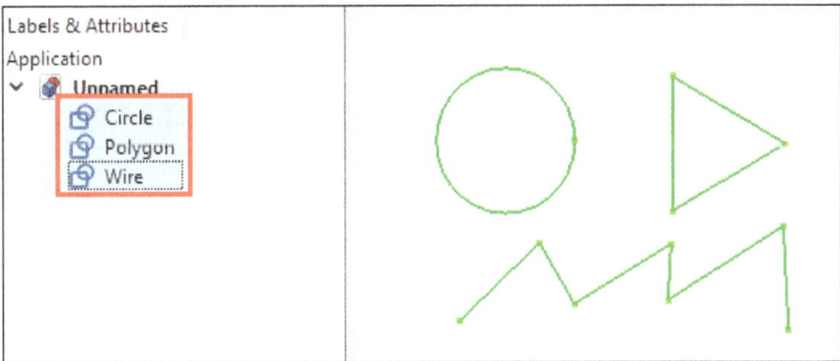

Figure-27. Selecting the objects to include

- Click on the **BuildingPart** tool from **Toolbar** in the **Arch** workbench; refer to Figure-28. The building part will be created in the Model tree view and all the selected objects will be included in the building part; refer to Figure-29.

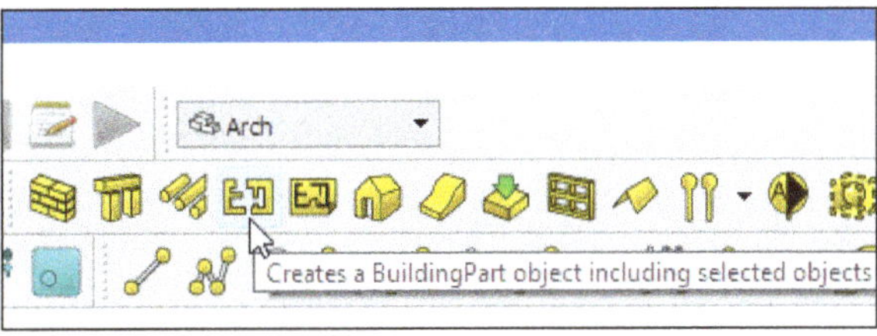

Figure-28. Building Part tool

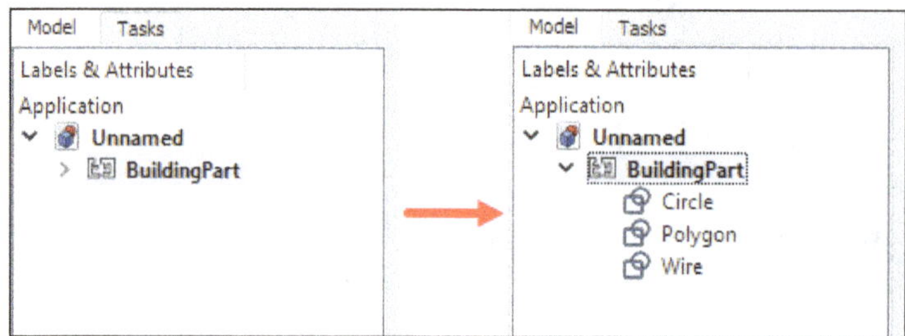

Figure-29. Objects included in building part

- If you want to add more objects to the building part then drag and drop the object to it or use **Add component** tool from the **Toolbar** which will be discussed later in this chapter.
- If you want to remove the objects then drag and drop the object out of it or use **Remove component** tool from the **Toolbar** which will be discussed later in this chapter.
- If you want to edit the properties of building part then select the building part created from Model tree view. The **Property editor** dialog will be displayed in the **Model** panel of **Combo View** with the parameters related to building part; refer to Figure-30.

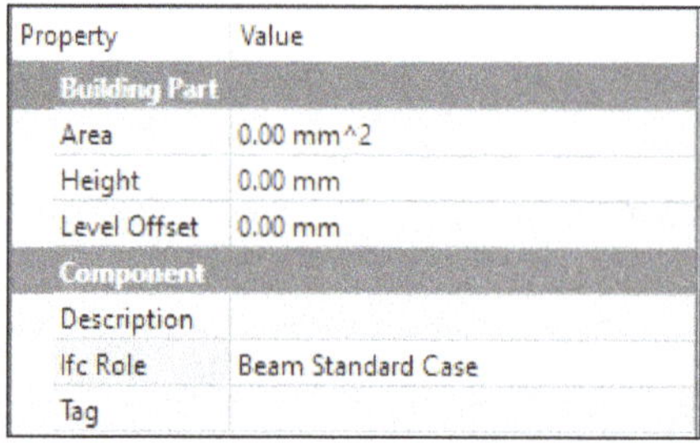

Property	Value
Building Part	
Area	0.00 mm^2
Height	0.00 mm
Level Offset	0.00 mm
Component	
Description	
Ifc Role	Beam Standard Case
Tag	

Figure-30. Property editor dialog with building part parameters

- Specify computed floor area of the building part in the **Area** edit box from **Building Part** section of the dialog.
- Specify desired height of the building part and of its children objects in the **Height** edit box.
- Enter desired value in **Level Offset** edit box. This value is added to the **Placement.Base.z** attribute of the building part to indicate a vertical offset without actually moving the object.
- Select desired IFC type of this object from **Ifc Role** drop-down in the **Component** section of the dialog.
- Specify desired description and tag for this object in the **Description** and **Tag** edit boxes of the dialog, respectively.
- On specifying the parameters in **Property editor** dialog, the properties of building part object will be modified.

Creating Floor

The **Floor** tool is a special type of FreeCAD group object that has a couple of additional properties particularly suited for building floors. The procedure to use this tool is discussed next.

- Select one or more objects from Model tree view or from the 3D view area which you want to include in a new floor; refer to Figure-31.

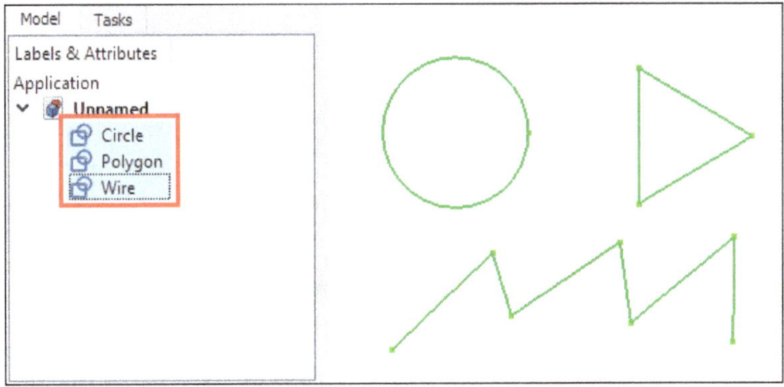

Figure-31. Selecting the objects to include in a floor

- Click on the **Floor** tool from **Toolbar** in the **Arch** workbench; refer to Figure-32. The Floor will be created in the Model tree view and all the selected objects will be included in the floor; refer to Figure-33.

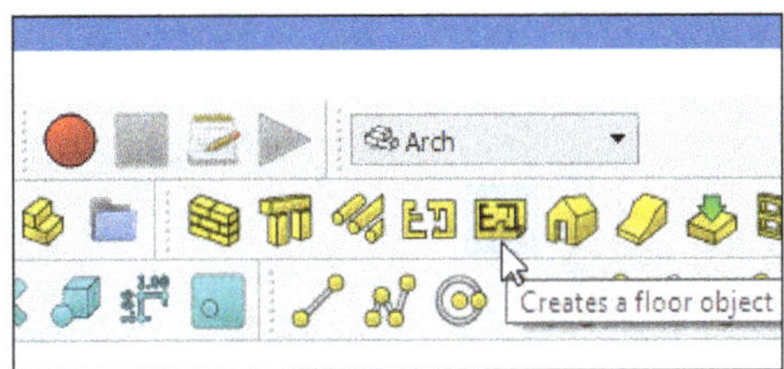

Figure-32. Floor tool

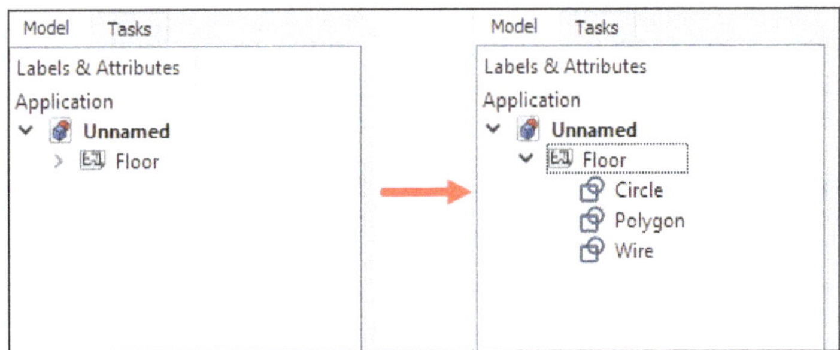

Figure-33. Objects included in a floor

- If you want to add more objects to the floor then drag and drop the object to it or use **Add component** tool from the **Toolbar** which will be discussed later in this chapter.
- If you want to remove the objects then drag and drop the object out of it or use **Remove component** tool from the **Toolbar** which will be discussed later in this chapter.
- The parameters in the **Property editor** dialog to edit the properties of floor object are same as discussed for previous tool.

Creating Building

The **Building** tool is a special type of FreeCAD group object particularly suited for representing a whole building unit. They are mostly used to organize your model by containing floor objects. The procedure to use this tool is discussed next.

- Select one or more objects from the Model tree view or from the 3D view area which you want to include in a building; refer to Figure-34.

Figure-34. Selecting the objects to include in a building

- Click on the **Building** tool from **Toolbar** in the **Arch** workbench; refer to Figure-35. The building will be created in the Model tree view and all the selected objects will be included in the building; refer to Figure-36.

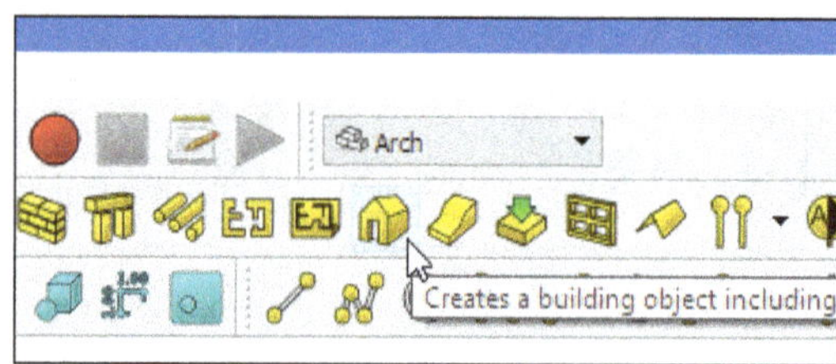

Figure-35. Building tool

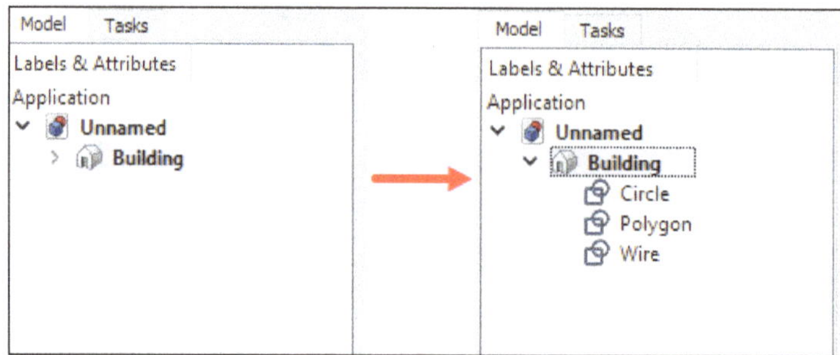

Figure-36. Objects included in building

- If you want to add more objects to the building then drag and drop the object to it or use **Add component** tool from the **Toolbar** which will be discussed later in this chapter.
- If you want to remove the objects then drag and drop the object out of it or use **Remove component** tool from the **Toolbar** which will be discussed later in this chapter.
- The parameters in the **Property editor** dialog to edit the properties of building object are same as discussed for previous tool.

Creating Site

The **Site** tool is a special object that combines properties of a standard FreeCAD group object and Arch objects. It is particularly suited for representing a whole project site or terrain. The procedure to use this tool is discussed next.

- Select one or more building objects from the Model tree view which you want to include in a site; refer to Figure-37.

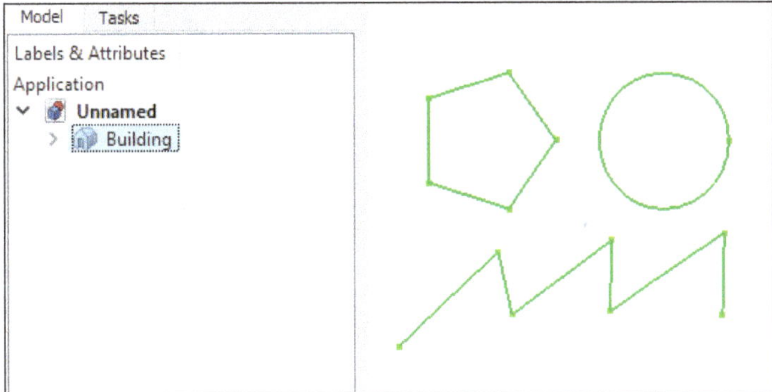

Figure-37. Selecting the building object

- Click on the **Site** tool from **Toolbar** in the **Arch** workbench; refer to Figure-38. The site will be created and the building object will be included in the site; refer to Figure-39.

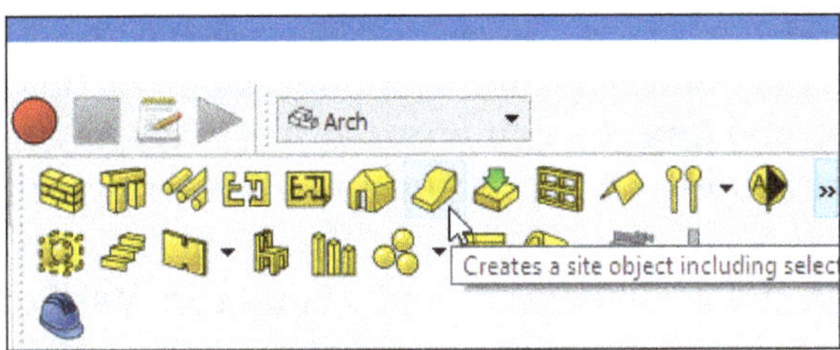

Figure-38. Site tool

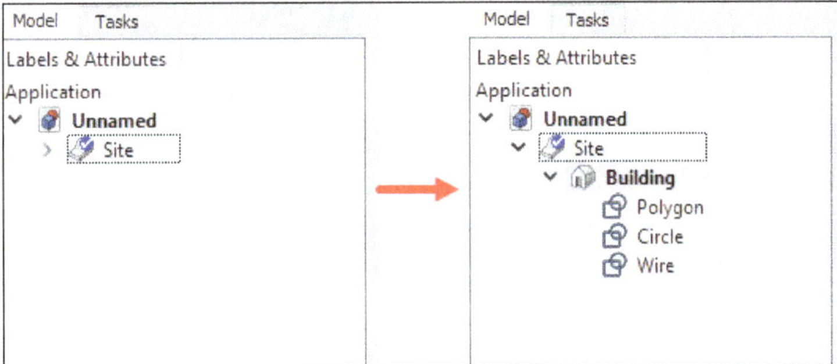

Figure-39. Building objects included in a site

- If you want to add more objects to the site then drag and drop the object to it or use **Add component** tool from the **Toolbar** which will be discussed later in this chapter.
- If you want to remove the objects then drag and drop the object out of it or use **Remove component** tool from the **Toolbar** which will be discussed later in this chapter.
- If you want to edit the properties of site object then select the site object created from the Model tree view. The **Property editor** dialog will be displayed in the **Model** panel of **Combo View** with the parameters related to site; refer to Figure-40.

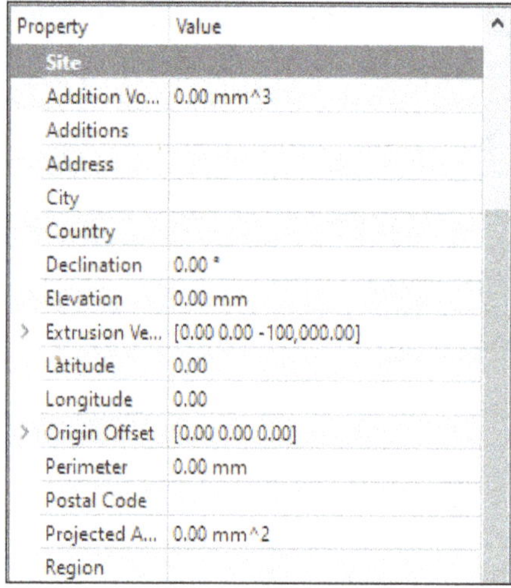

Figure-40. Property editor dialog with site parameters

- Enter desired value in **Addition Volume** and **Subtraction volume** edit boxes from **Site** section of the dialog to specify the volume of earth to be added to this terrain and to be removed from this terrain, respectively.
- Specify desired address, city, country, postal code, and region of this site in the **Address**, **City**, **Country**, **Postal code**, and **Region** edit boxes, respectively.
- Enter desired value in **Latitude** and **Longitude** edit boxes to specify the latitude and longitude of this site, respectively.
- Enter desired value in **x**, **y**, and **z** edit boxes from **Extrusion Vector** cascading menu to use the extrusion vector when performing boolean operations.
- Specify desired perimeter length of this terrain in the **Perimeter** edit box.
- Enter desired value in **Projected Area** edit box to specify area of the projection of this object onto the XY plane.
- Select desired option from **Remove Splitter** drop-down to specify whether to remove the splitters from resulting shape or not.
- Click on the **Terrain** button to specify base terrain of this site.
- Enter desired url in the **Url** edit box to show this site in a mapping website.
- On specifying the parameters in **Property editor** dialog, the properties of site object will be modified.

Adding an External Reference

The **Reference** tool allows you to place an object in the current document that copies its shape and colors from a part-based object stored in another FreeCAD file. If that FreeCAD file changes, the reference object is marked to be reloaded. The procedure to use this tool is discussed next.

- Click on the **Reference** tool from **Toolbar** in the **Arch** workbench; refer to Figure-41. The **External reference** dialog will be displayed in the **Tasks** panel of **Combo View**; refer to Figure-42.

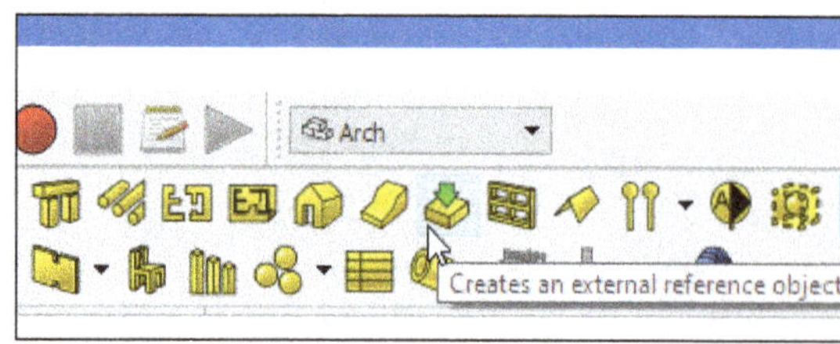

Figure-41. Reference tool

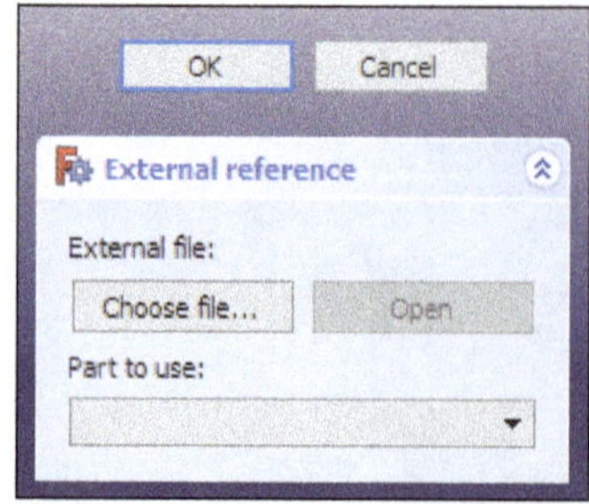

Figure-42. External reference dialog

- Click on the **Choose file** button from **External file** area of the dialog. The **Choose reference file** dialog box will be displayed; refer to Figure-43.

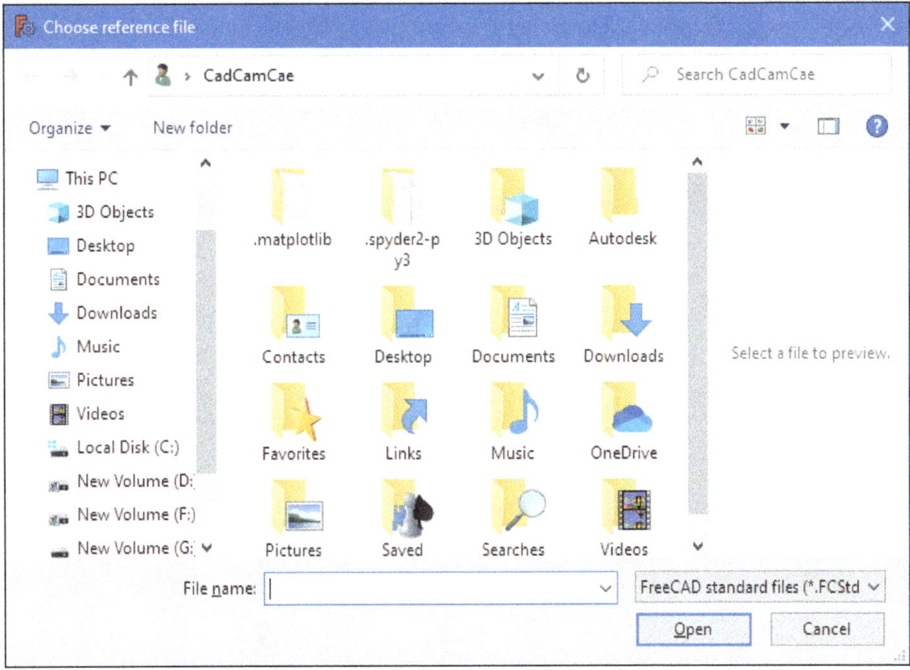

Figure-43. Choose reference file dialog box

- Select the existing FreeCAD file and click on **Open** button from the dialog box. The selected file will display in the **External file** area and the objects of selected reference file will be available in the **Part to use** drop down of **External reference** dialog; refer to Figure-44.

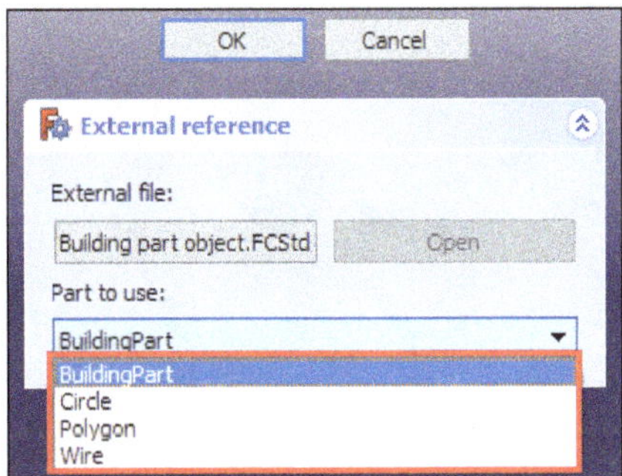

Figure-44. Objects of selected reference file

- Select one of the included part-based objects from the drop-down list and click on **OK** button from the dialog. The selected reference object will be added in the Model tree view; refer to Figure-45.

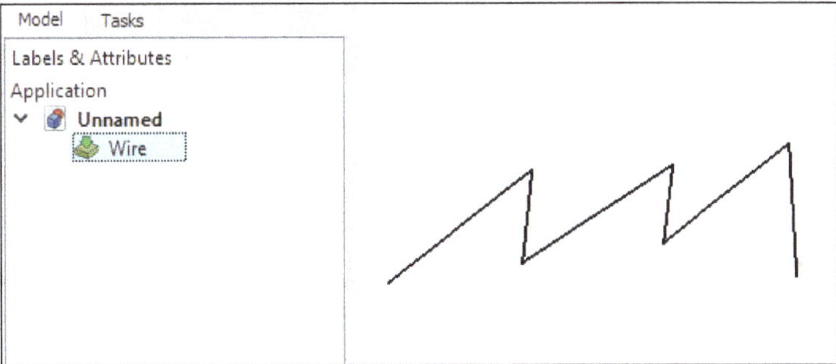

Figure-45. Reference object opened

- The reference object can be moved and rotated, the current position will be retained after reloading the object.
- If the original object gets moved in containing file, this movement will reflect in the reference object.
- If you want to open the containing file of reference object then select reference object and click **RMB** in the Model tree view. A shortcut menu will be displayed; refer to Figure-46.

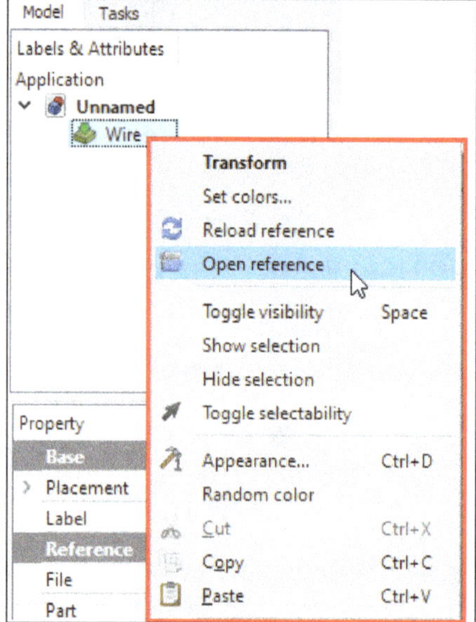

Figure-46. Shortcut menu

- Click on the **Open reference** option from the menu. The file containing the reference object will be opened; refer to Figure-47.

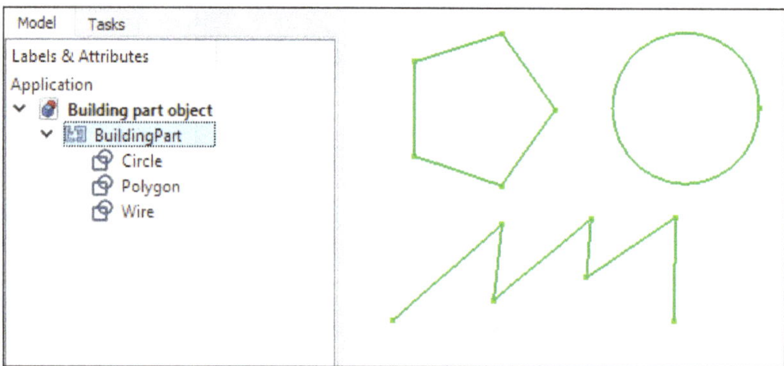

Figure-47. Reference object file

Creating Window

A **Window** is a base object for all kinds of embeddable objects such as windows and doors. The **Window** tool features several presets; this allows the user to create common types of windows and doors with certain editable parameters without the need for the user to create the base 2D objects and components manually. The procedure to use this tool is discussed next.

- Click on the **Window** tool from **Toolbar** in the **Arch** workbench; refer to Figure-48. The **Point** and **Window options** dialog will be displayed in the **Tasks** panel of **Combo View** along with the plus sign and structure of window in place of original cursor; refer to Figure-49 and Figure-50, respectively.

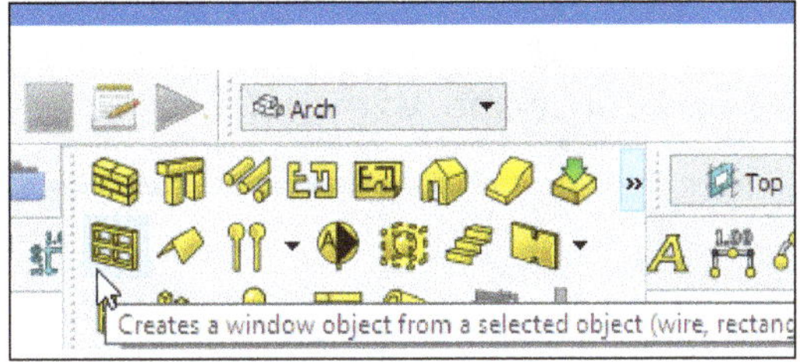

Figure-48. Window tool

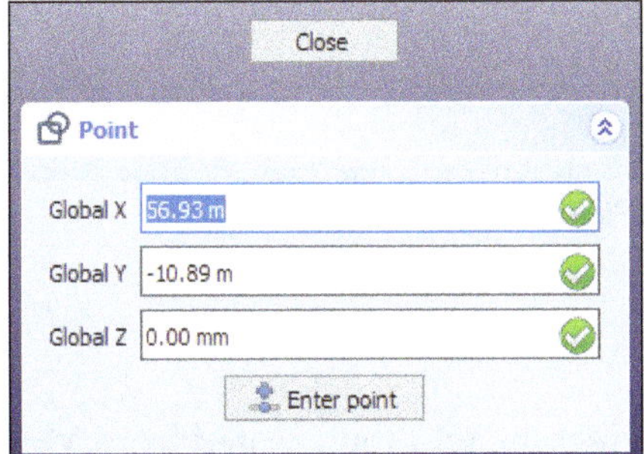

Figure-49. Point dialog

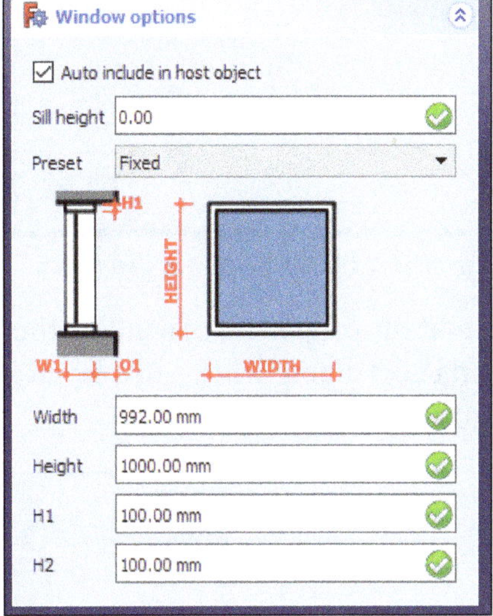

Figure-50. Window options dialog

- Select **Auto include in host object** check box to insert the window into host object on creation from **Windows options** dialog.
- Specify the height of sill for the window in the **Sill height** edit box of the dialog.
- Select desired preset for the window or door from **Preset** drop-down in the dialog.
- Specify desired width and height of the window in the **Width** and **Height** edit boxes of the dialog.
- After specifying all the parameters in the dialog, you will be asked to specify the location of point.
- Click in the 3D view area to specify the point or enter desired values for the x, y, and z coordinates in the **Global X**, **Global Y**, and **Global Z** edit boxes of the **Point** dialog, respectively.
- After specifying coordinates for the point in dialog, click on the **Enter point** button from the **Point** dialog. The window will be created; refer to Figure-51.

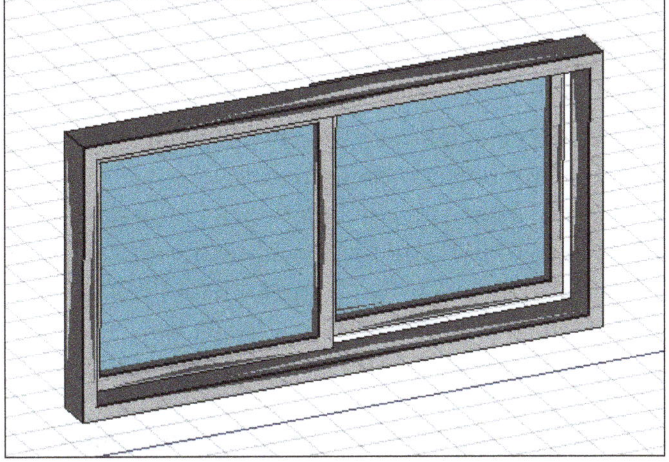

Figure-51. Window created

- If you want to get access, create, modify, or delete the components of a window then double click on the window created from the Model tree view. The **Window elements** and **Component** dialog will be displayed in the **Tasks** panel of **Combo View**; refer to Figure-52 and Figure-53, respectively.

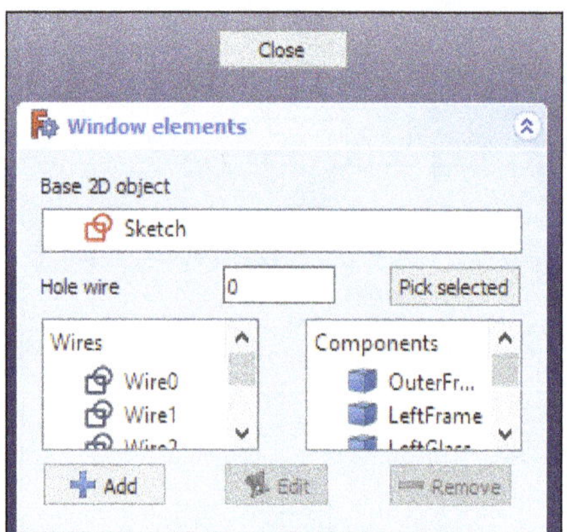

Figure-52. Window elements dialog

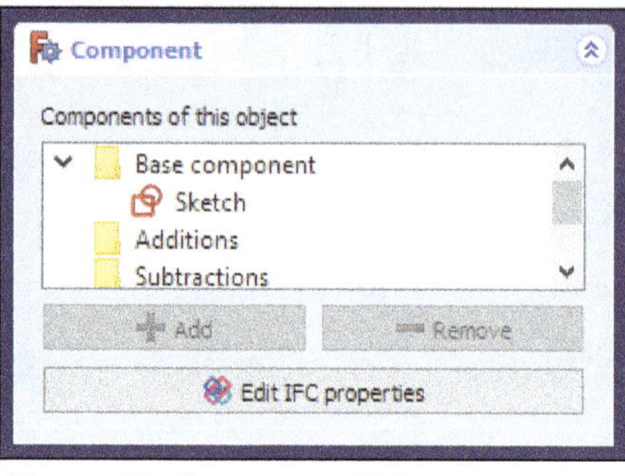

Figure-53. Component dialog

- If you want to edit the properties of window then select the window created from Model tree view. The **Property editor** dialog will be displayed in the **Model** panel of **Combo View** with parameters related to window; refer to Figure-54.

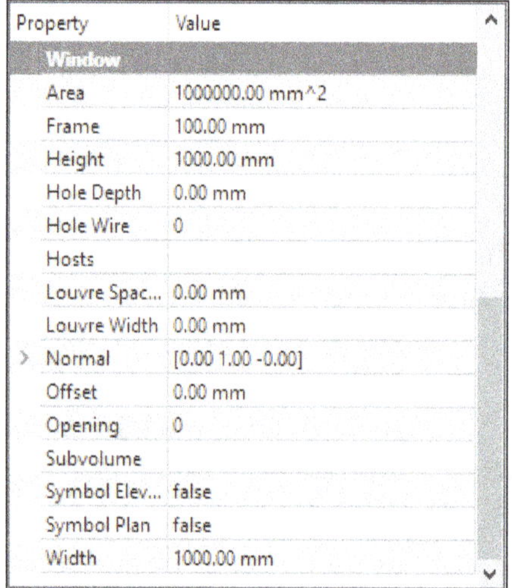

Figure-54. Property editor dialog with window parameters

- Specify desired height and width of the window in the **Height** and **Width** edit boxes from **Window** section of the dialog, respectively.
- Specify desired depth of the hole created by this window in its host object in the **Hole Depth** edit box.
- Enter desired value in **Host Wire** edit box to specify the number of wire from the base object that is used to create a hole in the host object of this window.
- Enter desired value in **Louvre Spacing** edit box to define the spacing between louvre elements.
- Enter desired value in **Louvre Width** edit box to define the size of the louvre elements.
- Enter desired value in **Opening** edit box to define opening mode of all the components and to define a hinge provided in them or in an earlier component in the list.
- Select desired value from **Symbol Elevation** drop-down to specify whether to show 2D opening symbol in elevation or not.
- Select desired value from **Symbol Plan** drop-down whether to show 2D opening symbol in plan or not.

- On specifying desired value in the **Property editor** dialog, the properties of window will be modified; refer to Figure-55.

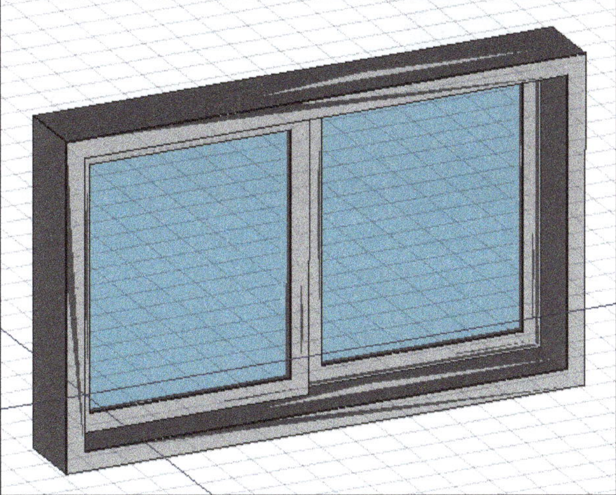

Figure-55. Window edited

Creating Roof

The **Roof** tool allows for the creation of a sloped roof from a selected wire. The created roof object is parametric keeping its relationship with the base object. The procedure to use this tool is discussed next.

- Select the wire from the Model tree view or from the 3D view area on which you want to create the roof; refer to Figure-56.

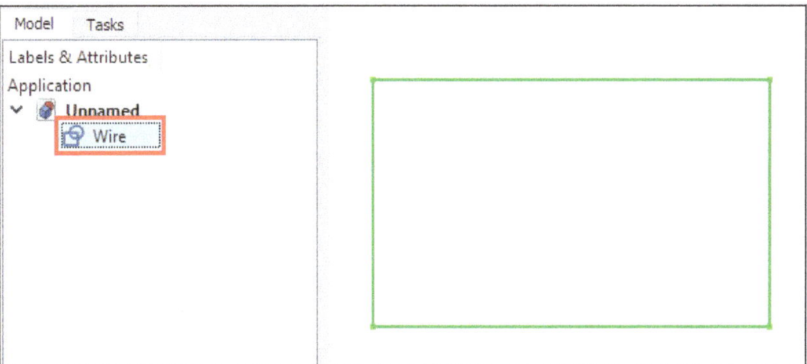

Figure-56. Selecting the wire to create roof

- Click on the **Roof** tool from **Toolbar** in the **Arch** workbench; refer to Figure-57. The default roof will be created; refer to Figure-58.

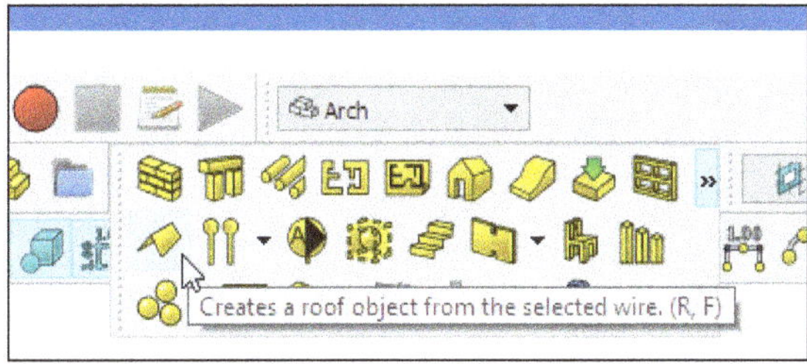

Figure-57. Roof tool

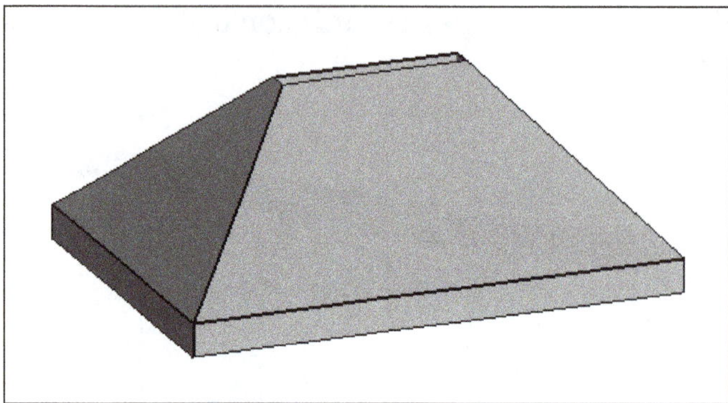

Figure-58. Default roof created

- If you want to edit the roof then select the roof created from the Model tree view and double-click on it. The **Roof** dialog will be displayed in the **Tasks** panel of **Combo View**; refer to Figure-59.

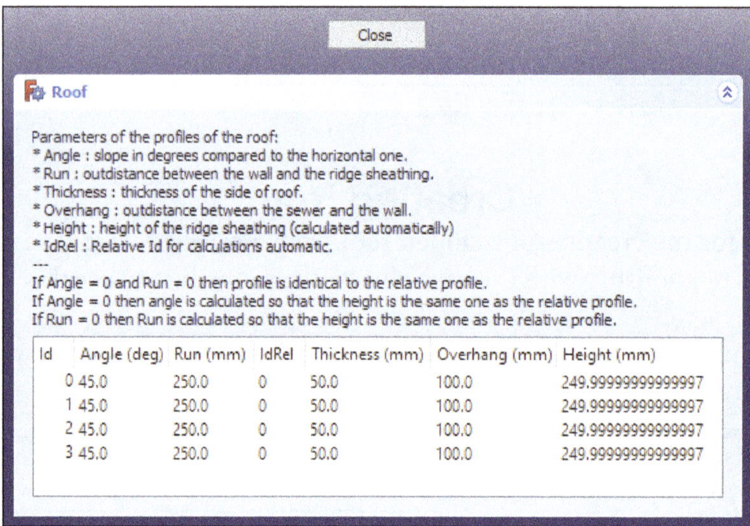

Figure-59. Roof dialog

- Specify desired values in the **Angle**, **Run**, **IdRel**, **Thickness**, **Overhang**, and **Height** edit boxes of the dialog. The roof will be edited.
- If you want to edit the properties of roof then select the roof created from the Model tree view. The **Property editor** dialog will be displayed in the **Model** panel of **Combo View** with the parameters related to roof object; refer to Figure-60.

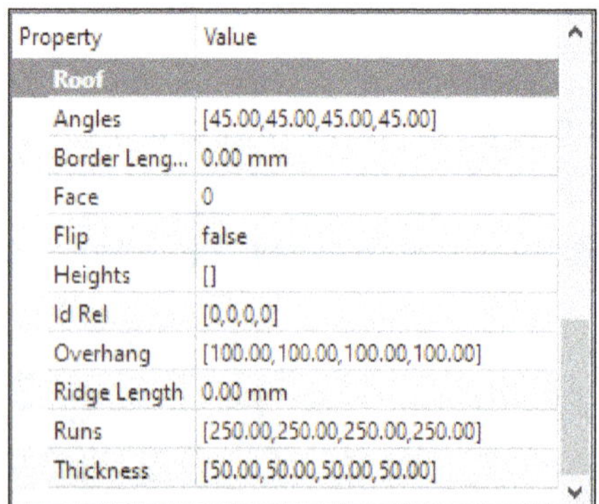

Figure-60. Property editor dialog with roof parameters

- Enter desired values in the **Angles** edit box to specify slope angle of the roof pane (an angle for each edge in the wire).
- Enter desired value in the **Face** edit box to specify face index of the base object to be used.
- Enter desired values in **Id Rel** edit box to specify relation Id of the slope angle of the roof.
- Enter desired values in **Overhang** edit box to specify overhang of the roof pane (an overhang for each edge in the wire).
- Enter desired values in **Runs** edit box to specify width of the roof pane (a run for each edge in the wire).
- Enter desired values in **Thickness** edit box to specify thickness of the roof pane (a thickness for each edge in the wire).
- On specifying parameters in the **Property editor** dialog, the properties of roof object will be modified.

Axis tools

The Axis tools allow you to place a series of axes in the current document.

Creating Axis

The **Axis** tool allows you to place a series of axes in the current document. The axes serve mainly as references to snap objects onto but can also be used together with **AxesSystems**, and can also be referenced by other Arch objects to create parametric arrays, for example, of beams or columns. **Grids** can also be used in place of axes. The procedure to use this tool is discussed next.

- Click on the **Axis** tool from **Axis tools** drop-down in the **Toolbar** of **Arch** workbench; refer to Figure-61. The axis will be created; refer to Figure-62.

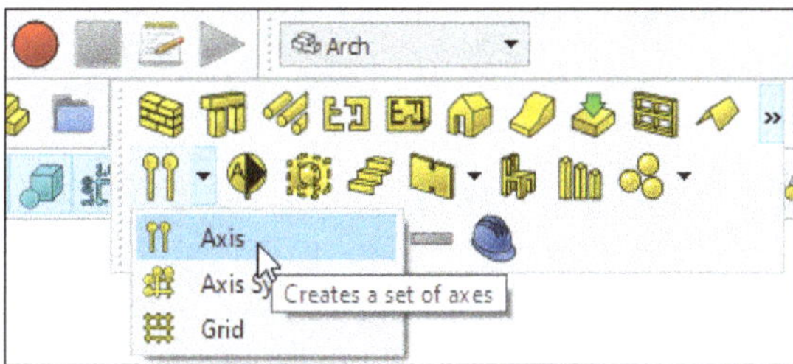

Figure-61. Axis tool

Figure-62. Axis created

- If you want to edit the parameters of axis then double-click on the axis created. The **Axes** dialog will be displayed in the **Tasks** panel of **Combo View**; refer to Figure-63.

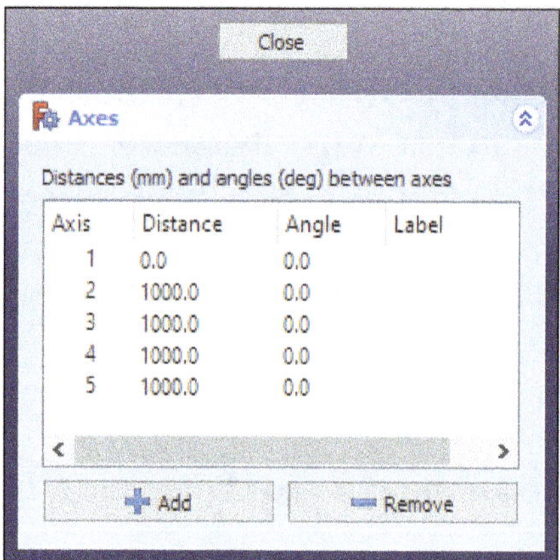

Figure-63. Axes dialog

- Click on **Add** button from the dialog to add the number of axis and if you want to remove the axis then select the axis from the dialog which you want to remove and click on **Remove** button.
- Specify the distance and angle of axis in the **Distance** and **Angle** edit boxes of the dialog, respectively.
- Click on **Close** button to close the dialog.
- If you want to edit the properties of axis then select the axis created from the Model tree view. The **Property editor** dialog will be displayed in the **Model** panel of **Combo View** with the parameters related to axis; refer to Figure-64.

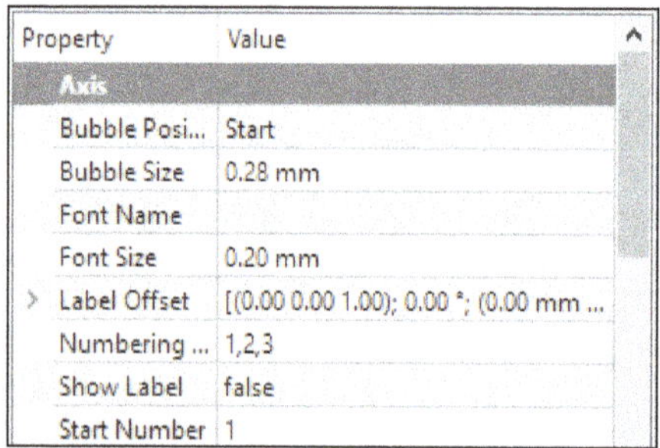

Figure-64. Property editor dialog with axis parameters

- Specify desired length of the axes in the **Length** edit box from **Axis** section of the dialog.
- Select desired option from **Bubble Position** drop-down to specify where the bubble is placed on the axis.
- Specify desired size of the axis bubbles in the **Bubble Size** edit box of the dialog.
- Specify desired font name to draw the bubble number and/or labels in the **Font Name** edit box.
- Specify desired size of the label text in the **Font Size** edit box of the dialog.
- Select desired option from **Numbering Style** drop-down to specify how the axes are numbered; 1, 2, 3, A, B, C, etc.
- Select desired option from **Show Label** drop down to turn the display of the label texts on/off.
- On specifying parameters in the **Property editor** dialog, the properties of axis will be modified.

Creating Axis System

The **Axis System** tool allows you to combine several axis to the document. This is useful to define the intersection points between the different axes. Arch objects can then use this system to duplicate their shape on the different intersection points. The procedure to use this tool is discussed next.

- Select the axes from the Model tree view or from the 3D view area; refer to Figure-65.

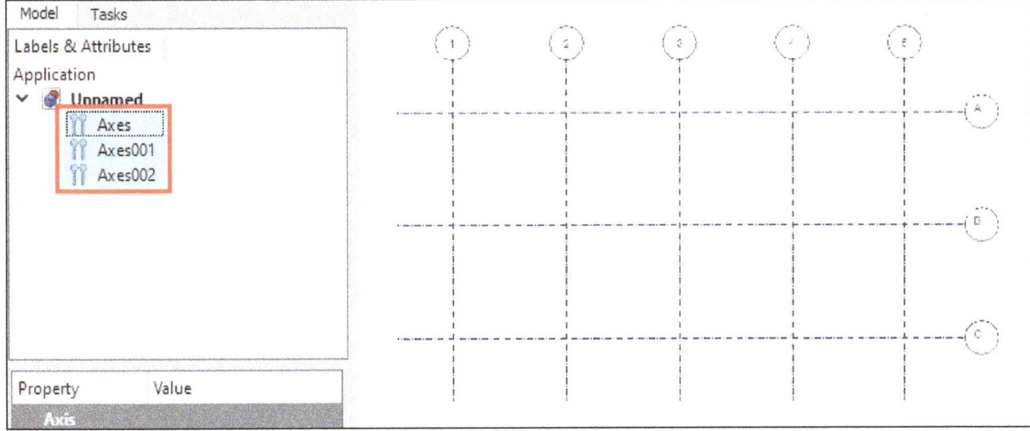

Figure-65. Selecting the axes

- Click on the **Axis System** tool from **Axis** tools drop-down in the **Toolbar** of **Arch** workbench; refer to Figure-66. The Axis system will be created and displayed in the Model tree view; refer to Figure-67.

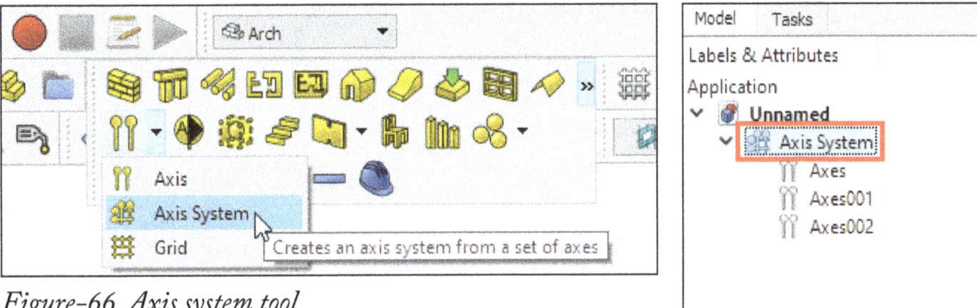

Figure-66. Axis system tool

Figure-67. Axis system created

- If you want to edit the parameters of axis system then double-click on the axis system created from the Model tree view. The **Axes** dialog will be displayed in the **Tasks** panel of **Combo View**; refer to Figure-68.

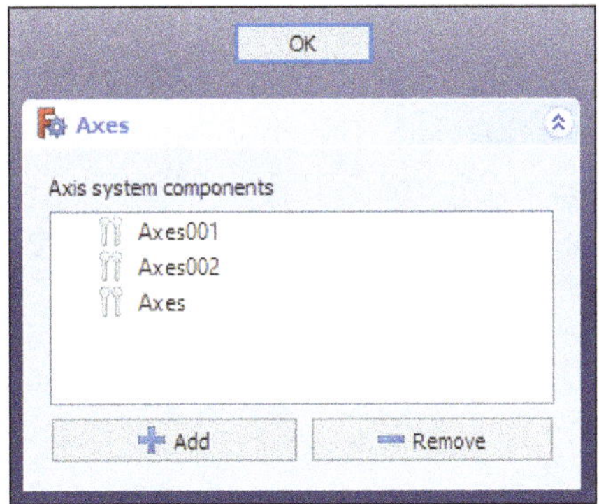

Figure-68. Axes dialog

- If you want to add more axes in the axis system then select the axis from the Model tree view and click on **Add** button from the **Axes** dialog and if you want to remove the axis then select the axis from **Axis system components** area of the dialog and click on **Remove** button from the dialog.
- Click on **OK** button to close the dialog.

Creating Grid

The **Grid** tool allows you to place a grid-like object in the document. This object is meant to serve as a base to build Arch objects that need a regular but complex frame, such as windows, curtain walls, column grids, railings, etc. The procedure to use this tool is discussed next.

- Click on the **Grid** tool from **Axis tools** drop-down in the **Toolbar** of **Arch** workbench; refer to Figure-69. The grid will be created and displayed in the Model tree view; refer to Figure-70.

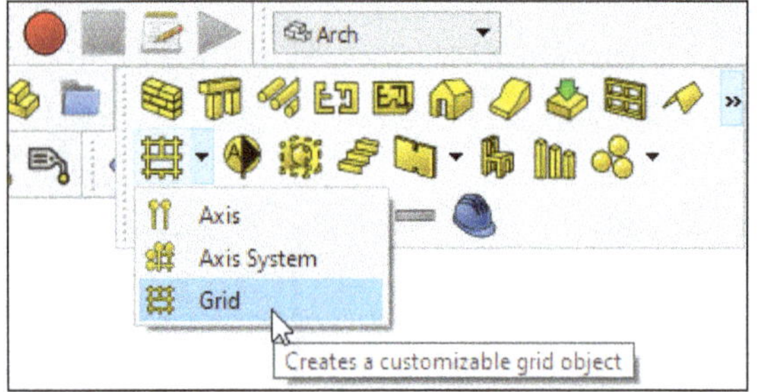

Figure-69. Grid tool

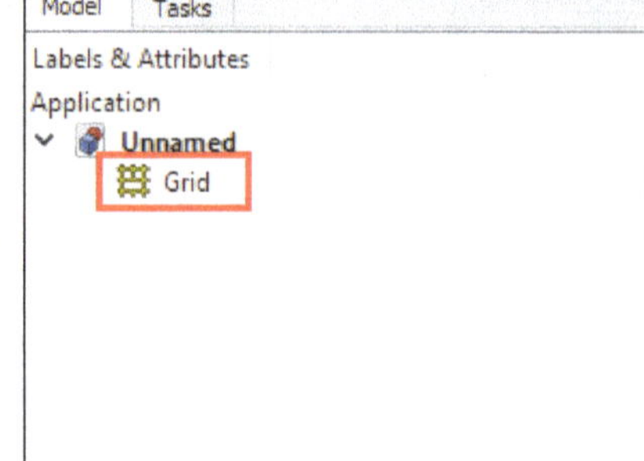

Figure-70. Grid created

- To edit the parameters of grid, double-click on the grid created from the Model tree view. The **Grid** dialog will be displayed in the **Tasks** panel of **Combo View**; refer to Figure-71.

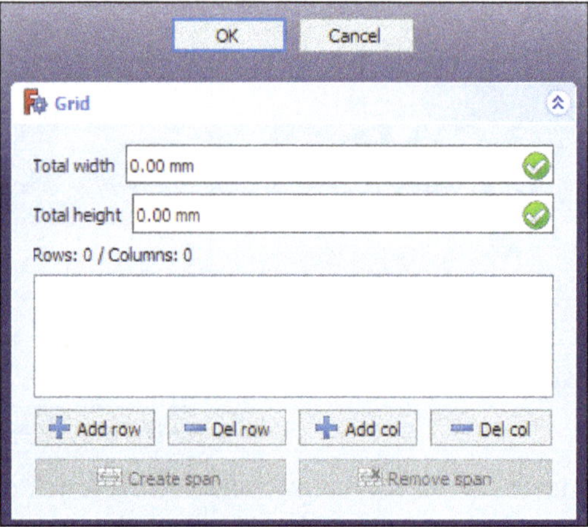

Figure-71. Grid dialog

- Specify desired width and height of the grid in the **Total width** and **Total height** edit boxes of the dialog, respectively.
- Click on the **Add row** button to add the number of rows in the grid and click on **Add col** button to add the number of columns in the grid. The rows and columns will be displayed in the **Rows/ Columns** area of the dialog; refer to Figure-72.

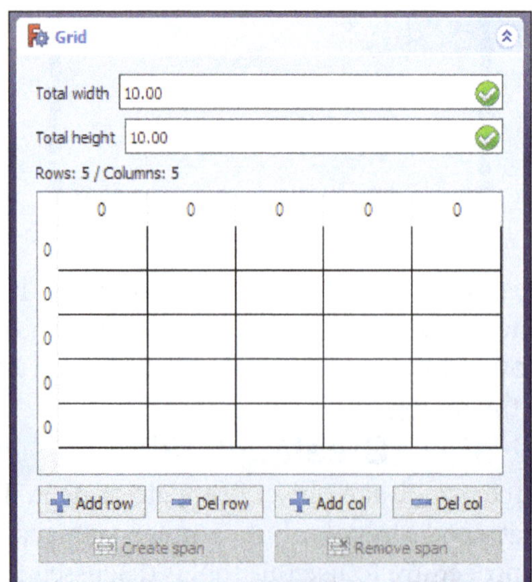

Figure-72. Grid dialog displaying rows and columns

- If you want to remove the row or column then select the row or column which you want to remove from the dialog and click on **Del row** or **Del col** buttons, respectively.
- Click on **OK** button from the dialog. The grid will be created; refer to Figure-73.

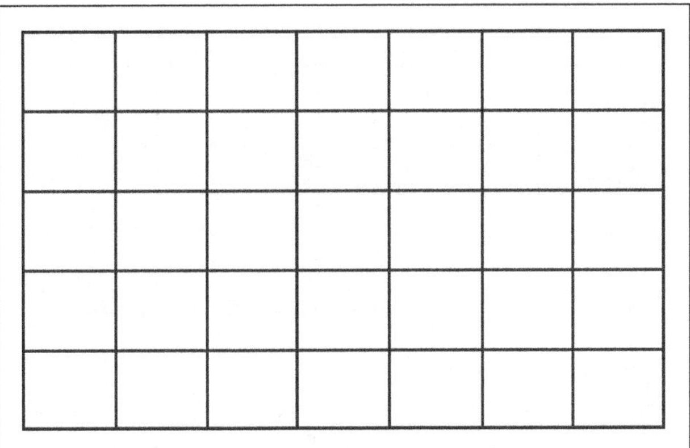

Figure-73. Grid created

- If you want to edit the properties of grid then select the grid created from the Model tree view. The **Property editor** dialog will be displayed in the **Model** panel of **Combo View** with the parameters related to grid; refer to Figure-74.

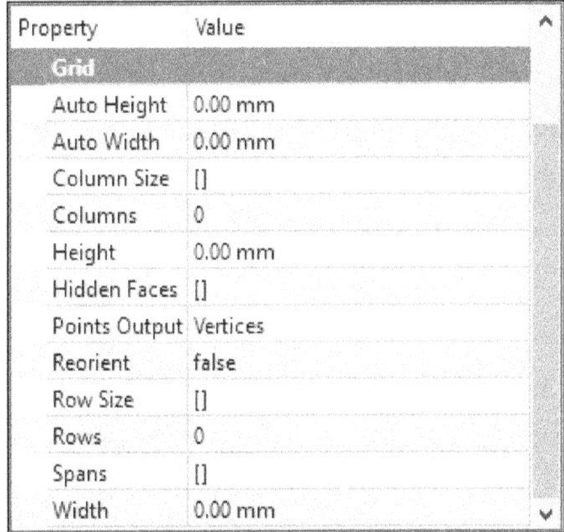

Figure-74. Property editor dialog with grid parameters

- Enter desired values in **Auto Height** and **Auto Width** edit boxes from **Grid** section of the dialog to create automatic row divisions and automatic column divisions, respectively.
- Specify desired size of columns and rows in the **Column Size** and **Row Size** edit boxes of the dialog, respectively.
- Specify desired number of columns and rows in the **Columns** and **Rows** edit boxes, respectively.
- Specify total height and total width of the grid in the **Height** and **Width** edit boxes, respectively.
- Enter desired value in **Hidden Faces** edit box to specify indices of faces to hide.
- Select desired option from **Points Output** drop-down to specify the type of 3D points produced by this grid object.
- Select desired option from **Reorient** drop-down to specify whether the grid must reorient its children along edge normals or not.
- On specifying parameters in the **Property editor** dialog, the properties of grid will be modified; refer to Figure-75.

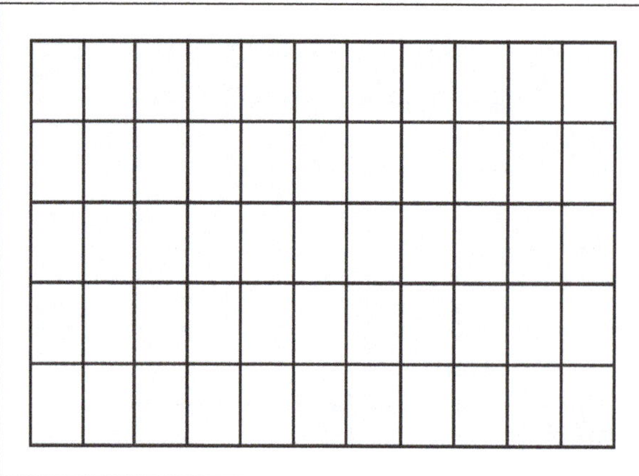

Figure-75. Grid edited

Creating Section Plane

The **Section Plane** tool is used to create section plane for generating section views. You can define the placement of section plane according to the current working plane and it can be relocated and reoriented by moving and rotating it, until it describes the 2D view you want to obtain. The Section plane object will only consider a certain set of objects, not all the objects of the document. The procedure to use this tool is discussed next.

- Set desired working plane to define where you want to place the section plane as discussed earlier.
- Select the object from the Model tree view or from the 3D view area which you want to include in the section view; refer to Figure-76.

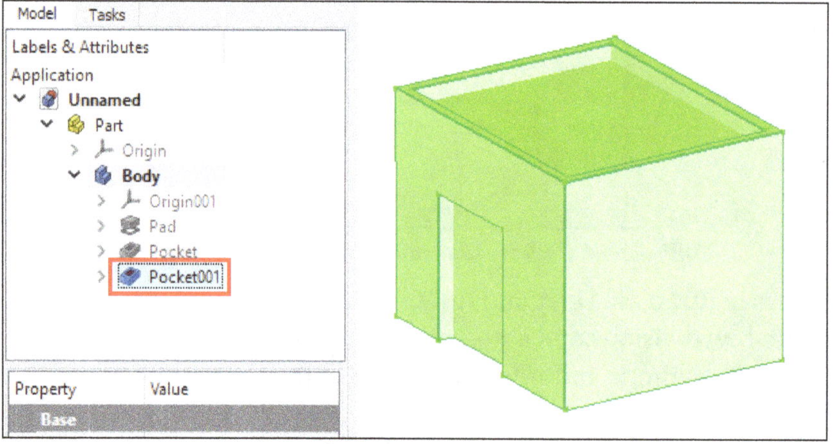

Figure-76. Selecting the object to include in section view

- Click on the **Section Plane** tool from **Toolbar** in the **Arch** workbench; refer to Figure-77. The section plane will be created; refer to Figure-78.

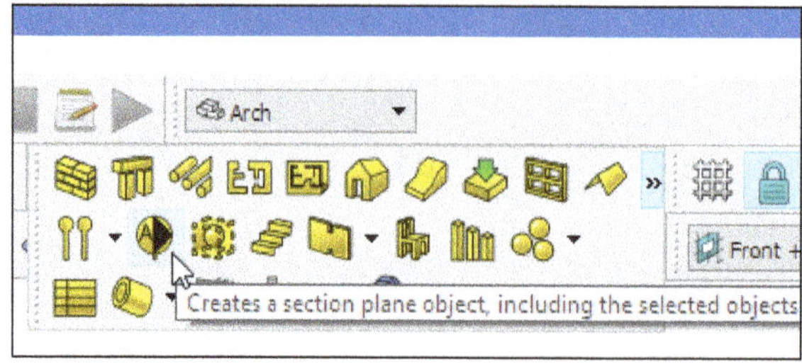

Figure-77. Section plane tool

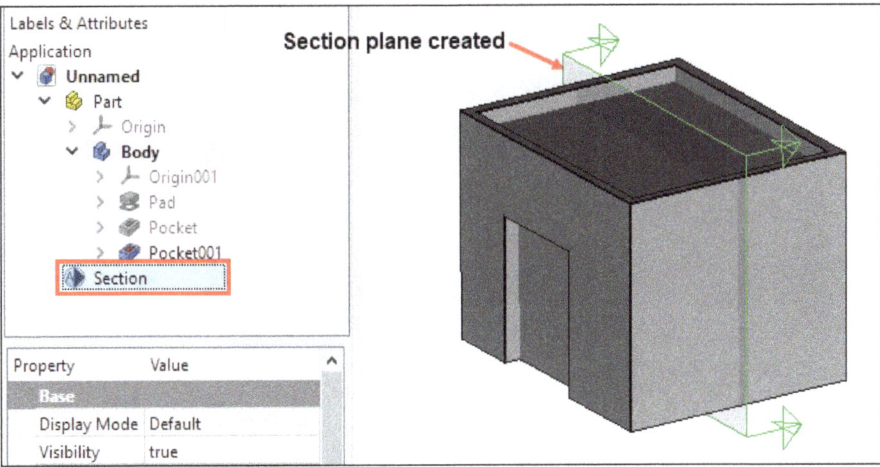

Figure-78. Section plane created

- You can move or rotate the section plane in desired position using **Move** tool as discussed earlier.
- If you want to edit the section plane then double-click on the section plane created from the Model tree view. The **Section plane settings** dialog will be displayed in the **Tasks** panel of **Combo View**; refer to Figure-79.

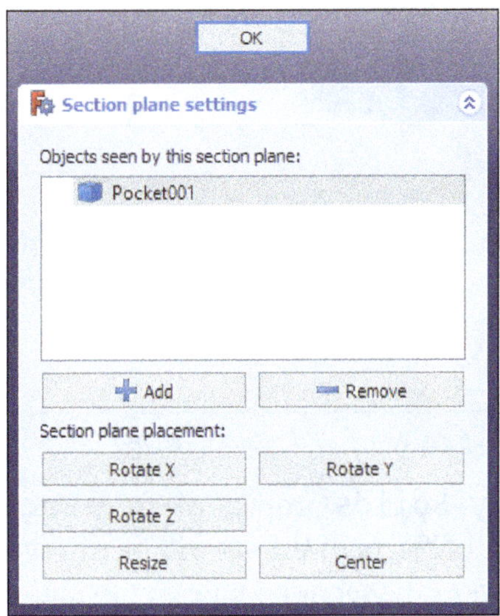

Figure-79. Section plane settings dialog

- The object selected for section plane will be displayed in the **Objects seen by this section plane** list box of the dialog.
- If you want to add more objects to be section then select the object from the Model tree view and click on **Add** button from the dialog.
- If you want to remove the objects then select the object from the dialog and click on **Remove** button.
- To rotate the section plane along x, y, or z direction then click on the **Rotate X**, **Rotate Y**, or **Rotate Z** button from **Section plane placement** area of the dialog, respectively.
- Click on the **Resize** button to modify the size of section plane.
- Click on the **Center** button to place the section plane at the center of object.
- Click on **OK** button to close the dialog.
- You can edit the properties of section plane using **Property editor** as discussed earlier.
- To create a section view of the object, select the section plane from the Model tree view or from the 3D view area and click on the **Shape 2D View** tool from the **Toolbar**. The 2D section view will be created; refer to Figure-80.

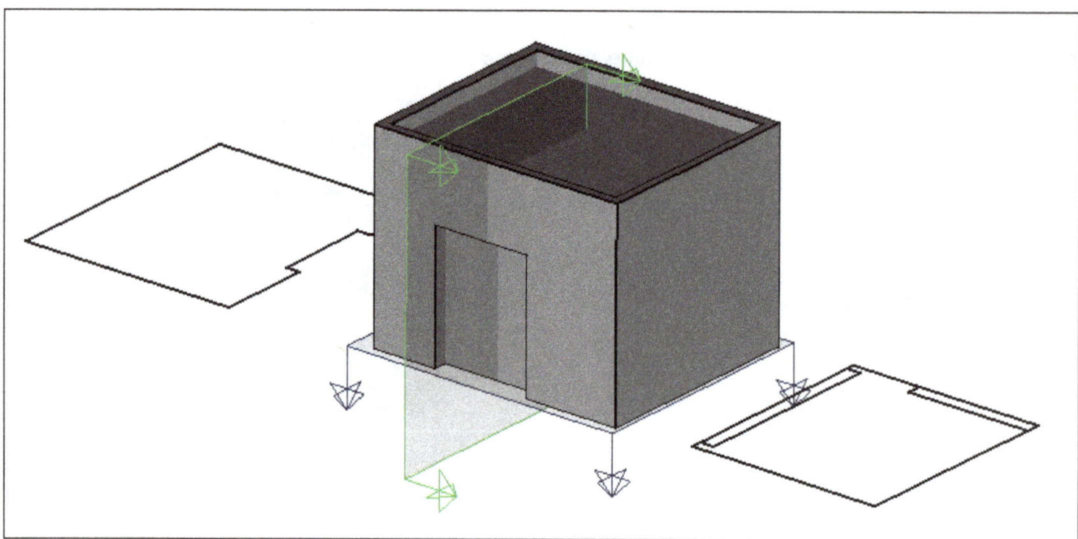

Figure-80. Section view created

- If you want to edit the properties of section then select the section plane created from the Model tree view. The **Property editor** dialog will be displayed in the **Model** panel of **Combo View** with parameters related to section plane; refer to Figure-81.

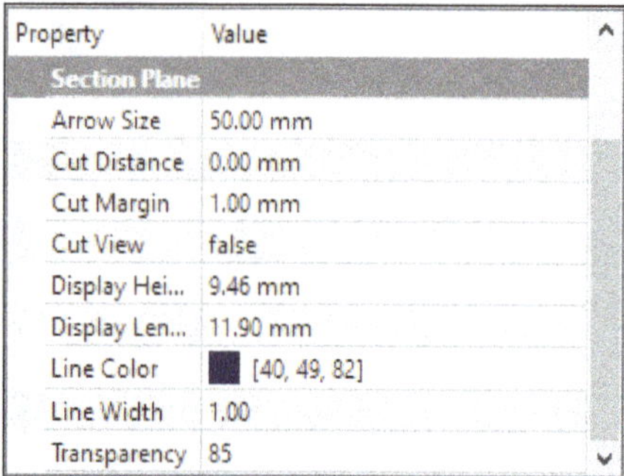

Figure-81. Property editor dialog with section plane parameters

- Select desired option from **Only Solids** drop-down from **Section Plane** section of the dialog to specify whether the non-solid objects in the set will be disregarded or not.
- Specify desired size of the arrows of the section plane gizmo in the 3D view area in the **Arrow Size** edit box of the dialog.
- Specify desired height and length of the section plane gizmo in the 3D view area in the **Display Height** and **Display Length** edit boxes, respectively.
- Select desired option from **Cut View** drop-down to specify whether the whole 3D view will be cut at the location of this section plane or not.
- On specifying parameters in the **Property editor** dialog, the properties of section plane will be modified.

Creating Space

The **Space** tool allows you to define an empty volume either by basing it on a solid shape or by defining its boundaries or a mix of both. The space object always defines a solid volume. The procedure to use this tool is discussed next.

- Select the existing solid object or faces on the boundary object from Model tree view or from the 3D view area which you want to create a space object; refer to Figure-82.

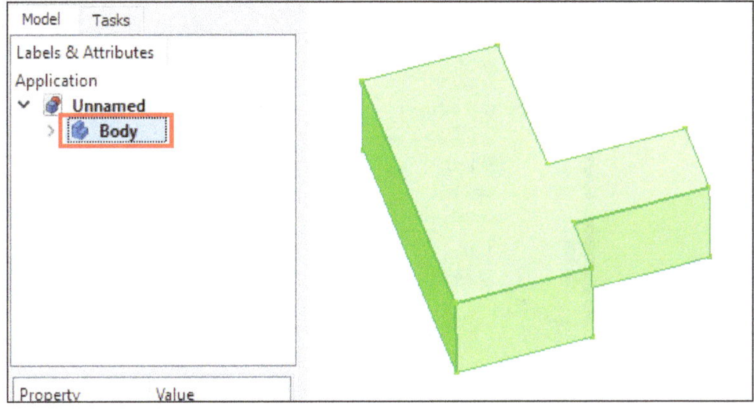

Figure-82. Selecting the solid object

- Click on the **Space** tool from **Toolbar** in the **Arch** workbench; refer to Figure-83. The space object will be created; refer to Figure-84.

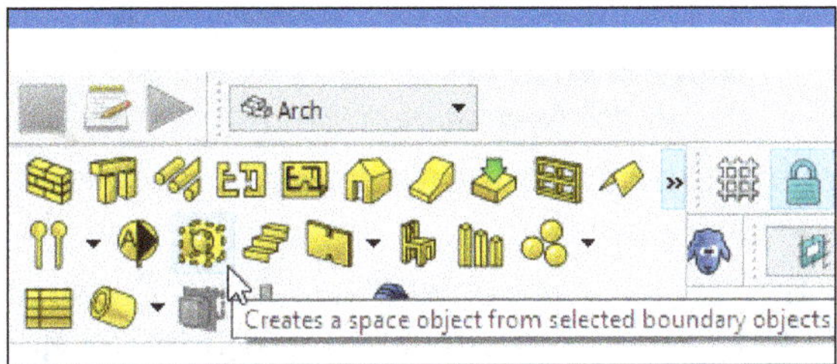

Figure-83. Space tool

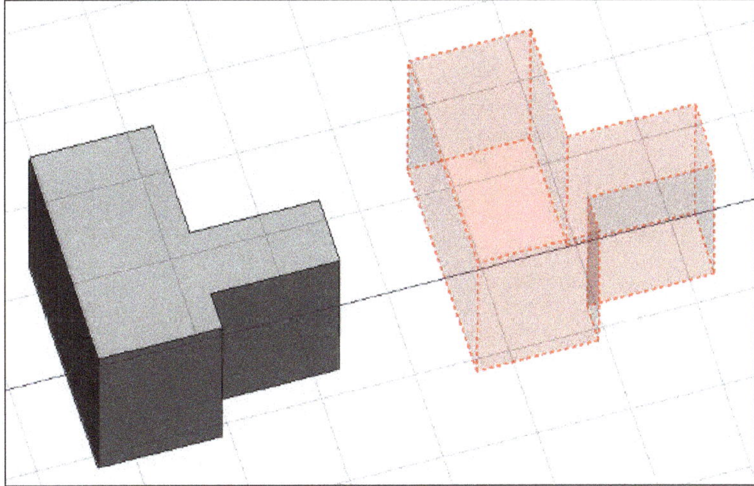
Figure-84. Space object created

- If you want to edit the space object then double-click on the space object created from the Model tree view. The **Component** dialog will be displayed in the **Tasks** panel of **Combo View**; refer to Figure-85.

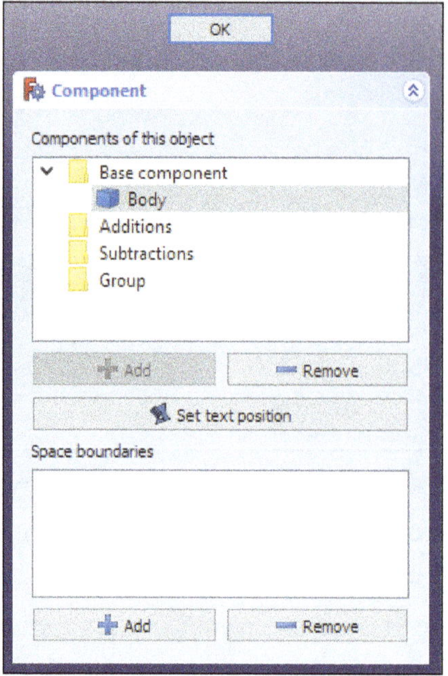

Figure-85. Component dialog

- Specify desired parameters from the **Component** dialog as discussed earlier.
- Click on **OK** button to close the dialog.
- If you want to edit the parameters of space object then select the space object created from Model tree view. The **Property editor** dialog will be displayed in the **Model** panel of **Combo View** with parameters related to space object; refer to Figure-86.

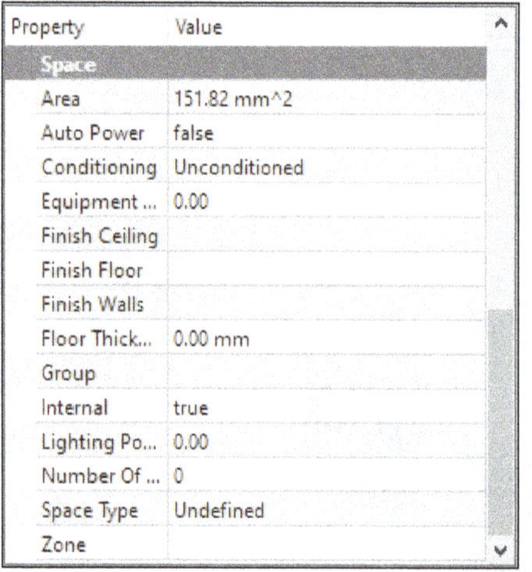

Figure-86. Property editor dialog with space object parameters

- Select desired option from **Auto Power** drop-down in the **Space** section of the dialog to specify whether equipment power will be automatically filled by the equipment included in this space or not.
- Select desired option from **Conditioning** drop-down to specify the type of air conditioning of this space.
- Enter desired value in **Equipment Power** edit box to specify the electric power needed by the equipment of this space in watts.
- Enter desired values in **Finish Ceiling**, **Finish Floor**, and **Finish Walls** edit boxes to specify finishing of ceiling, finishing of floor, and finishing of walls of this space, respectively.
- Specify desired thickness of the floor finish in the **Floor Thickness** edit box.
- Objects that are included inside this space, such as furniture will be displayed in the list box available by clicking on the **Group** button.

- Select desired option from **Internal** drop-down to specify whether this space is internal or external.
- Enter desired value in **Lighting Power** edit box to specify the electric power needed to light this space in watts.
- Specify desired number of people who typically occupy this space in the **Number Of People** edit box.
- Select desired type of this space from **Space Type** drop-down in the dialog.
- Specify desired number of decimals to use for calculated texts in the **Decimals** edit box.
- Specify desired size of the first line of text in the **First Line** edit box.
- Specify desired name, size, and color of the text in the **Font Name**, **Font Size**, and **Text Color** edit boxes of the dialog, respectively.
- Enter desired value in **Line Spacing** edit box to specify the space between lines of text.
- Select desired option from **Show Unit** drop-down to specify whether to show the unit suffix or not.
- Specify desired text to show in the **Text** edit box.
- Select desired option from **Text Align** drop-down to specify the justification of the text.
- Enter desired value in **x**, **y**, and **z** edit boxes from **Text Position** cascading menu to specify the position of text along x, y, and z directions, respectively.
- On specifying the parameters in **Property editor** dialog, the properties of space object will be modified.

Creating Stairs

The **Stairs** tool allows you to automatically build several types of stairs. Stairs can be built from scratch or from a straight line in which case, the stairs follow the line. If the line is not horizontal but has a vertical inclination, the stairs will also follow its slope. The procedure to use this tool is discussed next.

- Click on the **Stairs** tool from **Toolbar** in the **Arch** workbench; refer to Figure-87. The stairs will be created; refer to Figure-88.

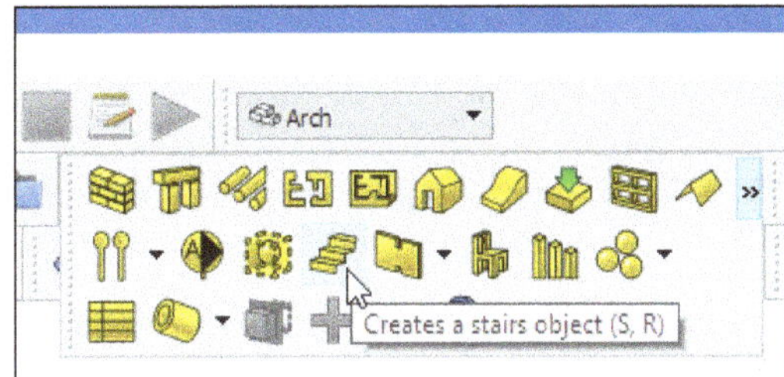

Figure-87. Stairs tool

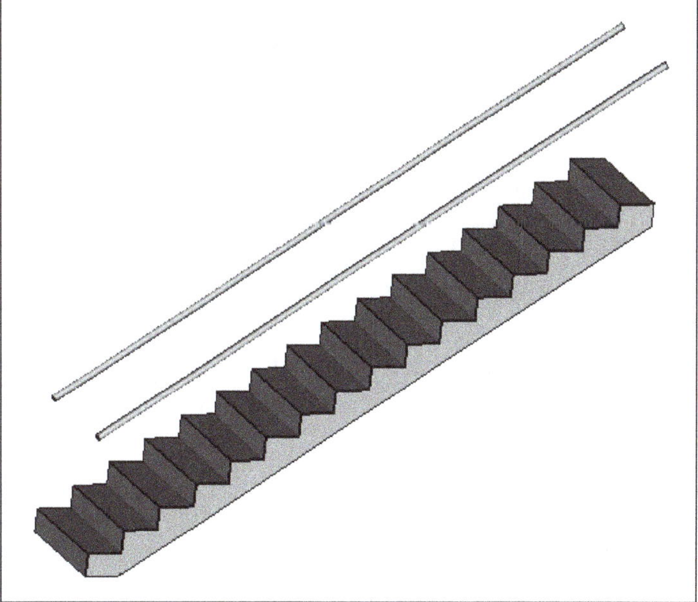

Figure-88. Stairs created

- If you want to edit the properties of stairs then select the stairs created from the Model tree view. The **Property editor** dialog will be displayed in the **Model** panel of **Combo View** with the options related to stairs.
- Select desired option from **Align** drop-down in the **Stairs** section of dialog to specify the alignment of these stairs on their baseline, if applicable.

- Specify desired height, length, and width of the stairs in the **Height**, **Length**, and **Width** edit boxes of the dialog, respectively.
- Specify desired size of nosing in the **Nosing** edit box from **Steps** section of the dialog.
- Specify number of steps (risers) in these stairs in the **Number of Steps** edit box.
- Specify desired height of the risers in the **Riser Height** edit box of the dialog.
- Specify depth of the treads in the **Tread Depth** edit box of the dialog.
- Specify thickness of the treads in the **Tread Thickness** edit box.
- Select desired type of landings from **Landings** drop-down in the **Structure** section of the dialog.
- Specify desired offset between the border of the stairs and the structure in the **Stringer Offset** edit box.
- Specify width of the stringers in the **Stringer Width** edit box of the dialog.
- Select desired type of structure of these stairs from **Structure** drop-down.
- Specify desired thickness of the structure from **Structure Thickness** edit box.
- Select desired type of winders from **Winders** drop-down in the dialog.
- On specifying the parameters in **Property editor** dialog. The properties of stairs will be modified; refer to Figure-89.

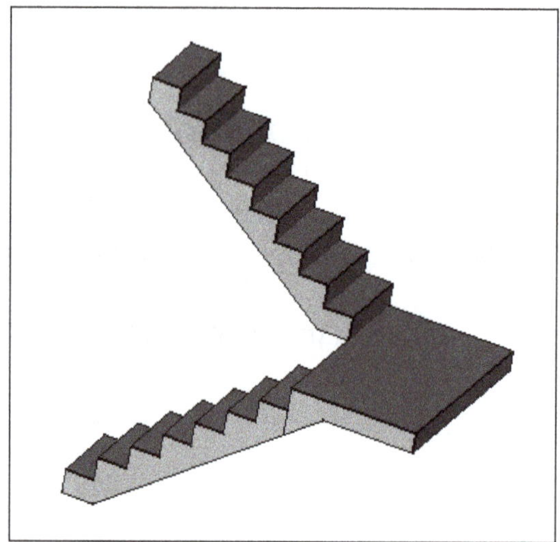

Figure-89. Stairs edited

Panel tools

The **Panel tools** allow you to build all kinds of panel-like elements.

Creating Panel

The **Panel** tool allows you to build all kinds of panel-like elements, typically for panel constructions but also for all kinds of objects that are based on a flat profile. The procedure to use this tool is discussed next.

- Click on the **Panel** tool from **Panel tools** drop-down in the **Toolbar** of **Arch** workbench; refer to Figure-90. The **Point** and **Panel options** dialog will be displayed in the **Tasks** panel of **Combo View** along with the plus sign and a structure of panel in place of original cursor; refer to Figure-91 and Figure-92. You will be asked to specify the location of panel.

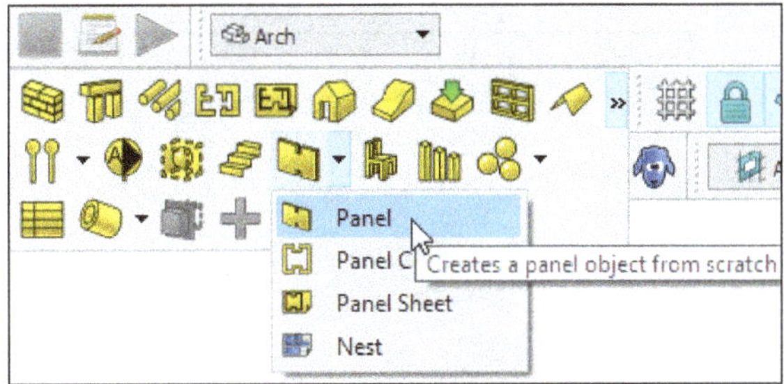

Figure-90. Panel tool

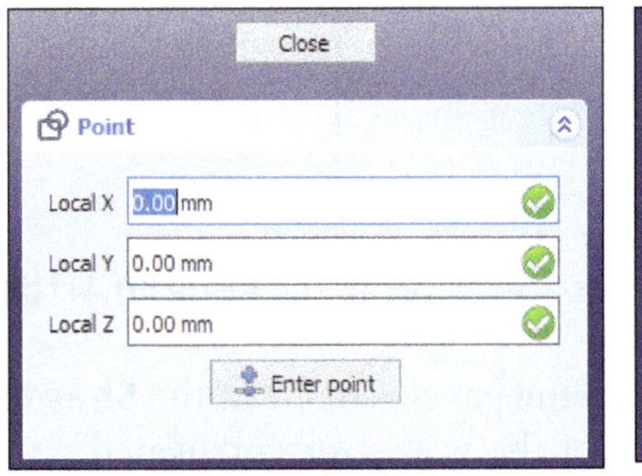

Figure-91. Point dialog

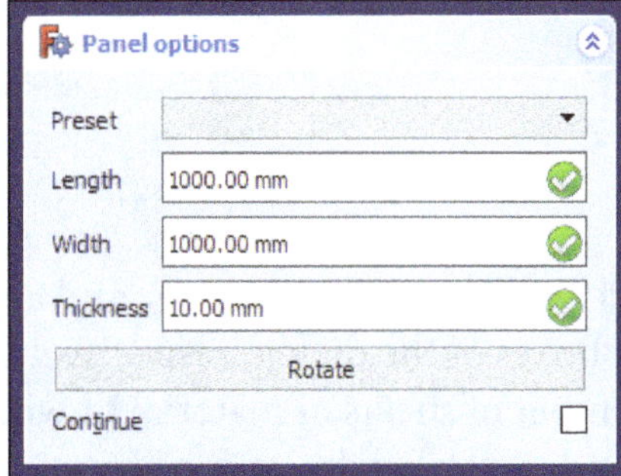

Figure-92. Panel options dialog

- Select desired preset of panel from **Preset** drop-down in the **Panel options** dialog.
- Specify desired length, width, and thickness of panel in the **Length**, **Width**, and **Thickness** edit boxes of the dialog, respectively.
- Click on **Rotate** button from the dialog to rotate the panel at desired location.
- After specifying parameters in the **Panel options** dialog, click in the 3D view area to specify the location of panel or enter desired values for x, y, and z coordinates in the **Local X**, **Local Y**, and **Local Z** edit boxes of the **Point** dialog, respectively.
- After specifying the coordinates in dialog, click on **Enter point** button. The panel will be created; refer to Figure-93.

Figure-93. Panel created

- If you want to edit the properties of panel then select the panel created from the Model tree view. The **Property editor** dialog will be displayed in the **Model** panel of **Combo View** with the parameters related to panel; refer to Figure-94.

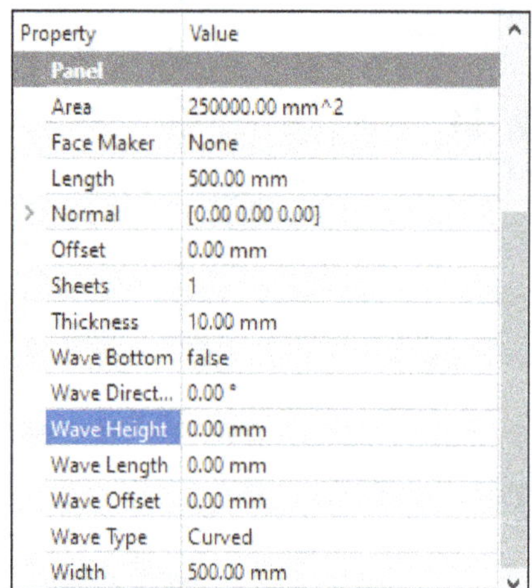

Figure-94. Property editor dialog with panel parameters

- Specify desired length, width, thickness, and area of the panel in the **Length**, **Width**, **Thickness**, and **Area** edit boxes of the dialog, respectively.
- Specify the number of sheets of material of which the panel is made in the **Sheets** edit box.
- Specify desired length, height, and orientation of the waves for corrugated panels in the **Wave Length**, **Wave Height**, and **Wave Direction** edit boxes of the dialog, respectively.
- Select desired type of the wave for corrugated panel from **Wave Type** drop-down in the dialog.
- Select desired option from **Wave Bottom** drop-down to specify whether the bottom wave of the panel is flat or not.
- On specifying the parameters in the **Property editor** dialog, the properties of panel will be modified; refer to Figure-95.

Figure-95. Panel edited

Creating Panel Cut

The **Panel Cut** tool creates a flat 2D view of a panel to be included in the 3D document of a panel sheet or directly exported to DXF. The procedure to use this tool is discussed next.

- Select one or more panel objects from Model tree view or from the 3D view area; refer to Figure-96.

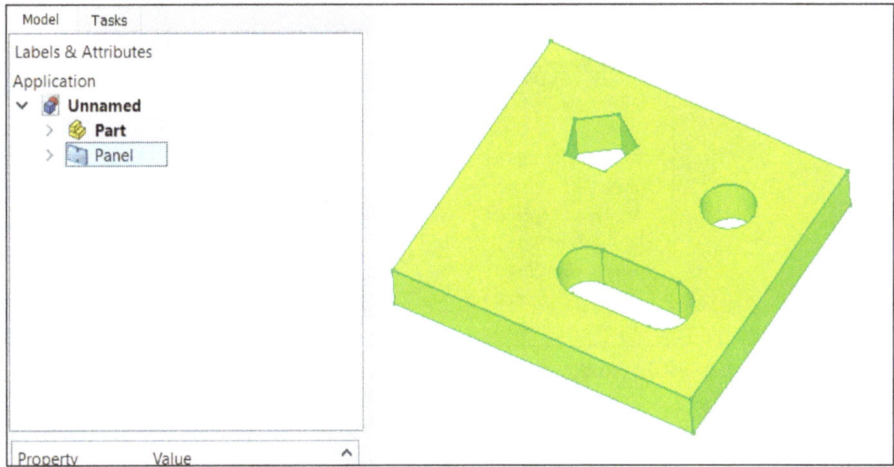

Figure-96. Selecting the panel object

- Click on the **Panel Cut** tool from **Panel tools** drop-down in the **Toolbar** of **Arch** workbench; refer to Figure-97. The panel cut object will be created; refer to Figure-98.

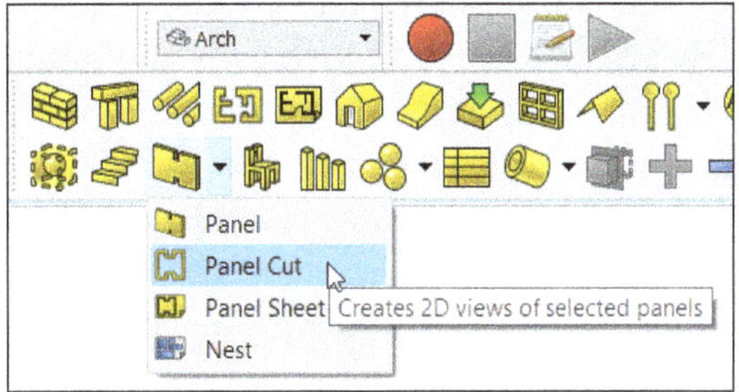

Figure-97. Panel cut tool

Figure-98. Panel cut object created

- If you want to edit the properties of panel cut then select the panel cut created from the Model tree view. The **Property editor** dialog will be displayed in the **Model** panel of **Combo View** with the parameters related to panel cut; refer to Figure-99.

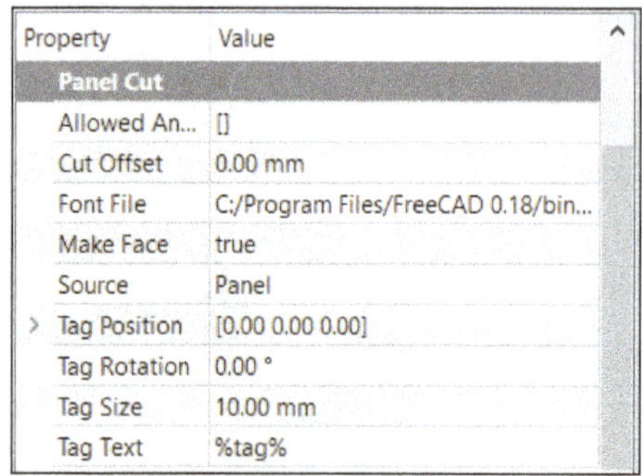

Figure-99. Property editor dialog with panel cut parameters

- Select desired font file to specify font of the tag text from **Font File** edit box in the **Panel Cut** section of the dialog.
- Select desired option from **Make Face** drop-down to specify whether the panel cut will be a face or a wire.
- Enter desired values in **x**, **y**, and **z** edit boxes from **Tag Position** cascading menu to specify the position of tag text along x, y, and z directions, respectively.
- Enter desired value in **Tag Rotation** edit box to specify the rotation of tag text.
- Specify desired size of the tag text in the **Text Size** edit box.
- Specify desired text to display in the **Tag Text** edit box of the dialog.
- Enter desired value in **Margin** edit box from **Arch** section to specify the margin that can be displayed outside the panel cut shape.
- Select desired option from **Show Margin** drop-down to specify whether to turn the display of the margin on/off.
- On specifying parameters in the **Property editor** dialog, the properties of panel cut will be modified.

Creating Panel Sheet

The **Panel Sheet** tool allows you to build a 2D sheet, including any number of panel cut objects or any other 2D object such as those made by the **Draft** workbench and **Sketcher** workbench. The Panel Sheet is typically made to layout cuts to be made by a CNC machine. These sheets can then be exported to a DXF file. The procedure to use this tool is discussed next.

- Select one or more panel cut objects or any other 2D objects from the Model tree view or from the 3D view area from which you want to create a panel sheet; refer to Figure-100.

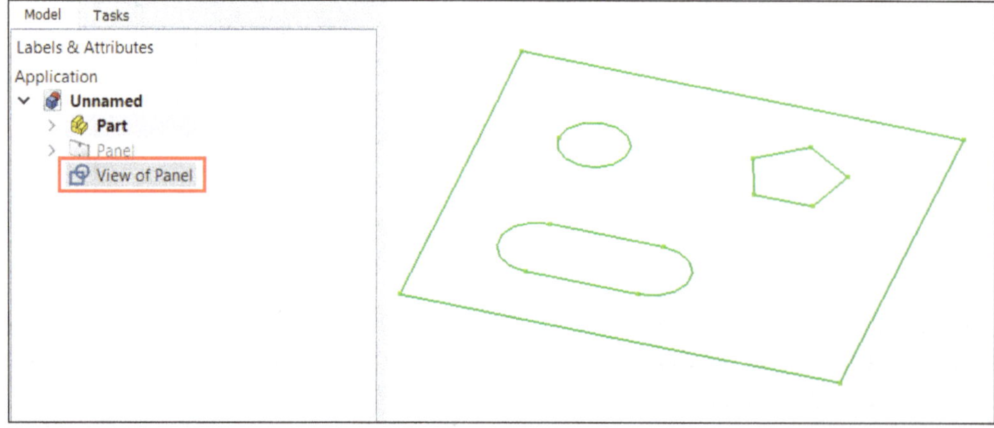

Figure-100. Selecting the panel cut object

- Click on the **Panel Sheet** tool from **Panel tools** drop-down in the **Toolbar** of **Arch** workbench; refer to Figure-101. A panel sheet of the object will be created; refer to Figure-102.

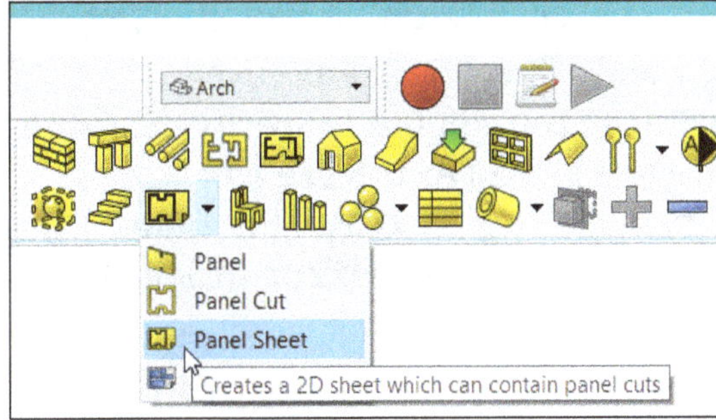

Figure-101. Panel Sheet tool

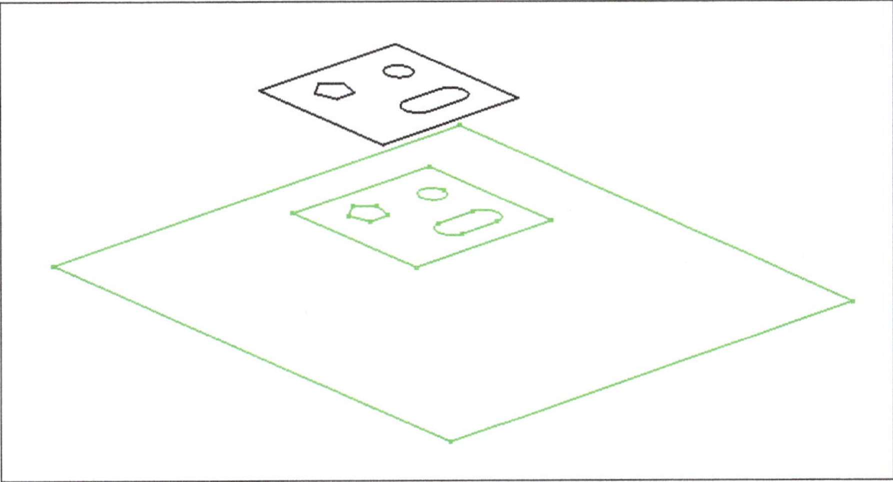

Figure-102. Panel sheet created

- If you want to edit the properties of panel sheet then select the panel sheet created from the Model tree view. The **Property editor** dialog will be displayed in the **Model** panel of **Combo View** with parameters related to panel sheet; refer to Figure-103.

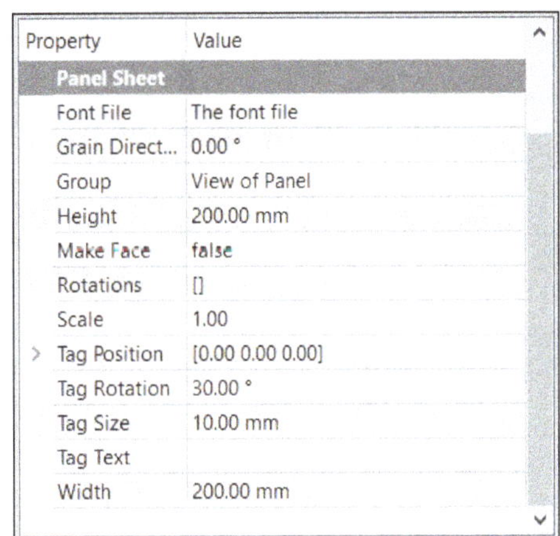

Figure-103. Property editor dialog with panel sheet parameters

- Select desired font file to specify font of the tag text from **Font File** edit box in the **Panel Sheet** section of the dialog.
- Enter desired value in **Grain direction** edit box which allows you to inform the main direction of the panel fiber.
- Specify desired height and width of the sheet in the **Height** and **Width** edit boxes of the dialog, respectively.
- Select desired option from **Make Face** drop-down to specify whether the panel cut will be a face or a wire.
- Enter desired value in **Scale** edit box to specify the scale applied to each panel view.

- Enter desired values in **x**, **y**, and **z** edit boxes from **Tag Position** cascading menu to specify the position of tag text along x, y, and z directions, respectively.
- Enter desired value in **Tag Rotation** edit box to specify the rotation of tag text.
- Specify desired size of the tag text in the **Text Size** edit box.
- Specify desired text to display in the **Tag Text** edit box of the dialog.
- On specifying parameters in the **Property editor** dialog, the properties of panel sheet will be modified.

Creating Nest

The **Nest** tool allows you to select a flat shape to be a container and a series of other flat shapes to be organized inside the space defined by the container shape. The procedure to use this tool is discussed next.

- Click on the **Nest** tool from **Toolbar** in the **Arch** workbench; refer to Figure-104. The **Nesting** dialog will be displayed in the **Tasks** panel of **Combo View**; refer to Figure-105.

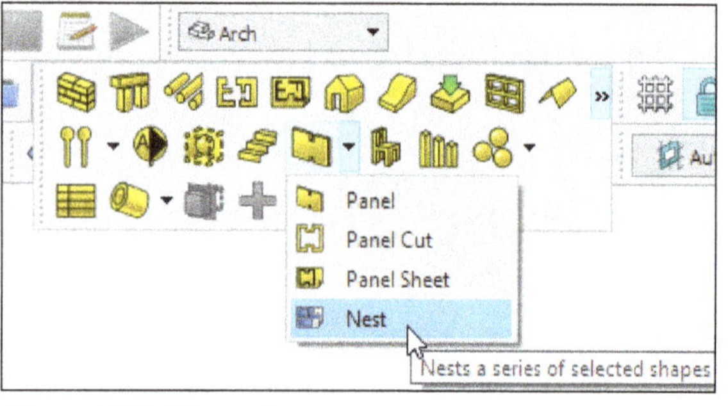

Figure-104. Nest tool

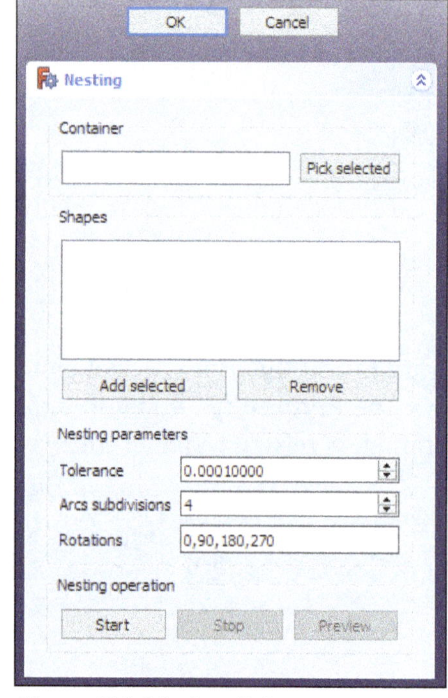

Figure-105. Nesting dialog

- Select a flat, rectangular object from Model tree view or from the 3D view area which is to be container; refer to Figure-106.

Figure-106. Selecting the rectangular object

- Click on **Pick selected** button from **Container** area of the dialog. The selected object will be displayed in the **Container** selection box.
- Select the other flat objects by holding the **CTRL** key that you want to place inside the container from Model tree view or from the 3D view area. Note that these objects must all be flat and in the same plane as the container.
- After selecting the objects, click on **Add selected** button to add it in the **Shapes** list box of the dialog.
- Specify desired tolerance value in the **Tolerance** edit box from **Nesting parameters** area.
- Enter desired value in **Arcs subdivisions** edit box to specify the number of segments to divide non-linear edges for calculations.
- Enter desired values in **Rotations** edit box to specify the angles to rotate the shapes.
- After specifying the parameters, click on **Start** button from **Nesting operation** area to start the calculation process.
- If you want to stop the calculation process then click on the **Stop** button.
- At the end of calculation process, click on the **Preview** button to create a temporary preview of the result.
- Click on **OK** button from the dialog to apply the result. The nest of the objects will be created; refer to Figure-107.

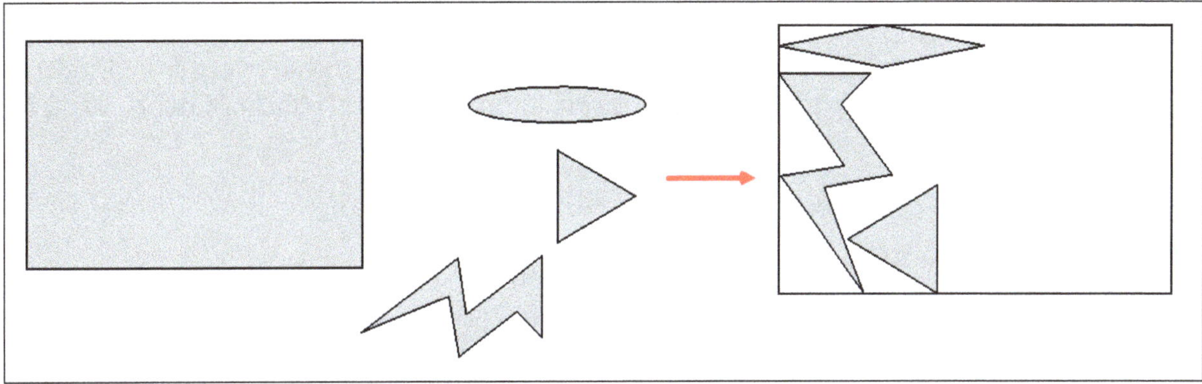

Figure-107. Nest created

Creating Equipment

The **Equipment** tool offers you a simple and convenient way to insert non-structural, standalone elements such as pieces of furniture, hydro-sanitary equipments, or electrical appliances to your projects. The procedure to use this tool is discussed next.

- Select the existing solid object or a mesh object from Model tree view or from the 3D view area; refer to Figure-108.

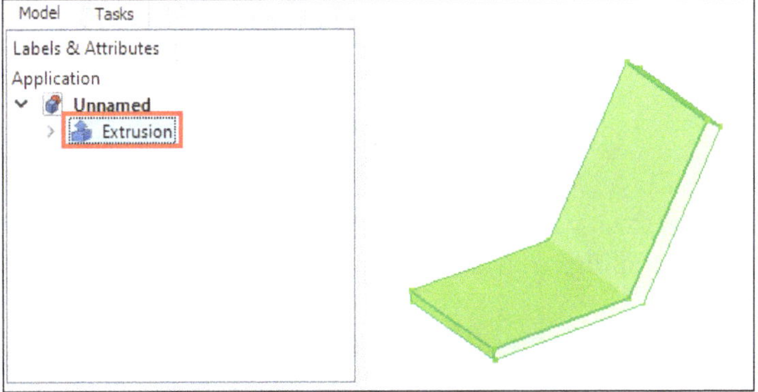

Figure-108. Selecting the existing solid object

- Click on the **Equipment** tool from **Toolbar** in the **Arch** workbench; refer to Figure-109. The non-structural element will be enclosed in an equipment object; refer to Figure-110.

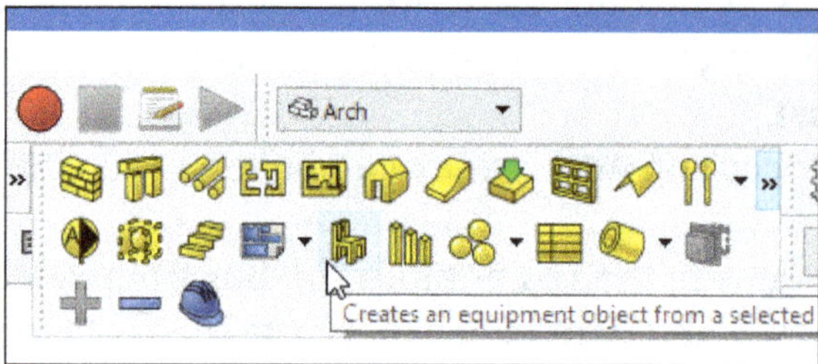

Figure-109. Equipment tool

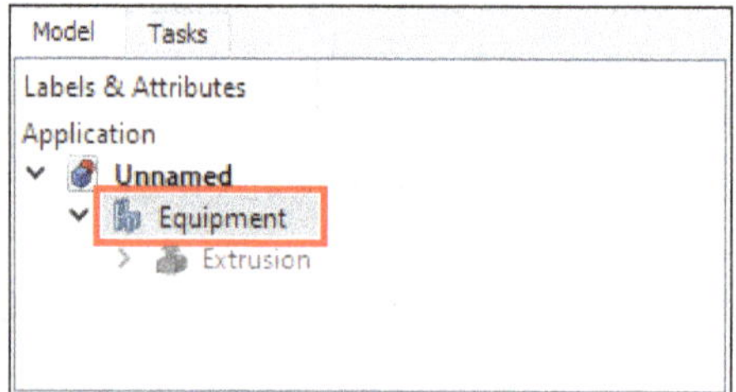

Figure-110. Equipment object created

- If you want to edit the properties of equipment then select the equipment created from Model tree view. The **Property editor** dialog will be displayed in the **Model** panel of **Combo View** with parameters related to equipment; refer to Figure-111.

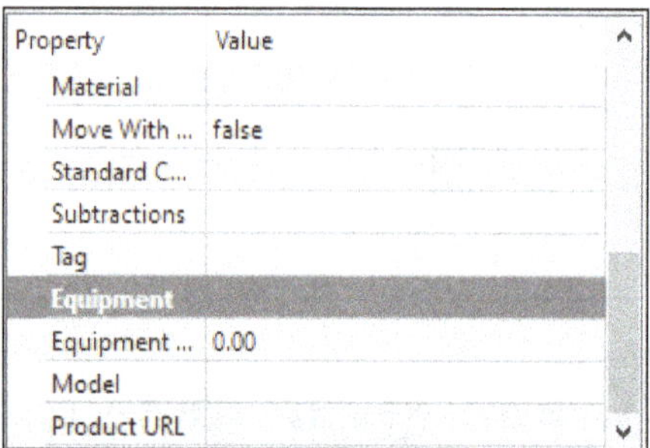

Figure-111. Property editor dialog with equipment parameters

- Enter desired value in **Equipment Power** edit box from **Equipment** section of the dialog to specify the electric power needed by the equipment of this model in watts.
- Specify desired description of the model for this equipment in the **Model** edit box.
- Enter desired URL of the product page where more information about this equipment can be found in the **Product URL** edit box of the dialog.
- On specifying parameters in the **Property editor** dialog, the properties of equipment will be modified.

Creating Frame

The **Frame** tool is used to build all kinds of frame objects based on a profile and a layout. The profile is extruded along the edges of the layout which can be any 2D object such as a sketch or a draft object. It is especially useful to create railings or frame walls. The procedure to use this tool is discussed next.

- First, create a layout object and a profile object from which you want to create a frame; refer to Figure-112.

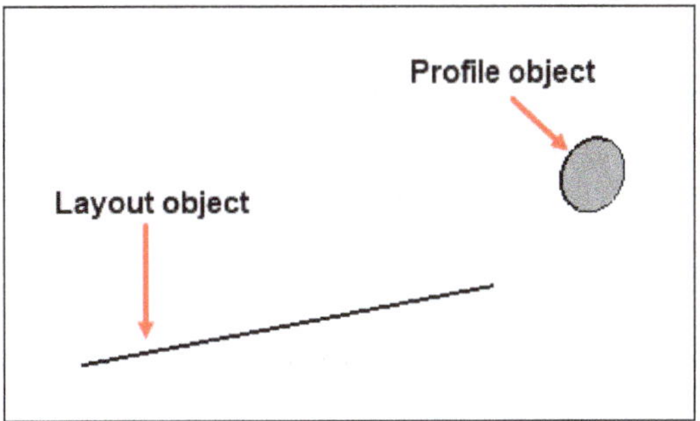

Figure-112. Creating the layout and profile object

- Select the created layout object and then while holding the **CTRL** key, select the profile object from the Model tree view or from the 3D view area.
- Click on the **Frame** tool from **Toolbar** in the **Arch** workbench; refer to Figure-113. The frame object will be created; refer to Figure-114.

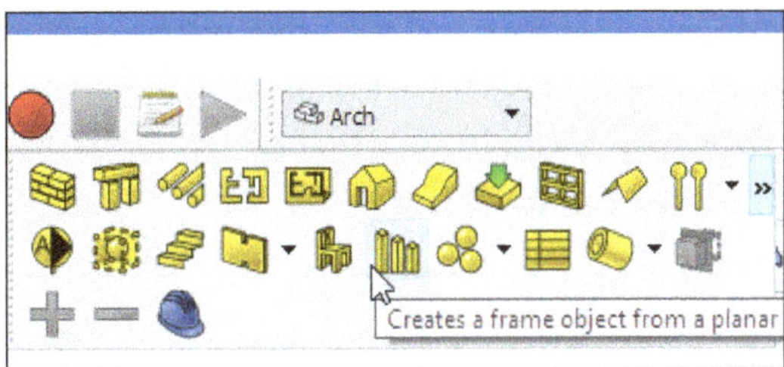

Figure-113. Frame tool

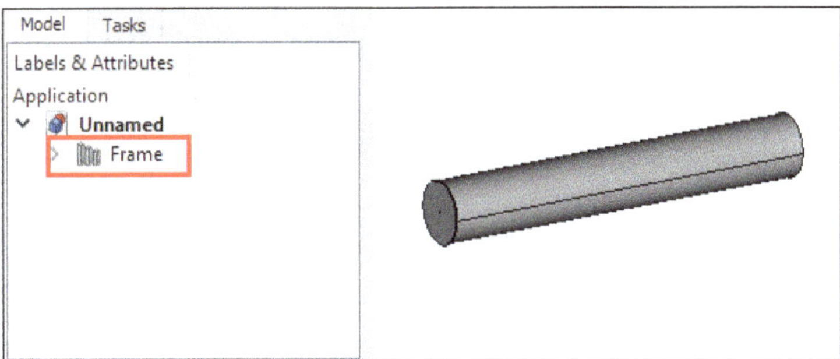

Figure-114. Frame object created

- If you want to edit the properties of frame object then select the frame object created from the Model tree view. The **Property editor** dialog will be displayed in the **Model** panel of **Combo View** with the parameters related to frame object; refer to Figure-115.

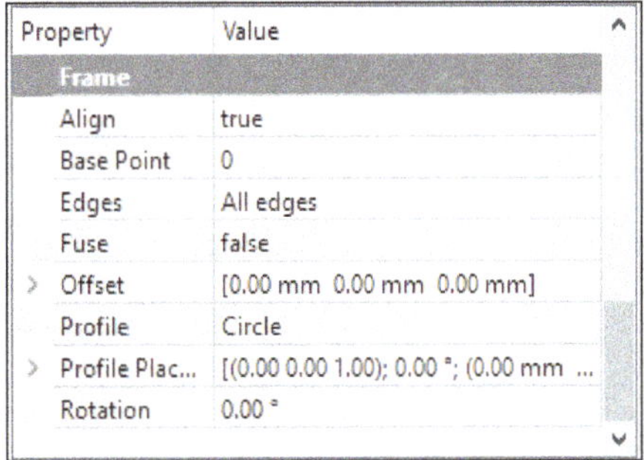

Figure-115. Property editor dialog with frame parameters

- Select desired option from **Align** drop-down in the **Frame** section of the dialog to specify whether the profile must be rotated to have its normal axis aligned with each edge or not.

- Select desired option from the **Edges** drop-down to specify the type of edges to be considered.
- Select desired option from **Fuse** drop-down to specify whether the geometry is fused or not.
- Enter desired values in **x**, **y**, and **z** edit boxes from **Offset** cascading menu to specify the distance between layout object and the frame object.
- The **Profile** edit box specifies the profile used to build this frame.
- Enter desired value in **Rotation** edit box to specify the rotation of the profile around its extrusion axis.
- On specifying parameters in the **Property editor** dialog, the properties of the frame object will be modified.

Material tools

The **Material tools** allow you to add materials to the active document.

Creating Material

The **Material** tool allows you to add materials to the active document and attribute a material to an Arch object. Materials are stored into a **Materials** folder in the active document. The procedure to use this tool is discussed next.

- Select one or more objects from the Model tree view or from the 3D view area to which you want to attribute a new material; refer to Figure-116.

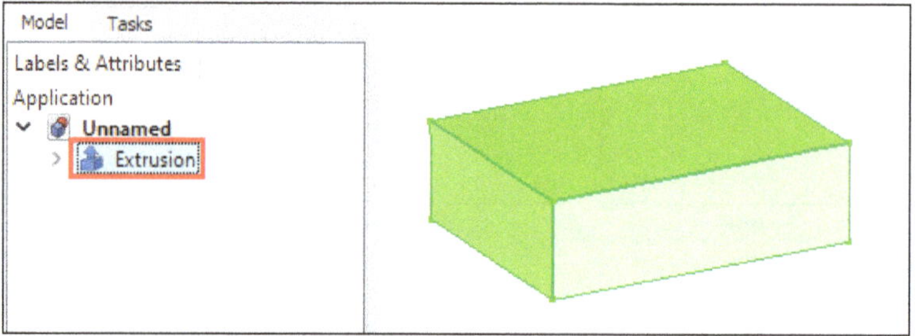

Figure-116. Selecting the object to attribute material

- Click on the **Material** tool from **Toolbar** in the **Arch** workbench; refer to Figure-117. The **Arch material** dialog will be displayed in the **Tasks** panel of **Combo View**; refer to Figure-118.

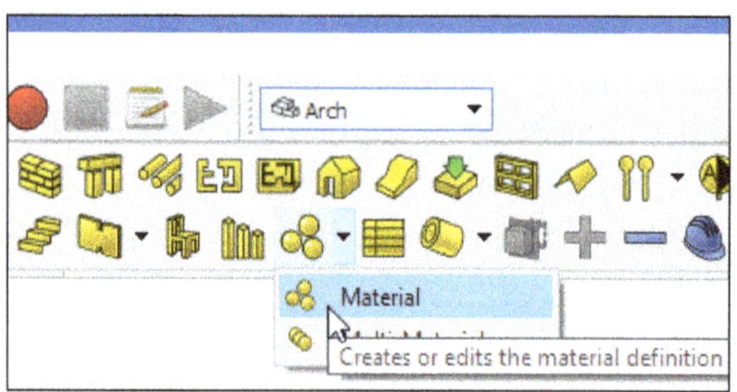

Figure-117. Material tool

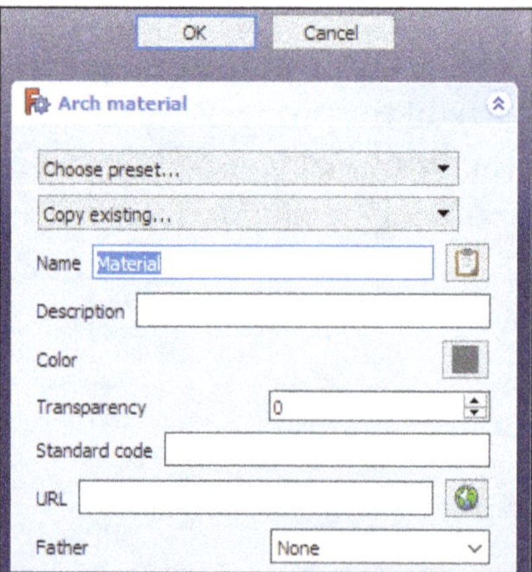

Figure-118. Arch material dialog

- Select one of the preset materials to be used from **Choose preset** drop-down in the dialog.
- Select desired option from **Copy existing** drop-down to copy the values from an existing material in the document.
- Specify desired name for the material in the **Name** edit box of the dialog.

- Click on the **Edit button** available next to the **Name** edit box. The **Material Editor** dialog box will be displayed which allows you to edit many additional properties and add your own custom ones; refer to Figure-119.

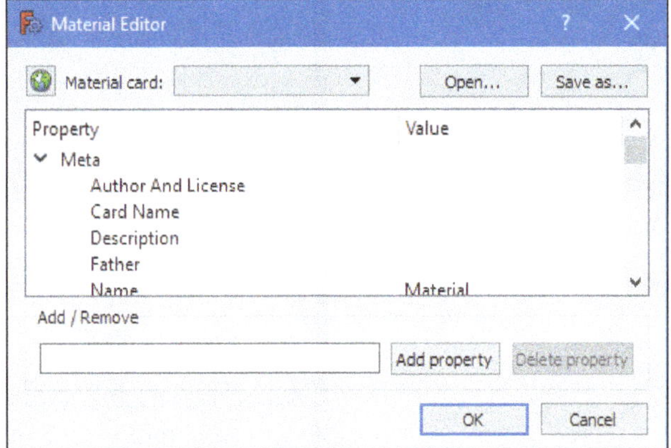

Figure-119. Material Editor dialog box

- Specify more detailed description of the material in the **Description** edit box of the **Arch material** dialog.
- Specify desired display color for the material which will be applied to all objects that use that material by clicking on the **Color** edit button.
- Specify desired transparency value for this material in the **Transparency** edit box.
- Specify a name and reference number of a specification system such as Masterformat or Omniclass in the **Standard code** edit box of the dialog.
- Enter desired url in the **URL** edit box where more information about the material can be found.
- Click on **OK** button from the dialog. The material will be applied to the object.

Creating Multi-Material

The **Multi-Material** tool defines a list of materials with their names and thickness values. This multi-materials list can then be added to an Arch object instead of a single Arch Material. The procedure to use this tool is discussed next.

- Select the object from the Model tree view or from the 3D view area which you want to attribute a new multi-material; refer to Figure-120.

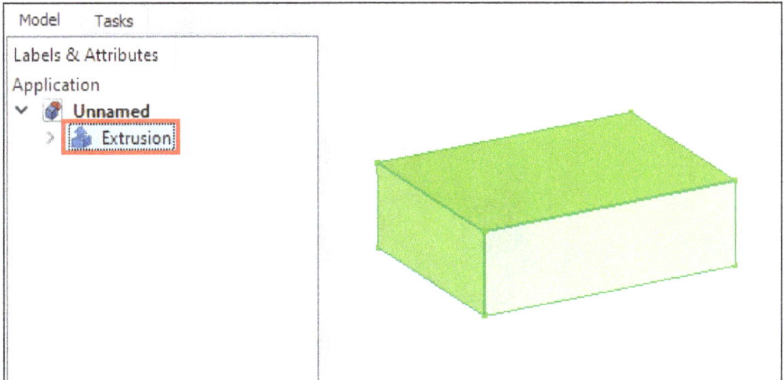

Figure-120. Selecting the object to attribute multi material

- Click on the **Multi-Material** tool from **Toolbar** in the **Arch** workbench; refer to Figure-121. The **Multimaterial definition** dialog will be displayed in the **Tasks** panel of **Combo View**; refer to Figure-122.

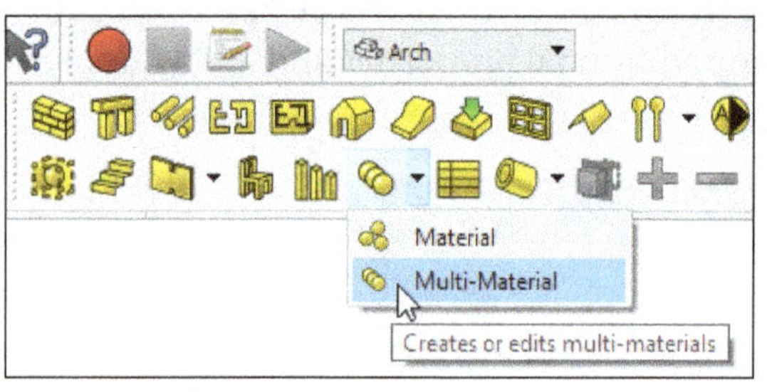

Figure-121. Multi Material tool

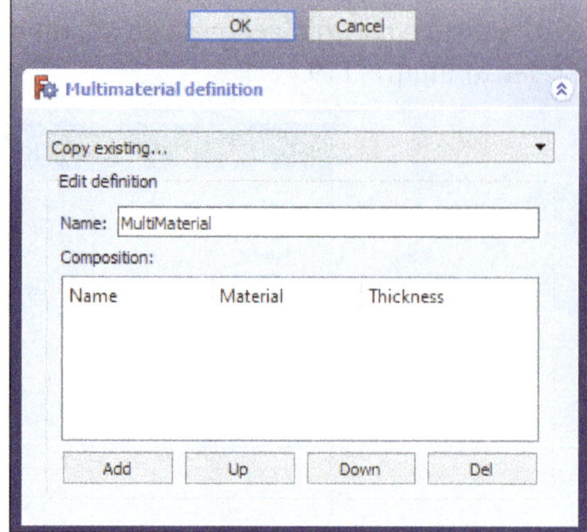

Figure-122. Multimaterial definition dialog

- Select desired option from **Copy existing** drop-down to copy the values from an existing material in the document.
- Specify desired name for the material in the **Name** edit box from **Edit definition** area of the dialog.
- The **Composition** list is the list of the different material layers that compose this multi-material.
- Click on the **Add** button to add a new layer to the list.
- Click on the **Up**, **Down**, and **Del** button to move a selected layer up, to move a selected layer down, and to delete a selected layer, respectively.
- Double-click on the layer name to edit the name of layer.
- Select desired material available in the same document from the **Material** drop-down.
- Enter desired thickness value in the **Thickness** edit box of the dialog.
- Click on **OK** button to close the dialog. The Multi-material will be applied to the object.

Creating Schedule

The **Schedule** tool allows you to create and automatically populate a spreadsheet with contents gathered from the model. The procedure to use this tool is discussed next.

- First, you need to create or open a FreeCAD document which contains some objects; refer to Figure-123.

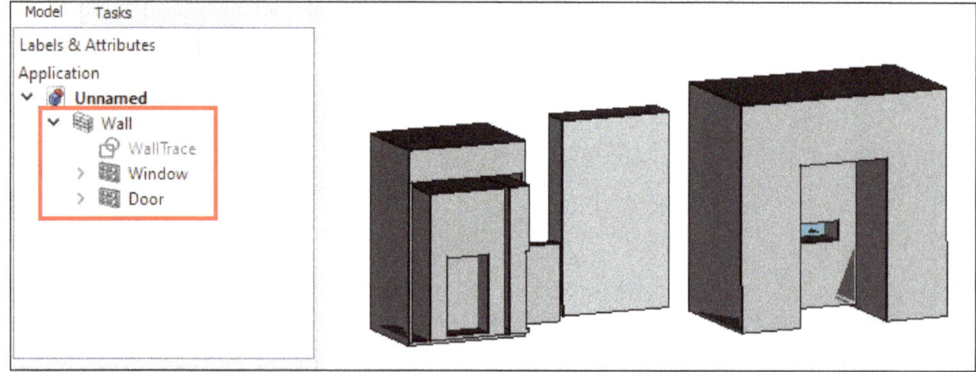

Figure-123. FreeCAD objects created

- Click on the **Schedule** tool from **Toolbar** in the **Arch** workbench; refer to Figure-124. The **Schedule definition** dialog will be displayed in the **Tasks** panel of **Combo View**; refer to Figure-125.

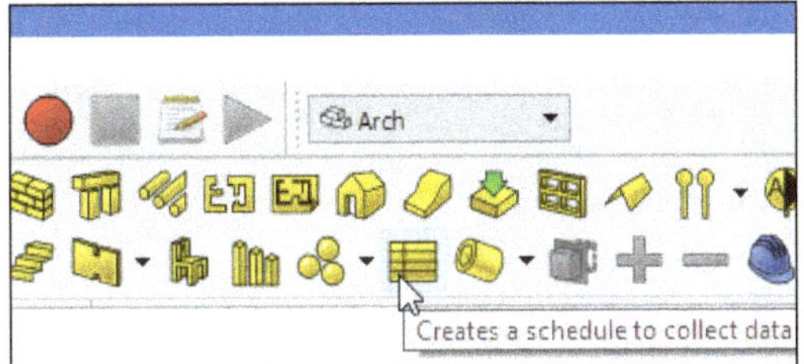

Figure-124. Schedule tool

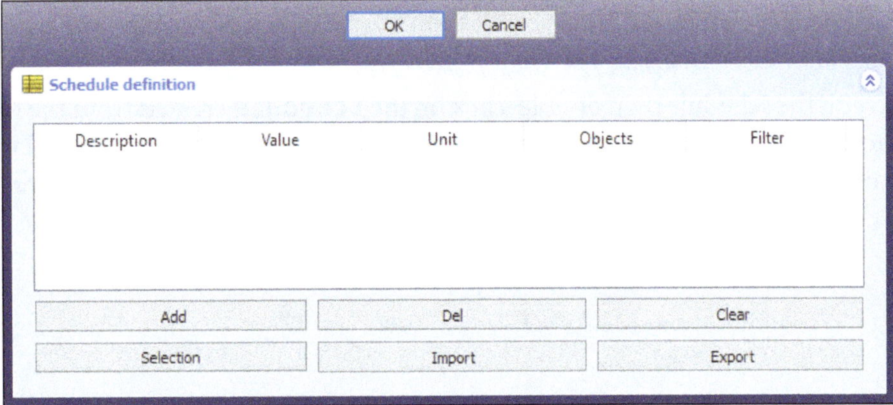

Figure-125. Schedule definition dialog

- Click on **Add** button to add a new line, click on **Del** button to delete the selected line, and click on **Clear** button to delete all the existing lines.
- Click on the **Selection** button to add objects currently selected in the document.
- Click on **Import** button to build the list created in another spreadsheet application and import that as a csv file here.
- Click on **Export** button to export the contents of Result spreadsheet to a csv file.
- Double-click each cell from that line in the dialog to specify desired values.
- Specify desired description for this operation in the **Description** edit box.
- Enter desired value in the **Value** edit box to retrieve for each object or to count the objects.
- Enter desired unit in the **Unit** edit box to express the resulting value.
- Specify the list of object names to be considered by this operation in the **Objects** edit box.
- Specify desired list of filters in the **Filter** edit box. Each filter is written in the form; filter:value.
- After specifying all the parameters, click on **OK** button from the dialog; refer to Figure-126. A new schedule object will be added to the document which contains a result spreadsheet; refer to Figure-127.

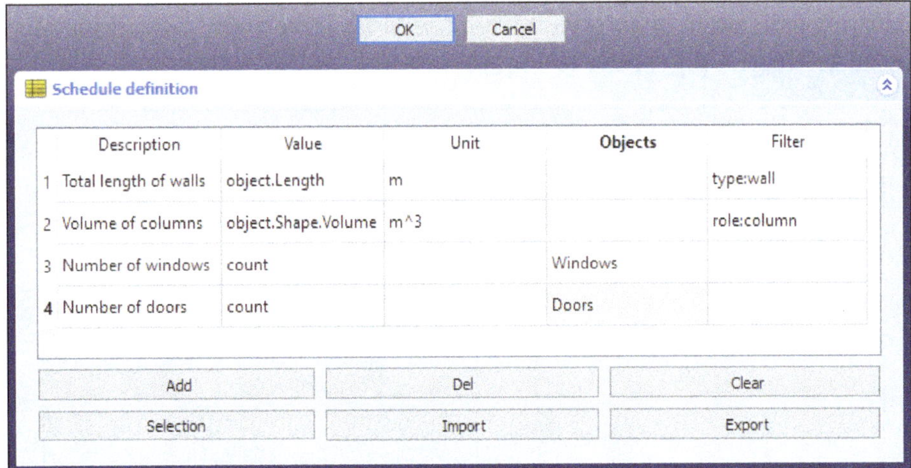

Figure-126. Schedule definition dialog with specified parameters

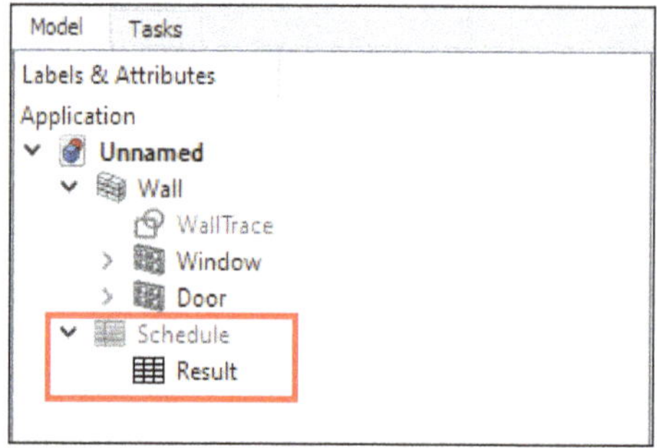

Figure-127. Schedule object added

- If you want to edit the schedule then double-click on the **Schedule** created from the Model tree view.
- If you want to get the results in spreadsheet itself then double-click on the **Result** created from the Model tree view. A new result file in spreadsheet will be opened; refer to Figure-128.

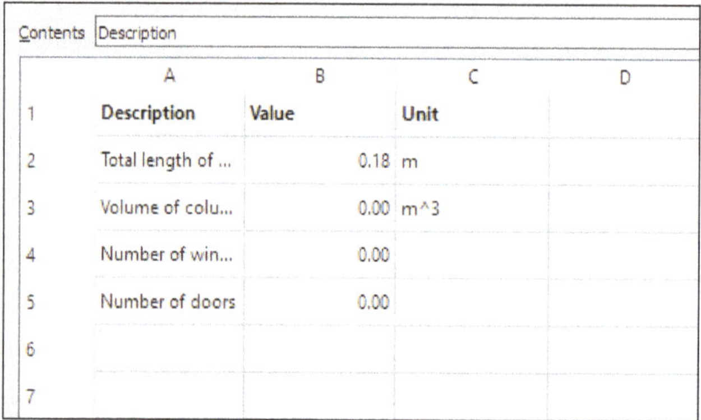

Contents: Description

	A	B	C	D
1	Description	Value	Unit	
2	Total length of ...	0.18	m	
3	Volume of colu...	0.00	m^3	
4	Number of win...	0.00		
5	Number of doors	0.00		
6				
7				

Figure-128. Result in spreadsheet opened

Pipe tools

The **Pipe** tools are used to create pipes from selected objects and to create corner connection between pipes. These tools are discussed next.

Creating Pipe

The **Pipe** tool allows you to create pipes from scratch or from selected objects. The selected objects must be part-based and contain only one open wire. The procedure to use this tool is discussed next.

- Select a linear line, wire, or an open sketch from the Model tree view or from the 3D view area with which you want to create a pipe; refer to Figure-129.

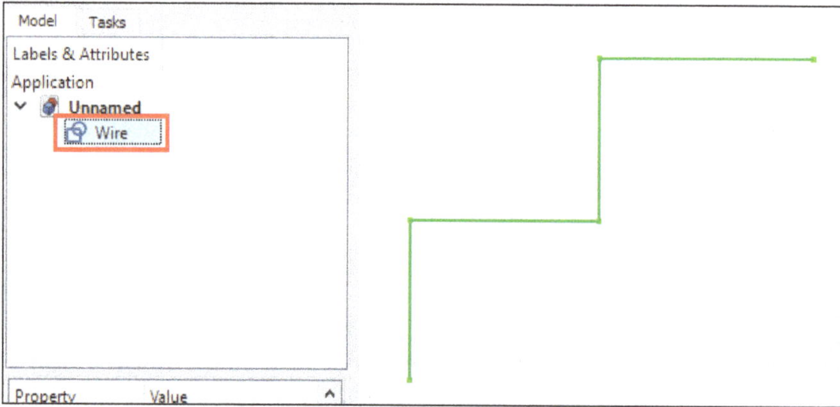

Figure-129. Selecting the wire to create pipe

- Click on the **Pipe** tool from **Pipe tools** drop-down in the **Toolbar** of **Arch** workbench; refer to Figure-130. The pipe will be created; refer to Figure-131.

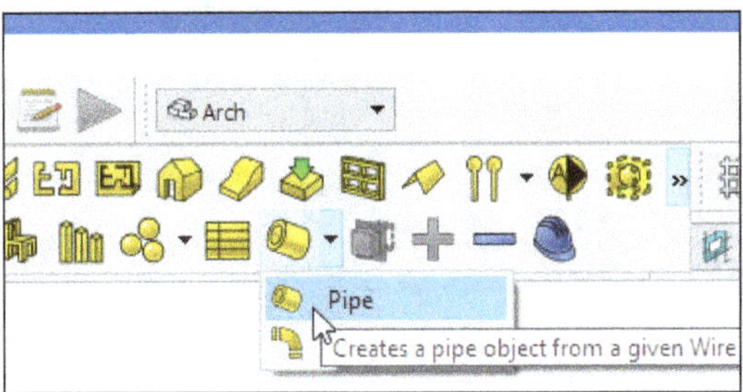

Figure-130. Pipe tool

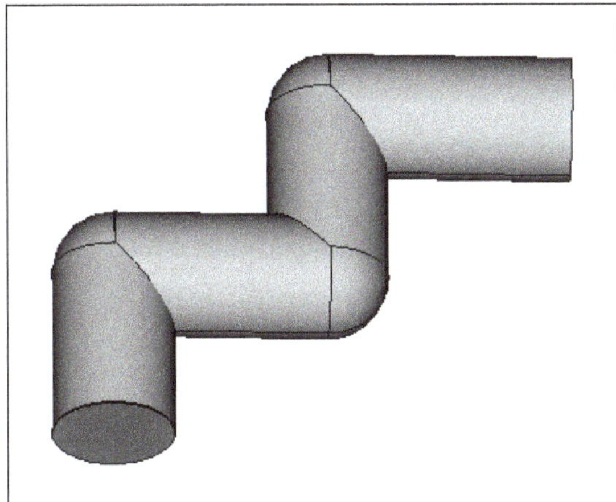

Figure-131. Pipe created

- If you want to edit the properties of pipe then select the pipe created from Model tree view. The **Property editor** dialog will be displayed in the **Model** panel of **Combo View** with parameters related to pipe; refer to Figure-132.

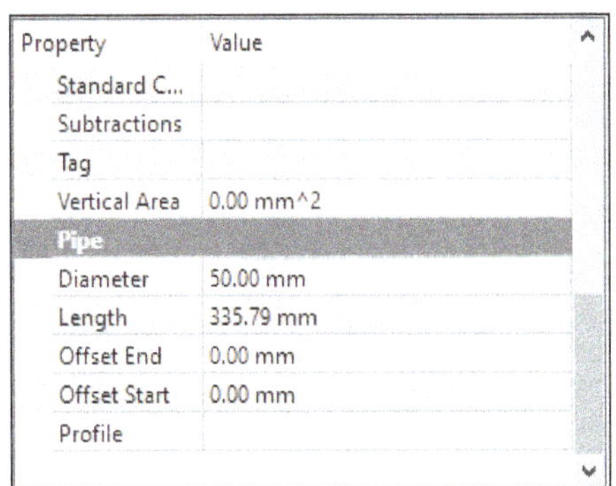

Figure-132. Property editor dialog with pipe parameters

- Specify desired diameter and length of pipe in the **Diameter** and **Length** edit boxes of the dialog, respectively.
- Specify desired offset from the end point and offset from the start point in the **Offset End** and **Offset Start** edit boxes, respectively.
- The **Profile** edit box specifies the base profile of this pipe. If profile is not specified then the pipe will be cylindrical.
- On specifying parameters in the **Property editor** dialog, the properties of pipe will be modified; refer to Figure-133.

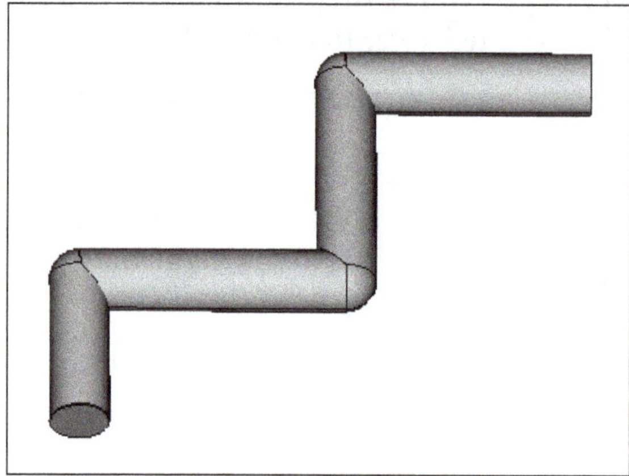

Figure-133. Pipe edited

Creating Pipe Connector

The **Pipe Connector** tool allows you to create corner or tee connection between two or three selected pipes. The procedure to use this tool is discussed next.

- Select two or three pipes from the Model tree view or from the 3D view area; refer to Figure-134.

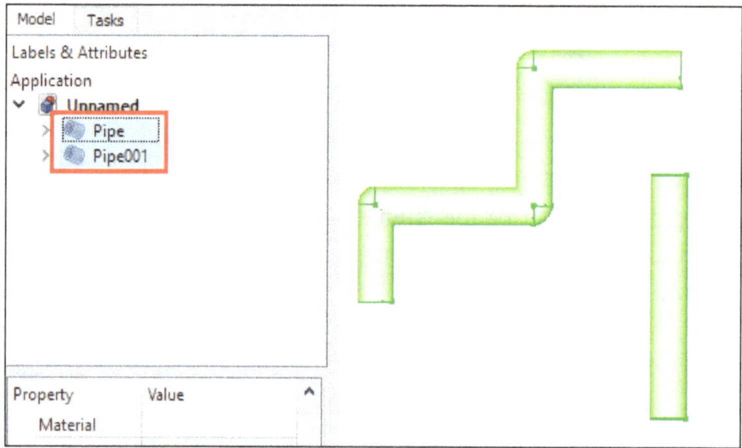

Figure-134. Selecting the pipes

- Click on the **Pipe Connector** tool from **Pipe tools** drop-down in the **Toolbar** of **Arch** workbench; refer to Figure-135. The pipe connector will be created; refer to Figure-136.

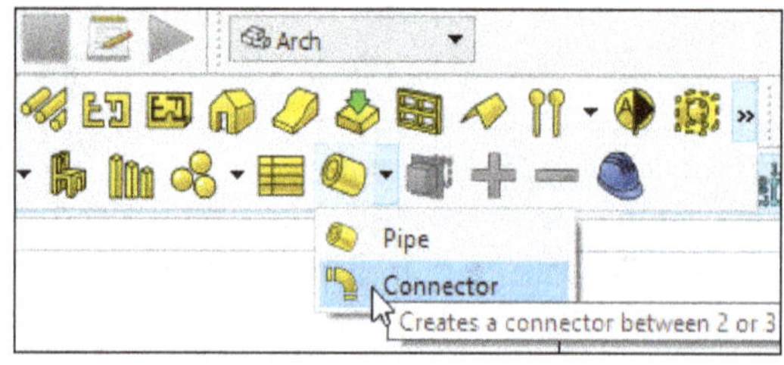

Figure-135. Pipe connector tool

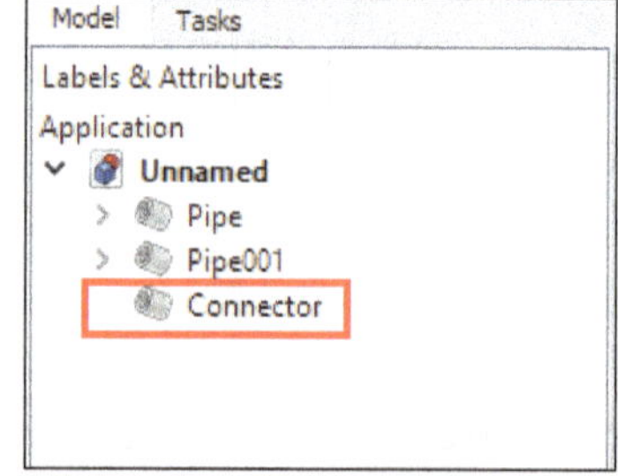

Figure-136. Pipe connector created

- You can edit the curvature radius of connector in the **Radius** edit box from **Pipe Connector** section of **Property editor** dialog.

Cutting the object with Plane

The **Cut Plane** tool allows you to cut an Arch object according to a plane. The procedure to use this tool is discussed next.

- First, select the object which you want to cut from the Model tree view or from the 3D view area and then by holding **CTRL** key, select face of the second object from 3D view area by which the first object will cut; refer to Figure-137.

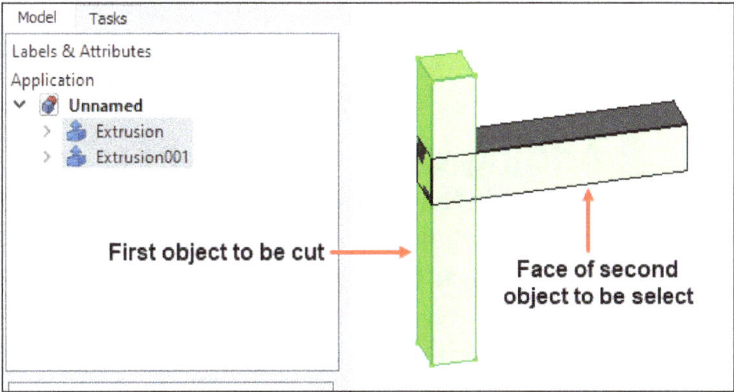

Figure-137. Selecting the objects to be cut

- Click on the **Cut Plane** tool from **Toolbar** in the **Arch** workbench; refer to Figure-138. The **Cut Plane** dialog will be displayed in the **Tasks** panel of **Combo View** along with the cut selection box; refer to Figure-139.

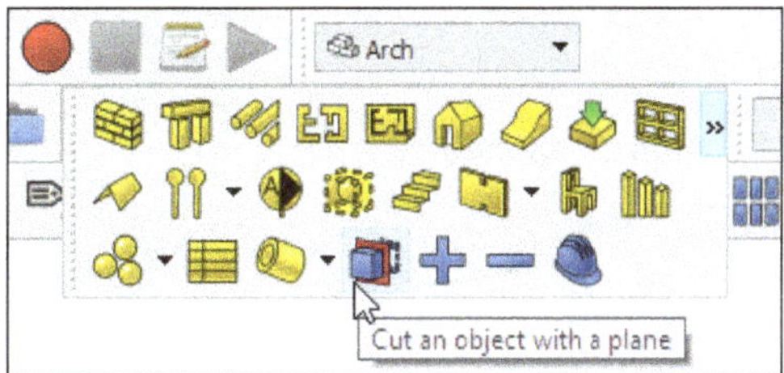

Figure-138. Cut Plane tool

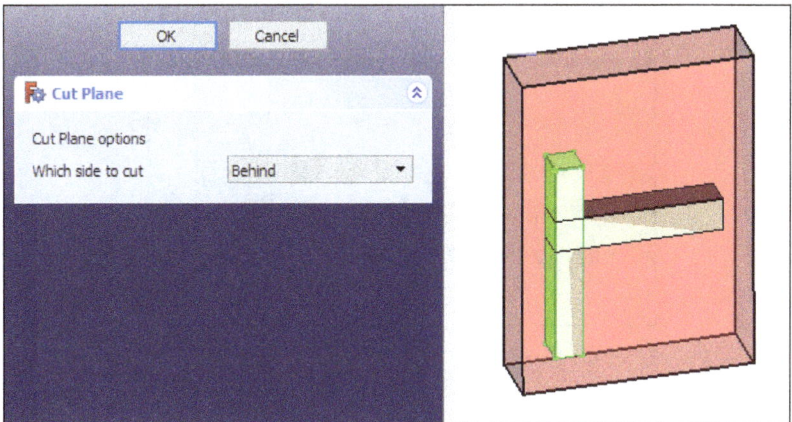

Figure-139. Cut Plane dialog with cut selection box

- Select **Behind** option to cut the object behind the normal face and select **Front** option to cut the object infront of normal face from **Which side to cut** drop-down in **Cut Plane options** section of the dialog.
- Click on **OK** button from the dialog. The object will be cut; refer to Figure-140.

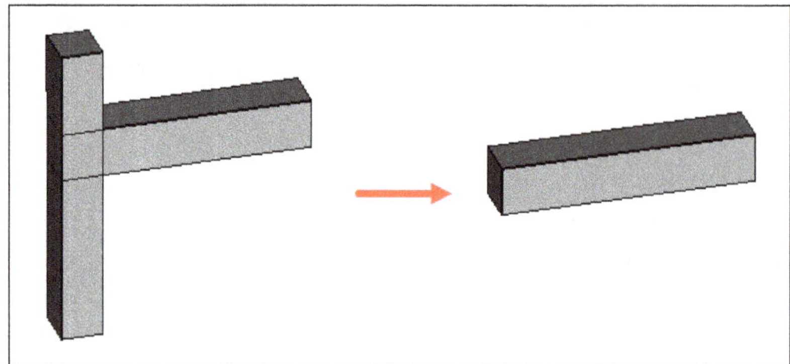

Figure-140. Object cut

Adding the Component

The **Add Component** tool allows you to add shape-based objects to an Arch component, such as a **Wall** or **Structure**. These objects then make part of the Arch component and allow you to modify its shape but keeping its base properties such as height and width. The procedure to use this tool is discussed next.

- First, select the objects from the Model tree view or from the 3D view area which you want to add to the Arch component and then by holding **CTRL** key, select the Arch component, such as a wall or structure ; refer to Figure-141.

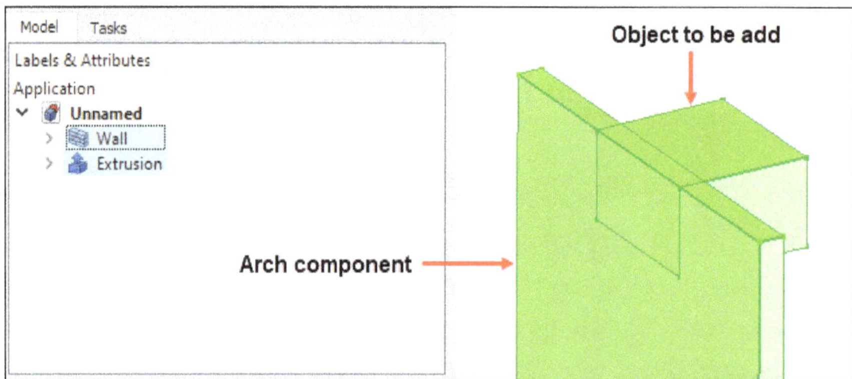

Figure-141. Selecting the object to add

- Click on the **Add component** tool from **Toolbar** in the **Arch** workbench; refer to Figure-142. The object will be added to the Arch component; refer to Figure-143.

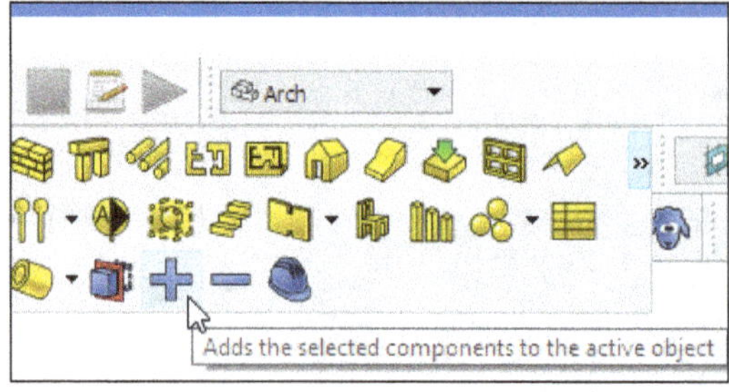

Figure-142. Add component tool

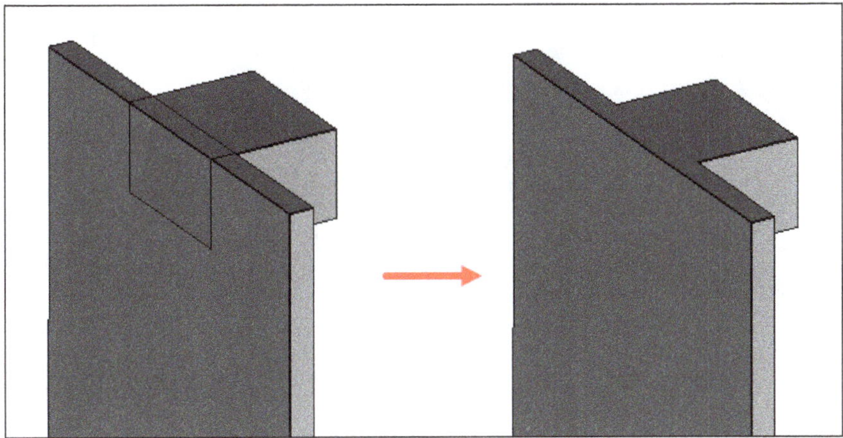

Figure-143. Object added

Removing the Component

The **Remove Component** tool allows you to remove a subcomponent from an Arch component, for example, remove a box that has been added to a wall. The procedure to use this tool is discussed next.

- First, select the object from the Model tree view or from the 3D view area which you want to remove from the Arch component and then by holding **CTRL** key, select the Arch component from the Model tree view or from the 3D view area; refer to Figure-144.

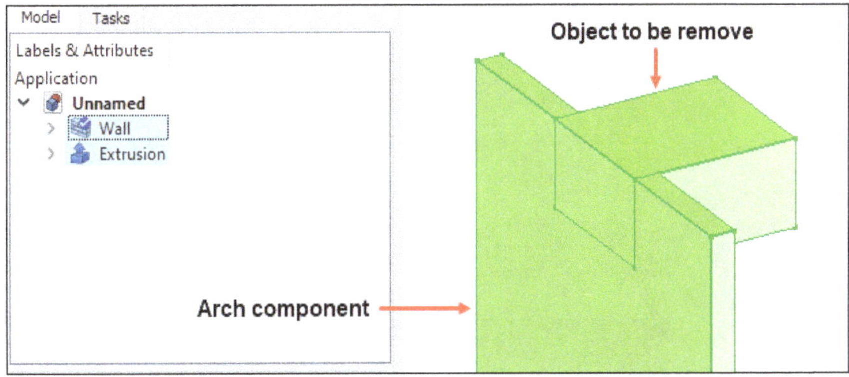

Figure-144. Selecting the object to remove

- Click on the **Remove component** tool from **Toolbar** in the **Arch** workbench; refer to Figure-145. The object will be removed; refer to Figure-146.

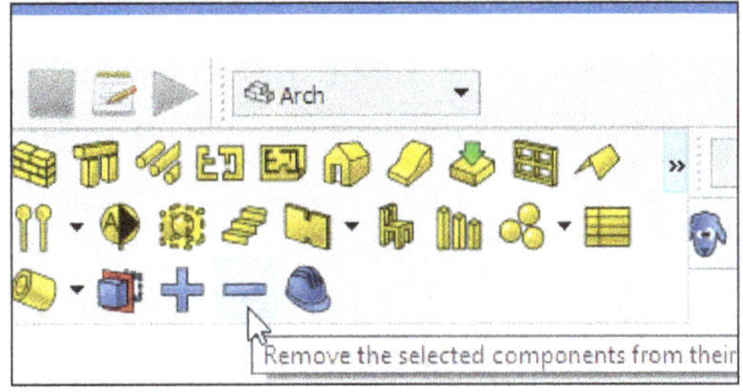

Figure-145. Remove component tool

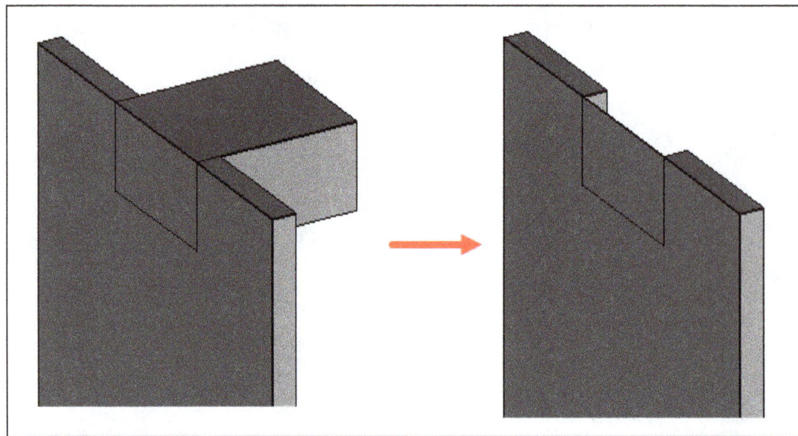

Figure-146. Object removed

Survey

The **Survey** tool enters a special surveying mode which allows you to quickly grab measurements and information from a model and transfer that information to other applications. The procedure to use this tool is discussed next.

- Click on the **Survey** tool from **Toolbar** in the **Arch** workbench; refer to Figure-147. The **Survey** dialog will be displayed in the **Tasks** panel of **Combo View**; refer to Figure-148.

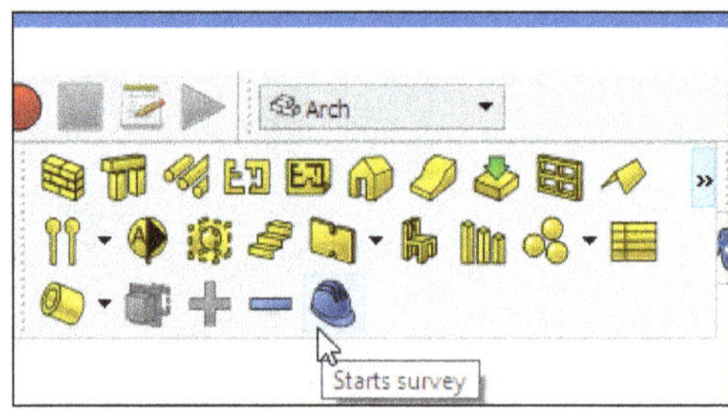

Figure-147. Survey tool

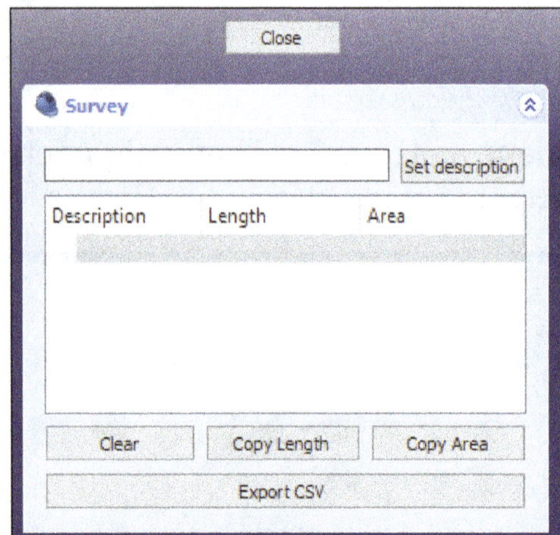

Figure-148. Survey dialog

- Click on the vertices, edges, faces of the object, or double-click on the whole object from the 3D view area to get the measurement. The label will be displayed showing the measurement of selected object in the 3D view area as well as in the **Survey** dialog; refer to Figure-149.

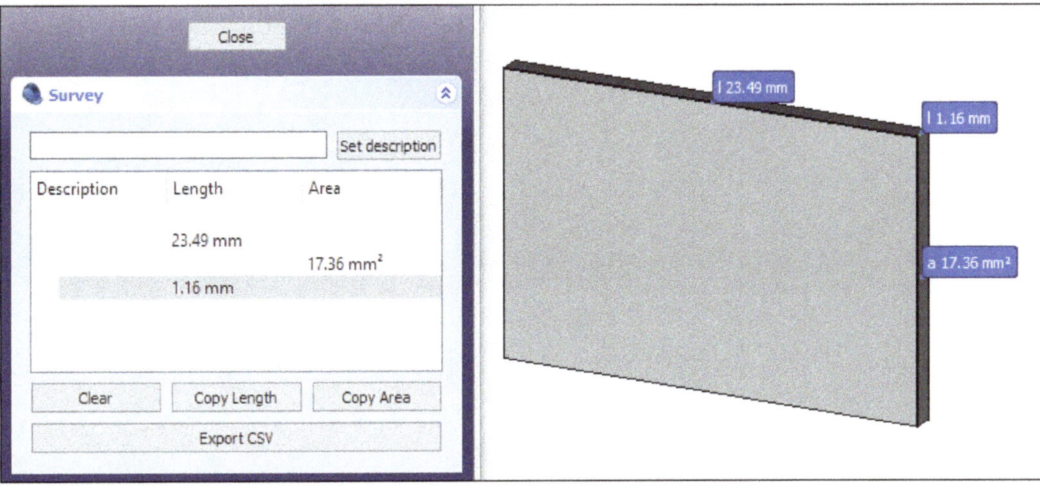

Figure-149. Measurement displayed

- If you want to add description for the measurement then select the measurement created from the dialog and specify desired description in the box available next to the **Set description** button.
- After specifying the description, click on the **Set description** button from the dialog. The description for that measurement will be added; refer to Figure-150.

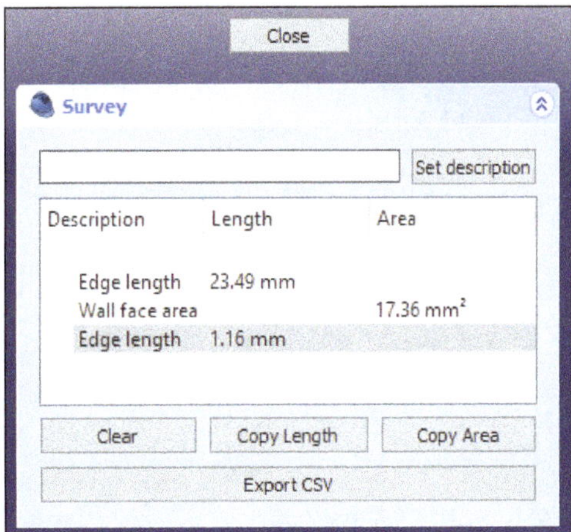

Figure-150. Description added

- Click on the **Export CSV** button from the dialog to export the contents of the dialog in csv file which can be opened in any spreadsheet application.
- Click on **Close** button to close the dialog.

For Student Notes

Chapter 8

Path Workbench

Topics Covered

The major topics covered in this chapter are:

- ***Introduction to Path Workbench***
- ***Create Jobs***
- ***Tool Manager***
- ***G-Code Generation***
- ***Basic Path Operations***
- ***Path Modification Tools***
- ***Path Utilities***
- ***Post-Processing***

INTRODUCTION

The **Path** workbench is used to generate machine codes for CNC machines using a FreeCAD 3D model; refer to Figure-1. Using these codes on CNC machines such as mills, lathes, laser cutters, or similar, you can machine real workpieces. These machine codes are also called G-codes.

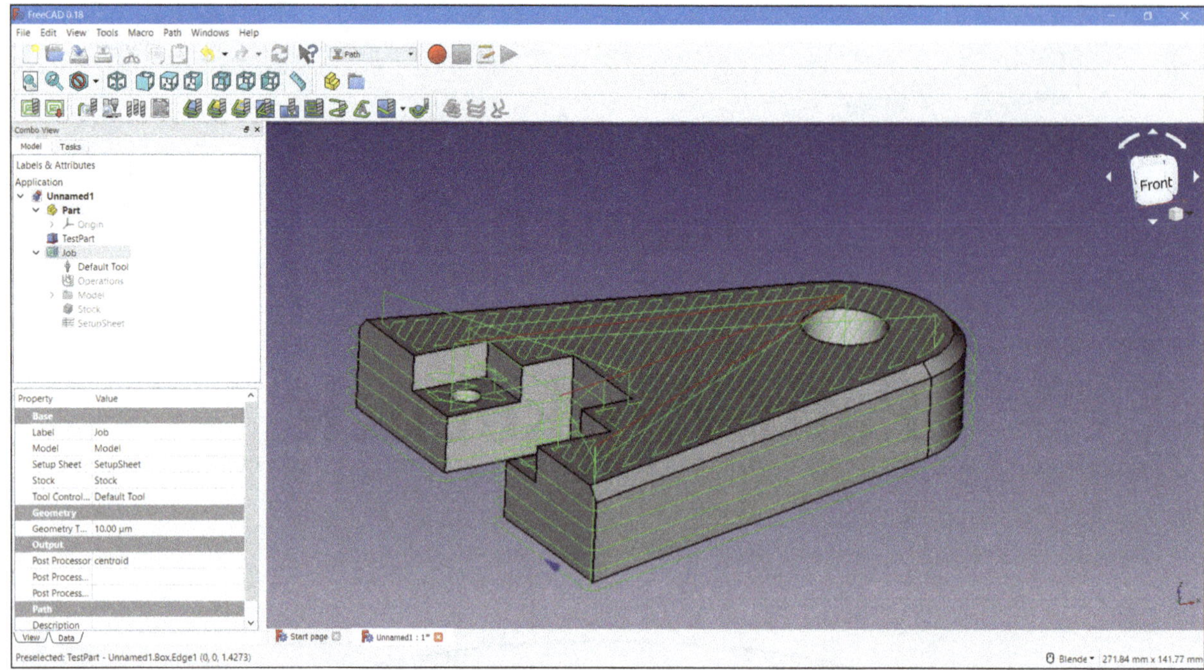

Figure-1. Path Workbench Overview

The FreeCAD Path workbench creates these machine instructions as follows:

- A 3D model of the part to be achieved after performing machining is created in modeling workbenches of FreeCAD.
- A Job is created in Path Workbench. This job contains all the information required to generate necessary G-Code for a CNC machine like Stock material, list of cutting tools to be used for machining with cutting parameters, machine characteristics, and so on.
- Tools are selected as required by the Job Operations.
- Machining paths are created using operations like contour and pocket. These path objects use internal FreeCAD G-Code dialect independent of the CNC machine.
- Machining paths are exported to G-codes matching to your machine. This step is called post processing uses a set of conversion codes called Post-processor which is specific to your machine.

STARTING PATH WORKBENCH

The **Path** workbench generates G-Code defining the paths required to mill the Project represented by the 3D model on the target mill in the Path Job Operations FreeCAD G-Code dialect which is later translated to the appropriate dialect for the target CNC controller by selecting the appropriate postprocessor.

- To start a new file in Path workbench, click on the **New** button from **File** menu and select **Path** workbench from **Switch between workbenches** drop-down in the **Toolbar**; refer to Figure-2. The tools related to **Path** workbench will be displayed in the **Toolbar**; refer to Figure-3.

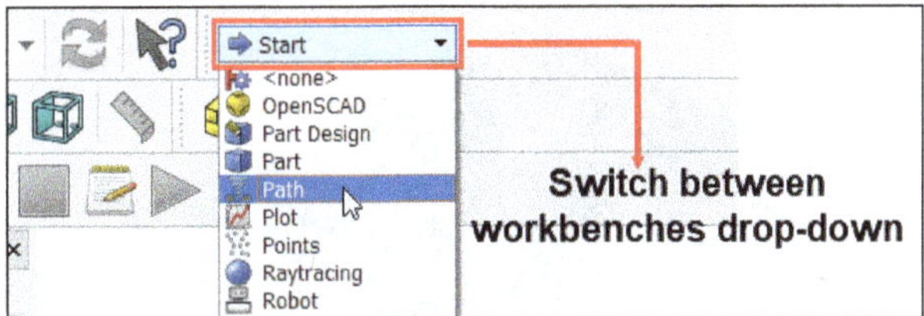

Figure-2. Path workbench

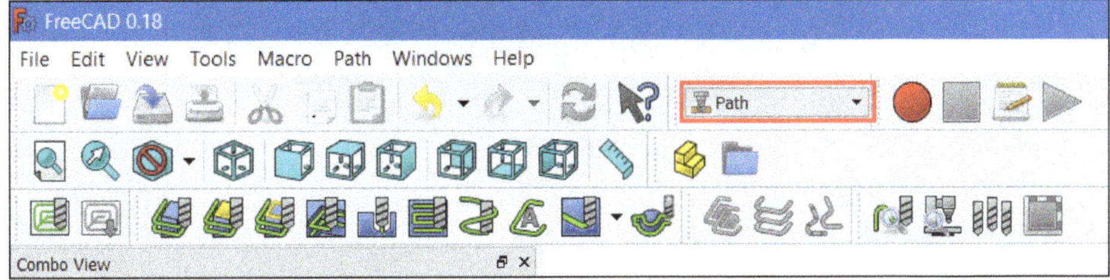

Figure-3. Path workbench tools

JOBS, TOOLS, AND GENERATING G-CODE

After creating model, the first important step for generating CNC program is setting up a job. In FreeCAD, job means defining machine, tool controllers, geometry orientation, and so on. The tools to perform job setup are available in the **Toolbar** of **Path** workbench; refer to Figure-4. These tools are discussed next.

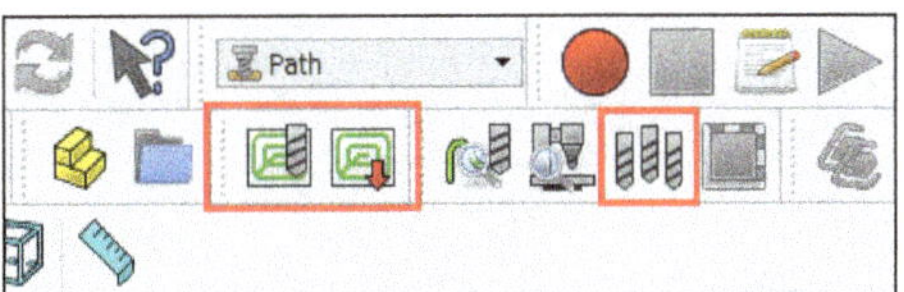

Figure-4. Tools for creating jobs and generating G codes

Creating Job

The **Job** tool is used to create a new machining job in the active document. The Job contains following information:

- A list of Tool-Controller definitions, specifying the geometry, Feeds, and Speeds for the Path Operations Tools.
- A Workflow sequential list of Path Operations.
- A Base Body—a clone used for offset.
- A Stock, representing the raw material that will be milled to get final model.
- A SetupSheet containing inputs used by the Path Operations, including static values and formulas.
- Configuration parameters specifying the output G-Code job's destination path, file name and extension, the Postprocessor—used to generate the appropriate dialect for the target CNC Controller, and so on.

The procedure to use this tool is discussed next.

- Open or create the 3D model for which you want to create the job; refer to Figure-5.

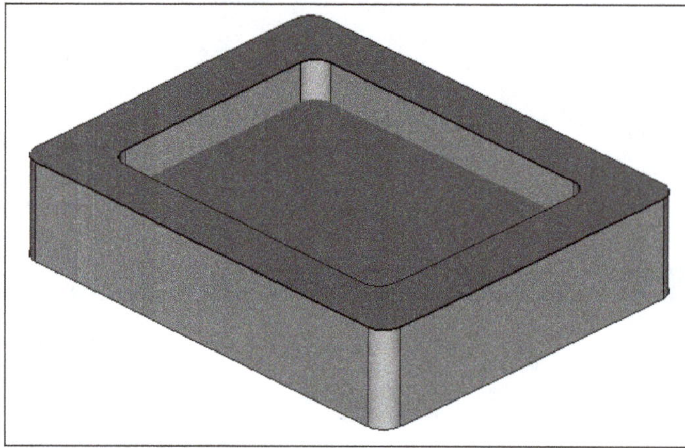

Figure-5. Model created

- Click on the **Job** tool from **Toolbar** in the **Path** workbench; refer to Figure-6. The **Create Job** dialog box will be displayed; refer to Figure-7. You will be asked to select the object.

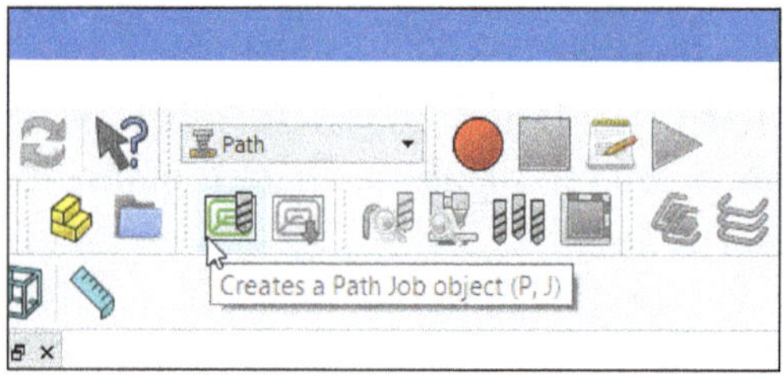

Figure-6. Job tool

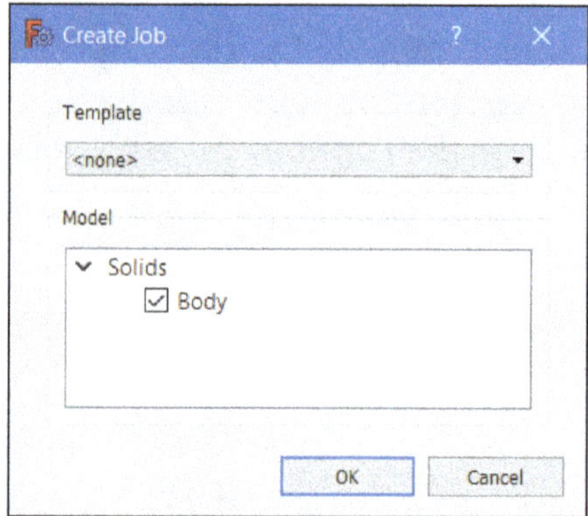

Figure-7. Create Job dialog box

- Select the object from **Model** list box in the dialog box for which you want to create a job and click on **OK** button from the dialog box. The **Job Edit** dialog will be displayed in the **Tasks** panel of **Combo View** with the **Setup** tab opened by default and the model will be enclosed by a bounding box; refer to Figure-8.

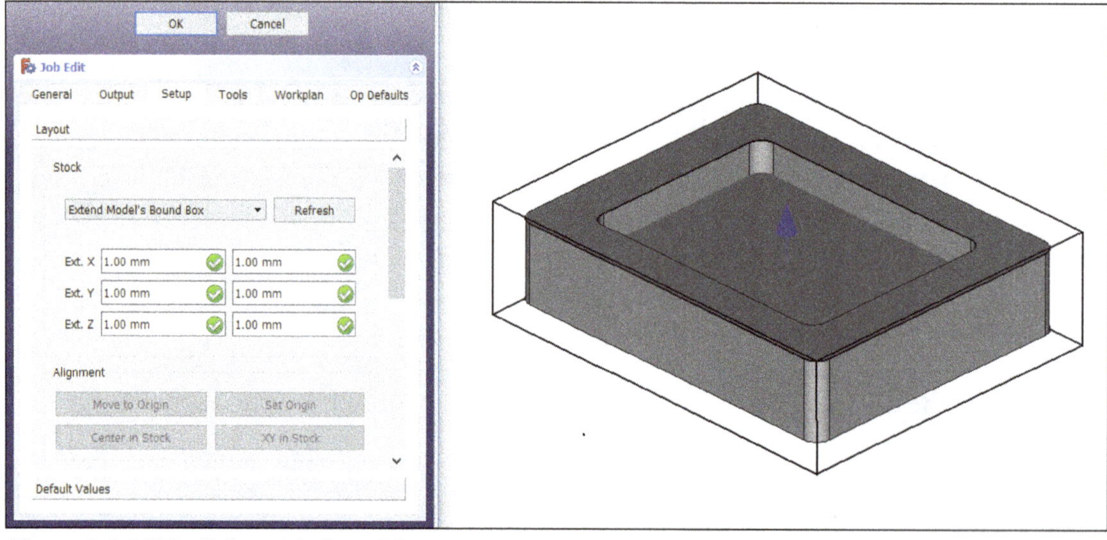

Figure-8. Job Edit dialog with the model

Defining Stock

- Select desired option from the drop-down in the **Stock** area of **Layout** section to define shape and size of the raw material to be machined. Select the **Create Box** option to create stock in box shape using parameters specified in **Length**, **Width**, and **Height** edit boxes; refer to Figure-9. Select the **Create Cylinder** option from the drop-down if you want to create a cylindrical stock using specified radius and height parameters; refer to Figure-10. Select the **Extend Model's Bound Box** option from the drop-down if you want to create stock by extending X, Y, and Z sides of base model by specified values; refer to Figure-11. Select the **Use Existing Solid** option from the drop-down to use existing solid currently available in the drawing area. On selecting this option, you can select any existing solid feature from the second drop-down of **Stock** area; refer to Figure-12.

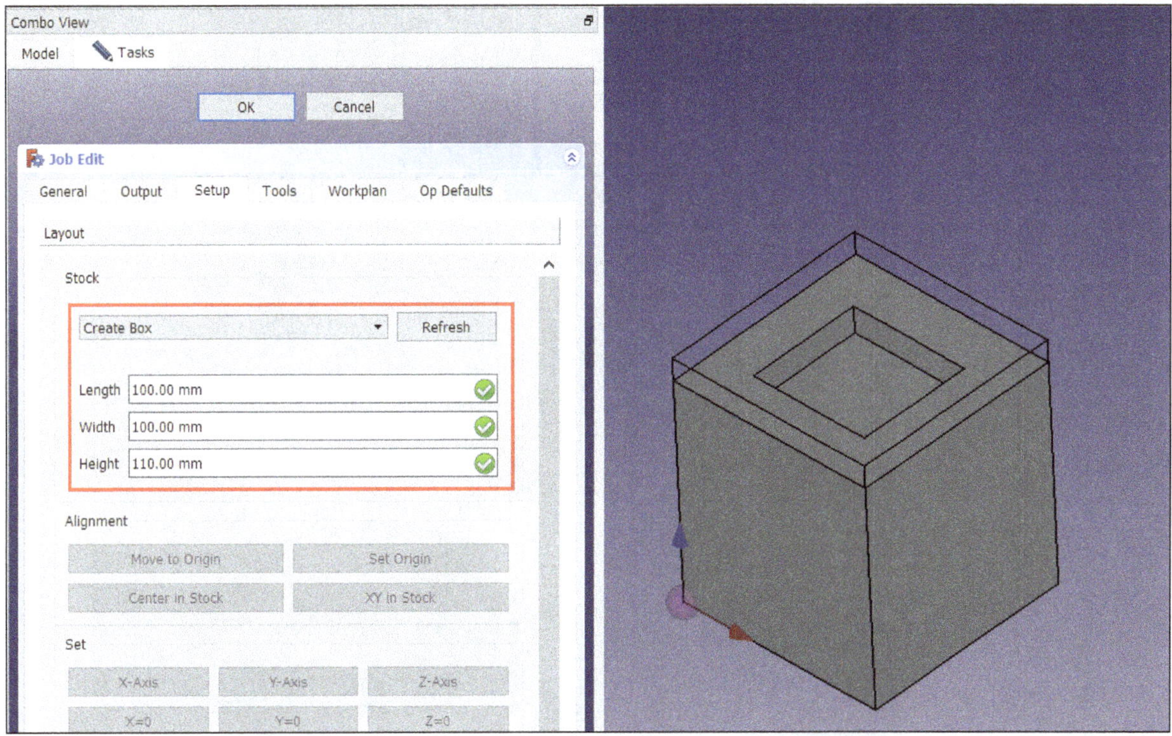

Figure-9. Create Box option for stock

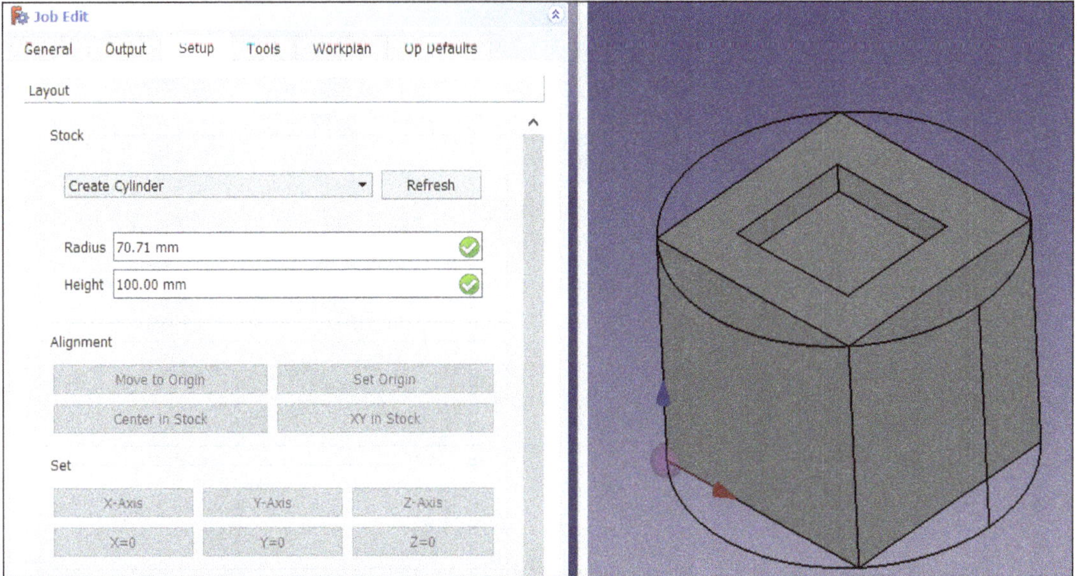

Figure-10. Create Cylinder option for stock

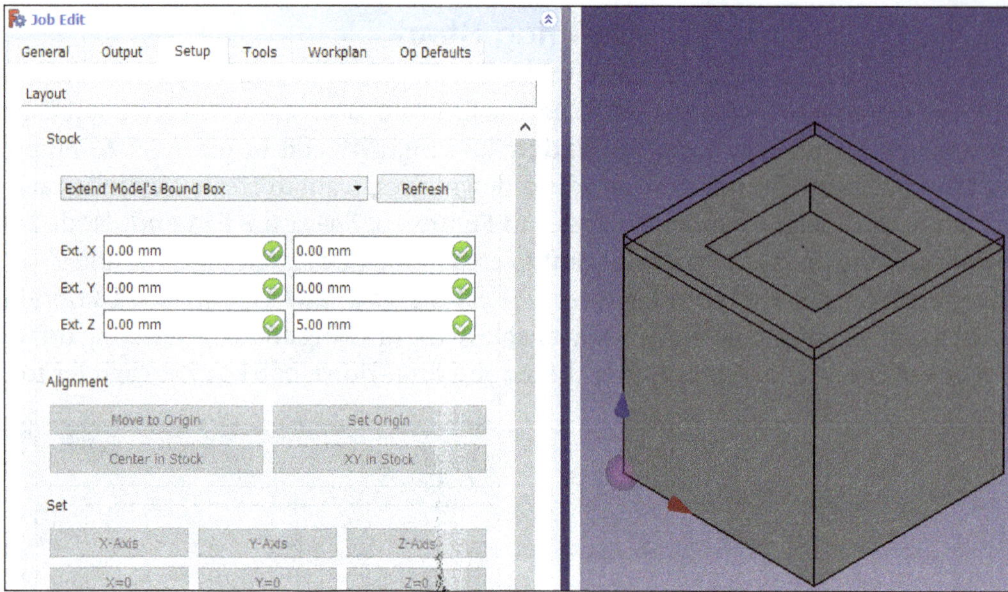

Figure-11. Extend model bound box option for stock

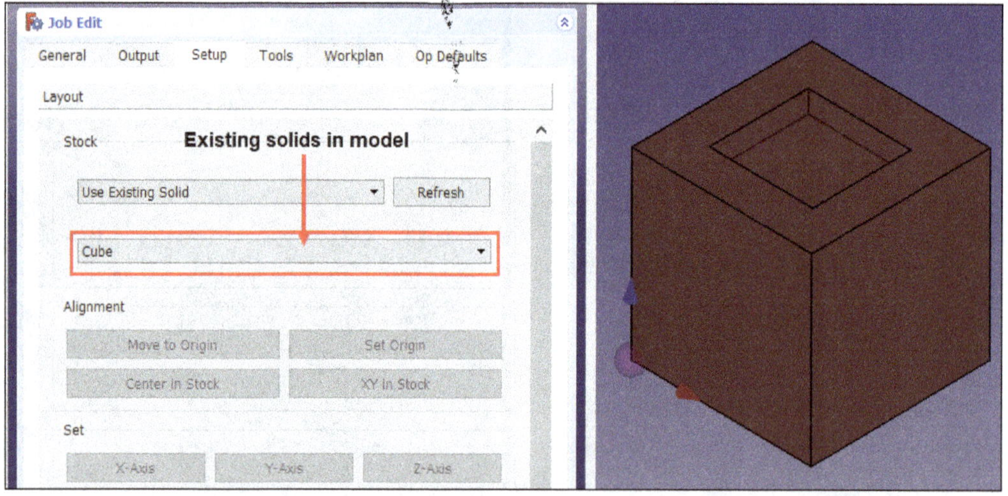

Figure-12. Using existing solid body

- Select a face of the model to specify alignment of model, the options in **Alignment**, **Set**, and other areas of dialog will become active; refer to Figure-13.

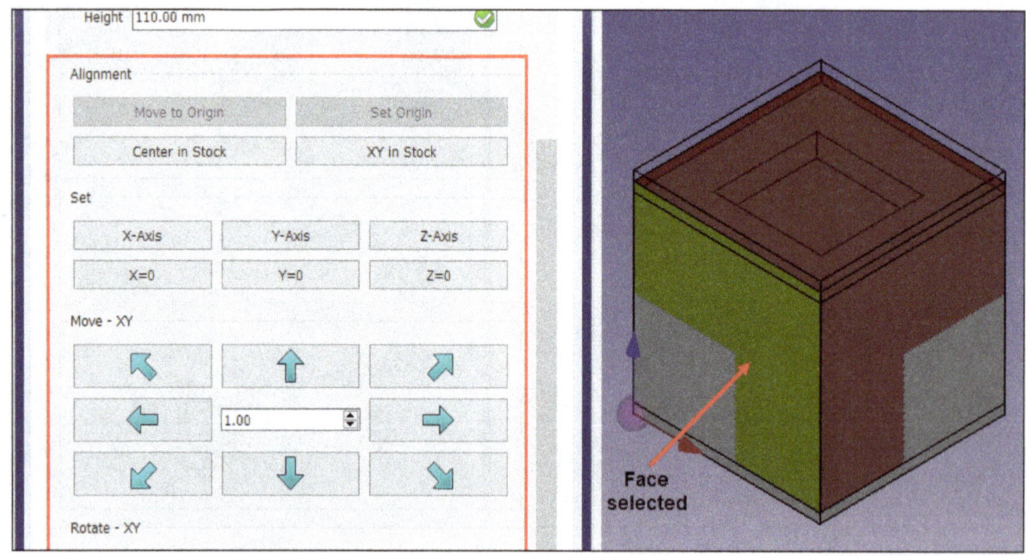

Figure-13. Alignment options for stock

- Select the **Center in Stock** button from the **Alignment** area to place the model in the center of stock. Select the **XY in Stock** button from the **Alignment** area to place the base of model within stock.
- Select desired button from the **Set** area to set X, Y, and Z as 0.
- Using the buttons in the **Move - XY** and **Rotate - XY** areas, you can move or rotate the model with respect to stock.

General Tab

- Click on the **General** tab from the dialog. The options in the **General** tab will be displayed; refer to Figure-14.

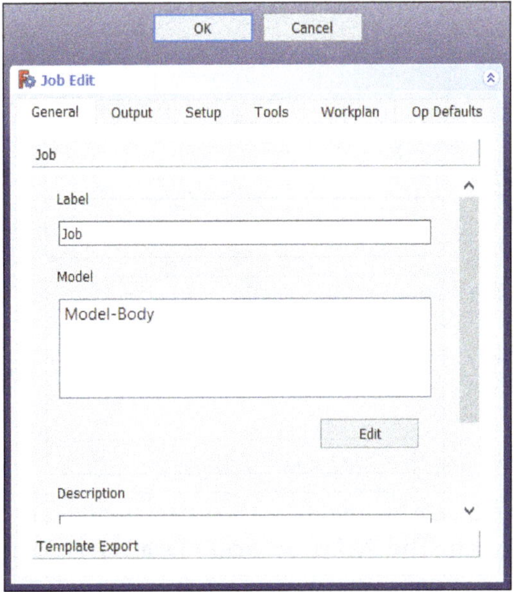

Figure-14. General tab of Job Edit dialog

- Specify desired name of job in the **Label** edit box from **Job** section of **General** tab.
- The **Model** edit box displays the base object which defines final shape of model for generating the paths of the job.
- If you want to add more models for creating a job then click on the **Edit** button below **Model** edit box. The **Model Selection** dialog box will be displayed; refer to Figure-15.

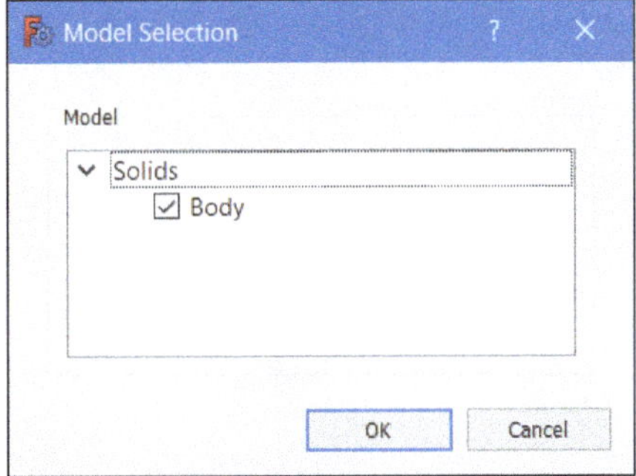

Figure-15. Model Selection dialog box

- Select desired model and click on **OK** button from the **Model Selection** dialog box. The model will be added in the **Job Edit** dialog.
- You can add some notes to the job in the **Description** edit box of the **General** tab of the dialog.
- Click on the **Template Export** button at the bottom in the dialog to export current specified parameters as a template. The options will be displayed as shown in Figure-16.

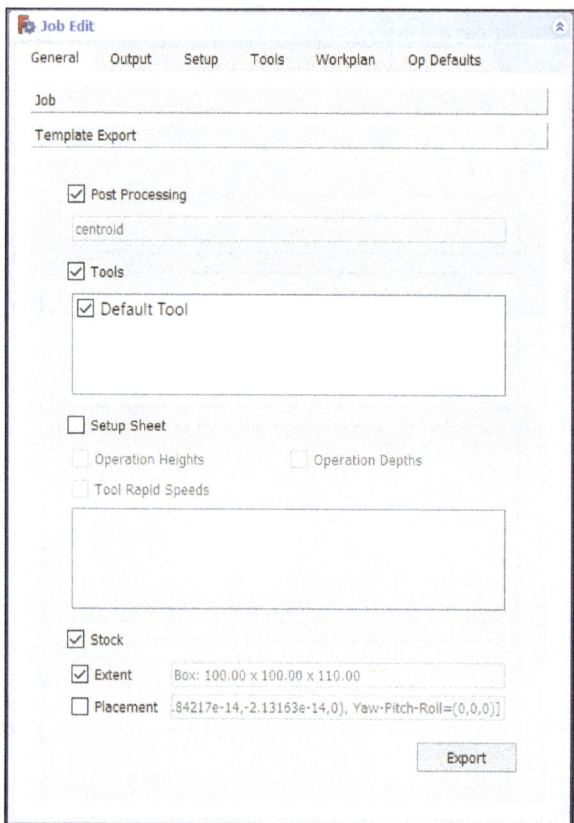

Figure-16. Template Export options

- Select check boxes from the dialog to specify which parameters are to be exported in the template and click on the **Export** button. The **Path - Job Template** dialog box will be displayed; refer to Figure-17.

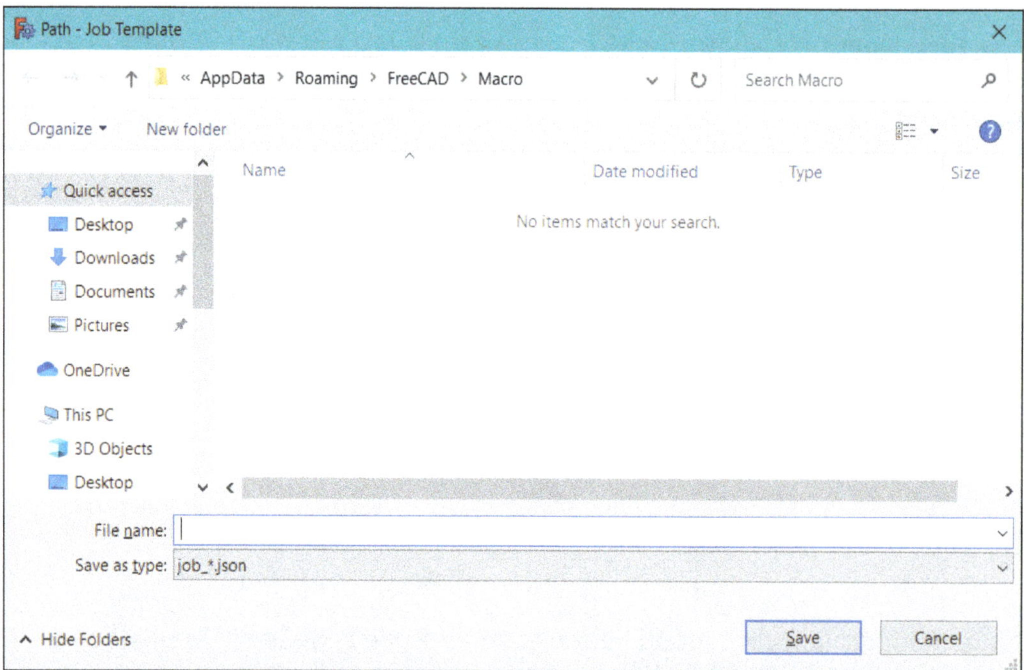

Figure-17. Path-Job Template dialog box

- Specify desired name for template in the **File name** edit box and save the template file at desired location.

Output Tab

- Click on the **Output** tab from **Job Edit** dialog. The options related to output tab will be displayed; refer to Figure-18.

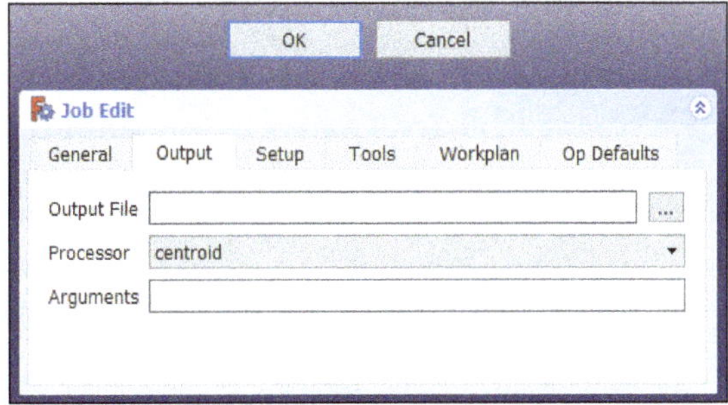

Figure-18. Output tab of Job Edit dialog

- Set desired name, extension, and the file path of the G-Code output in the **Output File** edit box of the **Output** tab.
- Select desired post processor for your machine from **Processor** drop-down.
- You can add desired arguments for the post processor as needed in the **Arguments** edit box of **Output** tab in the dialog.

Tools Tab

- Click on the **Tools** tab from **Job Edit** dialog to specify parameters related to cutting tools will be displayed in the dialog; refer to Figure-19.

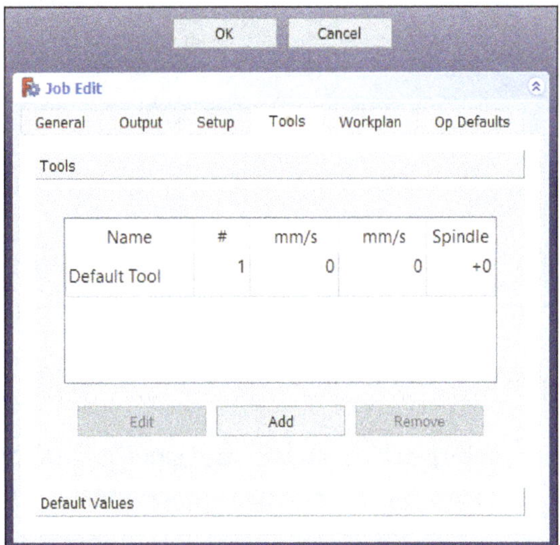

Figure-19. Tools tab of Job Edit dialog

- You can add the tool which you need for the operations of this job from **Tool Library** dialog box displayed by clicking on the **Add** button from **Tools** section of **Tools** tab; refer to Figure-20.

Figure-20. Tool Library dialog box

- The added tool will be displayed in the tool list box of **Tools** section in **Tools** tab of the **Job Edit** dialog.
- If you want to edit the tool then select the tool from the tool list and click on the **Edit** button from **Tools** section. The **Tool Controller Editor** dialog box will be displayed; refer to Figure-21.

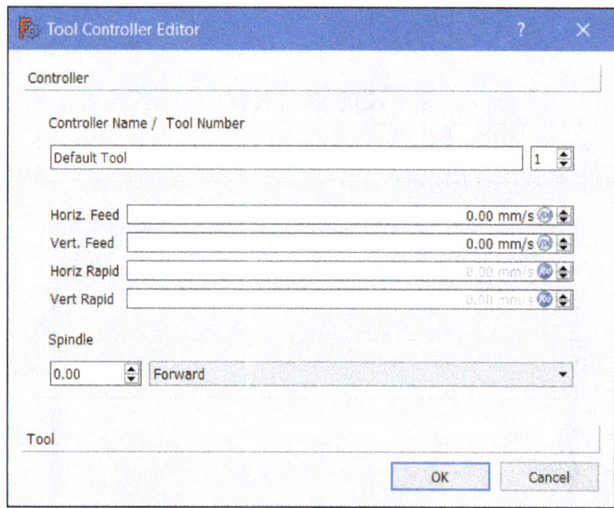

Figure-21. Tool Controller Editor dialog box

- After adding a tool, you can set/change the feed rate and spindle speed if you need a different feed rate in this job.
- Similarly, specify desired parameters in **Workplan** and **Op Defaults** tab of the **Job Edit** dialog.
- Click on **OK** button from **Job Edit** dialog. The Job will be created and displayed in the Model tree view; refer to Figure-22.

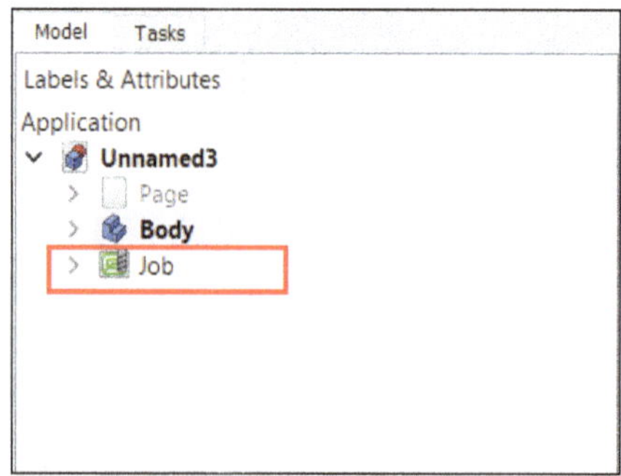

Figure-22. Job created

Tool Manager

The Tool Manager is used to create and manage cutting tools used in performing machining operations in the Path workbench. The procedure to use or manage cutting tools is discussed next.

- Click on the **Tool Manager** tool from **Toolbar** in the **Path** workbench; refer to Figure-23. The **Tool Library** dialog box will be displayed; refer to Figure-24. Note that the Tool Manager tool will be active in Toolbar only after you have created a job setup as discussed earlier.

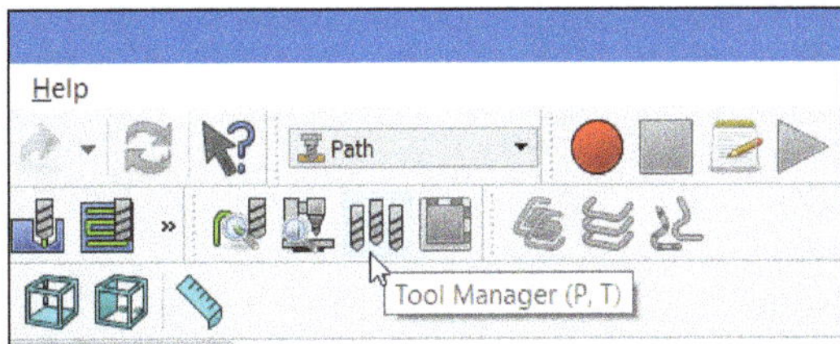

Figure-23. Tool Manager tool

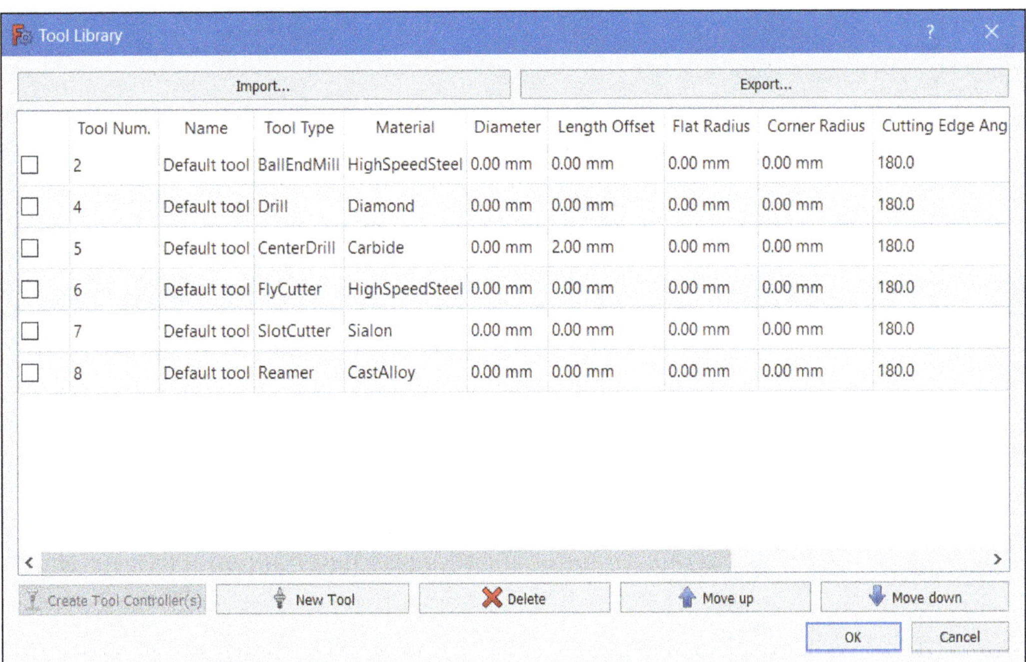

Figure-24. Tool Library dialog box

- Click on **Import** button from the dialog box to import a cutting tool table from an XML-file. The **Open tooltable** dialog box will be displayed; refer to Figure-25.

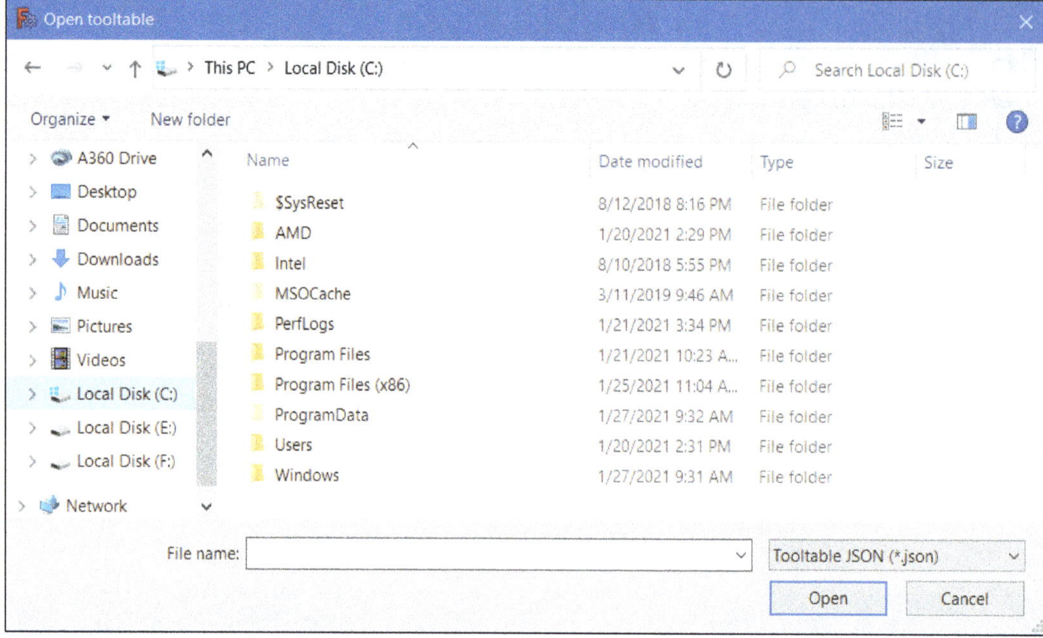

Figure-25. Open tooltable dialog box

- Select desired file to import and click on **Open** button from the dialog box. The cutting tool list of selected file will be displayed in the **Tool Library** dialog box.
- Click on **Export** button from the dialog box to export the tooltable to an XML-file. The **Save tooltable** dialog box will be displayed; refer to Figure-26.

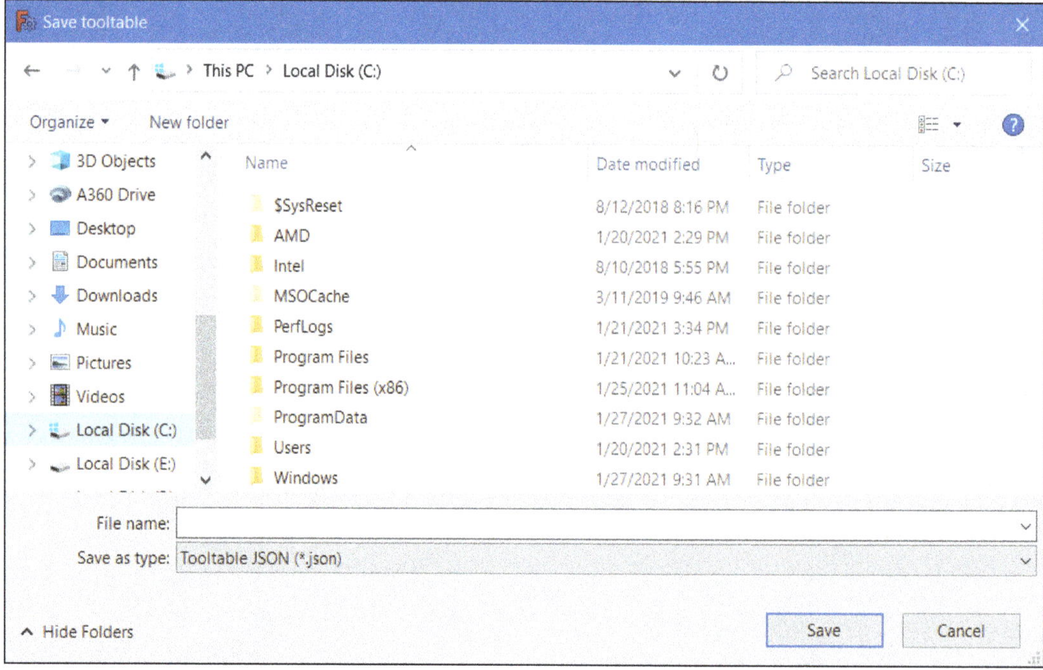

Figure-26. Save tooltable dialog box

- Specify desired name for file in the **File name** edit box and specify desired location to save the file.
- Click on **Save** button from the dialog box. The file will be saved.
- If you want to move a tool up or down in the tools list box then select the tool from the list box and click on **Move up** or **Move down** button, respectively.

- If you want to delete the tool from library then select the tool which you want to delete from the tools list box and click on **Delete** button.
- To create a cutting tool, click on **New Tool** button. The **Tool Editor** dialog box will be displayed; refer to Figure-27.

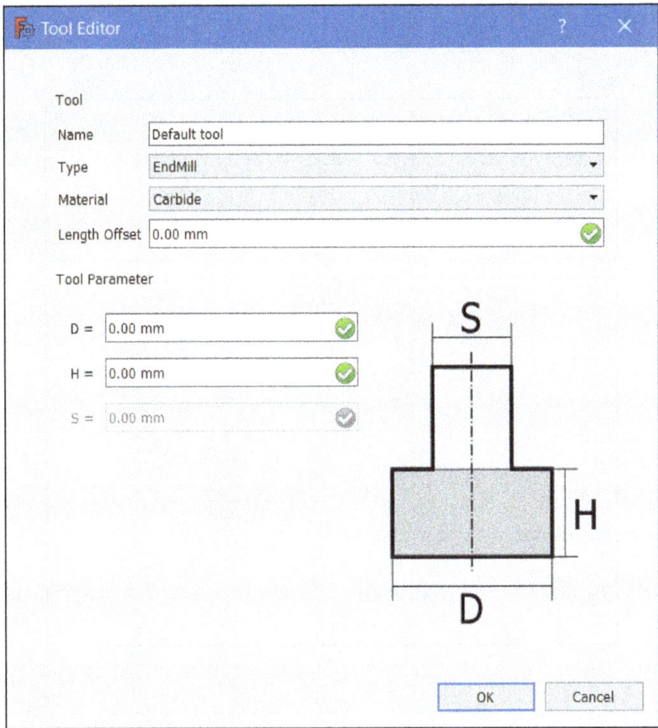

Figure-27. Tool Editor dialog box

- Specify desired parameters for the tool in the dialog box and click on **OK** button from the dialog box. The parameters specified for the tool will be applied and displayed in the library.
- To modify any cutting tool, double-click on the tool from the **Tool Library** dialog box. The **Tool Editor** dialog box will be displayed as discussed earlier.
- To create tool controller for the tool, select the check boxes for the tools in the **Tool Library** dialog box and click on **Create Tool Controller(s)** button from the dialog box.
- Click on **OK** button from the **Tool Library** dialog box.

Exporting Template

Exporting Job templates provide a convenient mechanism to save commonly used job definitions from within an existing Job. This facilitates the setup of future jobs that are largely similar by allowing job template import during the job creation process. The procedure to use this tool is discussed next.

- Select the Job created from the Model tree view and press **RMB** (Right Mouse Button). A shortcut menu will be displayed; refer to Figure-28.
- Click on **Export Template** tool from the shortcut menu. The **Job Template Export** dialog box will be displayed; refer to Figure-29.

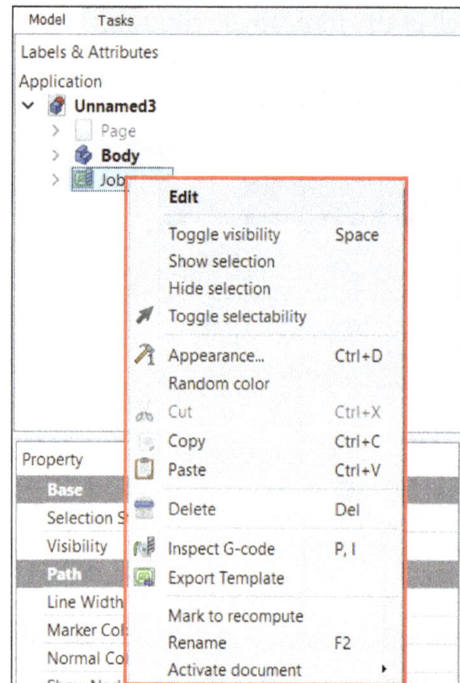

Figure-28. Shortcut menu displayed on selecting job

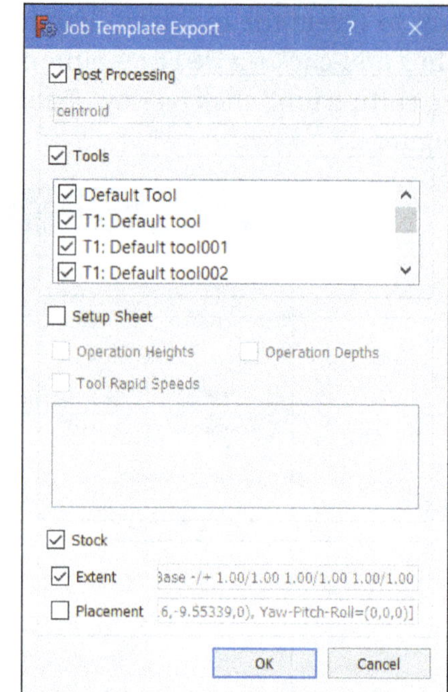

Figure-29. Job Template Export dialog box

- Select **Post Processing** check box from the dialog box to include all post processing settings in the template.
- Select the **Tools** check box to store tool controller definitions in the template.
- Select the check boxes of all the tool controllers which should be included in the template from **Tools** area of the dialog box.
- Select **Setup Sheet** check box to include values of the Setup Sheet in the template.
- Select **Operation Heights** check box from **Setup Sheet** area to include the default heights for operations in the template.
- Select **Operation Depths** check box from **Setup Sheet** area to include values of the setup sheet in the template.
- Select **Tool Rapid Speeds** check box to include the default rapid tool speeds in the template.
- Select **Stock** check box to include the creation of stock in the template. If a template does not include a stock definition, the default stock creation algorithm will be used.
- Select **Extent** check box to include the current size settings for the stock object in the template.
- Select **Placement** check box to store the current placement of the stock solid in the template.
- Click on **OK** button from the dialog box. The **Path-Job Template** dialog box will be displayed to save the template; refer to Figure-30.

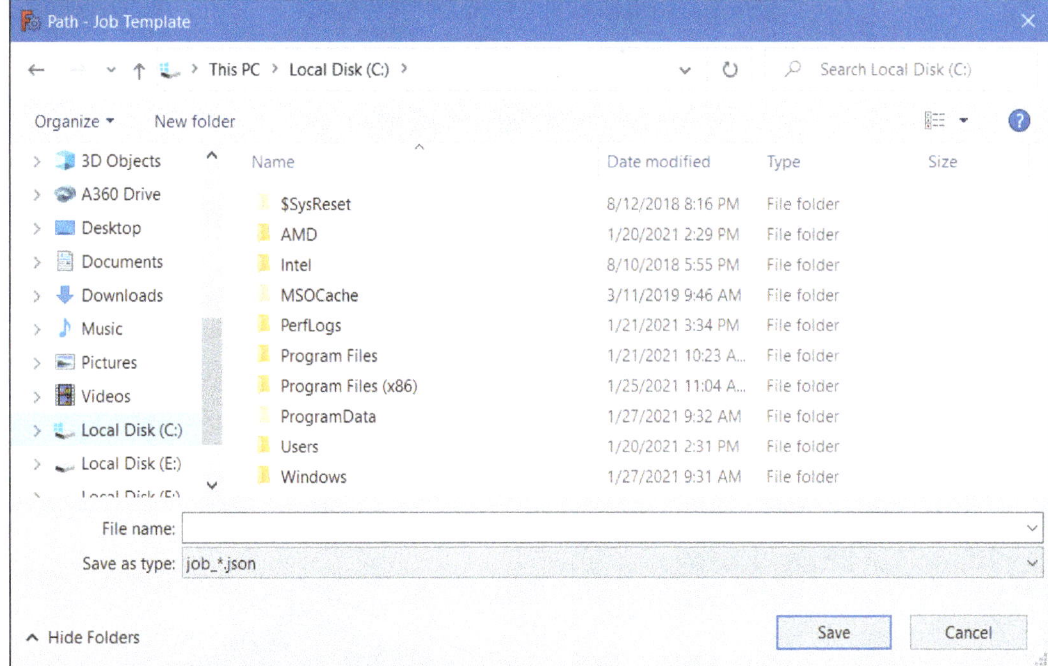

Figure-30. Path Job Template dialog box

- Specify name for the template file in the **File name** edit box and specify desired location to save the file.
- Click on **Save** button from the dialog box. The template will be saved.

BASIC PATH OPERATIONS

Basic path operations are Contour, Profile Faces, and Profile Edges operations used to remove material from the stock. These tools are available in the **Toolbar** of **Path** workbench; refer to Figure-31. These tools are discussed next.

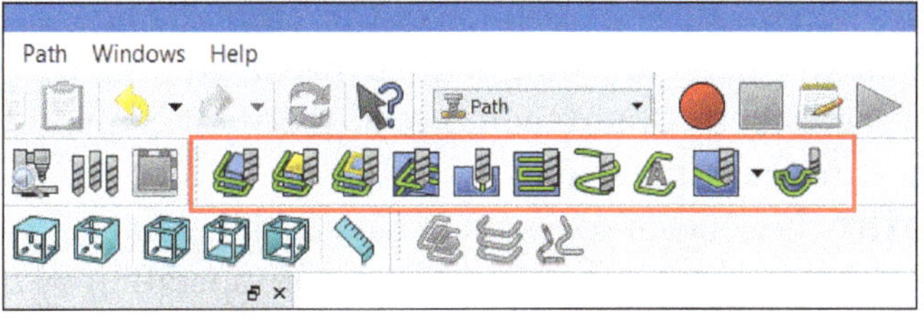

Figure-31. Basic Path Operations tool

Creating Contour Operation

The **Contour** tool creates a contour operation based on selected features of the model. It creates a simple external contour cut for 3D parts. The procedure to use this tool is discussed next.

- Click on the **Contour** tool from **Toolbar** in the **Path** workbench; refer to Figure-32. The **Choose a Tool Controller** dialog box will be displayed if you have not specified tool controller earlier for this operation while setting up the job; refer to Figure-33.
- Select desired tool from the drop-down and click on the **OK** button. The **Contour** dialog will be displayed in the **Tasks** panel of **Combo View** and the model will be enclosed within the bounding box; refer to Figure-34.

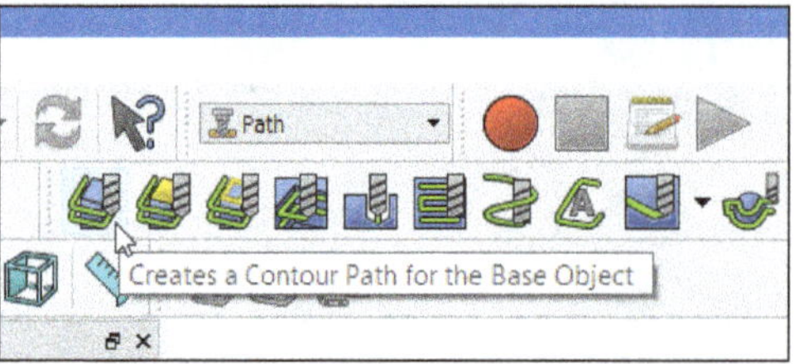

Figure-32. Profile tool

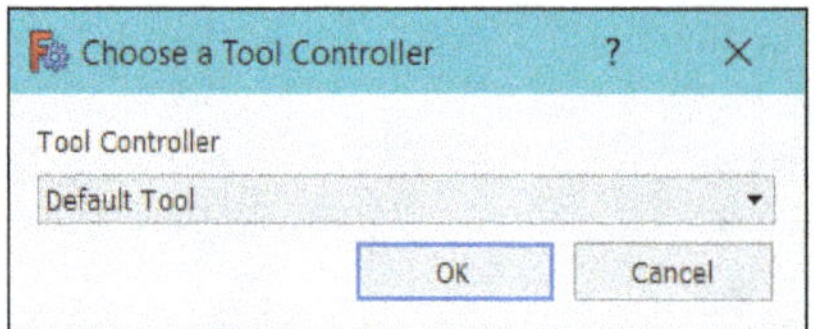

Figure-33. Choose a Tool Controller dialog box

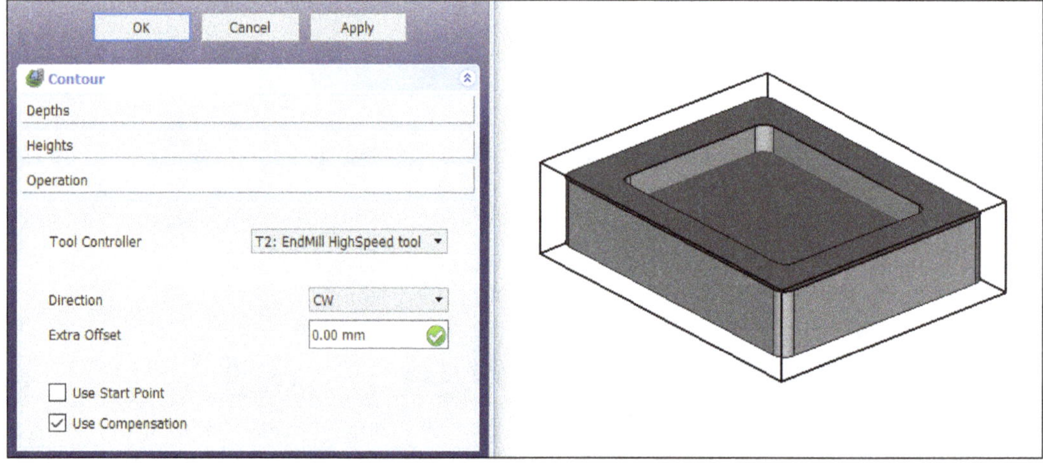

Figure-34. Contour dialog and the model enclosed within bounding box

- The tool controller specified while creating the Job for this operation will be displayed in the **Tool Controller** drop-down in the **Operation** tab of **Contour** dialog.
- Specify desired direction in which the profile is to be machined from **Direction** drop-down.
- Enter desired value in the **Extra Offset** edit box to specify the amount of extra material left by this operation in relation to the target shape.
- Select **Use Start Point** check box if you want to use starting point as exit point for this operation.
- Select **Use Compensation** check box to offset the profile operation by the tool radius. The offset direction is determined by the cut side.
- Enter desired values in **Start Depth** and **Final Depth** edit boxes in **Depths** tab of the dialog to specify start and final depth of tool for the operation, respectively.
- Enter desired value in **Step Down** edit box from **Depths** tab of the dialog to specify incremental step down of tool for each cutting pass of operation.
- Enter desired value in **Safe Height** edit box from **Height** tab of the dialog to specify the height above which Rapid motions are allowed.
- Enter desired value in **Clearance Height** edit box from **Height** tab of the dialog to specify the height needed to clear clamps and obstructions.
- Click on **Apply** button and then **OK** button from the dialog. The Contour operation path will be generated; refer to Figure-35.

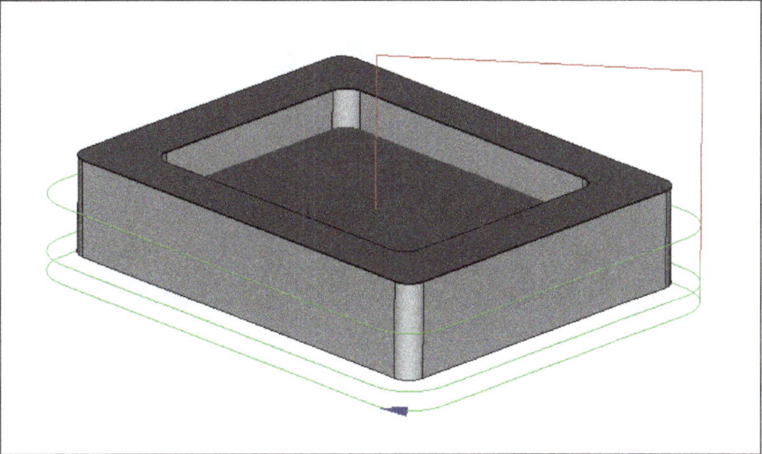

Figure-35. Contour operation path generated

Creating Profile Face Operation

The **Profile Face** tool creates a simple contour path for one or more selected faces of the model. The procedure to use this tool is discussed next.

- Click on the **Profile Face** tool from **Toolbar** in the **Path** workbench; refer to Figure-36. The **Profile Faces** dialog will be displayed in the **Tasks** panel of **Combo View** and the model will be enclosed within the bounding box; refer to Figure-37.

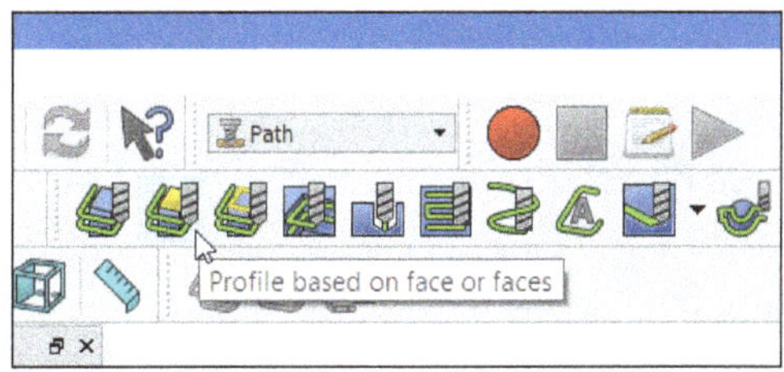

Figure-36. Profile Face tool

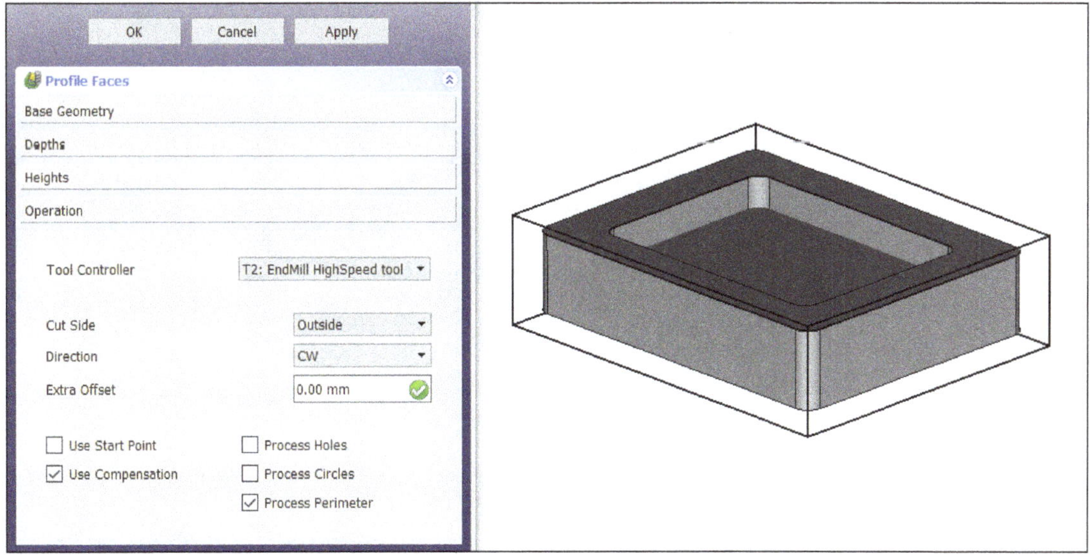

Figure-37. Profile Faces dialog and the model enclosed within bounding box

- Select one or more faces of the model in the 3D view area and click on **Add** button from **Base Geometry** tab of the dialog to add them to the list as the base geometries for this operation; refer to Figure-38.

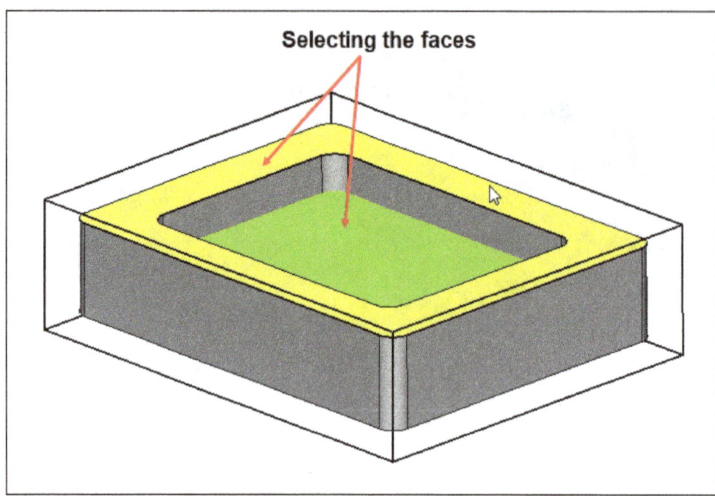

Figure-38. Selecting the faces for operation

- If you want to remove the geometries from the list then select the geometries from the list which you want to remove and click on **Remove** button.
- Click on the **Clear** button from the **Base Geometry** tab to clear all the geometries from the list.
- Select desired option from **Cut Side** drop-down in the **Operation** tab of the dialog to specify if the profile will be performed outside or inside the base geometry features.
- Select the **Process Holes** check box from **Operation** tab of the dialog if this profile operation will process holes in the base geometry.
- Select the **Process Circles** check box if you want this profile operation to be applied to cylindrical holes which normally get drilled.
- Select the **Process Perimeter** check box if this profile operation will process the outside perimeter of the base geometry shapes.
- Other parameters in this dialog have been discussed earlier.
- Click on **Apply** button and then **OK** button from the dialog. The Profile Face operation path will be generated; refer to Figure-39.

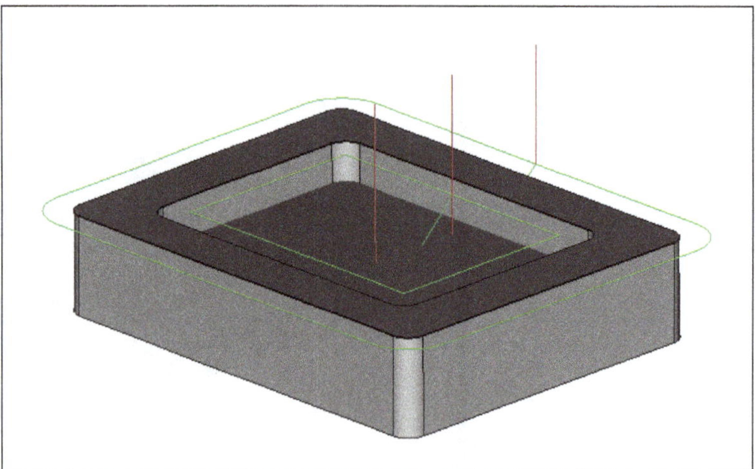

Figure-39. Profile face operation path generated

Creating Profile Edges Operation

The **Profile Edge** tool creates a simple contour path from selected edges. The procedure to use this tool is same as discussed for **Profile Face** tool. In case of Profile Edges operation, edges of the model will be selected in place of faces of the model as a base geometries to create the operation.

Creating Pocket Object

The **Pocket Shape** tool creates pocket machine operation for selected bottom faces or walls of one or more pockets in the model. The procedure to use this tool is discussed next.

- Click on the **Pocket Shape** tool from **Toolbar** in the **Path** workbench; refer to Figure-40. The **Pocket Shape** dialog will be displayed in the **Tasks** panel of **Combo View** and the model will be enclosed within the bounding box; refer to Figure-41.

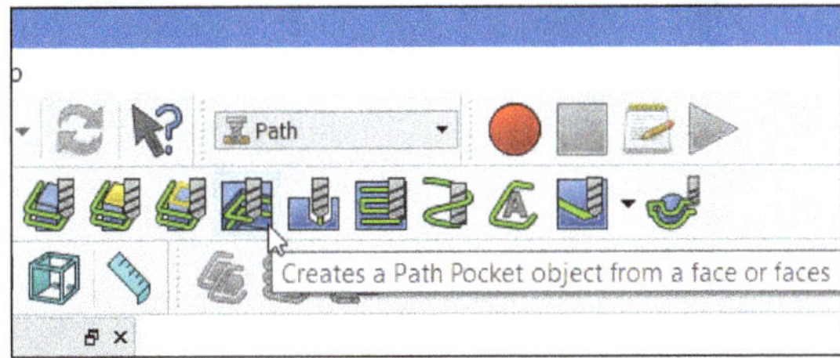

Figure-40. Pocket Shape tool

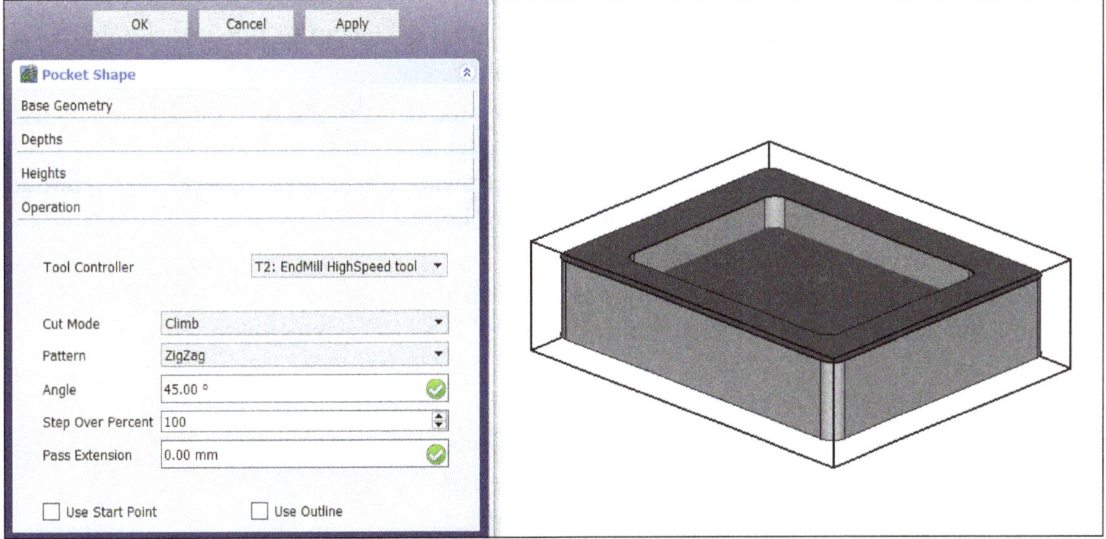

Figure-41. Pocket Shape dialog and the model enclosed within bounding box

- Select the bottom face or the walls of an object in the 3D view area for which you want to create pocket machining operation and click on **Add** button from **Base Geometry** tab of the dialog to add them to the list as the base geometries for this operation; refer to Figure-42.

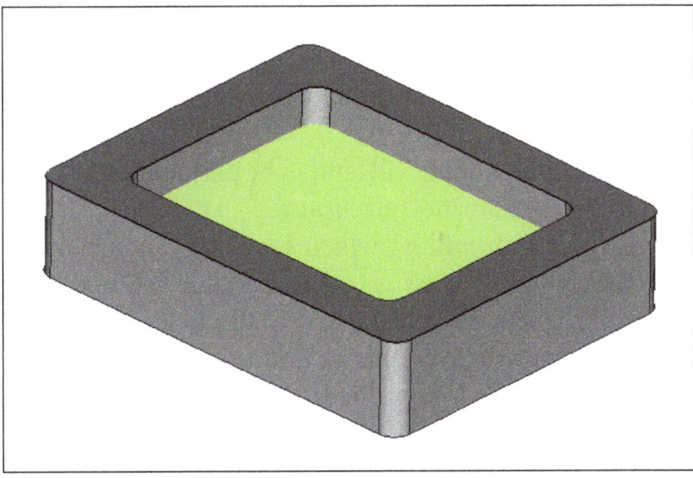

Figure-42. Selecting bottom face of the model

- Specify desired depth of the final cut of the operation in the **Finish Depth** edit box from **Depths** tab of the dialog. Specify the value 0 to perform finishing operation.
- Select desired cutting mode from **Cut Mode** drop down in the **Operation** tab of the dialog. The difference between conventional and climb cutting modes is shown in Figure-43.

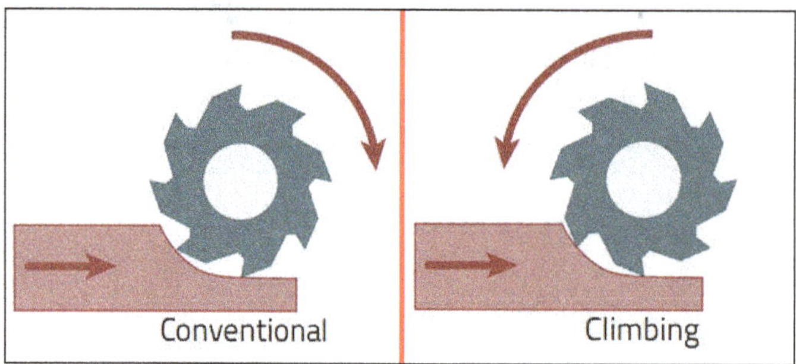

Figure-43. Conventional and climb cutting modes

Note: Conventional Milling V/S Climb Milling

Characteristics of Conventional Milling:

- The width of the chip starts from zero and increases as the cutter finishes slicing.
- The tooth meets the workpiece at the bottom of the cut.
- Upward forces are created that tend to lift the workpiece during face milling.
- More power is required to conventional mill than climb mill.
- Surface finish is worse because chips are carried upward by teeth and dropped in front of cutter. There's a lot of chip recutting. Flood cooling can help!
- Tools wear faster than with climb milling.
- Conventional milling is preferred for rough surfaces.
- Tool deflection during Conventional milling will tend to be parallel to the cut.

Characteristics of climb milling:

- The width of the chip starts at maximum and decreases.
- The tooth meets the workpiece at the top of the cut.
- Chips are dropped behind the cutter--less recutting.
- Less wear with tools lasting up to 50% longer.
- Improved surface finish because of less recutting.
- Less power required.
- Climb milling exerts a down force during face milling which makes work-holding and fixtures simpler. The down force may also help to reduce chatter in thin floors because it helps brace them against the surface beneath.
- Climb milling reduces work hardening.
- It can, however, cause chipping when milling hot rolled materials due to the hardened layer on the surface.
- Tool deflection during Climb milling will tend to be perpendicular to the cut, so it may increase or decrease the width of cut and affect accuracy.
- There is a problem with climb milling which is that it can get into trouble with backlash if cutter forces are great enough. The issue is that the table will tend to be pulled into the cutter when climb milling. If there is any backlash, this allows leeway for the pulling in the amount of the backlash. If there is enough backlash, and the cutter is operating at capacity, this can lead to breakage and potentially injury due to flying shrapnel.

Some worthwhile rules of thumb:

- When cutting half the cutter diameter or less, you should definitely climb mill (assuming your machine has low or no backlash and it is safe to do so!).

- Up to 3/4 of the cutter diameter, it doesn't matter which way you cut.

- When cutting from 3/4 to 1x the cutter diameter, you should prefer conventional milling.

- Select desired option from the **Pattern** drop-down in the **Operation** tab to specify the pattern in which the tool bit will move to clear the material. It is always better to select the pattern which resembles with walls of pocket.
- Enter desired value in the **Angle** edit box from **Operation** tab to specify the angle in which the pattern is applied.
- Enter desired value in **Step Over Percent** edit box to specify the amount by which the cutting tool will move laterally on each cutting pass in the pattern (specified in percent of the tool diameter).
- Specify desired distance value in the **Pass Extension** edit box from **Operation** tab by which, the facing operation will be extended beyond the boundary shape.
- Select **Use Outline** check box from the **Operation** tab if the operation uses the outline of the selected base geometry and ignores all holes and islands.
- Other parameters in the **Pocket Shape** dialog have been discussed earlier.
- Click on **Apply** button and then **OK** button from the dialog. The Pocket shape operation path will be generated; refer to Figure-44.

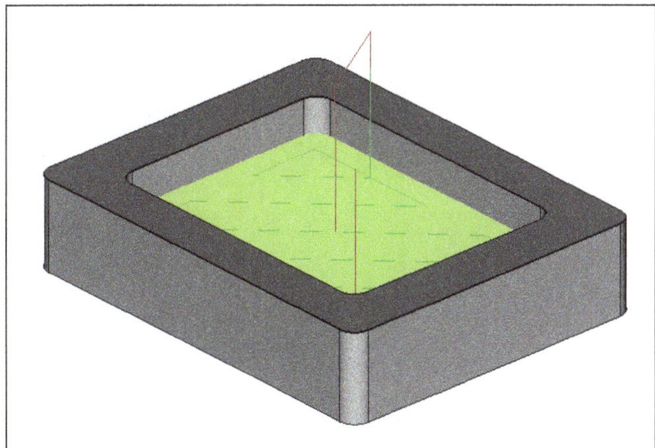

Figure-44. Pocket shape operation path generated

Performing Drill Operation

The **Drilling** tool generates a drilling operation in the job. The procedure to use this tool is discussed next.

- Click on the **Drilling** tool from **Toolbar** in the **Path** workbench; refer to Figure-45. The **Choose a Tool Controller** dialog box will be displayed if no controller has been selected for this operation during job setup; refer to Figure-46.

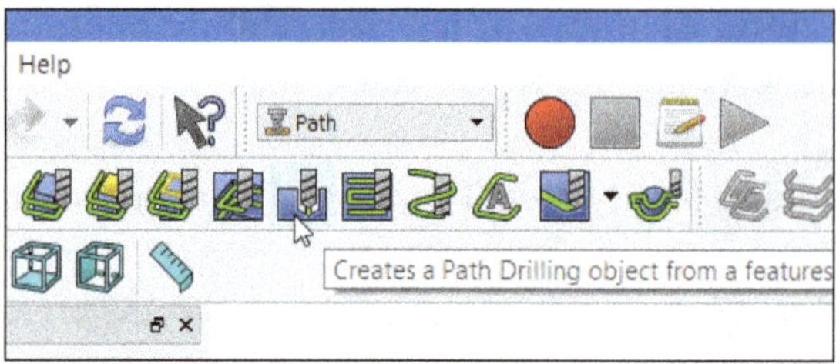

Figure-45. Drilling tool

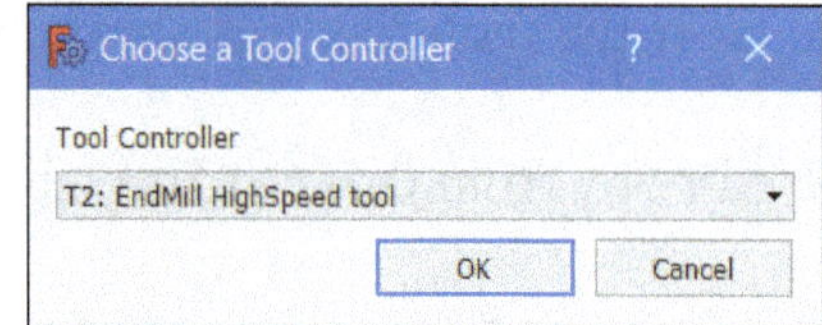

Figure-46. Choose a Tool Controller dialog box

- Select desired drilling tool from **Tool Controller** drop-down and click on **OK** button from the selection window. The **Drilling** dialog will be displayed in the **Tasks** panel of **Combo View** and the model will be enclosed within the bounding box; refer to Figure-47.

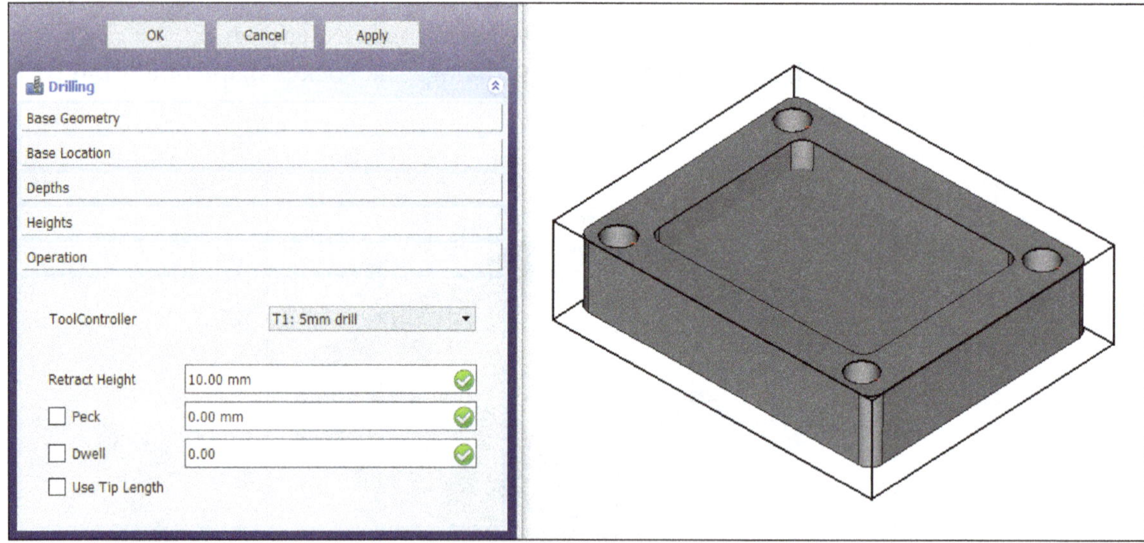

Figure-47. Drilling dialog and the model enclosed within bounding box

- Select the edges or/and faces of the hole feature of the model in the 3D view area on which you want to create the drilling operation; refer to Figure-48.

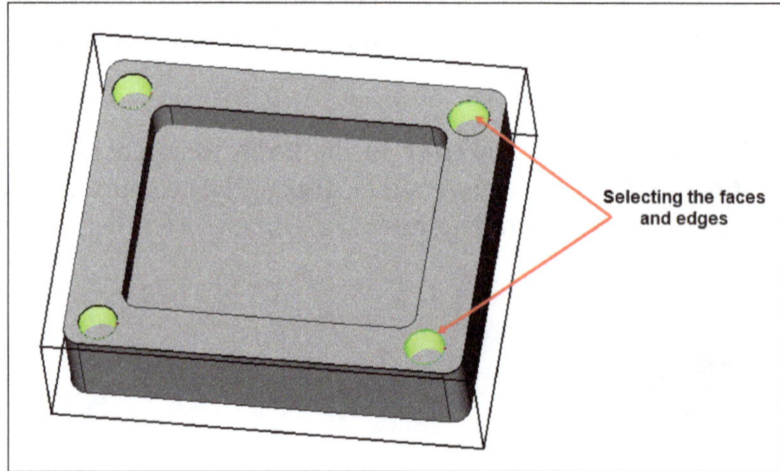

Figure-48. Selecting the faces and edges of hole feature

- After selecting the elements of the hole feature, click on **Add** button from the **Base Geometry** tab of dialog to add them in the selection list box.
- If you want to remove the element then select the element from the list box and click on **Remove** button from the **Base Geometry** tab.
- Click on the **Reset** button from the **Base Geometry** tab to delete all the current elements from the list and fills the list with all circular holes eligible for the drilling operation.
- Click on **Add** button from **Base Location** tab of the dialog. The direction edit boxes will be displayed in the tab and you will be asked to specify the locations.
- Click in the 3D view area at desired locations to specify the locations where you want to drill on the model or enter desired values in **Global X**, **Global Y**, and **Global Z** edit boxes to specify the drilling locations on the model.
- After specifying values in the edit boxes, click on **Save** button to save the drilling location in the list box or click on **Close** button to close the direction edit boxes.
- To remove the specified location, select the location from the list and click on **Remove** button from **Base Location** tab.
- Click on **Edit** button from **Base Location** tab to edit the specified location.
- Enter desired value in **Retract Height** edit box from **Operation** tab of the dialog to specify the incremental height where feed starts above the model surface.
- Select **Peck** check box and enter desired value in the edit box from **Operation** tab to specify incremental drill depth before retracting to clear chips in Peck cycle of drilling.
- Select **Dwell** check box and enter desired value in the edit box to specify the time of pause between peck cycles.
- Other parameters in the **Drilling** dialog have been discussed earlier.
- Click on **Apply** button and then **OK** button from the dialog. The drilling operation path will be generated; refer to Figure-49.

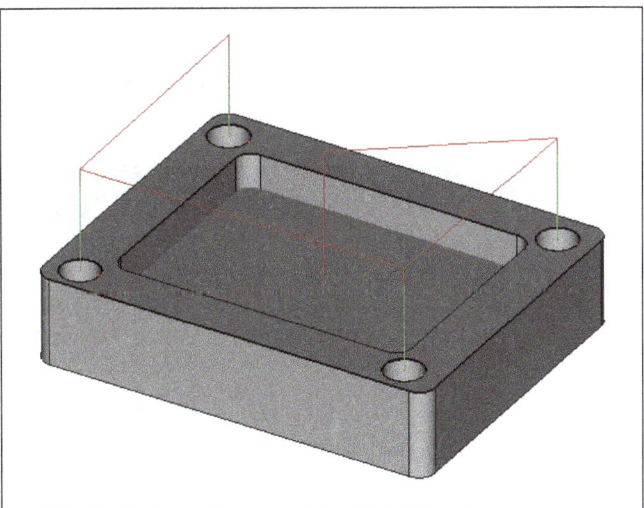

Figure-49. Drilling operation path generated

Creating Face Milling

The **Face Mill** tool creates toolpath to remove material from top flat face of the model. The procedure to use this tool is discussed next.

- Click on the **Face Mill** tool from **Toolbar** in the **Path** workbench; refer to Figure-50. The **MillFace** dialog will be displayed in the **Tasks** panel of **Combo View** and the model will be enclosed within the bounding box; refer to Figure-51.

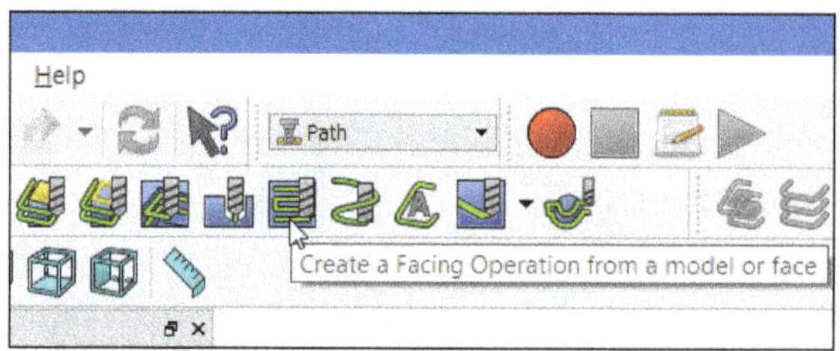

Figure-50. Face Mill tool

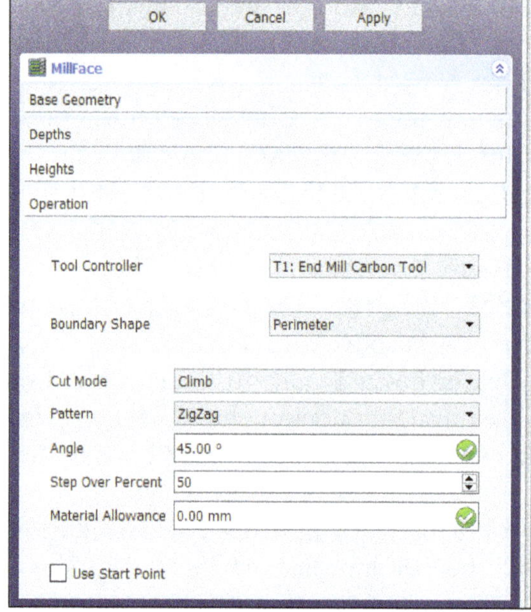

Figure-51. MillFace dialog

- Select one or more features of the model in the 3D view area and click on **Add** button from **Base Geometry** tab of the dialog to add them to the list as the base geometries for this operation; refer to Figure-52.

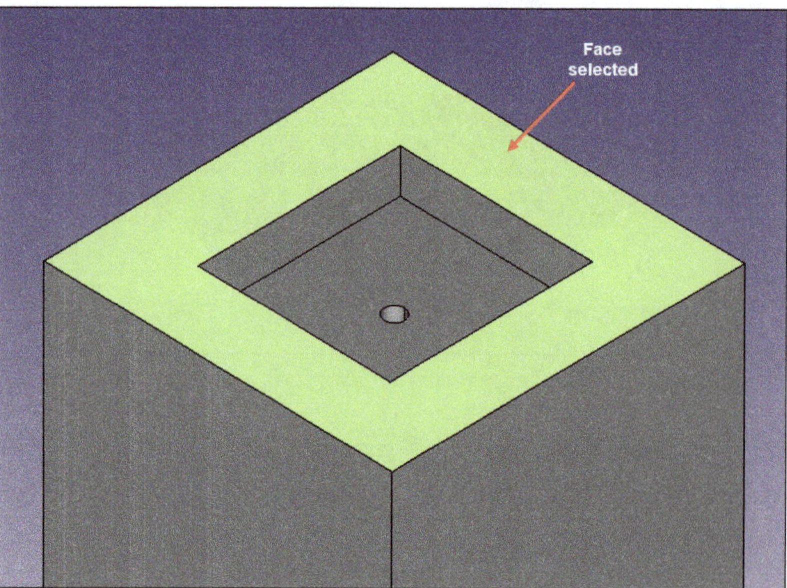

Figure-52. Selecting face of the model

- Enter desired value in **Material Allowance** edit box from **Operation** tab of the dialog to specify the amount of material that should be left by this operation in relation to the target shape.
- Other parameters in the **MillFace** dialog have been discussed earlier.

- Click on **Apply** button and then **OK** button from the dialog. The Face Mill operation path will be generated; refer to Figure-53.

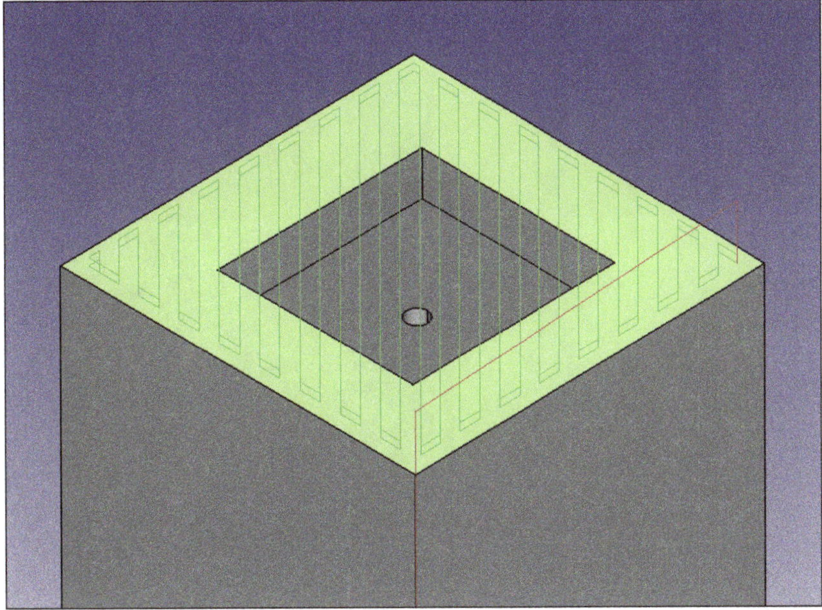

Figure-53. Face Mill operation path generated

Creating Helical Operation

The **Helix** tool appends a helical interpolation Operation to the Job. Clockwise Helix interpolation outputs (G2) G-Code commands. Counterclockwise outputs (G3) G-code commands. This operation is useful to machine circular vertical surfaces of the model. The procedure to use this tool is discussed next.

- Click on the **Helix** tool from **Toolbar** in the **Path** workbench; refer to Figure-54. The **Helix** dialog will be displayed in the **Tasks** panel of **Combo View** and the model will be enclosed within the boundary box; refer to Figure-55.

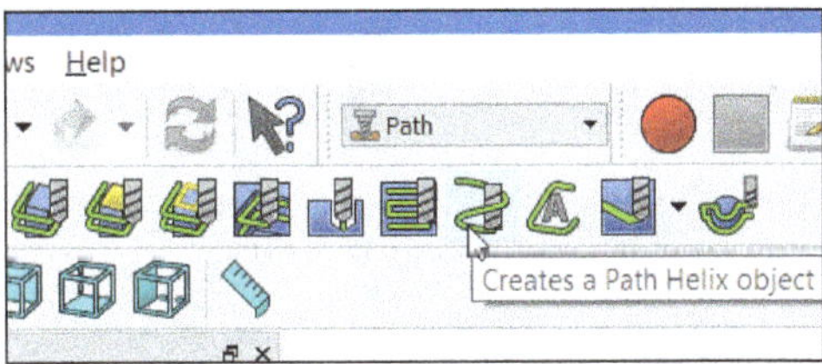

Figure-54. Helix tool

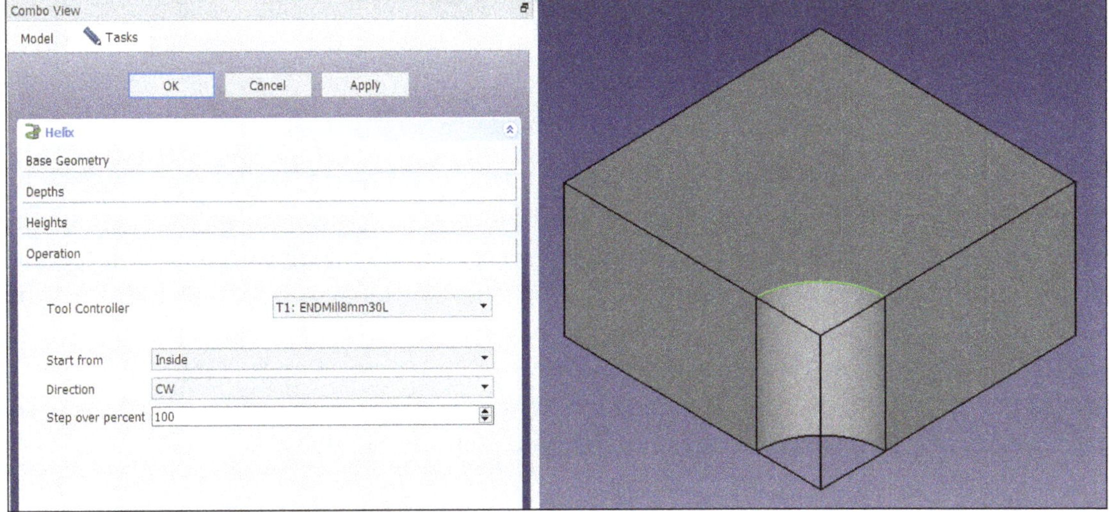

Figure-55. Helix dialog and the model enclosed within boundary box

- Select the edges of features to be machined by the helical operation and click on **Add** button from **Base Geometry** tab of the dialog to add them in the list as the base geometries for this operation.
- Specify appropriate values in the **Depths** section of dialog box; refer to Figure-56. Note that you can activate the files by clicking on the arrow buttons next to edit boxes.

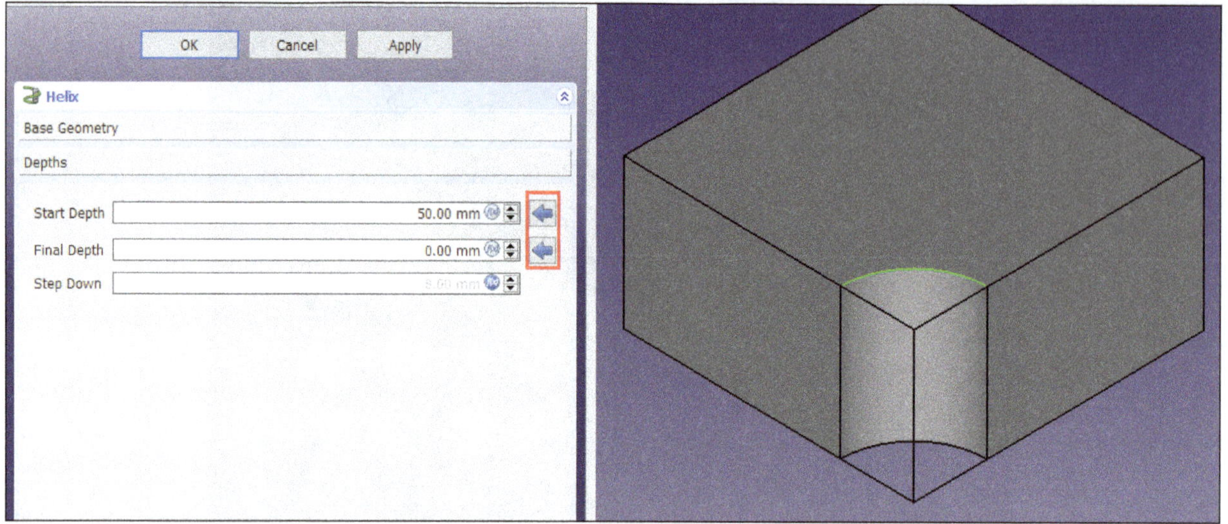

Figure-56. Specifying depths for helical toolpath

- Specify all the other parameters in the **Helix** dialog as discussed earlier.
- Click on **Apply** button and then **OK** button from the dialog. The helical operation path will be generated; refer to Figure-57.

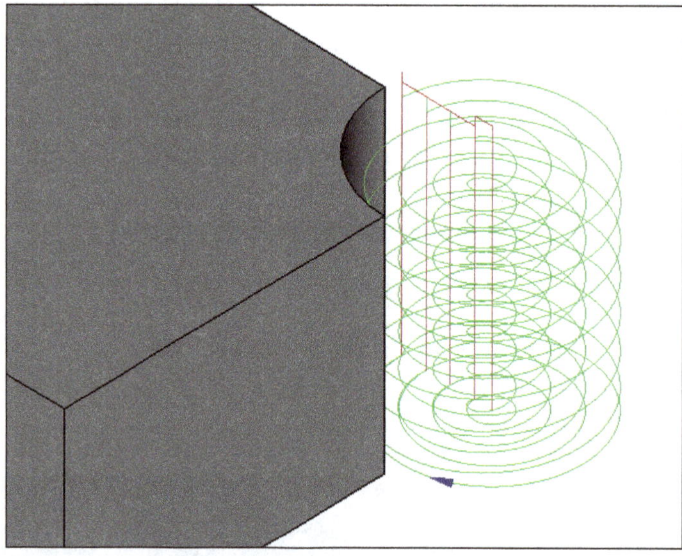

Figure-57. Helical operation path generated

Creating Adaptive Operation

The **Adaptive** tool uses an adaptive algorithm to create clearing and profiling paths that manage cutter engagement so that engagement and material removal never exceed a maximum value. This toolpath uses all the possible cutting strategies available FreeCAD for removing large stock of material. The procedure to use this tool is discussed next.

- Click on the **Adaptive** tool from **Toolbar** in the **Path** workbench; refer to Figure-58. The **Adaptive path operation** dialog will be displayed in the **Tasks** panel of **Combo View** and the model will be enclosed within the bounding box; refer to Figure-59.

Figure-58. Adaptive tool

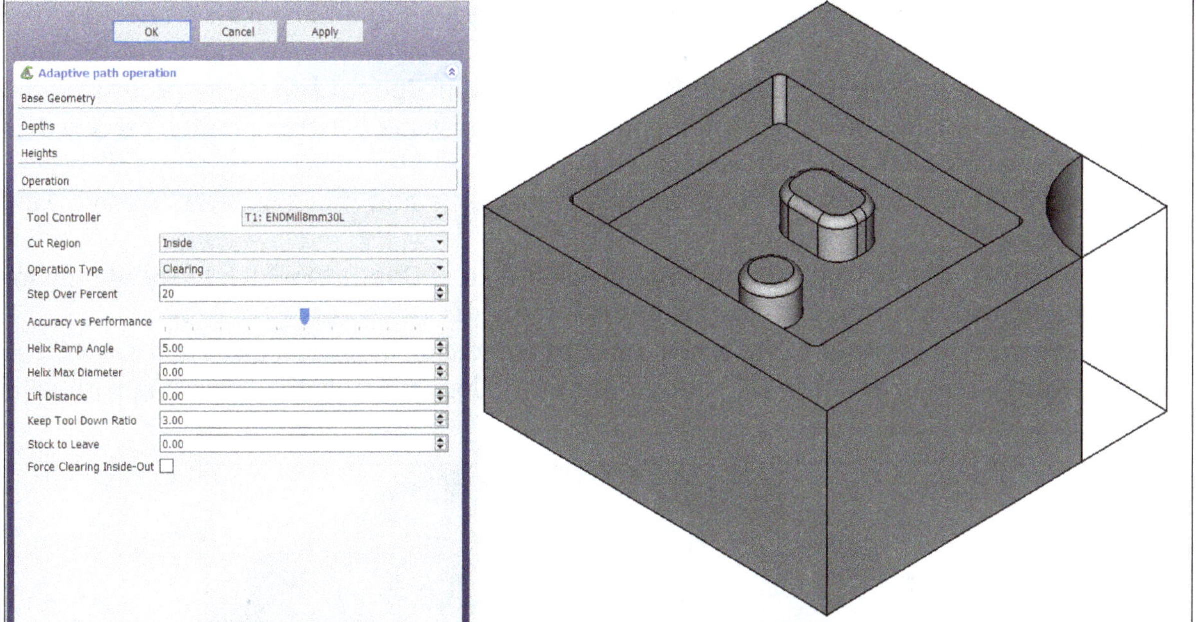

Figure-59. Adaptive path operation dialog and the model enclosed within bounding box

- Select one or more faces of the model in the 3D view area to be machined (generally base face) and click on **Add** button from **Base Geometry** tab of the dialog to add them to the list as the base geometries for this operation; refer to Figure-60.

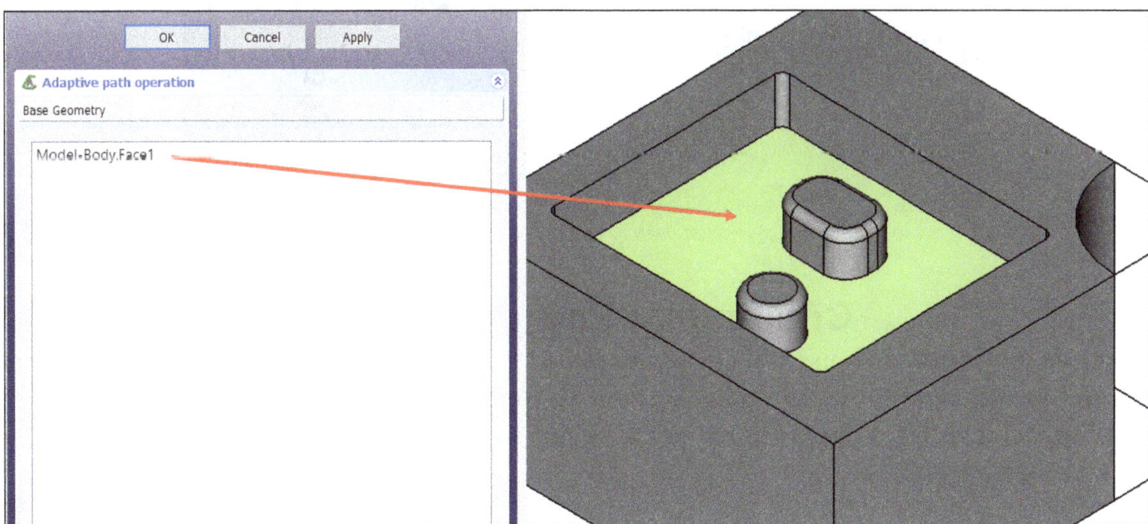

Figure-60. Selecting the face for adaptive operation

- Select desired option from **Cut Region** drop-down in the **Operation** tab to specify the cutting of model inside or outside of the selected shape.
- Select desired type of adaptive operation from **Operation Type** drop-down in the **Operation** tab of the dialog.

- Specify the percent of cutter diameter to step over on each pass in the **Step Over Percent** edit box of **Operation** tab.
- Slide the **Accuracy vs Performance** slider from **Operation** tab at desired position to influence the calculation performance vs stability and accuracy.
- Specify desired angle of the helix ramp entry in the **Helix Ramp Angle** edit box of **Operation** tab.
- Enter desired value in **Helix Max Diameter** edit box from **Operation** tab to specify the limit helix entry diameter. If limit larger than tool diameter or 0, tool diameter is used.
- Enter desired value in **Lift Distance** edit box from **Operation** tab to specify how much to lift the tool up during the rapid linking moves over cleared regions. If linking path is not clear, tool is raised to clearance height.
- Enter desired value in **Keep Tool Down Ratio** edit box from **Operation** tab to specify maximum length of keep tool down linking path compared to direct distance between point. If exceeded, link will be done by raising the tool to clearance height.
- Enter desired value in **Stock to Leave** edit box to specify how much material to leave (i.e. for finishing operation).
- Select the **Force Clearing Inside-Out** check box from **Operation** tab to specify the force plunging into material inside and clearing towards the edges.
- Other parameters in the **Adaptive path operation** dialog have been discussed earlier.
- Click on **Apply** button and then **OK** button from the dialog. The Adaptive operation path will be generated; refer to Figure-61.

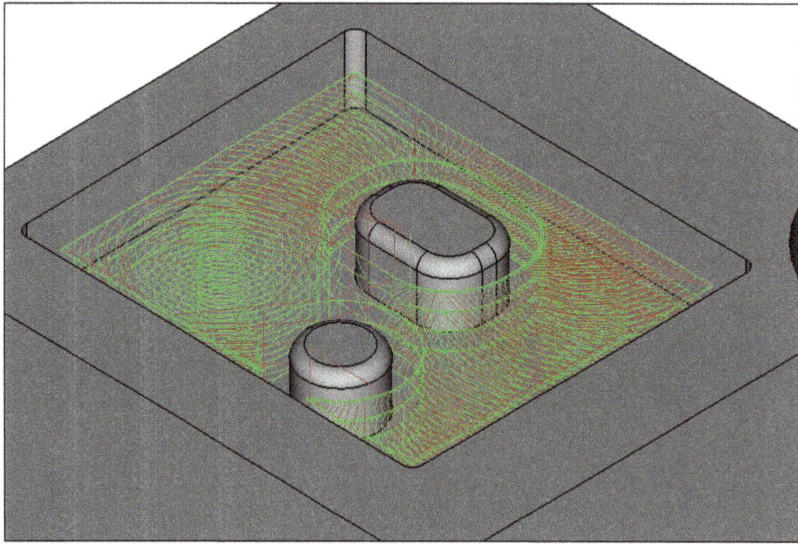

Figure-61. Adaptive operation path generated

Creating an Engraving Path

The **Engrave** tool is primarily used for engraving a **Draft ShapeString** onto a part like writing name on part. However, it may be useful for other kinds of 2D toolpaths. Note that if you want to engrave string on face of model then you will need a new job setup for string. The procedure to perform engraving of a string is discussed next.

- Create the string of text as desired on the face of model in **Draft** workbench.
- Click on the **Path Job** tool from the **Toolbar** in **Path** workbench. The **Create Job** dialog box will be displayed.
- Select the check box for **ShapeString** in **2D** node of list in the dialog box; refer to Figure-62 and click on the **OK** button. The **Job Edit** dialog will be displayed in the **Tasks** panel of **Combo View**.

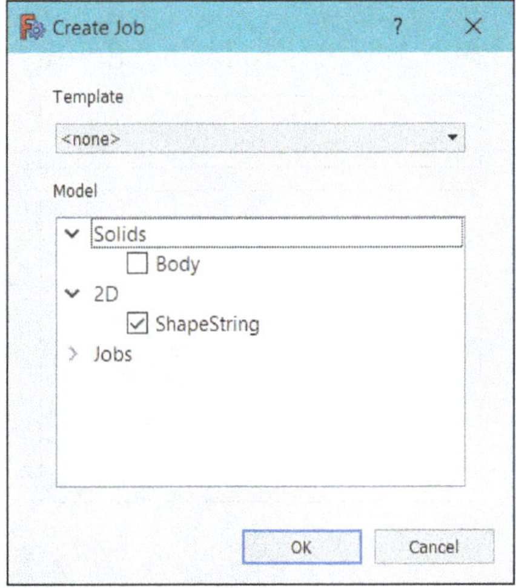

Figure-62. Shapestring selected for creating job

- Set the stock, cutting tool, and other parameters as discussed earlier and click on the **OK** button from the dialog box. A new job with name Job001 will be added in the **Model Tree**.
- Click on the **Engrave** tool from **Engraving Operations** drop-down in the **Toolbar** of **Path** workbench; refer to Figure-63. The **Choose a Path Job** dialog box will be displayed.
- Select the newly created job from the drop-down and click on the **OK** button. The **Choose a Tool Controller** dialog box will be displayed if there are multiple tools available. Select desired engraving tool from the drop-down and click on the **OK** button. The **Engrave** dialog will be displayed in the **Tasks** panel of **Combo View** and the model will be enclosed within the bounding box; refer to Figure-64.

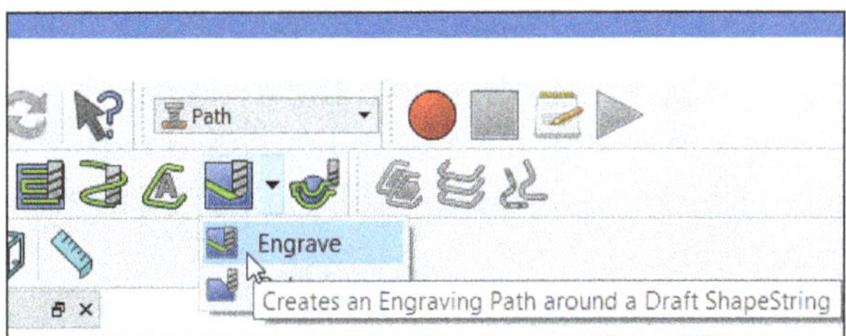

Figure-63. Engrave tool

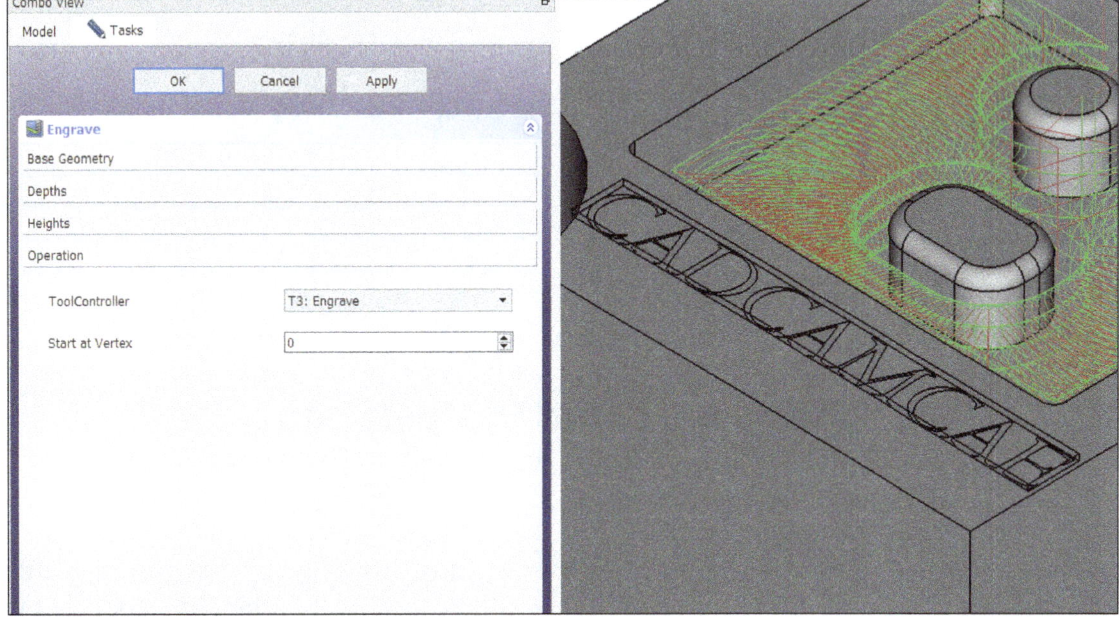

Figure-64. Engrave dialog and the model enclosed within bounding box

- Select faces of strings or every edge of the strings from the model in the 3D view area and click on **Add** button from **Base Geometry** tab of the dialog to add them to the list as the base geometries for this operation; refer to Figure-65.

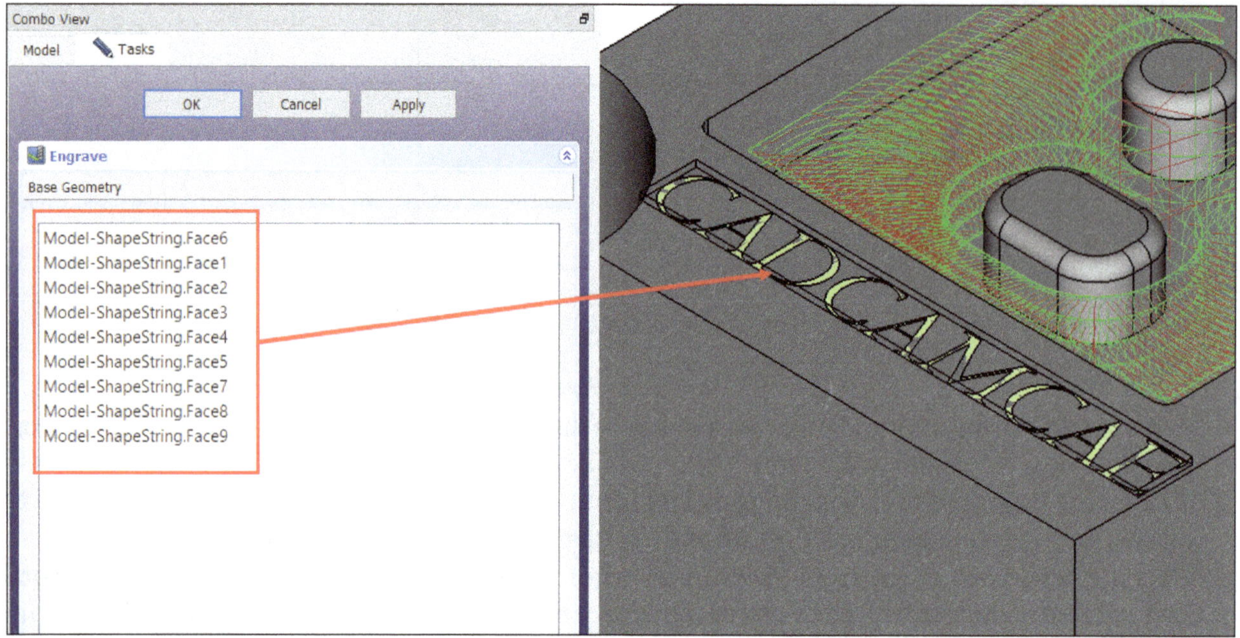

Figure-65. String faces selected

- Specify desired tool to be used for this operation from **ToolController** drop-down in the **Operation** tab of the dialog.
- Specify desired vertex number of the underlying shape string at which engraving should start in the **Start at Vertex** edit box of **Operation** tab in the dialog.
- Other parameters in the **Engrave** dialog have been discussed earlier.
- Click on **Apply** button and then **OK** button from the dialog. The Engraving operation path will be generated at specified depth; refer to Figure-66.

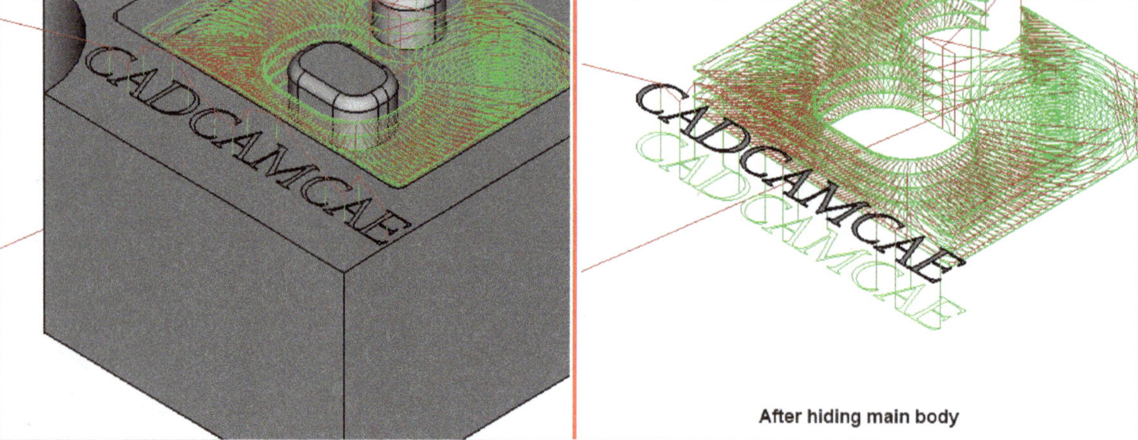

Figure-66. Engraving operation path generated

Creating Deburring Operation

The **Deburr** tool creates a deburr path along edges or around faces. The procedure to use this tool is discussed next.

- Click on the **Deburr** tool from **Engraving Operations** drop-down in the **Toolbar** of **Path** workbench; refer to Figure-67. The **Deburr** dialog will be displayed in the **Tasks** panel of **Combo View** and the model will be enclosed within the boundary box; refer to Figure-68.

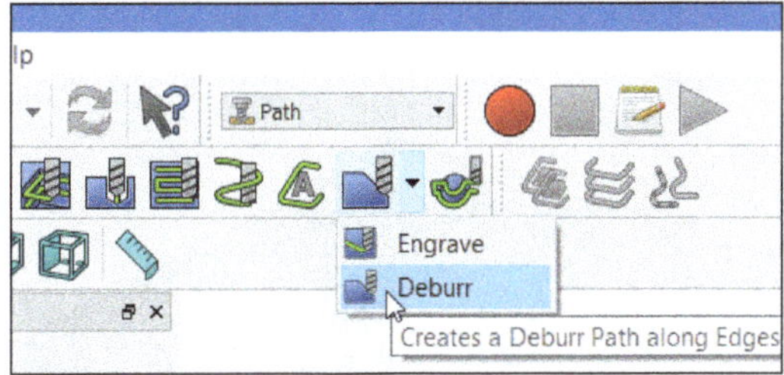

Figure-67. Deburr tool

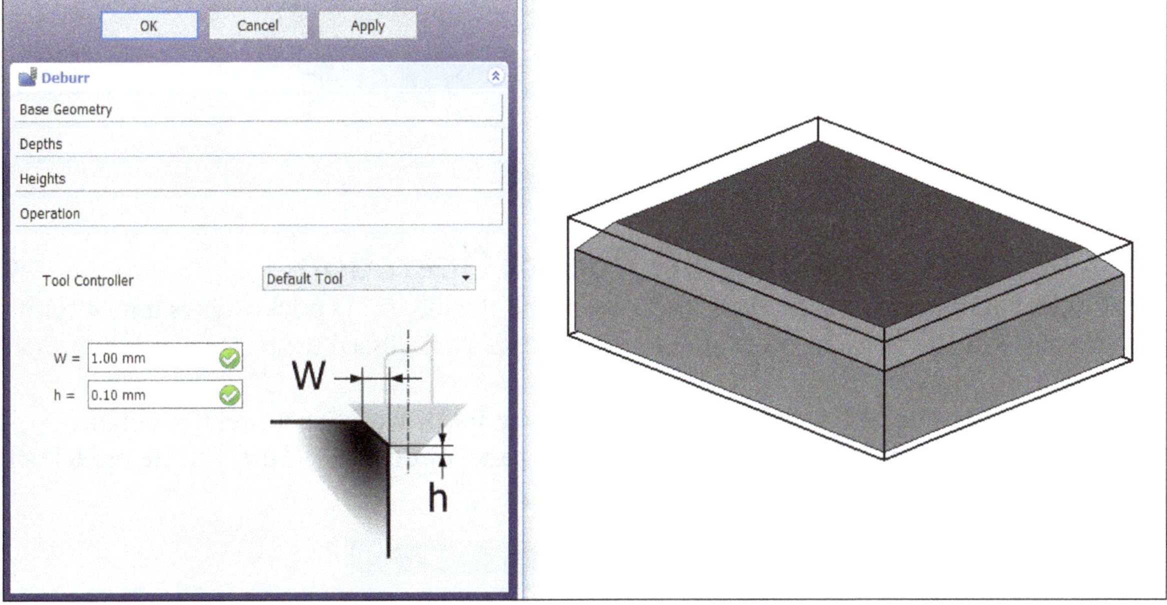

Figure-68. Deburr dialog and the model enclosed within bounding box

- Select the edges and/or faces of the model in the 3D view area on which you want to create the deburring operation and click on **Add** button from **Base Geometry** tab of the dialog to add them to the list as the base geometries for this operation; refer to Figure-69.

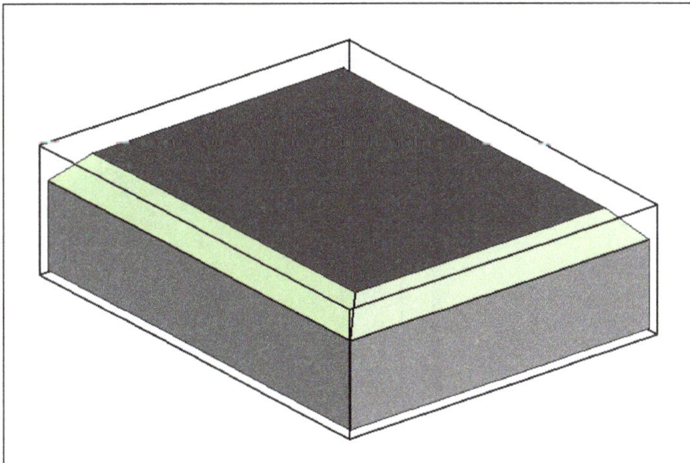
Figure-69. Selecting the faces of model for deburring

- Select desired tool to be used for this operation from **Tool Controller** drop-down in the **Operation** tab of dialog.
- Specify desired width of chamfer cut in the **W** edit box from **Operation** tab of the dialog.
- Specify desired extra depth of tool immersion in the **h** edit box from **Operation** tab of the dialog.
- Other parameters in the **Deburr** dialog have been discussed earlier.
- Click on **Apply** button and then **OK** button from the dialog. The Deburring operation path will be generated; refer to Figure-70.

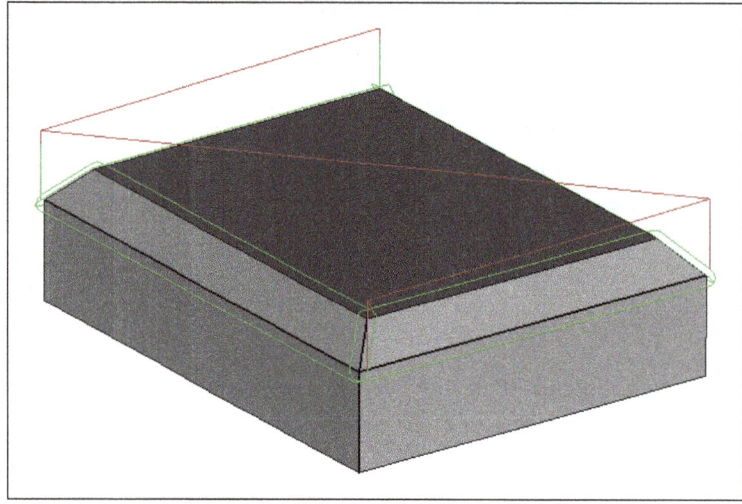

Figure-70. Deburr operation path generated

Creating 3D Pocket Operation

The **3D Pocket** tool inserts a path 3D Pocket object into the job. A 3D pocket takes into account the bottom surface of the pocket. The procedure to use this tool is discussed next.

- Click on the **3D Pocket** tool from **Toolbar** in the **Path** workbench; refer to Figure-71. The **Pocket 3D** dialog will be displayed in the **Tasks** panel of **Combo View** and the model will be enclosed within the bounding box; refer to Figure-72.

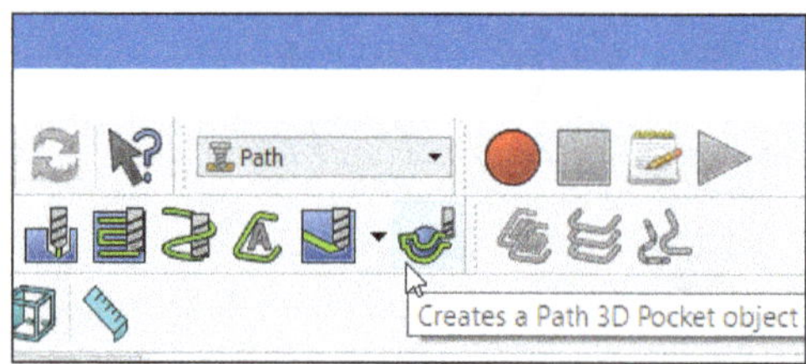

Figure-71. 3D Pocket tool

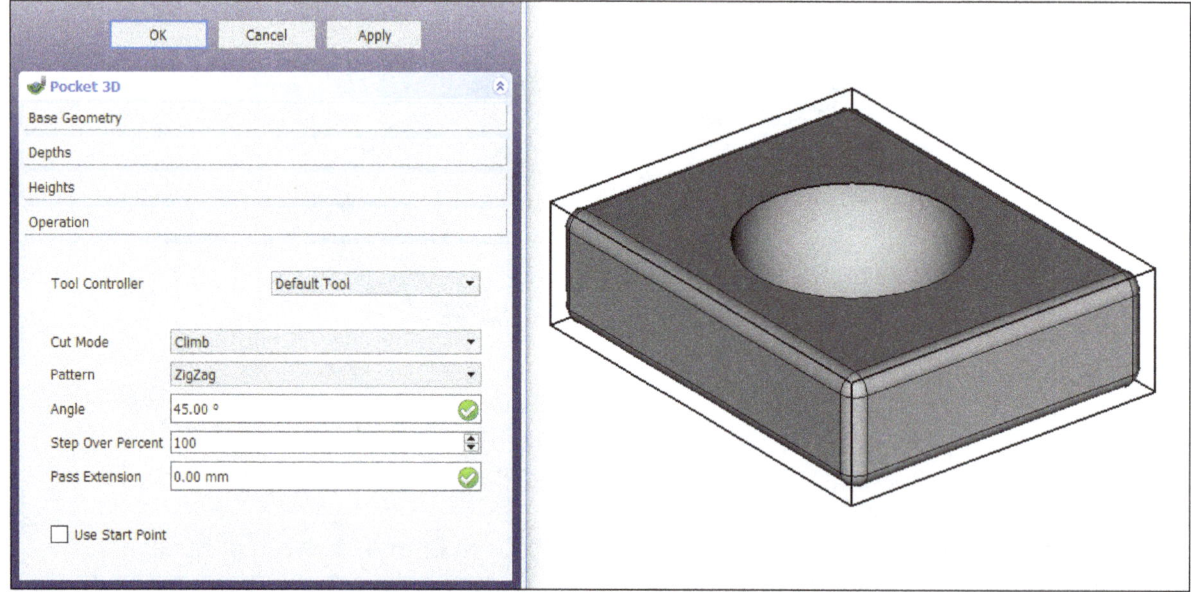

Figure-72. Pocket 3D dialog and the model enclosed within bounding box

- Select one or more faces from the model in the 3D view area and click on **Add** button from **Base Geometry** tab of the dialog to add them to the list as the base geometries for this operation; refer to Figure-73.

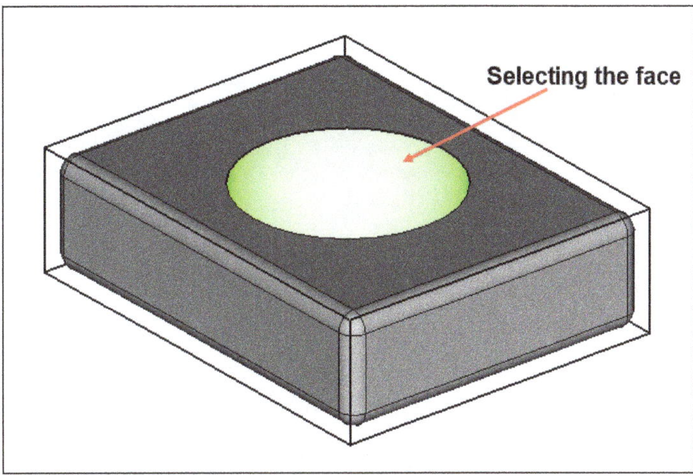

Figure-73. Selecting the face of model for 3D pocketing

- Specify all the parameters in the **Pocket 3D** dialog as discussed earlier.
- Click on **Apply** button and then **OK** button from the dialog. The 3D Pocket operation path will be generated; refer to Figure-74.

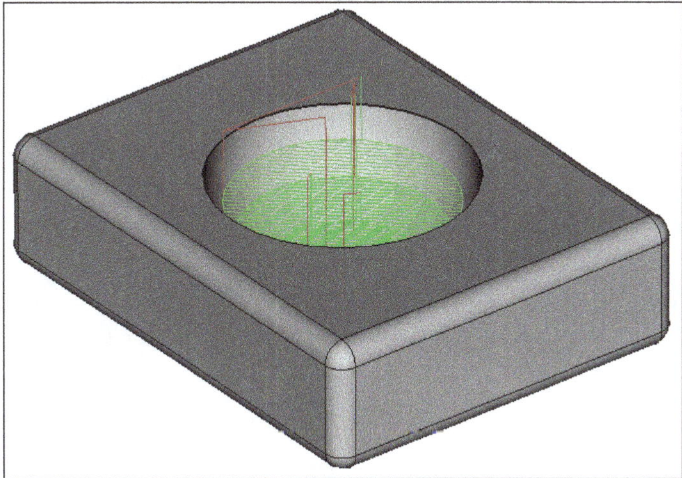

Figure-74. 3D Pocket operation path generated

PATH MODIFICATION TOOLS

There are three tools in **Toolbar** of **Path** workbench used to create multiple copies of the selected toolpaths. These tools are discussed next.

Creating Copy

The **Copy** tool creates a copy of selected path operation. The procedure to use this tool is discussed next.

- Select desired path operation from the Model tree view for which you want to create the copy or multiple copies; refer to Figure-75.

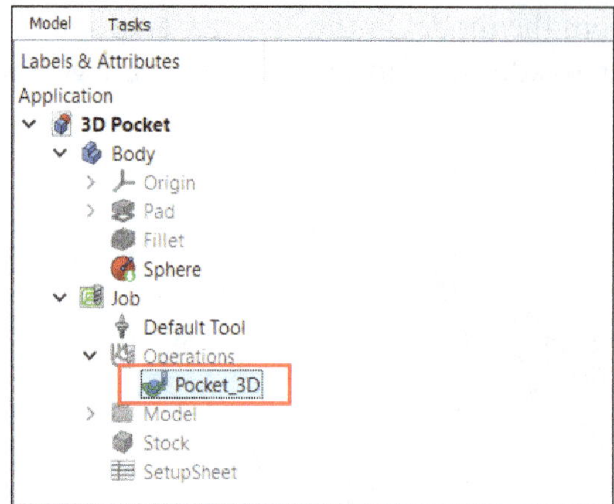

Figure-75. Selecting the path operation

- Click on the **Copy** tool from **Toolbar** in the **Path** workbench; refer to Figure-76. The copy of selected path operation will be created; refer to Figure-77.

Figure-76. Copy tool

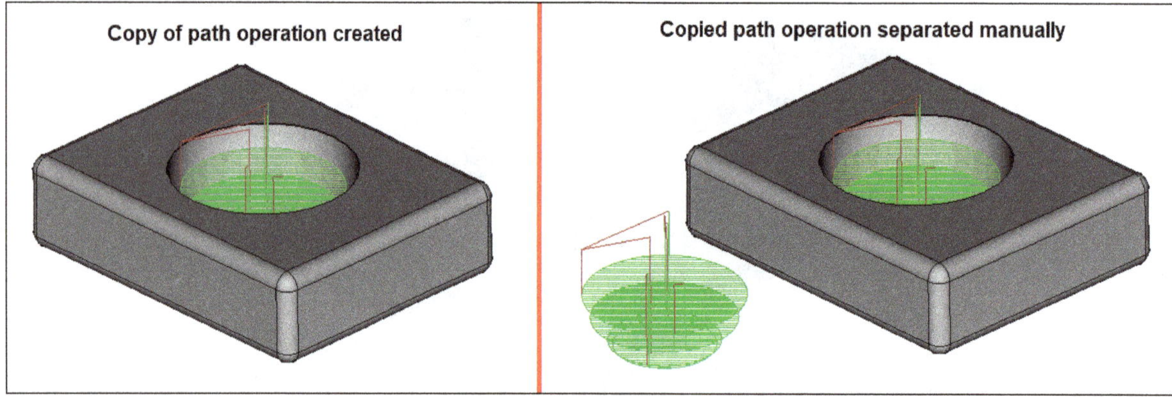

Figure-77. Copy of path operation created

- You can edit the properties of copied path operation using **Property editor** dialog as discussed earlier.

Creating Array

The **Array** tool creates a new path by duplicating another path several times at a certain interval distance. The procedure to use this tool is discussed next.

- Select desired path operation created from the Model tree view of which you want to create array; refer to Figure-78.

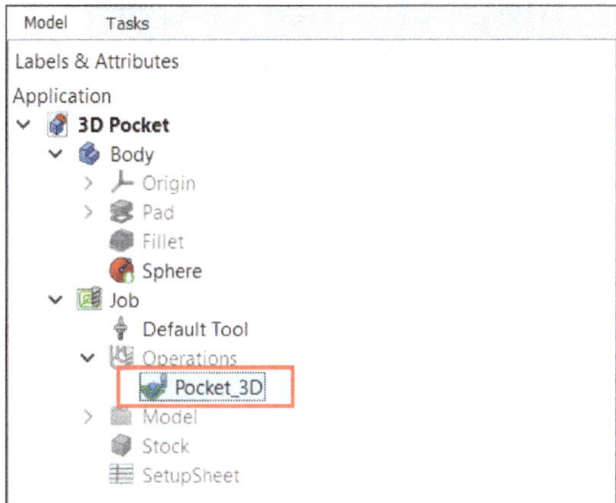

Figure-78. Selecting the path operation

- Click on the **Array** tool from **Toolbar** in the **Path** workbench; refer to Figure-79. The array of path operation will be created; refer to Figure-80.

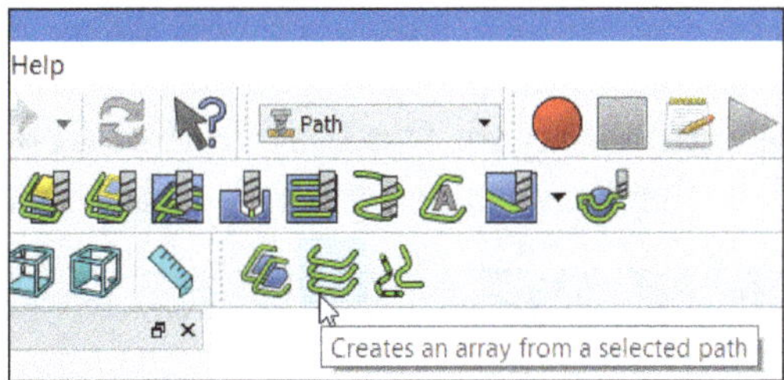

Figure-79. Array tool

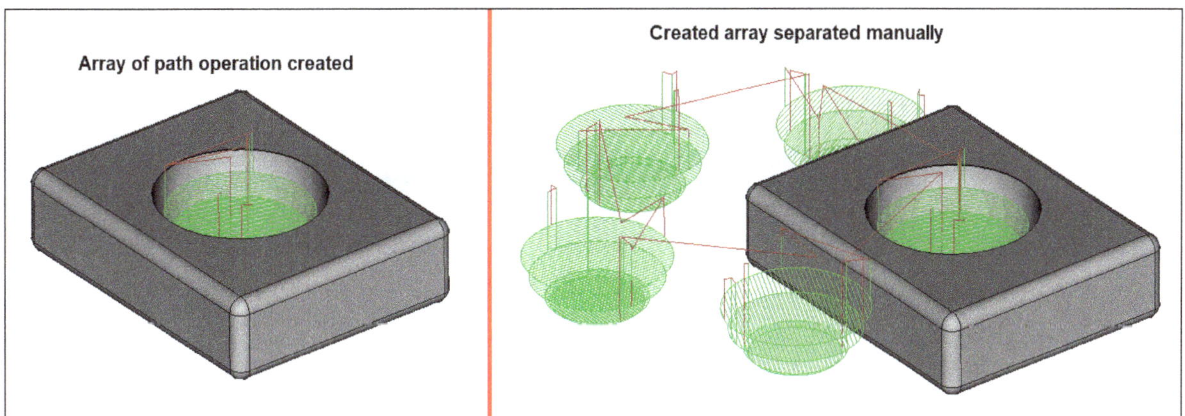

Figure-80. Array of path operation created

- You can edit the properties of array of path operation created from **Property editor** dialog as discussed earlier.

Creating Simple Copy

The **Simple Copy** tool creates a non-parametric copy of a given path. The procedure to use this tool is similar to discussed for **Copy** tool.

PATH UTILITIES

The tools in **Path Utilities** section of **Toolbar** are used to inspect, manage, and simulate toolpaths. These tools are discussed next.

Inspection of G-Code

The **G-Code Inspector** tool allows inspection of the internal FreeCAD G-code dialect contents of a Path Operation Object. The procedure to use this tool is discussed next.

- Select desired path operation from the Model Tree which you want to inspect for G-code; refer to Figure-81.

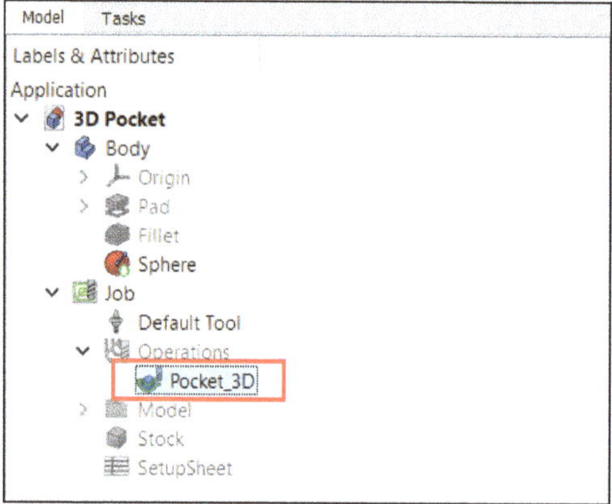

Figure-81. Selecting the path operation

- Click on the **G-Code Inspector** tool from **Toolbar** in the **Path** workbench; refer to Figure-82. The **FreeCAD** dialog box will be displayed showing the G-code of path operation; refer to Figure-83.

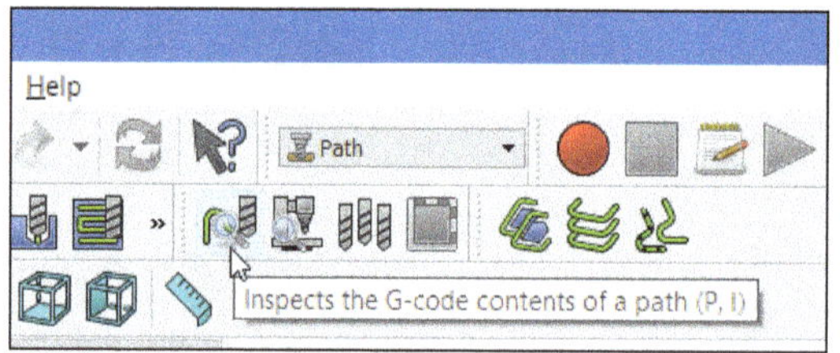

Figure-82. G-Code Inspector tool

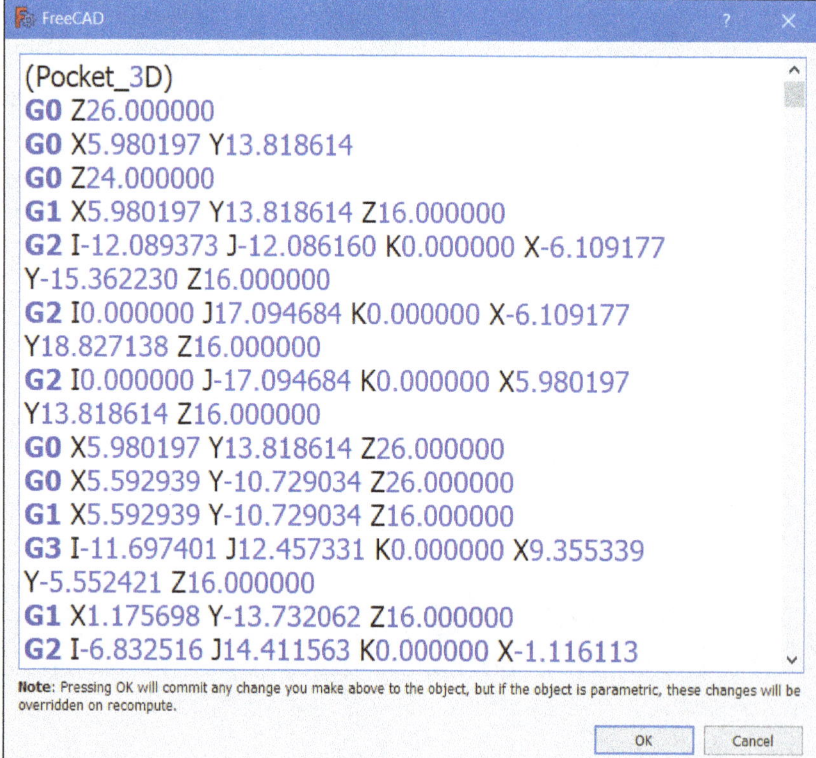

Figure-83. FreeCAD dialog box showing the G code

- Make any changes in the dialog box as desired and click on **OK** button from the dialog box to apply the changes.

Simulating the Path

The **Simulator** tool is used to simulate selected operations on 3D model along the G-Code paths, subtracting material from the stock, where the stock and tool overlap, providing visualization of the Job. This allows detection and isolation of errors prior to running the job on a mill. The procedure to use this tool is discussed next.

- Click on the **Simulator** tool from **Toolbar** in the **Path** workbench; refer to Figure-84. The **Path Simulator** dialog will be displayed in the **Tasks** panel of **Combo View** and the model will be enclosed within the stock, ready to be simulate; refer to Figure-85.

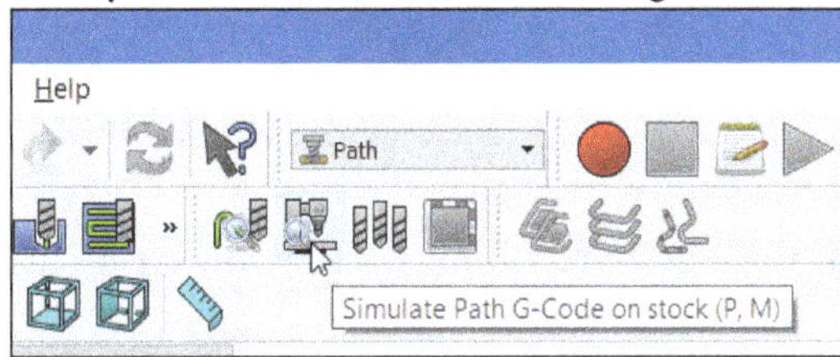

Figure-84. Simulator tool

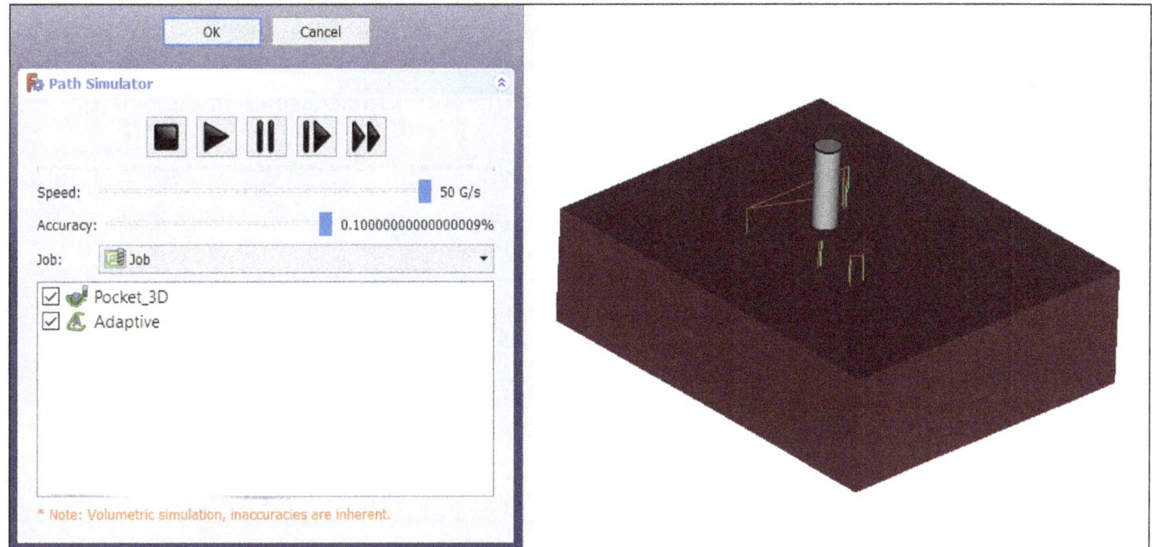

Figure-85. Path Simulator dialog and the model enclosed within stock

- All the operations created that are to be simulate will be selected by default in the **Operation list** area of the dialog.
- Deselect the operations which you do not want to be simulate from the **Operation list**.
- Select desired Job which you want to use as the basis of the simulation from **Job** drop-down in the dialog.
- Press the **Play** button to play or playback an animation of the operations.
- Press the **Stop** button to stop the animation.
- Press the **Pause** button to pause animation for troubleshooting purposes.
- Press the **Single-Step** button for slowing down the animation, this functionality helps troubleshooting and resolving specific cuts and/or movements.
- Press the **Fast-Forward** button to increase the speed substantially.
- Tune the speed of simulation by sliding the **Speed** slider from the dialog.
- Tune the accuracy of simulation by sliding the **Accuracy** slider from the dialog.
- After performing the simulation, the object will be displayed as shown in Figure-86.

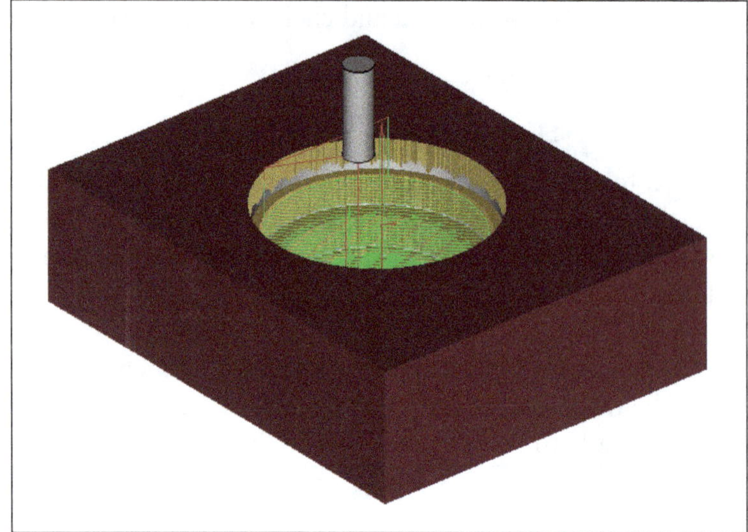

Figure-86. Object simulated

- Click on the **OK** button from the dialog to finish the simulation process.

POST-PROCESSING

The **Post Process** command exports the selected Path Jobs to a G-code file. Each CNC controller speaks a specific G-Code dialect-correct Postprocessor to translate the final output from the agnostic internal FreeCAD G-Code dialect. The procedure to use this tool is discussed next.

- Select the Path Job created from the Model tree view which you want to export; refer to Figure-87.

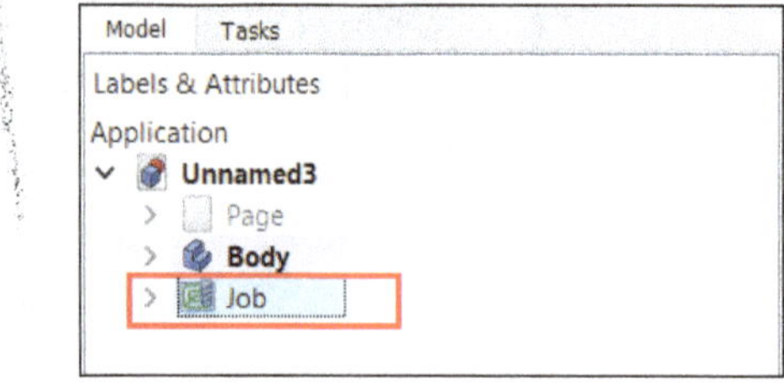

Figure-87. Job selected

- Click on the **Post Process** tool from **Toolbar** in the **Path** workbench; refer to Figure-88. The **Output File** dialog box will be displayed asking you to specify the file name and directory; refer to Figure-89.

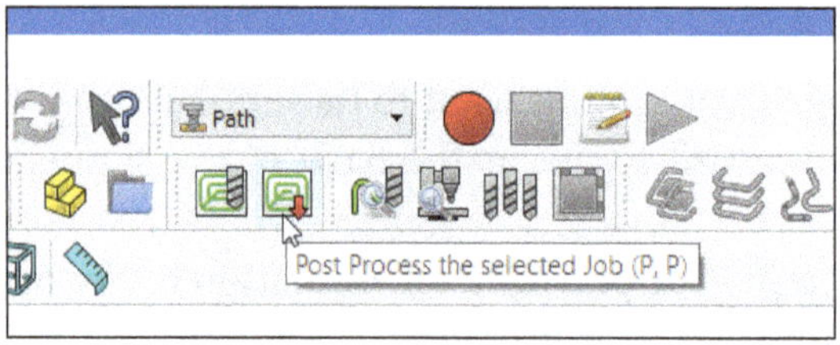

Figure-88. Post Process tool

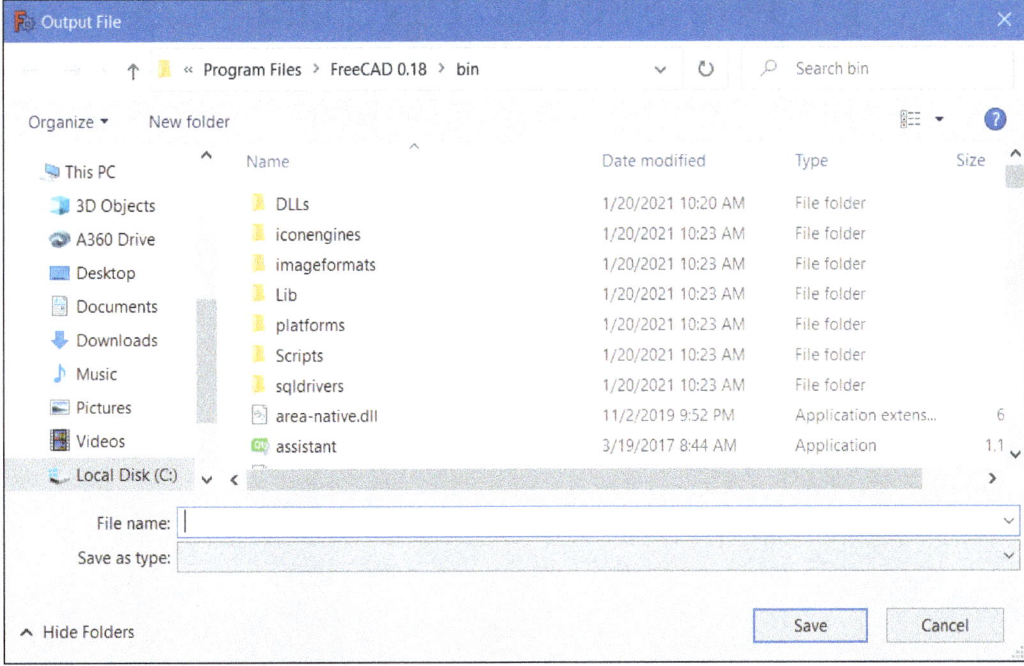

Figure-89. Output File dialog box

- Specify desired name of output file in the **File name** edit box and specify desired directory to save the file.
- Click on **Save** button to save the output file. The **FreeCAD** dialog box will be displayed showing the postprocessor properties of the job; refer to Figure-90.

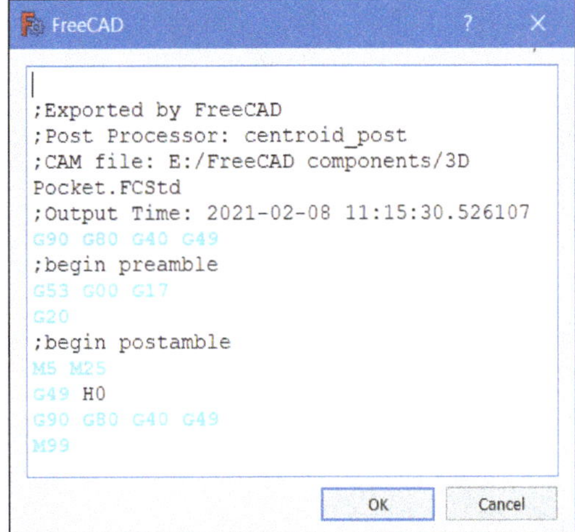

Figure-90. FreeCAD dialog box showing the post processor properties

For Student Notes

Chapter 9

FEM Workbench

Topics Covered

The major topics covered in this chapter are:

- ***Introduction to FEM Workbench***
- ***Starting FEM Workbench***
- ***Creating Analysis Container***
- ***Analysis Types in FreeCAD***
- ***Defining Material***
- ***Applying Constraints***
- ***Creating Geometry Elements***
- ***Meshing***
- ***Solvers***
- ***Post Processing***

INTRODUCTION TO FEM

The **FEM** Workbench provides a modern finite element analysis (FEA) workflow for FreeCAD. This means all the tools to perform an analysis are combined into one graphical user interface (GUI); refer to Figure-1.

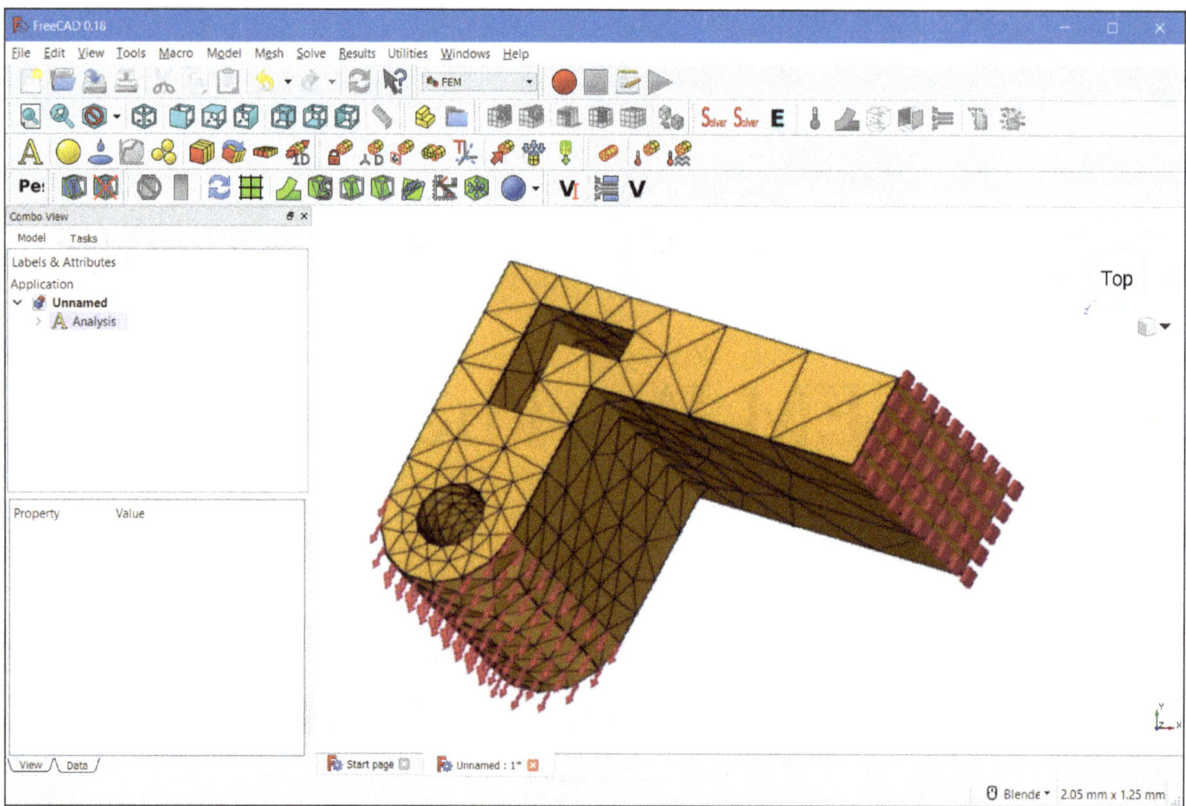

Figure-1. FEM Workbench Overview

The workflow for FEM is shown in Figure-2. The steps to carry out a finite element analysis are given next.

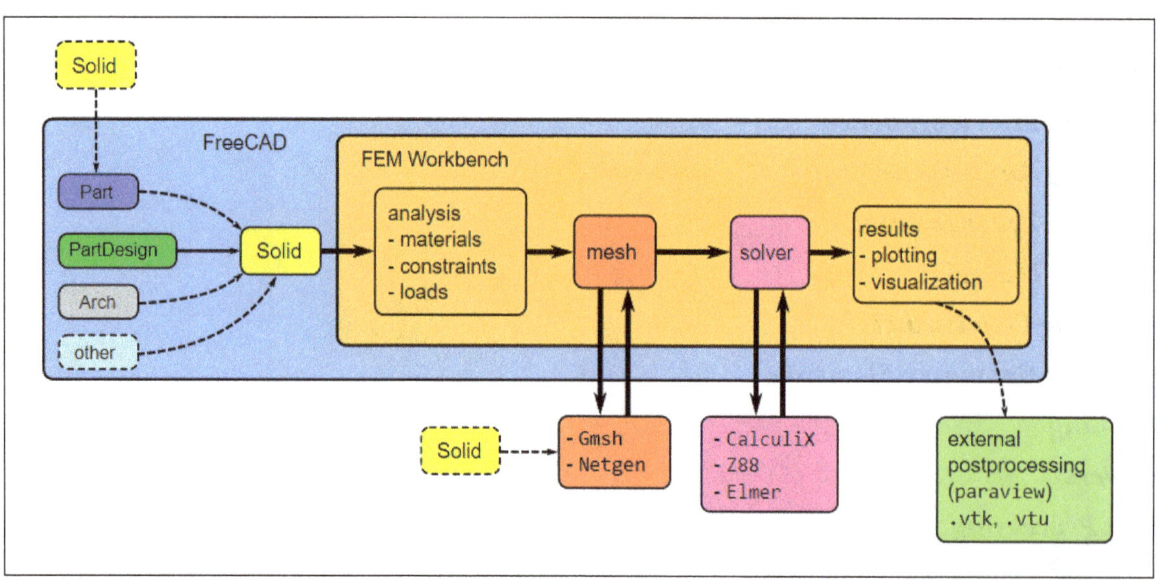

Figure-2. Workflow of FEM Workbench

1. **Preprocessing**: Setting up the analysis problem mainly on plain paper or in your mind.

2. **Modeling the geometry**: Creating the geometry with FreeCAD, or importing it from a different application.
3. **Creating an analysis**: This step is also called Defining Boundary Conditions. It involves adding simulation constraints such as loads and fixed supports to the geometric model. Adding materials to the parts of the geometric model. Creating a finite element mesh for the geometrical model, or importing it from a different application.
4. **Solving**: Running an external solver from within FreeCAD.
5. **Postprocessing**: Visualizing the analysis results from within FreeCAD, or exporting the results so they can be post processed with another application.

STARTING FEM WORKBENCH

The **FEM** Workbench calls two external programs to perform meshing of a solid object and perform the actual solution of the finite element problem.

- To start a new **FEM** file, click on the **New** button from **File** menu and select **FEM** workbench from **Switch between workbenches** drop-down in the **Toolbar**; refer to Figure-3. The tools related to **FEM** workbench will be displayed in the **Toolbar**; refer to Figure-4.

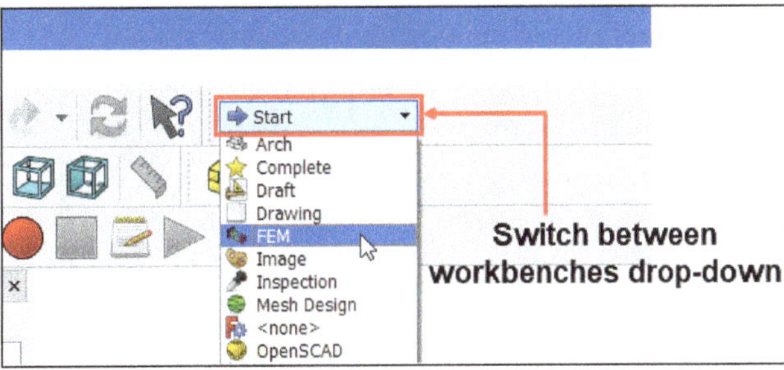

Figure-3. FEM Workbench

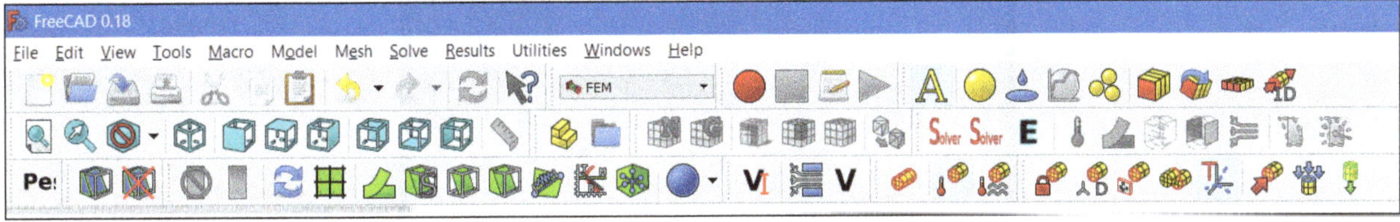

Figure-4. FEM Workbench tools

CREATING ANALYSIS CONTAINER

The **FEM** Analysis in FreeCAD could be seen as a Container that holds all objects of a Finite Element Analysis. It is mandatory to have an analysis container which holds all the needed objects. At least one of the following objects is needed for a mechanical analysis: material, fixed constraint, force constraint, or pressure constraint. The procedure to use this tool is discussed next.

- Click on the **Analysis Container** tool from **Toolbar** in the **FEM** workbench; refer to Figure-5. A new Analysis will be created and displayed in the Model tree view; refer to Figure-6.

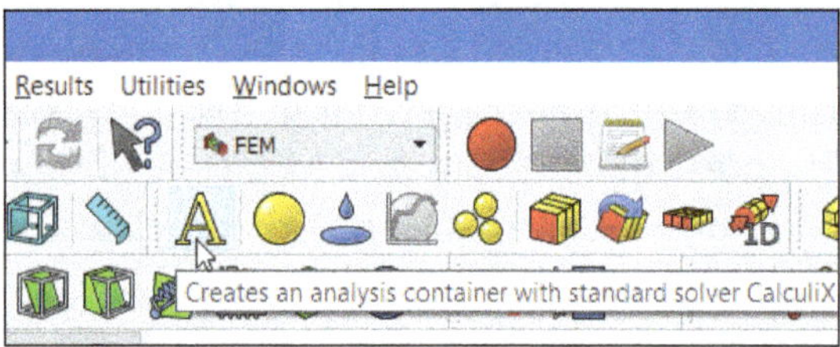

Figure-5. Analysis Container tool

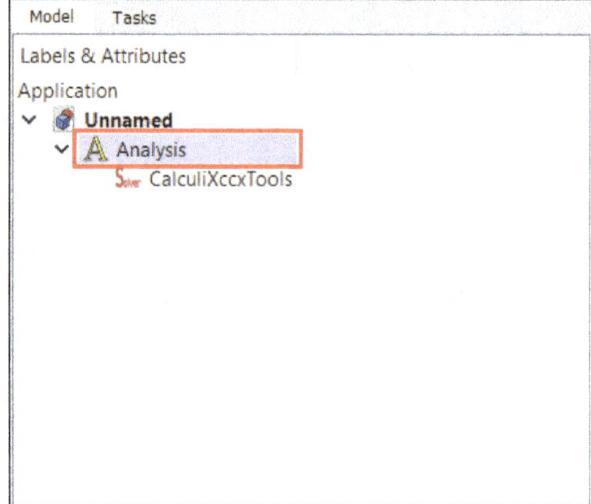

Figure-6. New Analysis created

- Expand the **Analysis** node and double-click on the **CalculiXccx Tools** element from the Model tree view. The **Mechanical analysis** dialog will be displayed in the **Tasks** panel of **Combo View**; refer to Figure-7.

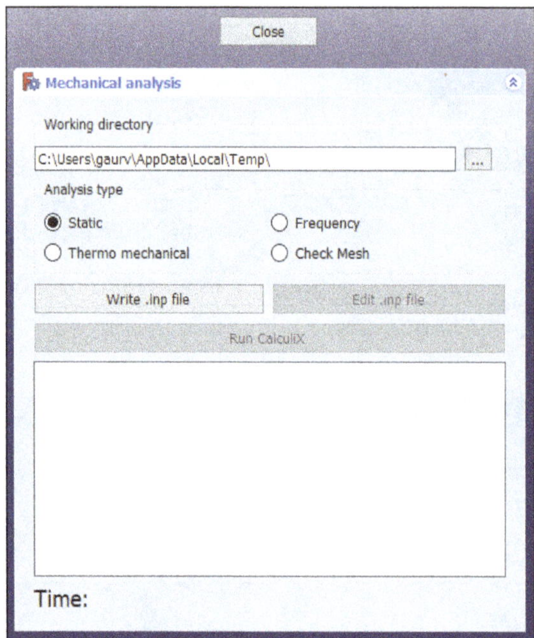

Figure-7. Mechanical analysis dialog

- Specify desired path to the working directory which will be used for CalculiX analysis files in the **Working directory** area of the dialog.
- Select desired radio button from **Analysis type** area of the dialog. Select the **Static** radio button from the dialog to apply the loads to a body due to which the body deforms and the effects of the loads are transmitted throughout the body in static environment. Select **Frequency** radio button from the dialog to design vibration isolation systems by avoiding resonance in specific frequency band. Select **Thermo mechanical** radio button from the dialog to check the distribution of heat over a body due to applied thermal loads. Select **Check Mesh** radio button to generate 3D tetrahedral solid elements, 2D triangular shell elements, and 1D beam elements for the model.
- Click on the **Write .inp file** button from the dialog to write CalculiX input file manually before running analysis.
- Click on the **Edit .inp file** button from the dialog to display and edit CalculiX input file manually before running analysis. In this case, it might be useful to use parameter "Split Input Writer = true".

- Click on the **Run CalculiX** button from the dialog to run the CalculiX input file.
- Click on **Close** button to close the dialog.

Some details about common analyses available in FreeCAD are given next.

Static Analysis

This is the most common type of analysis we perform. In this analysis, loads are applied to a body due to which the body deforms and the effects of the loads are transmitted throughout the body. To absorb the effect of loads, the body generates internal forces and reactions at the supports to balance the applied external loads. These internal forces and reactions cause stress and strain in the body. Static analysis refers to the calculation of displacements, strains, and stresses under the effect of external loads, based on some assumptions. The assumptions are as follows.

1. All loads are applied slowly and gradually until they reach their full magnitudes. After reaching their full magnitudes, load will remain constant (i.e. load will not vary against time).
2. Linearity assumption: The relationship between loads and resulting responses is linear. For example, if you double the magnitude of loads, the response of the model (displacements, strains, and stresses) will also double. You can make linearity assumption if:

- All materials in the model comply with Hooke's Law that is, stress is directly proportional to strain.
- The induced displacements are small enough to ignore the change in stiffness caused by loading.
- Boundary conditions do not vary during the application of loads. Loads must be constant in magnitude, direction, and distribution. They should not change while the model is deforming.

If the above assumptions are valid for your analysis then you can perform **Linear Static Analysis**. For example, a cantilever beam fixed at one end and force applied on other end; refer to Figure-8.

If the above assumptions are not valid then you need to perform the **Non-Linear Static analysis**. For example, an object attached with a spring being applied under forces; refer to Figure-9.

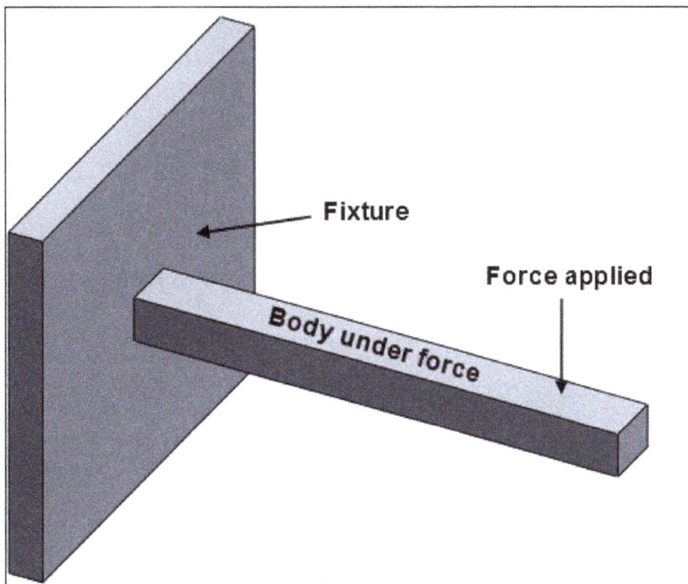

Figure-8. Linear static analysis example

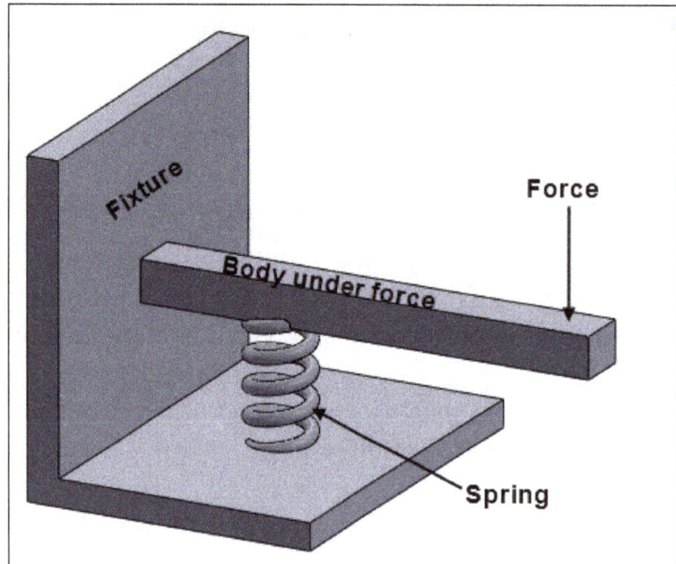

Figure-9. Non-linear static analysis example

Frequency Analysis (Modal Analysis)

By its very nature, vibration involves repetitive motion. Each occurrence of a complete motion sequence is called a "cycle." Frequency is defined as so many cycles in a given time period. "Cycles per seconds" or "Hertz". Individual parts have what engineers call "natural" frequencies. For example, a violin string at a certain tension will vibrate only at a set number of frequencies, which is why you can produce specific musical tones. There is a base frequency in which the entire string is going back and forth in a simple bow shape.

Harmonics and overtones occur because individual sections of the string can vibrate independently within the larger vibration. These various shapes are called "modes". The base frequency is said to vibrate in the first mode and so on up the ladder. Each mode shape will have an associated frequency. Higher mode shapes have higher frequencies.

The most disastrous kinds of consequences occur when a power-driven device such as a motor for example, produces a frequency at which an attached structure naturally vibrates. This event is called "resonance." If sufficient power is applied, the attached structure will be destroyed. Note that ancient armies, which normally marched "in step," were taken out of step when crossing bridges. Should the beat of the marching feet align with a natural frequency of the bridge, it could fall down. Engineers must design machines so that resonance does not occur during regular operation of machines. This is a major purpose of Modal Analysis. Ideally, the first mode has a frequency higher than any potential driving frequency. Frequently, resonance cannot be avoided, especially for short periods of time. For example, when a motor comes up to speed, it produces a variety of frequencies. So, it may pass through a resonant frequency.

Thermal analysis

There are three mechanisms of heat transfer. These mechanisms are Conduction, Convection, and Radiation. Thermal analysis calculates the temperature distribution in a body due to some or all of these mechanisms. In all three mechanisms, heat flows from a higher-temperature medium to a lower temperature one. Heat transfer by conduction and convection requires the presence of an intervening medium while heat transfer by radiation does not.

There are two modes of heat transfer analysis.

Steady State Thermal Analysis

In this type of analysis, we are only interested in the thermal conditions of the body when it reaches thermal equilibrium, but we are not interested in the time it takes to reach this status. The temperature of each point in the model will remain unchanged until a change occurs in the system. At equilibrium, the thermal energy entering the system is equal to the thermal energy leaving it. Generally, the only material property that is needed for steady state analysis is the thermal conductivity.

Transient Thermal Analysis

In this type of analysis, we are interested in knowing the thermal status of the model at different instances of time. A thermos designer, for example, knows that the temperature of the fluid inside will eventually be equal to the room temperature(steady state), but he is interested in finding out the temperature of the fluid as a function of time. In addition to the thermal conductivity, we also need to specify density, specific heat, initial temperature profile, and the period of time for which solutions are desired.

DEFINING MATERIAL FOR SOLID

The **Material Solid** tool specifies material properties for the model. The procedure to use this tool is discussed next.

- Click on the **Material Solid** tool from **Toolbar** in the **FEM** workbench; refer to Figure-10. The **FEM material** and **Geometry reference selector** dialog will be displayed in the **Tasks** panel of **Combo View**; refer to Figure-11 and Figure-12, respectively.

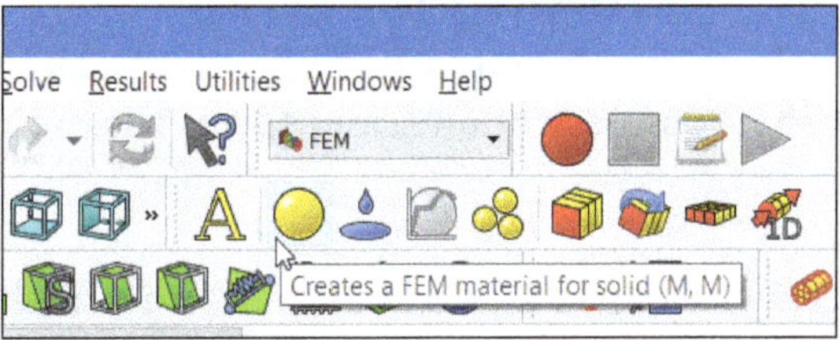

Figure-10. Material Solid tool

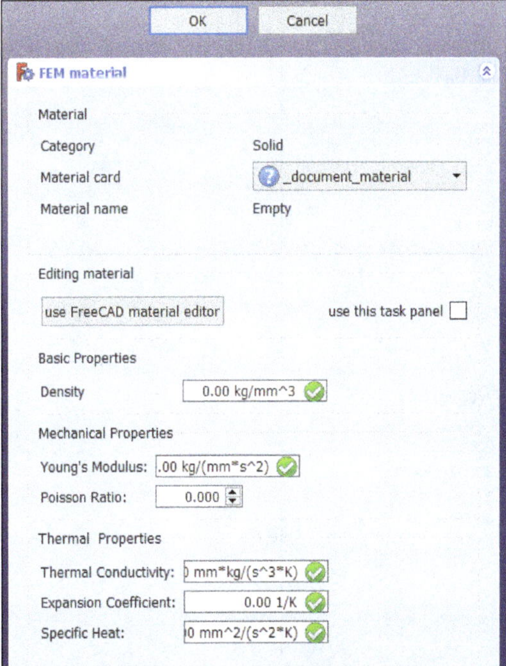

Figure-11. FEM material dialog

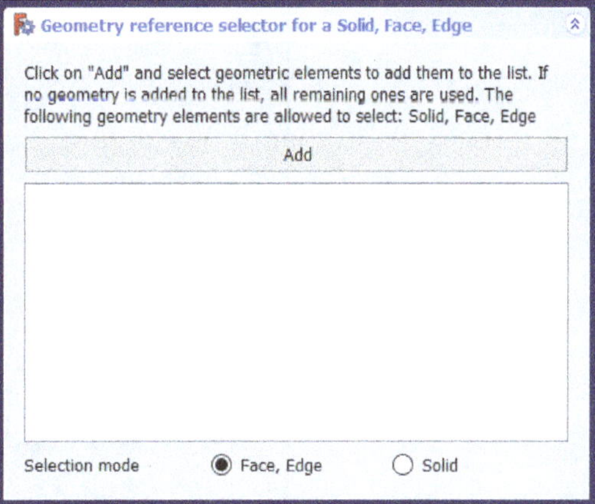

Figure-12. Geometry reference selector for a Solid Face Edge dialog

- Select desired material option from the **Material card** drop-down in **FEM material** dialog to be used as material for the model during analysis.
- Select desired radio button from **Selection mode** section of the **Geometry reference selector for a Solid, Face, Edge** dialog to define whether you want to apply material to single face/edge or solid object. Generally, the **Solid** radio button is used from this section. After selecting desired radio button, click on the **Add** button from the dialog and select the object to which you want to apply the material. The object will be added in the list; refer to Figure-13.

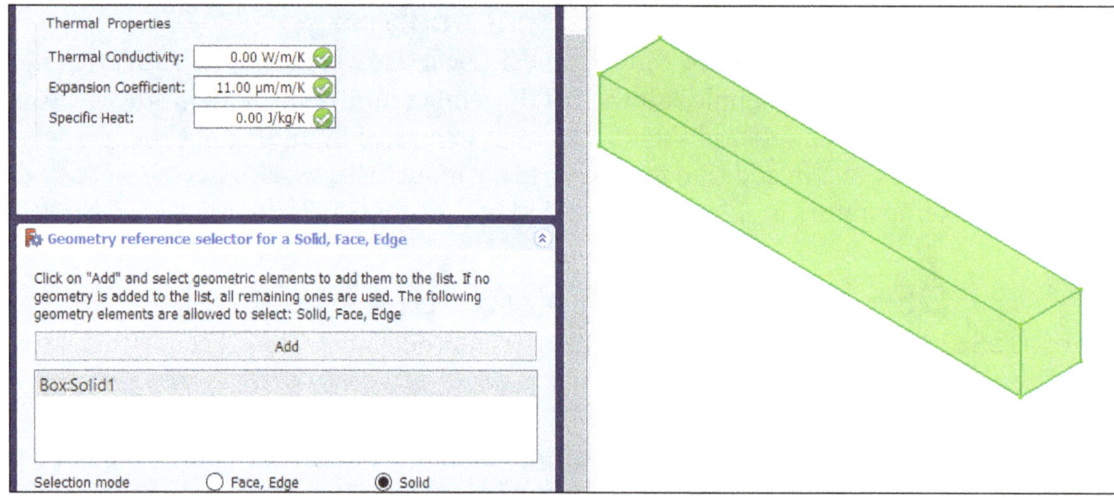

Figure-13. Object selected for applying material

Editing Material

The **use FreeCAD material editor** button in **FEM material** dialog is used to modify the parameters of material like density, Young's Modulus, and so on. The procedure to edit material is given next.

- Click on the **use FreeCAD material editor** button from the **FEM material** dialog in the **Tasks** panel of **Combo View**. The **Material Editor** dialog box will be displayed; refer to Figure-14.

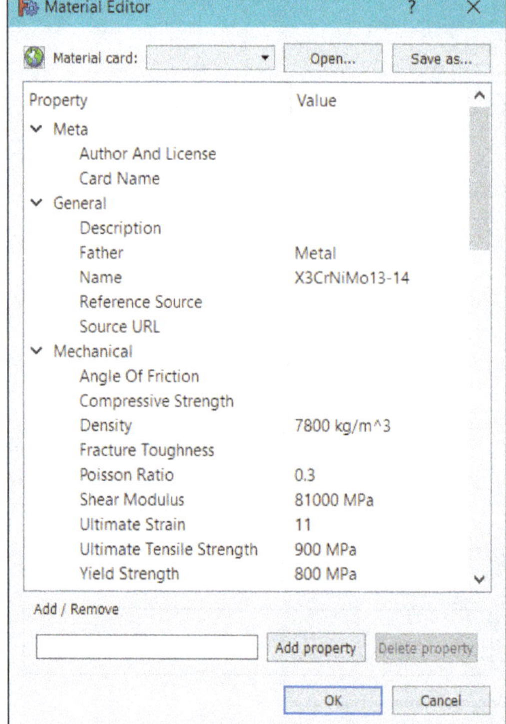

Figure-14. Material Editor dialog box

- Select desired material from the **Material card** drop-down to modify its parameter.
- Double-click in desired field of material to modify its value.
- If you want to add a new property for material then specify the name of property in the edit box of **Add/Remove** section in the dialog box and click on the **Add property** button. If you want to delete a property then select desired property from the dialog box and click on the **Delete property** button.
- You can open a new material by selecting the **Open** button and save current modified material properties by clicking on the **Save as** button.
- After setting desired parameters, click on the **OK** button from the dialog box to exit.

Note that you can modify the properties of material directly in **FEM material** dialog if you have selected the **use this task panel** check box in the dialog.

- Click on the **OK** button from the **Tasks** panel in **Combo View** to apply material.

After applying material to the model, the next step in FEM is to apply constraints to the model. Applying constraints is a part of defining boundary conditions. First, we will discuss the constraints which restrict the movements of model (Degree of Freedom) like fixing a face, allowing sliding of a face, and so on and then we will discuss the load constraints applied on model to check their effect.

APPLYING CONSTRAINTS

There are various toolbars in FEM workbench to apply constraints to the model; refer to Figure-15. You can also access the tools to apply constraints from **Model** menu in the FEM workbench; refer to Figure-16. The tools in these toolbars are discussed next.

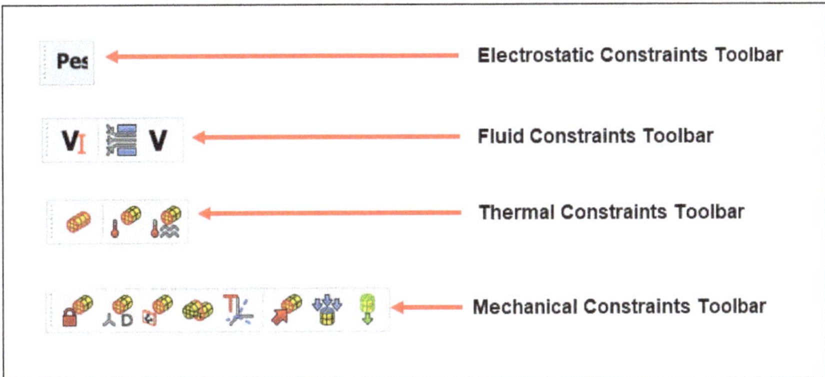

Figure-15. Constraint Toolbars

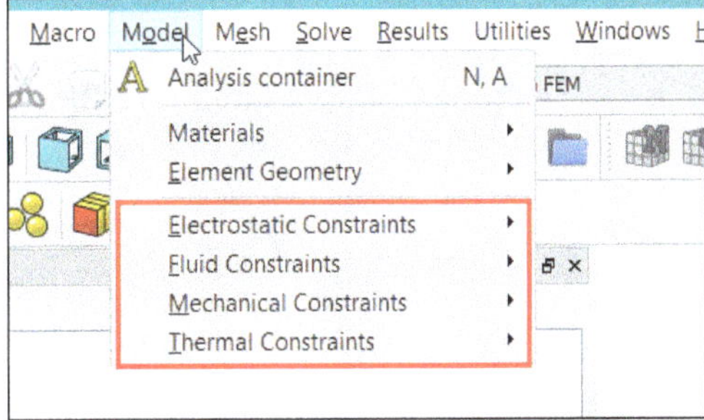

Figure-16. Constraint tools in Model menu

Mechanical Constraints

Mechanical constraints are used to stop physical movements of selected objects in specified directions. The tools to apply mechanical constraints are available in the **Mechanical Constraints Toolbar** and **Mechanical Constraints** cascading menu of **Model** menu as shown earlier. These tools are discussed next.

Applying Fix Constraint

The **Fix Constraint** tool in **Mechanical Constraint Toolbar** is used to stop movement of selected object in specified directions. The procedure to apply this constraint is given next.

- Click on the **Fix Constraint** tool from the **Mechanical Constraint Toolbar**; refer to Figure-17. The **FEM constraint parameters** dialog will be displayed in the **Tasks** panel of **Combo View**; refer to Figure-18.

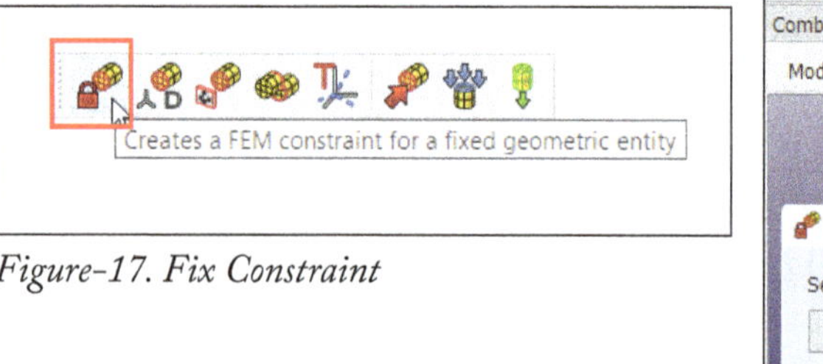

Figure-17. Fix Constraint

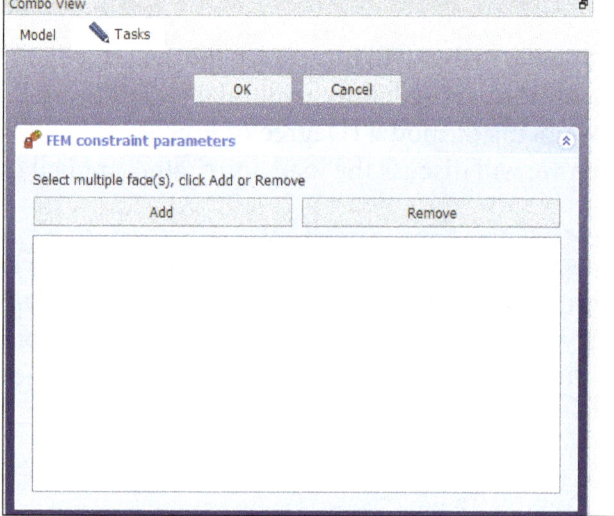

Figure-18. FEM constraint parameters dialog

- Select the face(s) whose movements are to be fixed and click on the **Add** button from the dialog box. Red boxes will be displayed attached to the selected faces showing that they are fixed; refer to Figure-19.

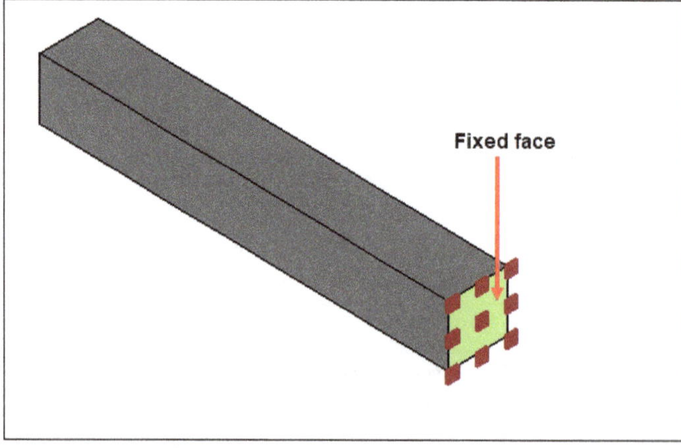

Figure-19. Fixed constraint applied to face

- After selecting desired faces, click on the **OK** button from the dialog to apply constraint.

Applying Displacement Constraint

The **Displacement Constraint** tool is used to specify the degree of freedom for selected objects in 3 linear and 3 rotation directions. The procedure to apply displacement constraint is given next.

- Click on the **Displacement Constraint** tool from the **Mechanical Constraints Toolbar** or **Model** menu. The **FEM constraint parameters** dialog will be displayed as shown in Figure-20.

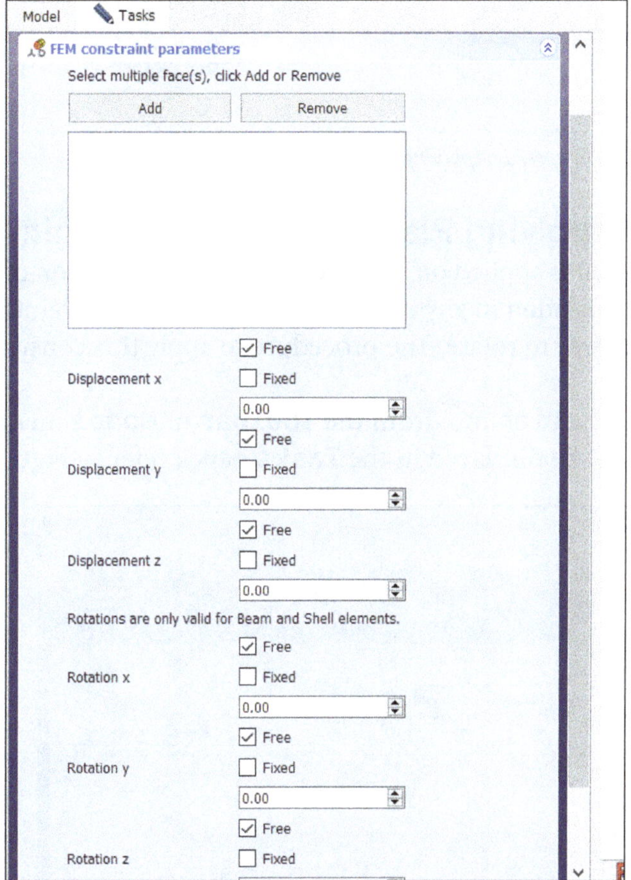

Figure-20. FEM constraint parameters dialog for displacement constraint

- Select the face(s) on which you want to apply constraint and click on the **Add** button from the dialog.
- Below the selection list box, there are six sections in the dialog box : **Displacement x**, **Displacement y**, **Displacement z**, **Rotation x**, **Rotation y**, and **Rotation z**. Select the **Free** check box for a section if you want the movement for respective direction to be free. You can also specify the limit up to which the displacement or rotation is allowed in edit box for that section. If you select the **Fixed** check box for a section then respective displacement or rotation will not be allowed. Note that rotation constraint can be applied for beam and shell elements only.
- After setting desired parameters, click on the **OK** button from the dialog to apply the constraint. The annotation for displacement constraint will be shown on the model; refer to Figure-21.

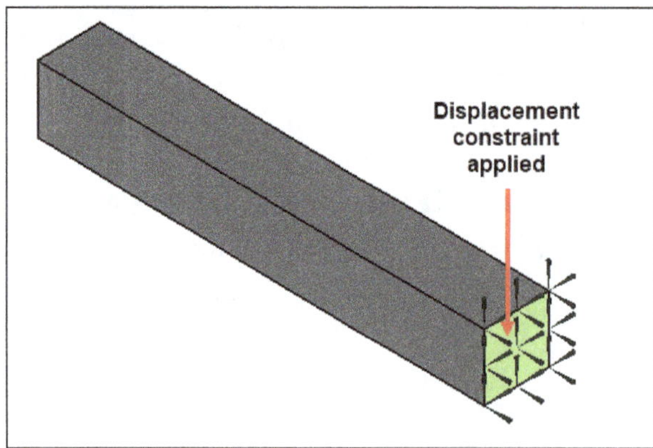

Figure-21. Applying displacement constraint

Applying Plane Rotation Constraint

The Plane Rotation constraint is applied on selected faces to allow rotation of model in plane parallel to selected face but restrict translation in any direction. This type of constraint is useful for fixing gears in an analysis so that they are free to rotate. The procedure to apply this constraint is given next.

- Click on the **Plane Rotation** tool from the **Toolbar** or **Model** menu. The **FEM constraint parameters** dialog will be displayed in the **Tasks** panel; refer to Figure-22.

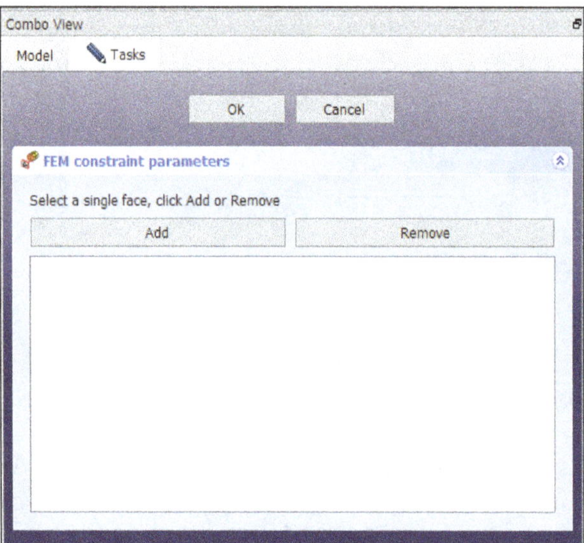

Figure-22. FEM constraint parameters dialog for plane rotation constraint

- Select the face on which you want to apply plane rotation constraint and click on the **Add** button from the dialog. The annotation of constraint will be displayed on the model face; refer to Figure-23.

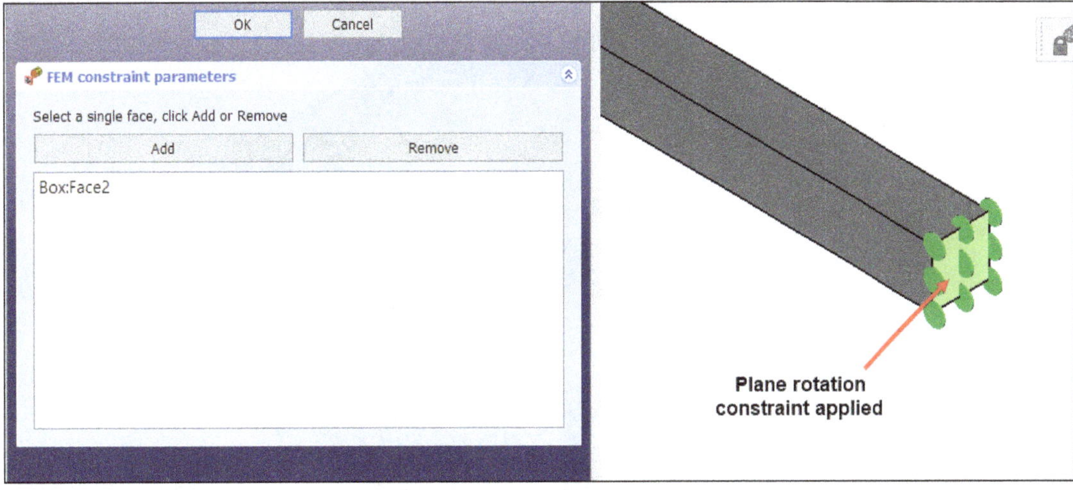

Figure-23. Applying plane rotation constraint

- After adding faces, click on the **OK** button from the dialog box to apply constraints and exit the dialog.

Applying Contact Constraint

The Contact constraint is applied to the model for making two selected faces in contact during the analysis. You can also specify the contact stiffness and friction coefficient between the faces when applying this constraint. This constraint is useful when you are working on an assembly of components. In assembly, you can define the parameters for contact between faces of two components. The procedure to apply this constraint is given next.

- Click on the **Contact Constraint** tool from the **Toolbar** or **Model** menu in **FEM** workbench. The **FEM constraint parameters** dialog box will be displayed as shown in Figure-24.

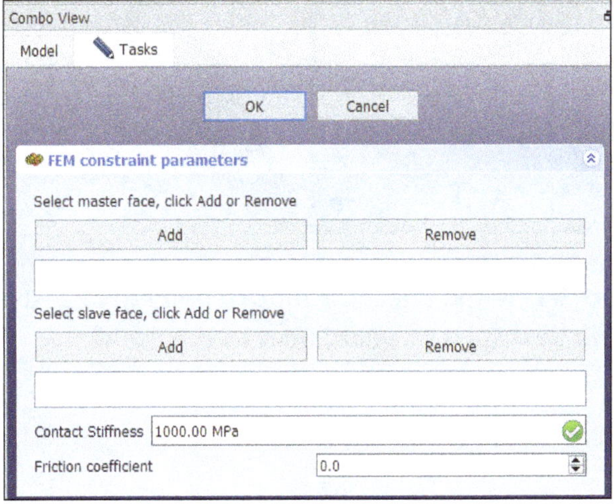

Figure-24. FEM constraint parameters dialog for contact constraint

- Select the face of first body to be used as master face and click on the **Add** button for **Select master face** section in the dialog; refer to Figure-25.

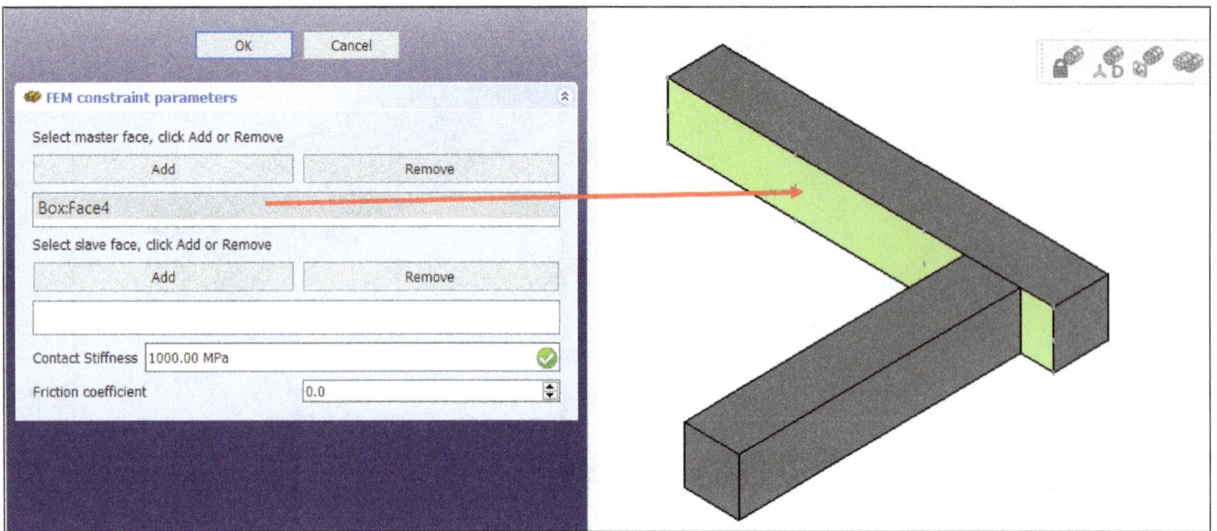

Figure-25. Master face selected

- Now, hide the master body by selecting any of its face and then pressing **SPACEBAR** from keyboard, so that you can select other face touching the master body.
- Select the slave face from other body and click on the **Add** button from **Select slave face** section of the dialog; refer to Figure-26.

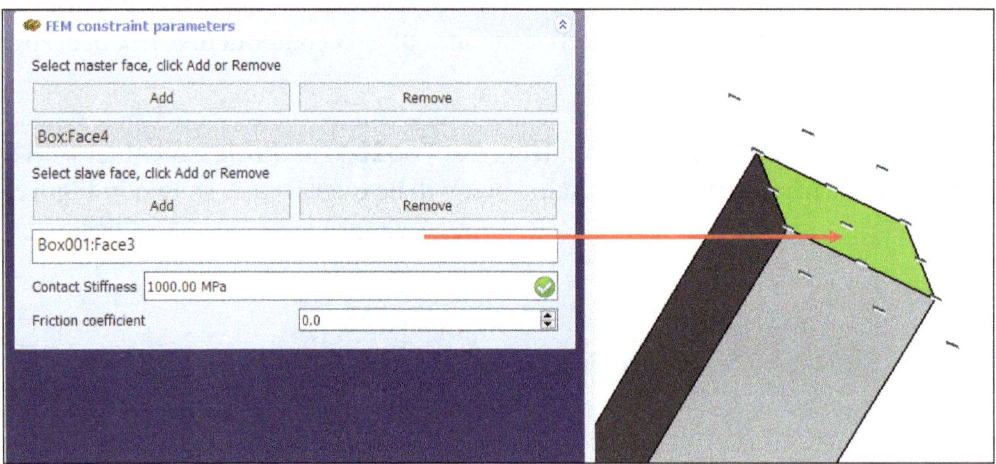

Figure-26. Slave face selected for constraint

- Open the **Model** panel in **Combo View** and select the hidden body to be displayed again and press **SPACEBAR**. The hidden body will be displayed again; refer to Figure-27.

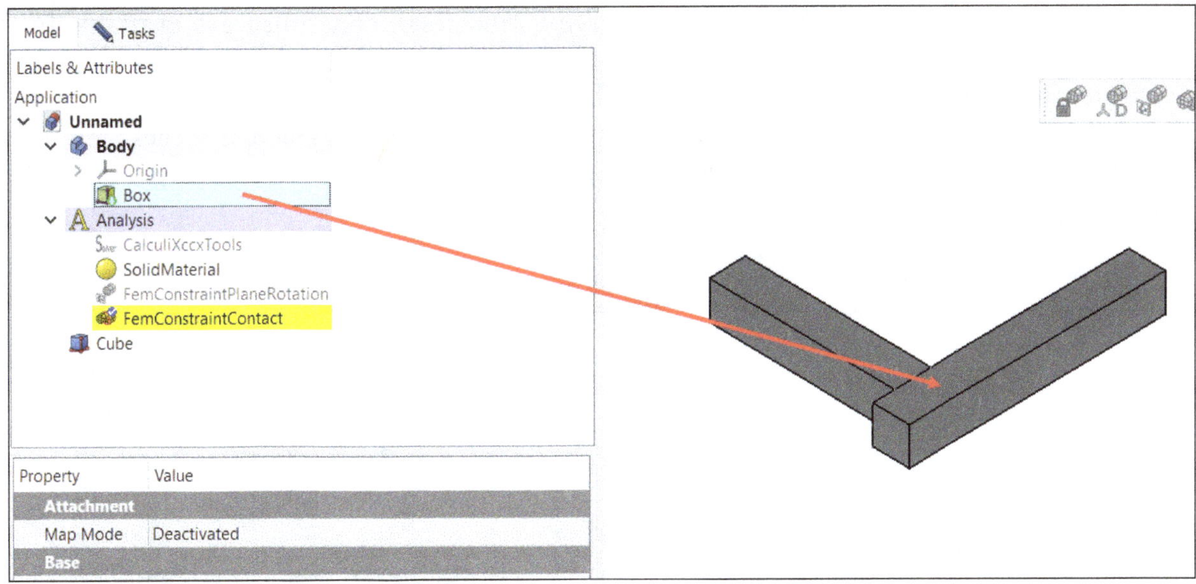

Figure-27. Displaying hidden body again

- Switch back to **Tasks** panel in the **Combo View** and specify desired value in **Contact Stiffness** edit box to define stiffness of contact point up to which the two faces will remain in contact under load.
- Click in the **Friction coefficient** edit box and specify desired value of friction coefficient between two faces.
- After setting desired parameters, click on the **OK** button from the dialog.

Applying Transform Constraint

The Transform constraint is used to transform coordinate system for faces on which displacement constraint is applied earlier. Using this constraint, you can modify the direction of displacement to desired orientation with respect to original coordinate system. The procedure to use this tool is given next.

- Click on the **Transform Constraint** tool from the **Toolbar** or **Model** menu in **FEM** workbench. The **FEM constraint parameters** dialog will be displayed in the **Tasks** panel of **Combo View**; refer to Figure-28.

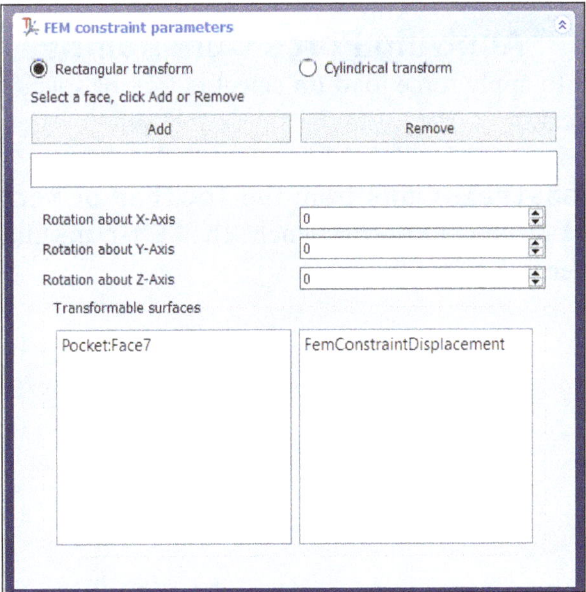

Figure-28. FEM constraint parameters dialog for transform constraint

- The surfaces/faces that can be used for applying transform constraint are available in the **Transformable surfaces** list box at the bottom in the dialog. Note that these are the faces on which displacement constraint has been applied.
- Select the face from model on which transform constraint is to be applied and click on the **Add** button from the dialog. The current orientation of coordinate system will be displayed on the face; refer to Figure-29.

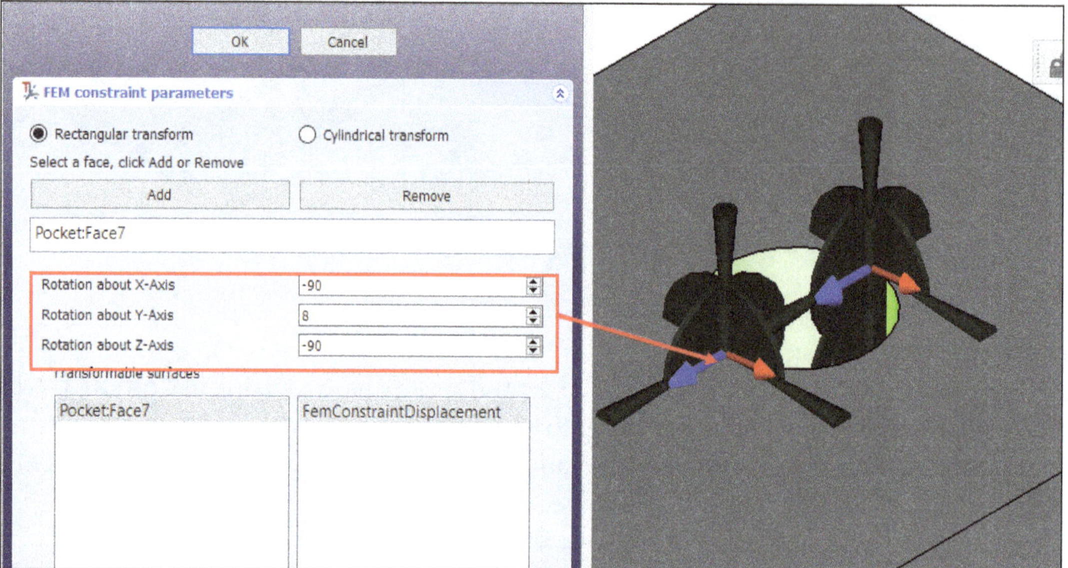

Figure-29. Orientation of coordinate system

- By default, the **Rectangular transform** radio button is selected, so you can apply rotation to each axis in Cartesian coordinate system. You can select the **Cylindrical** radio button to use Polar coordinate system for transformation.
- Set desired parameters in the **Rotation about X-Axis**, **Rotation about Y-Axis**, and **Rotation about Z-Axis** edit boxes. The coordinate system will orient according to the values specified in these edit boxes.
- Click on the **OK** button from the dialog to apply constraint.

Applying Force Constraint

The Force constraint is used to apply force load on selected face of specified value. The procedure to use this constraint is given next.

- Click on the **Force Constraint** tool from the **Toolbar** or **Mechanical Constraints** cascading menu of **Model** menu in **FEM** workbench. The **FEM constraint parameters** dialog box will be displayed; refer to Figure-30.

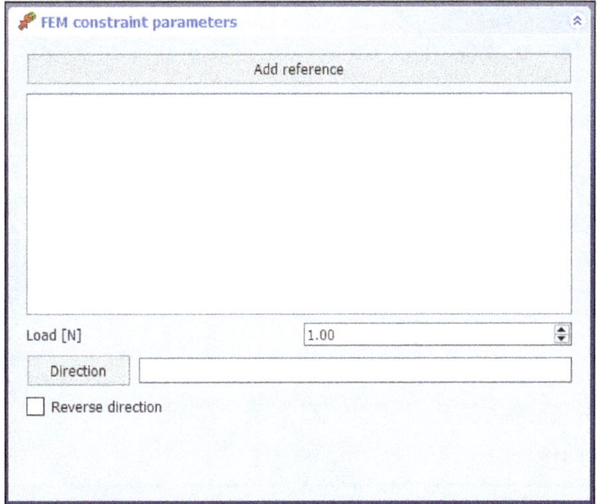

Figure-30. FEM constraint parameters dialog for force constraint

- Select desired face from the model on which you want to apply the load and specify desired load value in the **Area load** edit box.
- Select the **Reverse direction** check box to reverse direction of load.
- If you want to define desired direction for load then click on the **Direction** button and select desired reference; refer to Figure-31.

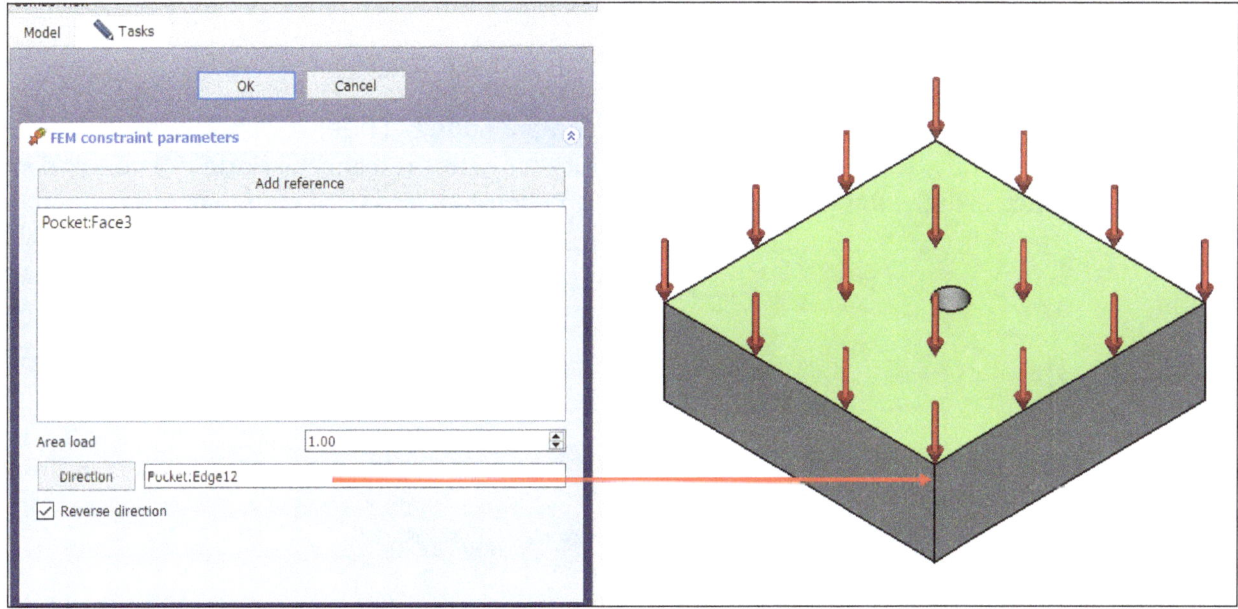

Figure-31. Defining direction for load

- After setting desired parameters, click on the **OK** button from the dialog.

Applying Pressure Load

The **Constraint Pressure** tool is used to define pressure load on selected face/surface. The procedure to use this tool is given next.

- Click on the **Constraint Pressure** tool from the **Toolbar** or **Model** menu in **FEM** workbench. The **FEM constraint parameters** dialog will be displayed; refer to Figure-32.

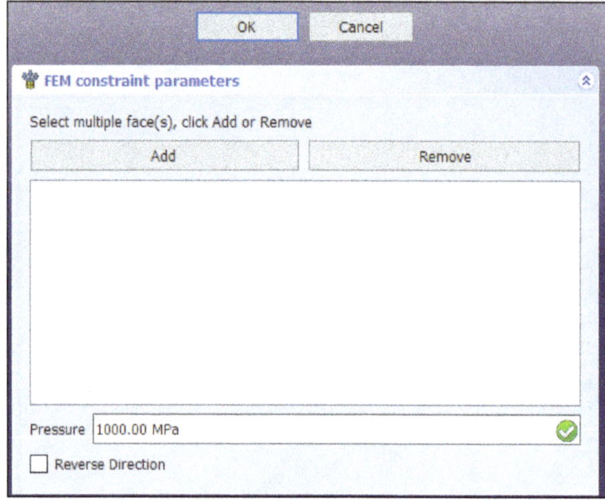

Figure-32. FEM constraint parameters dialog for pressure constraint

- Select the face(s) on which you want to specify pressure load and specify desired value in the **Pressure** edit box of the dialog.
- Specify the other parameters as discussed earlier and click on the **OK** button to apply load.

Applying Self Weight Constraint

The **Self Weight** constraint is applied on the model to check the effect of weight of the model in various situations. The procedure to use this tool is given next.

- Click on the **Constraint self weight** tool from the **Model** menu or from the **Toolbar**. The self weight constraint will be created and displayed in the **Model Tree**.
- If you want to modify the parameters of self weight then select it from the **Model Tree** and set desired parameters in the **Properties** box; refer to Figure-33.

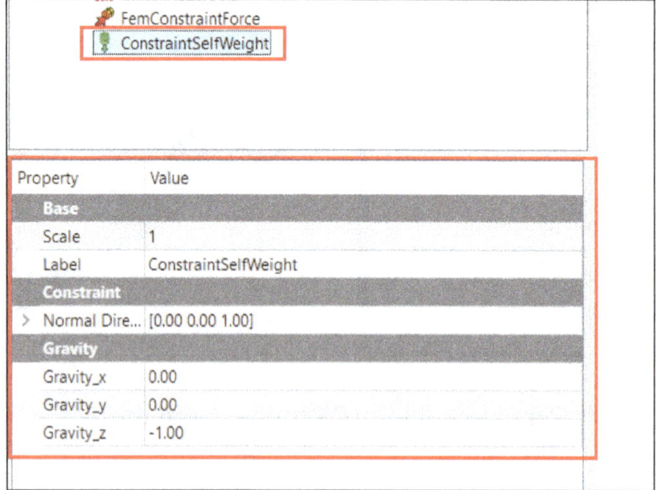

Figure-33. Self Weight properties

Applying Bearing Constraint

The Bearing constraint is used to apply constraints to selected faces for giving them bearing type load during analysis. This tool is not available by default in the Toolbar, so you need to access this tool from **Model** menu. The procedure to use this tool is given next.

- Click on the **Constraint bearing** tool from the **Mechanical Constraints** cascading menu of the **Model** menu in **FEM** workbench. The **FEM constraints parameters** dialog will be displayed as shown in Figure-34.

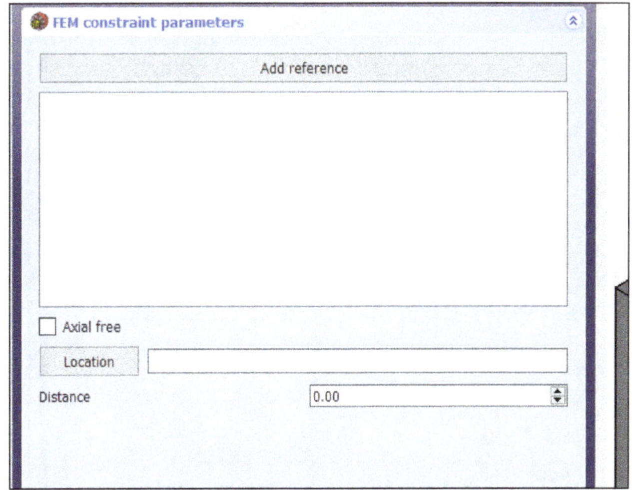

Figure-34. FEM constraint parameters dialog for bearing constraint

- Select the cylindrical face to which you want to apply bearing constraint.
- Select the **Axial free** check box if you want to allow translation of faces in direction parallel to cylindrical axis.
- Click on the **Location** button and select a point or line to define reference to be used for placing constraint.
- You can set the distance of constraint from selected point/line in the **Distance** spinner.
- After setting desired parameters, click on the **OK** button from the dialog. The annotation of constraint will be displayed on the model; refer to Figure-35.

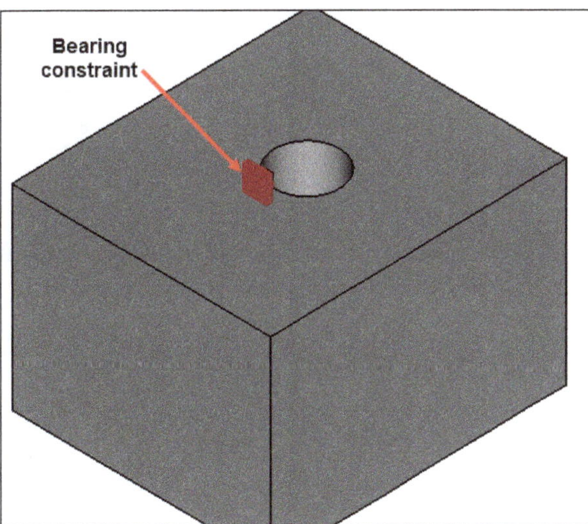

Figure-35. Bearing constraint

Applying Gear Constraint

The **Gear** constraint is used to restrict the rotation of selected face as if it is attached to a gear which needs specified load to be applied before it can rotate. In other words, a load equivalent to specified gear parameters will be applied on selected face/surface. The procedure to use this constraint is given next.

- Click on the **Constraint gear** tool from the **Mechanical Constraints** cascading menu of the **Model** menu. The **FEM constraint parameters** dialog will be displayed; refer to Figure-36.

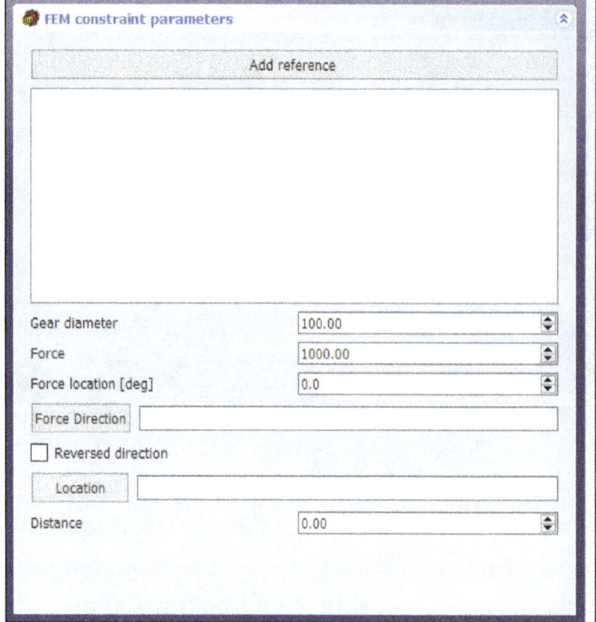

Figure-36. FEM constraint parameters dialog for gear constraint

- Select the cylindrical face to which you want to apply gear constraint. Preview of constraint will be displayed; refer to Figure-37.

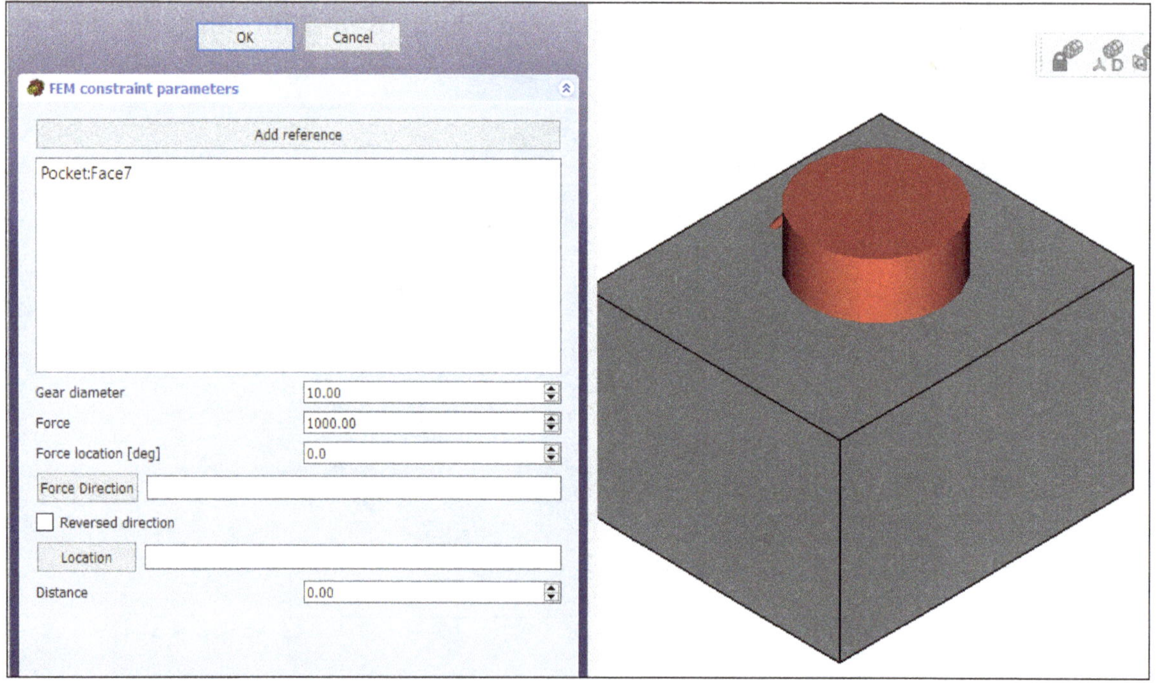

Figure-37. Preview of gear constraint

- Specify desired value in **Gear diameter** edit box to specify the diameter of gear to be used for constraining. This should represent actual gear attached to selected face.
- Specify desired parameters in the **Force** and **Force location[deg]** edit boxes to specify value and direction of force load on gear.
- Set the other parameters as discussed earlier and click on the **OK** button to apply the constraint.

Applying Pulley Constraint

The **Pulley** constraint is used to apply load on selected cylindrical face/surface equivalent to pulley load based on specified parameters. The procedure to apply this constraint is given next.

- Click on the **Constraint pulley** tool from the **Mechanical Constraints** cascading menu of **Model** menu in **FEM** workbench. The **FEM constraint parameters** dialog will be displayed as shown in Figure-38.

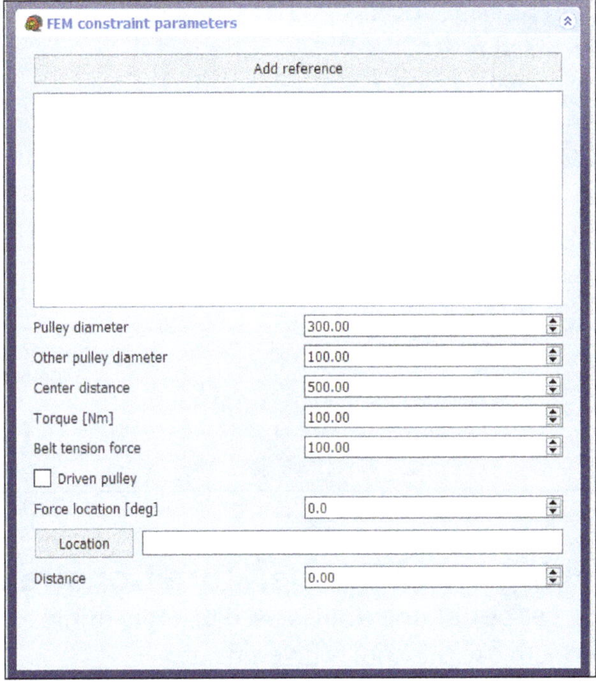

Figure-38. FEM constraint parameters dialog for pulley constraint

- Select the cylindrical face on which you want to apply pulley constraint (load). Preview of pulley constraint will be displayed; refer to Figure-39.

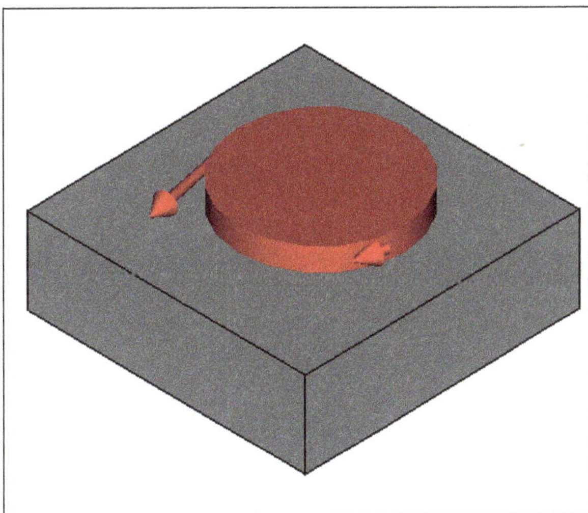

Figure-39. Preview of pulley constraint

- Set the parameters related to pulley in the edit boxes of dialog. Note that by default, selected cylindrical face is assigned as driving pulley. If you want to constraint selected face as driven pulley then select the **Driven pulley** check box from the dialog.
- After setting desired parameters, click on the **OK** button from dialog to apply constraint.

Thermal Constraints

The **Thermal** constraints are used to define various thermal conditions like temperature, heat flux, body heat source, and so on. The tools to apply thermal constraints are available in the **Thermal Constraints Toolbar** and **Thermal Constraints** cascading menu of the **Model** menu; refer to Figure-40. These tools are discussed next.

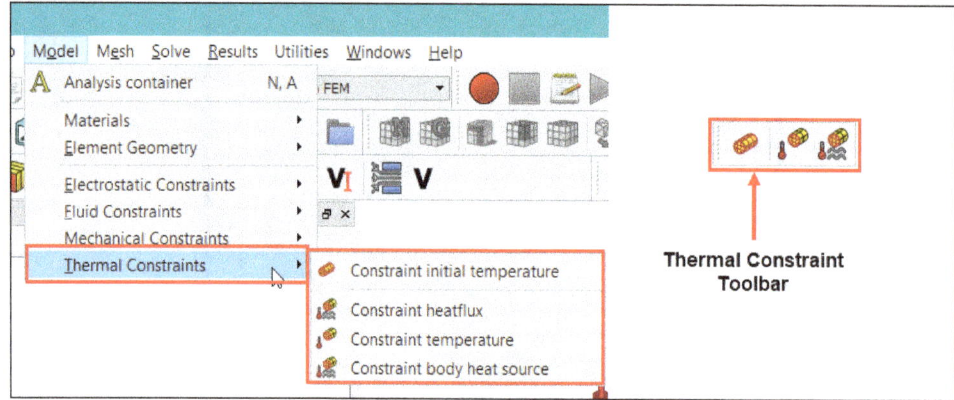

Figure-40. Thermal constraint tools

Defining Initial Temperature Constraint

The **Constraint initial temperature** tool is used to apply initial temperature to the model for thermal analysis. The procedure to use this tool is given next.

- Click on the **Constraint initial temperature** tool from the **Thermal Constraints Toolbar**. The **FEM constraint parameter** dialog will be displayed; refer to Figure-41.

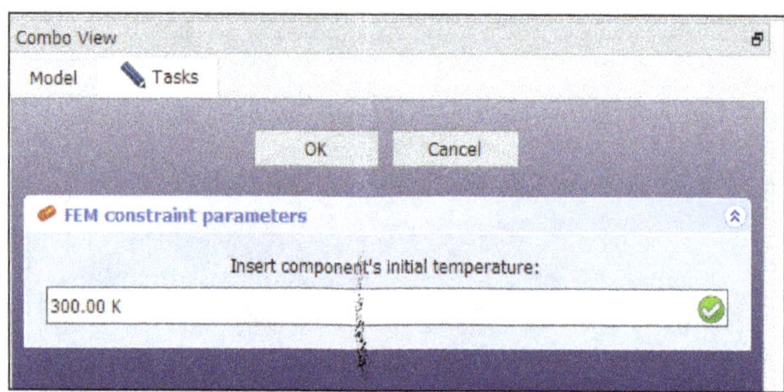

Figure-41. FEM constraint parameters dialog for initial temperature constraint

- Specify desired value of initial temperature in the edit box of dialog. Note that you need to specify temperature in Kelvin (K) unit.
- After specifying desired value, click on the **OK** button to apply constraint.

Applying Heat Flux Constraint

Heat flux is the rate of heat energy transfer from selected face(s). The **Constraint heatflux** tool in **Thermal Constraints** cascading menu of **Model** menu is used to define the heat flux for selected faces. The procedure to apply this constraint is given next.

- Click on the **Constraint heatflux** tool from **Thermal Constraints** cascading menu of **Model** menu. The **FEM constraint parameters** dialog will be displayed as shown in Figure-42.

- Select the **Surface Convection** radio button if you want to define heat convection rate for selected face/surface at specified temperature. Heat convection is the amount of heat passing through selected face/surface. Select the **Surface heat flux** radio button if you want to specify heat energy transfer rate.

- Specify the parameters as desired in the edit boxes of the dialog.
- Select the face on which you want to apply heat flux constraint and click on the **Add** button from the dialog. If you want to apply heat flux to multiple faces then you can select multiple faces while holding the **CTRL** key.
- After setting desired parameters, click on the **OK** button from the dialog. The constraint will be applied; refer to Figure-43.

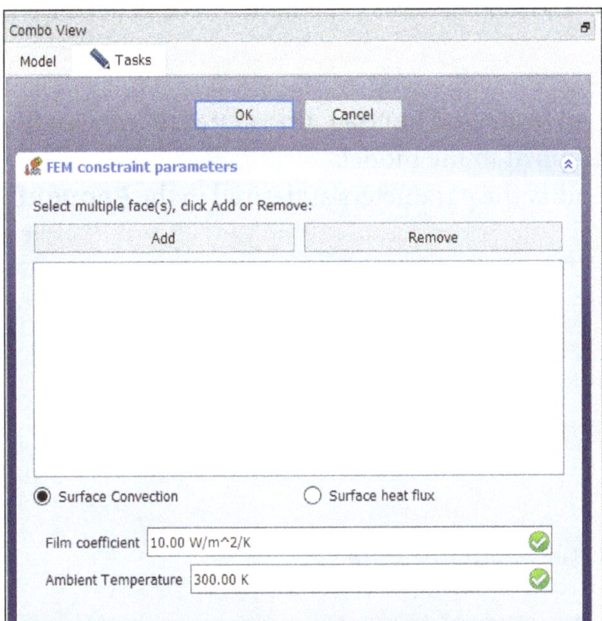

Figure-42. FEM constraint parameters dialog for heat flux constraint

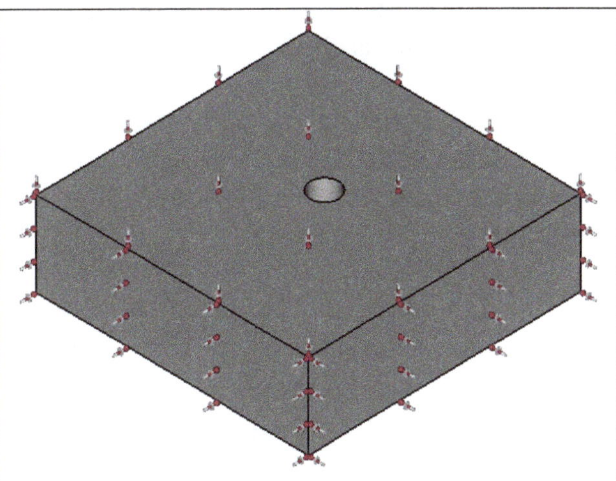

Figure-43. Heat flux constraint applied on faces

Applying Temperature Constraint

The **Constraint temperature** tool is used to apply fix temperature to selected faces. Applying fix temperature to a face makes that face a heat source for analysis. The procedure to apply this constraint is given next.

- Click on the **Constraint temperature** tool from the **Thermal Constraints** cascading menu in the **Model** menu. The **FEM constraint parameters** dialog will be displayed; refer to Figure-44.

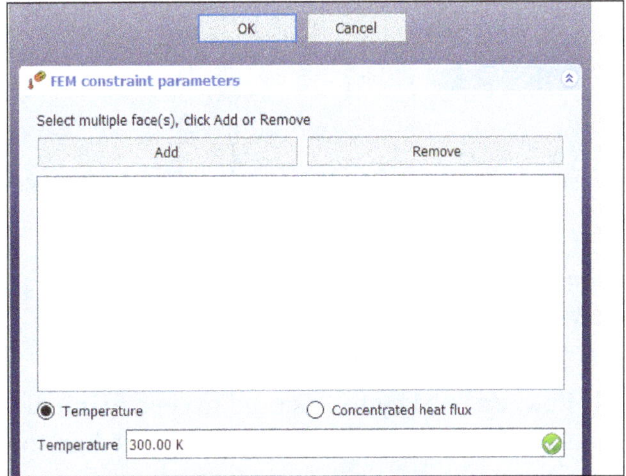

Figure-44. FEM constraint parameters dialog for thermal constraint

- Select the face(s) to which you want to apply thermal constraint and click on the **Add** button.
- Select the **Temperature** radio button to specify temperature of selected face(s).

- Select the **Concentrated heat flux** radio button from the bottom in the dialog if you want to specify total heat energy of the selected face.
- Specify desired value of temperature/heat flux in the edit box below the radio buttons.
- After setting desired parameters, click on the **OK** button from the dialog to apply the constraint.

Applying Body Heat Source Constraint

The **Body Heat Source** constraint is used to define the heat generated by model. The procedure to apply this constraint is given next.

- Click on the **Constraint body heat source** tool from the **Thermal Constraints** cascading menu of the **Model** menu. The constraint will be applied to the model.
- Select the constraint from the **Model Tree** and modify the parameters as desired in the **Property editor**; refer to Figure-45.

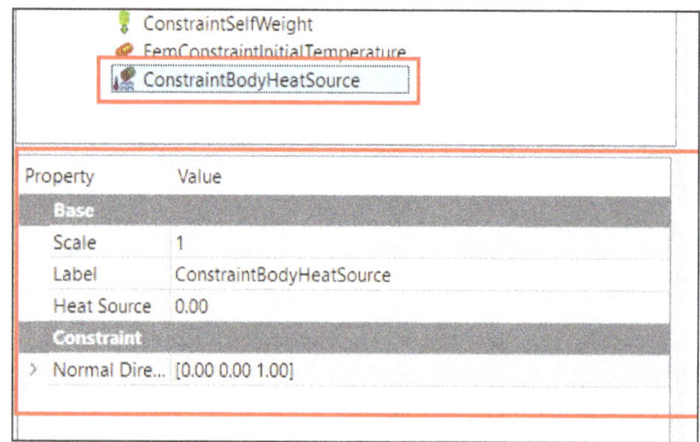

Figure-45. Property editor for body heat source constraint

- Specify desired value of heat in the **Heat Source** edit box. Click anywhere in the model area to exit the editing mode.

Fluid Constraints

The **Fluid** constraints are used to define boundary conditions for fluid flow in the model. These tools are available in the **Fluid Constraints Toolbar** and **Fluid Constraints** cascading menu of the **Model** menu; refer to Figure-46. These tools are discussed next.

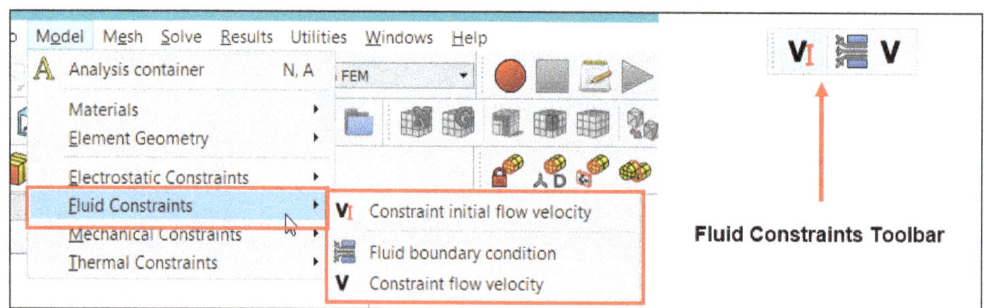

Figure-46. Fluid Constraint tools

Applying Initial Flow Velocity

The **Constraint initial flow velocity** tool is used to specify initial velocity for flow of fluid in the model. This constraint is useful for fluid flow analysis. The procedure to use this tool is given next.

- Click on the **Constraint initial flow velocity** tool from the **Fluid Constraints** cascading menu of the **Model** menu. The **Constraint Properties** dialog will be displayed; refer to Figure-47.

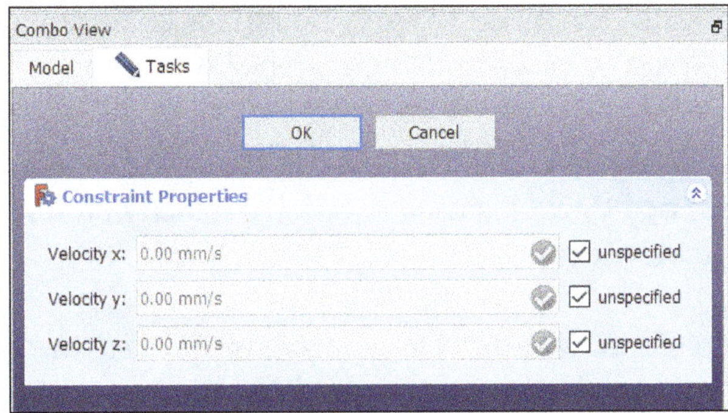

Figure-47. Constraint Properties dialog

- Clear the unspecified check box(es) for desired velocity direction(s) and specify the value(s) in respective edit box(es).
- After setting desired parameters, click on the **OK** button from the dialog to apply constraint.

Defining Fluid Boundary Conditions

The **Fluid boundary condition** tool is used to define boundary conditions for performing fluid flow analysis. The procedure to apply fluid boundary conditions is given next.

- Click on the **Fluid boundary condition** tool from the **Fluid Constraints** cascading menu of the **Model** menu. The **FEM constraint parameters** dialog will be displayed; refer to Figure-48.

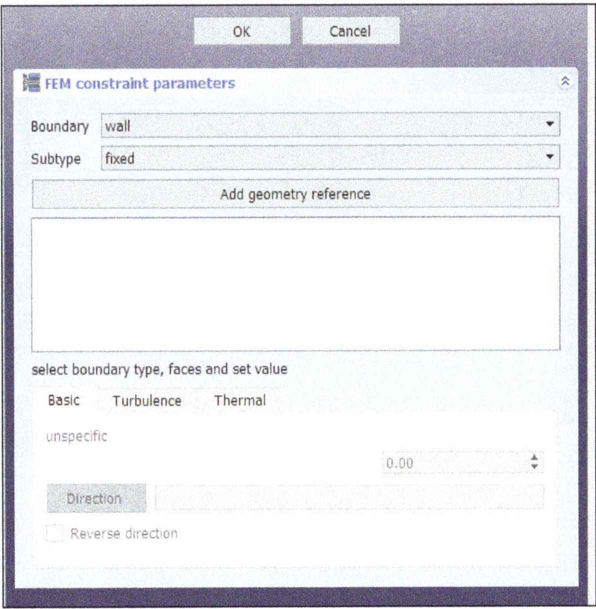

Figure-48. FEM constraint parameters dialog for fluid boundary conditions

- Select desired option from the **Boundary** drop-down in the dialog. Select the **inlet** option from the drop-down if you want to specify properties of fluid like pressure, velocity, flow rate, and so on at the inlet of fluid. Select the **wall** option from the drop-down if you want to specify physical conditions for wall in fluid domain like wall is fixed, slipping, partial slipping with specified slip ratio, moving at specified speed, rough, and so on. Select the **outlet** option from the drop-down if you want to specify fluid properties at the outlet like total pressure, static pressure, uniform velocity, or outflow parameter for fluid. Select the **interface** option from the drop-down if you want to specify symmetry parameters for selected face(s). Select the **freestream** option from the drop-down if you want to apply free fluid flow condition at selected face. The freestream condition is applicable at both inlet as well as outlet.
- After selecting desired option from the **Boundary** drop-down, select the related option from the **Subtype** drop-down below it.

- Select desired face(s) on which you want to apply boundary conditions.
- Specify the other parameters as desired and click on the **OK** button.

Applying Constraint Flow Velocity

The **Constraint Flow Velocity** tool is used to specify velocity of fluid through selected face. The procedure to use this tool is given next.

- Click on the **Constraint Flow Velocity** tool from the **Fluid Constraints** cascading menu of the **Model** menu. The **Select Faces/Edges/Vertexes** and **Constraint Properties** dialogs will be displayed in the **Tasks** panel of **Combo View**; refer to Figure-49.

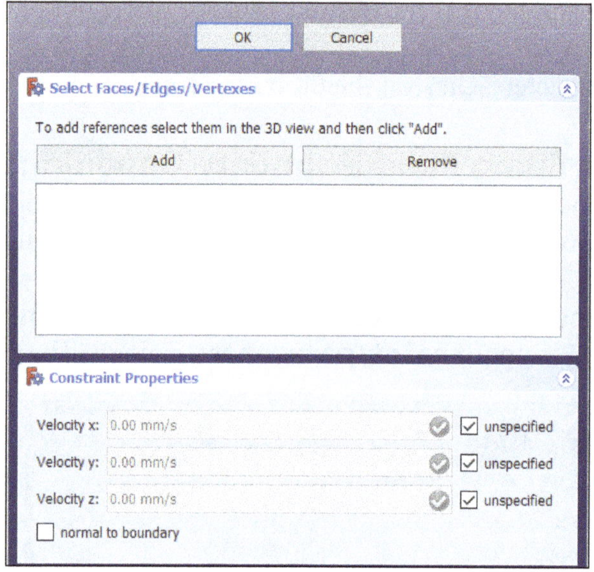

Figure-49. Constraint Properties and Select objects dialog

- Select the faces, edges, or vertices for which you want to define fluid velocity and click on the **Add** button.
- Set the other parameters as discussed earlier and click on the **OK** button.

Applying Electrostatic Potential Constraint

The Electrostatic Potential is used to specify electrical potential (voltage) at selected faces/edges/vertices. The procedure to use this tool is given next.

- Click on the **Constraint electrostatic potential** tool from **Electrostatic Constraints** cascading menu of the **Model** menu. The **Select Faces/Edges/Vertexes** and **Constraint Properties** dialogs will be displayed in the **Tasks** panel of **Combo View**.
- Select the faces/edges/vertices on which you want to apply electrostatic potential constraint and click on the **Add** button from the dialog.
- Clear the unspecified check box for **Potential** edit box and specify desired value of voltage.
- Select the **Potential Constant** check box from the dialog to keep voltage constant in analysis.
- After setting desired parameters, click on the **OK** button from the dialog to apply the constraint.

CREATING GEOMETRY ELEMENTS

In FreeCAD FEM, you can create 2D elements of specified thickness by using the tools in **Element Geometry** toolbar or **Element Geometry** cascading menu of the **Model** menu; refer to Figure-50. These tools are discussed next.

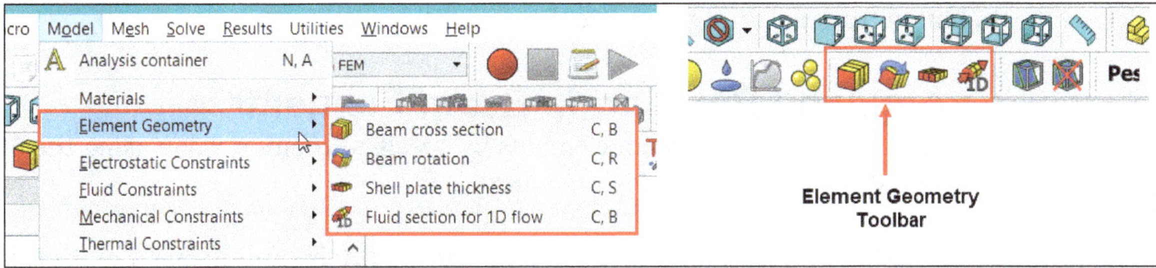

Figure-50. Element Geometry tools

Creating Beam Cross-section

The **Beam cross section** tool is used to create beam elements with different type of cross sections like circular beam, rectangular beam, and pipe beam. To create beam elements, you need to select reference edges from the model. These edges can be of a surface or a solid. The procedure to use this tool is given next.

- Click on the **Beam cross section** tool from the **Element Geometry** cascading menu of **Model** menu. The **Beam section parameters** and **Geometry reference selector for a Edge** dialogs will be displayed; refer to Figure-51.

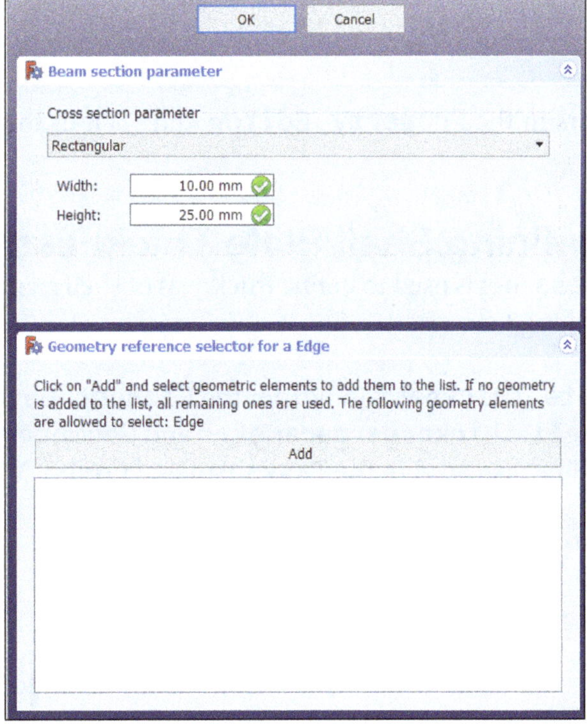

Figure-51. Beam section parameters and geometry selector dialogs

- Select desired option from the **Cross section parameter** drop-down in the dialog. Select the **Rectangular** option from the drop-down to create rectangular beam element. Select the **Circular** option to create circular rod elements. Select the **Pipe** option from the drop-down to create tube like elements.
- After selecting desired option from the drop-down, specify the related parameters in edit box(es) below the drop-down.
- After setting desired parameters in the dialog, click on the **Add** button from the **Geometry reference selector for an Edge** dialog and select the edge on which you want to create the element. Note that you need to click on **Add** button each time before you select an edge for creating element.

- Click on the **OK** button from the dialog to create the elements.

Applying Rotation to Beam Elements

You can change the default orientation of beam elements by specified rotation value using the **Beam rotation** tool. The procedure to use this tool is given next.

- Click on the **Beam rotation** tool from the **Element Geometry** cascading menu of the **Model** menu. The **ElementRotation1D** feature will be added in the **Model Tree**.
- Select the **ElementRotation1D** feature from **Model Tree**. The **Property Editor** dialog will be displayed; refer to Figure-52.

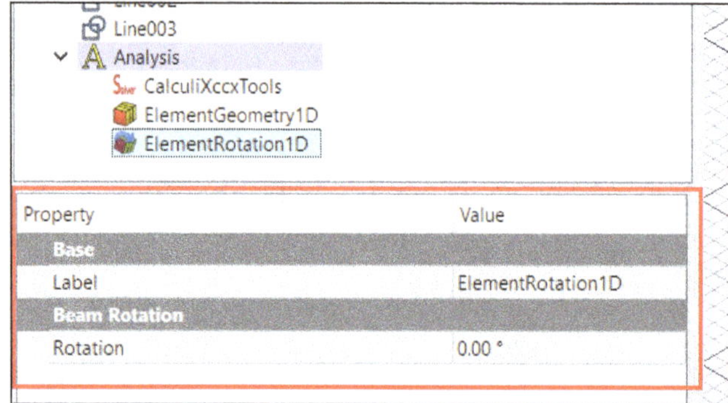

Figure-52. Beam Rotation Element properties

- Specify desired parameters in the **Property Editor** and click in the empty area of drawing to exit editing model.

Defining Shell Plate Thickness

The **Shell plate thickness** tool is used to define thickness of shell element based on selected face. The procedure to use this tool is given next.

- Click on the **Shell plate thickness** tool from the **Element Geometry** cascading menu of the **Model** menu. The **Shell thickness parameter** and **Geometry reference selector for a Face** dialogs will be displayed in the **Tasks** panel of **Combo View**; refer to Figure-53.

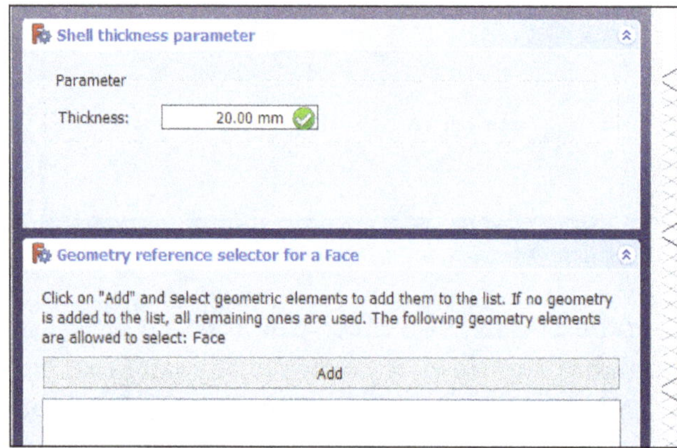

Figure-53. Dialogs for shell element

- Specify desired value of thickness for shell element in the **Thickness** edit box of the dialog.
- Click on the **Add** button from the dialog and select the face on which you want to create the shell element of specified thickness.
- After setting desired parameters, click on the **OK** button from the dialog to create the geometry.

Creating Fluid Section for 1D Flow

The **Fluid section for 1D flow** tool is used to create section in the model for fluid flow. The procedure to use this tool is given next.

- Click on the **Fluid section for 1D flow** tool from the **Element Geometry** cascading menu of the **Model** menu. The **Form** and **Geometry reference selector for a Edge** dialogs will be displayed; refer to Figure-54.

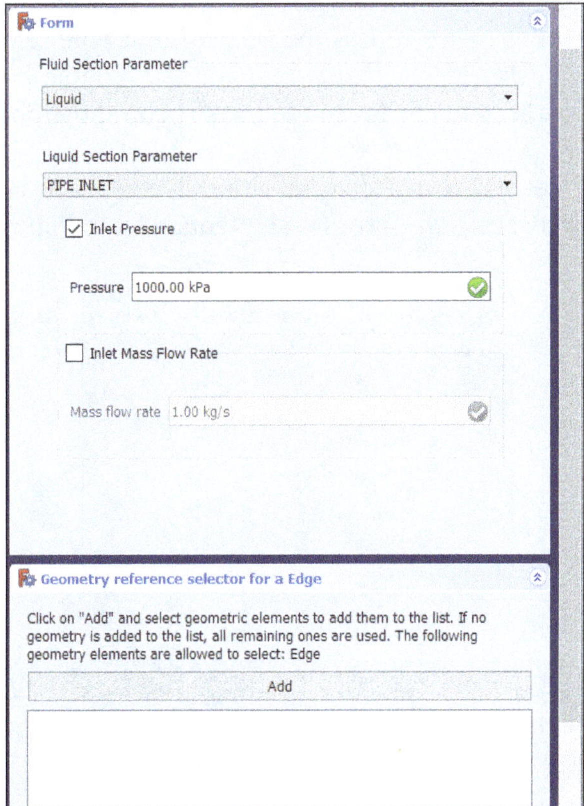

Figure-54. Dialogs for 1D fluid section

- Select desired option from the **Fluid Section Parameter** drop-down to define whether fluid section is for liquid, gas, or it is open for both liquid and gas. The other parameters will be displayed based on selected option.
- If you have selected **Liquid** option for fluid section then select desired option from the **Liquid Section Parameter** drop-down to specify whether the section is at inlet, outlet, diaphragm, or any other section of fluid domain.
- Specify the other parameter as needed and click on the **Add** button from the dialog. You will be asked to select an edge to define location of section.
- Select desired edge and click on the **OK** button from the dialog to create the section.

Note that the functionality to work with Beam elements has not been added yet up to version 18.0, so you will not be able to work with them in software.

MESHING

Meshing is the base of FEM (Finite Element Method) which is one of the methods used for FEA. Meshing divides the solid/shell models into elements of finite size and shape. These elements are joined at some common points called nodes. These nodes define the load transfer from one element to other element. Meshing is a very crucial step in design analysis. The automatic mesher in the software generates a mesh based on a global element size, tolerance, and local mesh control specifications. Mesh control lets you specify different sizes of elements for components, faces, edges, and vertices.

The software estimates a global element size for the model taking into consideration its volume, surface area, and other geometric details. The size of the generated mesh (number of nodes and elements) depends on the geometry and dimensions of the model, element size, mesh tolerance, mesh control, and contact specifications. In the early stages of design analysis where approximate results may suffice, you can specify a larger element size for a faster solution. For a more accurate solution, a smaller element size may be required.

Meshing generates 3D tetrahedral solid elements, 2D triangular shell elements, and 1D beam elements. A mesh consists of one type of elements unless the mixed mesh type is specified. Solid elements are naturally suitable for bulky models. Shell elements are naturally suitable for modeling thin parts (sheet metals), and beams and trusses are suitable for modeling structural members.

There are two plug-ins in FreeCAD to perform meshing: Netgen and Gmsh. The tools to perform meshing using these plug-ins are available in the **Mesh** menu; refer to Figure-55. The tools in this menu are discussed next.

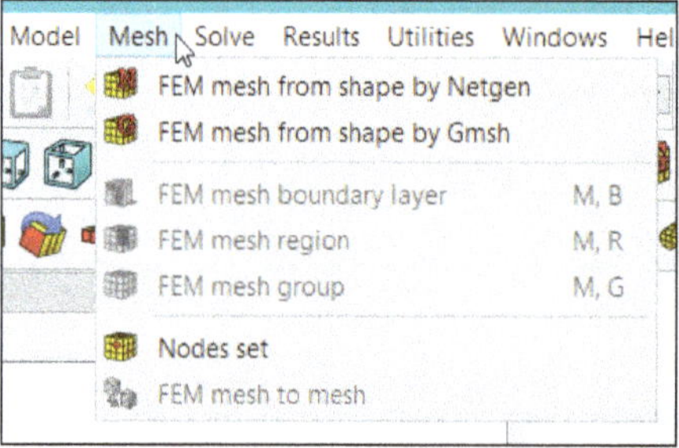

Figure-55. Mesh menu

Creating FEM mesh using Netgen Mesher

The **FEM mesh from shape by Netgen** tool is used to create mesh from selected shape(body) using Netgen plug-in. The procedure to use this tool is given next.

- Click on the **FEM mesh from shape by Netgen** tool from the **Mesh** menu after selecting the body. The **Tet Parameter** dialog will be displayed; refer to Figure-56.

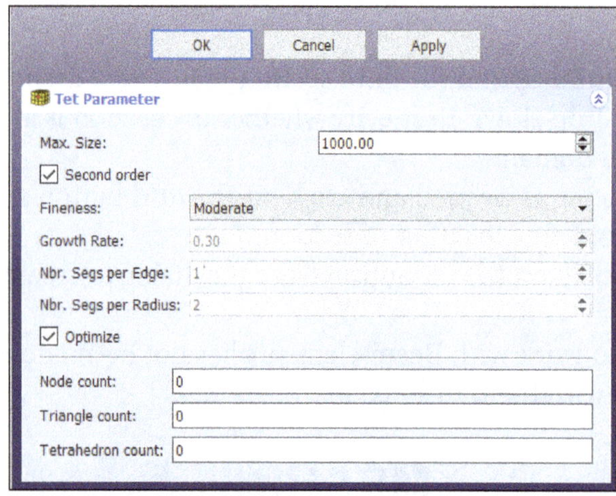

Figure-56. Tet Parameter dialog

- Specify desired value in **Max. Size** edit box to define maximum size of element that can be generated in current mesh. Note that this size can be length, width, or height of the element.
- Select the **Second order** check box if you want to generate elements in mesh with additional nodes that define curvature of segments; refer to Figure-57.

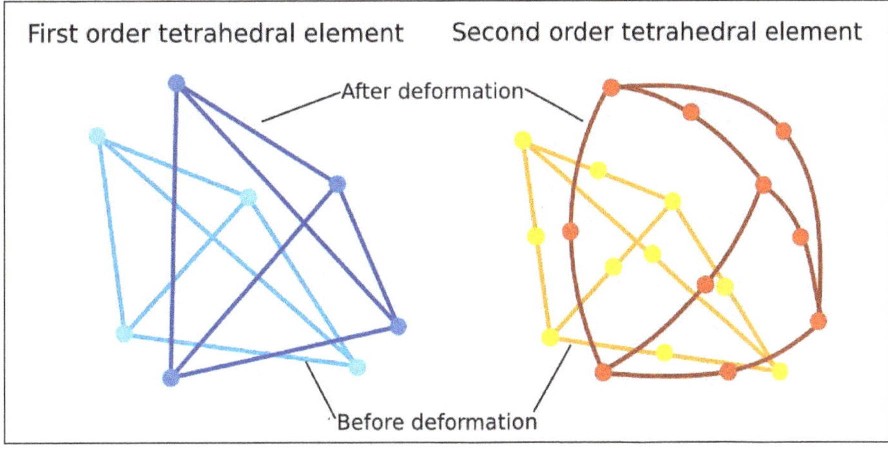

Figure-57. Difference between first order and second order mesh

- Select desired option from the **Fineness** drop-down to define smoothness in mesh elements at edges. A finer mesh gives better results of analysis but it also increases the computational time for analysis. Figure-58 shows same object mesh with both very coarse and very find mesh types.

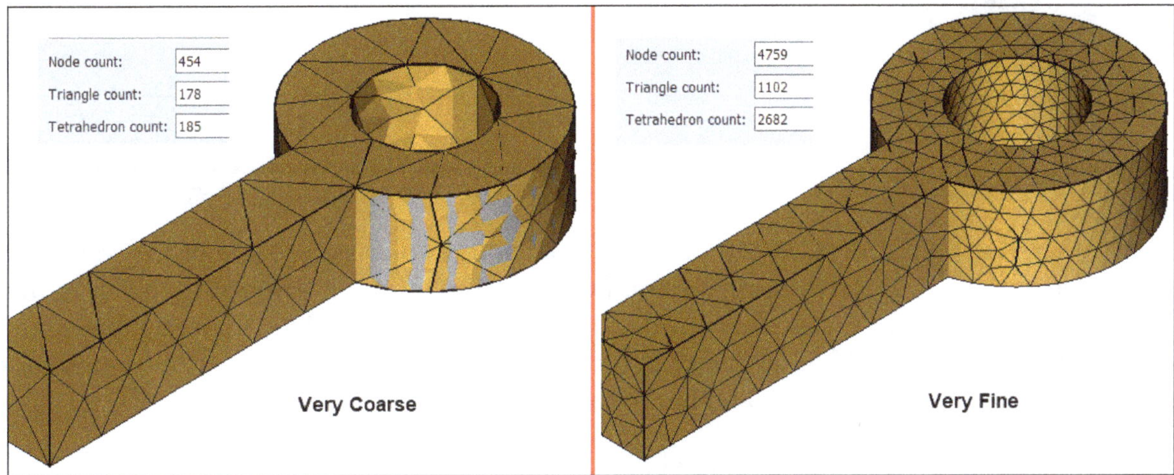

Figure-58. Mesh fineness

Note that a simple move from very coarse to very fine mesh can exponentially increase the number of elements as in our case. There is a certain threshold in meshing fineness after which you get a marginal accuracy at the loss of a lot of computing power and time. So, you should start with a moderate fineness before moving to finer side when performing analysis.

- If you want to manually specify the smoothness of your mesh then select the **UserDefined** option from the **Fineness** drop-down. The edit boxes below the drop-down will become active.
- Specify desired value in **Growth Rate** edit box to define increase in edge length of element when moving away from edges or complex regions of the model.
- Specify desired value in **Nbr. Segs per Edge** edit box to define number of segments in which the edges of model will be divided in mesh. Note that the mesh will be a compromise between maximum element size and number of segments per edge values. So, you may or may not get the definite number of segments per edge depending on the model.
- Specify desired value in **Nbr. Segs per Radius** edit box to define number of segments in circular/arc type edges.
- Select the **Optimize** check box to automatically increase number of elements near edges and complex regions.
- After setting desired parameters, click on the **OK** button from the dialog. The mesh will be generated; refer to Figure-59.

Creating FEM mesh using Gmsh Mesher

The **FEM mesh from shape by Gmsh** tool is used to create mesh using the Gmsh plug-in. The procedure to use this tool is given next.

- Click on the **FEM mesh from shape by Gmsh** tool from **Mesh** menu after selecting the body. The **FEM Mesh by Gmsh** dialog will be displayed; refer to Figure-60.

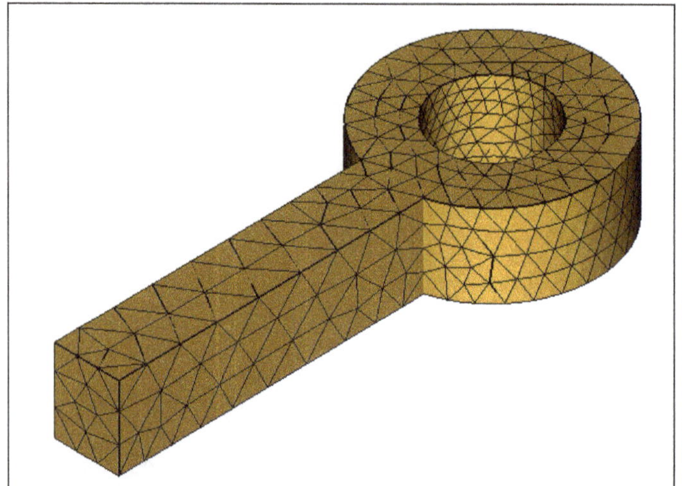

Figure-59. Mesh generated

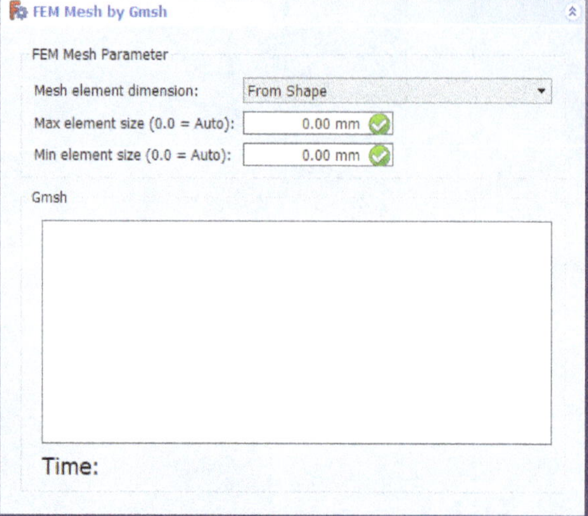

Figure-60. FEM Mesh by Gmsh dialog

- Select desired option from the **Mesh element dimension** drop-down to define what type of mesh elements are to be created. By default, **From Shape** option is selected in this drop-down, so dimension of element is automatically decided based on selected object. For example, for a 3D solid body, 3D elements will be generated in the mesh. You can manually specify the dimensions of element by selecting 1D, 2D, or 3D option from the drop-down.

The element types available in FreeCAD FEM are as follows:

1D Elements:

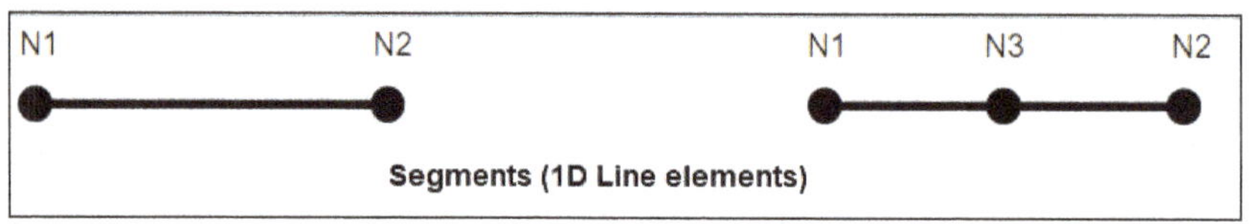

Figure-61. 1D elements

2D Elements:

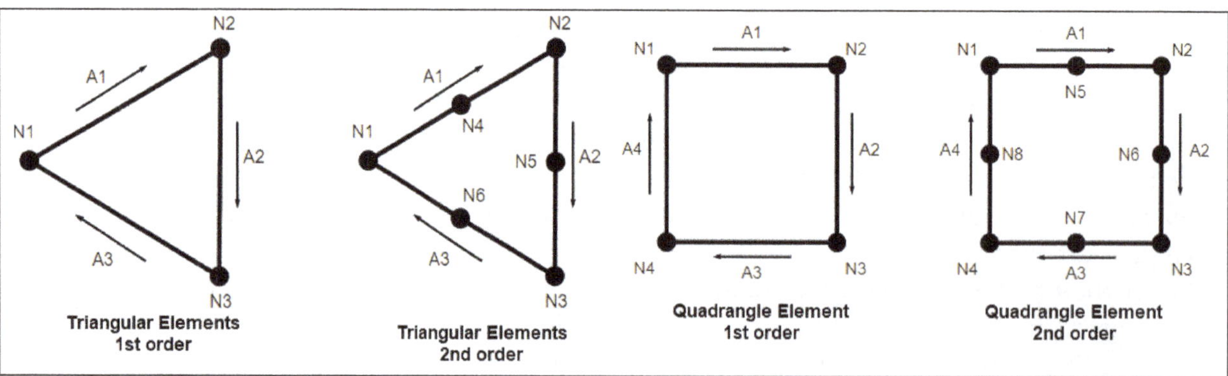

Figure-62. 2D elements

3D Elements:

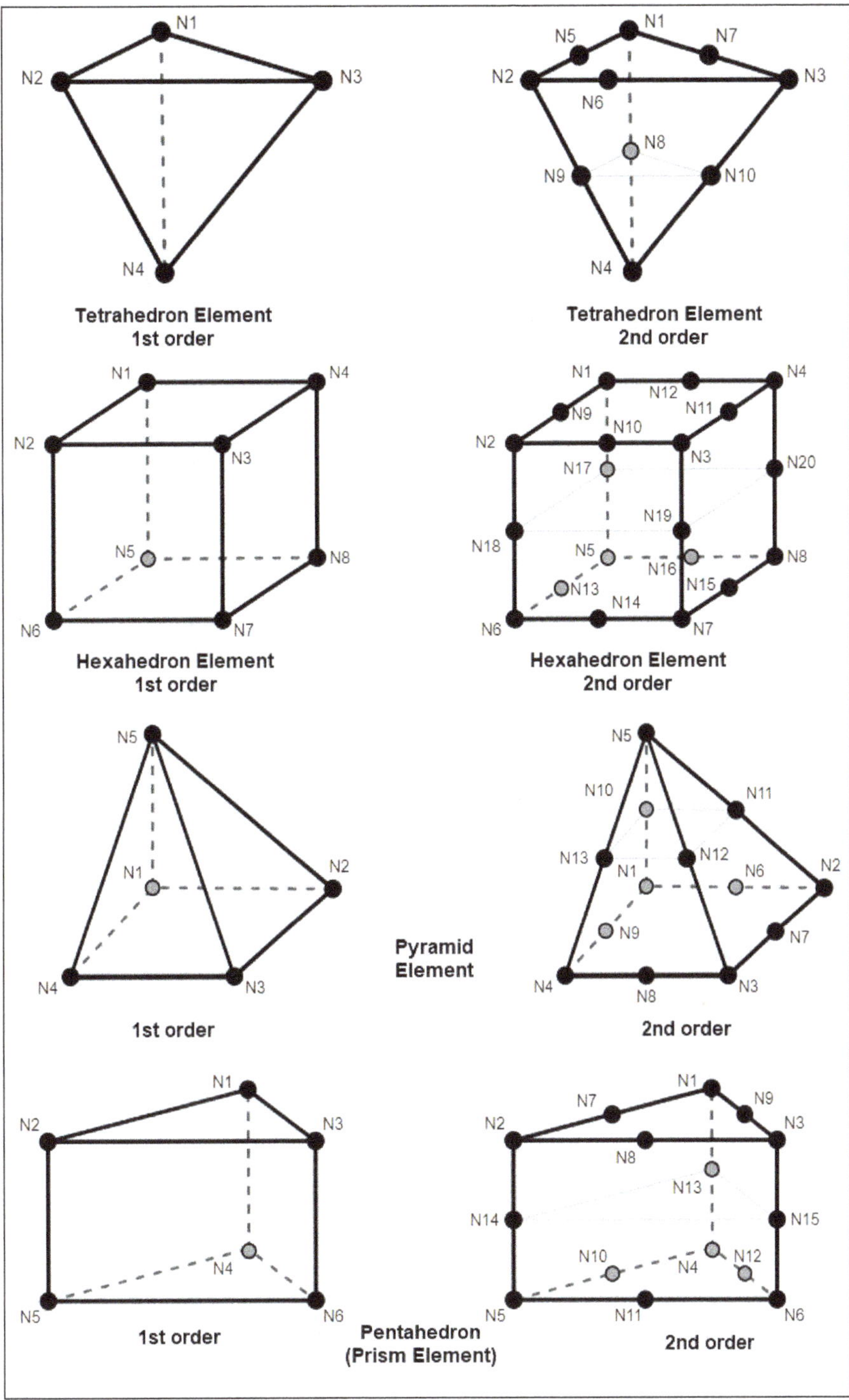

Figure-63. 3D elements

- Specify desired values in **Max element size** and **Min element size** edit boxes to define the maximum and minimum element size in mesh.
- After setting desired parameters, click on the **Apply** button to generate mesh.

Creating Boundary Mesh Layer

The **FEM mesh boundary layer** tool is used to create a layer of mesh with different element size from main mesh at the boundary of model. The procedure to use this tool is given next.

- Click on the **FEM mesh boundary layer** tool from the **Mesh** menu after selecting the **FEMMeshGmsh** feature (mesh created by Gmsh) from **Model Tree**. The **Mesh boundary layer settings** and **Geometry reference selector for a Solid, Face, Edge, Vertex** dialogs will be displayed; refer to Figure-64.

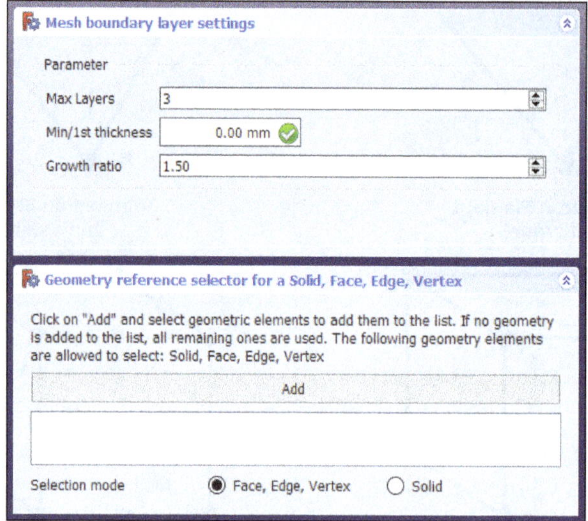

Figure-64. Mesh boundary layer settings and Geometry reference selector dialogs

- Specify desired value in the **Max Layers** edit box to define number of layers to be meshed by boundary layer settings.
- Specify desired value in the **Min/1st thickness** edit box to define the thickness of 1st layer in the mesh boundary.
- Specify desired value in **Growth ratio** edit box to specify the rate of increase/decrease in element size while moving away from boundary regions.
- Click on the **Add** button and select the solid/face/edge/vertex at which you want to apply boundary layer mesh.
- After setting desired parameters, click on the **OK** button from the dialog. The mesh will be created.

Create FEM mesh region

The **FEM mesh region** tool is used to create meshing for specific region of model with different element size from main mesh. This tool is useful for creating finer mesh in complex regions of model. The procedure to use this tool is given next.

- Click on the **FEM mesh region** tool from the **Mesh** menu. The **Mesh region** and **Geometry reference selector for a Solid, Face, Edge, Vertex** dialogs will be displayed; refer to Figure-65.
- Specify desired value of maximum element size for refined mesh in the **Max element size** edit box.
- Select desired radio button for **Selection mode** and add the geometry region to be refined.
- After setting desired parameters, click on the **OK** button from the dialog.

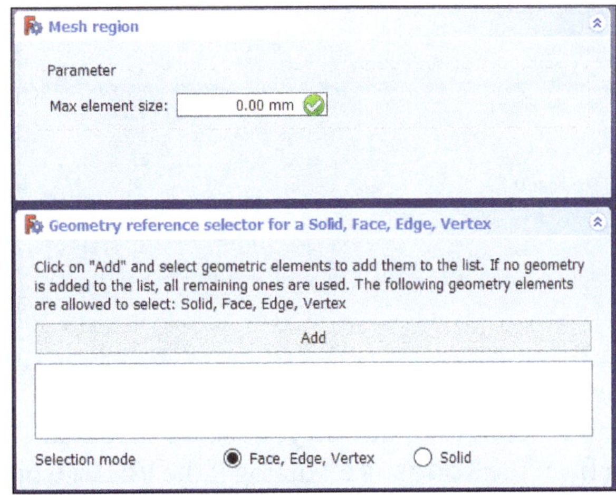

Figure-65. Mesh region and Geometry reference selector dialogs

Creating FEM mesh group

The **FEM mesh group** tool is used to create a group of mesh for exporting to other applications. The procedure to use this tool is given next.

- Click on the **FEM mesh group** tool from the **Mesh** menu. The **Mesh group** and **Geometry reference selector for a Solid, Face, Edge, Vertex** dialogs will be displayed; refer to Figure-66.

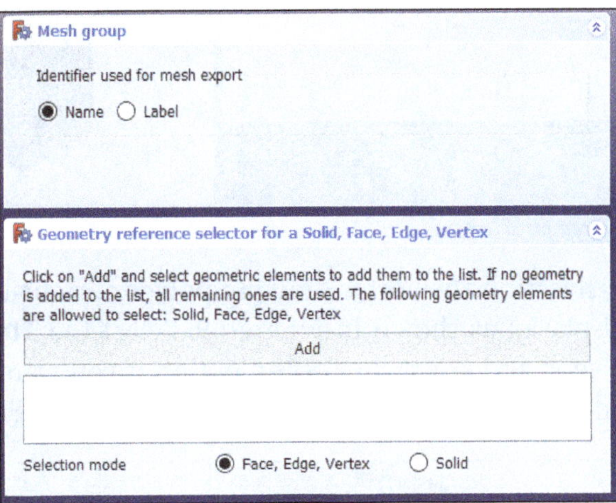

Figure-66. Mesh group dialogs

- Select desired radio button from **Mesh group** dialog to define whether name is used for exporting mesh or label.
- Select desired face, edge, vertex, or solid object to be added in mesh group. If no object is selected then all the unused objects will be selected for mesh group.
- After setting desired parameters, click on the **OK** button to create mesh group.

Creating Nodes set

The **Nodes set** tool is used to create a set of selected nodes for checking results of analysis later on those nodes. The procedure to use this tool is given next.

- Click on the **Nodes set** tool from the **Mesh** menu. The **TaskObjectName** and **Nodes set** dialogs will be displayed; refer to Figure-67.

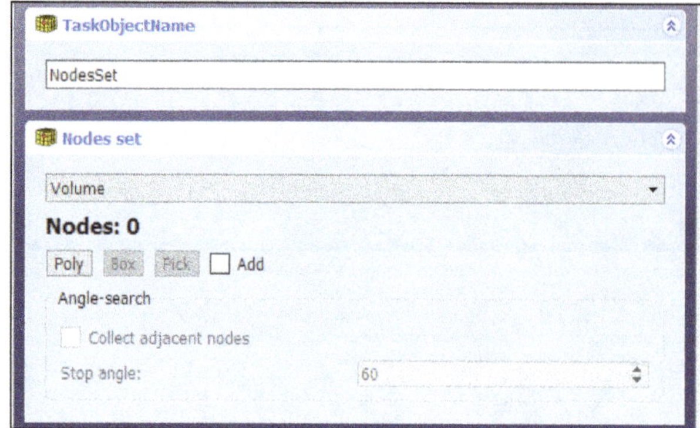

Figure-67. TaskObjectName and Nodes set dialogs

- Specify desired name of node set in the **TaskObjectName** edit box.
- Click on the **Poly** button from the **Nodes set** dialog if the **Volume** option is selected and click on the **Pick** button if the **Surface** option is selected in drop-down of this dialog. You will be asked to select the volume or surface based on your selection.
- Select desired volume or surface using the polygon or pick selection method respectively; refer to Figure-68.

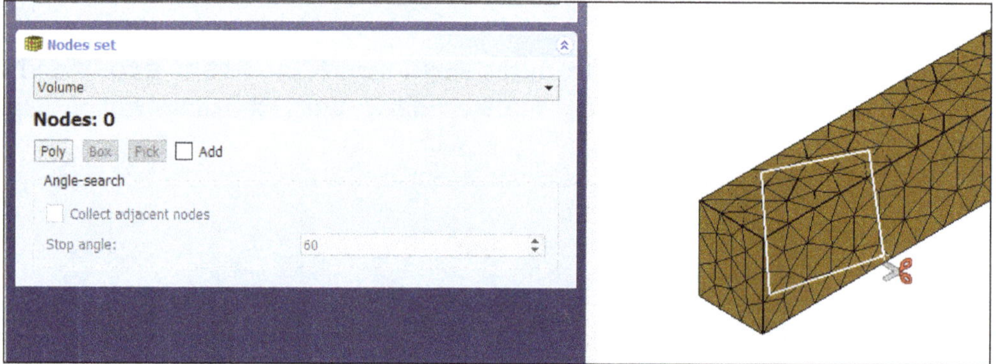

Figure-68. Polygon created for selection

- If you are using polygon selection then after drawing polygon, right-click in the drawing area. A shortcut menu will be displayed as shown in Figure-69. Select the **Inner** option to select all the nodes falling inside a polygon and select the **Outer** option if you want to select the nodes outside the polygon. Selected nodes will be displayed in green color; refer to Figure-70.

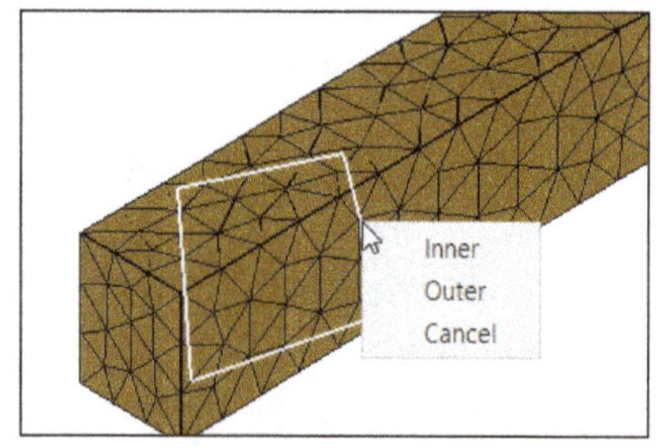

Figure-69. Shortcut menu for polygon selection

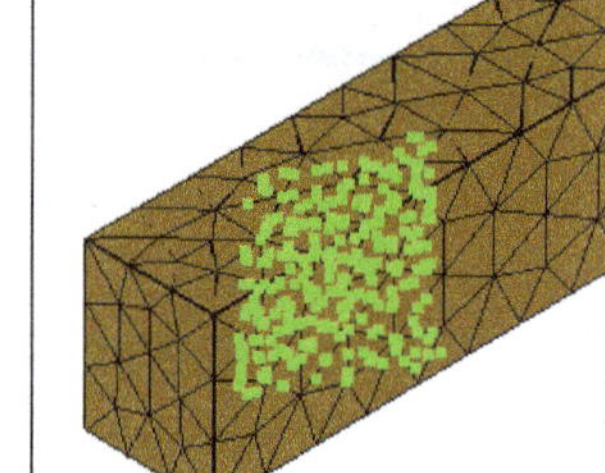

Figure-70. Selected nodes

- Click on the **OK** button from the dialog to create node set.

FEM Mesh to Polygonal Mesh Model Conversion

The Mesh created in FEM workbench is not a model mesh. It is representation of original model in the form of integration of small elements. You can convert the FEM mesh to a model mesh by using the **FEM mesh to mesh** tool in **Model** menu. The procedure to use this tool is given next.

- Click on the **FEM mesh to mesh** tool from the **Model** menu after selecting **FEMMeshGmsh** feature from the **Model Tree**. The FEM mesh will be converted to mesh model; refer to Figure-71.

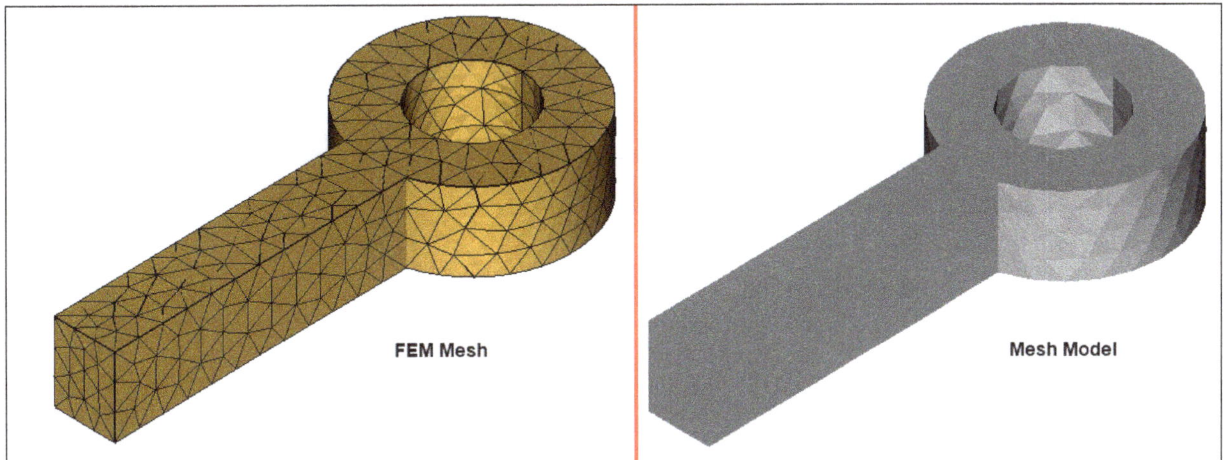

Figure-71. Mesh model conversion

SOLVERS

FreeCAD FEM has many analysis solvers available in **Solve** menu; refer to Figure-72. Some of these solvers are in experimental stage and some are specific to analysis type. Various solvers available in FreeCAD are discussed next.

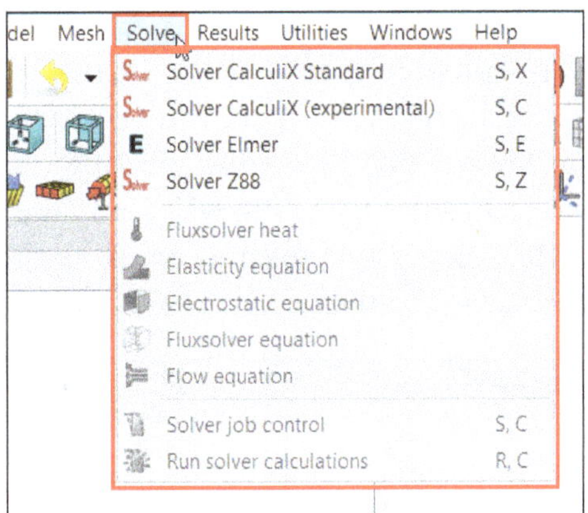

Figure-72. Solve menu

Calculix Standard Solver

CalculiX is a package designed to solve field problems. The method used by this solver is the finite element method. With CalculiX, Finite Element Models can be build, calculated, and post-processed. The solver is able to do linear and non-linear calculations. Static, dynamic, and thermal solutions can be obtained by this solver. Because the solver makes use of the abaqus input format, it is possible to use commercial pre-processors as well with this solver. The pre-processor of this solver is able to write mesh related data for nastran, abaqus, ansys, code-aster and for the free-cfd codes like duns, ISAAC and OpenFOAM. The procedure to use this solver is given next.

- Click on the **Solver CalculiX Standard** option from the **Solve** menu. The **CalculiXccxTools** feature will be added in the **Model Tree**. Note that when you start a mechanical analysis then **CalculiXccxTools** feature is added in the **Model Tree** by default.
- Double-click on the **CalculiXccxTools** feature. The **Mechanical analysis** dialog will be displayed in **Tasks** panel of **Combo View**; refer to Figure-73.

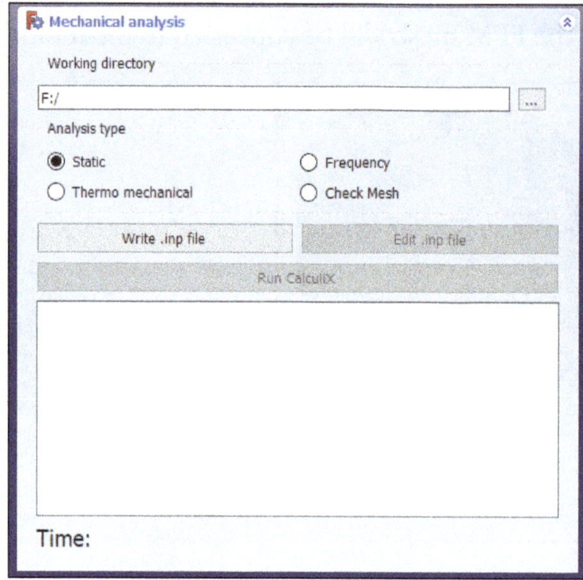

Figure-73. Mechanical analysis dialog

- Click on the **Write .inp file** button to create an input file for solver if you have specified all the other parameters of analysis like constraints, material, and so on in the model. This input file will tell the solver, what is to be calculated and what parameters are known.
- On clicking the **Write .inp file** button, the **Run CalculiX** button will become active and overview of input file writing process will be displayed; refer to Figure-74. Note that if there are parameters like material of body, constraints applied on the model, and loads which have not been defined yet then you will get an error message. Also, you can perform analysis on single body with one mesh feature. Multiple meshes are not allowed till writing this book in FreeCAD. Most of the time, mesh is major cause of error in solving the analysis, so make sure the mesh is created without error. Easiest way in FreeCAD to create a fine mesh is by using the **FEM mesh from shape by Netgen** tool if you are not familiar with advanced mesh terms.

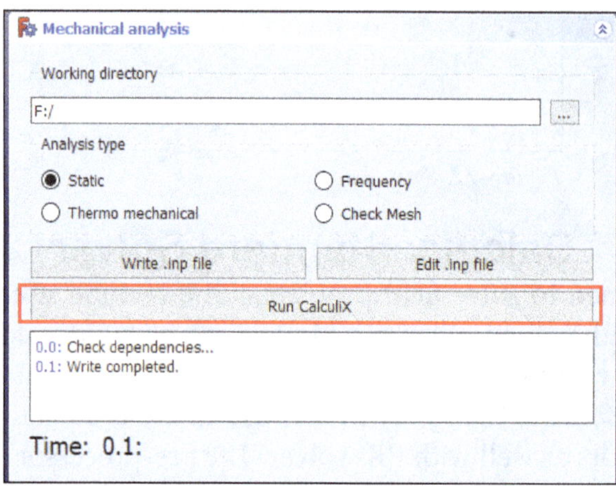

Figure-74. Input file writing process overview

- Click on the **Run CalculiX** button from the dialog. The resulting mesh will be generated if solver successfully runs. You may not get results if model is inconsistent, parameters specified are not feasible, or all the parameters are not defined.
- After running the analysis, click on the **Close** button to exit the dialog. The **Result mesh** feature will be added in the **Model Tree**; refer to Figure-75.

- To view results of analysis, select the CalculiX static results feature from the Model Tree and click on the **Show Results** option from the **Results** menu; refer to Figure-76. The **Show result** dialog will be displayed result mesh preview; refer to Figure-77.

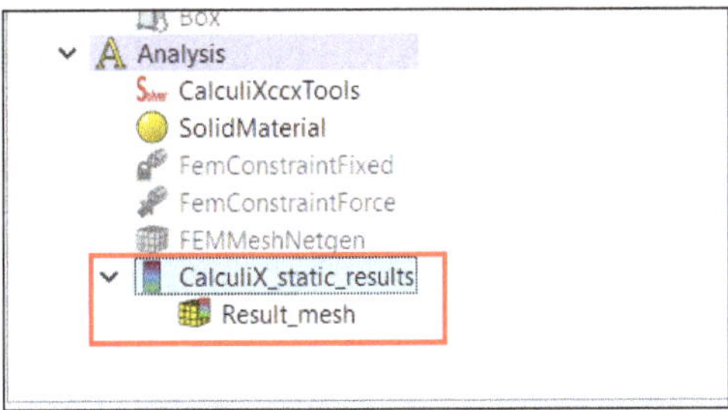

Figure-75. Result mesh feature generated

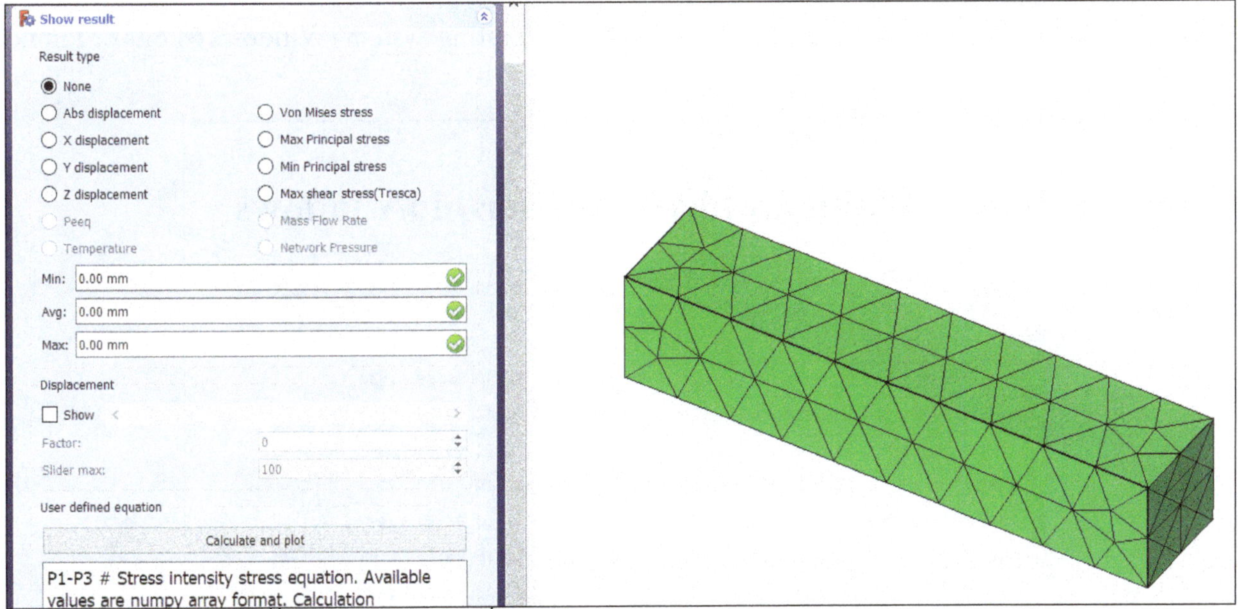

Figure-77. Show result dialog

- Select desired radio button from the **Result type** area of the dialog to check respective result.
- Select the **Show** check box from **Displacement** area of dialog and set desired magnification factor using slider to check displacement cause due to load.
- You can check the Minimum, Average, and Maximum value of selected result type from the **Min**, **Avg**, and **Max** edit boxes in the dialog.
- Click on the **Close** button from the top in the dialog to exit.

Calculix Experimental Solver

CalculiX Experimental solver is based on standard solver with some enhancements under development. The procedure to use this solver is similar to standard solver.

Elmer Solver

Elmer is an open source multi-physics simulation software mainly developed by CSC - IT Center for Science (CSC). Elmer development was started 1995 in collaboration with Finnish Universities, research institutes, and industry. After it's open source publication in 2005, the use and development of Elmer has become international.

Elmer includes physical models of fluid dynamics, structural mechanics, electromagnetism, heat transfer, and acoustics, for example. These are described by partial differential equations which Elmer solves by the Finite Element Method (FEM). **Note that in FreeCAD, you can use the Elmer Solver with mesh generated by using Gmsh Mesher only**. Before using this solver, you need to install the Elmer solver with its binary files which will be linked to FreeCAD GUI. Elmer requires two components to be interfaced with FreeCAD:

- ElmerGrid is the interface handling meshes.
- ElmerSolver is handling the computation.

The procedure to install this solver is given next.

- Go to the CSC binaries resources for Elmer: https://www.nic.funet.fi/pub/sci/physics/elmer/bin/ OR https://www.csc.fi/web/elmer/binaries
- Download and install the version best suited to your Operating System (Windows 64 bits or Linux); refer to Figure-78.

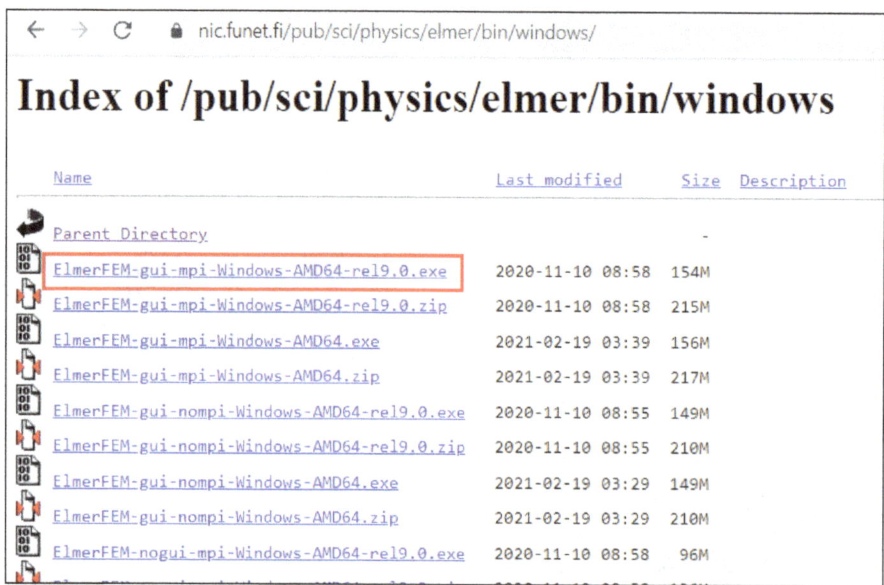

Figure-78. Elmer binaries for Windows

- After installing the Elmer solver, go to **Edit → Preferences → FEM → Elmer** in menu and link the correct path for both ElmerGrid and ElmerSolver; refer to Figure-79.

Now, we are ready to use the solver for performing analysis. The procedure to use this solver is given next.

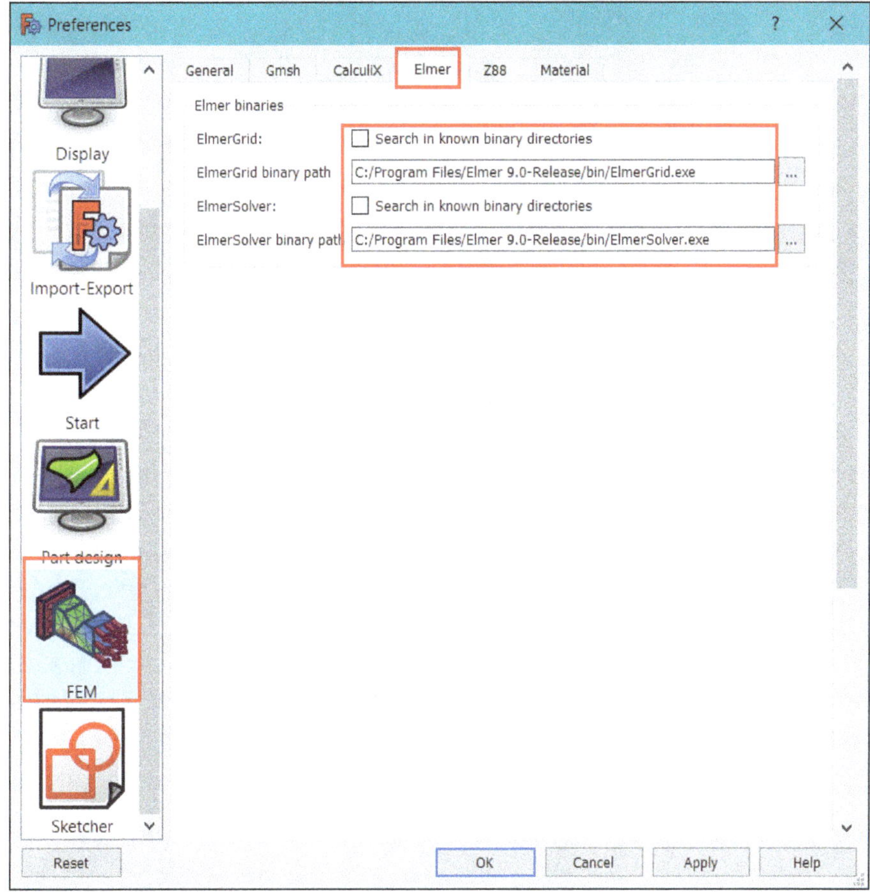

Figure-79. Linking Elmer with FreeCAD GUI

- Create the analysis setup including mesh create by Gmsh, thermal/mechanical/electrostatic/fluid loads applied on model, and material applied to model which has properties related to analysis; refer to Figure-80. After that, click on the **Solver Elmer** tool from the **Solve** menu. The **SolverElmer** feature will be added in the **Model Tree**.

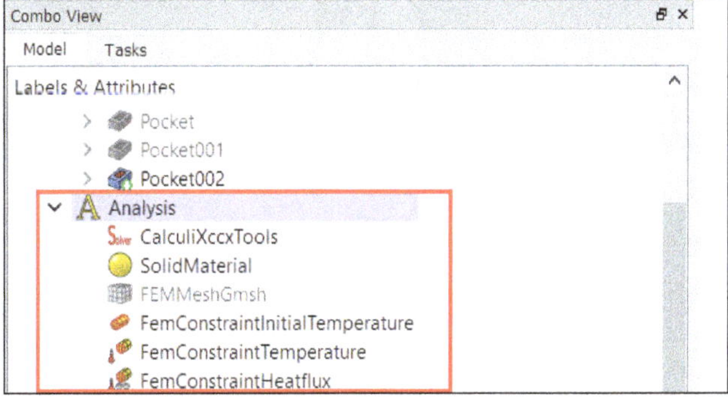

Figure-80. Features created before using Elmer solver for thermal analysis

For using Elmer solver, you need to create equations related to the analysis. There are five types of equations available in FreeCAD for different type of analyses; refer to Figure-81.

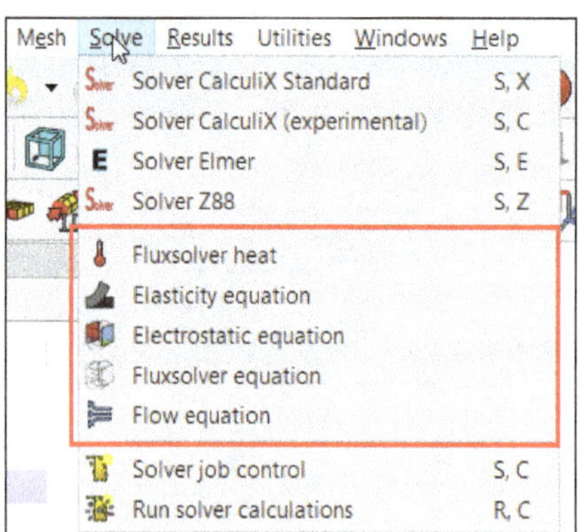

Figure-81. Equations available for Elmer solver

- Select the **Fluxsolver heat** option from the **Solve** menu to add thermal analysis equation. Select the **Elasticity equation** option from the **Solve** menu to add nonlinear analysis equation. Select the **Electrostatic equation** option to add electrostatic analysis equation. Select the **Fluxsolver equation** option to add magnetic flux analysis equation. Select the **Flow equation** option to add fluid dynamics analysis equation. In our case, we are using **Fluxsolver heat** option. On selecting this option, **Heat** feature will be added below **Elmer** solver in **Model Tree**.
- Select the feature from **Model Tree**. Properties of the equation will be displayed in the **Property Editor** dialog box; refer to Figure-82.

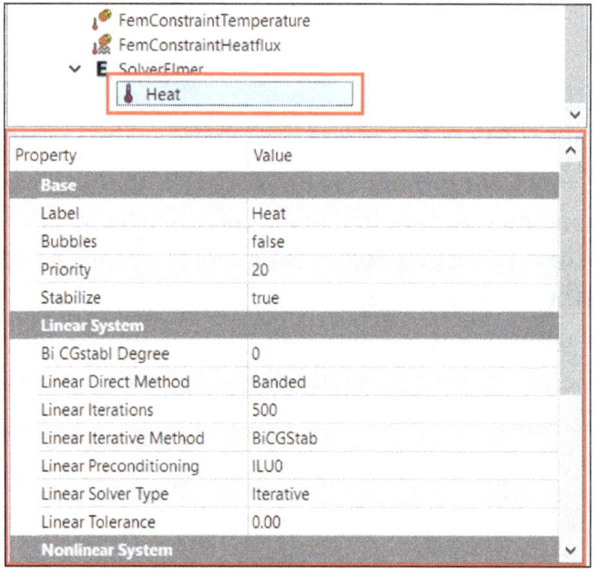

Figure-82. Properties for heat equation

- Specify desired parameters in the **Property Editor** dialog box. Various parameters for equation are given next:

--Base Section--

Label : Used to specify name of the equation.

Bubbles : Used to stabilize solution of equation depending on compressibility of model. If you are working on compressible fluids then you should set this value to True.

Priority : Used to define priority level of current equation with respect to other equations to be solved by Elmer solver. A higher value of priority means respective equation will be solved first.

Stabilize : Used to specify whether Stabilized finite element method will be used or RFB (Residual Free Bubble) stabilization will be used.

--Linear System--

BiCGstabl Degree : In numerical linear algebra, the biconjugate gradient stabilized method, often abbreviated as BiCGSTAB, is an iterative method developed by H. A. van der Vorst for the numerical solution of nonsymmetric linear systems. Used to solve iterative linear equations for 3D analysis cases in FreeCAD. Set desired value to define stabilization order for the solver.

Linear Direct Method : Select desired type of linear direct method for solving linear FEM equations. For stress analysis, a direct solver is used instead of an iterative solver. It is often difficult for the iterative solver to find a solution for a structure that contains parts with varying stiffness properties. Use the **Banded** option when matrix to be formed for equation is dense (In simple words, a small region is to be analyzed with minimum void areas). Use the **umfpack** option when problem is complex with load free matrices in equation. If you do not know what to select then go with **umfpack**, although it will take more resources but it will do the trick.

Linear Iterative Method : Select desired option from the drop-down to define which iterative method is to be used for solving linear equations if the problem becomes complex or the matrix of equation is large. Each iterative method has its pros and cons which you can browse on Internet.

Linear Preconditioning : Linear Preconditioning is used to form a matrix easily solvable by selected iteration method. Preconditioning is typically related to reducing a condition number of the problem. The options available in FreeCAD try to form Incomplete Lower unitriangular and Upper triangular (ILU) matrices for preconditioning.

Linear Solver Type : Used to define which type of solver will be used for solving current analysis problem. Select the **Direct** option if problem is small scale linear and select the **Iterative** option if problem is complex or large.

Linear Tolerance : Used to define the accuracy up to which solution is needed. The value specified here also acts as goal for convergence if true solution is not available.

--Nonlinear System--

Nonlinear Iterations : Used to specify the number of iterations to be performed for finding solution of problem.

Nonlinear Newton After Iterations : Used to specify when Newton method will be used for solving equations in terms of iterations. If a certain number of specified iterations have been solved then their solutions will be used as input for Newton method to further solve the problem.

Nonlinear Newton After Tolerance : Used to specify when Newton method will be introduced for solving equation in terms of tolerance. If specified tolerance in solution has been achieved then Newton method will further solve the equation for convergence.

Nonlinear Tolerance : Used to specify the tolerance allowed in final solution of problem.

Relaxation Factor : The relaxation factor is used to speed up convergence of iteration. You can assume relaxation factor as a value which removes the amount of error in solution of each iteration due to numerous unknown real conditions.

--Steady State--

Steady State Tolerance : Used to define the accuracy in results to be achieved after performing the analysis.

- After setting desired parameters in the **Property Editor** box, select the **SolverElmer** feature from the **Model Tree**. The options in **Property Editor** will be displayed as shown in Figure-83.

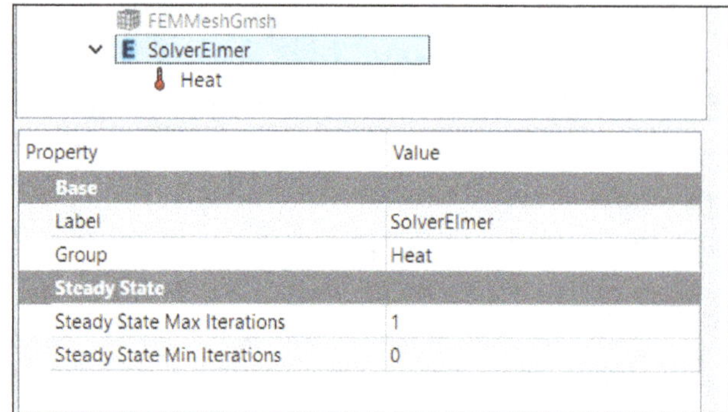

Figure-83. SolverElmer properties

- Specify the name and group for feature in the **Label** and **Group** fields, respectively.
- Specify desired values in the **Steady State Max Iterations** and **Steady State Min Iterations** fields to define maximum and minimum number of iterations to be solved when you are running coupled analysis (Coupled analysis involve multiple variables which are interdependent like temperature of body is function of current in analysis and current value changes with temperature of body as well). Generally, default value is **1** for max and **0** for min when you are solving for one variable only. If your analysis is coupled then specify desired value for iterations up to which you will be able to achieve solution of coupled equations.
- After setting desired solver parameters, double-click on the **SolverElmer** feature from **Model Tree**. The **Solver Control** dialog will be displayed in **Tasks** panel; refer to Figure-84.
- Click on the **Write** button from dialog to generate input file. Since, we are using external solver, so input file parameters will not be displayed in the dialog. If you want to check the input file then browse to the location of **Working Directory** edit box in the dialog and open the **case.sif** file in any Word processor program.
- Click on the **Run** button from the dialog after writing input file. If all the parameters specified are feasible then results will be generated with Output file; refer to Figure-85.

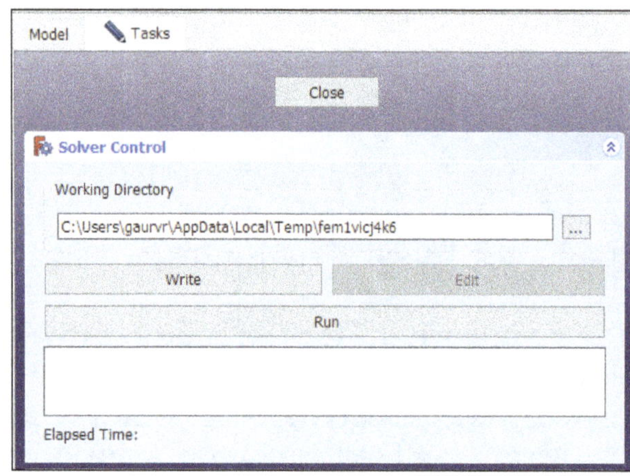

Figure-84. Solver Control dialog

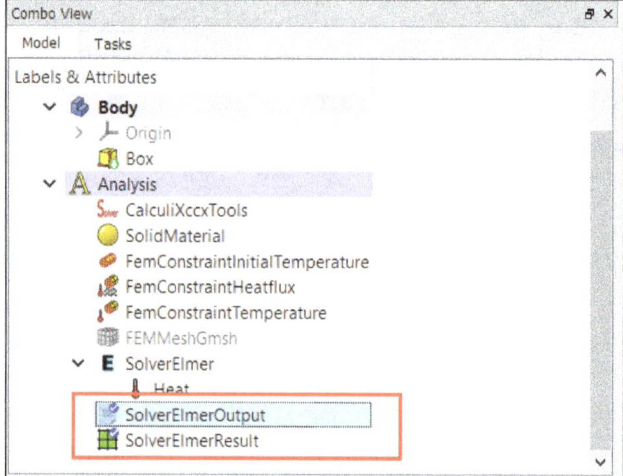

Figure-85. Elmer solver results generated

Note that sometimes, you may not get the results displayed in FreeCAD GUI. In that case, reach to the result file (*.vtu) of analysis in temporary folder of Elmer solver analysis and open it in ElmerGUI application which we have installed earlier.

ParaView is also a good post processor for analyzing results of analysis. You can download ParaView from https://www.paraview.org/download web page. Figure-86 shows the result of our current analysis in ParaView.

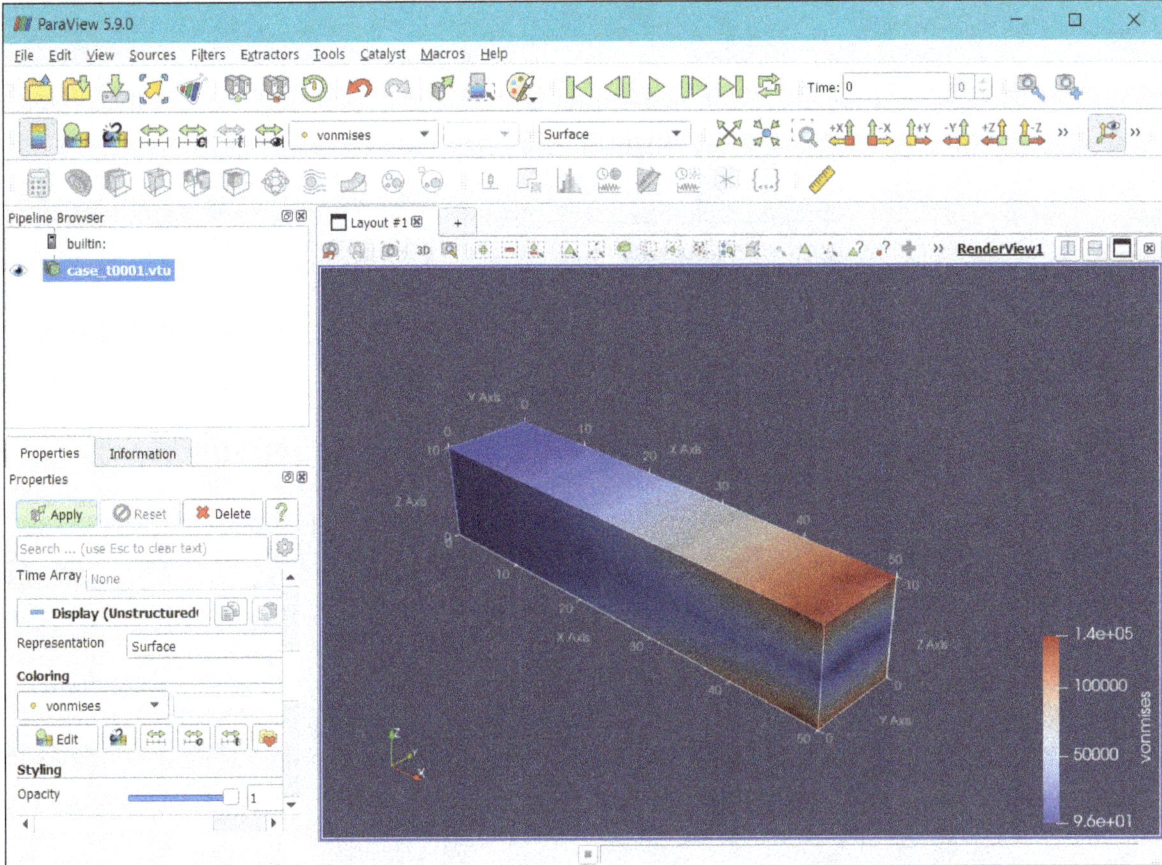

Figure-86. Result of Elmer analysis in ParaView

Similarly, you can use the other equations of Elmer Solver for performing related analyses.

Z88 Solver

The software was developed by Frank Rieg, a professor for engineering design and CAD at the University of Bayreuth. Originally written in FORTRAN 77, the program was ported to the programming language C in the early 1990s.
There are two programs for finite element analysis:

Z88OS is available as free software including the source code under the GNU General Public License. Due to the modular structure of the program and the open availability of the source code, it is possible to develop customized extensions, add-ons and several special case 2D and 3D continuum elements (e.g. anisotropic shell element).

Z88Aurora originally described the user interface of the Z88 finite element analysis program. After several additions and further development, it now comprises a significantly larger range of functionality than Z88OS. Z88Aurora is freeware, however the source code is not publicly available.
Z88Aurora's current version contains several computation modules:

In the case of linear static analyses, it is assumed that the result is proportional to the applied forces.

Nonlinear analyses are used for nonlinear geometries and nonlinear materials.

Using thermal and thermomechanical analyses, it is possible to not only compute results about temperature or heat currents, but also thermomechanical displacements and stresses.

By utilizing natural frequency simulation, natural frequencies and the resulting oscillations can be determined.

A contact module makes it possible to simulate interacting parts and assemblies. An integrated part management tool enables an effective handling of assemblies. There are options to simulate a glued connection or a friction-free connection and the contact discretization (type of contact: node-surface-contact, or surface-surface-contact), the mathematical imposition method (lagrange method, perturbed lagrange method, or penalty method), and the direction of contact stiffness (normal or tangential direction) can be changed via the contact settings. This module only supports tetrahedrons and hexahedrons with linear or quadratic shape functions. Additionally, the module is only available for linear mechanical strength analyses.

The procedure to use Z88 solver in FreeCAD is similar to using CalculiX solver which has been discussed earlier.

POST PROCESSING IN FREECAD FEM

The tools in **Results** menu of **FEM** workbench are used to perform post processing, so that you can analyze the results. The tools of this menu are active only when you performed the analysis and result of solver is selected in the **Model Tree**; refer to Figure-87. Various tools of this menu are discussed next.

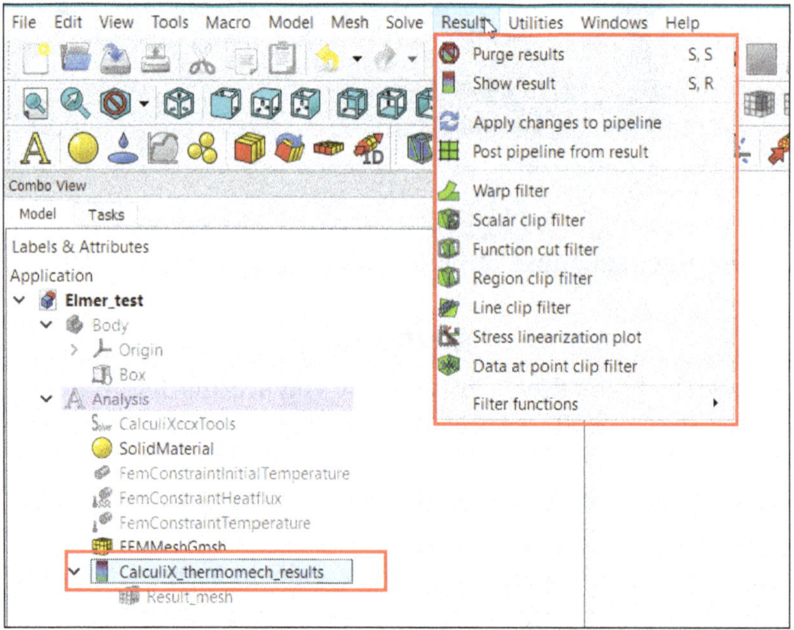

Figure-87. Results menu

Purging Results

The **Purge results** tool is used to delete the results generated by using **Show result** tool. Note that this tool is active only after you have used the **Show result** tool.

Showing Result

The **Show result** tool from the **Results** menu to generate analysis results. The procedure to use this tool has been discussed earlier.

Creating Post Processing Pipeline for Results

The **Post pipeline from result** tool is used to create post processing pipelines from the results. The procedure to use this tool is given next.

- Click on the **Post pipeline from result** tool from the **Results** menu after selecting solver's result feature from the **Model Tree**. The **Pipeline** feature will be added in the **Model Tree**.
- Double-click on this feature to modify its parameters. The **Result display options** dialog will be displayed; refer to Figure-88.

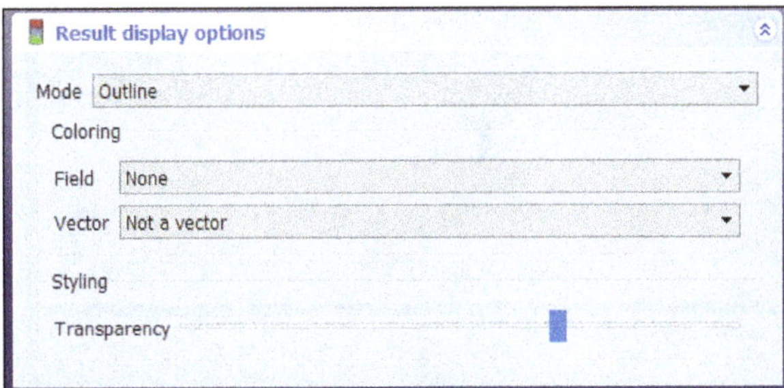

Figure-88. Result display options dialog

- Select desired option from the **Mode** drop-down to define what type of pipeline plot you want to generate using the result parameters. Select the **Outline** option to generate only border lines of the model. Select the **Nodes** option to generate colored nodes for analyzing results. Select the **Surface** option to generate result surface with multicolored annotations. Select the **Surface with Edges** option to generate result surface with wireframe edges. Select the **Wireframe** option to generate wireframe model of results. Select the **Wireframe (surface only)** option to generate wireframe model over the surface of model for analyzing results.
- Select desired option from the **Field** drop-down to define result parameter to be displayed over the pipeline plot. Preview of result will be displayed in the modeling area; refer to Figure-89.
- Select desired option from the **Vector** drop-down to define which component of vector is to be used for generating result model.
- Click on the **OK** button from the dialog to generate pipeline result.

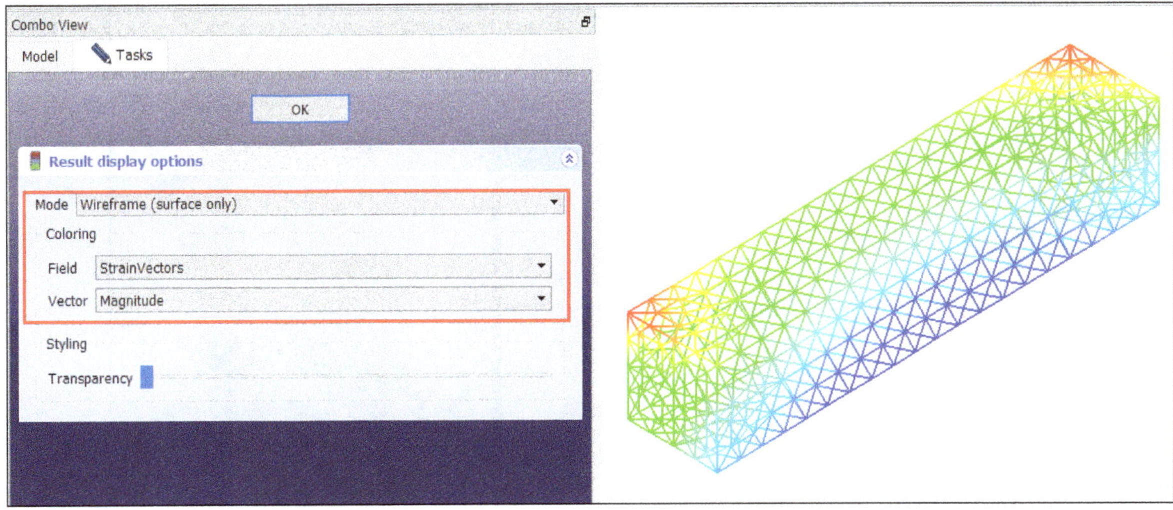

Figure-89. Preview of pipeline result

Applying Warp Filter

The **Warp filter** tool is used to distort the result model, so that higher values are further away from base model as compared to lower result values. The procedure to use this tool is given next.

- Click on the **Warp filter** tool from the **Results** menu after selecting **Pipeline** feature from the **Model Tree**. The **Warp options** and **Result display options** dialogs will be displayed; refer to Figure-90.

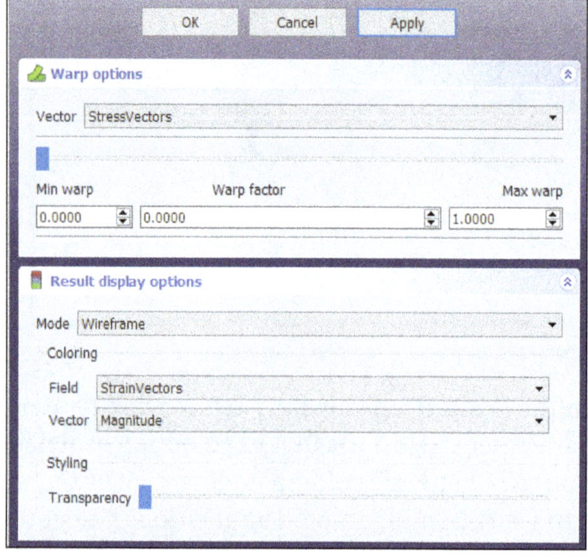

Figure-90. Warp options and Result display options dialogs

- Select desired option from the **Vector** drop-down to specify which result vector is to be displayed in warped model.
- Set the other parameters as discussed earlier in the **Result display options** dialog.
- Use the warp slider and warp related parameters in the **Warp options** dialog to check warped model of results; refer to Figure-91.

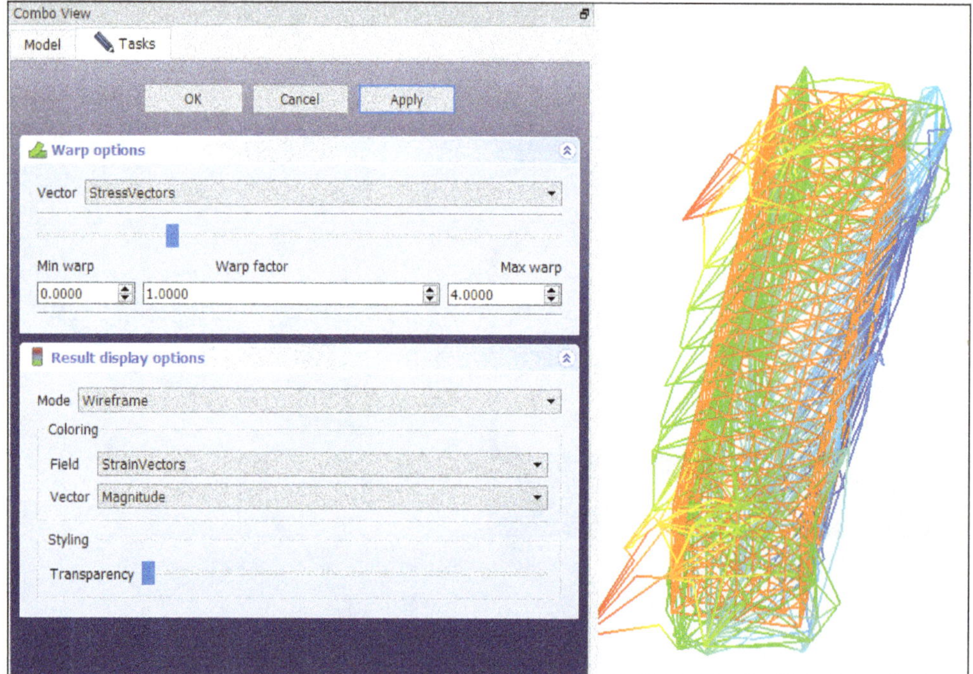

Figure-91. Warped model

- After setting desired parameters, click on the **OK** button from the dialog.

Applying Scalar Clip Filter

The **Scalar clip filter** tool is used to check internal section of model for result parameters. The procedure to use this tool is given next.

- Click on the **Scalar clip filter** tool from the **Results** menu after selecting **Pipeline** feature from **Model Tree**. The **Scalar clip options** and **Result display options** dialogs will be displayed.
- Set the parameters as discussed earlier in the dialogs and use the clipping slider for checking internal regions of model for analysis results; refer to Figure-92.

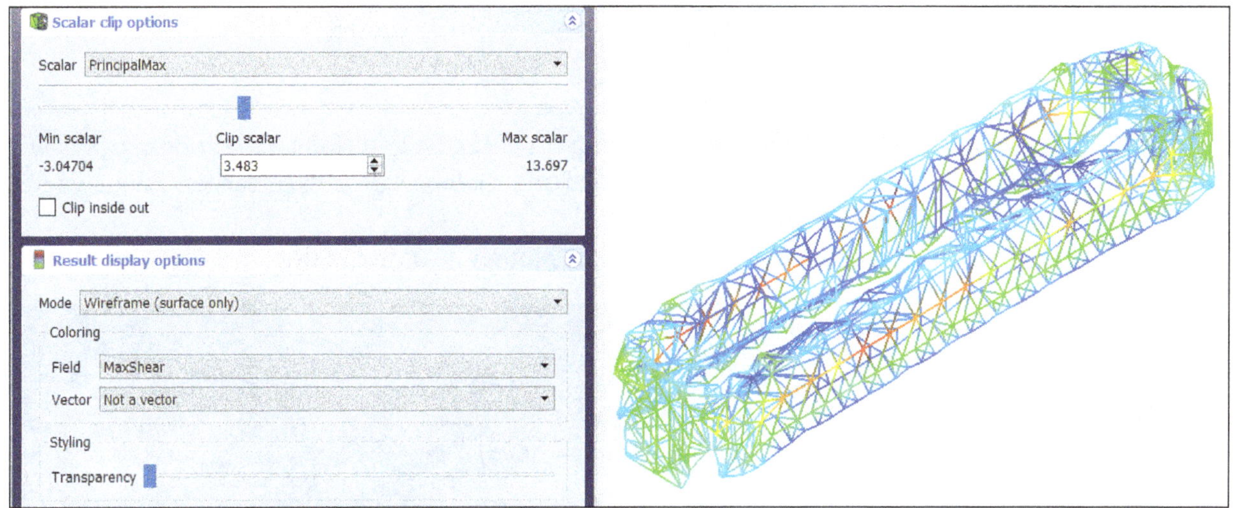

Figure-92. Pipeline results with scalar clipping

- Click on the **OK** button from the dialog to create the result model and exit dialog.

Applying Function Cut Filter

The **Function cut filter** tool is used to check the pipeline results of analysis along a plane or over the surface of a sphere. The procedure to use this tool is given next.

- Click on the **Function cut filter** tool from the **Results** menu after selecting the **Pipeline** feature from **Model Tree**. The **Function cut, choose implicit function** dialog will be displayed with **Result display options** dialog; refer to Figure-93.

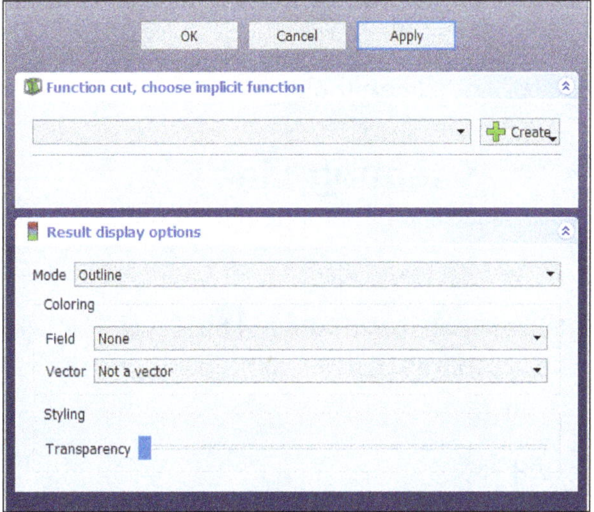

Figure-93. Function cut dialog

- Click on the **Create** button from the dialog and select desired option to define scope of result output geometrically. Select the **Plane** option if you want to display result on a plane intersecting with the model. Select the **Sphere** option if you want to check results on surface of a sphere intersecting with the model. We have selected **Plane** option in our case for tutorial.
- On selecting the **Plane** option, parameters related to location of plane will be displayed in the dialog; refer to Figure-94.

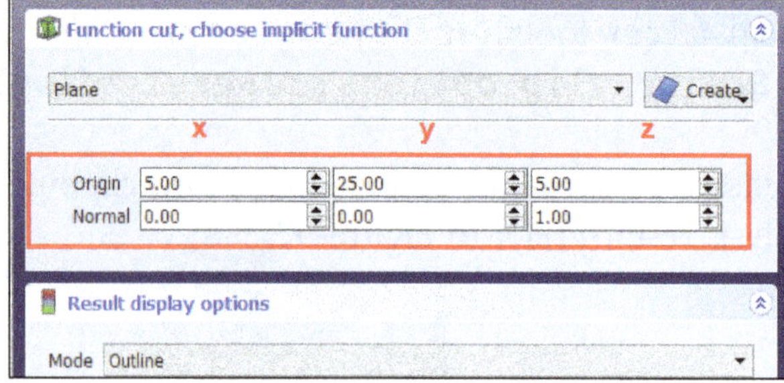

Figure-94. Parameters for plane option

- Set desired location and orientation for plane using the **Origin** and **Normal** edit boxes.
- Set desired options in the **Result display options** dialog as discussed earlier. Preview of the plot will be displayed; refer to Figure-95.
- Click on the **OK** button from the dialog to create the plot and exit the dialog.

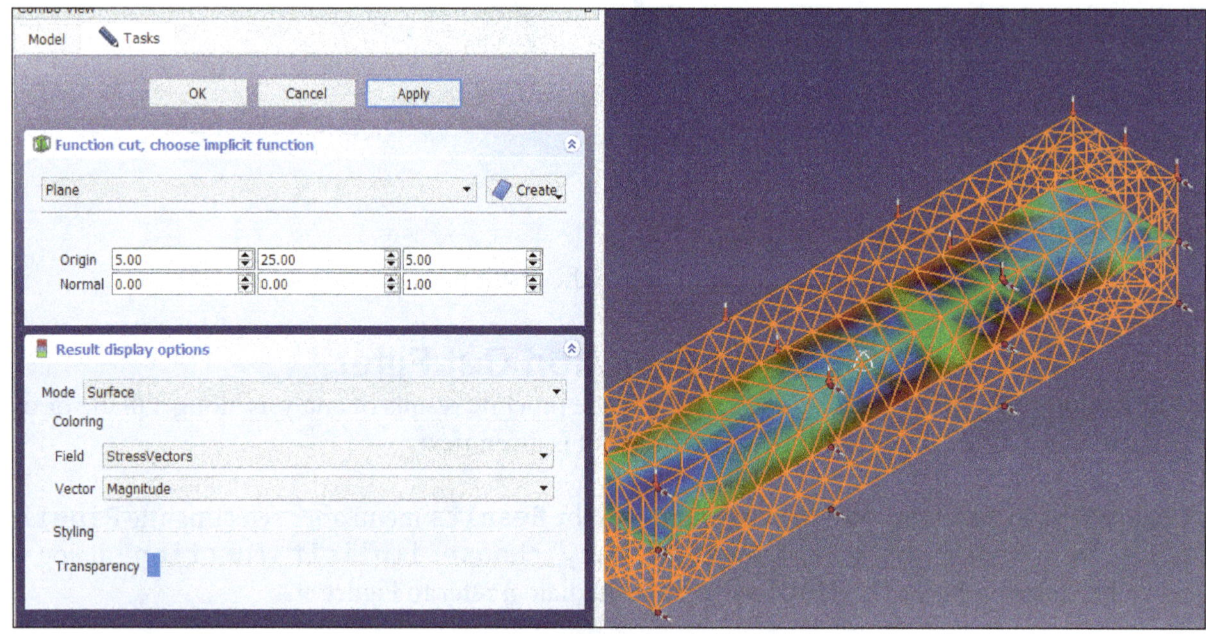

Figure-95. Preview of plot

Similarly, you can use **Region clip filter** tool.

Creating Line Plot

The **Line clip filter** tool is used to generate line plot of desired result parameter. The procedure to use this tool is given next.

- Click on the **Line clip filter** tool from the **Results** menu after selecting the **Pipeline** feature. The **Data along a line options** dialog will be displayed; refer to Figure-96.

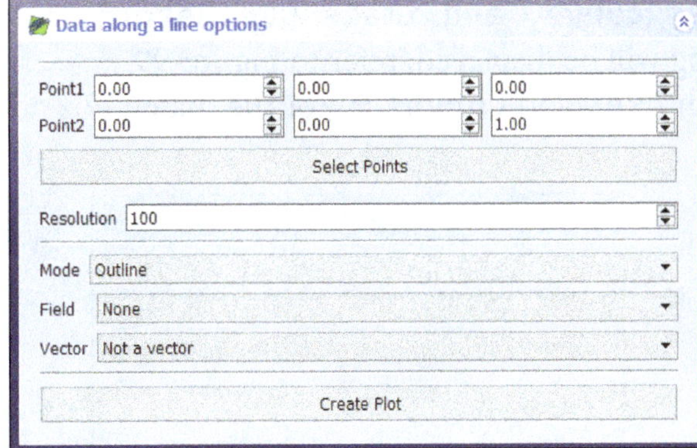

Figure-96. Data along a line options dialog

- Click on the **Select Points** button from the dialog and one by one select two points from model to define result output limits.
- Set the parameters as discussed earlier in **Resolution**, **Mode**, **Field**, and **Vector** drop-downs.
- Click on the **Create Plot** button to generate the plot. The plot will be created and preview of plot will be displayed in the model; refer to Figure-97.

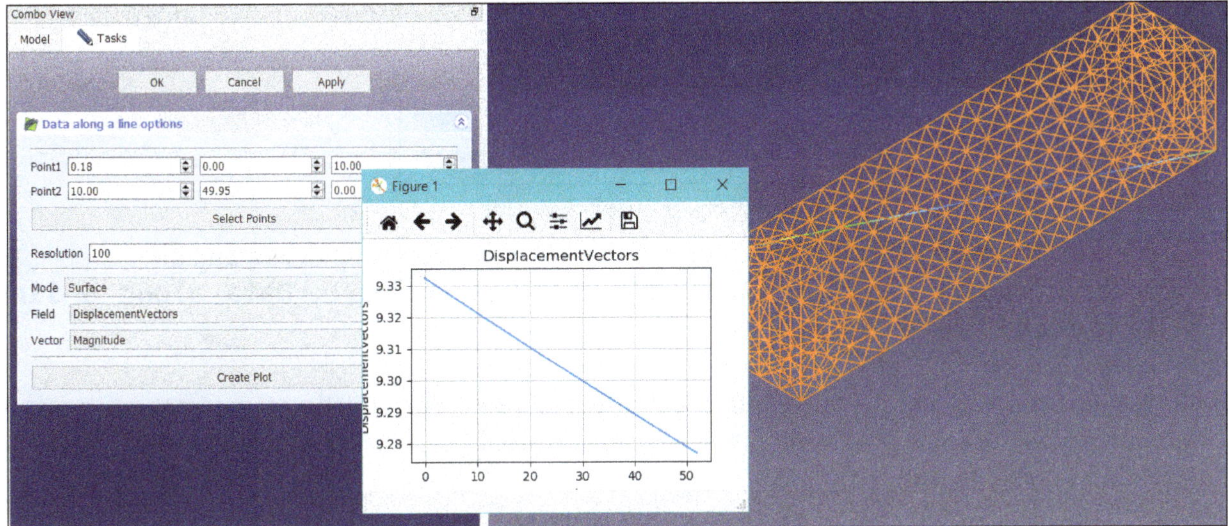

Figure-97. Line plot generated

- Using the **Zoom** button in the **Figure 1** dialog box, you can zoom into a specific region of plot by drawing a box.
- Using the **Pan** button you can move left/right/up/down in the plot to check values.
- The **Back** and **Forward** buttons are used to switch between previous and next views of plot.
- Click on the **Configure subplots** button to modify various spacings in the plot. On clicking this button, the **FreeCAD** dialog box will be displayed as shown in Figure-98. Set desired values and click on the **Close** button.

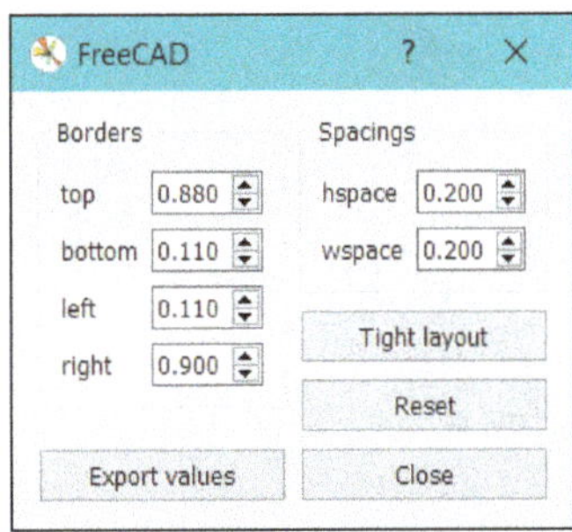

Figure-98. FreeCAD dialog box

- Click on the **Edit axes, curve, and image parameters** button from the dialog box. The **Figure options** dialog will be displayed; refer to Figure-99.
- Set desired parameters in **Axes** and **Curves** tab of the dialog box and click on the **OK** button to exit the dialog box.

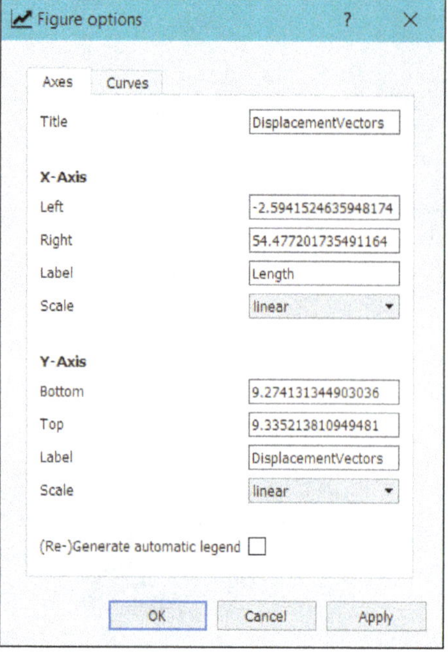

Figure-99. Figure options dialog box

- Click on the **Save** button from the **Figure 1** dialog to save current plot as a png file.
- Close the dialog after checking plot and click on the **OK** button from the **Data along a line** dialog to exit the tool.

Similarly, you can use other post processing tools in **Results** menu. Sometimes, you may get errors in postprocessing due to some programming needs of software. In such cases, your other post processor ParaView is available to help.

Note that till writing this book, FreeCAD FEM had good compatibility with Calculix Solver and its related analyses. We were not able to perform CFD analyses using Elmer solver. For CFD analyses, you can work with OpenFOAM solver externally instead.

Chapter 10

Tech Drawing and Image Workbench

Topics Covered

The major topics covered in this chapter are:

- ***Introduction to TechDraw Workbench***
- ***Starting New Drawing Page***
- ***Inserting Views***
- ***Creating Section Views and Detail Views***
- ***Applying Annotations***
- ***Clip Groups***
- ***Linking Dimensions of 3D Geometry***
- ***Exporting Drawing Pages***
- ***Image Workbench***

INTRODUCTION TO TECHDRAW WORKBENCH

The **TechDraw** workbench of FreeCAD is used to create technical drawings from the model earlier created in the software using other workbenches. To start creating technical drawings, select the **TechDraw** option from the **Switch between workbenches** drop-down in the **Toolbar**; refer to Figure-1. On selecting the TechDraw workbench, tools will be displayed as shown in Figure-2.

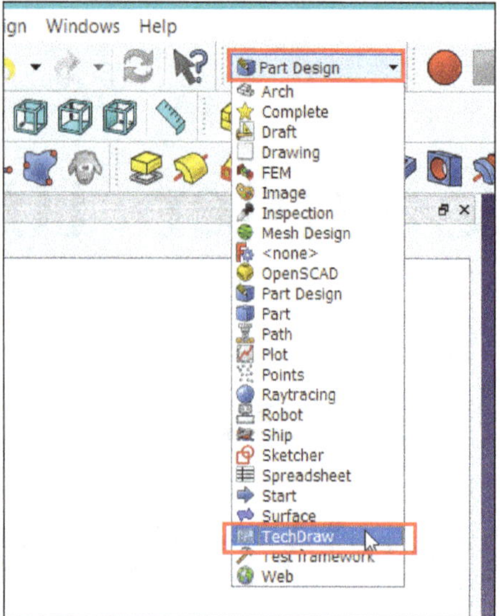

Figure-1. TechDraw workbench option

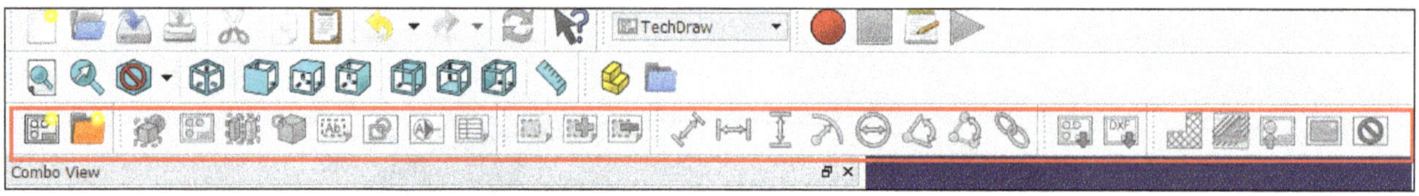

Figure-2. TechDraw toolbar

Various tools of this workbench are discussed next.

STARTING A NEW DRAWING PAGE

The **Insert new default page** tool is used to generate a drawing page using default drawing template. The procedure to use this tool is given next.

- Click on the **Insert new default page** tool from the **Toolbar** in **TechDraw** workbench. A new drawing page will be created and displayed in the interface; refer to Figure-3.

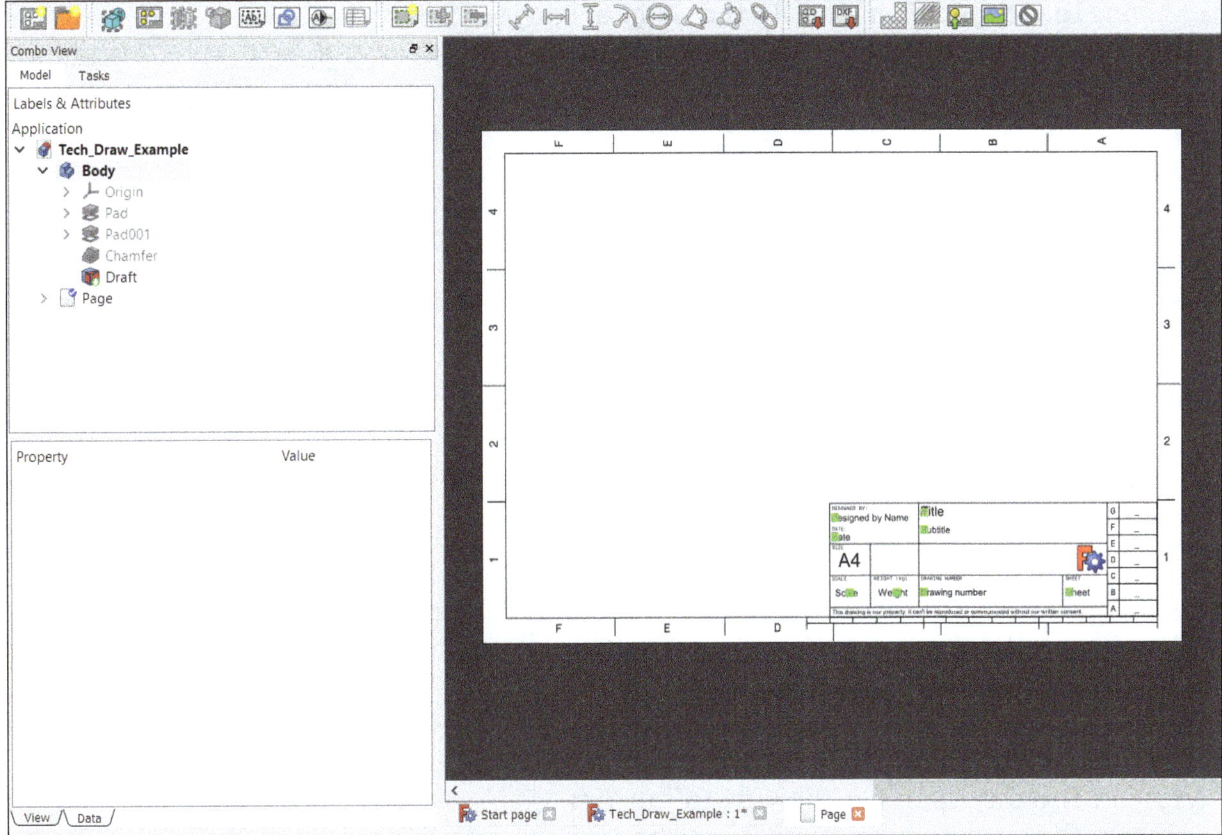

Figure-3. Drawing page generated

- Select the **Page** feature from the **Model Tree**. The properties of page will be displayed as shown in Figure-4.

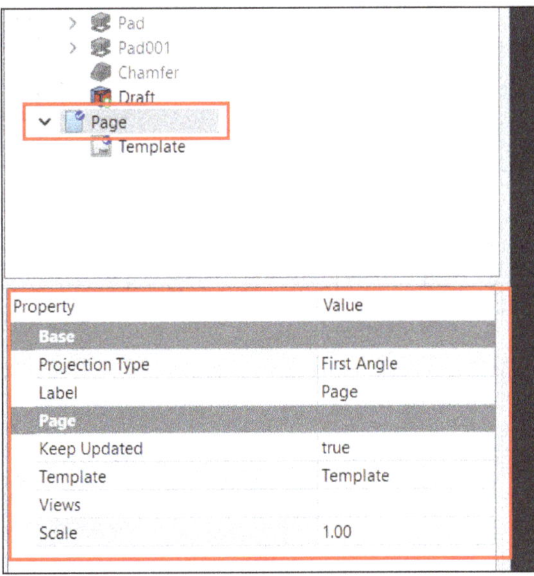

Figure-4. Properties for page feature

- Select desired option from the **Projection Type** field in **Property Editor** dialog box. Select the **First Angle** option from the field if you want to use First Angle projection for generating views. Select the **Third Angle** option from the field if you want to use Third Angle projection for generating views.

First Angle Projection and Third Angle Projection

Figure-5 shows an object with different view directions say, a, b, c, d, e, and f.

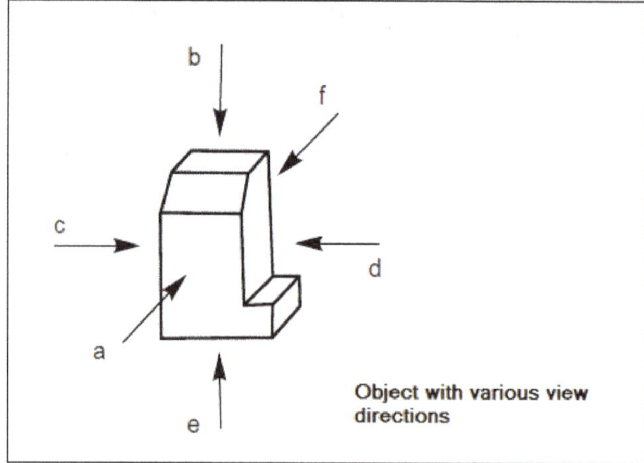

Figure-5. Object with view directions

Here,
1. View in the direction a = view from the front
2. View in the direction b = view from top
3. View in the direction c = view from the left
4. View in the direction d = view from the right
5. View in the direction e = view from bottom
6. View in the direction f = view from the back

In First Angle projection, these views are arranged as shown in Figure-6. In Third Angle projection, these views are arranged as shown in Figure-7.

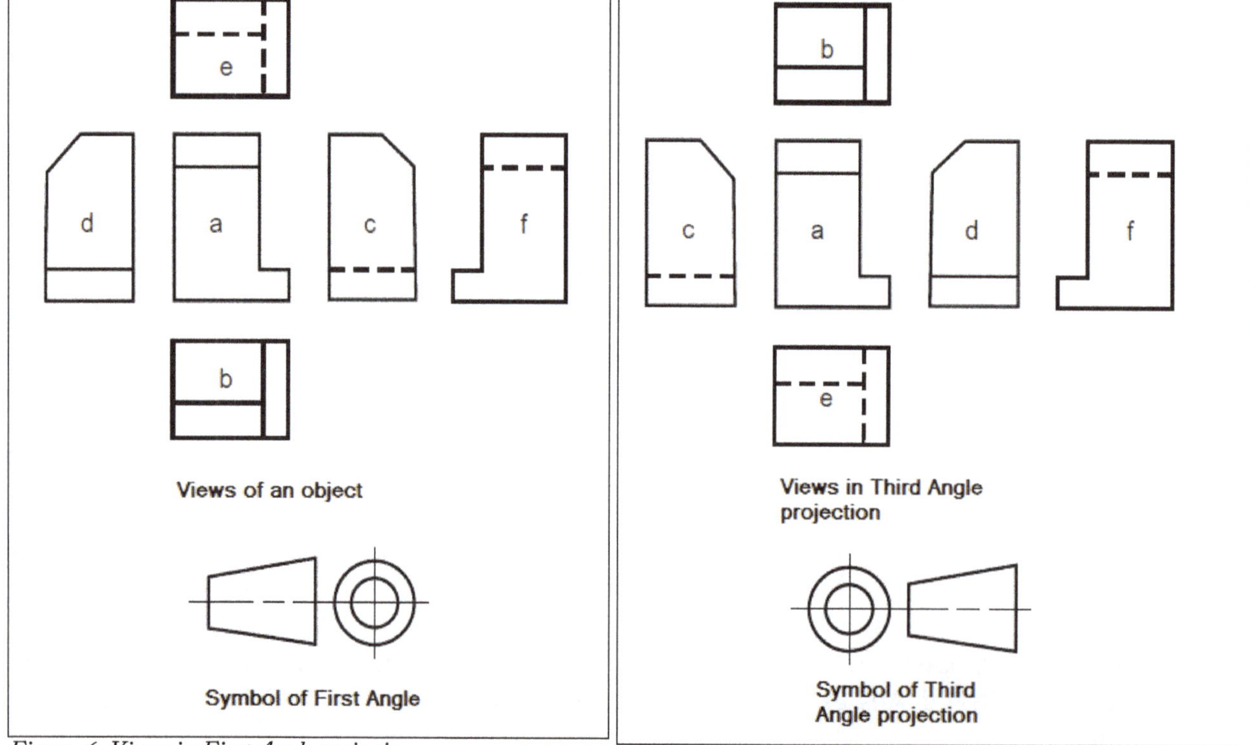

Figure-6. Views in First Angle projection

Figure-7. Views in Third Angle projection

- Specify desired name of page in the **Label** field of **Property Editor**.
- If you want to change template of drawing page and click on the **Template** link button from **Template** field of **Property Editor**. The **Template** feature will be selected in **Property Editor** and related options will be displayed; refer to Figure-8.
- Click on the **Browse** button in **Template** field of **Property Editor** and select desired template file.

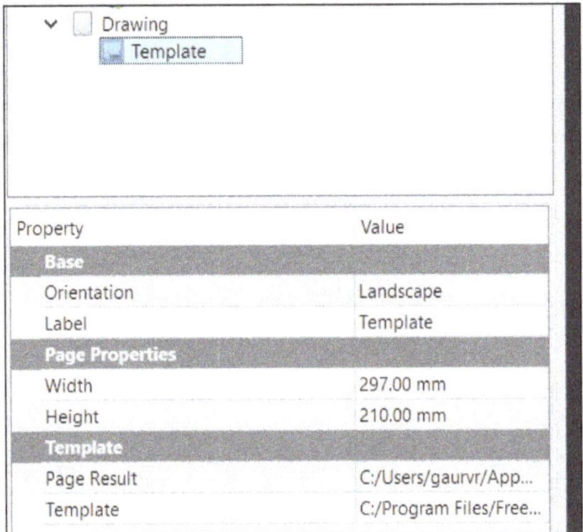

Figure-8. Properties of template

- You can specify the scale of drawing views in the **Scale** field of **Property Editor** for **Page**.

INSERTING NEW PAGE OF SELECTED TEMPLATE

The **Insert new Page using Template** tool is used to create new page using selected template. A template includes various features like size of page, orientation of page, scale of views, and so on. The procedure to use this tool is given next.

- Click on the **Insert new Page using Template** tool from the **Toolbar** or **TechDraw** menu of **TechDraw** workbench. The **Select a Template File** dialog box will be displayed; refer to Figure-9.

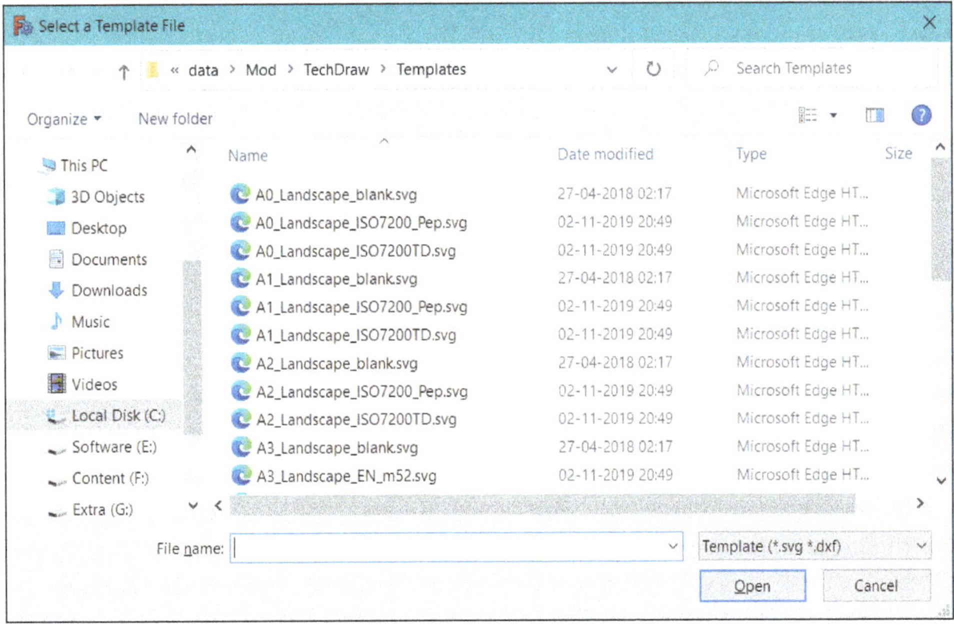

Figure-9. Select a Template File dialog box

- Select desired template from the dialog box and click on the **Open** button. The page will be created and displayed in the application; refer to Figure-10.

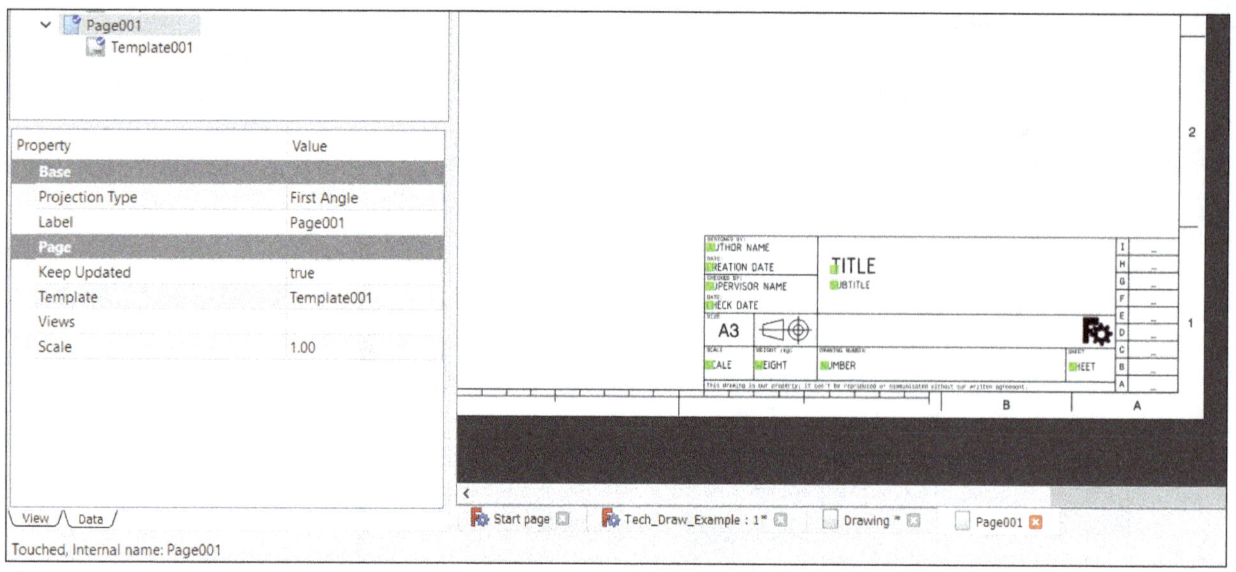

Figure-10. Page created by template

INSERTING VIEW IN PAGE

The **Insert View in Page** tool is used to insert current view of model in the page. The procedure to use this tool is given next.

- Select the tab of **Model** from the application menu and orient the model to desired view to be placed in the drawing; refer to Figure-11.

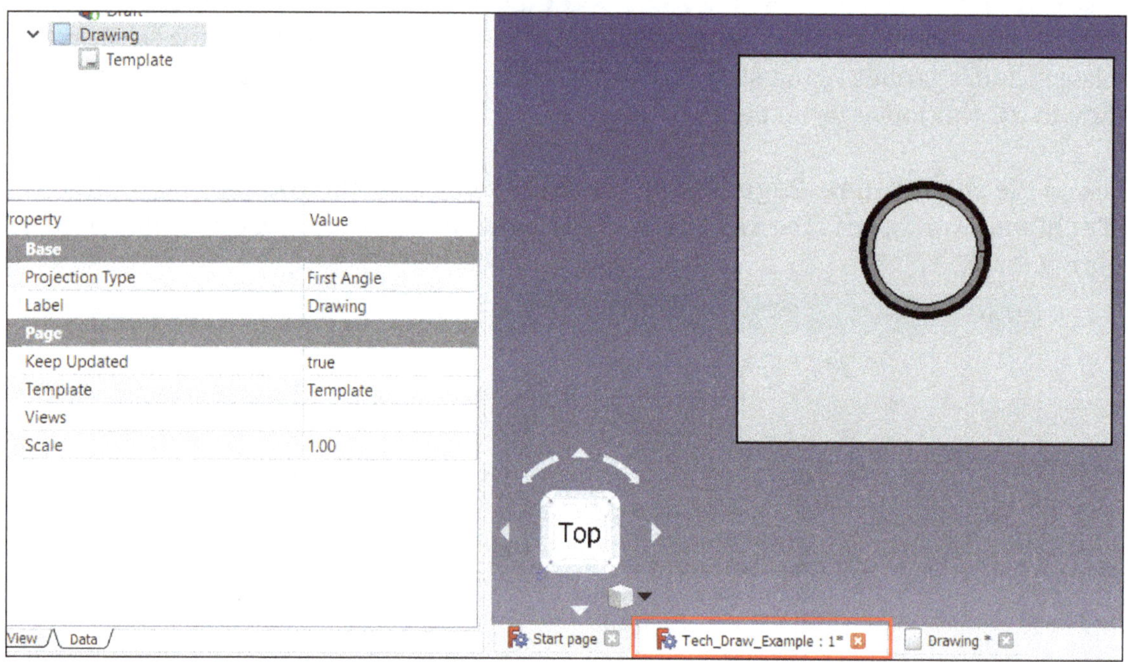

Figure-11. Model oriented for creating view

- Now, select the **Page** tab from the application window, select the model whose views are to be placed from **Model Tree**, and click on the **Insert View in Page** tool from the **Toolbar** or **TechDraw** menu. The current view of model will be placed in the page; refer to Figure-12. Note that the view generated has same orientation as defined in Figure-11.

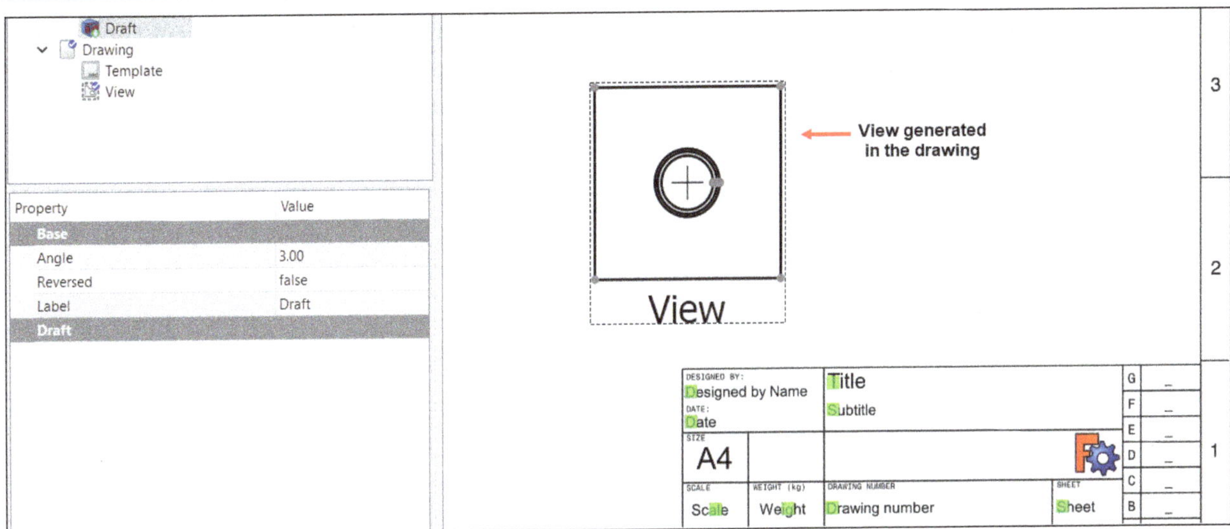

Figure-12. View generated in the drawing

- You can move this view by selecting the dotted boundary and dragging it to desired location.
- On selecting the view from **Model Tree**, various properties of the view are displayed in **Property Editor**; refer to Figure-13.

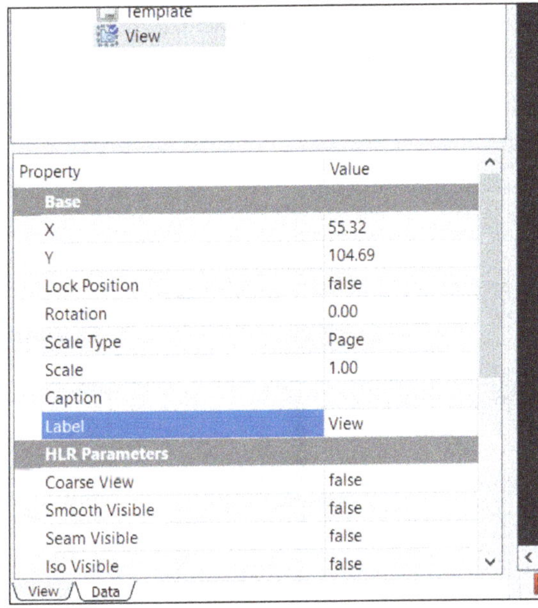

Figure-13. Properties of view

- Specify desired values in **X** and **Y** edit boxes to define location of view on page.
- Select **True** option for **Lock Position** field if you want to lock the position of view on page and do not want it to move.
- Specify desired value in **Rotation** field to rotate selected view and click on the **Recompute** button . The rotated view will be generated.
- Set desired (**True** or **False**) value for various display objects in **HLR Parameters** section of **Property Editor**.
- By default, orthographic views are generated from the model but if you want to generate perspective view of model then select the true option from **Perspective** field in the **Projection** section. You can also specify the Focus distance for perspective view in the **Focus** field.

INSERTING MULTIPLE LINKED VIEWS

The **Insert multiple linked views** tool is used to create multiple views (like front, left, and top views) of selected model object. The procedure to use this tool is given next.

- Click on the **Insert multiple linked views** tool from the **Toolbar** or select the **Insert Projection Group** tool from the **TechDraw** menu after selecting the model from **Model Tree**. The **Projection Group** dialog will be displayed with preview of base view; refer to Figure-14.

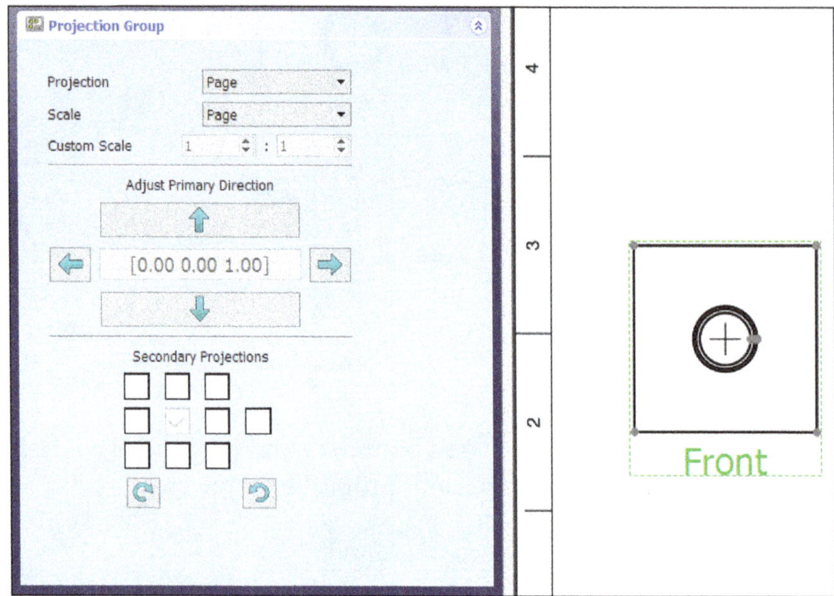

Figure-14. Projection Group dialog

- Select check boxes for views to be generated. Preview of views will be displayed accordingly in the drawing area; refer to Figure-15.

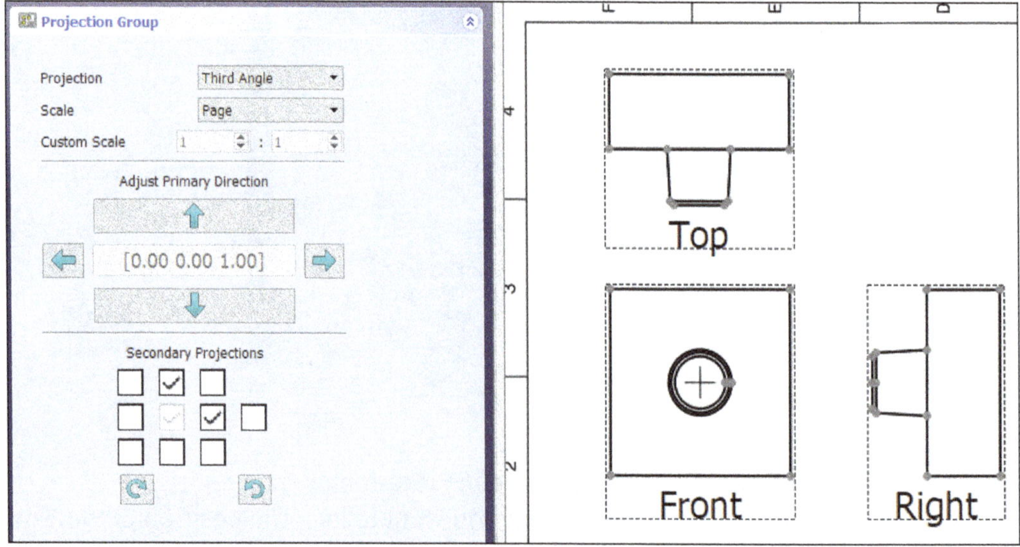

Figure-15. Preview of views

- Select desired option from the **Projection** drop-down to specify whether you want to use First Angle projection or Third Angle projection.
- Select desired option from the **Scale** drop-down to define whether you want to use Automatic scaling, Scale parameter set for page, or you want to specify custom scale values in the **Custom Scale** edit boxes. The views will be generated as per the scale option selected.
- After setting desired parameters, click on the **OK** button from the dialog to create the views.

CREATING SECTION VIEWS

The **Insert Section View in Page** tool is used to insert section view using selected view in the drawing area. The procedure to use this tool is given next.

- Click on the **Insert Section View in Page** tool from the **Toolbar** or **TechDraw** menu after selecting a view from the drawing area. The **Quick Section Parameters** dialog will be displayed; refer to Figure-16.

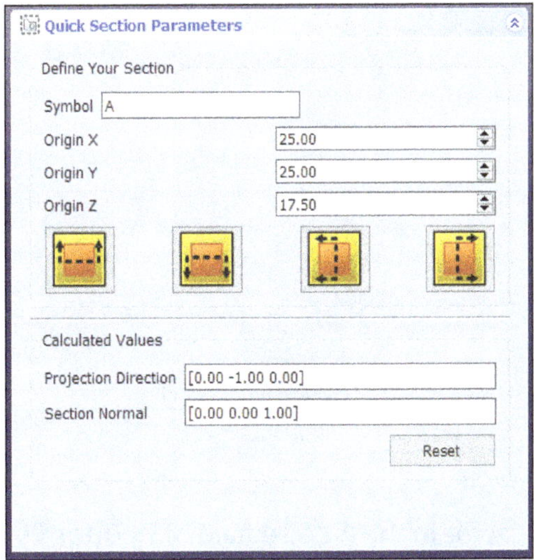

Figure-16. Quick Section Parameters dialog

- Select desired button from the dialog to specify the direction in which the section view will be generated. Preview of the section view will be displayed; refer to Figure-17.

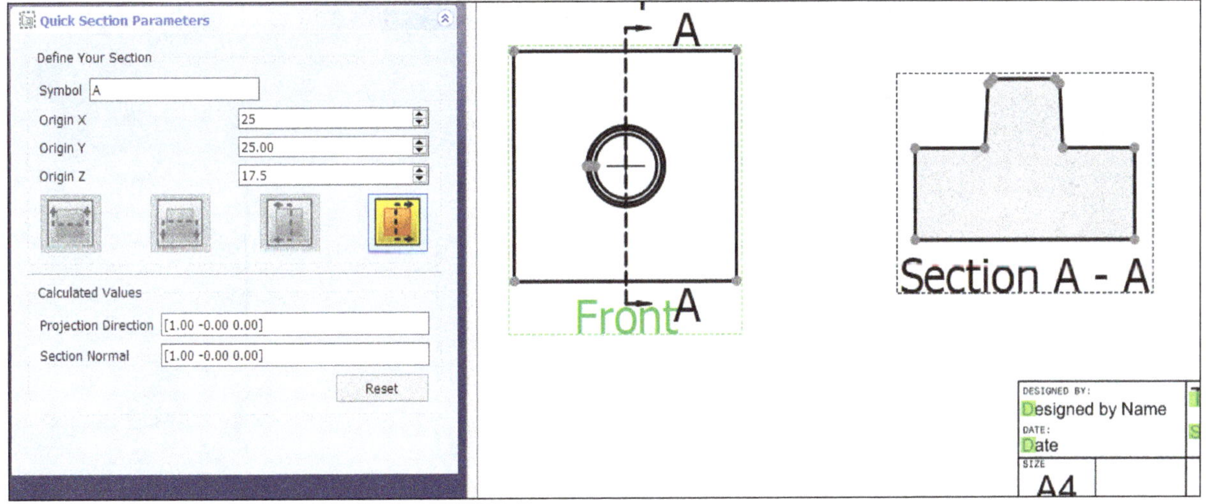

Figure-17. Preview of section view

- Click on the **OK** button from the dialog to create the section view.

INSERTING DETAIL VIEW

The **Insert Detail View** tool is used to create a magnified view of the model for defined region. The procedure to use this tool given next.

- Click on the **Insert Detail View** tool from the **Toolbar** or **TechDraw** menu after selecting a view. The detail view will be created; refer to Figure-18 and properties of detail view will be displayed in the **Property Editor**.

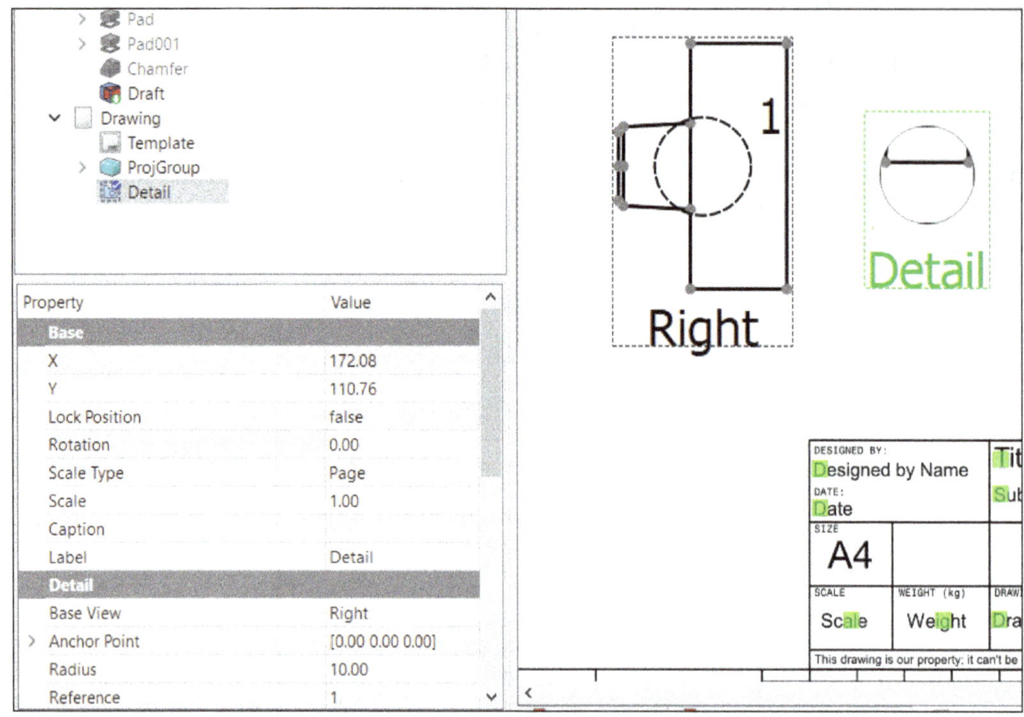

Figure-18. Detail view generated

- Specify desired value of scale factor in the **Scale** field of **Property Editor** to increase or decrease the size of detail view.
- Expand the **Anchor Point** field and specify desired location parameters in **x**, **y**, and **z** fields of **Property Editor** to specify location whose detail view will be generated with respect to center of selected main view; refer to Figure-19.
- Set the other parameters of **Property Editor** as discussed earlier and click in the empty area of drawing to exit.

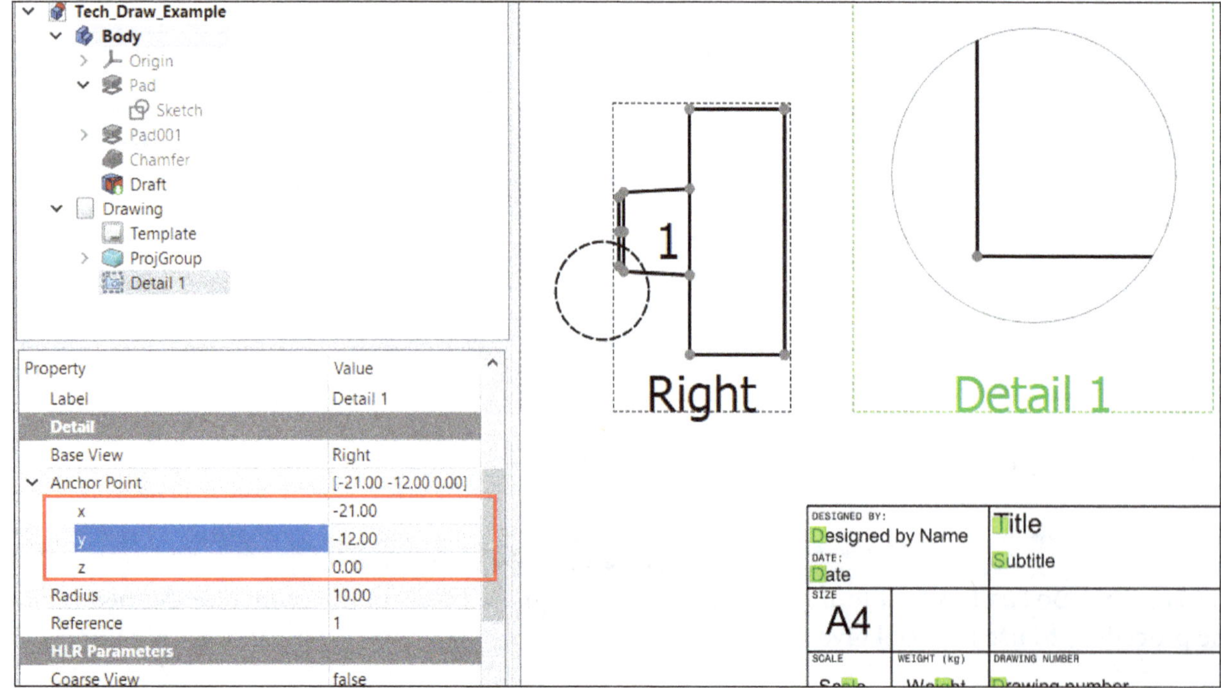

Figure-19. Setting location of detail view circle

INSERTING ANNOTATIONS

The **Insert Annotation** tool is used to insert desired text in the drawing view. The procedure to use this tool is given next.

- Click on the **Insert Annotation** tool from the **Toolbar** or **TechDraw** menu. The annotation will be placed in drawing area and properties of annotation will be displayed in the **Property Editor**; refer to Figure-20.

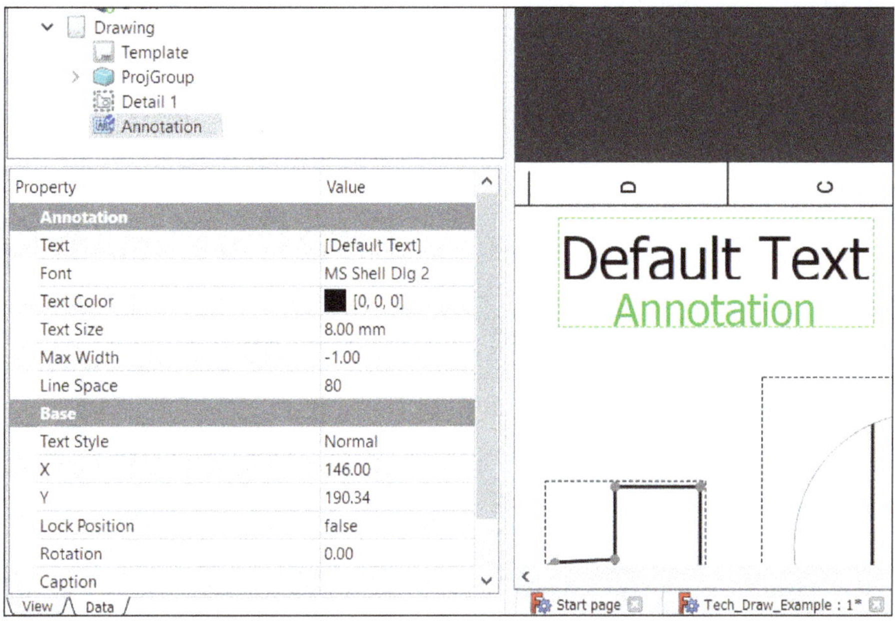

Figure-20. Annotation created

- Click in the **Text** field and then click on the **Browse** button. The **List** dialog box will be displayed; refer to Figure-21.

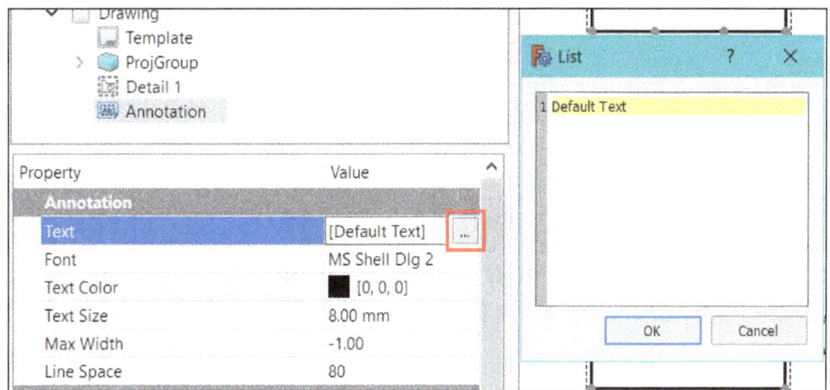

Figure-21. List dialog box

- Specify desired text in the edit box and click on the **OK** button from the dialog box. The annotation will be modified accordingly.
- Specify other parameters like font, text color, text size, and so on in the **Property Editor**.

INSERTING A VIEW OF DRAFT WORKBENCH OBJECTS

The **Insert a View of a Draft Workbench Objects** tool is used to create a view of objects in Draft workbench. The procedure to use this tool is given next.

- Click on the **Insert a View of a Draft Workbench Objects** tool from the **Toolbar** or **TechDraw** menu after selecting the Draft workbench objects to be inserted in the view. The draft views will be generated for selected objects; refer to Figure-22.

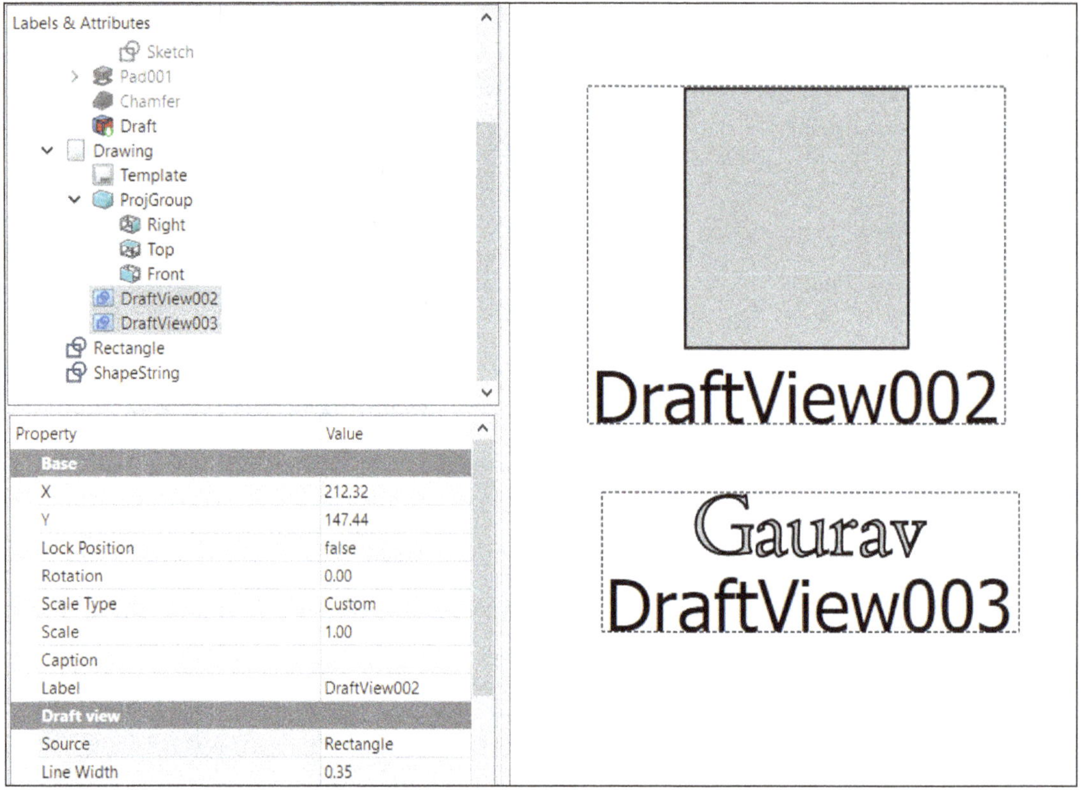

Figure-22. Draft views generated

- You can modify the parameters of views using the **Property Editor**.

The **Inserts a view of a Section Plane from Arch Workbench** tool is used to create views using section planes of **Arch** workbench. You can use this tool in the same way as discussed earlier.

The **Inserts a view of a selected spreadsheet** tool is used to create view using selected spreadsheet created in **Spreadsheet** workbench.

WORKING WITH CLIP GROUP

The **Insert Clip group** tool is used to insert a clipping group in the drawing. You can add or remove objects from the clip group by dragging them inside or out of the group. The procedure to use this tool is given next.

- Click on the **Insert Clip group** tool from the **Toolbar** or **TechDraw** menu. A clip group will be inserted in the drawing; refer to Figure-23.

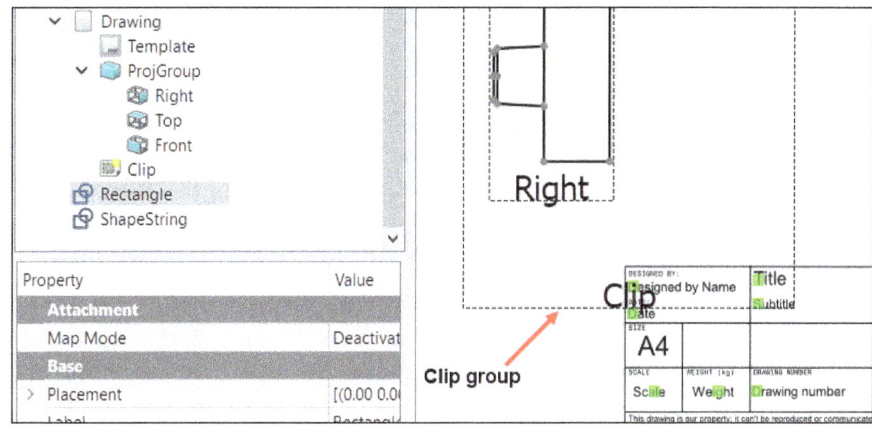

Figure-23. Clip group created

- You can drag the clip group to desired location as discussed for drawing views.
- Select the clip group recently created, select the object/view to be added in the clip group while holding the **CTRL** key and then click on the **Add a View to Clip group** tool from the **Toolbar** or **TechDraw** menu; refer to Figure-24. Selected view will be clipped and added in the clip group.

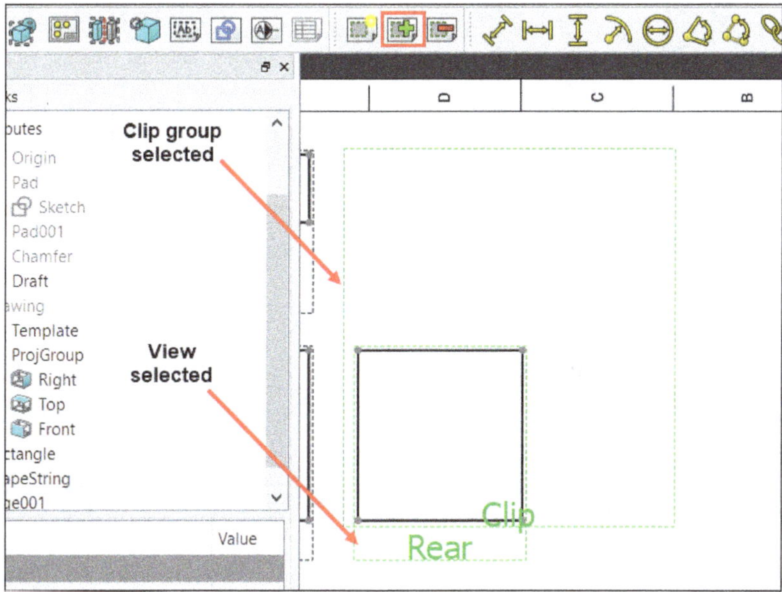

Figure-24. Adding view to clip group

- If you want to remove a view from clip group then select it from the clip group and click on the **Remove a View from Clip group** tool from the **Toolbar** or **TechDraw** menu. The view will be removed from clip group; refer to Figure-25.

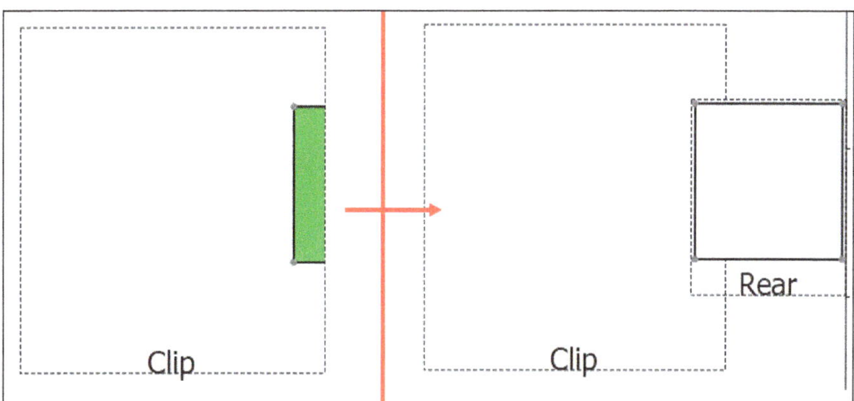

Figure-25. Removing from clip group

Note that you can use **CTRL+Z** and **CTRL+Y** shortcut keys to undo or redo previous operations.

INSERTING LENGTH DIMENSION

The **Insert a new length dimension** tool is used to apply a length dimension based on selected two vertices, an edge, or two edges. The procedure to use this tool is given next.

- After selecting two vertices, two edges (between which length dimension is to be created), or an edge to be dimensioned, click on the **Insert a new length dimension** tool from the **Toolbar** or **TechDraw** menu. The dimension will be generated; refer to Figure-26 and Figure-27.

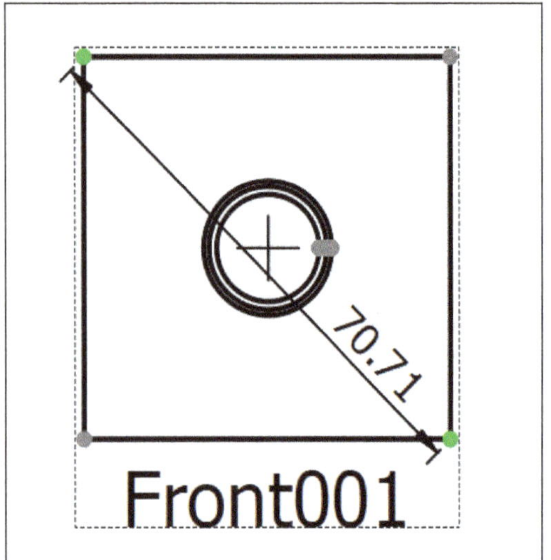

Figure-26. Dimension generated between two vertices

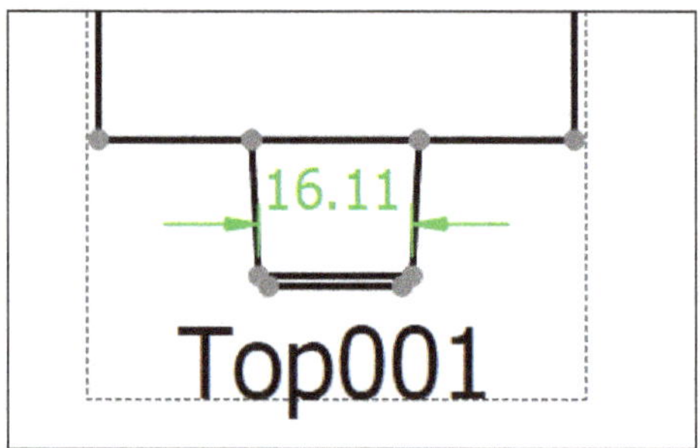

Figure-27. Dimension between two edges

Similarly, you can use other dimensioning tools to create horizontal distance dimension, vertical distance dimension, radius dimension, diameter dimension, angle dimension, and 3 point angle dimensions.

LINKING DIMENSION TO 3D GEOMETRY

The **Link a dimension to 3D geometry** tool is used to link selected dimension of **TechDraw** workbench to an edge or vertices of the 3D model. On doing so, the dimension in TechDraw changes as per the actual dimension in 3D model for same edge/vertices. The procedure to use this tool is given next.

- Create the dimension of edge/vertices that you want to be linked in **TechDraw** workbench page; refer to Figure-28.

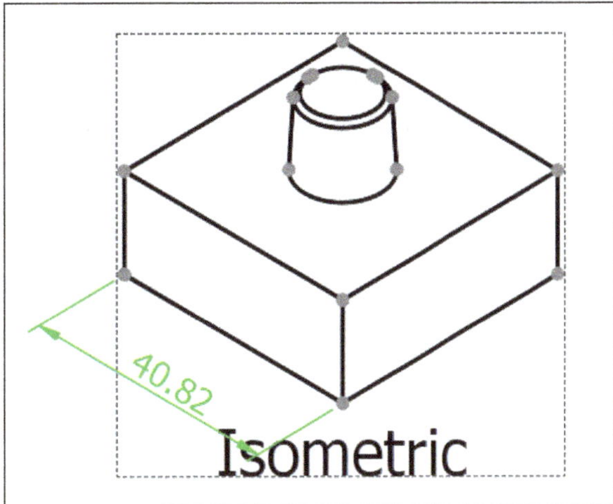

Figure-28. Dimension created in page

- Select the model tab at the bottom of drawing area and select the edge that you want to be linked to available dimensions along with page from Model Tree; refer to Figure-29.

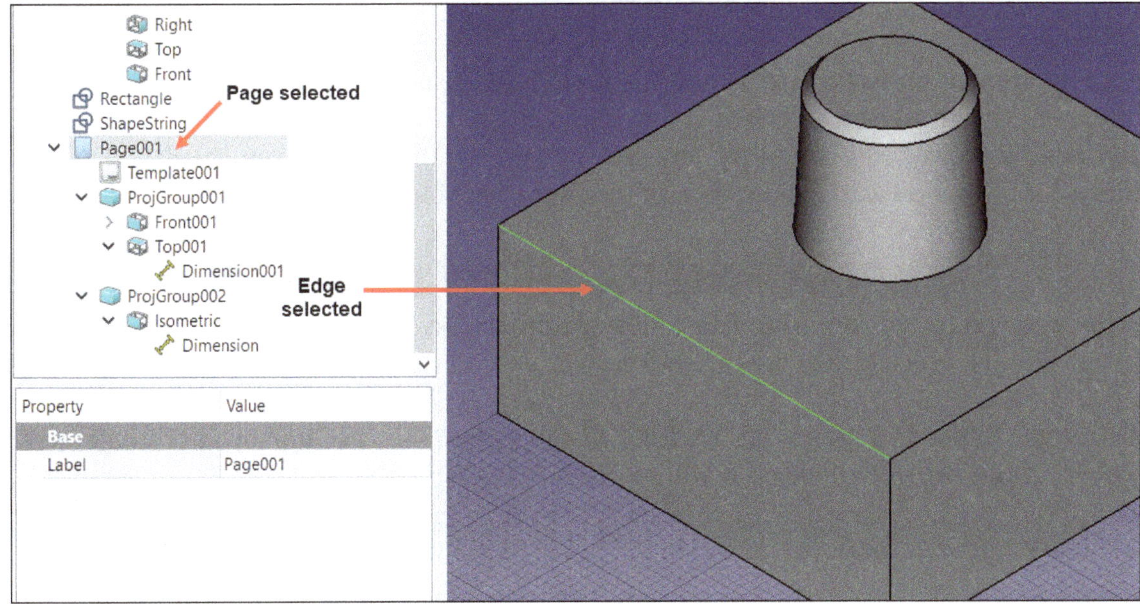

Figure-29. Page and edge selected for linking

- After selecting the edge and page, click on the **Link a dimension to 3D geometry** tool from the **Toolbar** or **TechDraw** menu. The **Link Dimension** dialog will be displayed; refer to Figure-30.

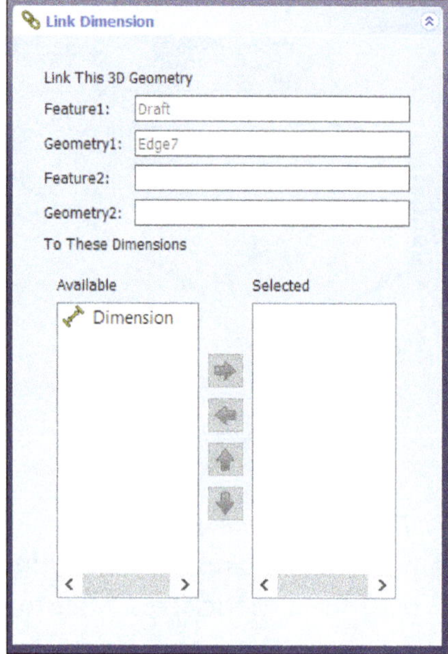

Figure-30. Link Dimension dialog

- Select the dimension to be linked from the **Available** area in the dialog and click on the **Add** button. The dimension will be linked to selected edge.
- After adding desired dimension for linking, click on the **OK** button from the dialog. The linked dimension will reflect actual dimension of the edge in 3D model.

EXPORTING PAGE TO SVG FILE

The **Export a page to an SVG file** tool is used to export selected technical drawing page as an SVG image file. The procedure to use this tool is given next.

- Click on the **Export a page to an SVG file** tool from the **Toolbar** or **TechDraw** menu. The **Export page as SVG** dialog box will be displayed; refer to Figure-31.

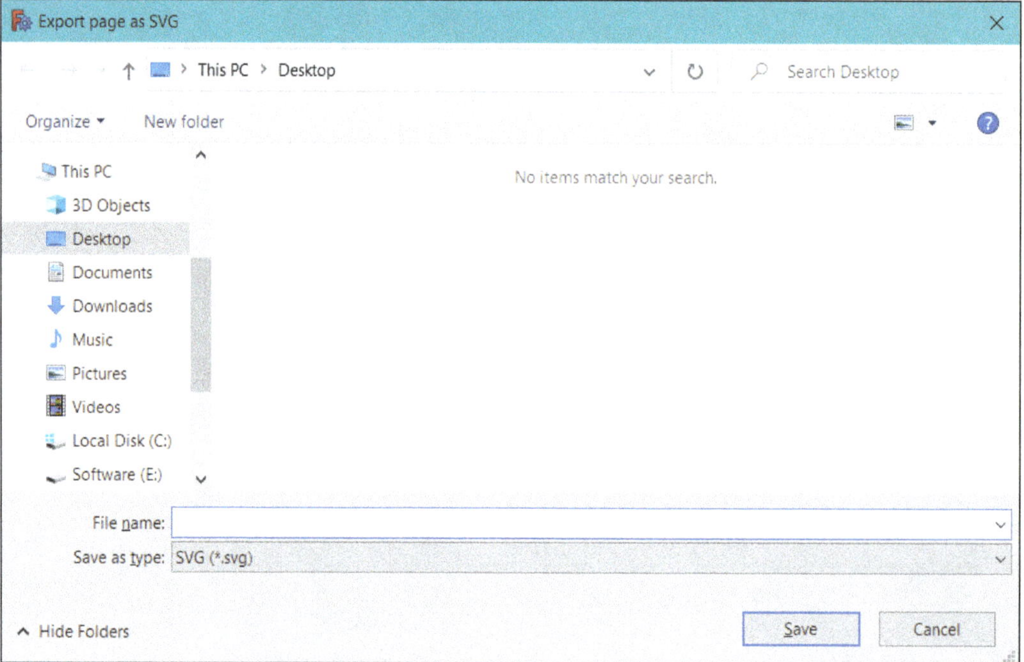

Figure-31. Export page as SVG dialog box

- Specify desired name of file to be exported in the **File name** edit box and click on the **Save** button to save the file. You can open this svg file in your default web browser.

EXPORTING PAGE AS DXF FILE

The **Export a page to a DXF file** tool is used to export current drawing page as DXF file generally used by CAD programs for generating engineering drawings. The procedure to use this tool is given next.

- Click on the **Export a page to a DXF file** tool from the **Toolbar** or **TechDraw** menu. The **Save Dxf File** dialog box will be displayed; refer to Figure-32.

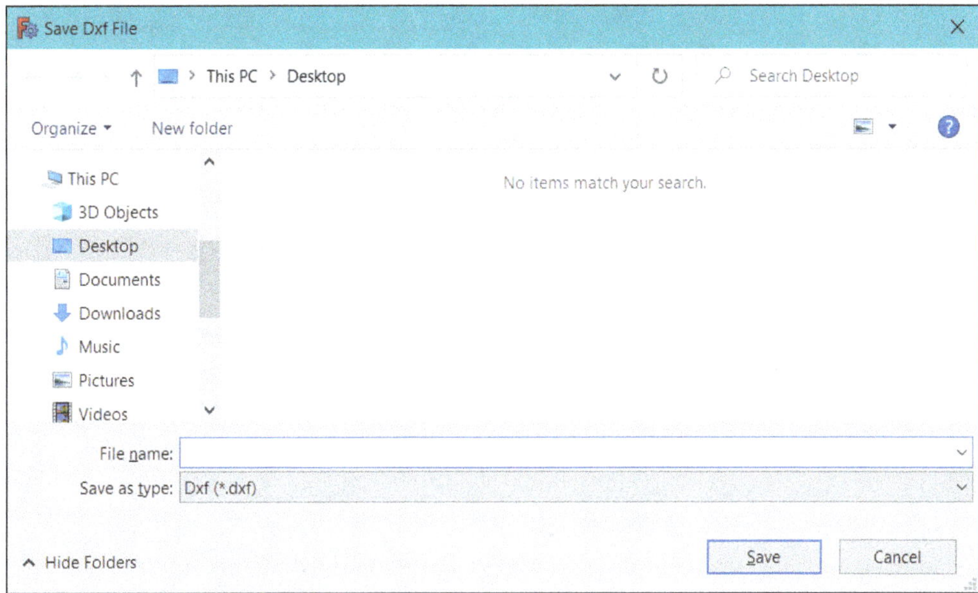

Figure-32. Save Dxf File dialog box

- Specify desired name of file in the **File name** edit box and click on the **Save** button from the dialog box. The file will be created. You can open this file using any Engineering drawing CAD software.

APPLYING HATCH PATTERN

The **Hatch a Face using image file** tool is used to apply hatch pattern in selected faces. The procedure to use this tool is given next.

- Select the face on which you want to apply hatch pattern and click on the **Hatch a Face using image file** tool from the **Toolbar** or **TechDraw** menu. The hatch pattern will be created; refer to Figure-33.

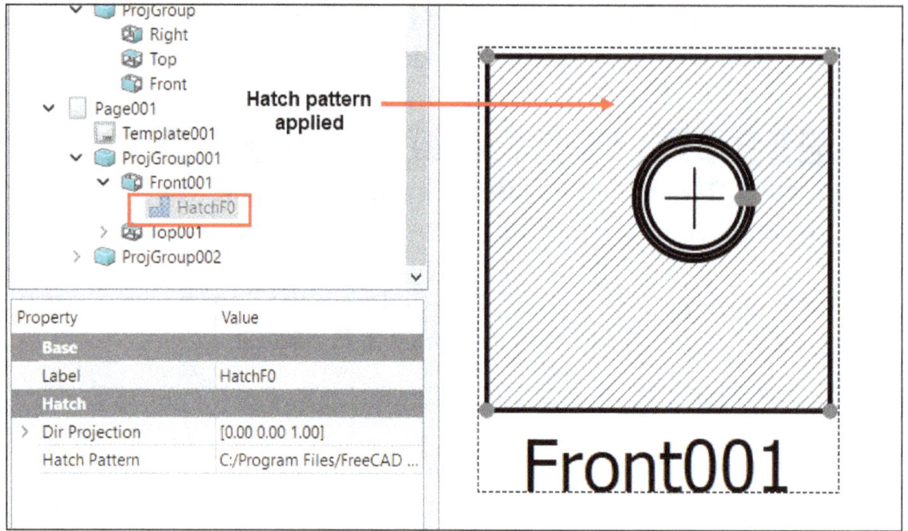

Figure-33. Hatch pattern applied

- Select the **HatchF0** feature from the **Model Tree** to modify parameters of hatch pattern.
- Click on the **Browse** button for **Hatch Pattern** field from **Property Editor**. The **Select a file** dialog box will be displayed; refer to Figure-34.

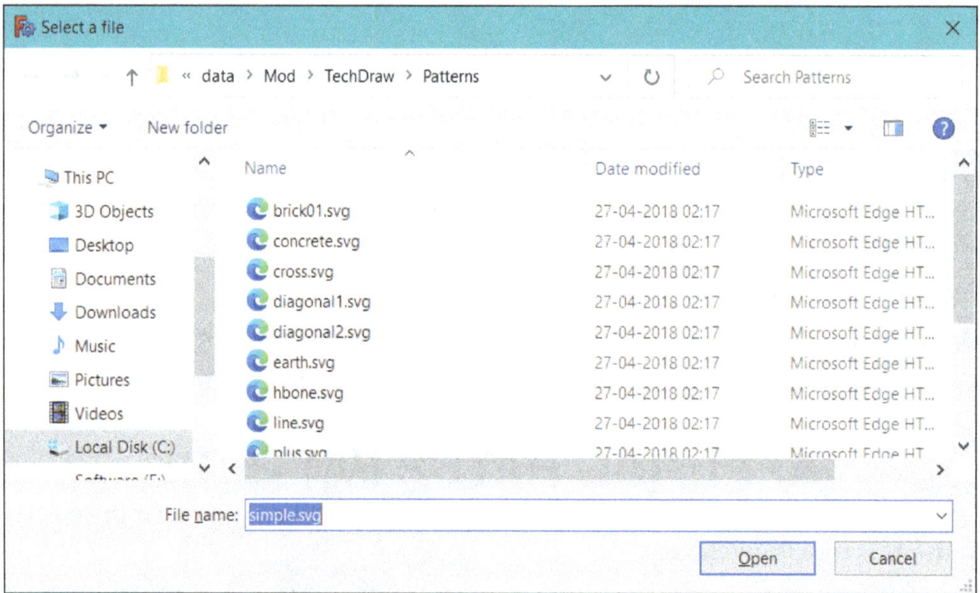

Figure-34. Select a file dialog box

- Select desired image file of hatch pattern from the dialog box and click on the **Open** button. The hatch pattern in drawing will be modified, accordingly.

APPLYING GEOMETRIC HATCH TO A FACE

The **Apply geometric hatch to a Face** tool is used to apply hatch based on specified geometric parameters. The procedure to use this tool is given next.

- Select a face and click on the **Apply geometric hatch to a Face** tool from the **Toolbar** or **TechDraw** menu. The **Apply Geometric Hatch to Face** dialog will be displayed; refer to Figure-35.

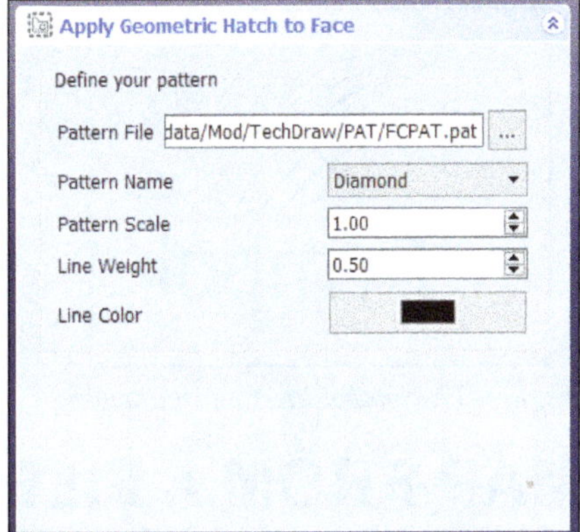

Figure-35. Apply Geometric Hatch to Face dialog

- Select desired option from the **Pattern Name** drop-down to define pattern to be created for hatching.
- Specify the scale, line weight, and line color for pattern in respective fields of the dialog.
- After setting desired parameters, click in the face on which you want to apply the pattern and click on the **OK** button from the dialog to create the pattern.

INSERTING SYMBOL FROM AN SVG FILE

The **Insert symbol from a svg file** tool is used to insert a symbol using a local svg file available. The procedure to use this tool is given next.

- Click on the **Insert symbol from a svg file** tool from the **Toolbar** or **TechDraw** menu. The **Choose an SVG file to open** dialog box will be displayed; refer to Figure-36.

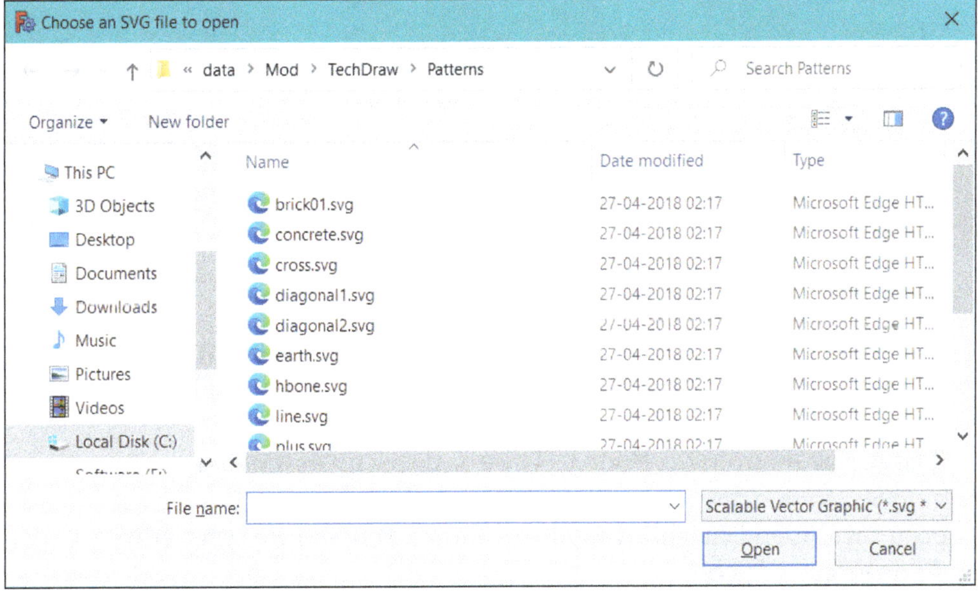

Figure-36. Choose an SVG file to open dialog box

- Select desired symbol file and click on the **Open** button. The symbol will be inserted at the center of the page; refer to Figure-37.

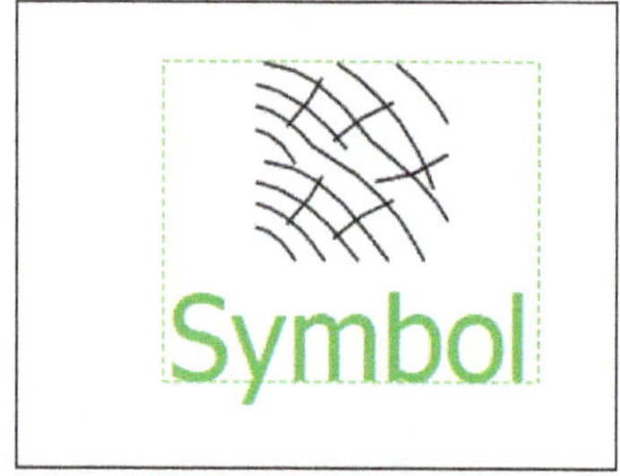

Figure-37. Symbol inserted in drawing

INSERTS A BITMAP FROM A FILE INTO A PAGE

The **Inserts a bitmap from a file into a Page** tool is used to insert an image file in the page. The procedure to use this tool is given next.

- Click on the **Inserts a bitmap from a file into a Page** tool from the **Toolbar** or **TechDraw** menu. The **Select an Image File** dialog box will be displayed; refer to Figure-38.

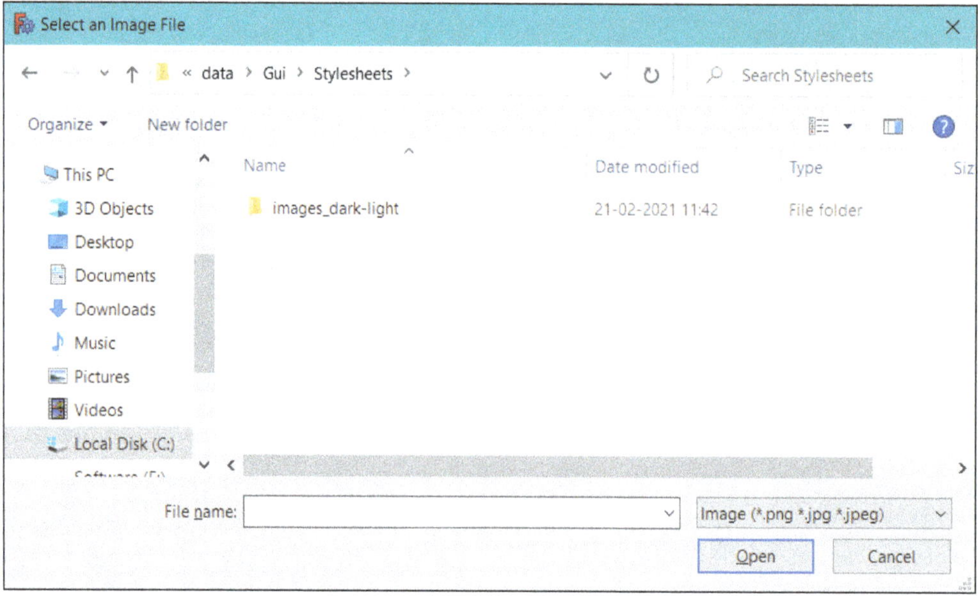

Figure-38. Select an Image File dialog box

- Select desired image file to be inserted in the page and click on the **Open** button. The selected image file will be inserted in the drawing. Note that only PNG and JPG/JPEG formats are supported for image insertion.
- Select the **Image** feature from the **Model Tree** and specify desired parameters in the **Property Editor**.
- You can specify width and height of image in the **Width** and **Height** edit boxes of the **Image** section in **Property Editor**.

TURNING VIEW FRAMES ON/OFF

The **Turn View Frames On/Off** tool is used to turn ON or turn OFF frames around the views in drawing page. Click on the **Turn View Frames On/Off** tool from **Toolbar** or **TechDraw** menu to toggle display of frames around the views.

IMAGE WORKBENCH

The **Image** workbench is used to place image file at any plane. Select the **Image** option from the **Switch between workbenches** drop-down to activate this workbench; refer to Figure-39.

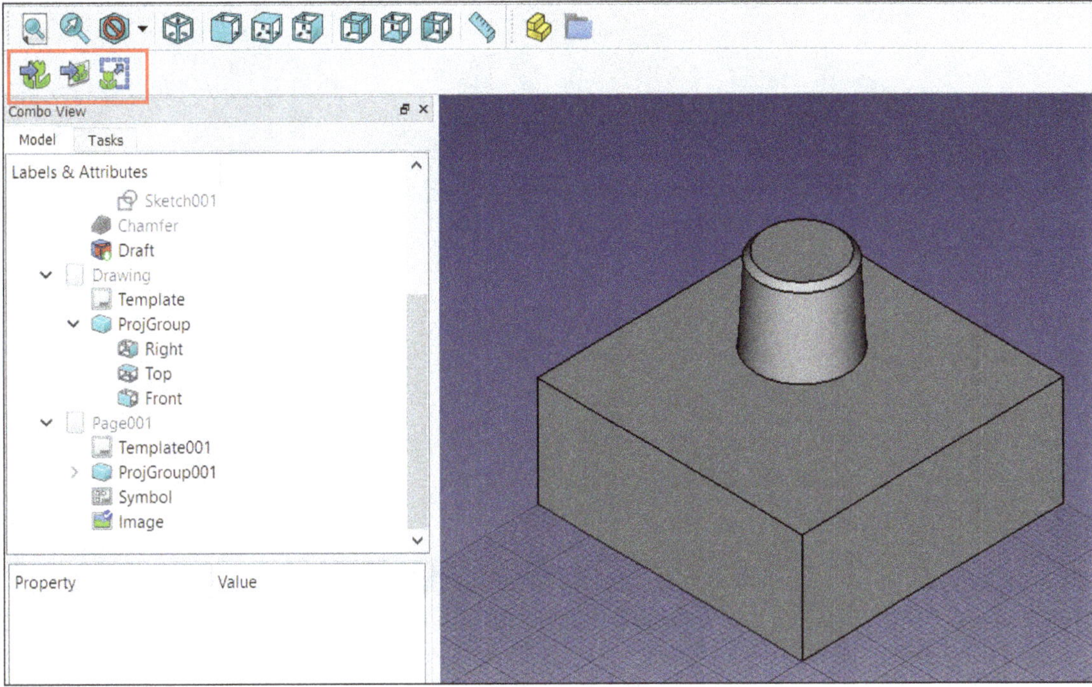

Figure-39. Image workbench

Various tools in this **Toolbar** are discussed next.

Opening Image in Image Viewer

The **Open image view** tool is used to open selected image in the **Image Viewer** app of FreeCAD.

- Click on the **Open image view** tool from the **Toolbar** of **Image** workbench. The **Choose an image file to open** dialog box will be displayed; refer to Figure-40.

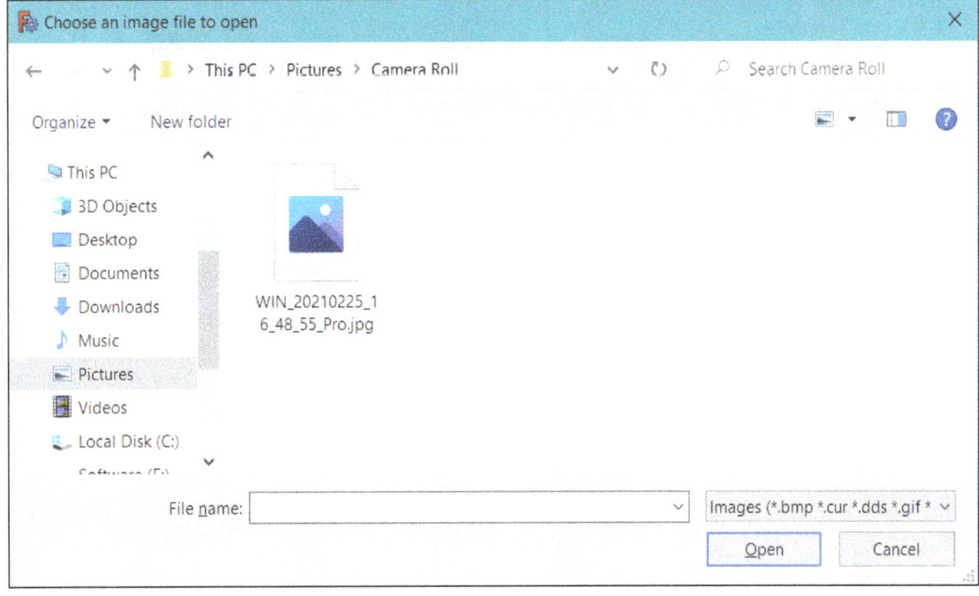

Figure-40. Choose an image file to open dialog box

- Select desired file from the dialog box and click on the **Open** button. The image file will be displayed in image viewer app; refer to Figure-41.

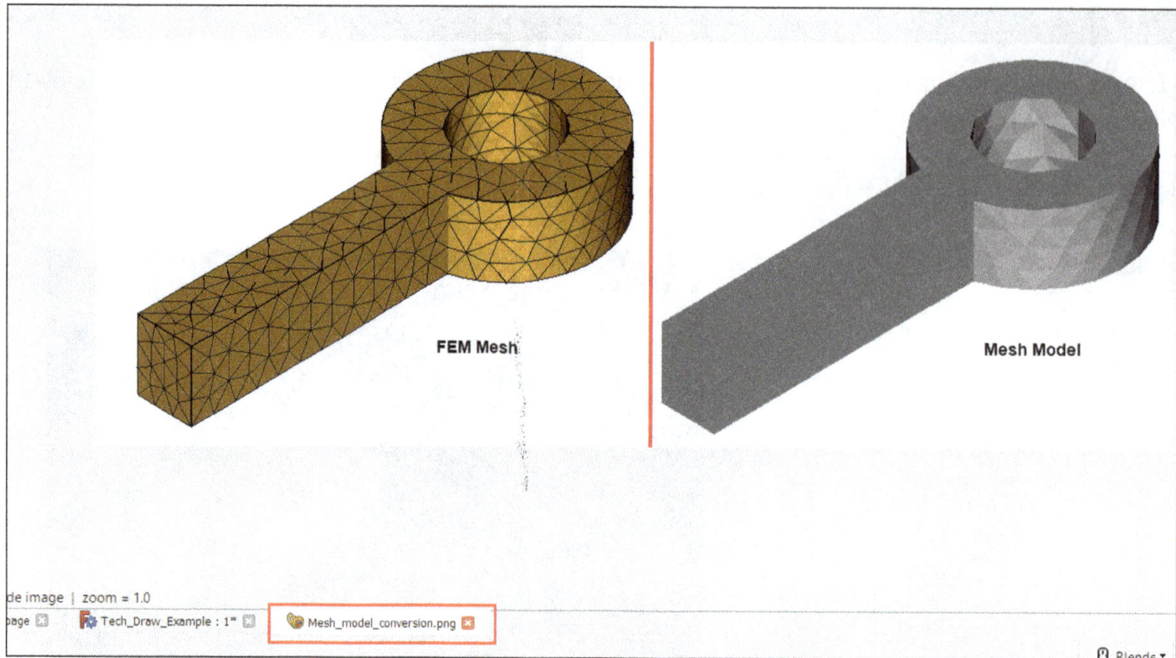

Figure-41. Image file open in Image Viewer app

Inserting Image in 3D Space

The **Create a planar image in the 3D space** tool is used to place an image on specified plane of model. The procedure to use this tool is given next.

- Click on the **Create a planar image in the 3D space** tool from the **Toolbar** in **Image** workbench. The **Choose an image file to open** dialog box will be displayed; refer to Figure-42.

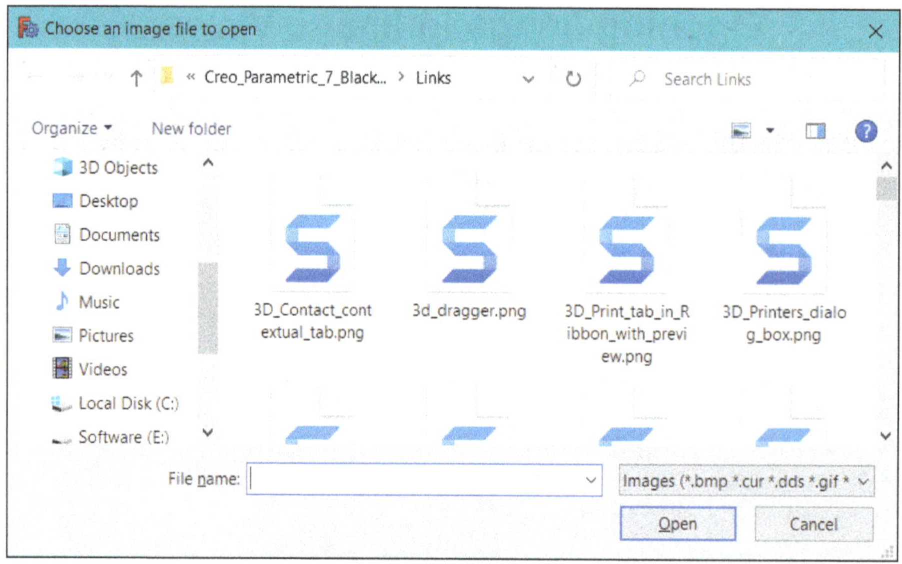

Figure-42. Choose an image file to open dialog box

- Select desired image file from the dialog box and click on the **Open** button. The **Choose orientation** dialog box will be displayed; refer to Figure-43.

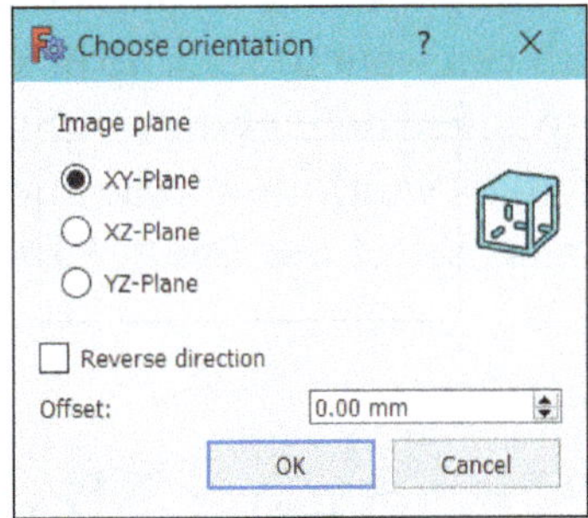

Figure-43. Choose orientation dialog box

- Select desired radio button from the **Image plane** area of the dialog box to specify the plane along which the image will be oriented while placing.
- You can reverse the direction of image placement by selecting the **Reverse direction** check box and you can specify the offset distance from selected plane in the **Offset** edit box.
- After setting desired parameters, click on the **OK** button from the dialog box. The image file will be placed on the selected plane; refer to Figure-44.

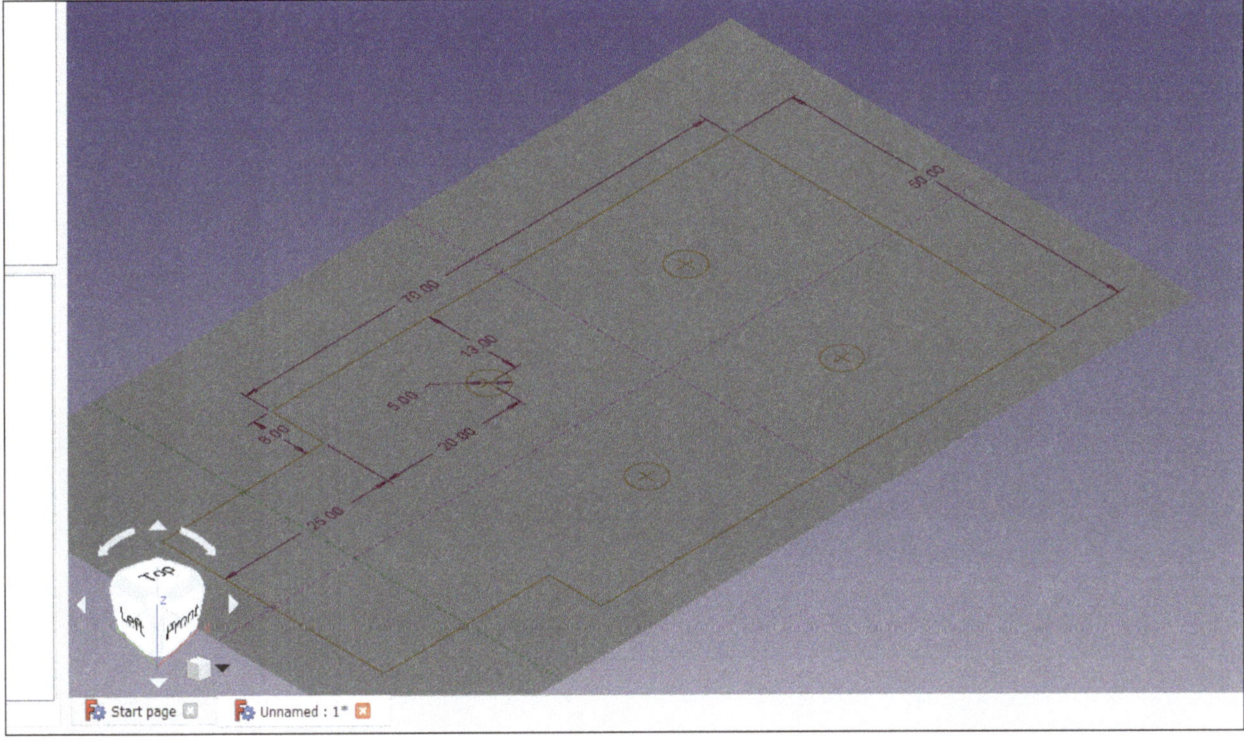

Figure-44. Image placed on plane

Applying Scale to Image on Plane

You can increase or decrease the size of an image file by using the **Scale** tool in **Toolbar**. The procedure to use this tool is given next.

- Click on the **Scales** tool from the **Toolbar** in **Image** workbench. The **Scale image plane** dialog box will be displayed; refer to Figure-45.

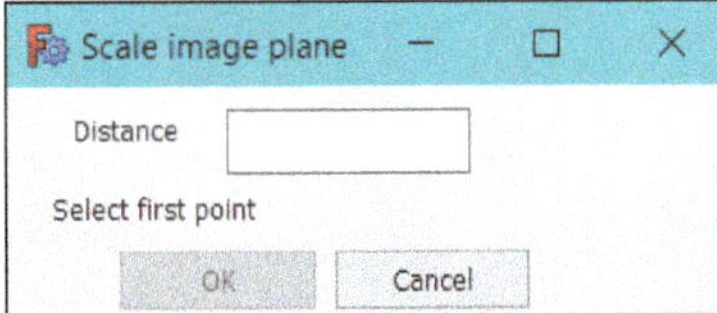

Figure-45. Scale image plane dialog box

- Click in the drawing area to specify first point and second point for scaling image.
- Specify the distance value up to which you want to scale up or scale down the image based on drawn line. In other words, the line earlier specified defines the length or width of image and distance value later specified defines what that length or width should be after scaling.
- After specifying desired distance value, select the **ImagePlane** from **Model Tree** to be scaled and click on the **OK** button.

FOR STUDENT NOTES

Chapter 11

Miscellaneous Workbench

Topics Covered

The major topics covered in this chapter are:

- ***Introduction to Surface Workbench***
- ***Creating Surfaces using Boundary Curves***
- ***Introduction to Mesh Workbench***
- ***Introduction to OpenSCAD Workbench***
- ***Introduction to Plot Workbench***
- ***Introduction to Points Workbench***

INTRODUCTION TO SURFACE WORKBENCH

The tools in **Surface** workbench are used to create surfaces using closed loop curve sections and boundary curves. Select the **Surface** option from **Switch between workbenches** drop-down in the **Toolbar**. The tools to create surfaces will be displayed in the **Toolbar**; refer to Figure-1.

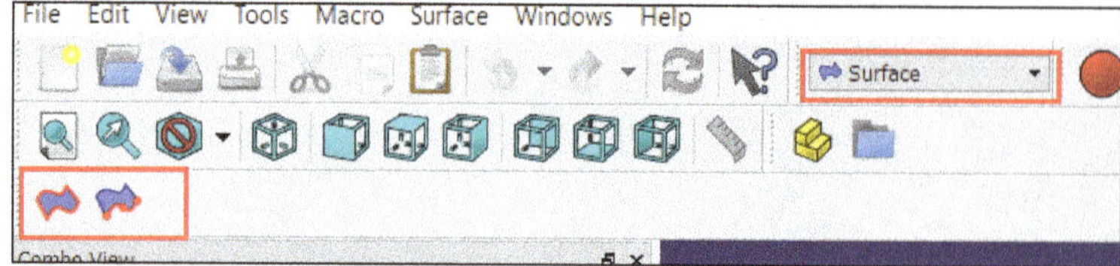

Figure-1. Tools in Surface workbench

Creating Fill Surface using Curves and Vertices

The **Filling** tool is used to create surface using selected curves and vertices. The procedure to use this tool is given next.

- Click on the **Filling** tool from the **Surface** menu or **Toolbar** in **Surface** workbench. The **Filling** dialog will be displayed; refer to Figure-2.

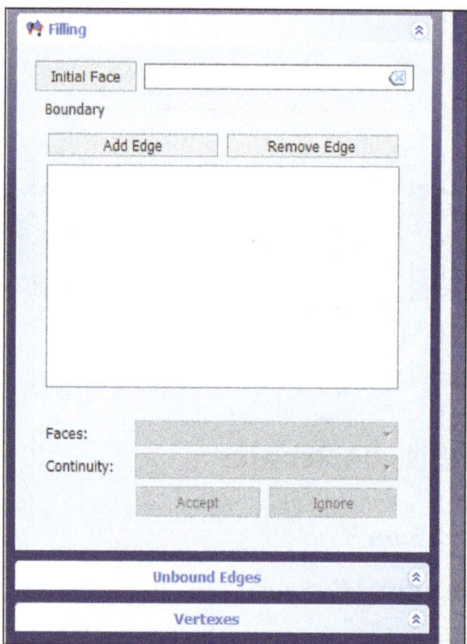

Figure-2. Filling dialog

- Click on the **Add Edge** button from dialog and select all the edges/curves forming a closed region for fill surface. Preview of the surface will be displayed; refer to Figure-3.

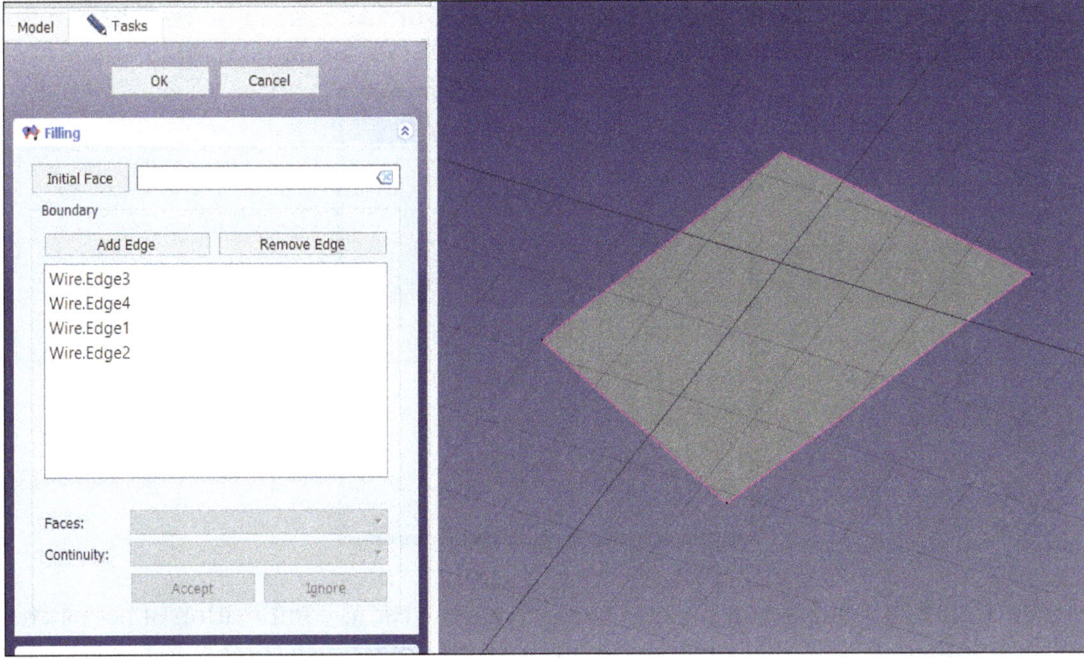

Figure-3. Preview of fill surface

- Expand the **Unbound Edges** dialog in the **Tasks** panel and click on the **Add Edge** button from this dialog to define the curves to be used as guide for creating surface.
- Select the curves from the drawing area. Note that the curves can be from sketch or draft workbenches. The curves may or may not be intersecting with the boundary curves; refer to Figure-4.

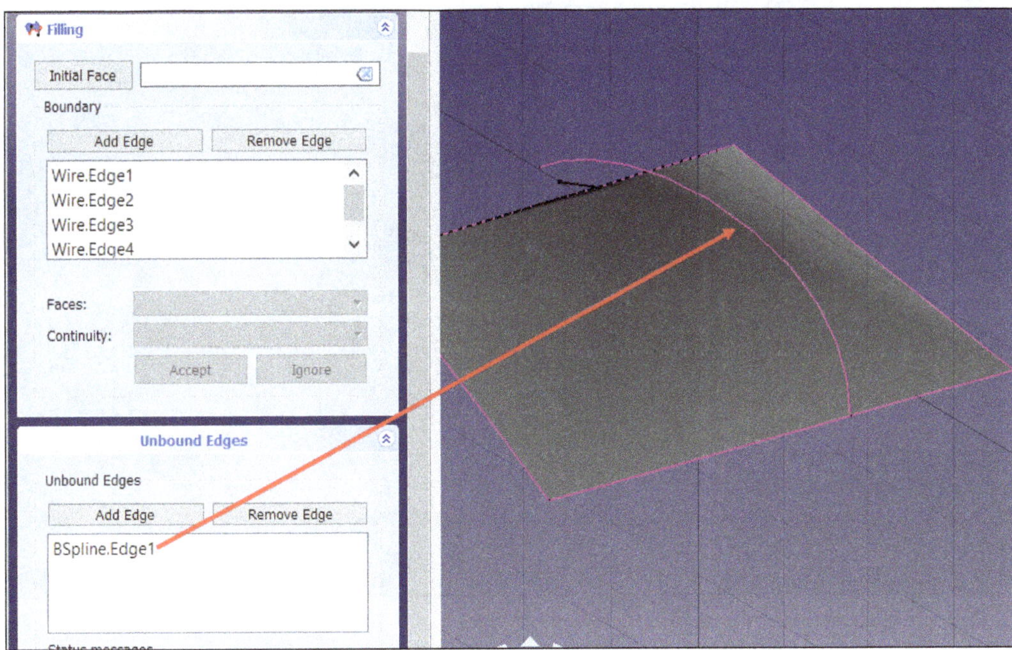

Figure-4. Unbound edge selected

- Similarly, you can use vertexes to refine shape of surface.
- After setting desired parameters, click on the **OK** button to create the surface.

Creating Surface using Boundary Curves

The **Fill Boundary Curves** tool is used to create surface using 2, 3, or 4 boundary edges. The procedure to use this tool is given next.

- Click on the **Fill boundary curves** tool from the **Surface** menu or **Toolbar** in **Surface** workbench. The **Surface** dialog will be displayed; refer to Figure-5.

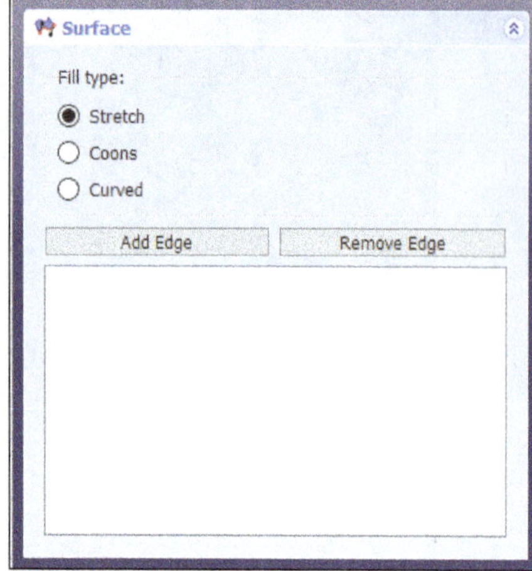

Figure-5. Surface dialog

- Select the **Stretch** radio button if you want to create surface as combination of flat patches.
- Select the **Coons** radio button if you want to surface with better finish compared to stretch but lesser finish as compared to curved.
- Select the **Curved** radio button if you want to create surfaces having finish following curvature of curves.
- After selecting desired radio button, click on the **Add Edge** button from the dialog and select two or more curve in succession. Preview of surface will be displayed; refer to Figure-6.

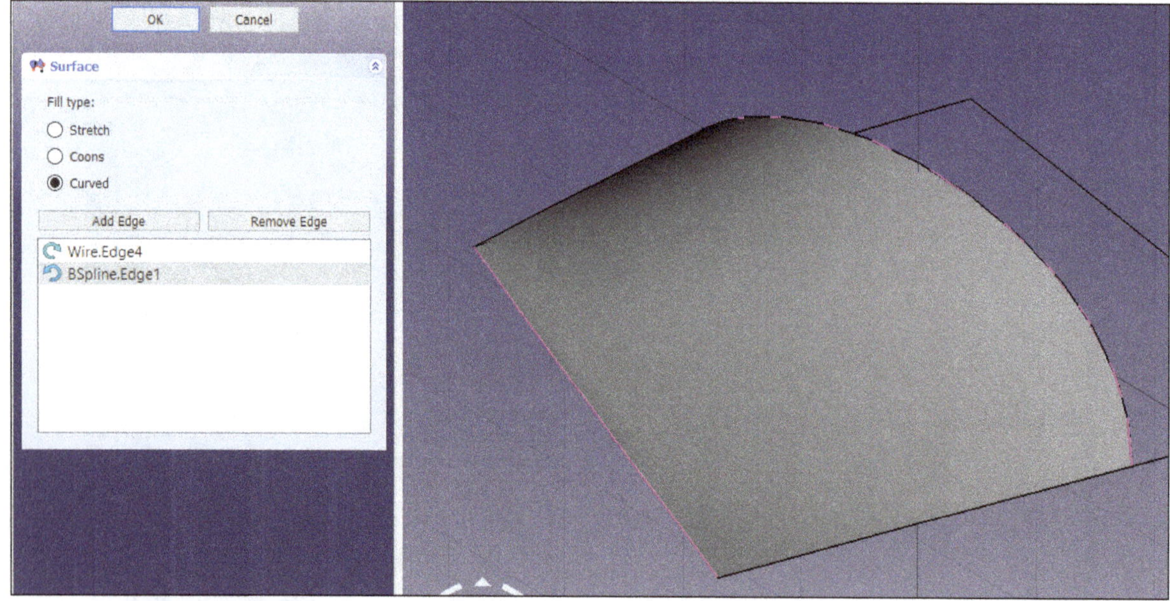

Figure-6. Preview of surface

- Click on the **OK** button from the dialog to create the surface.

Creating Surface by Extending a Face

The **Extend face** tool is used to create a surface by extending selected model face. The procedure to use this tool is given next.

- Select the face/surface you want to use for created extended surface and click on the **Extend face** tool from the **Surface** menu. Preview of extended surface will be displayed; refer to Figure-7.

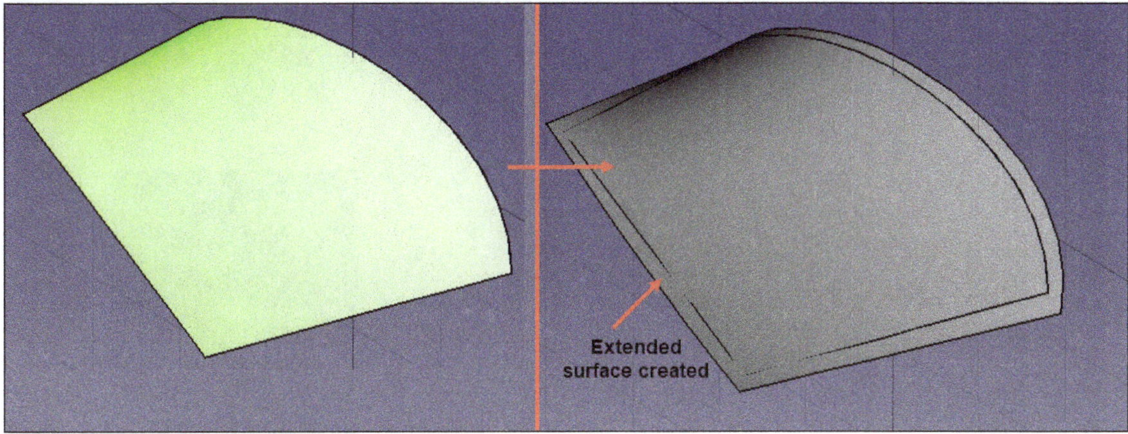

Figure-7. Extend face surface created

- Select the newly created surface from drawing area or **Model Tree** and specify desired parameters in the **Property Editor** to specify extension length in various directions; refer to Figure-8.

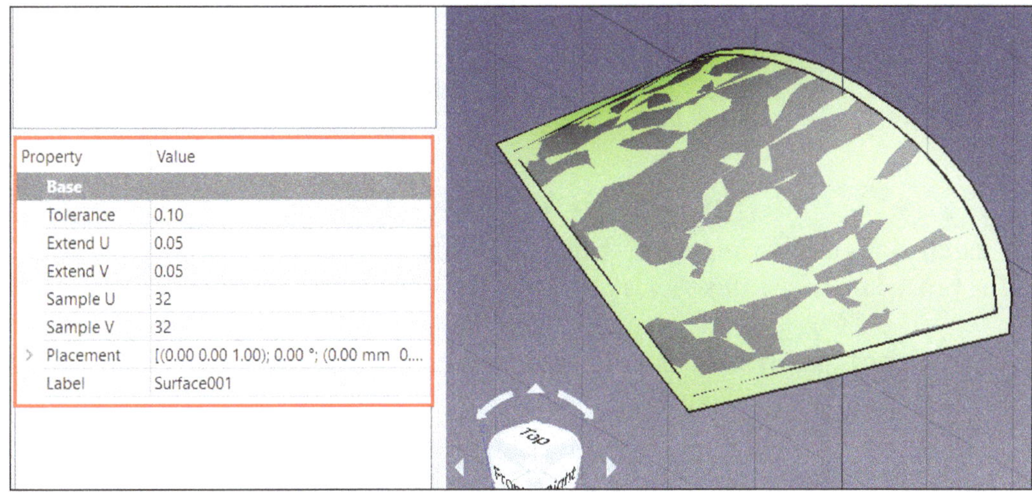

Figure-8. Properties of extended face surface

INTRODUCTION TO MESH WORKBENCH

The **Mesh** workbench of FreeCAD is used to import, create, and modify mesh objects generated in FreeCAD as well as other CAD software. Select the **Mesh** option from the **Switch between workbenches** drop-down to activate the **Mesh** workbench. The tools will be displayed as shown in Figure-9. Various tools in this workbench are discussed next.

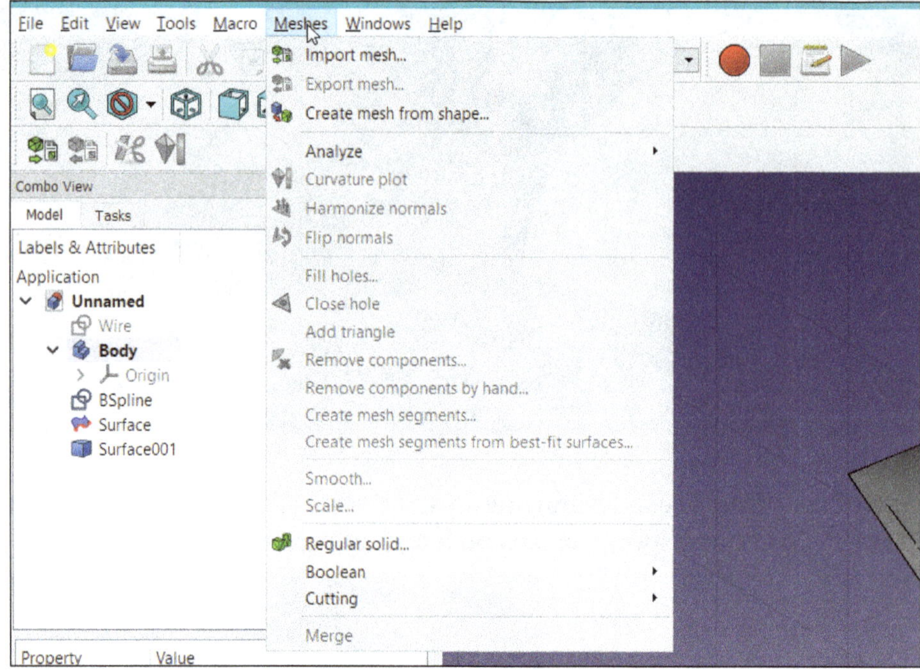

Figure-9. Mesh workbench tools

Importing Mesh Objects

The **Import mesh** tool is used to import mesh objects created in other CAD software like Alias, Inventor, and so on. The procedure to use this tool is given next.

- Click on the **Import mesh** tool from the **Meshes** menu or **Toolbar**. The **Import mesh** dialog box will be displayed; refer to Figure-10.

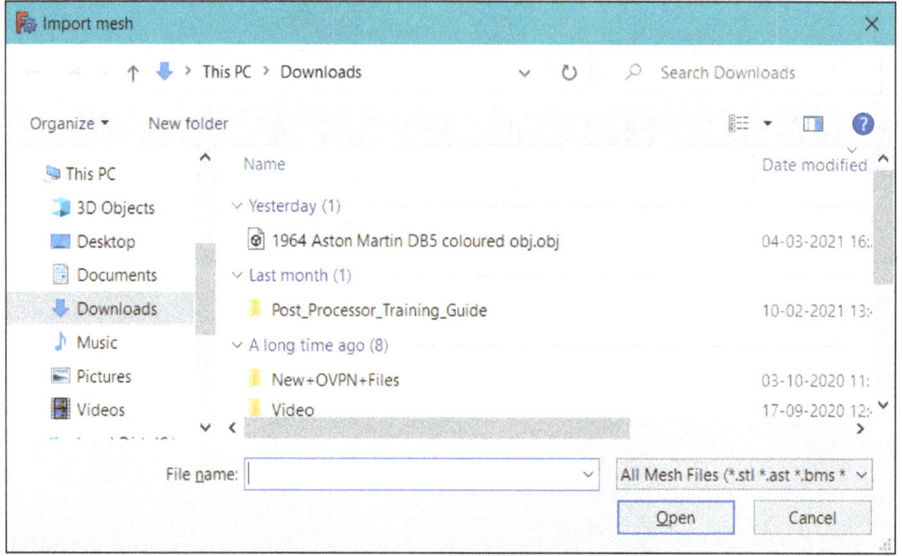

Figure-10. Import mesh dialog box

- Select desired mesh model file and click on the **Open** button. The mesh model will be displayed in the drawing area; refer to Figure-11.

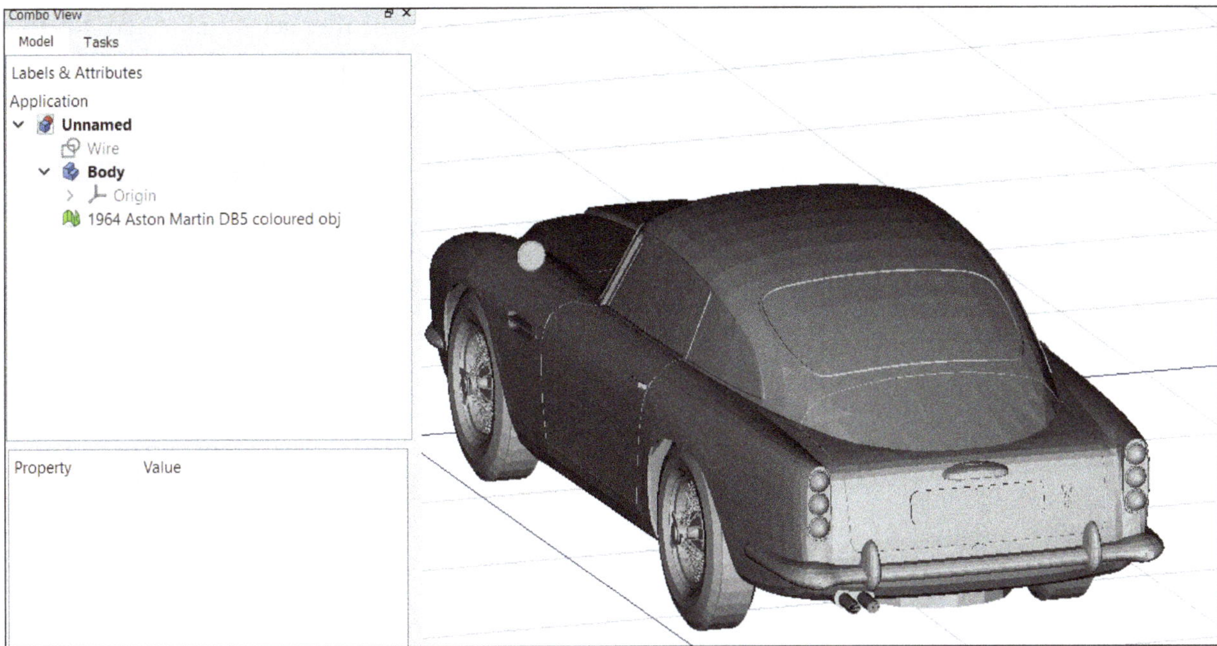

Figure-11. Mesh model imported

Exporting Mesh Model

The **Export mesh** tool is used to export mesh models created in FreeCAD for other CAD software. The procedure to use this tool is given next.

- Select the mesh model created in FreeCAD from **Model Tree** and click on the **Export mesh** tool from **Meshes** menu. The **Export mesh** dialog box will be displayed; refer to Figure-12.

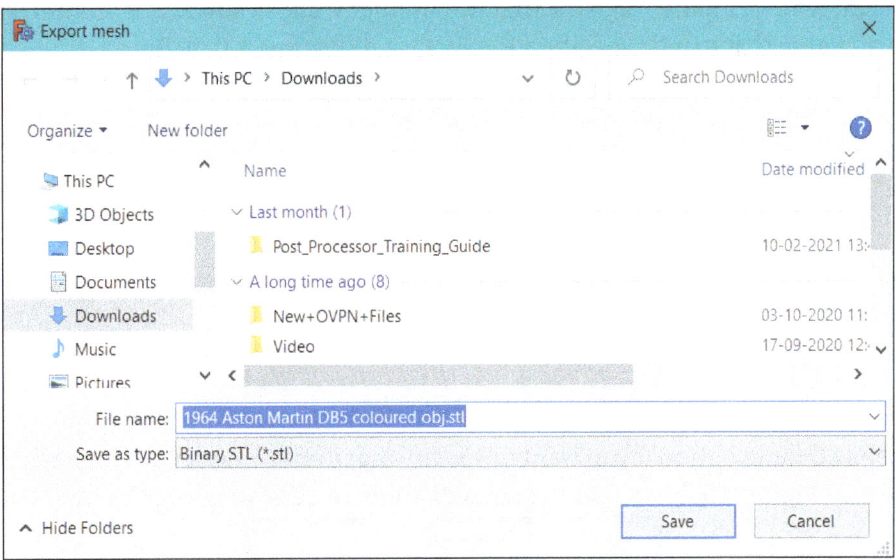

Figure-12. Export mesh dialog box

- Select desired format in which you want to export the file in **Save as type** drop-down.
- Specify desired name of file in the **File name** edit box and click on the **Save** button to export the file.

Creating Mesh from Shape

The **Create mesh from shape** tool is used to create mesh from selected solid body created using **Part** or **Part Design** workbench. The procedure to use this tool is given next.

- Click on the **Create mesh from shape** tool from **Meshes** menu after selecting solid body to be converted. The **Tessellation** dialog will be displayed; refer to Figure-13.

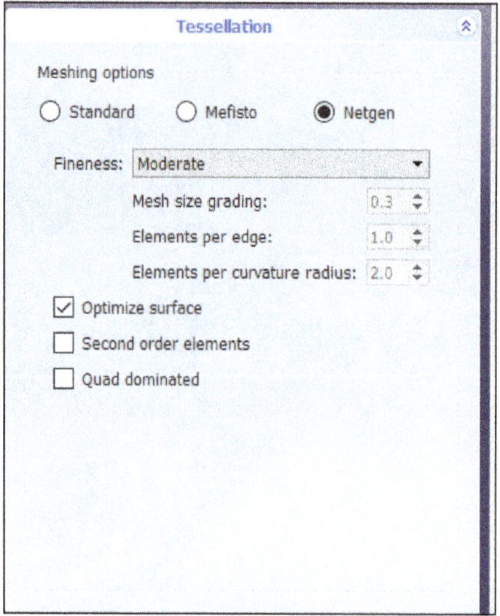

Figure-13. Tessellation dialog

- Select desired radio button from the **Meshing options** area to specify method to be used for generating mesh.
- Select the **Netgen** radio button to choose fineness of mesh from predefined mesh qualities; refer to Figure-13. You can select **Very coarse**, **Coarse**, **Moderate**, **Fine**, and **Very fine** to use presets of mesh quality. If you want to manually define mesh quality then select the **User defined** option and specify related parameters in the edit boxes below the drop-down.
- Select the **Standard** radio button if you want to specify surface deviation in the mesh to define quality of mesh; refer to Figure-14. Note that the smaller value of surface deviation gives finer mesh. The minimum value that can be specified for surface deviation is 0,001.

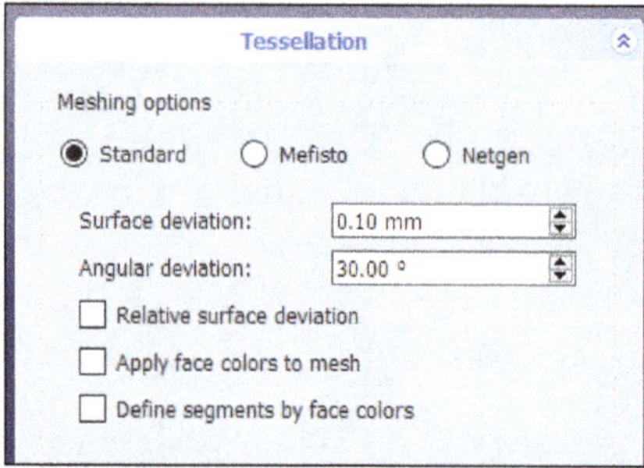

Figure-14. Standard mesh refinement

- Select the **Mefisto** radio button if you want to specify maximum edge length in mesh to define quality of mesh; refer to Figure-15. Note that the smaller value of edge length gives finer mesh.

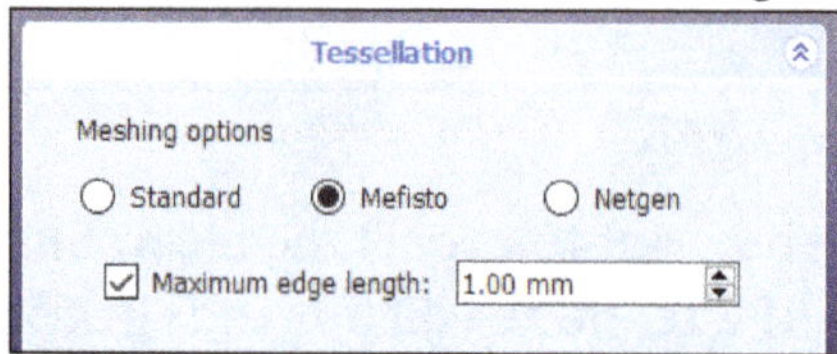

Figure-15. Mefisto mesh refinement

- Specify the other parameters as desired in the dialog and click on the **OK** button. The mesh model will be generated.

Mesh Analysis Tools

The tools in **Analyze** cascading menu of **Meshes** menu are used to analyze and repair various aspects of mesh model imported/created in FreeCAD; refer to Figure-16. Various tools in this menu are discussed next.

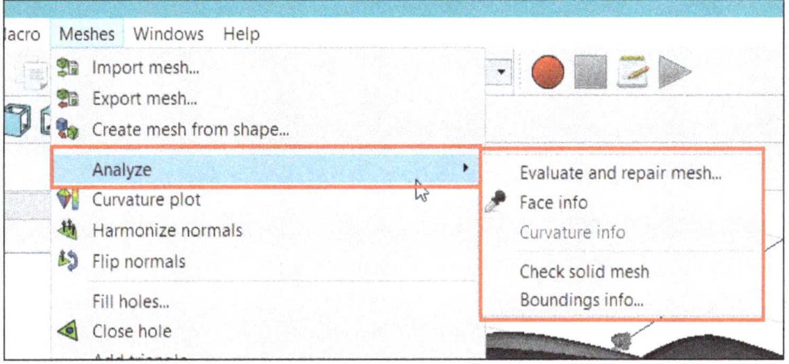

Figure-16. Analyze cascading menu

Evaluating and Repairing Mesh Model

The **Evaluate and repair mesh** tool is used to evaluate various aspects of mesh and repair them. The procedure to use this tool is given next.

- Click on the **Evaluate and repair mesh** tool from **Analyze** cascading menu of the **Meshes** menu. The **Evaluate & Repair Mesh** panel will be displayed; refer to Figure-17.
- There are various sections of panel to analyze various aspects of mesh model like orientation, duplicate faces, degenerated faces, and so on. Click on the **Analyze** button from the panel to check related results. If you want to analyze model based on all the analyses then click on the **Analyze** button from the **All above tests together** area of panel. The result of analyses will be displayed in respective sections of panel.
- Click on the **Repair** button from the panel to perform repairs in the model.

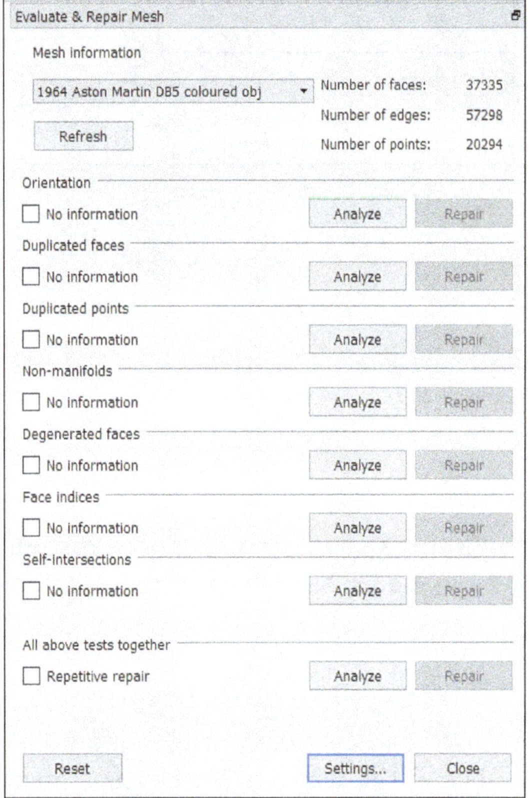

Figure-17. Evaluate and Repair Mesh panel

- Click on the **Settings** button from the panel to modify evaluation settings. The **Evaluation settings** dialog box will be displayed; refer to Figure-18.

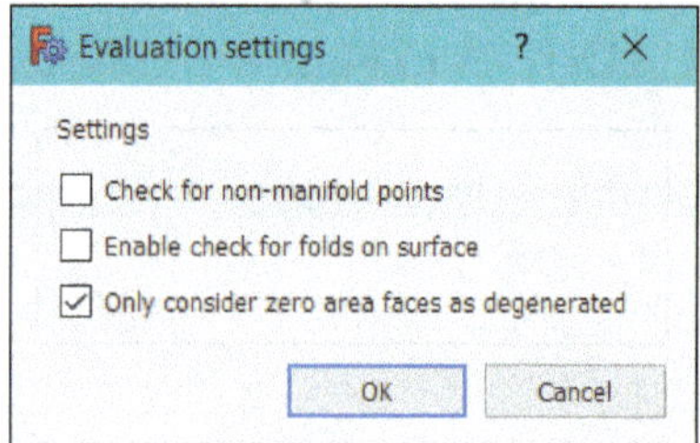

Figure-18. Evaluation settings dialog box

- Select the **Check for non-manifold points** check box to check mesh model for non-manifold points. Non-manifold points are locations of model where you cannot define precision of vertices.
- Select the **Enable check for folds on surface** check box to check for folds in the surface of mesh.
- Select the **Only consider zero area faces as degenerated** check box to define which faces are to be considered as degenerated.
- After setting desired parameters, click on the **OK** button.
- Click on the **Close** button from the panel to exit the panel.

Face Info

The **Face info** tool is used to check information of a face in mesh. Using this tool, you can index of selected face. The procedure to use this tool is given next.

- Click on the **Face info** tool from the **Analyze** cascading menu of **Meshes** menu. The cursor will change to pick face cursor.
- Click on desired faces of mesh model. Indexes of the faces will be displayed; refer to Figure-19.

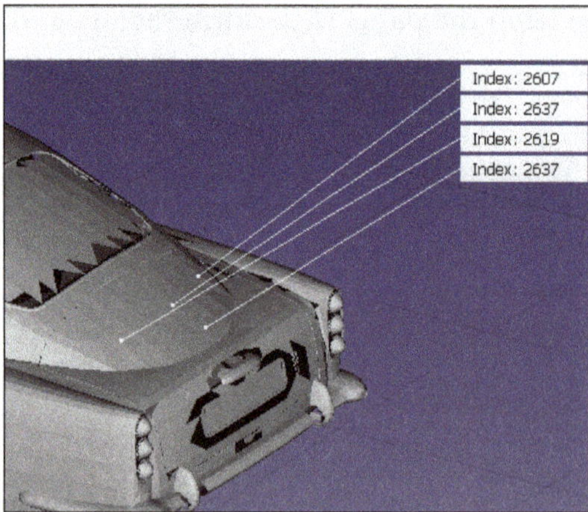

Figure-19. Indexes of faces displayed

- Right-click in the empty area of the drawing and select **Leave info mode** option from the shortcut menu.

Similarly, you can use the **Curvature info** tool of **Analyze** cascading menu in **Meshes** menu to check curvature information of mesh model. Note that this tool is active only when curvature plot has been generated.

Checking Solid Mesh

The **Check solid mesh** tool is used to check whether mesh is a solid mesh body or not. On clicking this tool, the **Solid Mesh** dialog box will be displayed showing information of mesh; refer to Figure-20.

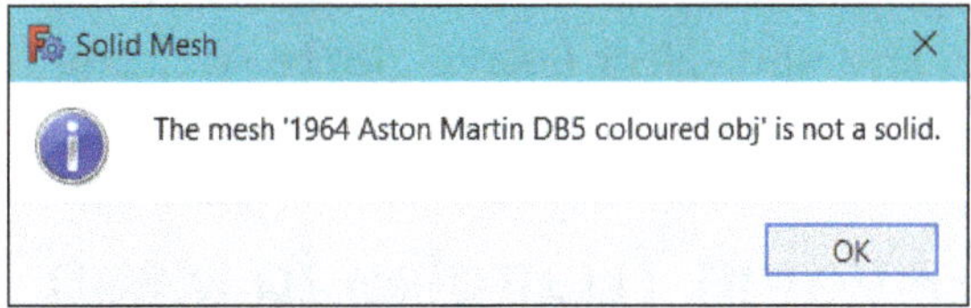

Figure-20. Solid Mesh dialog box

Checking Boundings Info

The **Boundings info** tool is used to check bounding parameters of the mesh body to define minimum and maximum point of the mesh; refer to Figure-21.

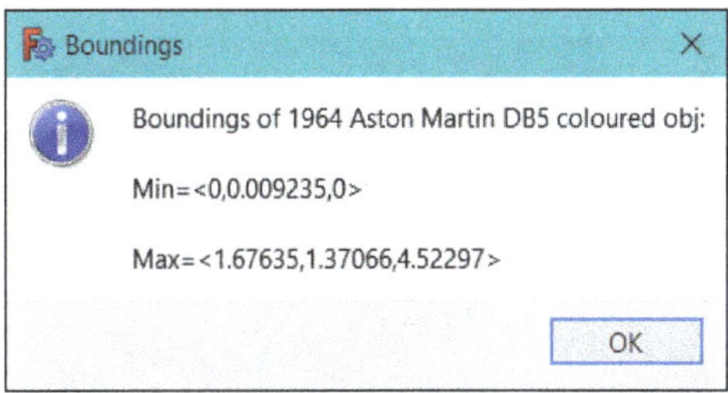

Figure-21. Boundings dialog box

Generating Curvature Plot

The **Curvature plot** tool is used to generate curvature plot for selected mesh model. Click on this tool after selecting mesh body, the curvature plot will be generated and displayed in the drawing area.

Harmonizing Normals

The **Harmonize normals** tool is used to make normals of faces in mesh share the same pattern. Click on the **Harmonize normals** tool from the **Meshes** menu after selecting the mesh body. All the normals of faces in mesh will point in same direction.

Flipping Normals of Mesh Faces

The **Flip normals** tool in **Meshes** menu is used to flip normal direction of mesh faces. After selecting mesh model, click on the **Flip normals** tool from the menu.

Filling Holes in Mesh

The **Fill holes** tool in **Meshes** menu is used to fill holes in the mesh based on specified edges. On clicking this tool, the **Fill holes** dialog box will be displayed; refer to Figure-22. Specify minimum number of edges to be considered for filling holes and click on the **OK** button. All the holes in mesh will be filled.

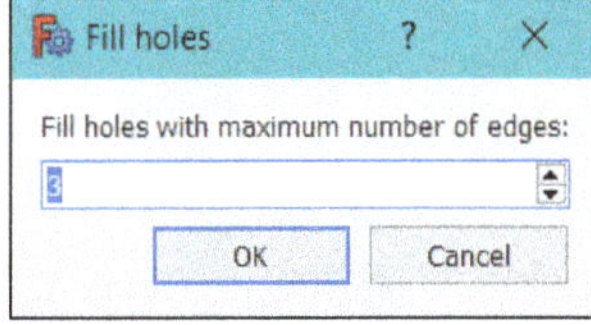

Figure-22. Fill holes dialog box

Closing Holes

The **Close hole** tool is used to close holes in selected faces of mesh. The procedure to use this tool is given next.

- Click on the **Close hole** tool from the **Meshes** menu. The cursor will change to face selection cursor.

- Click on the faces with empty holes/regions. The holes in selected faces will be filled.
- Right-click in the drawing area and select **Leave hole-filling mode** option to exit the hole closing mode.

Adding Triangles to Mesh

The **Add triangle** tool is used to add more triangles to the mesh body. On clicking this tool, you will be asked to select the points where more triangles for tessellation are to be added. One by one specify three points on the faces of mesh body to create a triangle. On selecting the points, a shortcut menu will be displayed; refer to Figure-23. Select the **Add triangle** option from shortcut menu. A triangle will be created in mesh.

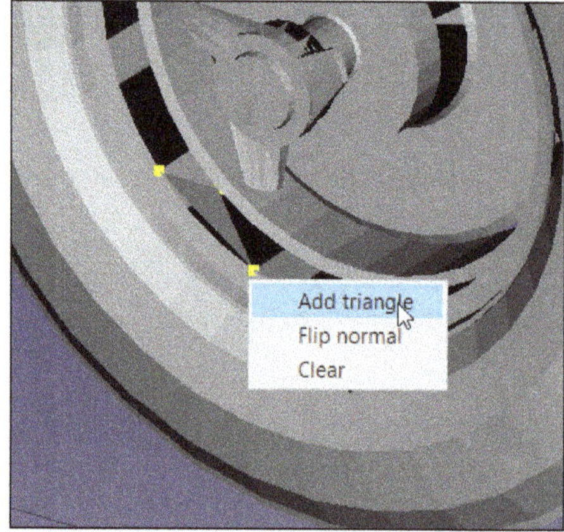

Figure-23. Shortcut menu for adding triangles

- Right-click in the drawing area and select **Finish** button to exit triangle creation mode.

Removing Components of Mesh

The **Remove components** tool is used to remove desired sections of the mesh model. The procedure to use this tool is given next.

- Click on the **Remove components** tool from the **Meshes** menu. The **Remove components** dialog will be displayed; refer to Figure-24.

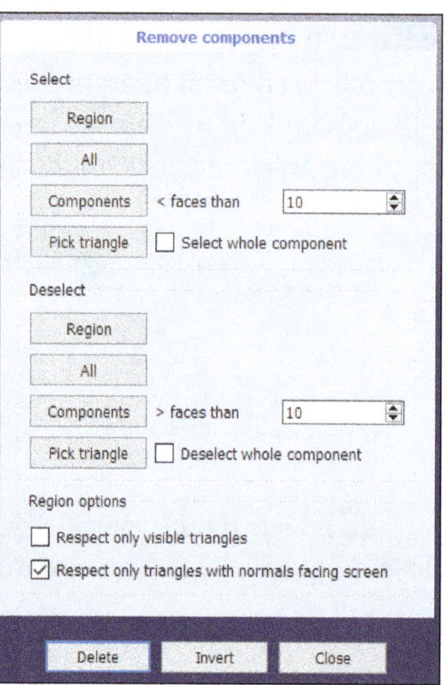

Figure-24. Remove components dialog

- Select the **Region** button from **Select** section and click & drag the cursor over faces of the mesh body to select them; refer to Figure-25. The faces in created region will get selected; refer to Figure-26.

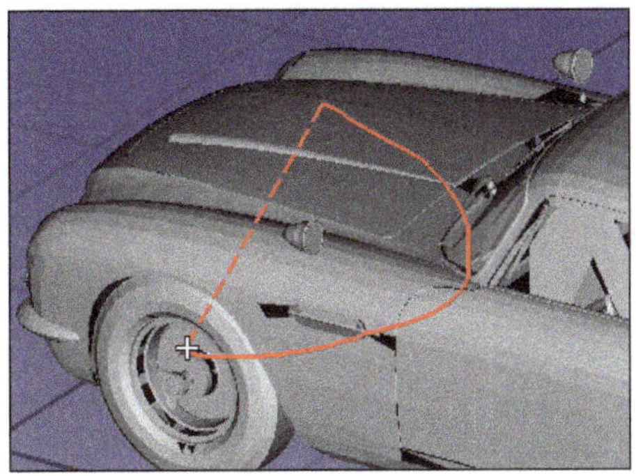

Figure-25. Selecting region on mesh body

Figure-26. Region selected

- If you want to select full body of mesh then select the **All** button. If you want to select specified number of faces from center of body then specify number of faces to be selected in the **< faces than** edit box and click on the **Components** button. If you want to individually select triangles from the mesh body then click on the **Pick triangle** button and select desired faces.
- If you want to deselect faces then you can deselect the faces in the same way from the **Deselect** section of dialog.
- Select the **Respect only visible triangles** check box to select or deselect visible triangles from the model.
- Select the **Respect only triangles with normals facing screen** check box to only select/deselect the triangles which have normals pointing towards the screen.
- After selecting desired faces, click on the **Delete** button to delete faces.

Removing Components by Hand

The **Remove components by hand** tool is used to remove selected faces of the mesh model. The procedure to use this tool is given next.

- Click on the **Remove components by hand** tool from the **Meshes** menu. The cursor will change to a hand sign for selection.
- Click on desired faces of mesh model to select them and right-click in the drawing area. A shortcut menu will be displayed; refer to Figure-27.

Figure-27. Shortcut menu for deleting faces

- Select the **Delete selected faces** option from the shortcut menu to delete selected faces. Select the **Clear selected faces** option to remove all faces from selection and start a new selection set. Select the **Leave removal mode** option to exit the face removal mode.

Creating Mesh Segments

The **Create mesh segments** tool is used to generate segments from the mesh model. The procedure to use this tool is given next.

- After selecting the mesh model, click on the **Create mesh segments** tool from the **Meshes** menu. The **Mesh segmentation** dialog will be displayed; refer to Figure-28.

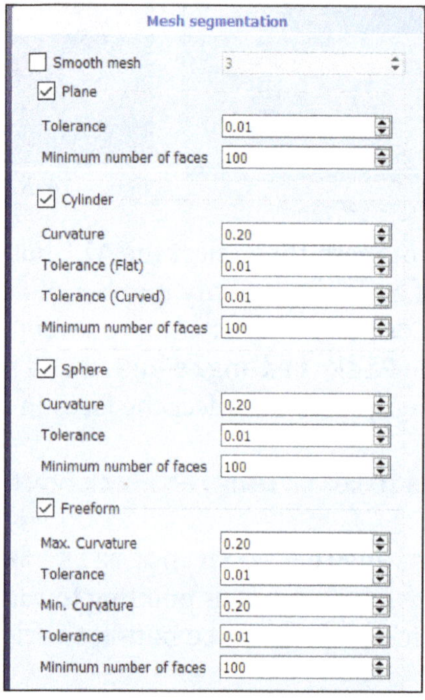

Figure-28. Mesh segmentation dialog

- Select check boxes for object types to be created in segmenting the mesh and specify related tolerances.
- After setting desired parameters, click on the **OK** button from the dialog. The segments will be created; refer to Figure-29.

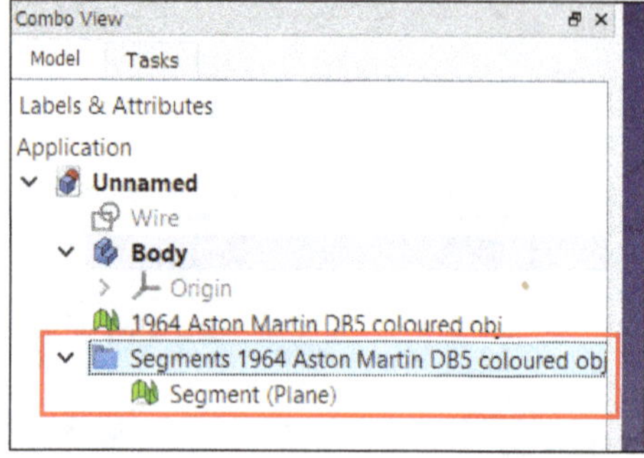

Figure-29. Segment created from mesh model

Similarly, you can use the **Create mesh segments from best-fit surfaces** tool to generate segments of mesh body.

Smoothening Mesh Model

The **Smooth** tool is used to smoothen selected mesh surface by specified degrees. The procedure to use this tool is given next.

- Select the mesh model from **Model Tree** or drawing area and then select the **Smooth** tool from the **Meshes** menu. The **Smoothing** dialog will be displayed; refer to Figure-30.

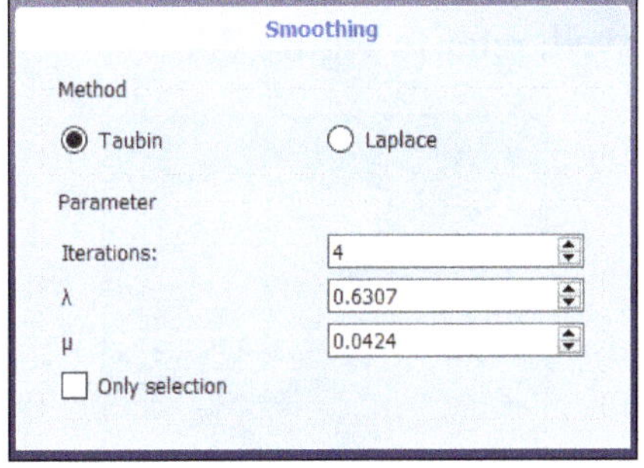

Figure-30. Smoothing dialog

- Select the **Taubin** radio button to use taubin smoothing and select the **Laplace** radio button to use Laplacian smoothing. Figure-31 shows difference between two smoothing methods.

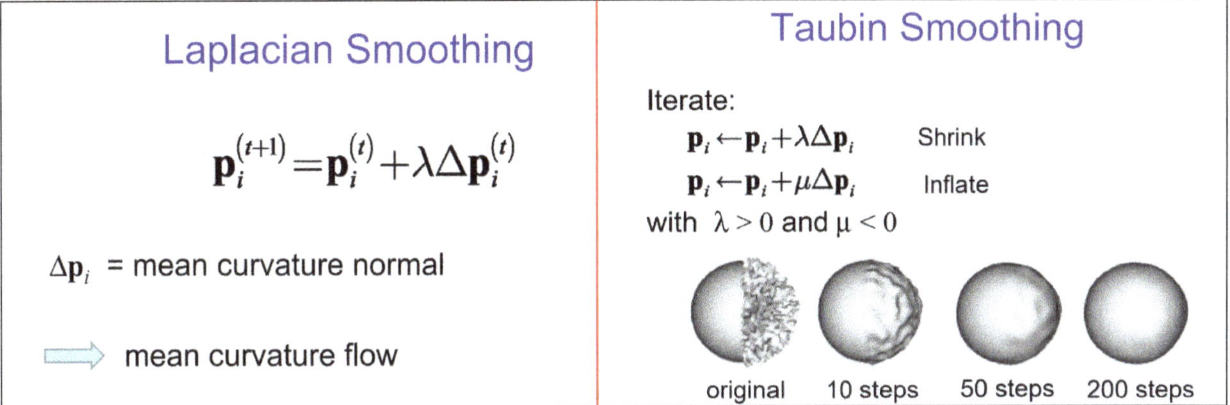

Figure-31. Smoothing methods

- Specify desired parameters in the dialog and click on the **OK** button to smoothen mesh model.

Scaling Mesh Model

The **Scale** tool is used to scale up/down the mesh model. The procedure to use this tool is given next.

- After selecting mesh model, click on the **Scale** tool from the **Meshes** menu. The **Scaling** dialog box will be displayed; refer to Figure-32.

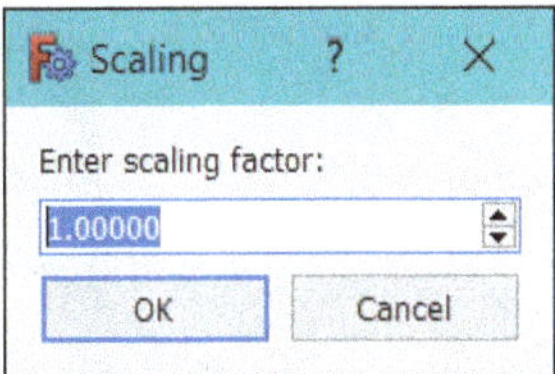

Figure-32. Scaling dialog box

- Specify desired value of scale factor in the edit box and click on the **OK** button. A value more than **1** will increase the size of mesh model and a value less than **1** but greater than **0** will decrease the size of mesh model.

Creating Regular Mesh Solids

The **Regular solid** tool is used to create regular mesh solids. The procedure to use this tool is given next.

- After selecting the mesh body, click on the **Regular solid** tool from **Meshes** menu. The **Regular Solid** dialog box will be displayed; refer to Figure-33.

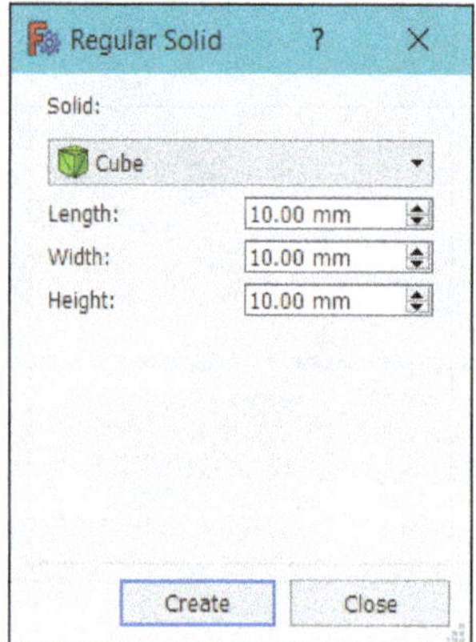

Figure-33. Regular Solid dialog box

- Select desired option from the **Solid** drop-down to define shape of solid and specify related parameters in the dialog box.
- After setting parameters, click on the **Create** button to create solid object at the origin.
- After creating solid, click on the **Close** button from the dialog box to exit.

Performing Boolean Operations on Meshes

The tools in **Boolean** cascading menu of **Meshes** menu are used to perform boolean operations like performing union, intersection, and difference of two mesh objects. Note that you need to install OpenSCAD using web link http://www.openscad.org/ and specify path for exe file of OpenSCAD in **Preferences** dialog box. The procedure to perform boolean operations is given next.

- Select two meshes from the **Model Tree** on which you want to perform boolean operations. The tools in **Boolean** cascading menu of **Meshes** menu will become active.
- Click on the **Union** tool from the cascading menu if you want to join two mesh bodies; refer to Figure-34.

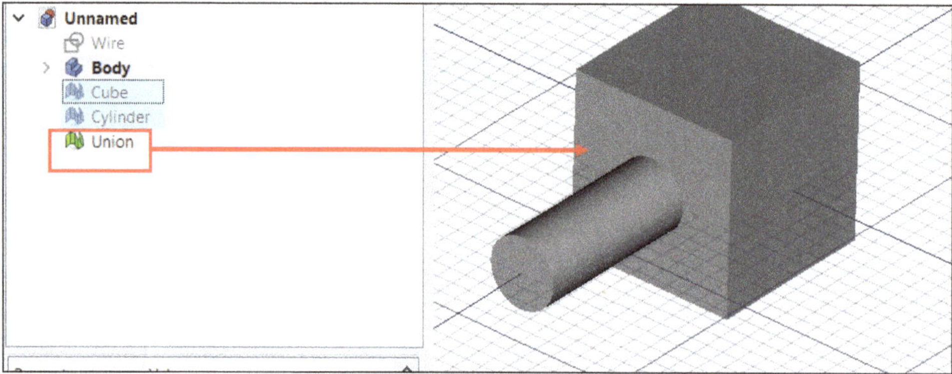

Figure-34. Union body created

- Click on the **Difference** tool to subtract 2nd selected object from 1st selected object. The **Difference** feature will be displayed; refer to Figure-35 .

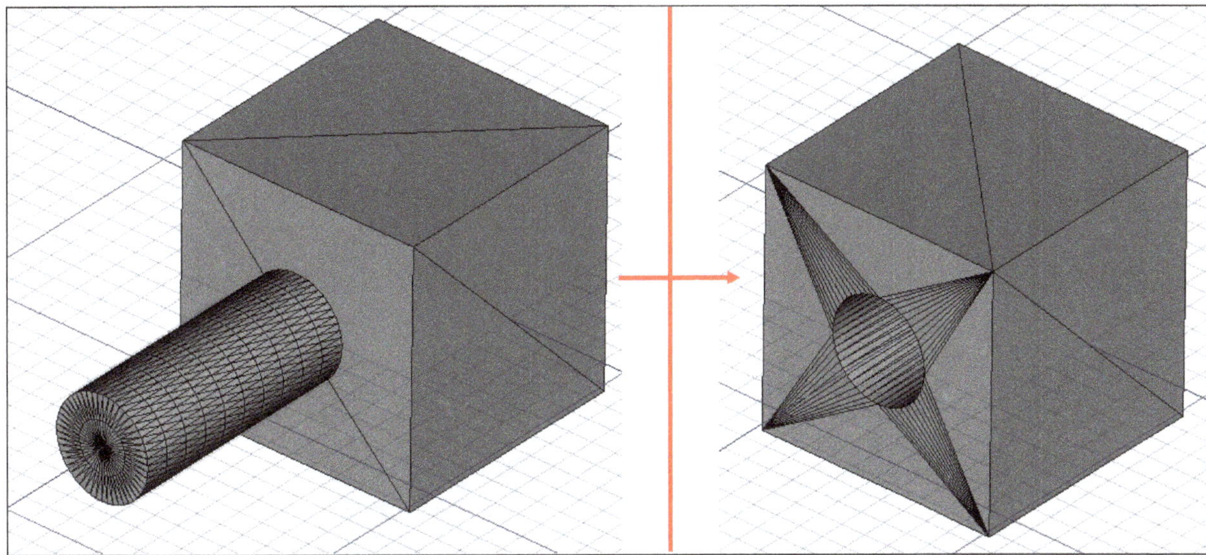

Figure-35. Difference feature

- Click on the **Intersection** tool from the **Boolean** cascading menu after selecting two mesh objects, the region common in both objects will be created as intersection object; refer to Figure-36.

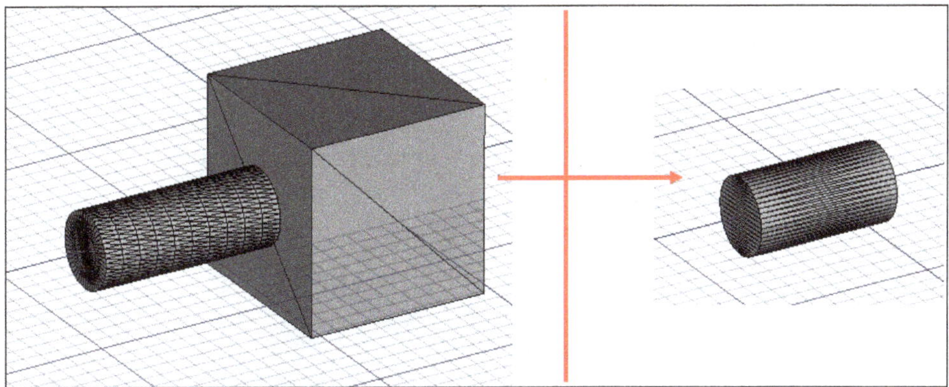

Figure-36. Intersection feature created

Mesh Cutting Operations

The tools in **Cutting** cascading menu of **Meshes** menu are used to cut mesh object using plane, section, or polygon selection; refer to Figure-37. Various tools in this cascading menu are discussed next.

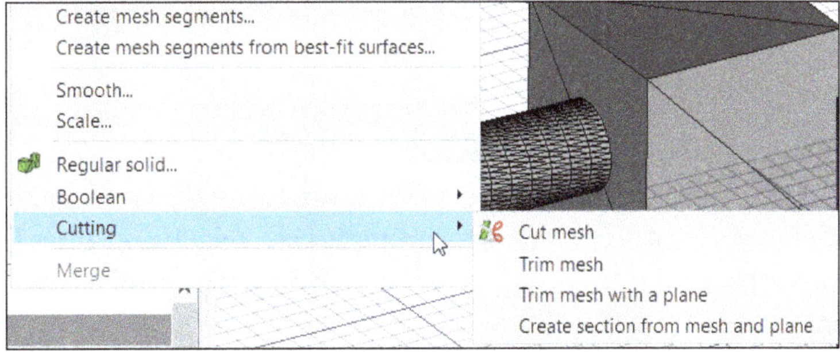

Figure-37. Cutting cascading menu

Cut Mesh Tool

The **Cut Mesh** tool is used to cut a mesh by drawing a polygon selection box. The procedure to use this tool is given next.

- Click on the **Cut Mesh** tool from the **Cutting** cascading menu of the **Meshes** menu after selecting the mesh body to be cut. The cursor will change to a scissor icon and you will be asked to draw a polygonal selection box.
- Draw a polygon on the body by specifying end points and right-click in the drawing area. A shortcut menu will be displayed; refer to Figure-38.

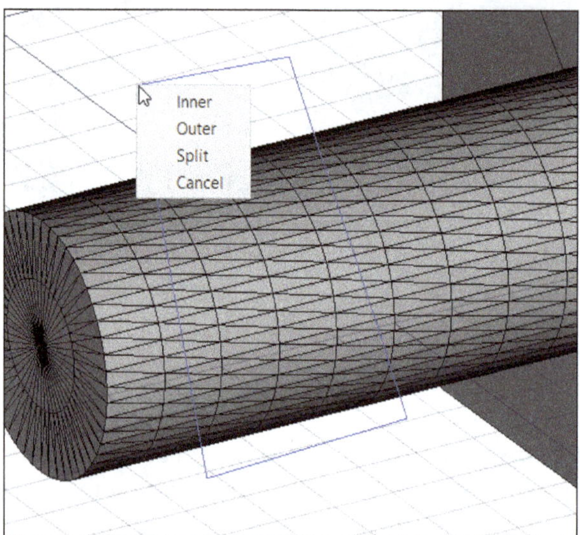

Figure-38. Shortcut menu for cutting

- Select the **Inner** option from shortcut menu to delete the mesh elements falling inside drawn polygon; refer to Figure-39. Select the **Outer** option from shortcut menu to delete mesh elements falling outside the selection polygon. Select the **Split** option from shortcut menu to delete the area falling inside selection polygon and create a new body. Select the **Cancel** option to abort the operation.

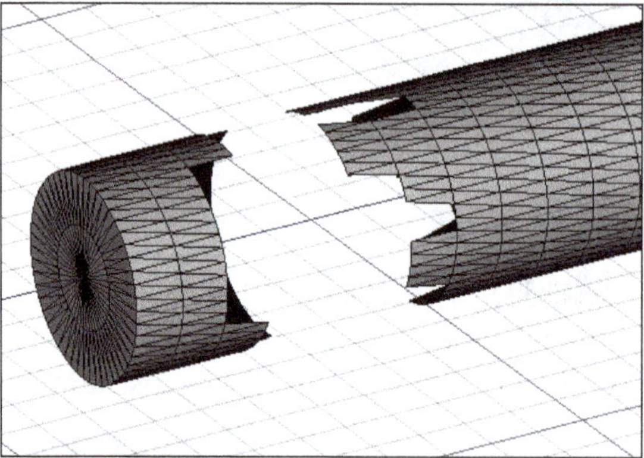

Figure-39. Cutting inner mesh elements

Trim Mesh Tool

The **Trim mesh** tool is used to trim a portion of mesh body using drawn selection polygon. Note that when you use **Cut mesh** tool then full elements falling inside selection polygon will be removed but when you use **Trim mesh** tool then these elements will also be trimmed at the boundary of selection polygon. The procedure to use **Trim mesh** tool is given next.

- Click on the **Trim mesh** tool from **Cutting** cascading menu of the **Meshes** menu after selecting the mesh body to be trimmed. You will be asked to draw a selection polygon.
- Create the polygon by specifying end points and right-click. A shortcut menu will be displayed.
- Select desired option from the shortcut menu to trim the mesh body; refer to Figure-40. The options in shortcut menu are same as discussed earlier.

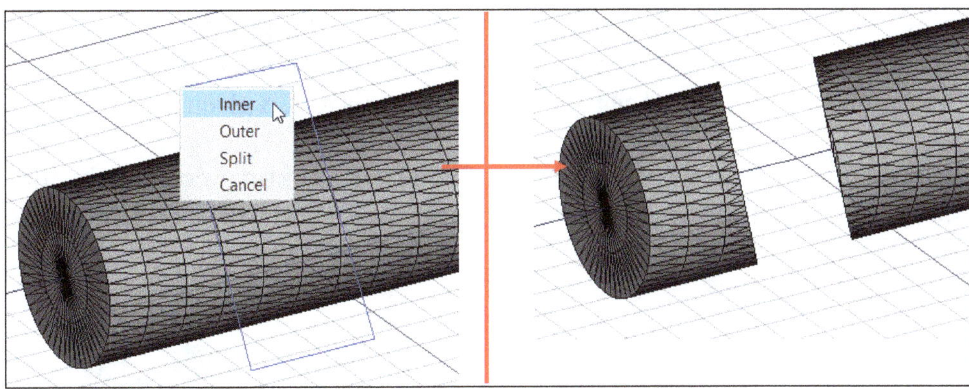

Figure-40. Trimming mesh

Trim mesh with a plane

The **Trim mesh with a plane** tool is used to trim mesh body using selected plane. Note that the plane must be created using the **Create primitives** tool of **Part** workbench; refer to Figure-41. The procedure to use this tool is given next.

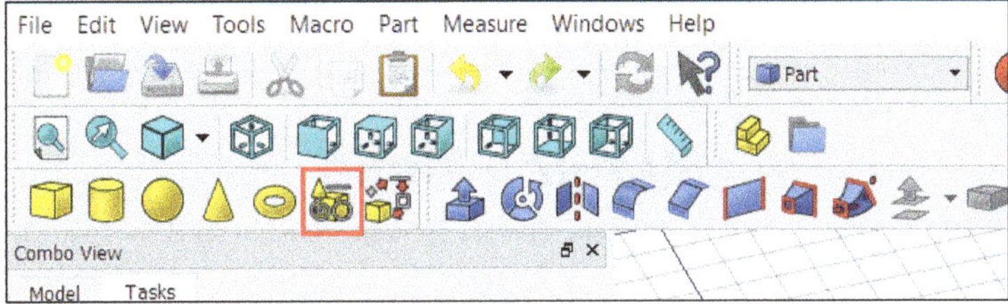

Figure-41. Create primitives tool

- Click on the **Trim mesh with a plane** tool from the **Cutting** cascading menu of the **Meshes** menu after selecting mesh body and an intersecting plane. The **Trim by plane** dialog box will be displayed; refer to Figure-42.

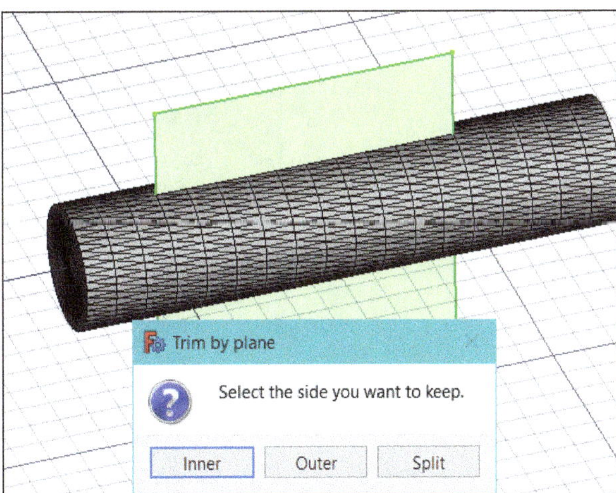

Figure-42. Trim by plane dialog box

- Select the **Inner** button from the dialog box to remove mesh body on the front side of plane and keep back side of mesh body.
- Select the **Outer** button from the dialog box to remove mesh body on the back side of plane and keep front side of mesh body.
- Select the **Split** button from the dialog box to split the mesh body into two pieces.

Create section from mesh and plane

The **Create section from mesh and plane** tool is used to generate section curves at the intersection of mesh body and plane. The procedure to use this tool is given next.

- After selecting plane and mesh body, click on the **Create section from mesh and plane** tool from the **Cutting** cascading menu of the **Mesh** menu. The section will be created; refer to Figure-43.

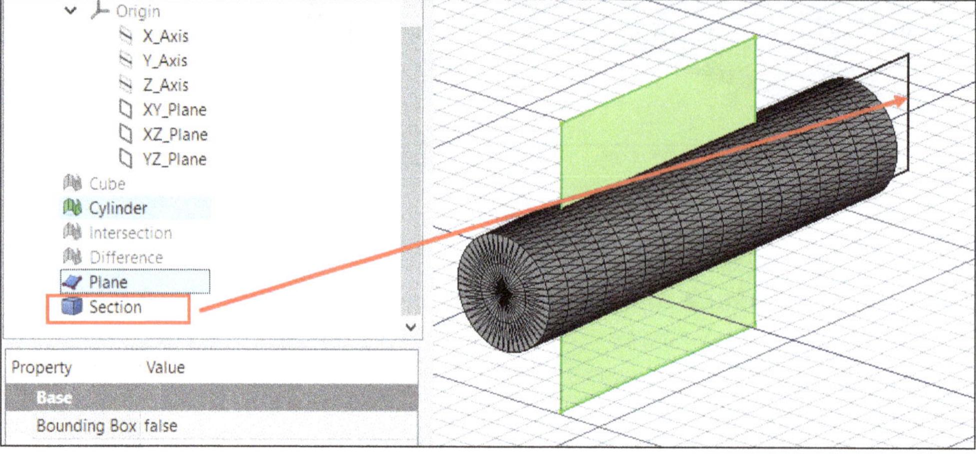

Figure-43. Section created

Merging Mesh

The **Merge** tool is used to combine two or more mesh bodies to form a single mesh. The procedure to use this tool is given next.

- Select the mesh bodies to be merged from the **Model Tree** and click on the **Merge** tool from the **Meshes** menu. The merged mesh body will be created; refer to Figure-44.

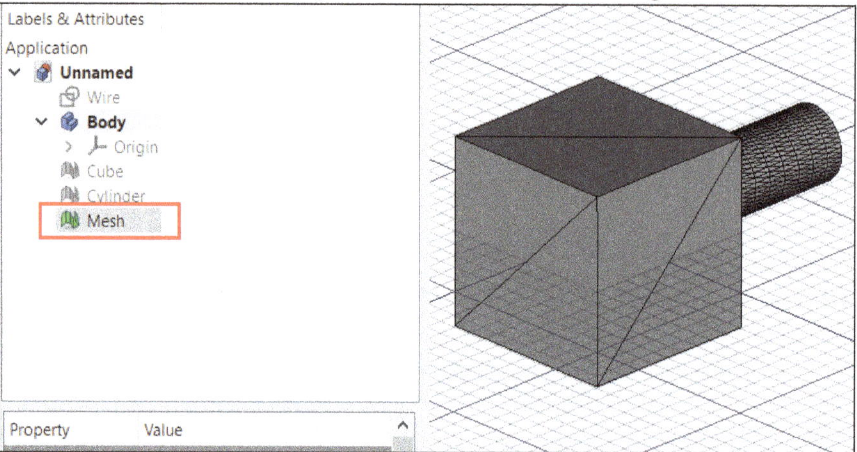

Figure-44. Mesh body created

POINTS WORKBENCH

The tools in the **Points** workbench are used to import, export, and cut point cloud using polygon. Select the **Points** option from **Switch between workbenches** drop-down in the **Toolbar**. The tools to work with point clouds will be displayed; refer to Figure-45.

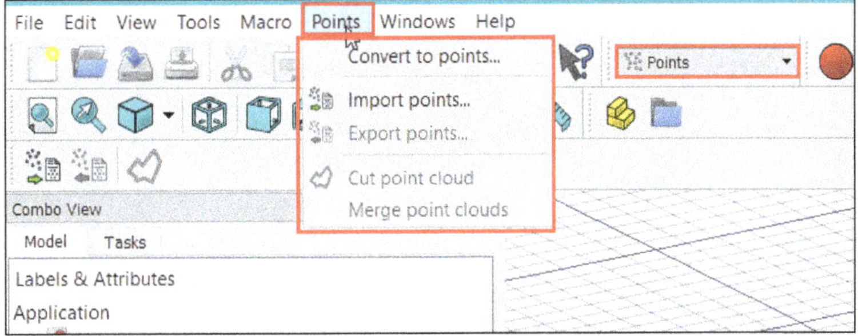

Figure-45. Points workbench

Converting Body to Points

The **Convert to points** tool is used to convert solid/surface/mesh bodies to a point cloud. The procedure to use this tool is given next.

- Select the solid/surface/mesh body to be converted to point cloud and click on the **Convert to points** tool from **Points** menu. The point cloud will be created; refer to Figure-46.

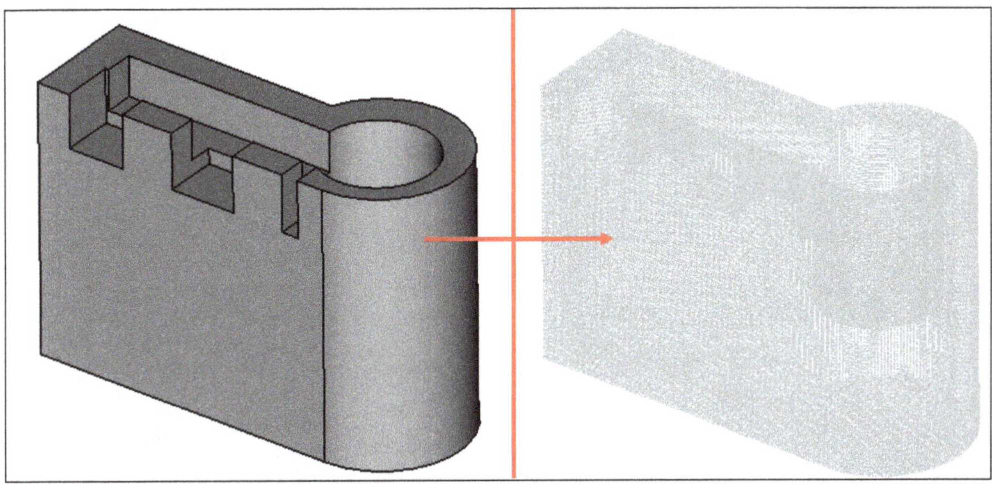

Figure-46. Point cloud created from solid body

Importing Point Cloud Data

The **Import points** tool is used to insert point cloud data from a non-native file. The procedure to use this tool is given next.

- Click on the **Import points** tool from the **Points** menu. The **Open** dialog box will be displayed; refer to Figure-47.
- Select the file with point cloud format like *.asc, *.pcd, or *.ply created from CAD software and click on the **Open** button. The file will open in the software; refer to Figure-48.

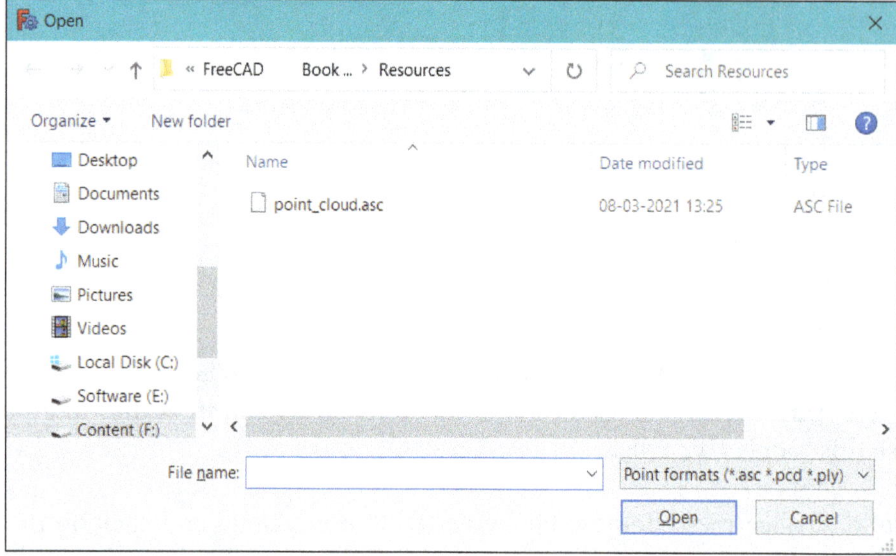

Figure-47. Open dialog box

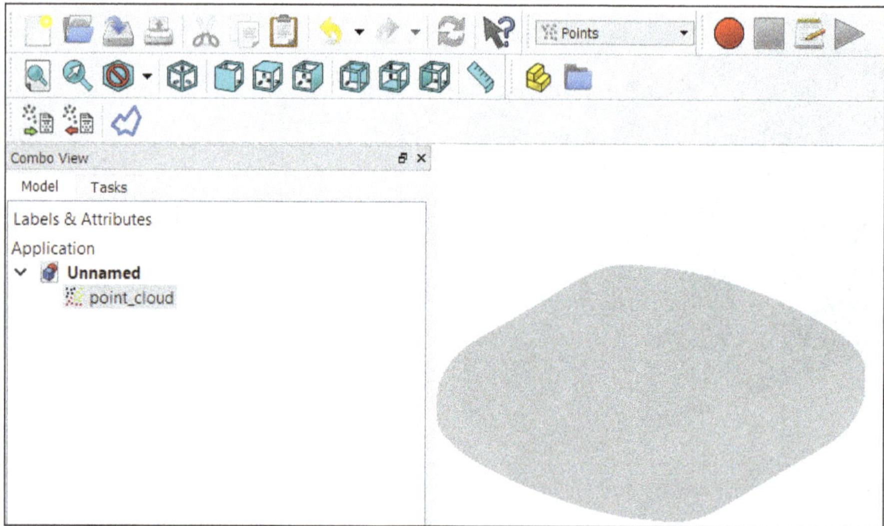

Figure-48. Point cloud file imported

Exporting Point Cloud Data

The **Export points** tool is used to export point cloud data created/modified in FreeCAD. The procedure to use this tool is given next.

- Click on the **Export points** tool from the **Points** menu after selecting the point cloud data model from the **Model Tree**. The **Save as** dialog box will be displayed; refer to Figure-49.
- Specify desired name of file in the **File name** edit box. You can specify desired point format with name or otherwise .asc format will be applied automatically.
- Click on the **Save** button from the dialog box to save the file at desired location.

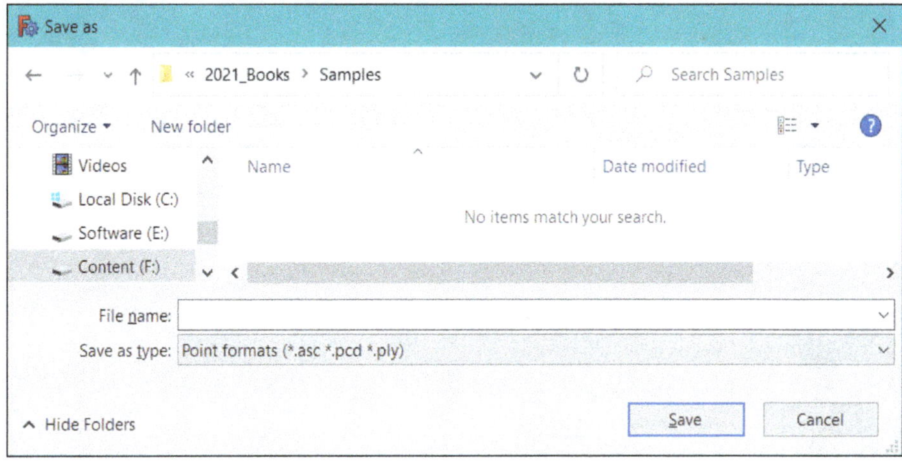

Figure-49. Save as dialog box

Cutting Point Cloud

The **Cut point cloud** tool is used to cut portion of point cloud as discussed earlier for mesh.

Similarly, you can use the **Merge point clouds** tool to join two or more point cloud models.

RAYTRACING WORKBENCH

The tools in **Raytracing** workbench are used to perform rendering of the model. To start **Raytracing** workbench, select the **Raytracing** option from the **Switch between workbenches** drop-down in the **Toolbar**; refer to Figure-50. Various tools in this workbench are discussed next.

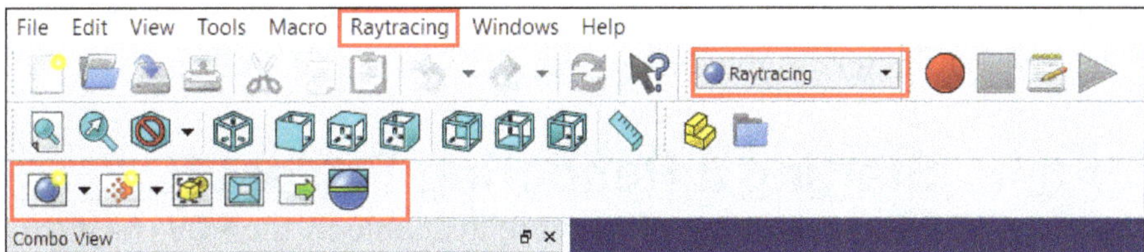

Figure-50. Raytracing workbench tools

Installing POV-Ray Executable

Before you perform rendering using FreeCAD, you need to install POV-Ray executable file. The procedure to download and install POV-Ray program is given next.

- Open your default web-browser and go to the link: http://www.povray.org/download.
- Click on the **Download Windows Installer** link button to download latest version of software.
- Double-click on the downloaded executable file and install it to desired location.
- Click on the **Preferences** option from **Edit** menu in **FreeCAD**. The **Preferences** dialog box will be displayed.
- Select the **Raytracing** option from left area of the dialog box. The options will be displayed as shown in Figure-51.

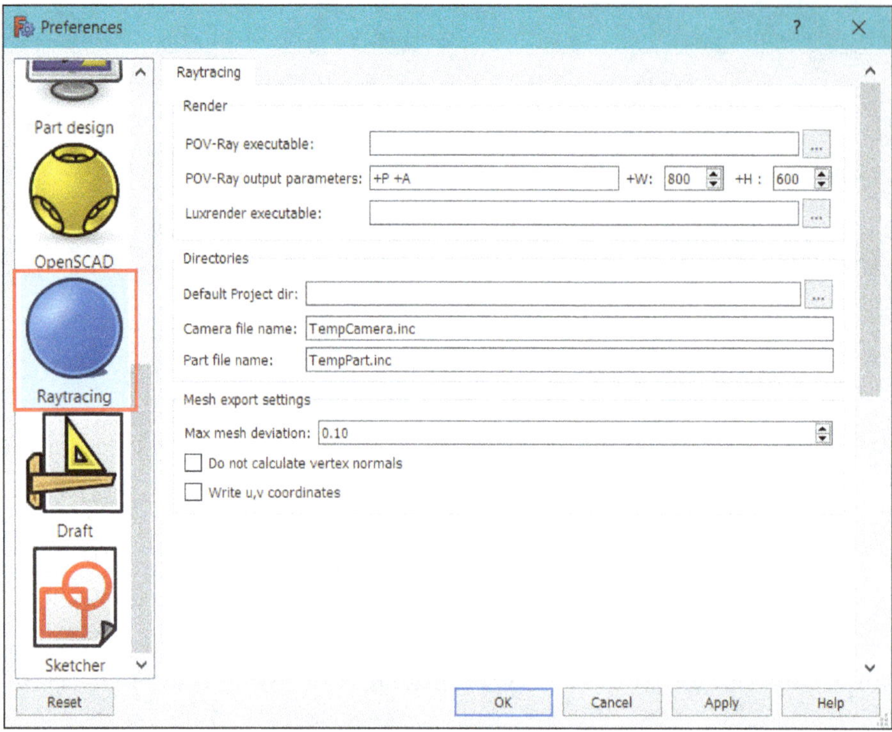

Figure-51. Raytracing options in Preferences dialog box

- Click on the **Browse** button next to **POV-Ray executable** edit box and select pvengine.exe file from the location C:\Program Files\POV-Ray\v3.7\bin if you have installed POV-Ray program at default location.
- You can install Luxrender application in the same way as discussed earlier for POV-Ray and set the executable location in **Luxrender executable** edit box of **Preferences** dialog box.
- After setting desired parameters in the dialog box, click on the **OK** button.

Starting a Rendering Project

The tools to start a new POV-Ray rendering project are available in the **POV-Ray** drop-down of the **Toolbar**; refer to Figure-52. The tools in this drop-down are discussed next.

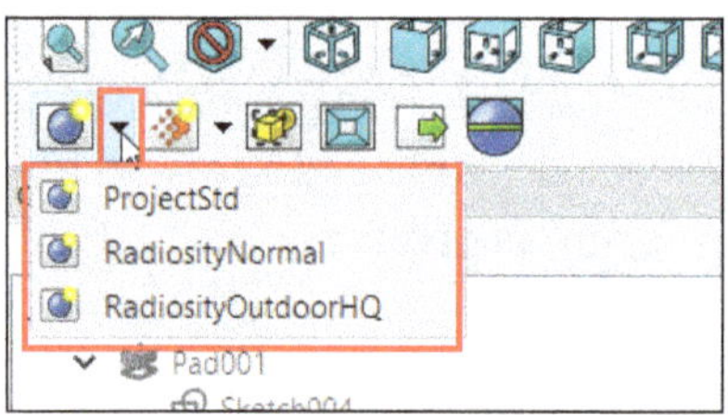

Figure-52. POV-Ray drop-down

- Click on the **ProjectStd** tool from the **POV-Ray** drop-down in the **Toolbar**. A new standard POV-Ray project will be added in the **Model Tree**. You can use the **RadiosityNormal** or **RadiosityOutdoorHQ** tools in the same way to start a new POV-Ray rendering project.
- If you have installed Luxrender application then you can start a new Luxrender project by using the tools in **New Luxrender project** drop-down; refer to Figure-53.

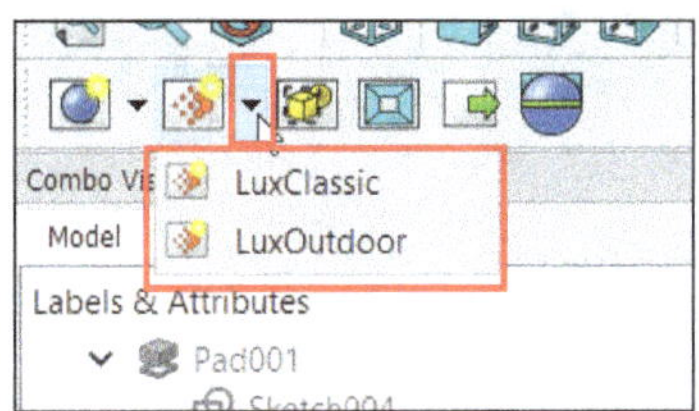

Figure-53. Luxrender drop-down

Adding Part to POV-Ray Project

After starting a new rendering project, the next step is to add parts to the Raytracing rendering project. The procedure to add a part in POV-Ray project is given next.

- Select the part you want to add in Raytracing project and select the project to which you want to add the part from **Model Tree**. After selecting both project and parts, click on the **Insert part** tool from the **Raytracing** menu or **Toolbar** in **Raytracing** workbench. The model will be added to project; refer to Figure-54.

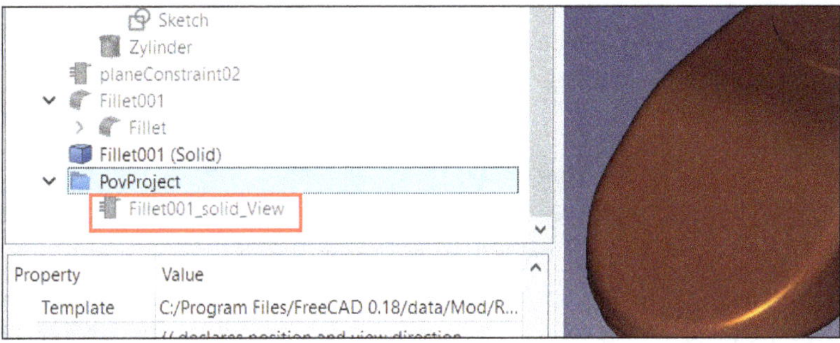

Figure-54. Model added to raytracing project

Resetting Camera

The **Reset Camera** tool is used to reset camera of rendering to match the current view. There are no parameters needed from your side when resetting camera. Click on the **Reset Camera** tool from the menu, the camera will set to current view automatically.

Exporting Raytracing project to a File

The **Export project** tool is used to export current raytracing project file for other rendering applications. The procedure to use this tool is given next.

- After setting raytracing project, click on the **Export project** tool from the **Raytracing** menu or **Toolbar**. The **Export page** dialog box will be displayed; refer to Figure-55.
- Specify desired name of project file in the **File name** edit box and save the file at desired location.

You can open this file in any raytracing renderer application like POV-Ray or Luxrender.

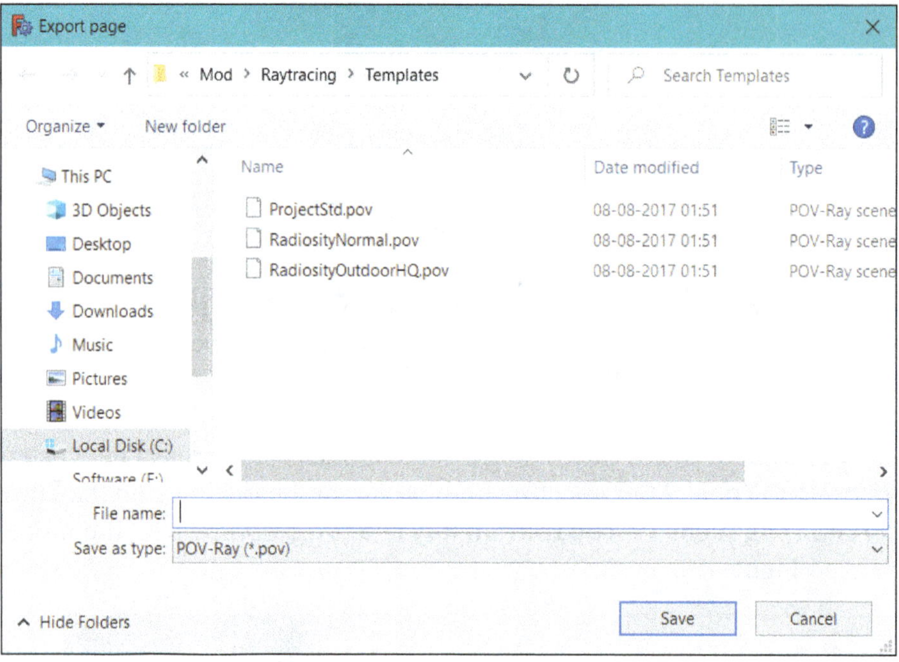

Figure-55. Export page dialog box

Performing Rendering

The **Render** tool is used to perform rendering using specified project parameters. Using this tool, generate an image file. The procedure to use this tool is given next.

- After setting all the parameters need for rendering project, select the project to be rendered from **Model Tree** and click on the **Render** tool from the **Toolbar** or **Raytracing** menu. The **Rendered image** dialog box will be displayed; refer to Figure-56.

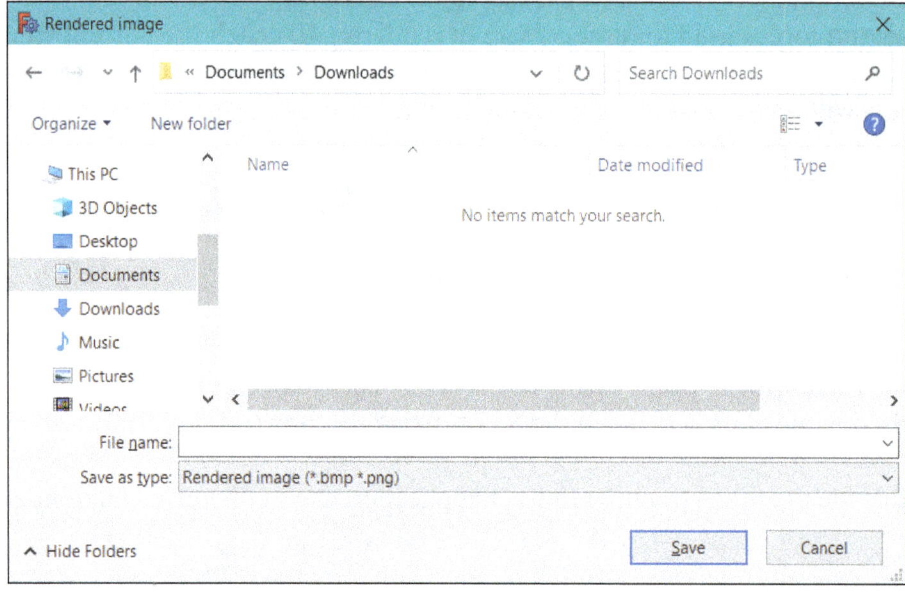

Figure-56. Rendered image dialog box

- Specify desired name in the **File name** edit box and click on the **Save** button. The rendered image will be generated at the specified location.

Note that if you get IO error when performing rendering then you need to export the project file and render it directly in POV-Ray application (make sure to run the application as administrator).

ROBOT WORKBENCH

The tools in **Robot** workbench are used to simulate path of robot movements. A robot is generally used for performing repeating tasks like welding an edge, tightening some screws, and so on which have predefined paths. The tools in **Robot** workbench are shown in Figure-57. Various tools of this workbench are discussed next.

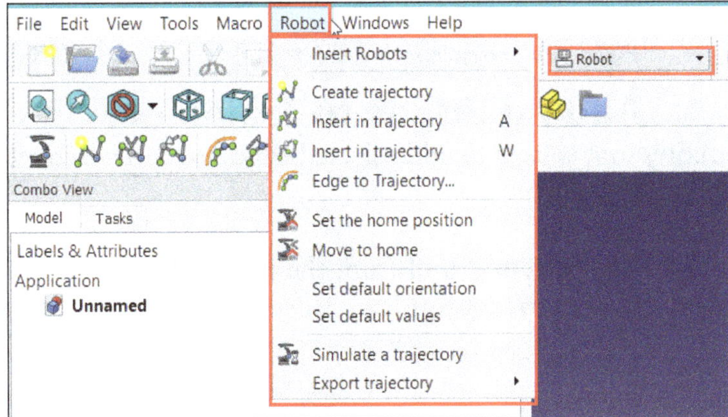

Figure-57. Robot workbench tools

Inserting Robot

The tools in **Insert Robots** cascading menu of **Robot** menu are used to insert different type of robots for path simulation; refer to Figure-58. Select desired robot from the menu for which you want to simulate the toolpath. The robot will be displayed in drawing area; refer to Figure-59.

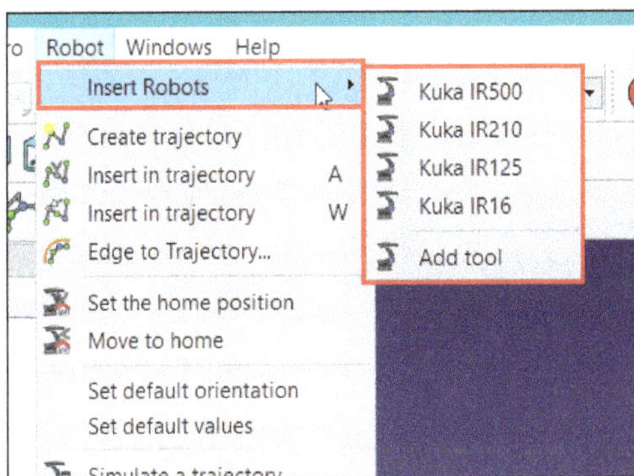

Figure-58. Insert Robots cascading menu

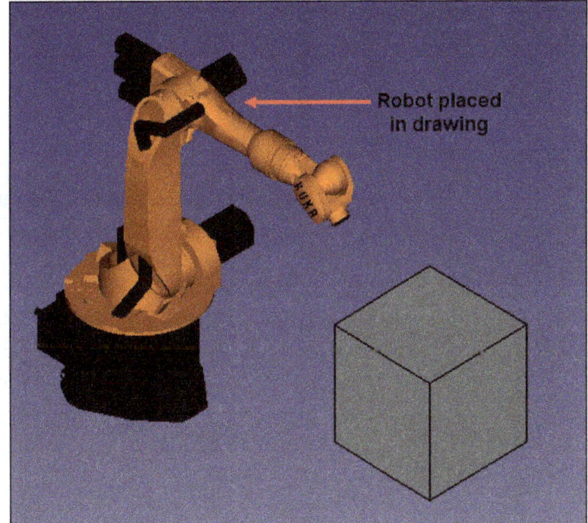

Figure-59. Robot placed in drawing

Creating Robot Trajectory

The **Create trajectory** tool is used to create an empty trajectory path for robot. Click on the **Create trajectory** tool from the **Robot** menu. An empty trajectory feature will be added in the **Model Tree**; refer to Figure-60.

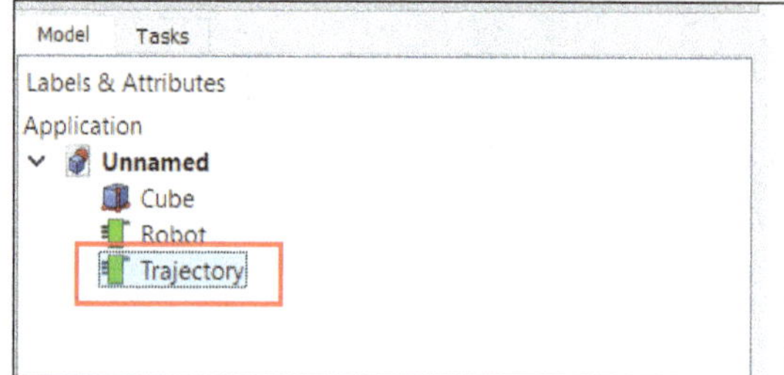

Figure-60. Empty trajectory added

Adding Edges to Trajectory

The **Edge to Trajectory** tool is used to add selected edges to the trajectory. The procedure to use this tool is given next.

- Click on the **Edge to Trajectory** tool from the **Robot** menu. The dialogs to add edges to trajectory will be displayed; refer to Figure-61.

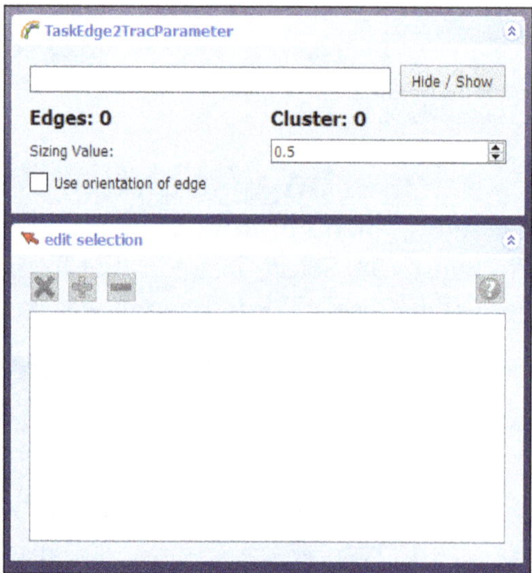

Figure-61. Dialogs for adding edges to trajectory

- Select the edges to be added in the trajectory while holding the **CTRL** key; refer to Figure-62.

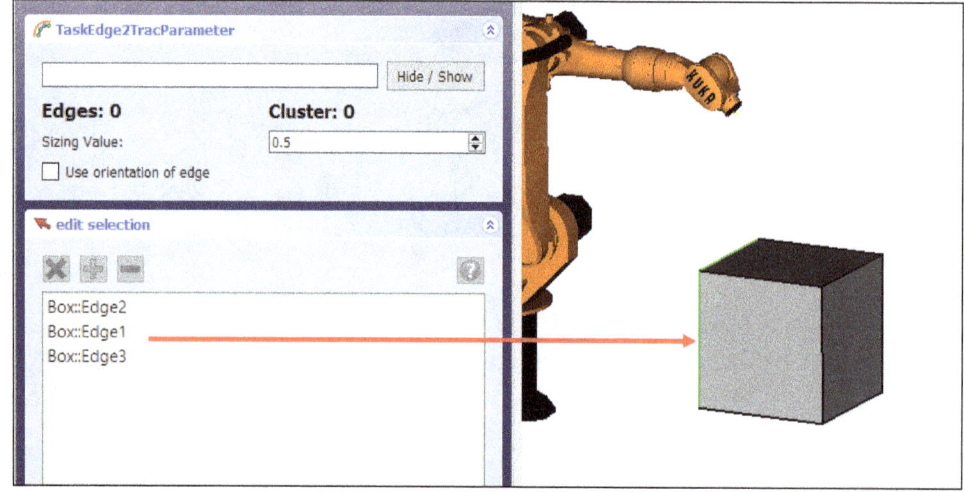

Figure-62. Edges selected for trajectory

- Click on the **Apply** button from the dialog after selecting edges and click on the **OK** button to add edges to trajectory.

Setting Default Orientation of Robot

The **Set default orientation** tool is used to set initial placement position and orientation of the selected robot. The procedure to use this tool is given next.

- Click on the **Set default orientation** tool from the **Robot** menu. The **Placement** dialog box will be displayed; refer to Figure-63.

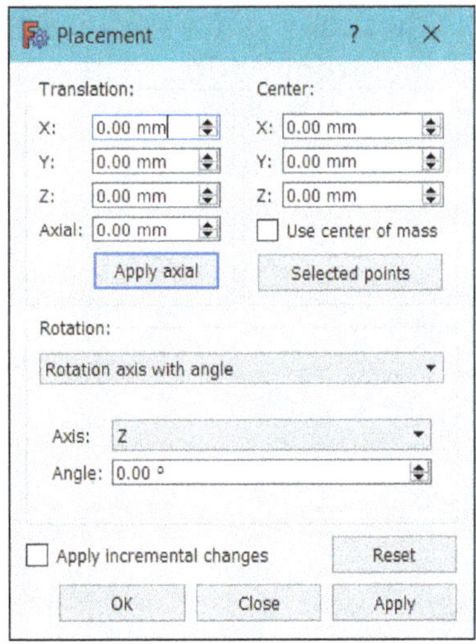

Figure-63. Placement dialog box

- Set desired parameters in the **Translation** area to move robot to desired location.
- Set desired values in **X**, **Y**, and **Z** edit boxes of **Center** area to define location of center of Robot for translation.
- In the **Rotation** area, you can rotate robot about selected axis in the **Axis** drop-down and define rotation angle in the **Angle** edit box.
- After setting desired changes, click on the **Apply** button to apply orientation and click on the **OK** button to exit the dialog box.

Setting Default Values for Robot

The **Set default values** tool is used to define default movement parameters for robot like velocity of movement, acceleration, and so on. Click on the **Set default values** tool from the **Robot** menu after selecting robot from **Model Tree**. The **Set default speed** dialog box will be displayed; refer to Figure-64. Specify desired value of speed and click on the **OK** button. The **Set default continuity** dialog box will be displayed; refer to Figure-65. Select the **False** option if you want robot movements to be in fragments. Select the **True** option if you want robot movements to be continuous. After selecting desired option, click on the **OK** button. The **Set default acceleration** dialog box will be displayed; refer to Figure-66. Specify desired value in the edit box for acceleration and click on the **OK** button.

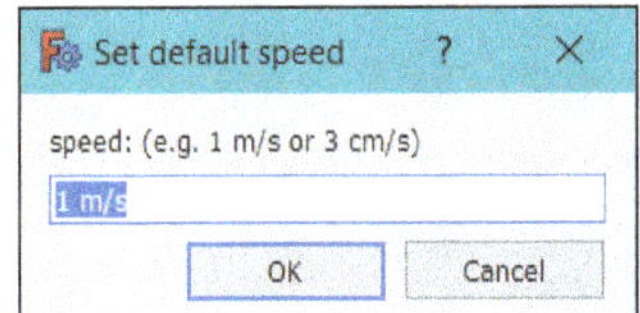

Figure-64. Set default speed dialog box

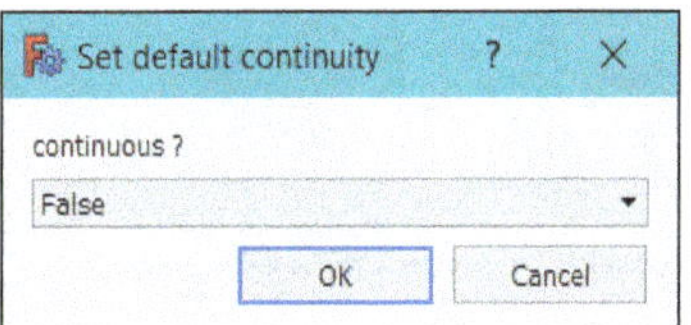

Figure-65. Set default continuity dialog box

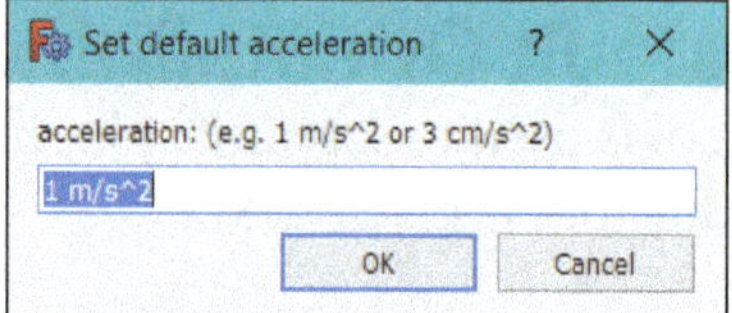

Figure-66. Set default acceleration dialog box

Adding Tool to Robot

To add a tool to Robot arm, first create the model of tool using **Part Design** workbench. Now, select the **Robot** and part model created as tool from **Model Tree**, switch to **Tasks** panel in the **Combo View** and click on the **Add tool** option from **Robot tools** dialog; refer to Figure-67. The selected body will be added as tool to robot arm; refer to Figure-68.

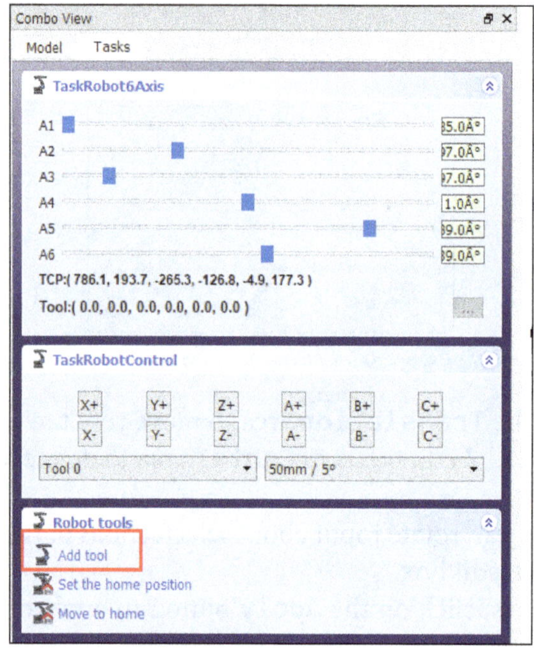

Figure-67. Add tool option

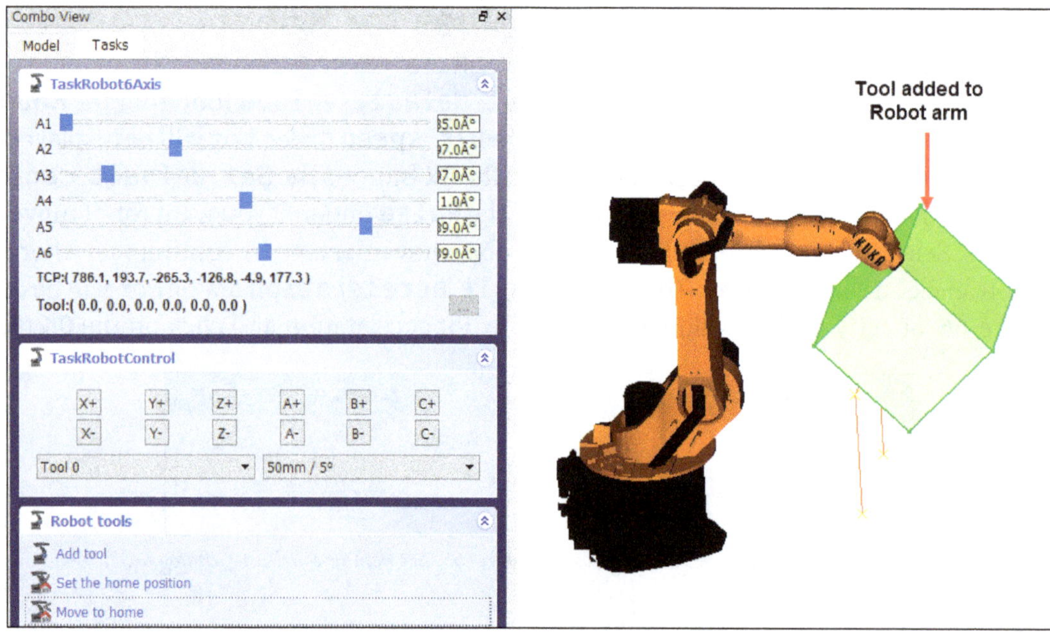

Figure-68. Tool added to robot arm

Simulating Robot Trajectory

The **Simulate a trajectory** tool is used to simulate movements of robot on the trajectory earlier created. Select the Robot and trajectory features from the **Model Tree** and click on the **Simulate a trajectory** tool from the **Robot** menu. Dialogs will be displayed in the **Tasks** panel to simulate robot trajectory; refer to Figure-69. You can move the slider in **Trajectory** dialog to check movement of robot arm or use the buttons to run simulation.

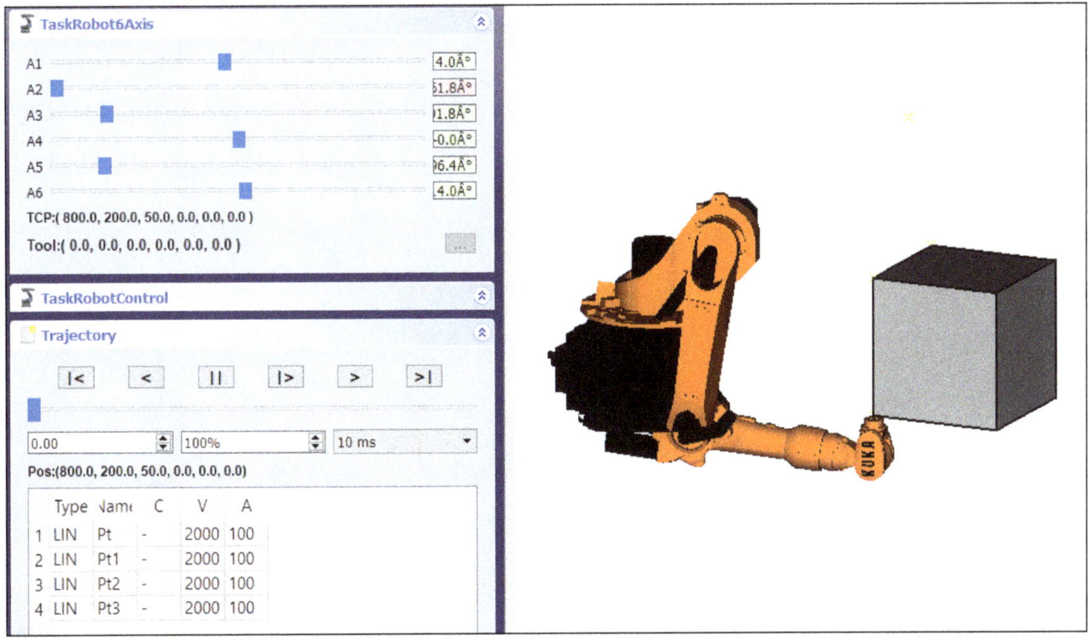

Figure-69. Dialogs for simulating robot

There are many other workbenches that can be used with FreeCAD to perform various tasks. You can download them and add them to the software using FreeCAD website links. We will try to discuss about more new workbenches in next editions of the book.

For Student Notes

Chapter 12

Miscellaneous Tool

Topics Covered

The major topics covered in this chapter are:

- ***Edit menu***
- ***View menu***
- ***Tools menu***
- ***Macro menu***

INTRODUCTION

In previous chapters, you have learned about various workbenches of FreeCAD. There are some tools in FreeCAD which are common for all the workbenches and are used to modify basic parameters of software like changing orientation of model, moving object in 3D space. These tools are available in various menus of FreeCAD which are discussed next.

EDIT MENU

The tools in **Edit** menu are used to perform common tasks like Undo, Redo, copy, paste, and so on; refer to Figure-1. These tools are discussed next.

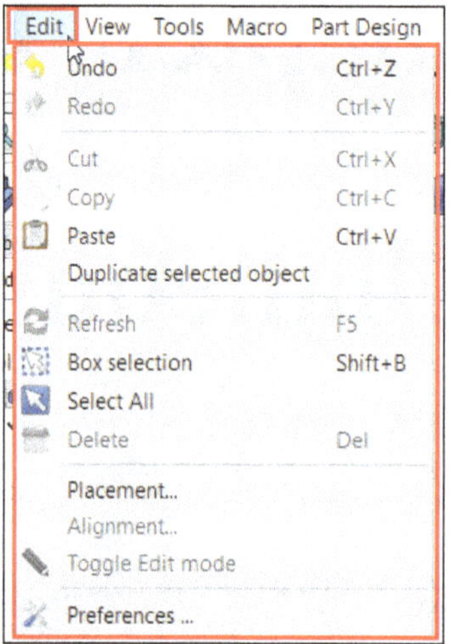

Figure-1. Edit menu

Undo Operation

The **Undo** tool in **Edit** menu is used to reverse the effect of any operation performed recently in the software. For example, you have drawn a line in FreeCAD sketcher and dragged it to a different location but now you want it to be back at its original location then you can use the **Undo** tool to do so. You can also use the shortcut key **CTRL+Z** to perform undo operation.

Redo Operation

The **Redo** tool in **Edit** menu is used to reverse the Undo operation. You can also use the shortcut key **CTRL+Y** to perform redo operation.

Cut, Copy, and Paste Operations

The **Cut** tool in **Edit** menu is used to copy selected objects in temporary memory of software and delete them from original location. You can also use **CTRL+X** shortcut keys for cut operation. The **Copy** tool in **Edit** menu is used to copy selected objects in temporary memory of software while keeping the original objects intact. You can also use **CTRL+C** shortcut keys for copy operation. After selecting **Cut** or **Copy** tool, click on the **Paste** tool from **Edit** menu to place copy of the objects in temporary memory. You can also use the **CTRL+V** shortcut keys for paste operation.

Duplicating Selected Objects

The **Duplicate selected objects** tool is used to generate a duplicate copy of selected objects. This tool is useful when creating assemblies using same components.

Refresh Operation

The **Refresh** tool is used to recompute any changes made in the model. Sometimes, changes made in the model do not reflect automatically in drawing area. For refreshing operation, you need to first mark an object for refreshing by selecting the object, right-clicking on it, and then selecting the **Mark to recompute** option from shortcut menu; refer to Figure-2. After marking an object to recompute, select the **Refresh** tool from **Edit** menu or press **F5** key from keyboard to update changes in the model.

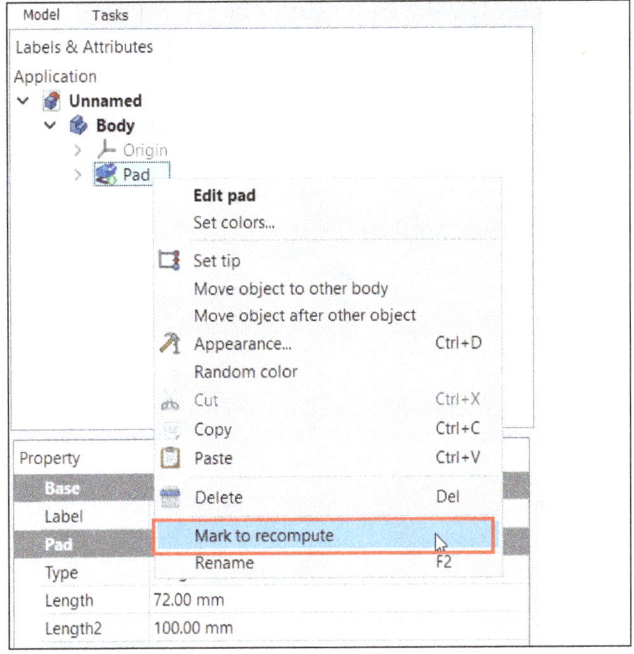

Figure-2. Mark to recompute option

Box Selection

The **Box Selection** tool is used to select objects which fall inside the boundaries of box selected created. To use box selection, select the **Box selection** tool from **Edit** menu and draw a box by dragging cursor; refer to Figure-3.

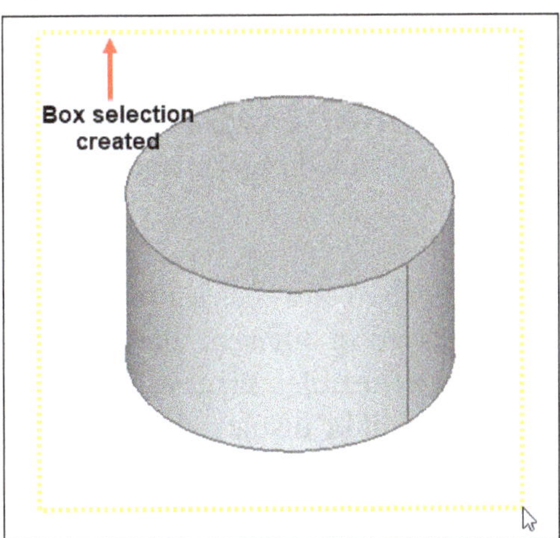

Figure-3. Box selection created

Select All

The **Select All** tool is used to select all the objects in drawing area.

Deleting Objects

The **Delete** tool in **Edit** menu is used to delete selected objects. You can select the objects from drawing area as well as **Model Tree**. You can also use the shortcut key **DELETE** from keyboard to delete selected objects.

Modifying Placement of Objects

The **Placement** tool in **Edit** menu is used to modify placement location and orientation of selected object. The procedure to do so is given next.

- Select the object whose placement is to be modified and click on the **Placement** tool from **Edit** menu. The **Placement** dialog will be displayed in **Tasks** panel of **Combo View**; refer to Figure-4.

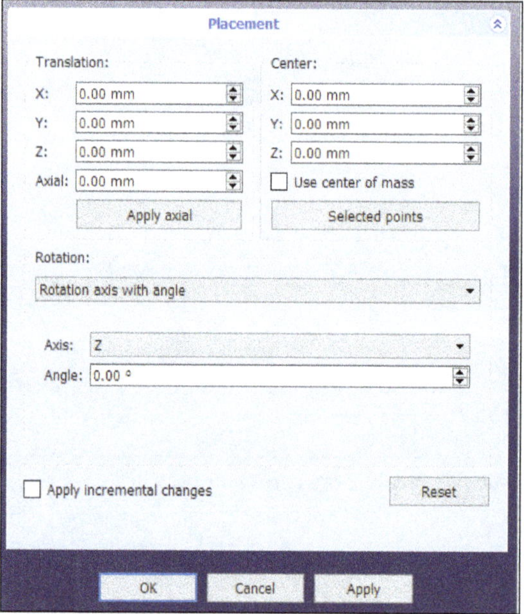

Figure-4. Placement dialog

The options in this dialog have been discussed earlier. Note that you may need to refresh the model after performing placement operation.

Alignment Operation

The **Alignment** tool is used to align two selected objects about selected points. The procedure to use this tool is given next.

- Select the two objects that you want to align and click on the **Alignment** tool from the **Edit** menu. The two selected objects will be displayed in different viewports of drawing area; refer to Figure-5.
- Select equal number of points on both the objects (movable as well as fixed object) and right-click in the drawing area. A shortcut menu will be displayed; refer to Figure-6.

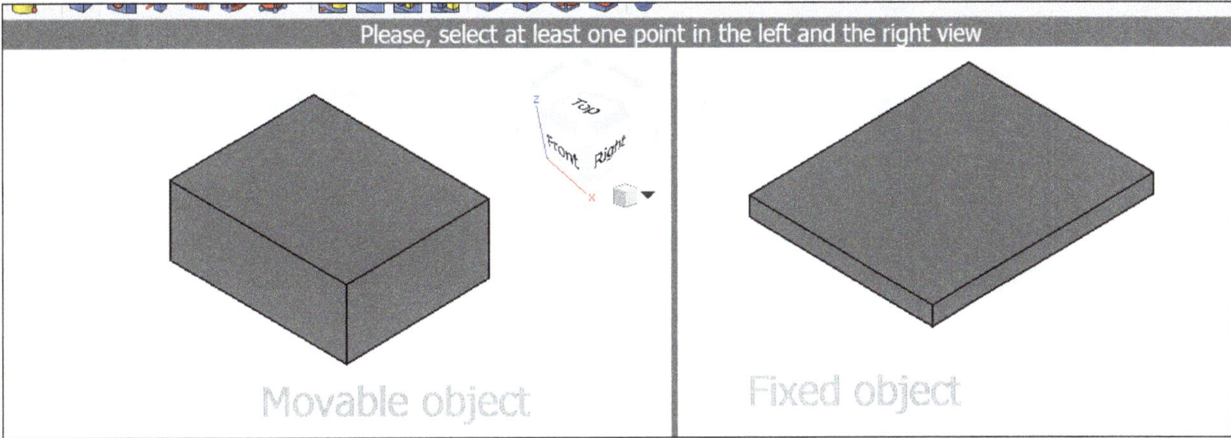

Figure-5. Objects displayed for alignment

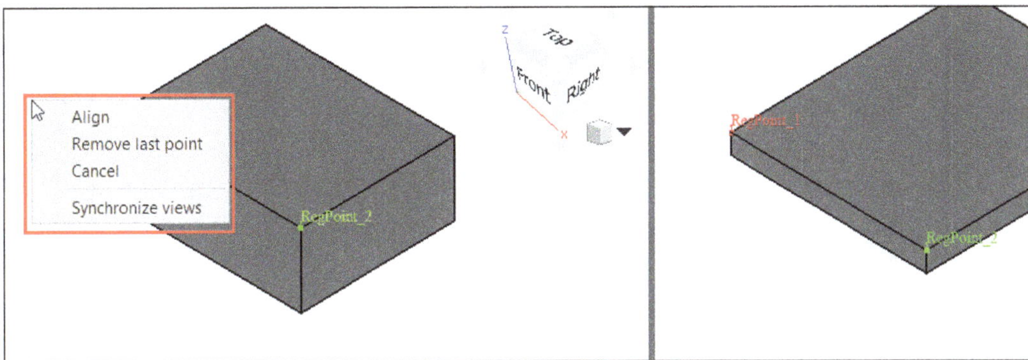

Figure-6. Shortcut menu for alignment

- Select the **Align** option from the shortcut menu to perform alignment operation. The two objects will be aligned as per the points selected on the bodies.

Activating Edit Mode

The **Toggle Edit mode** tool is used to modify a feature created in FreeCAD using its related dialog. The procedure to use this tool is given next.

- Select the feature that you want to be modified and click on the **Toggle Edit mode** tool from the **Edit** menu. The related dialog for modifying feature will be displayed. If you have selected an object/body which do not have any dialog associated with it then **Translation** and **Rotation** options will be displayed in the dialog; refer to Figure-7.

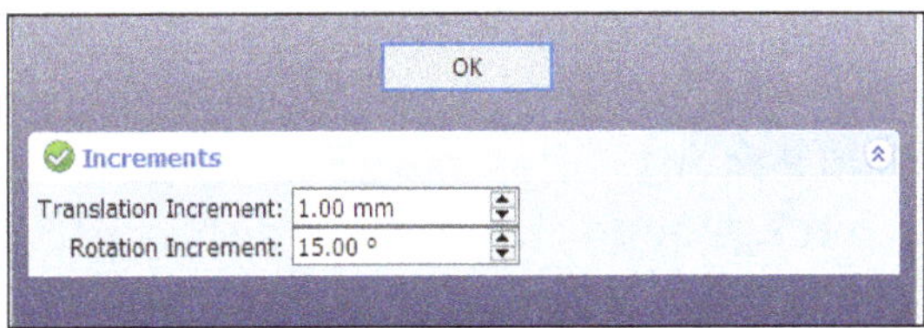

Figure-7. Increments dialog

- Set the parameters as desired and click on the **OK** button from the dialog.

Preferences

The **Preferences** tool is used to modify underlying parameters of FreeCAD application. The procedure to use this tool is given next.

- Click on the **Preferences** tool from the **Edit** menu. The **Preferences** dialog box will be displayed; refer to Figure-8.

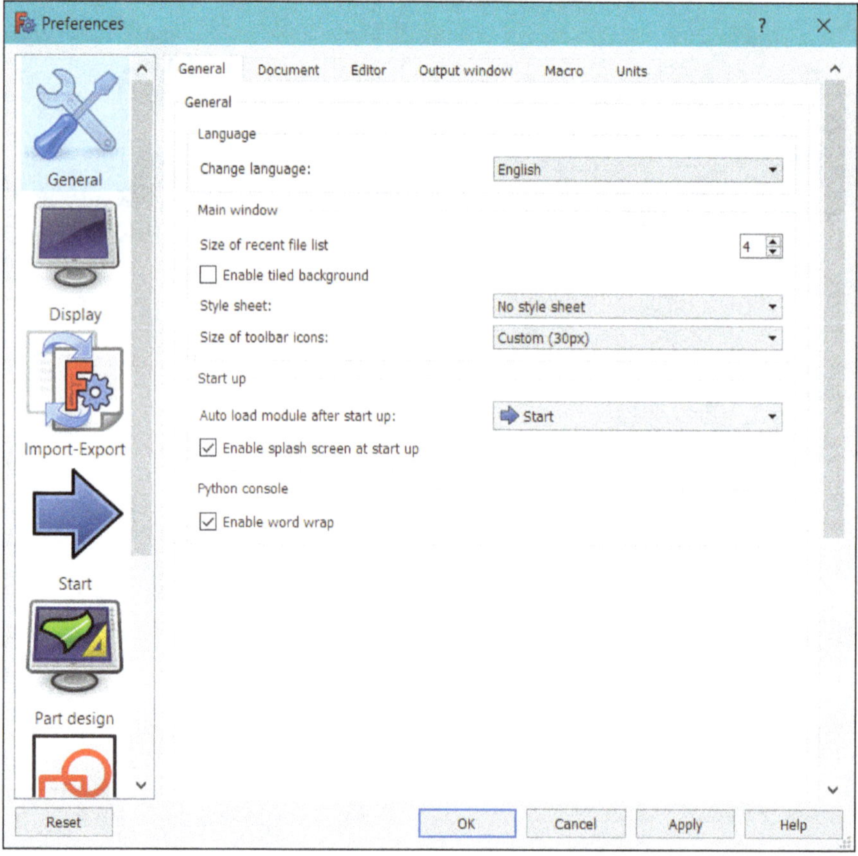

Figure-8. Preferences dialog box

The options in the dialog box are categorized in various pages like General page, Display page, Import-Export page, and so on. To modify options related to a category, select the respective page option from the left area of the dialog box. Some of the most common options of this dialog box are discussed next.

General Page

The options in **General** page are used to define general parameters for interface of application, documents, code editor, and so on. These options are discussed next.

General Tab

- Select desired option from the **Change language** drop-down to specify language used by application interface.
- Set desired value in the **Size of recent file list** edit box to specify number of recently worked files being displayed in **Recent files** cascading menu of **File** menu.
- Select the **Enable tiled background** check box to display background as tiled image.
- Select desired option from the **Style sheet** drop-down to define color and style of interface of application. For example, select the **Dark-blue** option to make background of application dark and selected tools/options highlighted in blue color.
- Select desired option from the **Size of toolbar icons** drop-down to define size of buttons in the toolbar.
- Select desired option from the **Auto load module after start up** drop-down to define which workbench will be active when you start the software.
- Select the **Enable splash screen at start up** check box to display splash screen showing modules being loaded in software when start the FreeCAD software.

- Select the **Enable word wrap** check box to automatically wrap codes in view area when sentences become long.

Document Tab

The options in **Document** tab are used to specify settings related to documents in FreeCAD; refer to Figure-9.

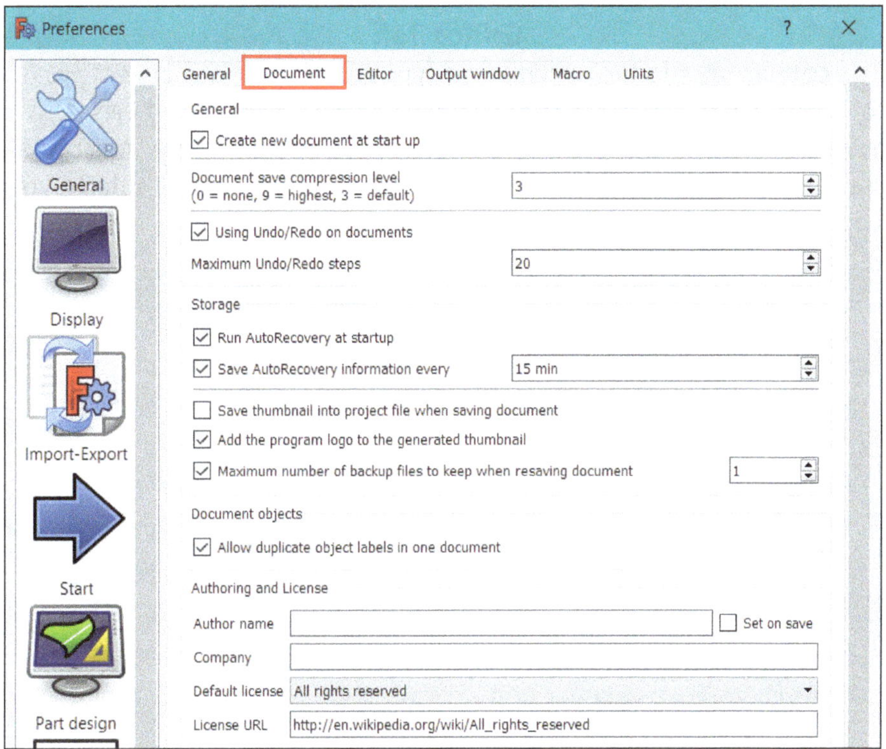

Figure-9. Document tab

- Select the **Create new document at start up** check box to automatically create a new document when you start the software.
- Set desired value in the **Document save compression level** edit box to define the compression level being used when saving the file. A higher value for compression will make saved files occupy less memory but it will also take longer to open the files in application. The minimum compression value is **0** and maximum compression value is **9**.
- Select the **Using Undo/Redo on documents** check box to allow undo/redo operations on the model. You can specify maximum number of undo/redo steps allowed in the **Maximum Undo/Redo steps** edit box.
- Select the **Run AutoRecovery at startup** check box to check for auto recovered documents at the startup.
- Select the **Save AutoRecovery information every** check box to automatically save file at regular interval specified in the edit box next to check box.
- Select the **Save thumbnail into project file when saving document** check box to add a thumbnail of model with project file while saving it.
- Select the **Add the program logo to the generated thumbnail** check box to also add logo of application with the thumbnail of model.
- Select the **Maximum number of backup files to keep when resaving document** check box to define the number of documents to be created as backup for a model.
- Select the **Allow duplicate object labels in one document** check box to allow duplicate labels of features in the **Model Tree** of a document.

- Specify desired values in **Author name** and **Company** edit boxes to define name of model author and designing company.
- If you select the **Set on save** check box then you will be asked to specify author name and company information while saving model.
- Similarly, you can specify license information in the **Default license** and **License URL** edit boxes.

Editor Tab

The options in **Editor** tab are used to specify font type and size for different types of texts in Python Console; refer to Figure-10. Select desired text type from **Display Items** list box and specify related parameters in the right area of dialog box. Select the **Enable line numbers** check box from the **Options** area of this tab to display line numbers in Python Console. Similarly, you can specify the indent sizes in **Indentation** area of the tab.

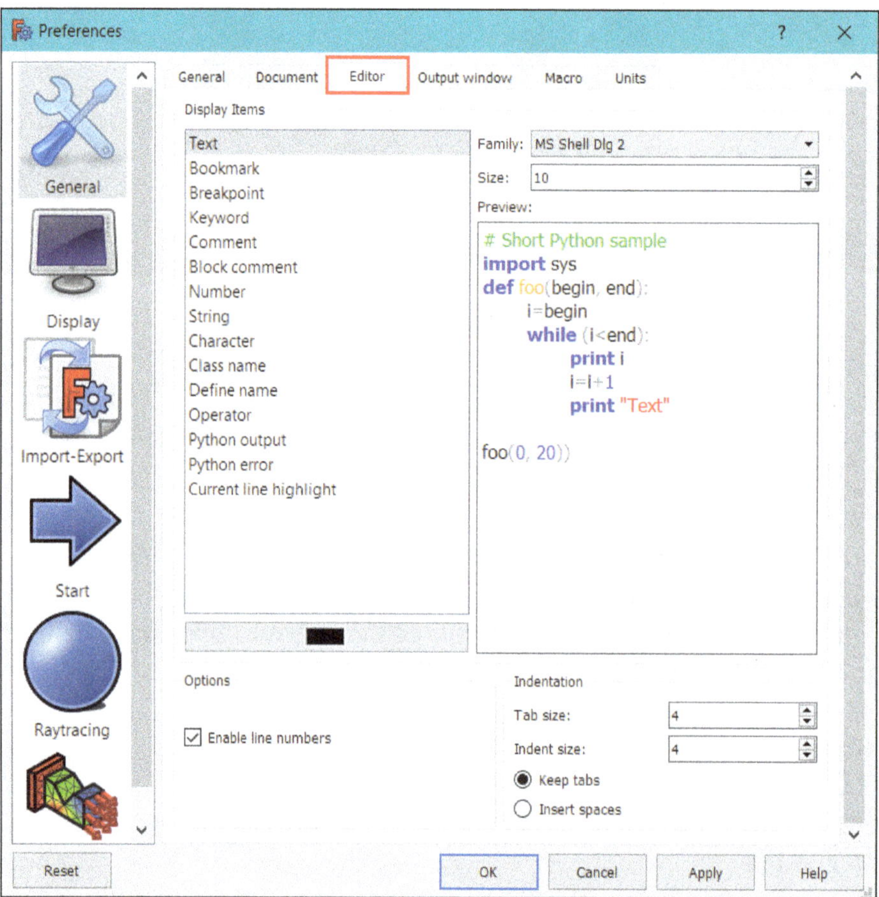

Figure-10. Editor tab

Output window Tab

The options in **Output window** tab are used to define how information will be displayed in the **Report view** panel; refer to Figure-11.

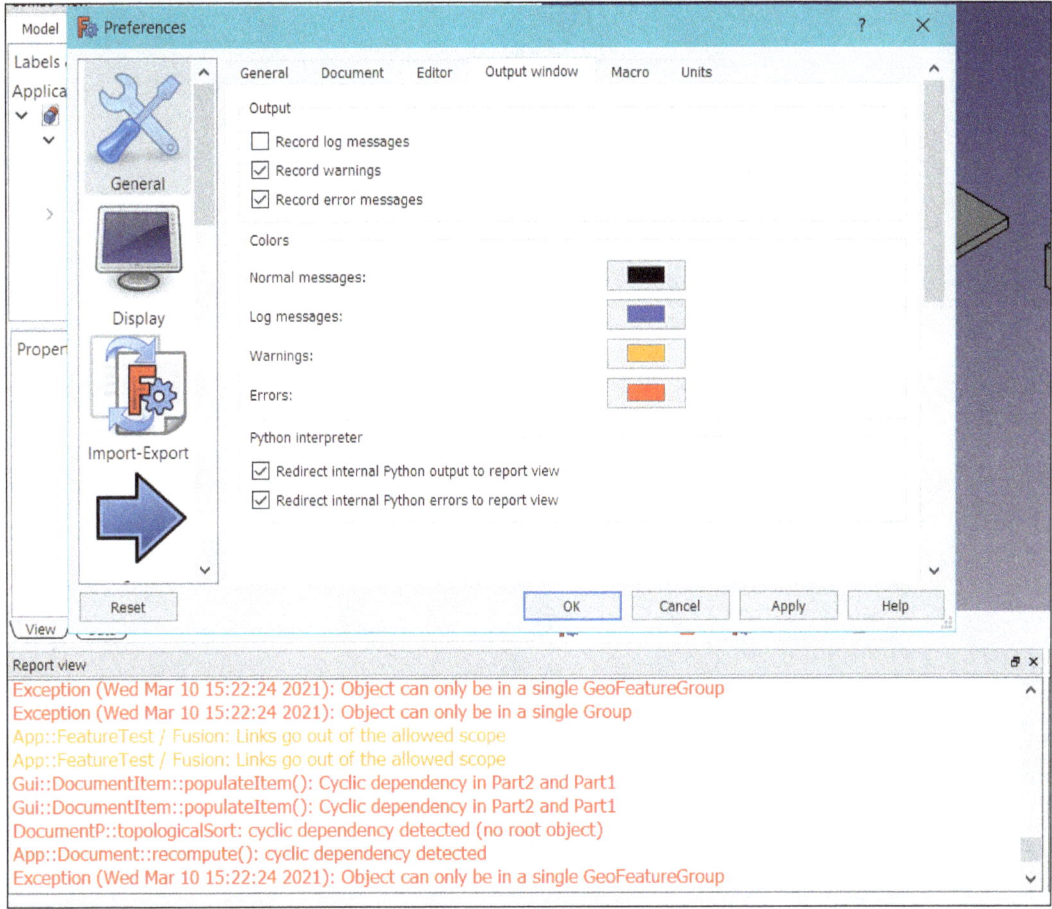

Figure-11. Output window tab

- Select the **Record log messages** check box to record the log messages displayed in **Report view** panel as a log file.
- Select the **Record warnings** check box to record warning messages in a file.
- Select the **Record error messages** check box to record error messages in a file.
- Set desired colors for Normal messages, Log messages, Warnings, and Errors in the **Colors** area of the dialog box.
- Select the check boxes from **Python interpreter** area of the tab to redirect outputs and errors of Python in **Report view** panel.

Macro Tab

The options in the **Macro** tab are used to define path and parameters for running macros in FreeCAD; refer to Figure-12.

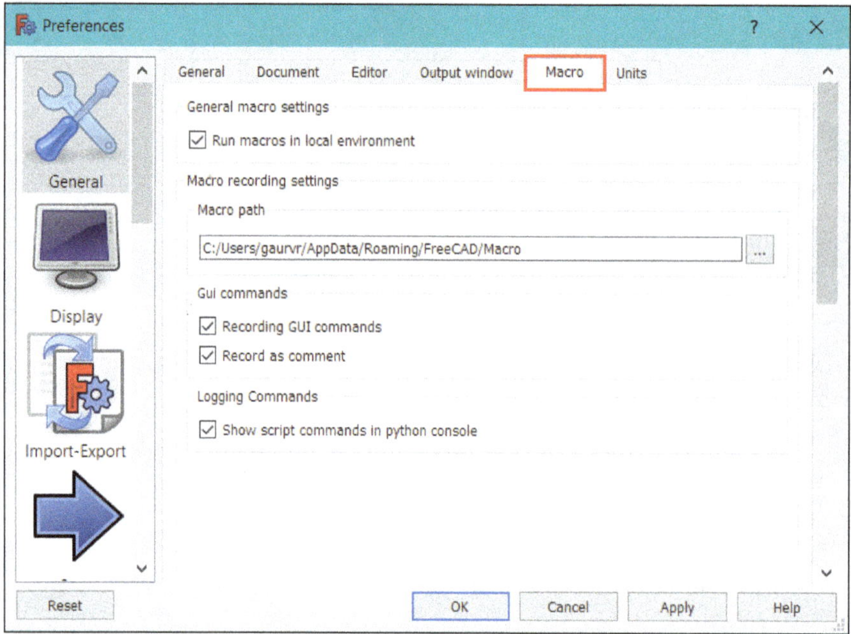

Figure-12. Macro tab

- Select the **Run macros in local environment** check box to run the macros internally in FreeCAD environment.
- Specify desired location in the **Macro path** area to define location where macro file will be saved and searched.
- Select check boxes from the **Gui commands** area to define whether graphic user interface commands and comments will be recorded in the macro.
- Select the **Show script commands in python console** check box to show scripts in python console when you activate a tool.

Units Tab

The options in the **Units** tab are used to specify unit system being used in FreeCAD for creating model; refer to Figure-13.

- Select desired option from the **User system** drop-down to define unit system used for creating model.

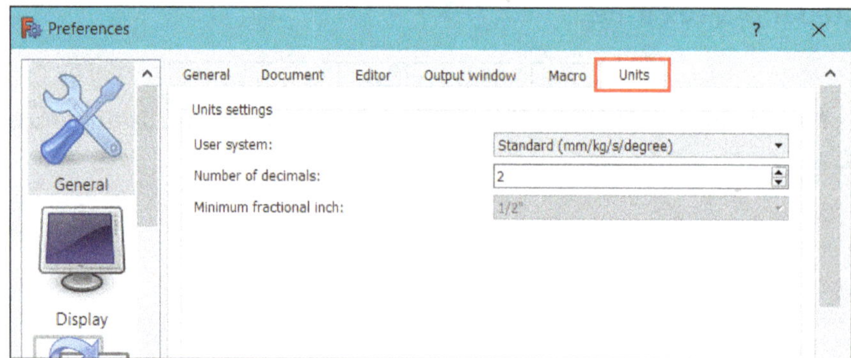

Figure-13. Units tab

- Specify desired value in the **Number of decimals** edit box to define precision up to which you want to specify values in FreeCAD.
- If you are using **Building US (ft-in/sqft/cuft)** unit system then you need to specify minimum fractional value allowed in the **Minimum fractional inch** drop-down.

Display Page

The options in the **Display** page of the dialog box are used to set parameters for display of various objects in the interface; refer to Figure-14.

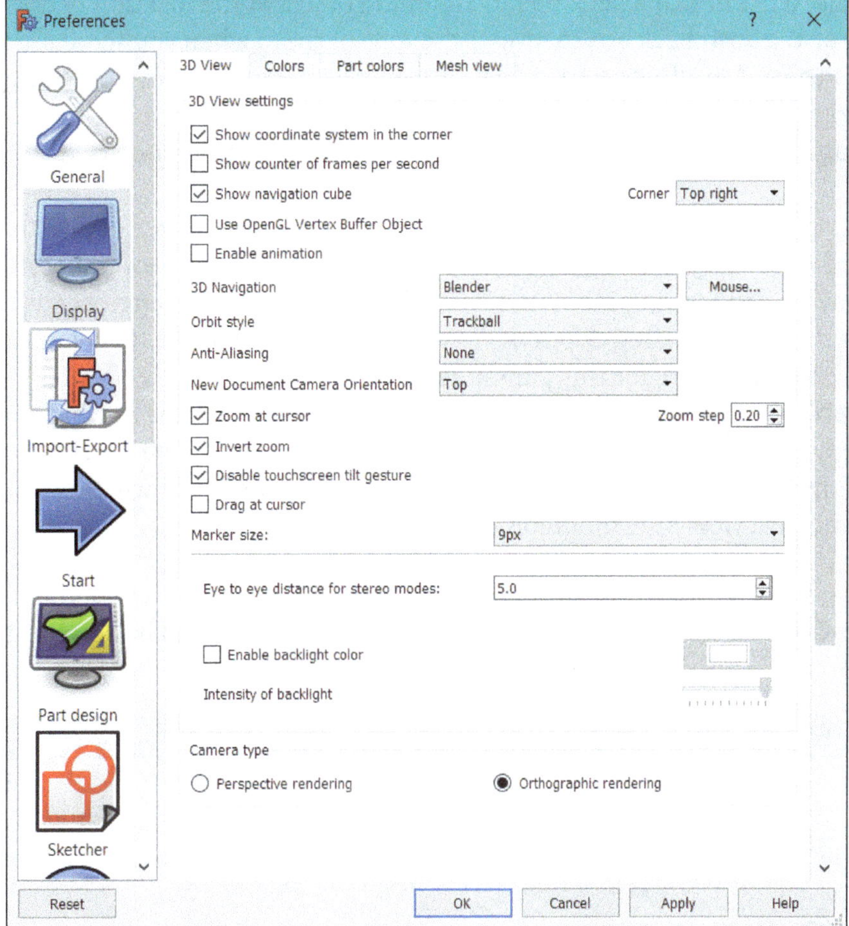

Figure-14. Display page

3D View Tab

- Select the **Show coordinate system in the corner** check box to display Coordinate system at the bottom right corner of the drawing area.
- Select the **Show counter of frames per second** check box to display a counter showing number of frames running in the software per second.
- Select the **Show navigation cube** check box to place 3D navigation cube for changing view. Select desired option from the **Corner** drop-down to define location of Cube in drawing area.
- Select **Use OpenGL Vertex Buffer Object** check box to use OpenGL buffer of your system for rendering objects in the drawing area.
- Select the **Enable animation** check box to animate the change of views in drawing area.
- Select desired option from the **3D Navigation** drop-down to specify what type of mouse gestures will be used in FreeCAD for rotating view, pan view, zoom view, or other orientation functions. You can find some common CAD application styles in this drop-down. Click on the **Mouse** button next to drop-down to check functions of different mouse buttons for selected style.
- Select desired option from the **Orbit style** drop-down to define how object will rotate in 3D orbit. Select the **Turntable** option from the drop-down to allow rotation of object as if it is placed on a turn table. Select the **Trackball** option to rotate the object freely in 3D orbit.

- Select desired option from the **Anti-Aliasing** drop-down to smoothen jagged edges of the model.
- Select desired option from the **New Document Camera Orientation** drop-down to define which orientation will be applied to the model by default when you start a new document.
- Select the **Zoom at cursor** check box to zoom the object at the location of cursor.
- Select the **Invert zoom** check box to reverse the style of zoom when you scroll up/down in drawing.
- Select the **Disable touchscreen tilt gesture** check box to disable tilting of screen when you perform pinch zooming.
- Select the **Drag at cursor** check box to perform view dragging operation from the position of cursor.
- Select desired option from the **Marker size** drop-down to define size of point/constraint marks.
- Specify desired value in **Eye to eye distance for stereo modes** edit box to define distance between two eye for enhancing 3D view of model. For a mechanical engineer, it hardly makes any difference but if you are creating different views of model for each eye then you need to specify the value carefully. Default value for this parameter is 5.
- Select the **Enable backlight color** check box to define the color in which backside of model will be highlighted and select desired color by using button next to the check box.
- Select desired radio button from the **Camera type** area to specify whether you want to render model in perspective view or orthographic view.

Colors Tab

The options in the **Colors** tab are used to modify colors for selection and background of drawing area; refer to Figure-15.

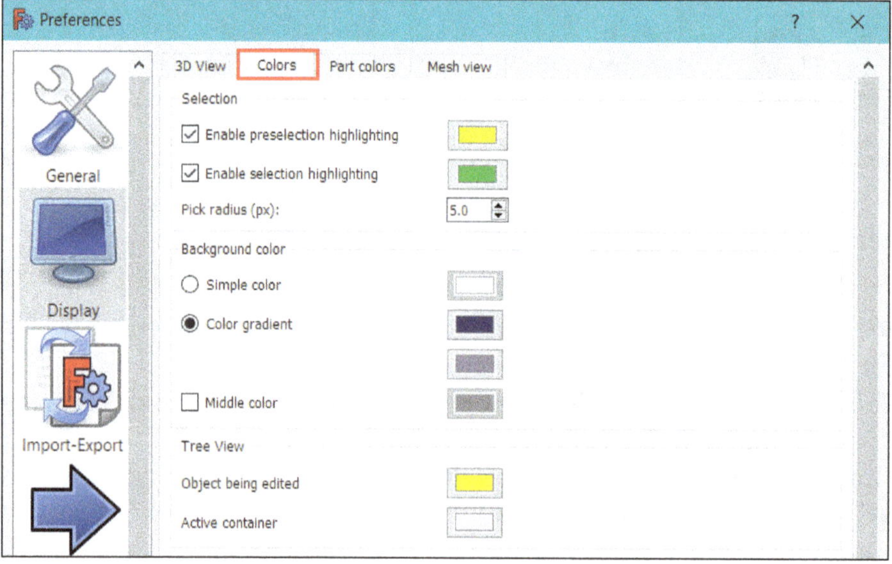

Figure-15. Colors tab

- Select the **Enable preselection highlighting** check box to highlight the object under cursor in drawing area. You can specify the color of highlighting by selecting the button next to the check box.
- Select the **Enable selection highlighting** check box to highlight the object selected from drawing area in color set next to the check box.
- Select the **Simple color** radio button from the **Background color** area to apply simple color in the background of drawing area. You can select desired color by using button next to the radio button.
- Select the **Color gradient** radio button to apply gradient of two colors in the background of drawing area.

- If you want to create a three color gradient background then select the **Middle color** check box.
- Similarly, you can set colors for objects in **Model Tree** by using options in the **Tree View** area of the dialog box.

The options of **Part colors** tab and **Mesh view** tab can be modified in the same way. After setting desired parameters, click on the **OK** button from the dialog box to apply changes.

VIEW MENU

The tools in the **View** menu are used to create and manage views of drawing area. You can also display/hide various elements of user interface of software by using the tools in this menu; refer to Figure-16. Some of the important tools in this menu have been discussed earlier in Chapter 1. The other tools in this menu are discussed next.

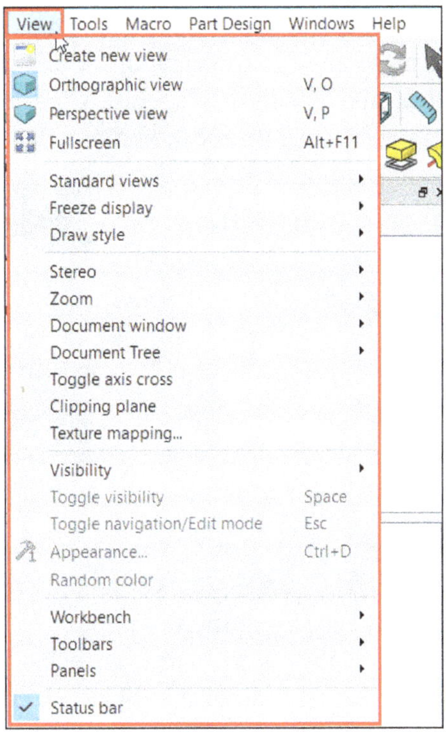

Figure-16. View menu

Stereoscopic Views

The tools in the **Stereo** cascading menu are used to display the model stereoscopic color style; refer to Figure-17.

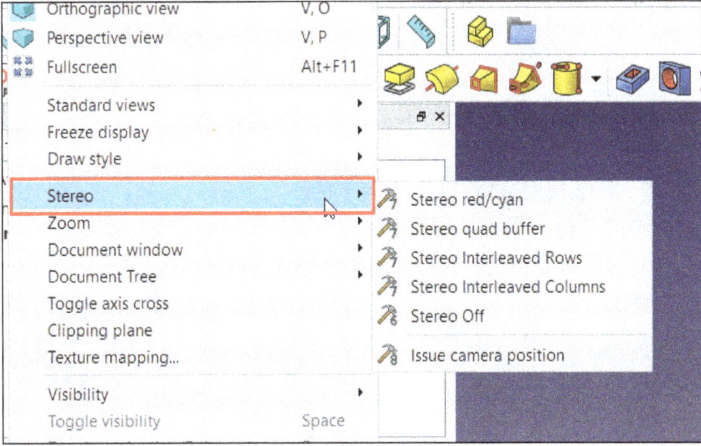

Figure-17. Stereo cascading menu

- Select the **Stereo red/cyan** option from the cascading menu to apply red and cyan color filters on the model, so that stereoscopic 3D effect can be achieved when viewing model using 3D glasses.
- Select the **Stereo quad buffer** option from the cascading menu to switch to Quad buffer stereo display mode if supported by your hardware. Quad-buffered stereo mode provides separate left and right eye frame buffers. It allows the window display in stereo while all other windows functions appear as normal.
- Select the **Stereo Interleaved Rows** or **Stereo Interleaved Columns** option to apply stereo color filters alternatively to every pixel row or column in the display, respectively.
- Select the **Stereo Off** option to stop stereoscopic display of model.
- Select the **Issue camera position** tool from the cascading menu to display current position of camera in 3D space. Note that the parameters will be displayed in **Report view** panel.

Zoom Options

The options in the **Zoom** cascading menu are used to zoom in/out and zoom specific objects; refer to Figure-18. Select the **Zoom In** option to enlarge the view of model and select the **Zoom Out** option to diminish the view of model. If you want to zoom on a specific object then select the **Box zoom** option and create a selection box around the object to be zoomed at.

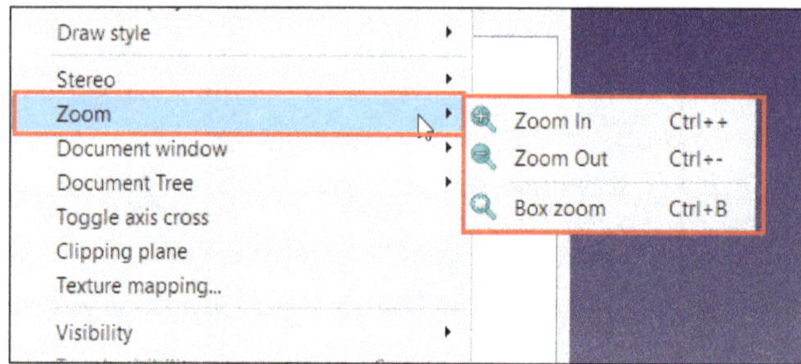

Figure-18. Zoom cascading menu

Document Windows Options

The options in the **Document window** cascading menu are used to dock, undock, and full screen the document windows where models are displayed; refer to Figure-19. Select the **Undocked** option from menu to make document window free floating in interface. Select the **Docked** option to make document window attached as tile in interface. Select the **Fullscreen** option to display document window on full screen. You can also use **F11** shortcut key to enter and exit full screen mode.

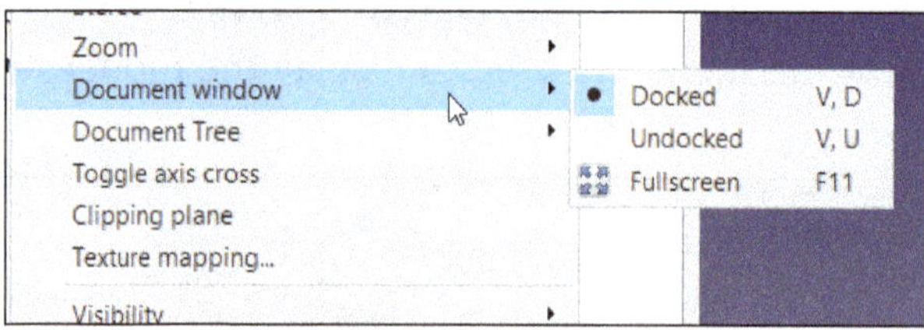

Figure-19. Document window cascading menu

Document Tree Options

The options in the **Document Tree** cascading menu are used to modify display parameters for document tree; refer to Figure-20. Select the **Single Document** option from cascading menu to display features of current document in **Model Tree**. Select the **Multi Document** option from the cascading menu to display features of all the documents currently open in **Model Tree**. Select the **Collapse/Expand** option from the cascading menu to expand or collapse all the feature nodes displayed in **Model Tree**.

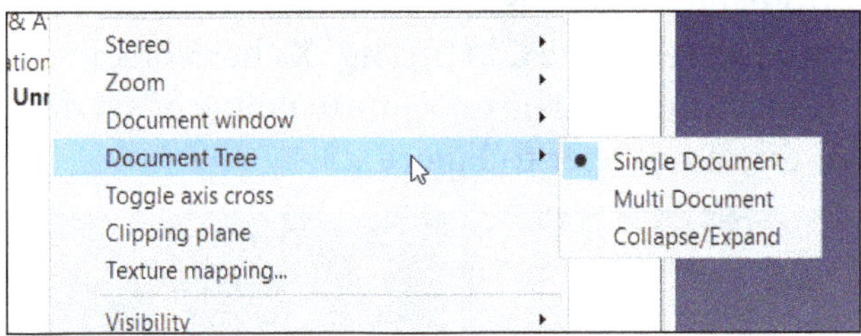

Figure-20. Document Tree cascading menu

Toggle Axis Cross

The **Toggle Axis Cross** option in **View** menu is used to toggle global coordinate system On/Off. When the Axis Cross is On, it will be displayed at the center of drawing area as shown in Figure-21.

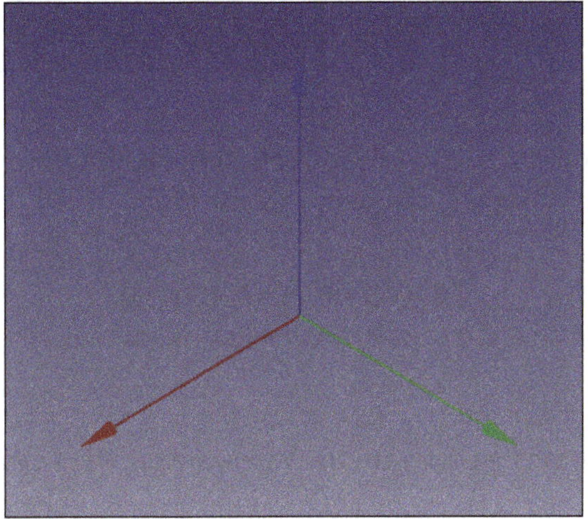

Figure-21. Axis cross

Clipping Plane

The **Clipping plane** tool in **View** menu is used to clip the model for checking interior of model. The procedure to use this tool is given next.

- Click on the **Clipping plane** tool from the **View** menu. The **Clipping** dialog box will be displayed; refer to Figure-22.

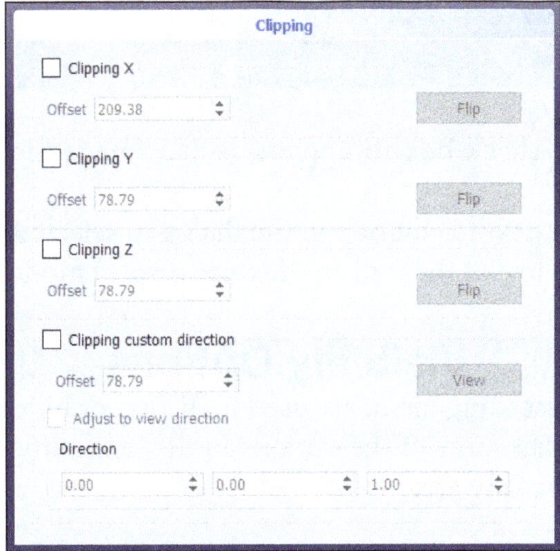

Figure-22. Clipping dialog box

- Select desired check box from the **Clipping** dialog to define in which direction you want to perform clipping of model. For example, select the **Clipping X** check box to clip model along X direction.
- After selecting check box, specify desired offset value to define offset distance of clipping plane from origin parallel to selected direction; refer to Figure-23.

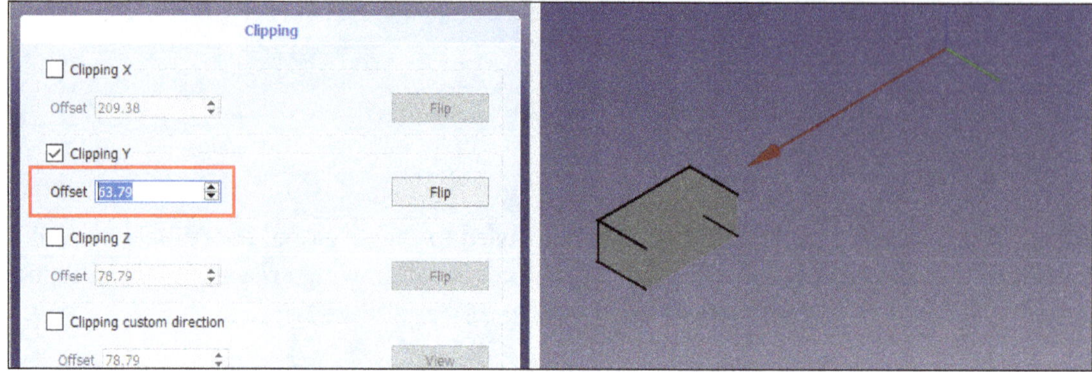

Figure-23. Clipping a model

- Using the **Flip** button, you can reverse the direction of clipping.
- Click on the **Close** button from the dialog to exit the tool.

Texture Mapping

The **Texture mapping** tool is used to apply texture on the model. The procedure to use this tool is given next.

- Click on the **Texture mapping** tool from the **View** menu. The **Texture** dialog will be displayed; refer to Figure-24.

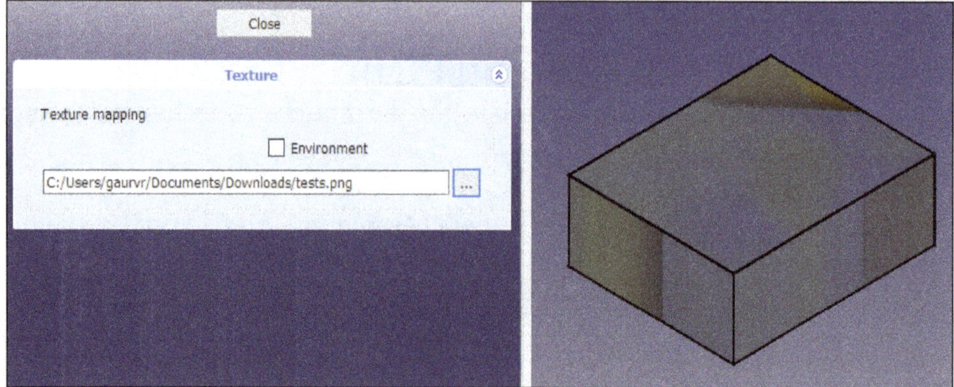

Figure-24. Texture dialog

- Select the **Environment** check box to apply selected image file on all the faces of model as environment.
- Click on the browse button next to edit box in the dialog to selected desired texture image file.
- Click on the **Close** button to exit the tool and texture display mode.

Visibility Options

The options in **Visibility** cascading menu are used to display or hide various objects in the drawing area; refer to Figure-25. Note that some of the options in this cascading menu are available when you have selected an object from drawing area.

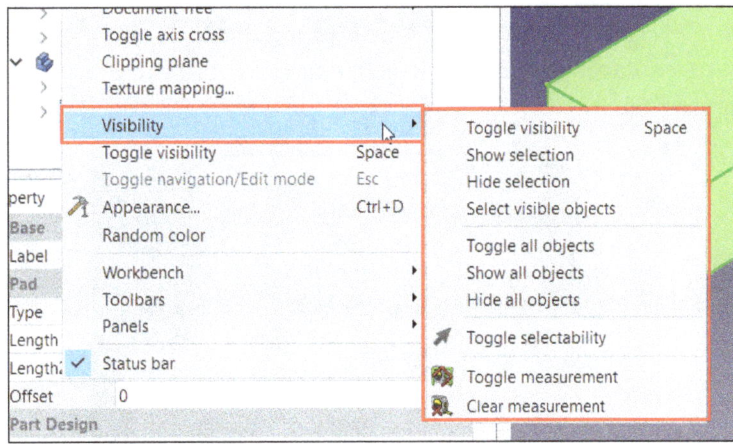

Figure-25. Visibility cascading menu

- Select the **Toggle visibility** option from the **Visibility** cascading menu to hide selected component if it is visible and display the component if it is hidden in drawing area. You can also use the **SPACEBAR** key to toggle visibility. Note that although an object is hidden in drawing area, you can still select it from **Model Tree**. Hidden objects are displayed as washed out Black & White icons in the **Model Tree**.
- Select the **Show selection** option from the cascading menu to display selected object in drawing area.
- Select the **Hide selection** option from the cascading menu to hide selected object in drawing area.
- Select the **Select visible objects** option to select all the objects currently visible in drawing area.
- Select the **Toggle all objects** option from the **Visibility** cascading menu to hide currently visible objects and display currently hidden objects.
- Select the **Show all objects** option from the cascading menu to display all the objects of **Model Tree** in drawing area.
- Select the **Hide all objects** option from the cascading menu to hide all the objects of **Model Tree** from drawing area.
- Select the **Toggle selectability** option to make currently selected objects unselectable in drawing area. Note that you can still select those objects from **Model Tree**. If you have made an object unselectable then you can use this tool again to make it selectable.
- Similarly, you can use the **Toggle measurement** and **Clear measurement** options to display or hide measurement annotations in drawing area.

The **Toggle navigation/Edit mode** tool is used to switch the editing mode of a tool temporarily for rotation/reorientation of model. Till FreeCAD version 0.18, the tool was not working as intended, so we have not discussed the procedure here.

The **Appearance** tool is used to apply different colors and appearances to the model. This tool has been discussed earlier.

Apply Random Colors to Model

The **Random color** tool is used to apply random colors to the faces of selected model. After selecting model, click on the **Random color** tool from the **View** cascading menu.

Workbench Options

The options in **Workbench** cascading menu are used to switch between various workbenches. You can also use the **Switch between workbenches** drop-down of toolbar to select desired workbench.

Showing/Hiding Toolbars

The options in **Toolbars** cascading menu are used to display/hide various toolbars from the interface of software; refer to Figure-26.

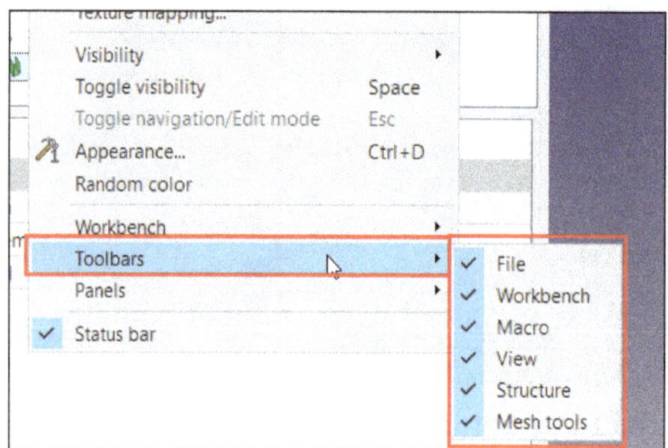

Figure-26. Toolbars cascading menu

Select desired option from the cascading menu to display respective toolbar. Note that toolbars for options which have tick mark before them are showing in interface.

Showing/Hiding Panels

The options in the **Panels** cascading menu are used to display/hide various panels of the interface like report view, panel view, selection view, and so on; refer to Figure-27.

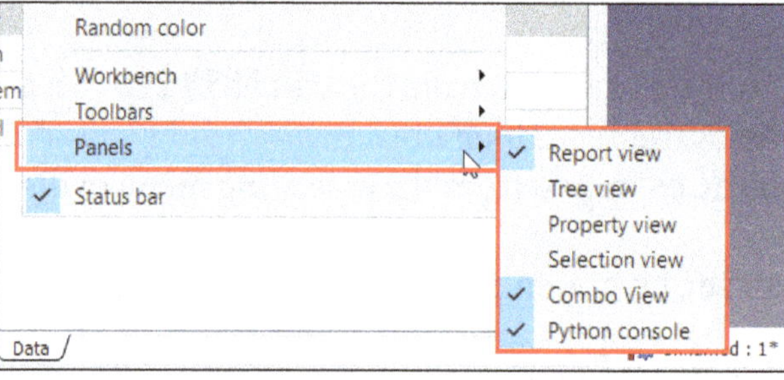

Figure-27. Panels cascading menu

Select desired option from **Panels** cascading menu to display respective panel in interface.

Status Bar

The **Status Bar** is used to display information like coordinates, mouse style, and so on. Select the **Status Bar** option from **View** menu to display **Status Bar** in interface.

MACRO MENU

The options in **Macro** menu are used to create and manage macros of FreeCAD. Macros are set of instructions for software to automate repetitive tasks; refer to Figure-28. For example, if you want to perform scale up of 10 sketches one by one then you can create a macro to do so after recording for just one sketch. Various tools of this menu are discussed next.

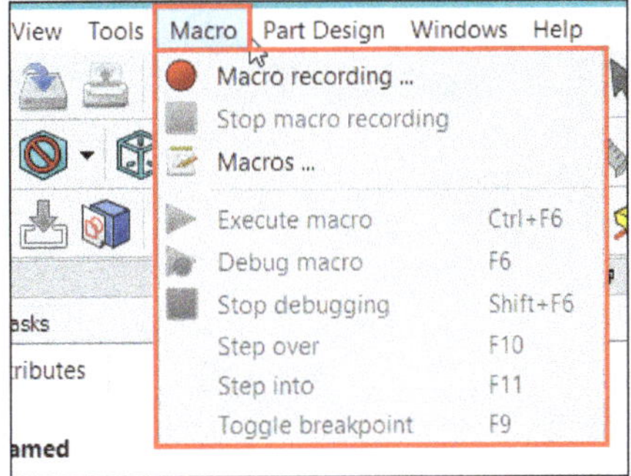

Figure-28. Macro menu

Recording Macro

The **Macro recording** tool in **Macro** menu is used to record sequence of operations performed in FreeCAD. The procedure to use this tool is given next.

- Click on the **Macro recording** tool from the **Model** menu. The **Macro recording** dialog box will be displayed; refer to Figure-29.

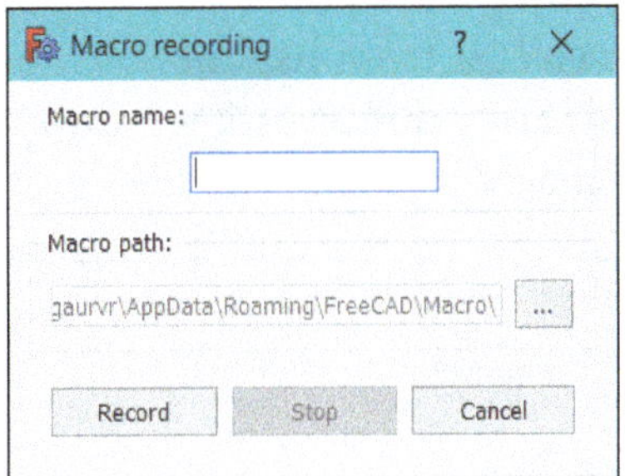

Figure-29. Macro recording dialog box

- Specify desired name for macro in the **Macro name** edit box.
- Click on the **Browse** button for **Macro path** edit box to specify location for saving macro and click on the **Record** button. The macro recording mode will start.
- Perform desired operations in the drawing area to be recorded and then click on the **Stop macro recording** tool from **Macro** menu. The macro will be created and saved at earlier specified location.

Executing and Managing Macros

The **Macros** tool in **Macro** menu is used to execute, edit, delete, or perform other tasks for macros. The procedure to use this tool is given next.

- Click on the **Macros** tool from the **Macro** menu. The **Execute macro** dialog box will be displayed; refer to Figure-30, with the list of macros available in location specified in **User macros location** edit box.

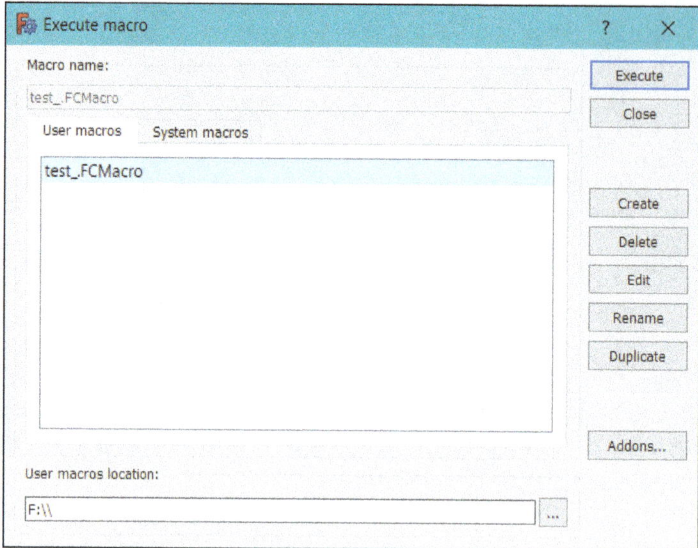

Figure-30. Execute macro dialog box

- Select desired macro from the list and click on the **Execute** button to run the macro.
- If you want to edit a macro then select it from list and click on the **Edit** button in the **Execute macro** dialog box. The **Macro Editor document** window will be displayed; refer to Figure-31. Use the Python coding to modify macro and save the file.

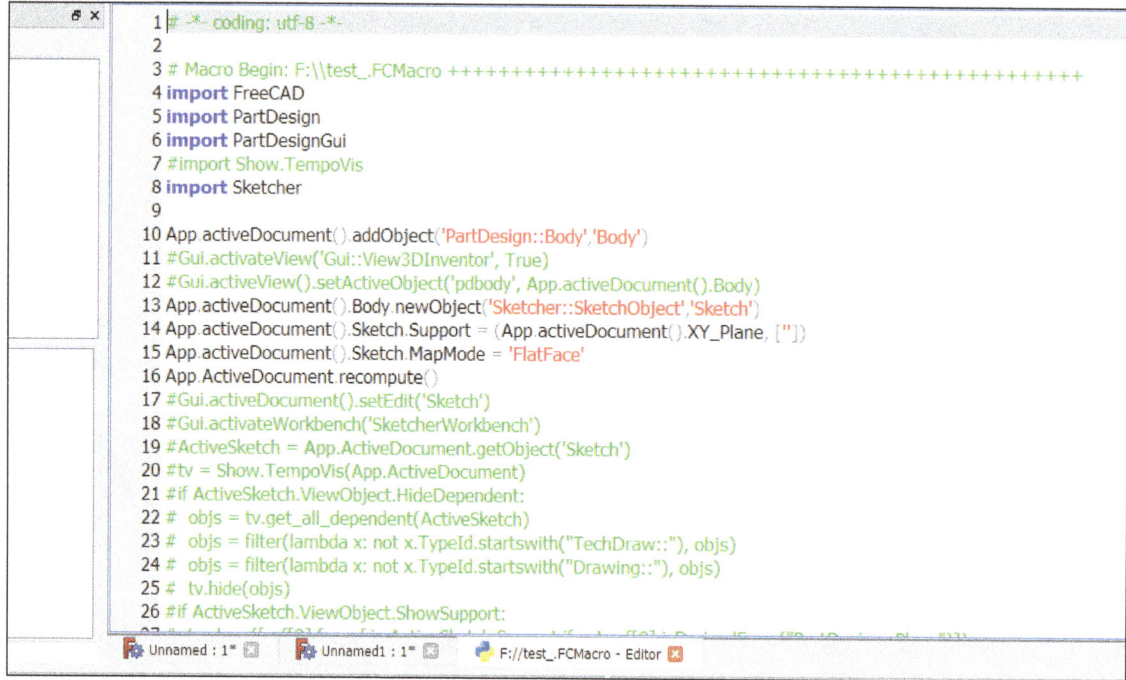

Figure-31. Macro Editor dialog box

- You can use the **Delete**, **Rename**, and **Duplicate** buttons in dialog box to delete, rename, or create duplicate copy of selected macro, respectively.
- Click on the **Addons** button to import macros available online. The **Addon manager** dialog box will be displayed. Click on the **Macros** tab to check available macros; refer to Figure-32. Select desired macro from list and execute or Install/update as desired.

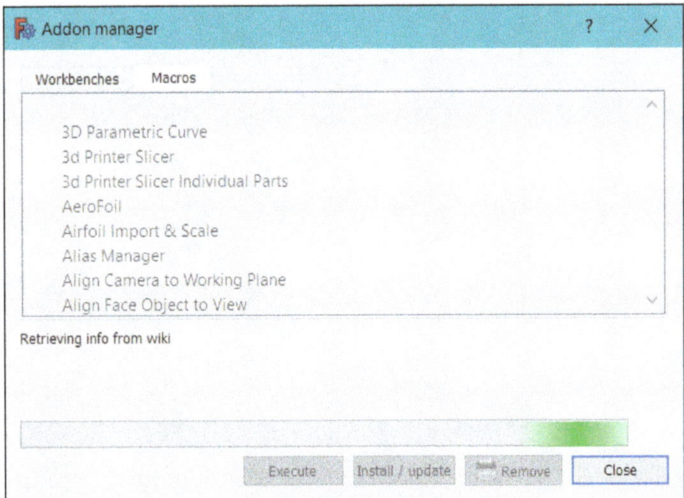

Figure-32. Macros tab in Addon manager

- Click on the **Close** button to exit dialog boxes.

For Student Notes

Index

J

K

L

M

N

O

P

R

S

T

U

V

W

Z

OTHER BOOKS BY CADCAMCAE WORKS

Autodesk Inventor 2021 Black Book

Autodesk Revit 2021 Black Book

Autodesk Fusion 360 Black Book (V 2.0.10027)

AutoCAD Electrical 2021 Black Book
AutoCAD Electrical 2020 Black Book

SolidWorks 2021 Black Book
SolidWorks 2020 Black Book
SolidWorks 2019 Black Book

SolidWorks Simulation 2021 Black Book
SolidWorks Simulation 2020 Black Book

SolidWorks Flow Simulation 2021 Black Book
SolidWorks Flow Simulation 2020 Black Book

SolidWorks Electrical 2021 Black Book
SolidWorks Electrical 2020 Black Book

Mastercam X7 for SolidWorks 2014 Black Book
Mastercam 2017 for SolidWorks Black Book

Creo Parametric 7.0 Black Book
Creo Parametric 6.0 Black Book
Creo Parametric 5.0 Black Book

Creo Manufacturing 4.0 Black Book

Autodesk CFD 2018 Black Book

Basics of Autodesk Inventor Nastran 2020
Basics of Autodesk Inventor Nastran 2021

ETABS 2016 Black Book
ETABS 2018 Black Book

Solid Edge 2021 Black Book

www.ingramcontent.com/pod-product-compliance
Lightning Source LLC
Chambersburg PA
CBHW061328190326
41458CB00011B/3930
9781774590263